YALE UNIVERSITY

MRS. HEPSA ELY SILLIMAN

MEMORIAL LECTURES

THE LATE CENOZOIC GLACIAL AGES

edited by Karl K. Turekian

New Haven and London, Yale University Press, 1971

Library of Congress catalog card number: 70–140540.
International standard book number: 0–300–01420–1.

Set in Caledonia type,
and printed in the United States of America by
Connecticut Printers, Inc., Hartford, Connecticut.

Distributed in Great Britain, Europe, and Africa by
Yale University Press, Ltd., London; in Canada by
McGill-Queen's University Press, Montreal; in Mexico
by Centro Interamericano de Libros Académicos,
Mexico City; in Australasia by Australia and New
Zealand Book Co., Pty., Ltd., Artarmon, New South
Wales; in India by UBS Publishers' Distributors Pvt.,
Ltd., Delhi; in Japan by John Weatherhill, Inc.,
Tokyo.

THE SILLIMAN FOUNDATION LECTURES
On the foundation established in memory of
Mrs. Hepsa Ely Silliman, the President and
Fellows of Yale University present an annual
course of lectures designed to illustrate the
presence and providence of God as manifested in
the natural and moral world. It was the belief
of the testator that any orderly presentation of
the facts of nature or history contributed to this
end more effectively than dogmatic or polemical
theology, which should therefore be excluded
from the scope of the lectures. The subjects are
selected rather from the domains of natural
science and history, giving special prominence to
astronomy, chemistry, geology, and anatomy.
The present work constitutes the forty-third
volume published on this foundation.

Contents

Until the middle of this century the ice caps in Greenland and Antarctica were commonly considered to be ephemeral relicts of the ice ages recorded on the North American and European continents. In this context it was reasonable to look for the limits of the most recent ice age in the Pleistocene stratigraphic record of Europe and America. In most minds the boundary between the Pleistocene epoch and the Pliocene epoch was also the index of the beginning of this latest ice age experienced by the earth. Within the confines of this definition, problems such as the number of major ice advances during the Pleistocene and the exact stratigraphic indicator and the absolute date of the Pliocene-Pleistocene boundary were of paramount importance.

Recently this parochial view of the Pleistocene ice ages has been shattered by discoveries from the study of deep-sea sediments and the geophysics of the ocean bottom coupled with detailed investigations of the glacial history of Antarctica. It is now clear that ice has been on the continents in large enough quantities to form ice caps at least since the Miocene and probably even earlier.

Confronted with the accumulating evidence supporting the idea of ice on the continents prior to any artificially constructed Pleistocene-Pliocene boundary, it is evident that we must now think about the Late Cenozoic glacial ages instead of merely the Pleistocene glacial ages. But as the key to the past is in the present, so this extension could only have come from the detailed analysis of the Pleistocene record.

In this book the emphasis is on the Late Cenozoic glacial ages. A description of the worldwide tectonic events on land and sea during this period provides the context in which to set the possibly changing climatic condition. The technique for understanding the significance of palaeontologic variation in deep-sea cores developed for the Pleistocene can be extended to the older deep-sea sediments being recovered under the auspices of the National Ocean Sediment Coring Program by the *Glomar Challenger*. The nuances of ice advances in North America and Europe can be related by analogy and by isotope-dated stratigraphic correlation to events in Antarctica.

The palynological data provide a method of extending climatic inferences from the more easily understood Pleistocene to progressively older and older samples from lake deposits.

Understanding of the controls on sea level, the present-day calcium carbonate deposition in the ocean, the mechanism controlling the oxygen isotopic composition of ice caps, and the fluctuation of the intensity of particles and photons from the sun and their effect on the earth must come principally from the study of the Pleistocene before models for the whole of the Late Cenozoic can be constructed.

The study of mammalian evolution during the Late Cenozoic ice ages now must consider a larger period during which the apparently rapid, climate-dictated, adjustments of the ice ages occurred. Of course, the evolution of man himself took place under the changing climatic conditions of the Late Cenozoic.

The last few years have been pivotal in this transformation from the concept of the Pleistocene ice ages to the Late Cenozoic glacial ages. It seemed a proper time to bring together men of various disciplines who would focus on this change in perspective. The retirement of Richard Foster Flint, Henry Barnard Davis Professor of Geology at Yale, provided a particularly appropriate time for such a convocation; a conference on the Late Cenozoic Glacial Ages was convened under the sponsorship of the Silliman Lectures of Yale University in December 1969 to honor him. It was singularly appropriate that we coupled such a conference with Richard Foster Flint. He has been one of the most important guides through the Pleistocene, and as we seek to project the insights gained from the study of the Pleistocene into the Late Cenozoic glacial ages, he stands at the threshold beckoning the Friends of the Pleistocene to a broader horizon.

K. K. T.

New Haven, Connecticut
1970

ACKNOWLEDGMENTS

Without the active cooperation of Brian J. Skinner, Chairman of the Department of Geology and Geophysics, the conference which gave rise to this volume could not have been as successful as it was. The strong professional and personal interest in the production of this book by Mrs. Anne Wilde of Yale Press has been evident to all the contributors and the editor. Mrs. Ellen T. Drake, with extraordinary skill deriving from her strong scientific interests, has produced an uncommonly comprehensive index.

Contributors

Richard L. Armstrong, Department of Geology and Geophysics, Yale University, New Haven, Connecticut 06520

Allan W. H. Bé, Lamont-Doherty Geological Observatory, Columbia University, Palisades, New York 10964

Walter William Bishop, Department of Geology, Bedford College, University of London, London N.W. 1, England

Arthur L. Bloom, Department of Geological Sciences, Cornell University, Ithaca, New York 14850

Wallace S. Broecker, Lamont-Doherty Geological Observatory, Columbia University, Palisades, New York 10964

H. B. Clausen, Physical Laboratory II, H. C. Ørsted Institute, University of Copenhagen, Denmark

Paul E. Damon, Laboratory of Isotope Geochemistry, School of Earth Sciences, University of Arizona, Tucson, Arizona 85721

W. Dansgaard, Physical Laboratory II, H. C. Ørsted Institute, University of Copenhagen, Denmark

George H. Denton, Department of Geological Sciences, University of Maine, Orono, Maine 04473

K. O. Emery, Woods Hole Oceanographic Institution, Woods Hole, Massachusetts 02543

Cesare Emiliani, School of Marine and Atmospheric Sciences, University of Miami, Miami, Florida 33149

D. B. Ericson, Lamont-Doherty Geological Observatory, Columbia University, Palisades, New York 10964

Maurice Ewing, Lamont-Doherty Geological Observatory, Columbia University, Palisades, New York 10964

William R. Farrand, Department of Geology and Mineralogy, University of Michigan, Ann Arbor, Michigan 48104

C. C. Flerow, Paleontological Museum, Academy of Sciences of the U.S.S.R., Moscow V-71, U.S.S.R.

Kenneth Hunkins, Lamont-Doherty Geological Observatory, Columbia University, Palisades, New York 10964

John Imbrie, Department of Geological Sciences, Brown University, Providence, Rhode Island 02906

S. J. Johnsen, Physical Laboratory II, H. C. Ørsted Institute, University of Copenhagen, Denmark

Nilva G. Kipp, Department of Geological Sciences, Brown University, Providence, Rhode Island 02906

Kazimierz Kowalski, Polish Academy of Sciences, Institute of Systematic Zoology, Krakow, Poland

C. C. Langway, Jr., U.S. Army Cold Regions Research and Engineering Laboratory, Hanover, New Hampshire 03755

Barrie C. McDonald, Geological Survey of Canada, Ottawa, Canada

Guy Mathieu, Lamont-Doherty Geological Observatory, Columbia University, Palisades, New York 10964

H. W. Menard, Institute of Marine Resources and Scripps Institution of Oceanography, University of California at San Diego, La Jolla, California 92037

Hiroshi Niino, Tokai University, Tokyo, Japan

Neil D. Opdyke, Lamont-Doherty Geological Observatory, Columbia University, Palisades, New York 10964

Stephen C. Porter, Quaternary Research Center, University of Washington, Seattle, Washington 98105

Minze Stuiver, Departments of Geological Sciences and Zoology, Quaternary Research Center, University of Washington, Seattle, Washington 98105

Beverly Sullivan, United States Geological Survey, Washington, D.C. 20242

T. van der Hammen, Afdeling Palynologi, University of Amsterdam, Amsterdam 4, Netherlands

T. A. Wijmstra, Afdeling Palynologi, University of Amsterdam, Amsterdam 4, Netherlands

G. Wollin, Lamont-Doherty Geological Observatory, Columbia University, Palisades, New York 10964

H. E. Wright, Jr., Department of Geloogy, University of Minnesota, Minneapolis, Minnesota 55455

W. H. Zagwijn, Afdeling Palynologi, University of Amsterdam, Amsterdam 4, Netherlands

H. W. Menard

1. THE LATE CENOZOIC HISTORY OF THE
PACIFIC AND INDIAN OCEAN BASINS

Despite increasingly intense oceanographic exploration, data about the deep sea floor are still relatively rare. Thus it would be difficult to attempt a meaningful synthesis of the late Tertiary history of the Pacific and Indian ocean basins if a unifying hypothesis did not exist. Such a hypothesis has been formulated during the last few years by the gradual development and enlargement of some ideas of Hess (1962) and Dietz (1961) concerning sea floor spreading and its effects at continental margins. Wilson (1965) developed the importance and consequences of crustal continuity by emphasizing that, for an earth of constant surface area, as much crust is destroyed at oceanic trenches as is created at midocean ridges. Moreover, the regions of creation and destruction of crust must be connected by faults with horizontal motion which he called transforms.

These ideas were grossly modified by the discovery that the surface of the earth consists of very large tectonic plates which, although moving, are essentially rigid. Wilson proposed that the North Pacific was part of a tectonic plate bounded at the trailing edge by the East Pacific rise crest, at the leading edge by the Aleutian, Kurile, Japan, and Marianas trenches, and with the San Andreas and parallel faults forming the connecting transforms. A theorem of Euler states that if one rigid spherical surface moves over another, two points remained fixed. For example, the points over the poles would not change if the spinning earth had a plastic cover. Consequently, the motion of a rigid crustal plate on the surface of the earth can be described in terms of a rotation about a point on the globe although not necessarily on the plate. Geometry alone shows that every point on the rigid moving plate moves in an arc of a small circle relative to the point of rotation or "Euler pole." McKenzie and Parker (1967) showed that the slip vectors determined by first motion of earthquakes bordering the North Pacific tectonic plate followed small circles relative to a point near Hudson Bay. Further analysis of slip vectors subsequently confirmed that other suspected tectonic plates are at present rotating around Euler poles (Isacks et al. 1968). Moreover, the earthquakes at the trailing edges of plates are tensional, those at the leading edges are compressional, and those on transform faults are strike slip.

The distribution and motion of earthquakes show that large, rigid tectonic plates now exist and that almost all the deformation of the earth occurs at the interacting edges of the plates. However, earthquakes tell us nothing of the past. Morgan (1968) developed the analysis that demonstrates the persistence of large moving plates for periods of at least 100 million years. The trailing edges of the plates are offset in many steps by ridge–ridge transform faults which produce the distinctive linear topography of a fracture zone. Inasmuch as they are transforms, the motion along the faults is horizontal and follows a small circle relative to the Euler pole of relative motion between the two spreading plates. Morgan plotted great-circle "perpendiculars" to fracture zones on various suspected plates and found that they intersect at a point, and thus rotation of a large rigid plate has occurred. The duration of motion is given by considerations of continental drift but also, and with more resolution, by the sea floor spreading which occurs at the trailing edges of plates. The Vine-Matthews mechanism (1963) records the history of the reversals of the magnetic field in the dikes which form much of the volcanic layer of the oceanic crust. If the separating plates are rigid, the spreading rate varies according to the cosine of the latitude relative to the Euler pole. Morgan demonstrated that this relationship exists in several places and thus independently confirmed the long persistence of the rigidity of large moving plates.

Subsequent developments have been numerous and reassuring, with the result that the following paradigm of plate tectonics has emerged and gained almost universal acceptance.

1. The surface of the earth is broken into large, rigid plates, most of which have been moving for at least 100 million years. However, new plates are formed from time to time and others are destroyed by being consumed at the leading edge.

2. The plates are uplifted at the trailing edge because of thermal and perhaps phase change expansion of the mantle. A pair of adjacent plates spreading apart thus forms a midocean ridge.

3. The plates, initially high at the trailing edge, begin to sink as they age and drift. This continues at a decreasing rate, but even the oldest oceanic crust appears to be sinking at about 20 meters per million years.

4. While the plates are created at the trailing edge they are simultaneously destroyed at the leading edge. Globally, creation and destruction must balance, but this is not necessarily true of each plate because all the plates are in relative motion. Moreover, the rate of destruction varies with the cosine of the latitude of the Euler pole for any two colliding plates.

5. LePichon (1968) made the first global synthesis of plate tectonics by computing relative motions where they have not been observed. He found that the mode of plate deformation at the leading edges is a function of the rate at which the leading edges come together. If less than 6 cm per year, the plates buckle and form high, folded mountain ranges. If the rate is greater, one of the leading edges

plunges into the mantle to form the Benioff zone of intermediate and deep-focus earthquakes (Isacks et al. 1968) which is accompanied by a deep trench and usually an island arc.

6. The plates are very thin at the trailing edge where they solidify, but they gradually thicken to about 50–100 km within a few tens of millions of years as they drift and sink. The increasing thickness can be usefully equated to an isotherm which can be calculated from models of heat flow (Langseth et al. 1966; McKenzie 1967). The cold plate eventually plunges back into the mantle at an island arc and gradually warms and thins again (Griggs, personal communication, 1970) until it vanishes. The slope of the Benioff zone is inversely proportional to the distance from the Euler pole for the two converging plates (Luyendyk 1970). This suggests that the leading edge plunges at a relatively constant rate and thus that the slope is roughly a function of the rate of convergence.

7. Very little happens within the plates compared to the edges. Volcanism is intense where it occurs, as in the Hawaiian Islands, but such places are few (Menard 1969a). Otherwise, an almost total absence of earthquakes implies little deformation except for minor adjustments accompanying general subsidence.

Given this paradigm we can reconstruct the history of the Pacific and Indian ocean basins in late Tertiary time.

CREATION OF CRUST

The area of new oceanic crust created during the last ten million years is 2.6×10^7 km^2, which is equivalent to 5 percent of the area of the earth and 9 percent of the area of ocean basins (Fig. 1). This borders on the spectacular and certainly affirms that crustal creation has been unusually intense. At such a rate the oceanic crust would be renewed in only 110 million years—only half the probable rate at which the existing crust is formed. The distribution of new crust is by no means uniform. In a general way, it increases from north to south in the Atlantic but the total amount is relatively small. It increases from the Red Sea to south of Australia in the Indian Ocean basin and the amount is roughly double that of the Atlantic. The Pacific, however, is the principal locus of new crust because of the very fast spreading in the south equatorial region. Both the East Pacific rise (Pitman et al. 1968) and the Chile rise (Herron and Hayes 1969) contain wide bands of new crust. Likewise a patch of young crust occurs between Fiji and the New Hebrides Islands (C. Chase, personal communication, 1970) where the oceanic crust is elevated and the heat flow is high (Sclater and Menard 1967).

The lithosphere created at spreading centers has an average section of 1 km of volcanic rock, 2–5 km of oceanic crustal layer, and a few kilometers of rigid, cooling lithosphere. As the new material ages, the thickness of the volcanic layer remains relatively constant but that of the other layers gradually increases for some tens of millions of years (McKenzie 1967; Shor et al., in press). Eventually the oceanic layer becomes uniformly about 5 km thick, and the lithosphere reaches

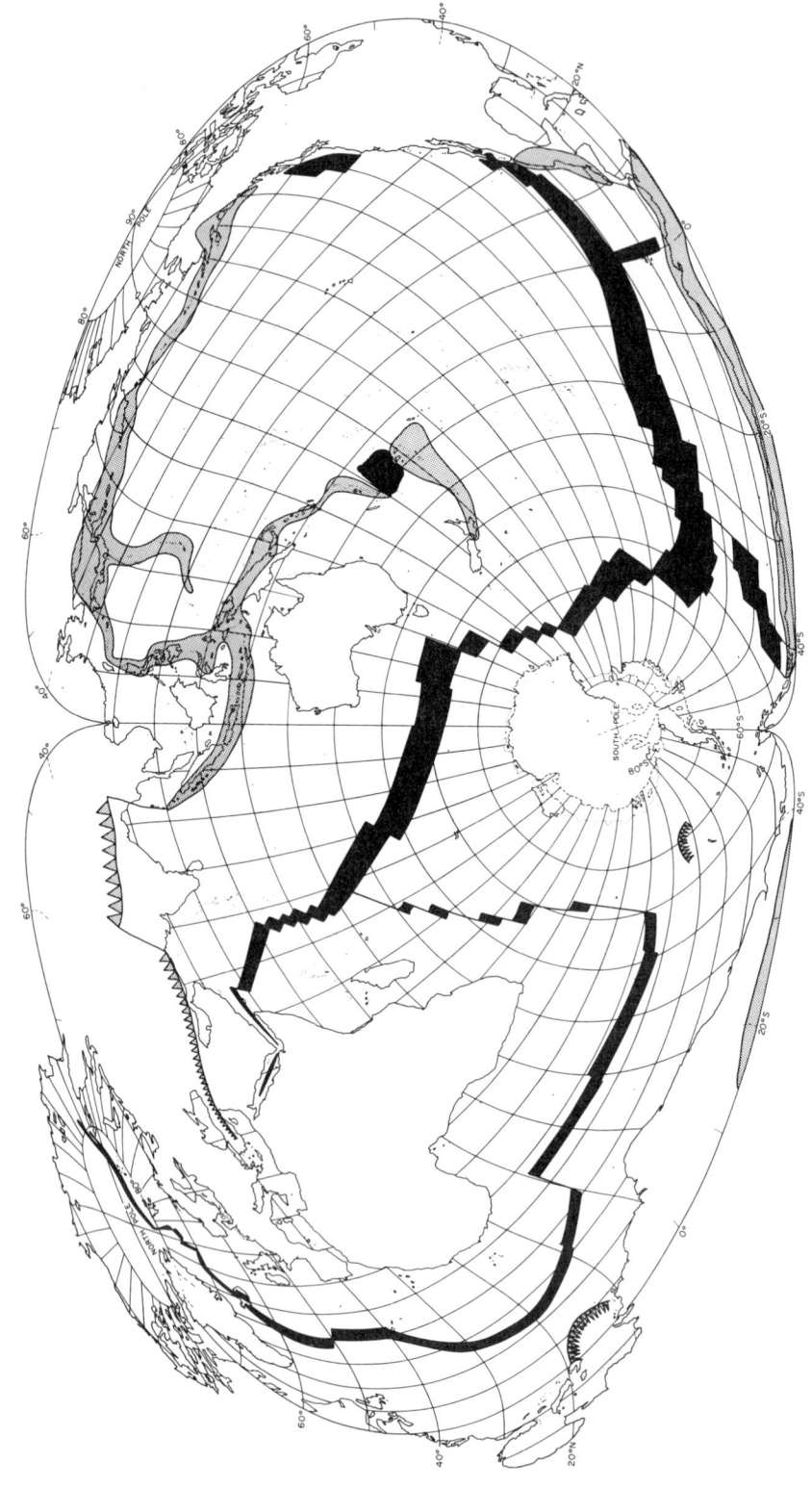

Fig. 1. Equal area map showing crust created during the last 10 million years (black). The shaded region is where the crust is restored to the mantle at island arcs and young mountains. The area of the shaded region corresponds to the distribution of earthquake epicenters deeper than 100 km and presumably is the extent of the plunging crust. The area of new crust and of crust being destroyed is about equal.

a thickness of about 50 km or even as much as 100 km. The lower layers accumulate on the bottom of the moving plate rather than at the trailing edge, but the volume of material can be computed as if it all formed in the same place.

Dredged samples of the volcanic layer show it to be tholeiitic basalt with a distinctive and highly uniform composition (Engel 1963; Engel et al. 1965). The thickness of lava in any spreading center seems to be inversely proportional to the spreading rate, with the result that the discharge is constant (Menard 1967). The total discharge for all spreading centers is about 4 km³ per year, which is quadruple the 1 km³ per year estimated on much less reliable data for subaerial volcanoes (Kuenen 1950). For the last 10 million years the volume of new volcanic layer is by chance also four times the total volume of all identifiable continental flood basalts of all ages (Verhoogen 1946). These figures confirm the fact that oceanic volcanism is strongly dominant over continental (Menard 1964). We shall consider the additional increment of volcanism on moving oceanic islands in a later section, but this plate edge volcanism alone strongly suggests a higher intensity than the average for the past. For example, a discharge of 4 km³ per year would yield the total volume of the continents in little more than a third of geological time, or much faster than previously estimated (Wilson 1951).

The composition of the oceanic layer is still conjectural. Either partially serpentinized periodotic mantle (Hess 1962) or basalt derived from a pyrolite mantle (Ringwood and Green 1966) seems possible. The remarkable uniformity of the seismic velocities of 6.81 ± 0.16 (Shor et al., in press) and the equally uniform thickness of the older crust appear to accord most reasonably with a layer with a simple composition and a resistance to alteration. The original thickness of this layer is greater where spreading is fast; that is, it is directly proportional to spreading or just the opposite of the volcanic layer. However, the oceanic layer ultimately reaches a uniform thickness regardless of spreading rate, and thus we can calculate the discharge of this layer to be 5 km times the area of new crust, which gives an average rate of 13 km³ per year for the last 10 million years. Once again, this rate is far greater than would have been expected in earlier paradigms of the history of the earth. This accumulation would yield the volume of the continents in only 540 million years. Either plate motion was less intense during most of geological time or else most of this material returns to the mantle along the Benioff zone.

The lithosphere presumably is merely cold and rigid material, with the composition of the mantle, and thus the discharge of material introduces no problems. It is of interest to calculate whether the unexpectedly rapid flux of surface materials indicates a correspondingly rapid overturn of the mantle. The discharge of lithosphere is 130 km³ per year according to the same arguments introduced for the oceanic layer. The volume of the mantle is about 840×10^9 km³ and thus a single overturn would require 6.4×10^9 years, or more than the age of the earth. The plate tectonics and continental drift paradigm does not require that the mantle has overturned even once.

DESTRUCTION OF CRUST

If the earth is not expanding, the area of crust created is exactly balanced by that destroyed. LePichon (1968) showed that the creation of crust during Tertiary time has not been radially symmetrical, with the result that the earth would be lopsided if crust had not been consumed. This demonstrates that expansion alone cannot explain plate tectonics or even sea floor spreading, but it does not eliminate a small but perhaps important expansion occurring simultaneously with the motion of large rigid plates. The paradigm does not include expansion but merely because it does not seem necessary on a small scale and because large-scale expansion is ruled out by other considerations (Menard 1966).

Nevertheless, it is always worthwhile to seek an independent measure of the crust destroyed during the last 10 million years to see if it balances that created or whether destruction is significantly less, which would suggest expansion. The down-dip length of the Benioff zone is roughly proportional to the rate of crustal convergence in the island arc (Isacks et al. 1969) and can be explained by convergence at a constant rate for 10 million years. On this hypothesis, we need merely measure the area of the Benioff zone from the epicenter plots of Barazangi and Dorman (1969) and apply a correction for dip to obtain the area of crust destroyed. The area of the Benioff zone projected to the surface is roughly 2.0×10^7 km², or three-quarters of the area created. If a dip of 45° is assumed, the area of the Benioff zone is 2.9×10^7 km², or somewhat larger than the area created at spreading centers. It appears that the identifiable lithosphere thrust down into the mantle is reasonably equivalent to that created during the last 10 million years. This means that Figure 1 shows not only the locus of crustal creation but probably also of crustal destruction and thus the total flux of material.

The crust generated at the trailing edge of the India plate is more extensive than that destroyed at the leading edge in the island arcs of Indonesia and New Britain and the mountains of New Guinea. Consequently, the island arcs and mountains are migrating to the northeast, and part of the extension in the Indian Ocean is balanced by convergence in the North Pacific. The edge phenomena between the Pacific and Indian plates should show a complex history of changes in direction as the boundary migrates.

The destruction of crust under South America is comparable to that under the island arcs fringing Asia and is capable of absorbing all the crust created in the Atlantic plus about a third of that on the East Pacific rise. The remainder of the East Pacific rise crust is balanced by destruction in the northern and western Pacific where the leading edge of the Pacific plate converges on Asia.

The composition of material destroyed is about the same as that created except for the addition of a few hundred meters to a kilometer of pelagic sediment. The composition differs according to latitude, with carbonates predominating near the equator and silicates in high latitudes. Presumably, as the lithosphere plunges and is reheated, the lower-melting components migrate upward to produce andesitic volcanoes in island arcs (Ringwood and Green 1966). Variations of the composition of andesitic magmas may be related to the composition of the sediments of the adjacent sea floor, but this has not yet been demonstrated.

EVOLUTION OF SEA WATER

On the evidence of its seismic velocity, Hess (1962) estimated that the oceanic layer is 70 percent serpentinized peridotite and thus contains 25 percent water by volume. He assumed ocean spreading at 1 cm per year and crust destroyed at the same rate in island arcs. Destruction involves reheating which drives out water which then migrates to the surface, yielding 0.4 km³ per year. The spreading rate he used produces an ocean of about the present volume in 4×10^9 years, which suggested a process with uniform intensity during geological time. The new measured spreading rates present quite a different picture. According to his model, the discharge of water is 3.3 km³ per year, which would produce the ocean in only 420 million years. Put in different terms, the same rate of generation for only the last 100 million years would yield 24 percent of the volume of the oceans. Only four such tectonic episodes could occur in the history of the earth.

The oceans may have been produced by degassing of the mantle at a constant rate (Rubey 1951), but the constancy is little more than a convenient assumption. Revelle (1955) argued that 25 percent of the oceans may have accumulated in the last 100 million years in order to account for the depth of central Pacific guyots. Additional data on paleobathymetry suggest that half the sinking in the region may have been localized but that the volume of the ocean may have increased by 15 percent (Menard 1969b). Even so, the last 100 million and the last 10 million years may have seen far more intense degassing and ocean accumulation than the geological average.

The conjectural evidence for a large increase in the volume of the oceans is in quantitative agreement with the oceanic crustal model proposed by Hess. This is suggestive but does not eliminate the possibility that the water reaches the ocean directly by volcanism at spreading centers without being stored in the crust. Thus a significant increase in ocean volume does not require a serpentinized crust.

MIDPLATE VOLCANISM

Within the deep ocean basins are numerous active and recently active volcanoes which lie on and are drifting with otherwise inactive central parts of tectonic plates. The outpourings of these volcanoes do not compare in intensity with those at the plate edges, but they cannot be neglected with regard to factors which might influence late Tertiary volcanism. They disgorge CO_2 and ash directly into the atmosphere, whereas almost all plate edge volcanism is under kilometers of water and is virtually ash-free. Thus an exceptional intensity of island volcanism might have influenced the greenhouse effect or the transparency of the air out of all proportion to the volume of the lava. Was the island volcanism more intense than normal in the Pleistocene?

Within the oceanic plates, away from the edges, are 22 major active volcanic islands and 84 inactive but still extant volcanic islands (Menard and Ladd 1963). The latter presumably would be eroded away and sink below the sea in less than 10 million years after they become inactive. This suggests a total of roughly 100 ocean basin islands active during late Tertiary time. This surely is more intense volcanism than normal or else the ocean basins would be choked with large volcanoes instead of the small ones that predominate. Nevertheless, it is difficult to estimate the amount of volcanism above normal. In the Pacific about one-quarter of the large groups of volcanoes, including all ages from 100-million-year guyots to young islands, are now active. Volcanoes that originate near plate edges commonly grow relatively slowly and are active for tens of millions of years. Those within the plates, however, may grow extremely rapidly and can hardly endure so long (Menard 1969a). Thus the present midplate volcanism probably began no earlier than late Tertiary time and is of an intensity previously identified only in very late Mesozoic–very early Tertiary time in the Pacific.

The same arguments can be advanced for the other ocean basins. A large fraction of the volcanic groups within the plates was active during late Tertiary time. Many of these are not very far from the edges of the Africa and America plates and probably originated at the edges and then drifted with the plates. Oceanic exploration is by no means complete, but the physiographic diagrams of the Atlantic and Indian ocean basins (Heezen and Tharp 1961, 1964, 1968) show nothing like the same number of drowned ancient islands on the outer flanks of the ridges as now exist on the younger crust in the center. Once again, without offering a quantitative measure, this suggests that volcanism within the basins has been more intense than usual during the last 10 million years.

Loss of Volcanoes into Trenches

The creation of active new volcanoes is balanced by the loss of inactive and generally ancient volcanoes into trenches and down along the Benioff zone. Crudely visualized, a large guyot, perhaps 4 km high, projects like a giant wart above the smooth sea floor plunging down into a trench. Regardless of the geometry it is difficult to see how the volcano can avoid being sheared off from the plunging crust. Being of relatively low density the volcano should initially tend to clog the trench. This is an effect observable in many places in the western Pacific where the continuity of the trenches is broken by both large and small inactive submarine volcanoes (Figs. 2, 3). The paradigm of plate tectonics requires a continuous plate edge and thus the volcanoes are mere superficial plugs which conceal an underlying trench.

Two things might occur. The low-density volcanoes could continue to collect as a scum over the descending maelstrom of the convergence zone and gradually obscure the trench. Alternatively, they may gradually be broken up and pulled down along the Benioff zone. The former possibility would result in the occurrence of a Benioff zone without a trench—which is not observed. The fracturing of

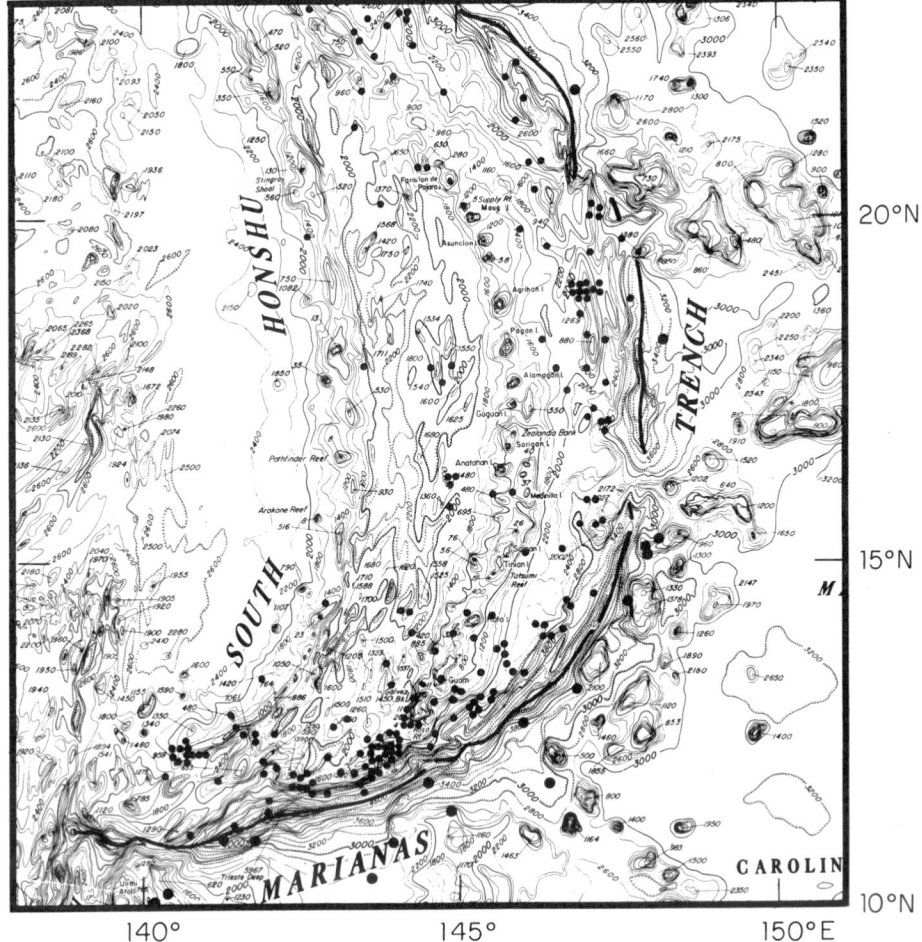

Fig. 2. Bathymetry surface and earthquakes along the Marianas trench. Earthquakes in the Pacific tectonic plate east of the trench axis tend to be concentrated where volcanoes are moving into the trench (after Gutenberg and Richter 1949).

volcanoes in trenches is also unobserved, but the pattern of shallow earthquakes suggests it is occurring. Very few shallow earthquakes lie east of the axis of the Marianas trench, for example, and most are on the sides of volcanoes impinging on the trench (Fig. 2).

POSITION OF CONTINENTS

The continents have drifted during the last 10 million years, but plate tectonics gives us only the relative motion between them and not the absolute motion compared to a fixed geographical frame of reference such as the pole. Nevertheless, the relative motions amount to hundreds up to more than a thousand kilometers, and thus the surface distribution may have been changed enough to influence whatever conditions favor glaciation.

Ideally, the motion of the plates relative to the equator or north pole can be

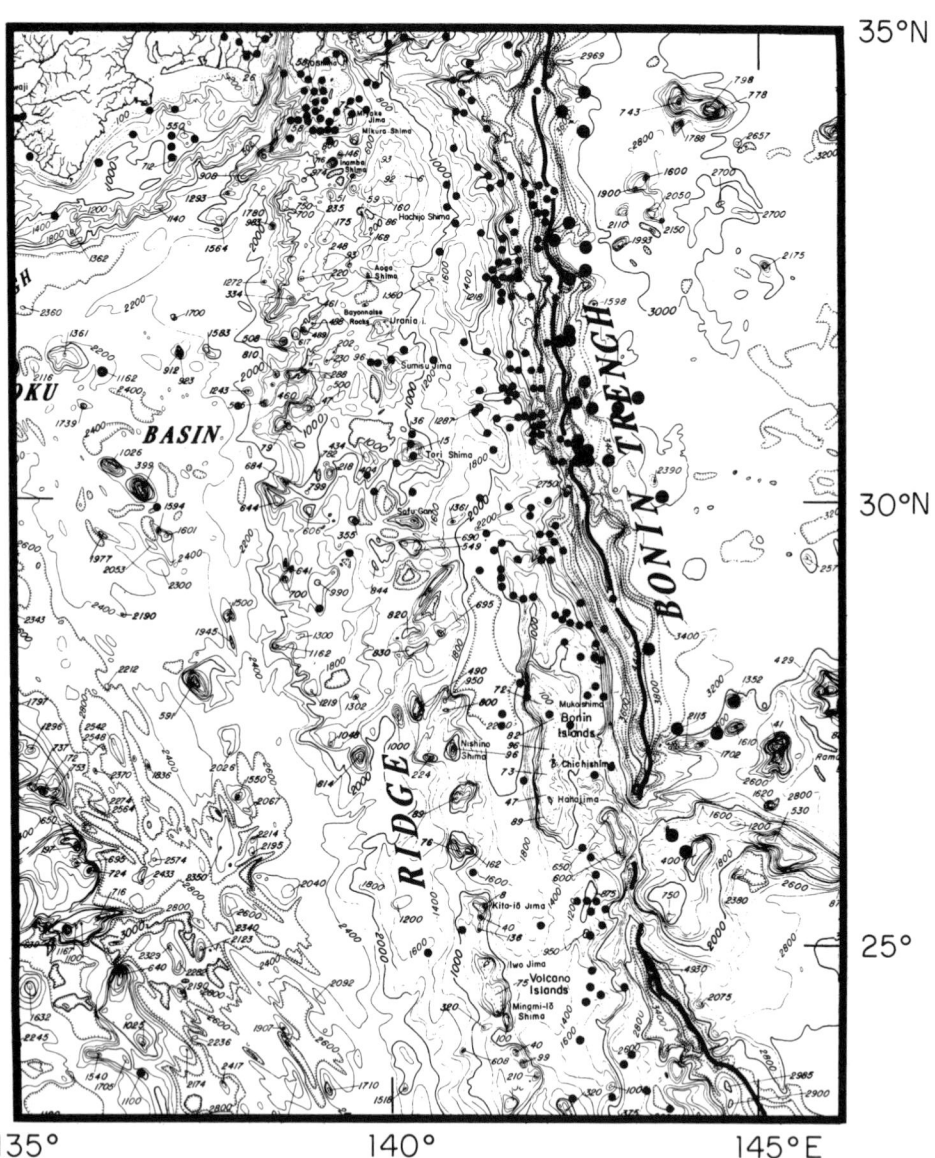

Fig. 3. Bathymetry and surface earthquakes along the Bonin trench. Earthquakes tend to be concentrated in the Pacific plate where large volcanoes are intersecting the trench.

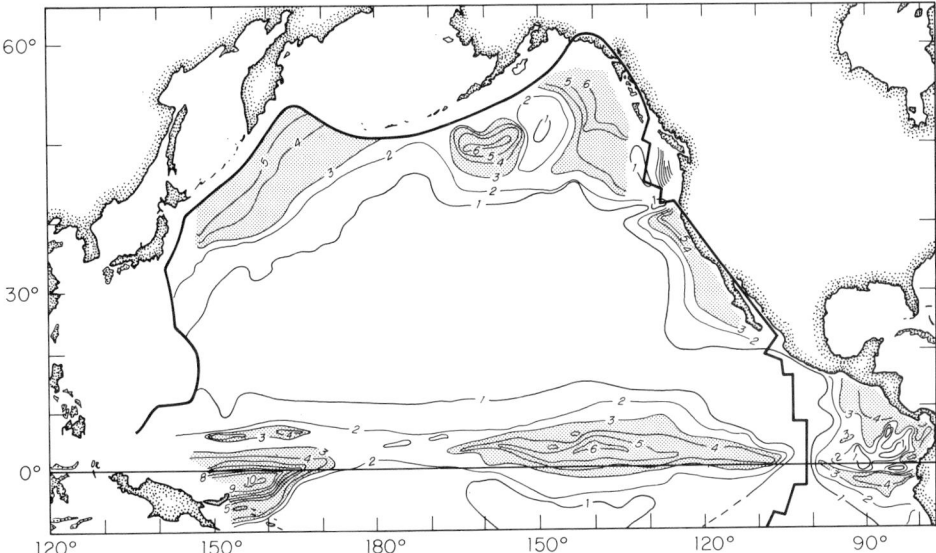

Fig. 4. Isopach map of sub-bottom reflectors in the North Pacific, redrawn from Ewing et al. (1968); thicknesses in hundreds of meters, assuming sound velocity in sediment is the same as in sea water. The equatorial ridge of pelagic sediment is north of the equator in the Pacific plate and south of it to the south of the Galapagos Islands.

deduced from paleontology or paleoclimatology or paleomagnetism, but for a mere 10 million years this is difficult. The most promising criterion appears to be the motion of the Pacific plate relative to the equator as indicated by the belt of intense pelagic sedimentation. The approximate locus of the belt is given by sub-bottom profiling (Ewing et al. 1968). It lies well north of the equator on the west side of the East Pacific rise but is south of the equator between the rise crest and South America (Fig. 4). The preliminary results of the National Ocean Sediment Coring Program (planned by JOIDES), indicate that the locus of maximum sedimentation has moved north during late Tertiary time (C. C. von der Borch, personal communication, 1970). The best fit for the combined JOIDES stratigraphy and sub-bottom profiling thicknesses is a simple northern rotation at 1.9×10^{-7} degrees per year around an Euler pole at the equator and 95° W longitude.

  If this preliminary result is correct, the relative motions of all the continents can be calculated by deconvoluting the motions determined by LePichon (1968). The motion of the Pacific plate, which is more than a third of the earth's surface, is in a geographical framework because it is relative to the equator. Thus all the latitudinal motions calculated by deconvolution are also in this geographical framework; that is, they are motions relative to the pole and equator. On the other hand, the longitudinal motions are not fixed in any such framework and are merely relative to the longitude of the Pacific plate which may itself be changing. Fortunately, longitude is itself arbitrary and of no consequence in geology except for the relative motion between continents.

With the Pacific plate rotating slowly north, the Antarctic rotates clockwise with a resultant northward motion at longitude 120°E. Australia, which is separating from the Antarctic in the general region of longitude 120°E, also moves north and even faster than that part of Antarctica. South America and Africa move almost entirely longitudinally. North America moves generally southeast and Eurasia rotates in such manner that Spain moves generally east and Kamchatka moves northeast. Thus the great continental mass of Asia has gradually encroached on the Arctic basin, and Alaska and the Canadian archipelago have moved away from the basin. The net result has been an increase in the continental area in very high latitudes.

Conclusions—Marine Geology and Glaciation

Among the many factors that may have contributed to the formation of late Tertiary glaciers are crustal displacements, changes in ocean currents, variations in atmospheric CO_2 and turbidity, and the formation of highlands (Flint 1957). Information regarding all these factors is given by marine geology and plate tectonics. Crustal displacements, for example, have occurred and it appears that the area of land in very high latitudes has increased. This may have favored cooling in Asia which may have affected global atmospheric circulation. Likewise, the continuing opening of the North Atlantic has surely influenced the pattern of ocean currents there and in the Arctic Ocean.

Volcanism has been unusually intense during late Tertiary time both at the trailing edges of plates and within the plates. Plate motion has been unusually intense for a much longer period. This requires the destruction of lithosphere to have been intense during that time and implies intense volcanism in island arcs. In sum, volcanism at the leading and trailing edges of plates has been relatively intense for all of the Tertiary and an additional increment of volcanism has occurred within the plates during late Tertiary time. Volcanoes are the source of the CO_2 and fine ash which may have affected glaciation.

Although not certain, it appears that the volume and depth of the oceans have increased significantly during the late Tertiary. This requires some related vertical motion of the continents in order to maintain the observed freeboard. The continuing erosion of the exposed continents also requires a continuing rise of the continents. Neither of these phenomena necessitates or even suggests an overcompensation whereby the continents rise to their present exceptional height and perhaps facilitate glaciation. However, the process and motion within the mantle, which elevate or accompany the elevation of continents, are obscure. Is the necessary influx of mantle hotter than the material it replaces? If so, the intensity of plate motion and volcanism in the late Tertiary may also be accompanied by a flux of unusually hot mantle under continents, with a resultant thermal expansion and an overcompensating rise. This mechanism is entirely conjectural, but it should not obscure the fact that large vertical motions occur under the oceans and are somehow balanced by continental motion of proportional scale.

The quantitative significance of these factors can be evaluated only by specialists in glacial geology and the history of the Pleistocene epoch. I have attempted here only to focus their attention on some promising aspects of glacial research which have recently been exposed by marine geology.

Summary. Marine geology guided by the paradigm of plate tectonics suggests that late Tertiary time has been marked by intense volcanism, continental drift, and vertical motion of continents and the sea floor. All these factors may have influenced the onset of glaciation.

REFERENCES

Barazangi, M., and J. Dorman, World seismicity maps compiled from ESSA, Coast & Geodetic Survey, Epicenter Data, 1961–1967, *Bull. Seism. Soc. Am., 59*, 369, 1969.

Dietz, R. S., Continent and ocean basin evolution by spreading of the sea floor, *Nature, 190*, 864, 1961.

Engel, A. E. J., Geologic evolution of North America, *Science, 140*, 143, 1963.

Engel, A. E. J., C. G. Engel, and R. G. Havens, Chemical characteristics of oceanic basalts and the upper mantle, *Bull. Geol. Soc. Am., 76*, 719, 1965.

Ewing, J., M. Ewing, T. Aitken, and W. J. Ludwig, North Pacific sediment layers measured by seismic profiling, *Geophys. Monograph 12*, edited by Leon Knopoff et al., p. 147, American Geophysical Union, Washington, D.C., 1968.

Flint, R. F., *Glacial and Pleistocene Geology*, 553 pp., Wiley, New York, 1957.

Gutenberg, B., and C. F. Richter, *Seismicity of the Earth and Associated Phenomena*, 310 pp., Princeton University Press, 1949.

Heezen, B. C., and M. Tharp, Physiographic diagram of the South Atlantic Ocean, *Geol. Soc. Am.*, New York, 1961.

Heezen, B. C., and M. Tharp, Physiographic diagram of the Indian Ocean, *Geol. Soc. Am.*, New York, 1964.

Heezen, B. C., and M. Tharp, Physiographic diagram of the North Atlantic Ocean, revised, *Geol. Soc. Am.*, New York, 1968.

Herron, E. M., and D. E. Hayes, A geophysical study of the Chile Ridge, *Earth Planetary Sci. Letters, 6*, 77, 1969.

Hess, H. H., History of Ocean Basins, in *Petrologic Studies*, 599 pp., edited by A. E. J. Engel et al., Geol. Soc. Am., New York, 1962.

Isacks, B., J. Oliver, and L. R. Sykes, Seismology and the new global tectonics, *J. Geophys. Res., 73*, 5855, 1968.

Isacks, B., L. R. Sykes, and J. Oliver, Focal mechanisms of deep and shallow earthquakes in the Tonga-Kermadec region and the tectonics of island arcs, *Bull. Geol. Soc. Am., 80*, 1443, 1969.

Kuenen, Ph. H., *Marine Geology*, 568 pp., Wiley, New York, 1950.

Langseth, M. G., Jr., X. LePichon, and M. Ewing, Crustal Structure of the mid-ocean ridges, 5. Heat flow through the Atlantic Ocean floor and convection currents, *J. Geophys. Res., 71*, 5321, 1966.

LePichon, X., Sea-floor spreading and continental drift, *J. Geophys. Res., 73*, 3661, 1968.

Luyendyk, B. P., Geophysical profiles over the northwestern African continental margin (abstract), *Am. Geophys. Union*, 1970.

McKenzie, D. P., Some remarks on heat flow and gravity anomalies, *J. Geophys. Res., 72,* 6261, 1967.

McKenzie, D. P., and R. L. Parker, The North Pacific: An example of tectonics on a sphere, *Nature, 216,* 1276, 1967.

Menard, H. W., *Marine Geology of the Pacific,* 271 pp., McGraw-Hill, New York, 1964.

Menard, H. W., Sea floor relief and mantle convection, in *Physics and Chemistry of the Earth,* edited by L. H. Ahrens et al., vol. 6, p. 315, Pergamon, New York, 1966.

Menard, H. W., Sea-floor spreading, topography and the second layer, *Science, 157,* 923, 1967.

Menard, H. W., Growth of drifting volcanoes, *J. Geophys. Res., 74,* 4827, 1969a.

Menard, H. W., Elevation and subsidence of oceanic crust, *Earth Planetary Sci. Letters,* 6, 257, 1969b.

Menard, H. W., and H. S. Ladd, Oceanic islands, seamounts, guyots and atolls, in *The Sea,* edited by M. N. Hill, vol. 3, p. 365, Interscience, New York, 1963.

Morgan, W. J., Rises, trenches, great faults and crustal blocks, *J. Geophys. Res., 73,* 1959, 1968.

Pitman, W. C., III, E. M. Herron, and J. R. Heirtzler, Magnetic anomalies in the Pacific and sea floor spreading, *J. Geophys. Res., 73,* 2069, 1968.

Revelle, R., On the history of the oceans, *Marine Res. (Sears Found. Marine Res.), 14,* 446, 1955.

Ringwood, A. E., and D. H. Green, An experimental investigation of the gabbro-eclogite transformation and some geophysical implications, *Tectonophysics, 3,* 383, 1966.

Rubey, W. W., Geologic history of sea water, *Bull. Geol. Soc. Am., 62,* 1111, 1951.

Sclater, J. G., and H. W. Menard, Topography and heat flow of the Fiji Plateau, *Nature, 216,* 991, 1967.

Shor, G. G., Jr., H. W. Menard, and R. W. Raitt, Structure of the Pacific Basin, in *The Sea,* vol. 4, in press.

Verhoogen, J., Volcanic heat, *Am. J. Sci., 244,* 745, 1946.

Vine, F. J., and D. H. Matthews, Magnetic anomalies over oceanic ridges, *Nature, 199,* 947, 1963.

Wilson, J. T., On the growth of continents in *The Papers and Proceedings of the Royal Society of Tasmania, 1950,* p. 85, Hobart, Tasmania, 1951.

Wilson, J. T., A new class of faults and their bearing on continental drift, *Nature, 207,* 343, 1965.

Paul E. Damon

2. THE RELATIONSHIP BETWEEN LATE CENOZOIC VOLCANISM AND TECTONISM AND OROGENIC-EPEIROGENIC PERIODICITY

First of all we have seen hardly anything of the earth's crust below a depth of 2 km; secondly, only one third of the globe is open to geological investigation —the rest is ocean; thirdly, a large portion of the continents is covered by shallow water or alluvial deposits, and of the remaining fraction only a very small portion is really well known.

L. U. De Sitter *in* Structural Geology, *1956, p. 483*

This digression from discussion of the familiar ocean basins to the mysterious continents may serve to emphasize that large elevated regions of the continents and the ocean basins may be produced by the same bulges of the mantle. The origin of rises may be determined by studying plateaus. Unfortunately we know even less about plateaus than about rises.

H. W. Menard *in* Marine Geology of the Pacific, *1964, p. 152*

These quotations from De Sitter in 1956 and Menard in 1964 serve to remind us of the giant steps—"the great leap forward"—taken by geology during the last decade or so. The once inaccessible ocean basins now seem familiar, and some of us must return to dry land once again and take a fresh look at the "mysterious" continents. Many questions remain to be answered, such as: Is orogeny periodic, episodic, or continuous? What is the relationship between continental volcanism and continental drift? What is the relationship between orogenic and epeirogenic movements on the continents and sea floor spreading? From my vantage point on the continental flank of the East Pacific rise here in Arizona, I would like to address myself to these questions, particularly, as they relate to the Late Cenozoic.

Is Continental Orogeny Periodic, Episodic, or Continuous?

On the subject of orogenic periodicity there are almost as many opinions as there are geologists. However, without attempting to list all plausible permutations and combinations, it will suffice to list only the point of view of certain major protagonists and their disciples.

1. During the Phanerozoic eon there were four major revolutions caused by periodic melting of the substratum. The inter-revolutionary periods were subject to minor disturbances resulting from partial melting and tidal creep of the substratum (Joly 1930).

2. Phanerozoic earth history consists of aperiodic, episodic, orogenic phases of very short duration (ca. 300,000 years) followed by anorogenic periods of very long duration (Stille 1924, 1940).

3. Superposed on a general tectonic unrest two major periodicities may be discerned—a longer periodicity of about 300 m.y. and a shorter periodicity of about 60 m.y. (Umbgrove 1947; Rutten 1949).

4. Phanerozoic earth history consists of essential uniformity in tectonism and volcanism with sporadic and not very great fluctuations in intensity. However, on a regional scale (e.g. western U.S.A.) plutonism may be catastrophic (Gilluly 1949, 1963).

It is safe to say that between them, Rutten and Gilluly demolished Stille's theory (item 3), which had reigned supreme for almost two decades. The existence of orogenic episodes of such short duration followed by long quiescent episodes is not a tenable point of view. When I arrived in Arizona in 1957, Gilluly's ideas (item 4) reigned supreme. The burden of proof fell on anyone who attempted to defend another proposition. Nevertheless, after seven years of geochronologic reseach in southwestern United States, I began, with the help of my students, to formulate an alternative point of view (Damon et al. 1964; Damon and Bikerman 1964; Damon and Mauger 1966).

The stimulus for the alternative point of view resulted from a statistical treatment of K-Ar dates for silicic volcanism and hypabyssal plutonism within the Basin and Range Province (Fig. 1). It can be seen that silicic magmatism during late Cretaceous and Cenozoic time, when plotted on a 5-m.y. class interval, resolves itself into two quasi-Gaussian distributions with a standard deviation of about 7.5 m.y. Furthermore, the two distributions peak almost exactly at the Cretaceous-Paleogene and Paleogene-Neogene boundaries. This result was not sought after, nor did it appear to be a mere coincidence, inasmuch as classical geologists had always sought to place boundaries at significant, observable "breaks" in the geologic record.

The suggested cause of these hypothetical discontinuities has been best expressed by Umbgrove (1947) in his classical treatise entitled *The Pulse of the Earth*. Briefly stated, diastrophism accompanied by silicic magmatism results in mountain building and a regression of the seas. Climatic deterioration thus ensues. Consequently, evolution is accelerated as a result of climatic stress and the

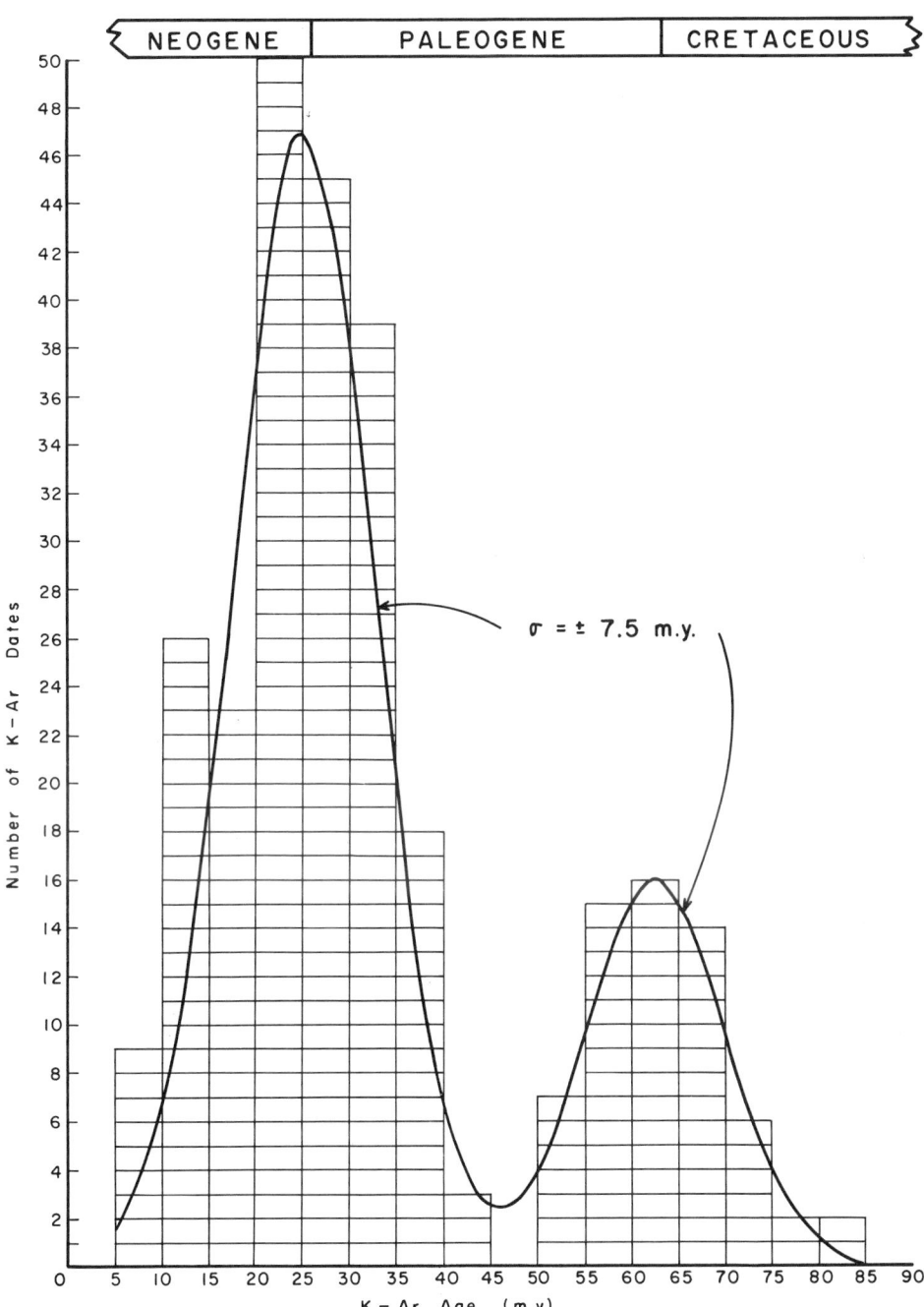

Fig. 1. Current version of histogram published previously by Damon and Mauger (1966) and Damon (1968b). Data are taken from Appendix I of Eastwood (1970) plus additional data from Armstrong (1970) and Damon et al. (1969).

destruction of floral and faunal habitats. Damon and Mauger (1966; see also Damon 1968a,b) suggested that the Laramide orogeny was a good example of Umbgrove's pulse of the earth. Recently, Kauffman and Kent (1968) have demonstrated a twofold increase in the rate of evolution of ammonites and other invertebrates during regressive, as compared to transgressive, stratigraphic sequences during the Cretaceous. Thus, even relatively minor fluctuations have measurable evolutionary effects.

At the time the original histogram was constructed, insufficient reliable data were available to extrapolate the K-Ar dating curve to earlier times and so, as an expedient, the area of transgression of the epicontinental seas onto the present outline of North America was used as an orogenic indicator. We obtained the areal data by planimeter measurements, using Schuchert's (1955) paleogeographic maps for want of a more up-to-date equivalent. The data for the Mesozoic and Cenozoic are presented graphically in Figure 2. Again, much to our surprise, each of the classical North American orogenies is represented by distinct regressions of the epicontinental seas. Subsequently, Evernden and Kistler (1970) have demonstrated the occurrence in California and western Nevada of five epochs of pre-Laramide, Mesozoic magma generation and emplacement (10–15 m.y. duration). Evernden and Kistler then correlated their magmatic episodes with the graph in Figure 2 and arrived at the following conclusion: "Our data require a refinement of this interpretation since the beginning of each Mesozoic plutonic episode is approximately coincident with a peak which marks the beginning of a temporary reversal of the general transgression curve" (p. 77 of preprint). These intrusive epochs and their ages are given in Table 1. A similar pattern has been observed for the Mesozoic batholithic rocks of British Columbia, Canada (Gabrielse and Reesor 1964). In order to explain the relationship in Figure 2, Grasty (1967) constructed a simple diastrophic model involving two "continents," one deformed and one undeformed. He concluded from this model (p. 5) that "an orogenic event is necessarily associated with a eustatic fall of sea level and that the depth of the epicontinental seas is a sensitive indicator of these events." It is now evident that the major features of Umbgrove's concept of the pulse of the earth are valid.

Encouraged by the success attained by the use of transgression of the epi-

Table 1. Intrusive Epochs in California and Western Nevada

(After Evernden and Kistler 1970)

M.Y.	Geologic age	Intrusive epoch
90–79	Late Cretaceous	Cathedral Range
117–104	Early Cretaceous	Huntington Lake
148–132	Late Jurassic	Yosemite
180–160	Early and middle Jurassic	Inyo Mountains
210–195	Middle and late Triassic	Lee Vining

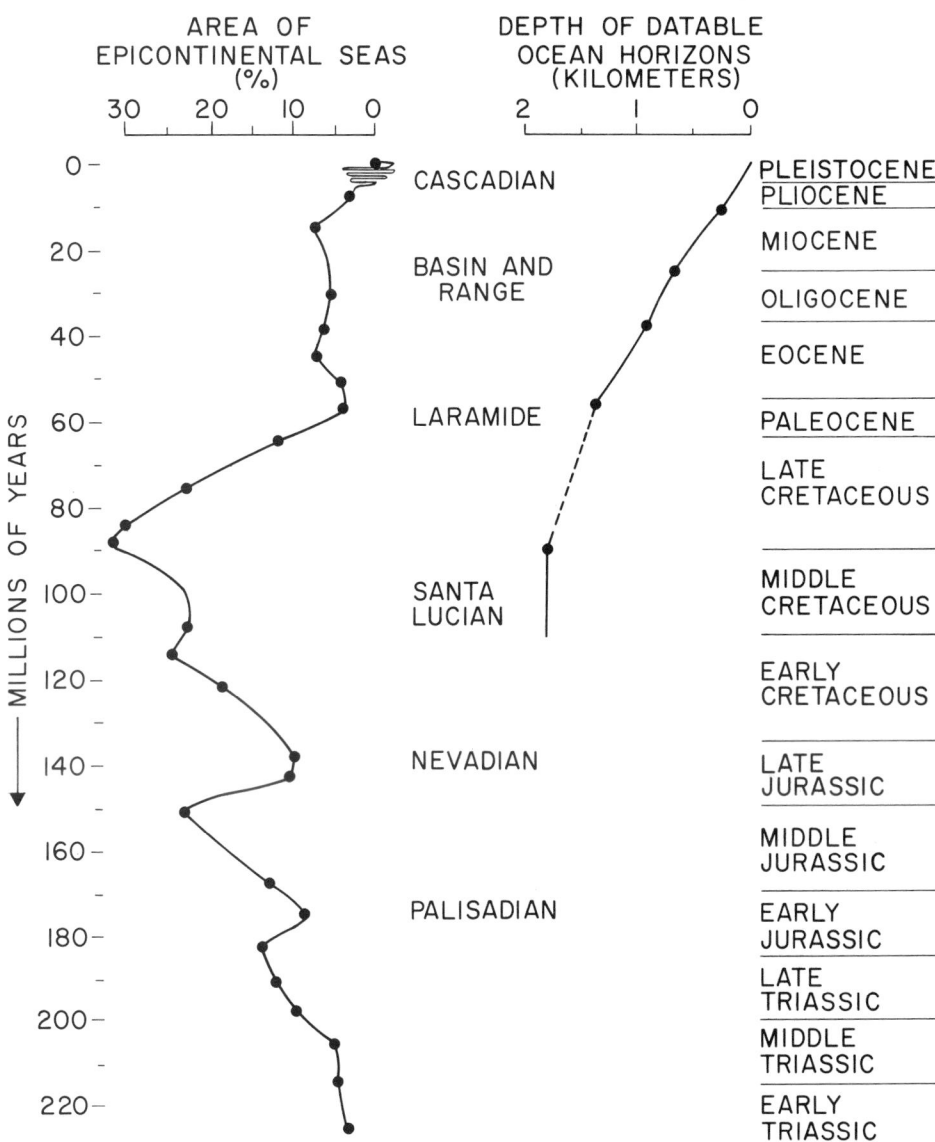

Fig. 2. Extent of the epicontinental seas and depth to datable ocean horizons. Area of epicontinental seas measured in units of percent of the present North American land area covered (including also Central America, northern Colombia, and Greenland, but excluding Hudson Bay). Depth of datable horizons data were from Menard and Ladd (1963); after Damon and Mauger (1966).

continental seas onto the present outline of North America as a measure of orogeny, I have made planimeter measurements for all of Schuchert's paleogeographic maps. The measurements were plotted as a histogram, with a 10-m.y. class interval to minimize correlation problems and overcome the sparsity of data which would result for class intervals of shorter duration (Fig. 3). In addition to the Mesozoic and Cenozoic orogenies previously noted (Fig. 2), the three classical Paleozoic North American orogenies (Taconic, Acadian, and Appalachian) show up as distinct regressions. Several other distinct regressions of nearly as great intensity are also indicated. These correspond to Caledonian and Hercynian movements in Europe.

In order to distinguish broad epeirogenic movements from the orogenic epochs, I passed a smooth curve through the transgression peaks and measured the extent of regression from the smoothed curve (Fig. 4). Stille's orogenies are

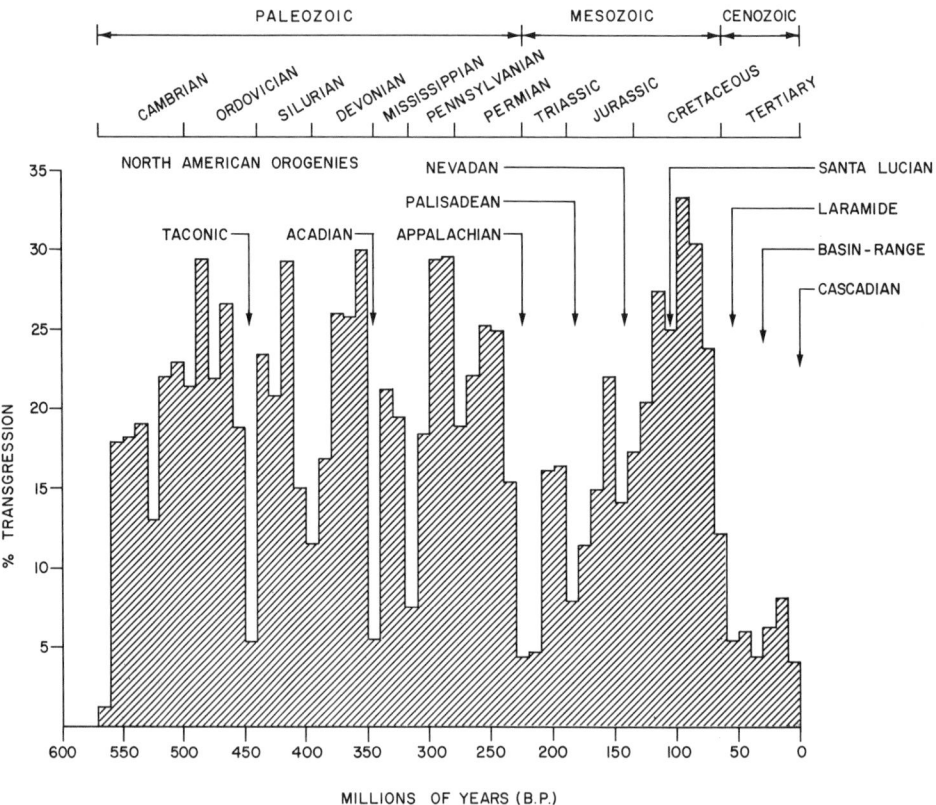

Fig. 3. Extent of epicontinental seas during the Phanerozoic as measured from Schuchert's (1955) paleographic maps. Area measured is the same as for Figure 2. Geological Society of London time scale (Harland et al. 1964) is used in this and following figures.

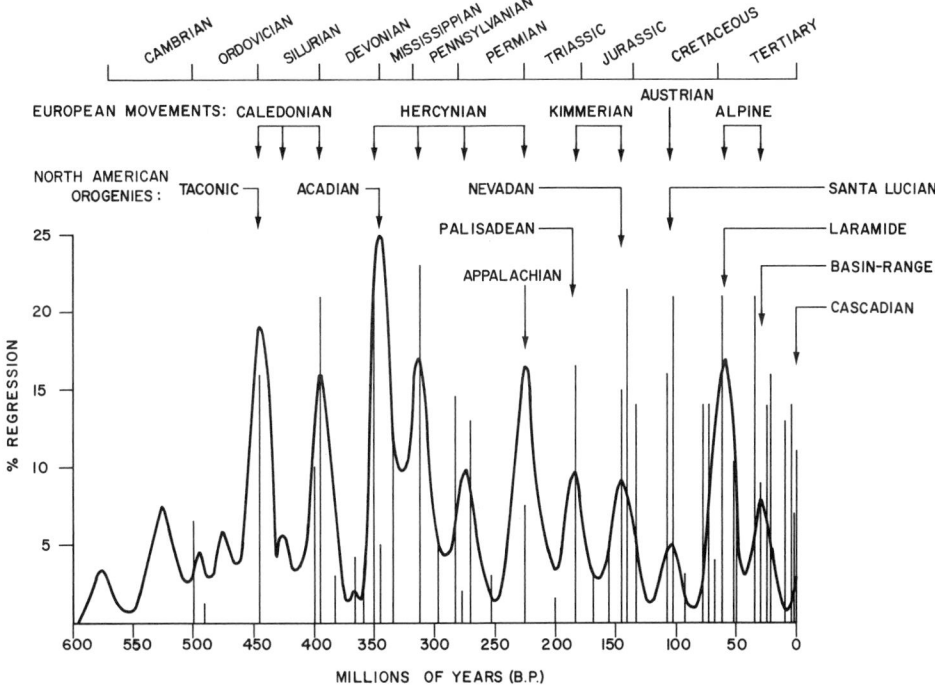

Fig. 4. Orogenic events measured as percent regression from a smooth curve passed through the peaks of marine transgression.

shown as straight vertical lines. Each of the main peaks can be correlated with classical major orogenies. Typically, several of Stille's orogenies fall under each peak but few occur during the quiescent periods. Apparently, most of Stille's orogenies are minor or regional episodes within more extensive events. The most pronounced peak corresponds to the Acadian orogeny. That this is no mere co-incidence is evident from the analysis of Lyons and Faul (1968), of northern Appalachian geology in which they conclude (p. 315) that "The overridingly important event was the Acadian Orogeny."

Interestingly enough, if the orogenic pulses are approximated by Gaussian curves, the standard deviation is about 9 m.y. which is close to the value of ±7.5 m.y. observed for Laramide and mid-Tertiary magmatism within the Basin and Range Province. Clearly, the orogenic pulses are of more than regional significance and occur periodically. However, their apparent intensities differ by an order of magnitude. The relationship between apparent intensity and actual intensity of orogeny is a function of epeiric sea and continental shelf bathymetry, a subject which I intend to treat quantitatively in a subsequent paper.

The ages of the boundaries between geologic periods and the ages of regressive minima are presented in Table 2. There appears to be a good correspondence in most cases. The boundary between the Triassic and Jurassic is a notable ex-

Table 2. Period Boundaries and Regression Minima

Boundary between periods	Age of boundary	Age of regression minima
Precambrian–Cambrian	570	570–580
		520–530
Cambrian–Ordovician	500	490–500
		470–480
Ordovician–Silurian	430–440	440–450
		420–430
Silurian–Devonian	395	390–400
Devonian–Mississippian	345	340–350
Mississippian–Pennsylvanian	320	310–320
Pennsylvanian–Permian	280	270–280
Permian–Triassic	225	220–230
Triassic–Jurassic	190–195	175–185
Jurassic–Cretaceous	136	135–145
		100–110
Cretaceous–Paleogene	65	50–60
Paleogene–Neogene	26	20–30

Table 3. Duration of Periods and Orogenic Cycles

	Duration	S.D.	Spread
Periods	49.5	15.5	25–71
Orogenic cycles	35.9	10.7	20–50

ception. Where minima do not correspond to geologic boundaries, e.g. within the Cambrian, Ordovician, and Cretaceous, the geologic periods are more than 60 m.y. in duration. The Silurian is an exception to this rule.

The durations of periods and orogenic cycles are compared in Table 3. The orogenic cycles are 36 ± 11 (s.d.) m.y. in duration. The geologic periods are 50 ± 16 (s.d.) m.y. long. The three long periods account for the greater duration of geologic periods as compared to the orogenic cycles. Very recently, Reso (1969, p. 187) has suggested that one of these periods, the Cretaceous, should properly be divided into two: "The paleontologic record indicates that the mid-Cretaceous biological changes are as great as those which distinguish most systematic boundaries. Thus the evidence suggests that according to the methods employed to establish systems, the Cretaceous time interval should be regarded as comprising two Periods." Reso places this break at the base of the Cenomanian, which closely corresponds to the Santa Lucian regressive minimum in Figure 2.

Thus, to summarize my answer to the question posed at the beginning of this chapter: (1) Orogenies are periodic and can be approximated by Gaussian distributions for the intensities of diastrophism and magmatism; (2) the standard deviation of an orogenic pulse is about 8 m.y.; (3) the period of cycles is 36 ± 11 m.y.; (4) the apparent intensities of orogenic pulses vary by an order of magnitude; and (5) the rules used to define geologic boundaries typically resulted in these boundaries being fixed near the peak of orogenic pulses.

EPEIROGENY VERSUS OROGENY

In order to remove the periodic orogenic oscillations, the data of Figure 3 were subjected to a standard harmonic analysis, dropping all but five terms to emphasize the epeirogenic oscillations of sea level (Fig. 5). The overall trend is a rise in sea level during Cambrian and Ordovician time followed by oscillation around a sea level which was about 20 percent greater than at present during the remainder of the Paleozoic. Thereafter, sea level falls to a minimum in early Mesozoic time followed by a rise until a maximum is reached during the Turonian stage of the Cretaceous. Following the Turonian maximum, sea level continues to fall until the present condition of extreme regression is attained. There is no evidence for an overall change in the above-water area of the North American continent during the Phanerozoic era.

If we accept the model of Mitchell and Reading (1969) in which continents oscillate between rises, an interesting explanation of the main features of post-

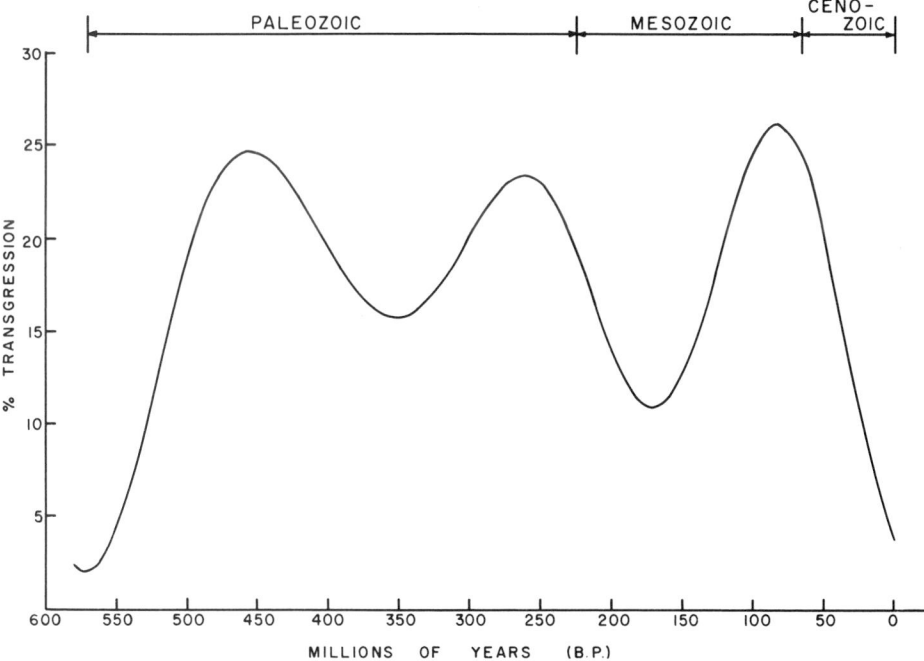

Fig. 5. Epeirogenic movements separated from orogenic movements by standard harmonic analysis, dropping all but five terms.

Paleozoic North American geology results. This is summarized by the cartoons in Figure 6. It is assumed in Figure 6 that the mid-Atlantic rise and East Pacific rise have been permanent features of the earth's crust (mega-undations: Van Bemmelen 1964) during Mesozoic time and have not moved relative to each other. It is also assumed that these mega-undations affect the warping of continents and sea floor in a similar way (see Damon and Mauger 1966, Fig. 5; or Damon 1968b, Fig. 22).

According to this model, during early Mesozoic time, eastern North America

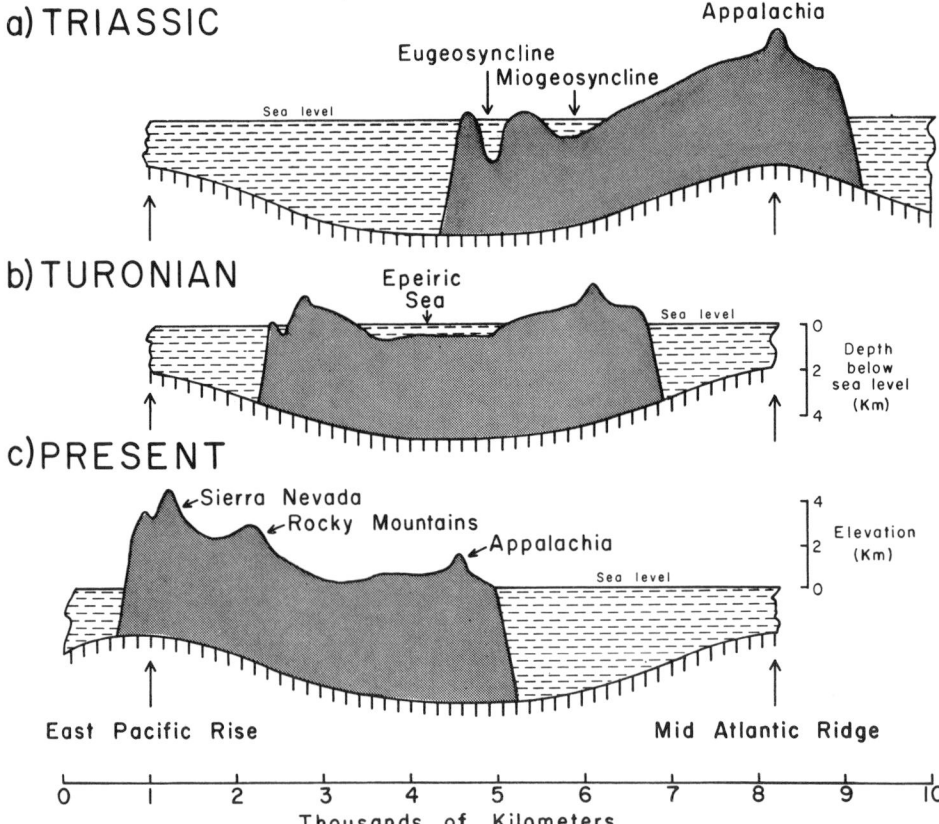

Fig. 6. The migration of North America from the mid-Atlantic rise to the East Pacific rise. The lower line represents the configuration of the sea floor which would be produced by the rises in the absence of a continent. The undulation resulting from causes in the mantle also affects the continents. Only that portion of the continent which rises above the sea floor configuration is shown. Eastern and western North America are 180° out of phase during the epeirogenic cycle. An intermediate position of maximum transgression was reached during the Turonian epoch when the North American continent was centered between the two rises.

had overridden the mid-Atlantic rise (Fig. 6a). Consequently, in accord with the geologic record, eastern North America was emergent beyond the present coastline and in many ways was similar to western North America today (Moore 1958). For example, Appalachia was a mountainous area with intermontane basins such as occur in Nevada and the Great Valley of California today, and products of erosion were carried far to the west. On the other hand, there was marine submergence of the Pacific border which extended eastward as far as Arizona and Idaho.

Following this, the North American continent moved westward down the slope of the mid-Atlantic mega-undation and the epeiric seas continued to transgress until Turonian time (Fig. 6b) when the North American continent was midway in its course between the two mega-undations. At this time, one-third of the present continent was covered by the epeiric seas of the Rocky Mountain's geosyncline (Figs. 3, 5).

Continued westward movement of North America up the slope of the mega-undation comprising the East Pacific rise resulted in the emergence of western North America with the continuous retreat of the epeiric seas except for temporary and minor transgressions following the Laramide and mid-Tertiary orogenies (Fig. 6c). The westward movement resulted in the destruction of an Atlantic-type geosyncline (nomenclature is that of Mitchell and Reading 1969) and the creation of an Andean-type geosyncline. The Andean type of geosyncline was then also destroyed as western North America overrode the crest of the East Pacific rise (Hamilton 1969). During this time, an Atlantic-type geosyncline continued to develop off the east coast of North America (Drake 1966; Mitchell and Reading 1969). Western North America and eastern North America are 180° out of phase during the oscillation. Western North America is now emergent, eastern North America relatively submerged. As a consequence, Cretaceous strandline sediments of Turonian age may be found at an elevation of several kilometers on the Colorado plateau but are buried beneath the sea off the Atlantic coast.

The average rate of westward movement required to complete the passage between the two extreme positions (Figs. 6a, c) is about 2 cm per year, which is equal to the average rate of sea floor spreading as discussed in the following section. Because the slope of the East Pacific rise is about 1 meter per kilometer, we might expect large areas of western North America, such as the Colorado plateau, to be elevated at an average rate of 2 cm per 1000 years, as suggested by Damon and Mauger (1966), while sea mounts sink at about the same rate (Fig. 2).

Thus, the epeirogenic movements of North America may be explained without apparent inconsistency as the result of the oscillation of the American block (see Fig. 6; LePichon 1968) between the two great mega-undations comprising the mid-Atlantic and East Pacific rises. Such epeirogenic cycles involve vast crustal displacements occurring over hundreds of millions of years, whereas orogenic movements (meso-undations according to the classification of Van Bemmelen 1964) are much more restricted in time and space.

Is Sea Floor Spreading Continuous or Discontinuous?

Three papers are most frequently cited for evidence supporting periodic or episodic sea floor spreading during Mesozoic and Cenozoic time (Ewing and Ewing 1967; Ewing et al. 1968; LePichon 1968). According to LePichon (p. 3693):

> Three main episodes of spreading are recognized—late Mesozoic, early Cenozoic, and late Cenozoic. The beginning of each cycle of spreading is marked by the reorganization of the global pattern of motion. A correlation is made between slowing of spreading at the ends of the two previous cycles and paroxysms of orogenic phases. This history of spreading follows closely one advocated by Ewing *et al.* (1968) to explain the sediment distribution.

However, the most recent results from the deep sea drilling project of the National Ocean Sediment Coring Program (Percival 1969) seem to indicate that sea floor spreading during the entire Cenozoic in the South Atlantic has been quasicontinuous at an average rate of 2 cm per year for movement on one limb (half-value).

Plio-Pleistocene spreading rate measurements in the South Atlantic vary from about 1.5 cm per year to 2.25 cm per year for the half-value, whereas measurements of the equivalent rate in the North Atlantic vary between about 1 cm and 1.5 cm per year (LePichon 1968). These rates are significantly less than the rate of movement (5.6 cm per year) deduced by Yeats (1968) for the eastward migration of the East Pacific rise relative to the North American continent during Miocene time. According to Yeats, the eastward movement terminated about 12 m.y. ago near the end of the Miocene epoch.

No doubt there will be more to be said on this subject in the near future. Perhaps sea floor spreading during the Cenozoic has been continuous but with at least minor changes in rate. However, Yeats's evidence for episodic westward drift of the North American continent relative to the East Pacific rise appears to be quite convincing.

Is There a Relationship between Continental Orogeny and Continental Drift?

LePichon (1968) and Dott (1969) have suggested a correlation between orogeny and major movements of lithospheric plates. As stated by Dott (p. 874):

> According to this "new global tectonics" we should expect closer worldwide relationships among continents and between the sea floors and continents than we had reason to suspect before (Morgan 1968; LePichon 1968). All movements of lithosphere plates should be so interrelated that "no mid-oceanic ridge can be understood independently of the others" (LePichon 1968). If true, then students of continental structure and history should be in a position to shed considerable light upon less-known sea floor history; well-known land areas should provide tests or refinements of postulated spreading histories.

With this perspective in mind, I intend now to suggest a relationship between the Basin and Range orogeny and the westward migration of North America relative to the East Pacific rise.

In Figure 7, K-Ar dates for andesitic to rhyolitic volcanic rocks and hypabyssal plutons from the Basin and Range Province are plotted on a 2-m.y. class interval. When more than one date is available for a correlatable rock unit, an average date is given which has been weighted according to the precision of individual analyses. Dates for Late Cenozoic basalts are plotted as a separate histogram in the insert on Figure 7. Several features of the histogram are of particular interest. (1) Magmatism of the Basin and Range orogeny follows a pronounced quiescent period and is terminated by a period of magmatic quiescence (see also Fig. 1). (2) There are distinct magmatic episodes within the orogeny, which are of at least regional extent, reminiscent of Stille's orogenies, but of longer duration than had been suggested by Stille (see Fig. 4). (3) Basaltic volcanism waxes as silicic volcanism wanes (field data convincingly support the radiometric data upon

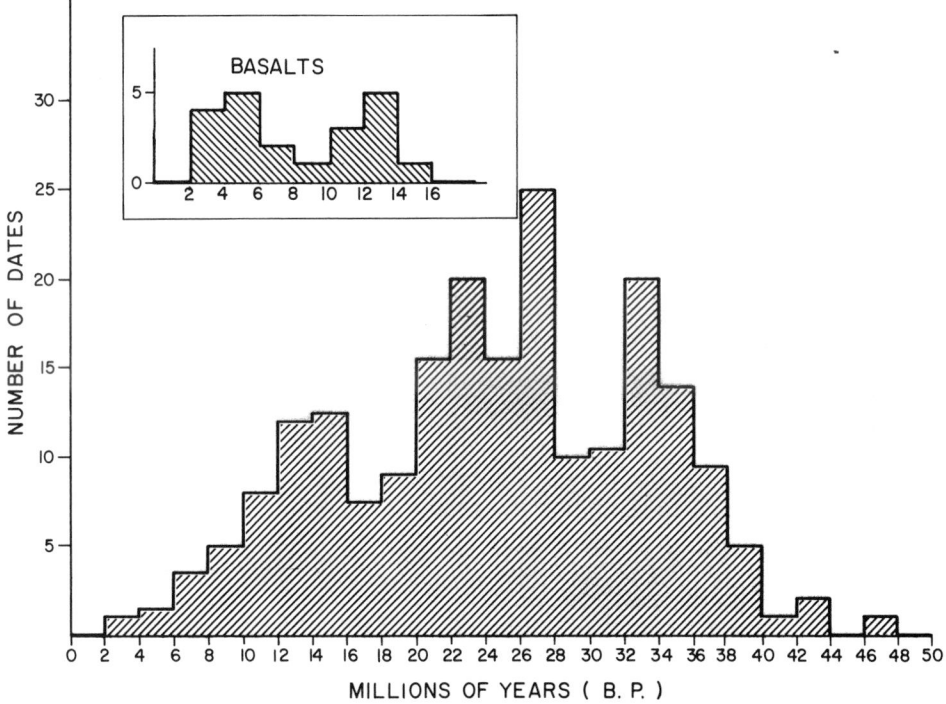

Fig. 7. Histogram for Basin and Range orogeny magmatism plotted on a 2-m.y. class interval. No attempt was made to collect data for the last 2 m.y., and so this interval should be ignored. Data are taken from Appendix I of Eastwood (1970) plus data from Ozima et al. (1967), Damon et al. (1969), and Armstrong (1970). Histogram in insert is for late basalts only. The primary histogram is for andesitic to rhyolitic volcanics and hypabyssal plutons.

which point 3 is based). (4) There is a sudden increase in basaltic volcanism when, according to Yeats (1968), the East Pacific rise terminates its eastward movement relative to the North American continent.

There is an interesting correlation between age of volcanism and initial Sr^{87}/Sr^{86} ratio. The andesitic through rhyolitic volcanic rocks which are older than 20 m.y., from southeastern Arizona, have initial ratios varying between 0.7067 and 0.7096, in contrast to late Miocene basalts which have initial ratios slightly less than 0.704 (Bikerman 1967; Percious 1968; Eastwood 1970). The very low initial ratios for the young basalts seem to indicate that very little contamination, if any, could have occurred during the transit of the basaltic magmas to the surface. Thus, the crust in late Miocene and Pliocene time constituted a relatively brittle, inert, and therefore presumably a relatively cool conduit system which allowed basaltic magmas to be transported through it without contamination or hybridization. This was not the case during the preceding more intense period of mid-Tertiary magmatism.

Figure 8 illustrates some major aspects of the present condition of the crust and mantle underlying western North America. The silicic upper crust is underlain by a more mafic lower crust, and below that is the anomalous upper mantle which is a region of relatively low seismic velocities. Basalt dikes are shown within the Basin and Range Province and on the margin of the Colorado plateau. As previously mentioned, these dikes contain relatively uncontaminated basalt with low Sr^{87}/Sr^{86} ratios which is believed to have been derived from the man-

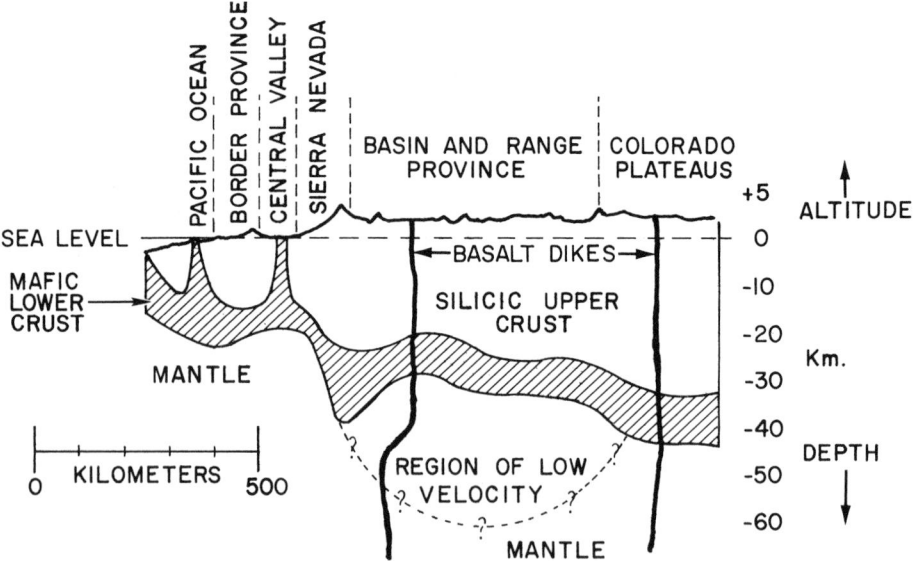

Fig. 8. Cross section through crust and upper mantle of the western states, based on papers by Pakiser (1963) and Pakiser and Zietz (1965).

tle (Eastwood 1970). The magnetic high under the central valley of California (Grantz and Zietz 1960; Zietz et al. 1969) is interpreted as a now solidified upwelling of mafic magma from the lower crust. It occurs within the general region of the easternmost migration of the East Pacific rise (Yeats 1968).

Figure 9 illustrates a hypothetical cross section through the crust and upper mantle at the beginning of the Neogene (Oligocene-Miocene boundary) during the peak of the Basin and Range orogeny. The crust and mantle were at least partly molten. The locus of the present anomalous mantle is interpreted as a crust–mantle mix (Cook 1965). The asthenosphere had migrated upward into the lithosphere (nomenclature is that of Isacks and Oliver 1968). Large parts of the Basin and Range Province were still elevated above the Colorado plateau so that drainage was to the east and northeast from the present Basin and Range Province onto the Colorado plateau (McKee et al. 1967). A Benioff zone plunged eastward from a trench off the west coast (Gilluly 1969; Hamilton 1969) feeding marine sediments and basalt into the western part of the region of crust–mantle mix. The initial Sr^{87}/Sr^{86} ratios of lavas extruded during that time are predictable from Armstrong's (1968) "model for the evolution of strontium and lead isotopes in a dynamic earth," which involves the processes implied in Figure 9.

What Is the Relationship between Rate of Epeirogenic Uplift and Drift?

If, as seems evident, the North American continent has migrated westward relative to the East Pacific rise and maintains its existence during the migrations as a crustal bulge originating in the mantle, then there should be a simple relationship between the rate of drift and the rate of uplift. As previously suggested, if the rate

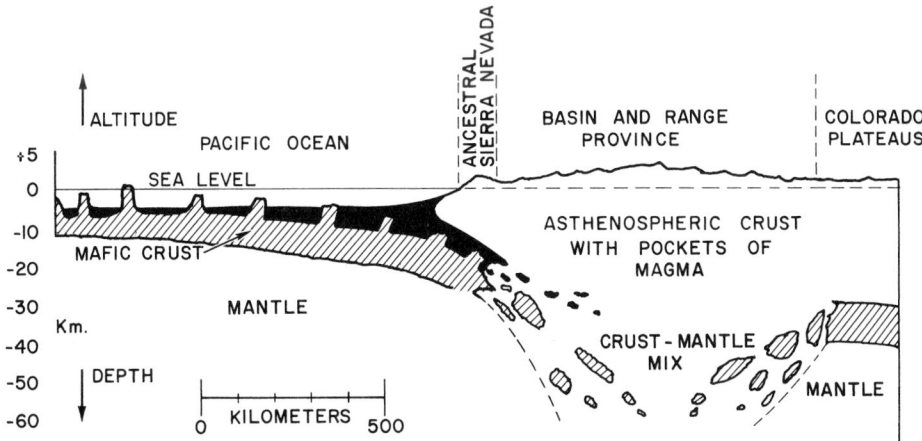

Fig. 9. Hypothetical cross section through the crust and upper mantle 26 m.y. ago at the beginning of the Neogene. A Benioff zone is supposed to feed marine sediments and volcanic rocks into what is now the low-velocity region. The interpretation of the low-velocity zone as the result of a crust–mantle mix follows a suggestion of Cook (1965).

of westward drift is 2 cm per year and the slope of the rise is 1 meter per kilometer, then the rate of uplift would be 2 cm per 1000 years.

However, the rate of westward drift should certainly depend on the state of the continental crust and upper mantle. A plastic asthenospheric crust would present less resistance to movement than a rigid lithospheric crust. If we assume that: (1) the rate of westward drift is proportional to the plasticity of the crust; (2) the rate of silicic magmatism above some cutoff point as given in Figure 7 is a measure of crustal plasticity; (3) the cutoff point for drift occurs at the level of silicic magmatism existing at about 10 m.y. ago, when, according to Yeats (1968), drift ceased and, as suggested in this paper, the continental crust became a relatively cool and brittle conduit system allowing basalt dikes to be transported through the crust without appreciable contamination; (4) the slope of the East Pacific rise is 1 meter per kilometer (1/1000); and (5) the average rate of drift above the cutoff point is 5.6 cm per year as shown by Yeats (1968); then, from this model, the graph of Figure 10 can be derived.

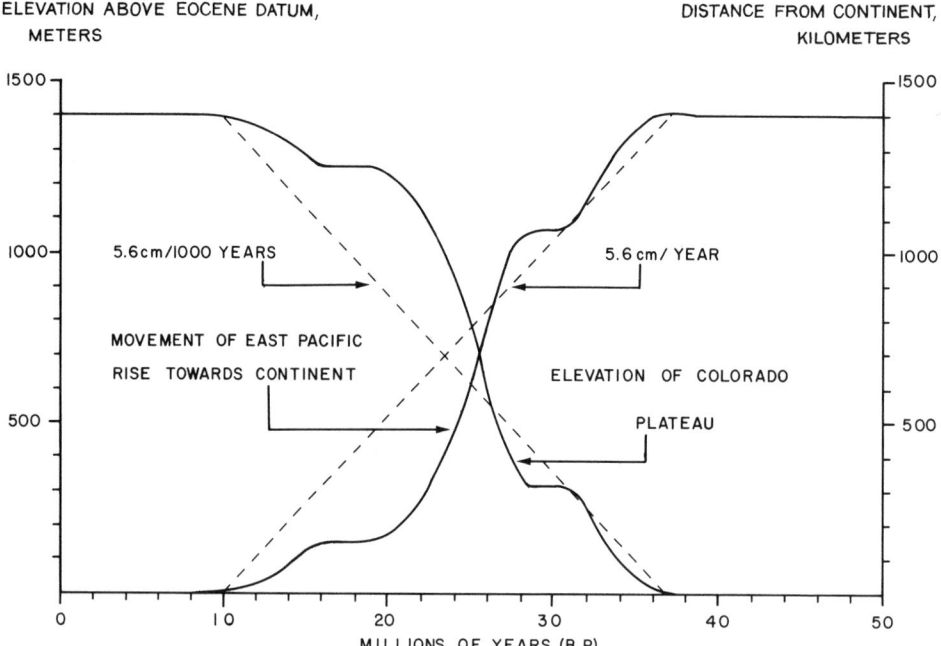

Fig. 10. Movement of the East Pacific rise relative to the continent and uplift of the Colorado plateau. Kilometers from continent refers to the distance of the center of spreading from the present latitude and longitude of San Diego, California. The elevation of the Colorado plateau is measured in meters above its altitude in late Eocene time when it was already well above sea level. The most rapid movement of the rise and elevation of the Colorado plateau occurred, according to the model described in the text, during early Neogene time.

During Eocene time, the center of spreading was about 1400 km east of the shoreline near San Diego, California. In early Oligocene time, westward migration of the continent relative to the center of spreading began and continued at a varying rate until late Miocene time when it reached its present site east of the Peninsular Ranges (Yeats 1968). The maximum rate of migration, 13 cm per year, occurred during early Miocene time. At that time, the Colorado plateau was rising at a rate of 13 cm per 1000 years. This rapid uplift of the Colorado plateau in early Miocene time appears to be in accord with the geologic record. According to Hunt (1956, p. 77):

> During early Miocene time, as block faulting progressed in the Basin and Range Province, the Colorado Plateau probably became well defined as a structural unit and reached an altitude distinctly higher than the basins, though perhaps not so high as the ranges. Epeirogenic uplift of the Plateau began at this stage, and as the Plateau broke away from and was raised higher than the basins to the west, south, and southeast, large areas on the Plateau must have begun draining to those basins. At this point, aggradation on the Plateau ended and degradation began.

Prior uplift of the Colorado plateau, which had been covered by epeiric seas during Turonian time, would also have occurred, according to this model, in Laramide time. If we assume that magmatism was as intense during Laramide time as it was in mid-Tertiary time, but that erosion has removed most of the volcanic rocks of that age, then the average rate of uplift of the Colorado plateau was about 3 cm per 1000 years. The rate of uplift calculated by Damon and Mauger (1966) should be less because of erosion—hence their figure of 2 cm per 1000 years for the average rate of uplift.

This model suggests the following chronology of geologic events affecting the Colorado plateau subsequent to the maximum extension of the epeiric seas in Turonian time. Epeirogenic uplift of the plateau began in Santonian time (about 78 m.y. ago) and reached a maximum rate of uplift at the beginning of the Cenozoic era. Epeirogenic uplift ceased in middle Eocene time (about 48 m.y. ago) at which time the average elevation of the plateau was roughly 1.4 km. Erosional degradation occurred during late Eocene time, reducing the Colorado plateau to an average elevation of about 1 km. Epeirogenic uplift followed during Oligocene and Miocene time when the plateau reached a maximum average elevation of 2.4 km in late Miocene time and has been subjected to erosional degradation since then. This, I believe, is in reasonable accord with the geologic record of the Colorado plateau as described by Hunt (1956).

CONCLUSIONS

Proof of the power of a scientific theory, such as the atomic theory, is its effectiveness in helping the scientist correlate vast amounts of data and its potency in fathering new concepts, new relationships, and new tests of its validity. Certainly, the new global tectonics has proven its potency. So many ideas come to mind stimulated by this rapidly developing theory that one scarcely has the time or energy to explore them fully. Fortunately for the reader, most of the ideas which now come to this author's mind are beyond the scope of the paper and so I will conclude by briefly summarizing a few conclusions which must, as always in scientific work, remain tentative.

1. Orogenies are periodic and can be approximated by Gaussian distributions for the apparent intensities of diastrophism and magmatism. The standard deviation of an orogenic pulse is about 8 m.y. The period of cycles is 36 ± 11 m.y. The intensities of orogenic pulses vary by an order of magnitude.

2. The rules used to define geologic boundaries typically resulted in the boundaries being fixed near the peak of orogenic pulses. Certain long geologic periods, such as the Cretaceous, could appropriately be divided into two geologic periods.

3. The model of Mitchell and Reading (1969), in which continents oscillate between rises, provides an effective explanation of the main features of post-Paleozoic North American geology.

4. Epeirogenic uplift of continents is also a periodic phenomenon related to continental orogeny. During orogenic pulses the asthenosphere migrates upward into the lithosphere, and the resulting plastic crust allows continental drift to proceed more rapidly.

5. The rates of sea floor spreading and continental drift are measured in centimeters per year. Vertical epeirogenic movements proceed at centimeters per thousand years. The rate of vertical epeirogenic movement is determined by the rate of sea floor spreading and the slope of rises (approximately 1 meter per kilometer).

6. The late Cenozoic tectonic state of western North America is not typical of past states. It is at one extreme of an epeirogenic cycle during which it overrode the East Pacific rise. Consequently, it has been epeirogenically upwarped to an unusually high average altitude. Furthermore, the western states are entering a new period of orogeny, the Cascadian orogeny which will become more intense and spread eastward.

Summary. A statistical treatment of late Mezozoic and Cenozoic magmatic events coupled with analysis of paleogeographic data is used to demonstrate that orogenesis is periodic with a wavelength of 36 ± 11 m.y. The intensities of diastrophism and magmatism can be represented by a Gaussian distribution with a standard deviation of ± 8 m.y. Anorogenic periods are not of very long duration relative to orogenic periods, as suggested by Stille (1924, 1940). The intensities of orogenic pulses vary by an order of magnitude. The Acadian orogeny was the most intense event occurring during Phanerozoic time, as suggested by Lyons and Faul (1968).

Epeirogenic uplift of continents is also a periodic phenomenon related to continental orogeny. During orogenic pulses, the athenosphere migrates upward into the lithosphere, and the resulting plastic crust allows continental drift to proceed more rapidly. The rate of vertical epeirogenic displacement is determined by the rate of continental drift and the slope of rises ($\sim$1 meter per kilometer). Consequently, it may be measured in centimeters per thousand years as compared to centimeters per year for continental drift.

The Late Cenozoic tectonic state of western North America is not typical of its past history. It is at one extreme of an epeirogenic cycle and at an unusually high average altitude. Furthermore, the Cascadian orogeny is only in its beginning stages and, if the past is an adequate key to the future, it should become more intense and spread eastward.

Acknowledgments. It is a pleasure to thank Mr. William K. Smith and Professor John R. Sturgul for formulating the computer program used to obtain the graph in Figure 5. I am also grateful to Professor Donald E. Livingston for editing the manuscript.

This work was supported by U.S. Atomic Energy Commission contract AT(11-1)-689, and the State of Arizona.

References

Armstrong, R. L., A model for the evolution of strontium and lead isotopes in a dynamic earth, *Rev. Geophys.*, 6, 175, 1968.

Armstrong, R. L., Geochronology of the Tertiary igneous rocks, eastern Basin and Range Province, western Utah, eastern Nevada, and vicinity, U.S.A., *Geochim. Cosmochim. Acta*, 1970, in press.

Bikerman, M., Isotopic studies in the Roskruge Mountains, Pima County, Arizona, *Bull. Geol. Soc. Am.*, 78, 1029, 1967.

Cook, K. L., Rift system in the Basin and Range Province, in *The World Rift System,* edited by T. N. Irvine, p. 246, Geological Society of Canada Paper 66, Queen's Printer, Ottawa, Canada, 1965.

Damon, P. E., The relationship between terrestrial factors and climate, *Meteorol. Monographs, 8,* 106, 1968a.

Damon, P. E., Potassium-argon dating of igneous and metamorphic rocks with applications to the Basin ranges of Arizona and Sonora, in *Radiometric Dating for Geologists,* edited by E. I. Hamilton, and R. M. Farquhar, p. 1, Interscience, New York, 1968b.

Damon, P. E., and M. Bikerman, Potassium-argon dating of post-Laramide plutonic and volcanic rocks within the Basin and Range Province of Southeastern Arizona and adjacent areas, *Ariz. Geol. Soc. Dig., 7,* 63, 1964.

Damon, P. E., and R. L. Mauger, Epeirogeny-orogeny viewed from the Basin and Range Province, *Trans. AIME, 235,* 99, 1966.

Damon, P. E., R. L. Mauger, and M. Bikerman, K-Ar dating of Laramide plutonic and volcanic rocks within the Basin and Range Province of Arizona and Sonora, Proceedings of the 22d International Geological Society, entitled *Cretaceous-Tertiary Boundary Including Volcanic Activity,* pt. 3, p. 45, New Delhi, India, 1964.

Damon, P. E., and Associates, Correlation and Chronology of Ore Deposits and Volcanic Rocks, Annual Progress Report No. COO-689-120 to *Research Division, U.S. Atomic Energy Commission,* 1969.

De Sitter, L. U., *Structural Geology,* 552 pp., McGraw-Hill, New York, 1956.

Dott, R. H., Jr., Circum-Pacific late Cenozoic structural rejuvenation: Implications for sea floor spreading, *Science, 166,* 874, 1969.

Drake, C. L., Recent investigations on the continental margin of eastern United States, in *Continental Margins and Island Arc.* edited by W. H. Poole, p. 33, Geological Survey of Canada Paper 66-15, Queen's Printer, Ottawa, Canada, 1966.

Eastwood, R. L., A geochemical-petrological study of mid-Tertiary volcanism in parts of Pima and Pinal Counties, Arizona, Ph.D. dissertation, University of Arizona, Tucson, Ariz. 1970.

Evernden, J. F., and R. W. Kistler, Chronology of emplacement of Mesozoic batholithic complexes in California and western Nevada, 91 pp., preprint of *U.S., Geol. Surv., Profess. Papers, 623,* 1970.

Ewing, J., and M. Ewing, Sediment distribution on the mid-ocean ridges with respect to spreading of the sea floor, *Science, 156,* 1590, 1967.

Ewing, M., J. I. Ewing, R. Leydon, and T. Aitken, Seismic reflection profiler results in the Pacific, in *The Crust and Upper Mantle of the Pacific Area, Geophys. Monograph 12,* American Geophysical Union, Washington, D.C., 1968.

Gabrielse, H., and J. E. Reesor, Geochronology of plutonic rocks in two areas of the Canadian Cordillera, in *Geochronology in Canada,* edited by F. Fitz Osborne, p. 96, University of Toronto Press, Canada, 1964.

Gilluly, J., Distribution of mountain building in geologic time, *Bull. Geol. Soc. Am., 60,* 561, 1949.

Gilluly, J., The tectonic evolution of the western United States, *Quart. J. Geol. Soc. London, 119,* 133, 1963.

Gilluly, J., Oceanic sediment volumes and continental drift, *Science, 166,* 992, 1969.

Grantz, A., and I. Zietz, Possible significance of broad magnetic highs over belts of moderately deformed sedimentary rocks in Alaska and California, in Geological Survey Research 1960, *U.S., Geol. Surv., Profess. Papers, 400–B,* p. B342, 1960.

Grasty, R. L., Orogeny, a cause of world-wide regression of the seas, *Nature, 216,* 779, 1967.

Hamilton, W., Mesozoic California and underflow of Pacific mantle, *Bull. Geol. Soc. Am., 80,* 2409, 1969.

Harland, W. B., et al., *The Phanerozoic Time Scale,* edited by W. B. Harland, A. Gilbert Smith, and B. Wilcock, 458 pp., *Quart. J. Geol. Soc. London,* Suppl., *1205,* Burlington House, London, 1964.

Hunt, C. B., Cenozoic geology of the Colorado Plateau, *U.S., Geol. Surv., Profess. Papers, 279,* 99 pp., U.S. Government Printing Office, Washington, D.C., 1956.

Isacks, B., and J. Oliver, Seismology and the new global tectonics, *J. Geophys. Res., 73,* 5855, 1968.

Joly, J., *The Surface-History of the Earth,* 211 pp., Oxford at the Clarendon Press, England, 1930.

Kauffman, E. G., and H. C. Kent, Cretaceous biostratigraphy of western interior United States, paper presented at the annual meeting of the Geological Society of America, Mexico City, Mexico, November 1968.

LePichon, X., Sea-floor spreading and continental drift, *J. Geophys. Res., 73,* 3661, 1968.

Lyons, J. B., and H. Faul, Isotope geochronology of the northern Appalachians, in *Studies of Appalachian Geology,* edited by E. Zen, W. S. White, J. B. Hadley, and J. B. Thompson, Jr., p. 305, Interscience, Wiley, New York, 1968.

McKee, E. D., et al., *Evolution of the Colorado River,* edited by E. D. McKee, R. F. Wilson, W. J. Breed, and C. S. Breed, 67 pp., Museum of Northern Arizona, Flagstaff, Ariz., 1967.

Menard, H. W., *Marine Geology of the Pacific*, 271 pp., McGraw-Hill, New York, 1964.

Menard, H. W., and H. S. Ladd, Oceanic islands, seamounts, guyots, and atolls, in *The Sea*, p. 365, Interscience, New York, 1963.

Mitchell, A. H., and H. C. Reading, Continental margins, geosynclines, and ocean floor spreading, *J. Geol.*, 77, 629, 1969.

Moore, Raymond C., *Introduction to Historical Geology*, 656 pp., McGraw-Hill, New York, 1958.

Morgan, W. J., Rises, trenches, great faults, and crustal blocks, *J. Geophys. Res.*, 73, 1959, 1968.

Ozima, M., M. Kono, I. Kaneoka, H. Kinoshita, K. Kobayashi, T. Nagata, E. E. Larson, and D. Strangway, Paleomagnetism and potassium-argon ages of some volcanic rocks from the Rio Grande Gorge, New Mexico, *J. Geophys. Res.*, 72, 2615, 1967.

Pakiser, L. C., Structure of the crust and upper mantle in the western United States, *J. Geophys. Res.*, 68, 5747, 1963.

Pakiser, L. C., and I. Zietz, Transcontinental crustal and upper-mantle structure, *Rev. Geophys.*, 3, 505, 1965.

Percious, J. K., Geochemical investigation of the Del Bac Hills volcanics, Pima County, Arizona, M.Sc. thesis, University of Arizona, Tucson, Ariz., 1968.

Percival, S. F., Jr., Paleontologic evidence for sea floor spreading in the south Atlantic-Leg III (Dakar-Rio de Janeiro) of the deep sea drilling project, paper presented at the annual meeting of the Geological Society of America, Atlantic City, N.J., 10–12 November 1969.

Reso, A., The mid-Cretaceous paleontological break, paper presented at the annual meeting of the Geological Society of America, Atlantic City, N.J., 10–12 November 1969.

Rutten, L. M. R., Frequency and periodicity of orogenetic movements, *Bull. Geol. Soc. Am.*, 60, 1755, 1949.

Schuchert, C., *Atlas of Paleogeographic Maps of North America*, 177 pp., Wiley, New Jersey, 1955.

Stille, H., *Grundfragen der vergleichenden Tektonik*, 443 pp., Gebruder Borntraeger, Berlin, 1924.

Stille, H., *Einfuhrung in den Bau Amerikas*, 717 pp., Gebruder Borntraeger, Berlin, 1940.

Umbgrove, J. H. F., *The Pulse of the Earth*, 358 pp., Martinus Nijhoff, The Hague, Holland, 1947.

Van Bemmelen, R. W., The evolution of the Atlantic mega-undation (causing the American continental drift), *Tectonophysics*, 1, 385, 1964.

Yeats, R. S., Southern California structure, seafloor spreading, and history of the Pacific Basin, *Bull. Geol. Soc. Am.*, 79, 1693, 1968.

Zietz, I., P. C. Bateman, J. E. Case, M. D. Crittenden, Jr., A. Griscom, E. R. King, R. J. Roberts, and G. R. Lorentzen, Aeromagnetic investigation of crustal structure for a strip across the western United States, *Bull. Geol. Soc. Am.*, 80, 1703, 1969.

W. Dansgaard, S. J. Johnsen, H. B. Clausen,
and C. C. Langway, Jr.

3. CLIMATIC RECORD REVEALED BY THE CAMP CENTURY ICE CORE

In recent years it has become obvious that the deposits on the two large ice sheets in Antarctica and Greenland contain paleoclimatic information in the form of varying isotopic composition of the ice (Dansgaard 1954; Epstein and Sharp 1959; Dansgaard et al. 1969). The index of concentration, δ, of oxygen-18 and deuterium (Craig 1961) in high-latitude precipitation is mainly determined by its temperature of formation (Dansgaard 1954, 1964; Picciotto et al. 1960). This causes seasonal variations of δ in accumulated snow and ice (Epstein and Sharp 1959; Dansgaard et al. 1960; Gonfiantini et al. 1963) as well as long-term variations due to climatic changes (Dansgaard et al. 1969). Increasing δ's upward in the core indicate warming climate, and decreasing δ's indicate cooling climate in the course of time. Since other parameters influence the isotopic composition, one must bear in mind that a vertical δ profile in a high polar glacier should not be strictly interpreted as a paleotemperature record (Dansgaard et al. 1969). This reservation is particularly relevant in the case of temperate glaciers, where refreezing meltwater and isotopic exchange between water and snow complicate the situation (Arnason 1969).

Compared with the paleoclimatic isotope studies on calcareous species in deep-sea cores (Emiliani 1966), O^{18} investigations of high polar ice cores involve the advantage that one can, practically speaking, go as far in details as desirable, the only limitation being obliteration of isotopic gradients by diffusion in solid ice, which is a very slow process (Johnsen, in preparation). Furthermore, ice cores from high polar regions offer a continuous sedimentary sequence, which is not necessarily true for deep-sea cores. However, none of the radioactive ice-dating methods—C^{14} (Scholander et al. 1962; Oeschger et al. 1967), Si^{32} (Dansgaard et al. 1966; Clausen et al. 1967), Pb^{210} (Goldberg 1963; Crozaz and Langway 1966), tritium (Aegerter et al. 1969; Ambach et al. 1969)—can be applied on the older parts of deep ice cores, if only for the reason that the half-lives of the radioactive elements are too short.

Hitherto the most extensive study of an ice core (Dansgaard et al. 1969) was carried out on that recovered in 1966 from Camp Century on the North Greenland ice sheet, 225 km east of Thule (Hansen et al. 1966). The 1390-meter ice core reaches from the surface to bedrock. Some 7500 core increments were cut in a continuous sequence and analyzed for O^{18}. The main series of samples described in this paper were cut to represent 10 to 20 years per sample in the time range 0 to 800 years B.P., 50 years between 1000 and 12,000 B.P., and 200 years between 12,000 and 157,000 B.P., to judge from the time scale evaluated in Dansgaard and Johnsen (1969) and outlined in the next section.

I. DATING THE ICE CORE

To interpret the climatic history revealed by the δ measurements it is necessary to establish a time scale for the core. In principle it is possible to use the seasonal δ variation measured on small (10 cm³) samples of the core and count the summer maxima and winter minima downward from the surface to obtain, as in tree rings, an absolute chronology. However, not only would this procedure require an enormous number of measurements, it also would not take us very far back in time because molecular diffusion in the solid ice gradually obliterates the short-term oscillations that remain after firnification (Johnsen, in preparation). Consequently, the most obvious possibility is to calculate the age–depth relationships of the ice core by developing a physical model that incorporates (1) the generally accepted glacier flow theory (as suggested by the thin curves in Fig. 1) and, (2) reasonable assumptions concerning the parameters that influence it. As shown in section II the δ measurements enable us to estimate the validity of the assumptions and even indicate corrections of the resulting time scale.

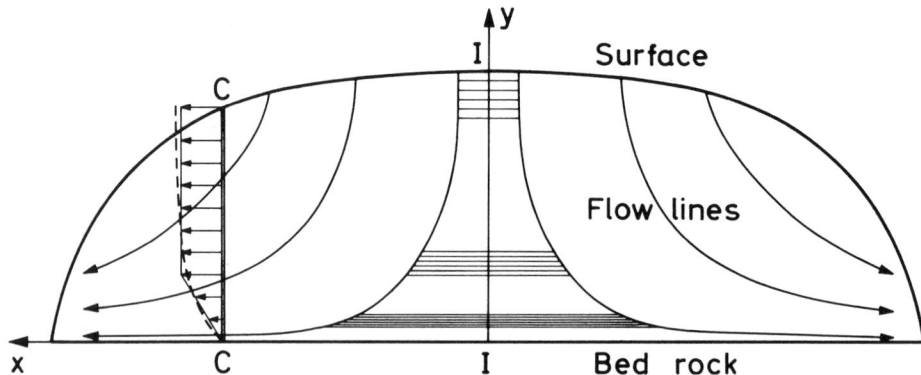

Fig. 1. Vertical cross section of an ice sheet resting on a horizontal subsurface. Ice particles deposited on the snow surface will follow lines that travel closer to the base the farther inland the site of deposition. An ice mass formed around the divide (I–I) will be plastically deformed (thinned) with depth as suggested by the lined areas. The dashed curve along the vertical ice core (C–C) shows the calculated horizontal velocity profile V_x (Weertman 1968b). The horizontal arrows along C–C show the adopted approximation to V_x (Dansgaard et al. 1969).

Figure 1 shows a vertical section in an ice sheet resting on horizontal bedrock. The ice divide is denoted by I–I. The amount of ice deposited at the surface around location I is buried by succeeding snowfalls and sinks into the ice sheet; at the same time the annual accumulation layers become thinner by plastic deformation. The core from Camp Century (C–C) contains ice formed between C and I under environmental conditions essentially similar to those at C. The surface velocity at C is 3.3 m per year (Mock 1968), and therefore even a 15,000-year-old (deep) section of the ice core originated less than 50 km farther inland.

We shall first consider the classical Nye model (1951, 1957, 1959), which assumes that the horizontal velocity component V_x is uniform along the core. This concept implies sliding on the bottom. For an ice sheet in mass balance the continuity equation is

$$\frac{\lambda_H}{\tau} \cdot x = V_X \cdot H$$

λ_H being the net accumulation in meters of ice, τ the period (1 year), and H the thickness of the ice sheet. The equation of incompressibility gives

$$\frac{\partial V_x}{\partial x} + \frac{\partial V_y}{\partial y} = 0$$

If λ is the thickness of a deformed annual layer deposited t years ago, we have $\lambda = -V_y \cdot \tau$, and

$$\frac{\partial V_x}{\partial x} = -\frac{\partial V_y}{\partial y} = \frac{\lambda_H}{H\tau}$$

As $V_y = 0$ for $y = 0$, we get

$$V_y = \int_0^{-\lambda} \frac{-\lambda}{\tau} dV_y = -\frac{\lambda_H}{H\tau} \int_0^y dy \qquad \lambda = \frac{\lambda_H}{H} y$$

Hence,

$$t = \int_H^y \frac{dy}{V_y} = \frac{H\tau}{\lambda_H} \ln \frac{H}{y}$$

This is a time scale for the core which, in fact, is applicable in the upper layers of the ice sheet. We have only to introduce $\lambda_H = 0.35 \pm 0.04$ meter of ice, which is the mean value for Camp Century over the last 100 years (Crozaz and Langway 1966). However, the present temperature at the bedrock interface is $-13°C$ (Hanson and Langway 1966). The ice is therefore frozen to the bottom, and V_x must be zero for $y = 0$, invalidating Nye's model at great depths.

In the model described below, we use a more realistic horizontal velocity profile by integrating Glen's law:

$$\dot{\epsilon} = k\sigma^n \qquad V_x = \int_0^y k\sigma^n \, dy$$

$\dot{\epsilon}$ and σ being the shear strain rate and the shear stress, and k and n being experimentally determined constants that depend on the temperature. Using the measured temperature profile along the Camp Century core, one gets the result shown by the dashed curve along C–C in Figure 1 (Weertman 1968a). Applying the approximation shown by the full curve along C–C (V_x uniform down to $y = h = 400$ m) and following a similar procedure as with the Nye model, we end up with an age-versus-depth relationship of

$$
t = \begin{cases} \dfrac{(2H-h)\tau}{2\lambda_H} \ln \dfrac{2H-h}{2y-h} & h \leqslant y \leqslant H \\[2em] t_h + \dfrac{(2H-h)\tau}{\lambda_H} \left(\dfrac{h}{y} - 1 \right) & 0 < y \leqslant h \end{cases} \tag{1}
$$

shown nomographically in Figure 2. The equations have been used with $H = 1370$ instead of 1390-meter ice thickness at Camp Century to account for the lower densities in the upper firn layers, and with $\lambda_H = 0.35$ meter of ice, which is the mean for the last 100 years. Indeed, one century is only a tiny fraction of the time range of interest here, and the λ_H value mentioned need not be representative. In fact, λ_H is a function of t, and it is impossible to find the relationship between these two parameters if only for the reason that a given annual layer has undergone plastic thinning, which depends upon the temperatures it has been exposed to since deposition. Most probably, λ_H was considerably lower than 0.35 meter during the glaciation, but on the other hand this is more or less counteracted by the lower strain rates and higher viscosity of the ice which slows the plastic thinning of the layers. Superimposed upon the temporal variations of λ_H there is a time-dependent geographical variation, which influences the stratigraphy in the ice core because the various increments were not originally deposited on the same location. Consequently, there are many reasons for considering the proposed time scale as only a first approximation to absolute chronology.

II. The Climatic Record

In a preliminary study (Dansgaard et al. 1969) the δ's of the ice core increments plotted against the t values calculated from equation 1 showed a fairly good agreement with the main features of climatic records known from other studies (pollen, ice sheet retreats and advances, deep-sea cores) and dated by C^{14}, Pa^{231} and Th^{230}. When the continuous δ record was extended to span the entire glaciation (Dansgaard et al., in press), it became obvious that equation 1 led to a better approximation to absolute chronology when used with a λ_H lower than 35 cm.

In this work we shall go a step farther by considering individually the three core intervals (meters below surface):

1. 0–283
2. 0–1150
3. 0–1373

In order to obtain the best possible approximation to absolute chronology, we

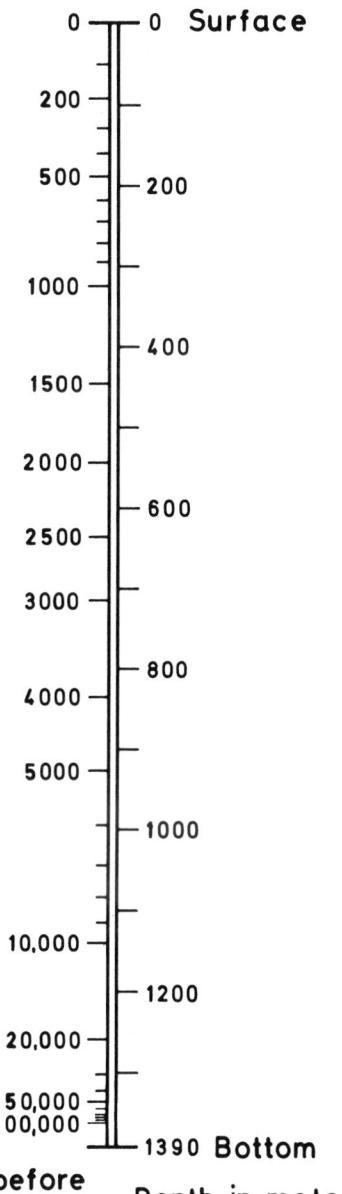

Years before
present

Depth in meters

Fig. 2. A depth–age nomograph for the
Camp Century ice core (1390 meters long)
(Dansgaard et al. 1969).

shall use equation 1 with the measured $\lambda_H = 35$ cm only in the first interval that spans approximately the last eight centuries. In the second interval, the original time scale will be adjusted to bring correspondence between observed δ oscillations and similar absolutely dated variations in the C^{14} concentration in the atmosphere. Finally, in the third interval we shall apply a continuously varying correction to the preliminary time scale.

Core Depths 0 to 283 Meters ($1107 \leqslant y \leqslant 1370$ m)

This upper part of the core spans the interval 0 to 780 years B.P. according to equation 1, when used with $\lambda_H = 0.35$ m, $H = 1370$ m, and $h = 400$ m. The isotope study (Johnsen et al., 1970) was based on samples cut from the core to represent generally 10 to 20 years each. The δ values are plotted to the right in Figure 3 as a function of t. The hatched areas depict relatively warm periods ($\delta > -$ 29 per mil), e.g. the well-known climatic optimum around 1930 and the warm periods in the middle of the eighteenth and sixteenth centuries. Low δ's (< -29 per mil) reflect relatively cold periods, e.g. in the 1820s. The "little ice age" from 1600 to 1740 (pack ice is known to have surrounded Iceland completely in 1690), and the generally cool conditions in the fifteenth century that suggest large amounts of pack ice, is a contributory reason for the fatal breaking off of relations between Iceland and the Norse settlements in Greenland.

On the face of it, one might estimate ten δ maxima indicated by arrows between 1240 and 1930, corresponding to a climatic oscillation with a period of approximately 63 years. This would be in fairly good agreement with the 66-year period suggested by studies of the Greenland fauna (Vibe 1967). However, the step curve expectedly reflects systematic oscillations superimposed by accidental deviations that do not recur regularly. The cause for short time intervals with accidentally high (low) δ's could, e.g., be a series of years with unusually high (low) amounts of isotopically heavy summer snow, and a low δ value could be due to high volcanic activity with abundant dust in the atmosphere and consequent low influx of solar radiation. Thus, the step curve calls for a more objective interpretation, as may appear by using the Fourier transform[1] on all available δ data. The

---

1. Periodicities in continuous time series of the parameter $\delta(t)$, $t_1 \leqslant t \leqslant t_2$, are detected by calculating the Fourier transform

$$F = \int_{t_1}^{t_2} \delta(t)e^{-i\omega t}\, dt$$

where $\omega = 2\pi/T$ is the cyclic frequency and T the period in question. Plotting F^2 against $1/T$ gives a so-called power spectrum of $\delta(t)$. A relatively high peak in the plot indicates a tendency of the $\delta(t)$ data to oscillate at the frequency of the peak. Minor and/or irregular peaks reflect either chance periodicity induced by noise present in the data or they may be due to a "D.C. component" in the data, i.e.

$$\int_{t_1}^{t_2} \delta(t)\, dt \neq 0$$

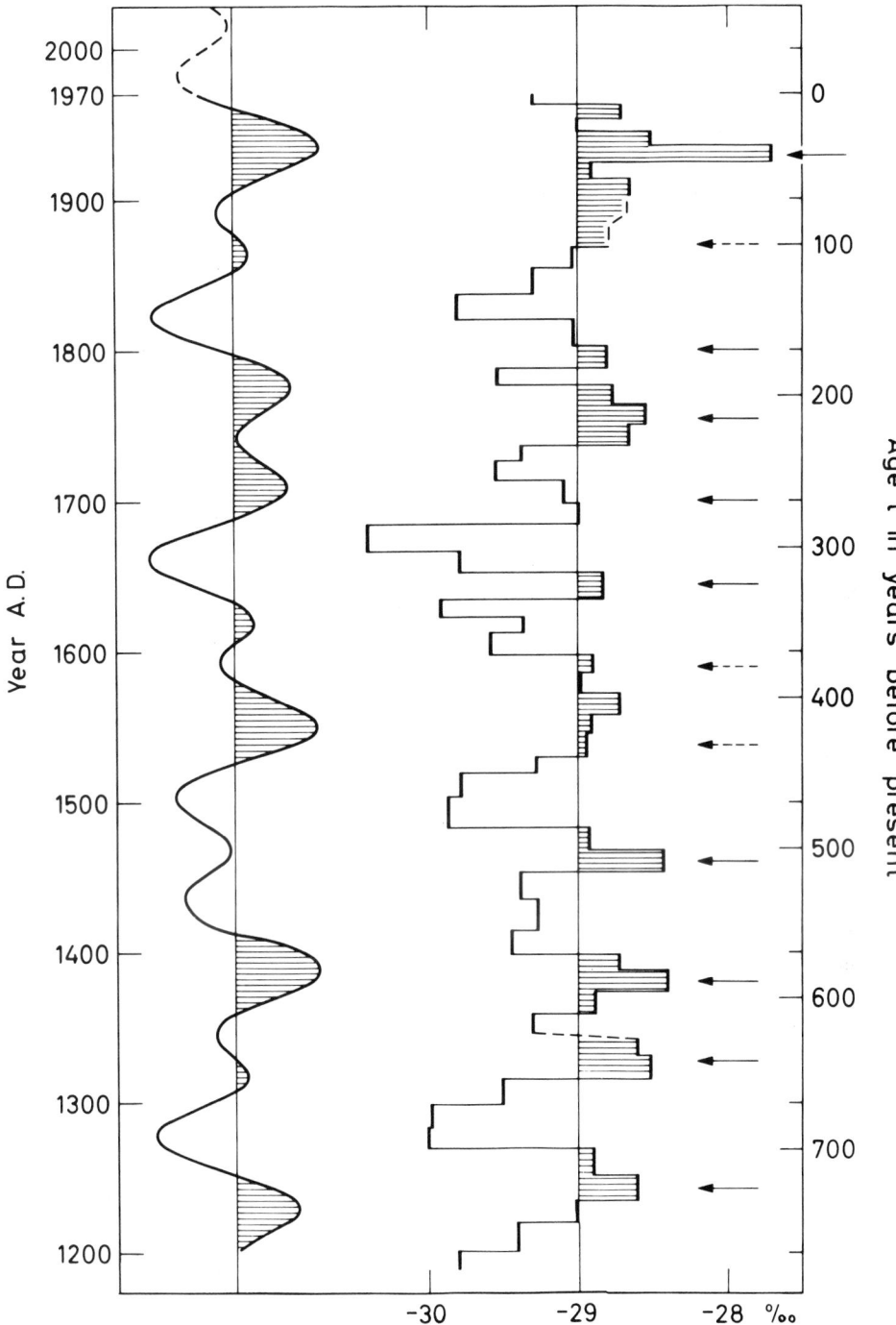

Fig. 3. Right: Oxygen-18 values of increments of an ice core from Camp Century plotted against time t since the deposition of the ice. The hatched areas correspond to relatively warm periods (Johnsen et al. 1970). Left: A synthesis of the two harmonics that dominate the step curve, judged from the spectral analysis shown in Figure 4. The dashed extrapolation suggests the probable future climatic development as a continued cooling through the next one or two decades, followed by a warming trend toward a new climatic optimum around the year 2015 A.D.

resulting power spectrum (Fig. 4) shows two dominant peaks that correspond to periods of 78 years and 181 years (the time scale assumed to be correct).

The spectral analysis also gives the amplitudes and the phases of the oscillations involved. This enables us to reproduce the δ curve by combining the two dominant harmonics (see left part of Fig. 3). The smoothed curve depicts the main trends of the step curve as far as the medium frequencies are concerned, but of course it neglects long-term as well as short-term variations such as noise (and the 11-year sunspot period, the climatic influence of which is detectable in very detailed analysis).

The 78-year-period has been noticed before as oscillations in the length of the sunspot cycle (Gleissberg 1944; Schove 1955). Maximum length (∼ 12 years) occurred around 1662, 1728, 1816, and 1895 A.D. (all close to δ minima, cf. the smoothed curve in Fig. 3), and minimum length (∼ 10 years) occurred around 1706, 1770, 1850, and 1930 (all close to δ maxima). In addition, relatively long time intervals with short sunspot cycles occurred at 1560 to 1590, 1750 to 1790,

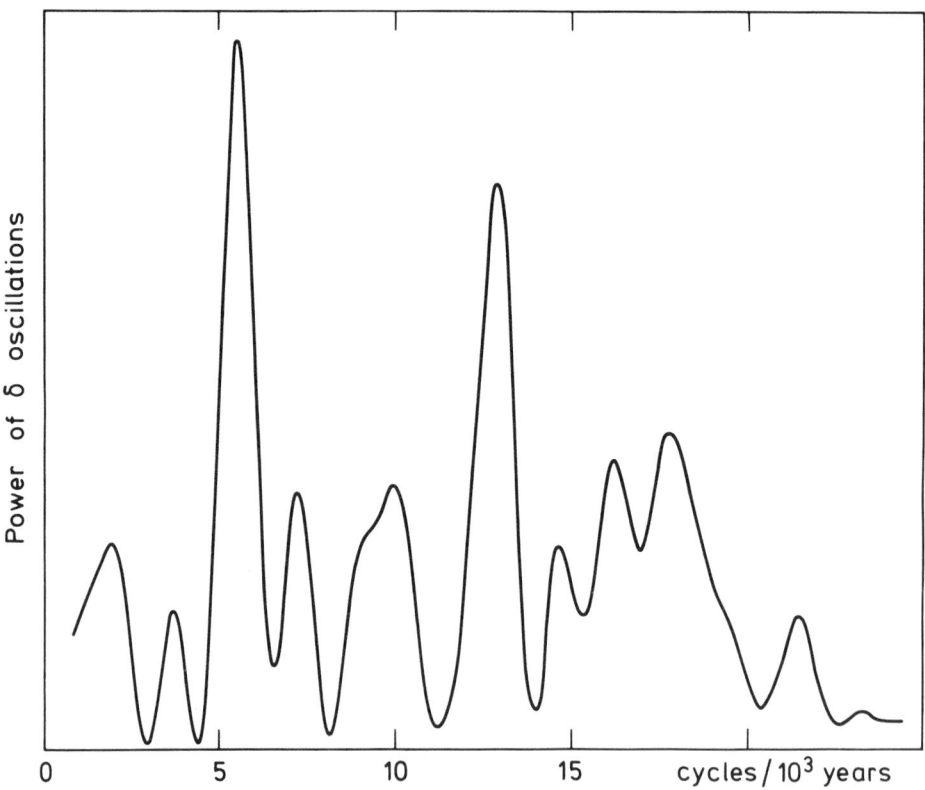

Fig. 4. Fourier power spectrum of the step curve in Figure 3. The two dominating peaks correspond to climatic oscillations with periods of 78 and 181 years ± 5 percent (Johnsen et al. 1970).

and 1900 to 1950 (Schove 1955). These intervals essentially coincide with the three last maxima in the 180-year component of the δ curve. Thus, the 180-year and the 78-year periods seem to originate from changing conditions on the sun.

Only the medium frequencies are of interest for a first approximation prognosis of the δ record within the coming 30 or 50 years. Therefore, it makes sense to extrapolate the smoothed curve, the more so as it can be done quite objectively. The dashed extrapolation to the upper left in Figure 3 suggests the most probable future development of the δ curve as a continued decrease in the next one or two decades followed by an increasing tendency (toward a maximum between 2010 and 2020). This trend remains essentially unchanged when more than the two most dominant harmonics are considered for the synthesis of the smoothed curve.

It should be emphasized that, like the δ curve of the past, the future curve will be influenced by "accidental" events and possibly by modern effects due to pollution of the atmosphere by dust and carbon dioxide. If we neglect the latter rather unknown effects, there is a more than 85 percent probability that the δ curve will keep on the "cold" side of the −29 per mil line in the next 10 to 20 years, to judge from a comparison between the two curves in Figure 3.

Core Depths 296 to 1150 Meters ($240 \leqslant y \leqslant 1094$ m)

This larger part of the core spans the time back to 10,000 B.P. in the preliminary time scale (equation 1). However, the measured mean λ_H (0.35 m) during the last century is hardly representative for a period two orders of magnitude longer, even though some compensation for changes in λ_H may be expected, as pointed out in section I.

At present, the only way to check the time scale is to determine by measurement whether all the layers in the core really are as thick as they would be if a steady state in a broad sense had prevailed (i.e. time-independent rate of accumulation, temperature, flow pattern, location of the ice divide, etc.). This technique must be based upon a periodic isotopic variation that can be followed far back in time (down to great core depths) and that can be correlated to independent measurements on an absolute time scale.

In the search for a suitable climatic oscillation the Fourier transform was applied on 180 δ values, each representing 50 years in the time interval 1000–10,000 B.P. according to equation 1 (the last millennium was disregarded because, unfortunately, part of the core increment representing the interval 800–1000 B.P. is missing and part is in poor physical condition due to a new drilling technique being applied at the corresponding depth). Sliding subintervals of generally 4000 years were treated separately in the range of periods $300 \leqslant T \leqslant 500$ years.[2] The influence of the generally decreasing δ's toward the late glacial was eliminated by

2. The sliding subinterval technique is required in cases where the total interval is many times longer than the period looked for. Treating the total interval as a whole might suppress real and persistent oscillations, if the time scale is not absolutely linear. Furthermore, the sliding subinterval technique involves the advantage that spurious peaks induced by noise will not be persistent in the whole series of spectra.

using the δ deviations $(\Delta\delta)$ from a smoothed curve instead of the δ's themselves.[3] Each of these spectra, shown in Figure 5, contains up to three peaks, but only the dominating peak is persistent through the entire Holocene. The corresponding period varies only $\pm$ 3 percent around the mean value $T = 350$ years, which indicates that the *variation* of the thickness of the layers in the upper 1100 meters of the core is in accordance with the flow model, a truly constant period assumed in the corresponding climatic variation.

The 350-year period in the δ curve may be correlated with the 405-year period in the C^{14} concentration in the atmosphere that appears in C^{14} measurements on tree rings (Suess 1970) dated on an absolute time scale.[4] If this correlation is real, it suggests that a λ_H value 15 percent lower than that previously applied might bring our time scale closer to absolute chronology in the Holocene. Further support for this suggestion is rendered by an antiphase correlation between another peak at $T = 2100$ years in the δ spectrum and a ca. 2400-year period in the C^{14} data.

In Figure 6 the δ's measured on the upper 1150-meter core length have been plotted against t calculated from equation 1 with a λ_H value 15 percent lower than the measured 0.35 meter. Back to 12,300 B.P. in the corrected time scale, all δ's exceeding a smoothed curve are set off in black. The smoothed curve follows $\delta = -29$ per mil back to 8000 B.P., from which point it follows the generally decreasing tendency of the δ's as we approach the time of the glaciation.

Going backward from zero B.P. we notice that, in general, the last millennium has apparently been colder than the preceding one. The first half of the third and fourth millennia B.P. were cold. The postglacial climatic optimum is depicted by an almost continuous series of high δ's from 4100 back to 8000, with two extremes close to 5000 and 6000 B.P.

3. This technique also acts to eliminate a "D.C. component" in the data (cf. footnote 1). A mean, or trend curve, is calculated as

$$\bar{\delta}(t) = \int_{t-\Delta t}^{t+\Delta t} \delta(\tau) \cdot w(t-\tau)\, d\tau$$

$w(t)$ being a symmetrical, normalized weight function defined in the interval $-\Delta t \leqslant t \leqslant \Delta t$. In order to obtain values of $\bar{\delta}(t)$ even at the end of the interval $t_1 \leqslant t \leqslant t_2$, outside which no measured values are available, $\delta(t)$ must be extended to the interval $-\Delta t + t_1 \leqslant t \leqslant t_2 + \Delta t$ by assigning appropriate constants in the extended interval.

$\bar{\delta}(t)$ is called the convolution of $\delta(t)$ and the kernel $w(t)$. It should contain no oscillations in the frequency range of interest. This is ensured by using $w(t) = \sin(\omega_c t)/\omega_c t$ as a kernel, which cuts off effectively all cyclic frequencies $> \omega_c$. The difference $\Delta\delta(t) = \bar{\delta}(t) - \delta(t)$, therefore, does contain all the cyclic frequencies of the $\delta(t)$ curve higher than ω_c, and $\int_{t_1}^{t_2} \Delta\delta(t)\, dt$ is close to zero.

4. A possible explanation for this correlation is that high (low) solar activity causes relatively warm (cold) conditions at high latitudes and, thereby, high (low) δ's. At the same time, plasma emitted from the sun cuts off the cosmic radiation to a higher (lower) degree than normal, which, in turn, causes lower (higher) rate of C^{14} production in the stratosphere.

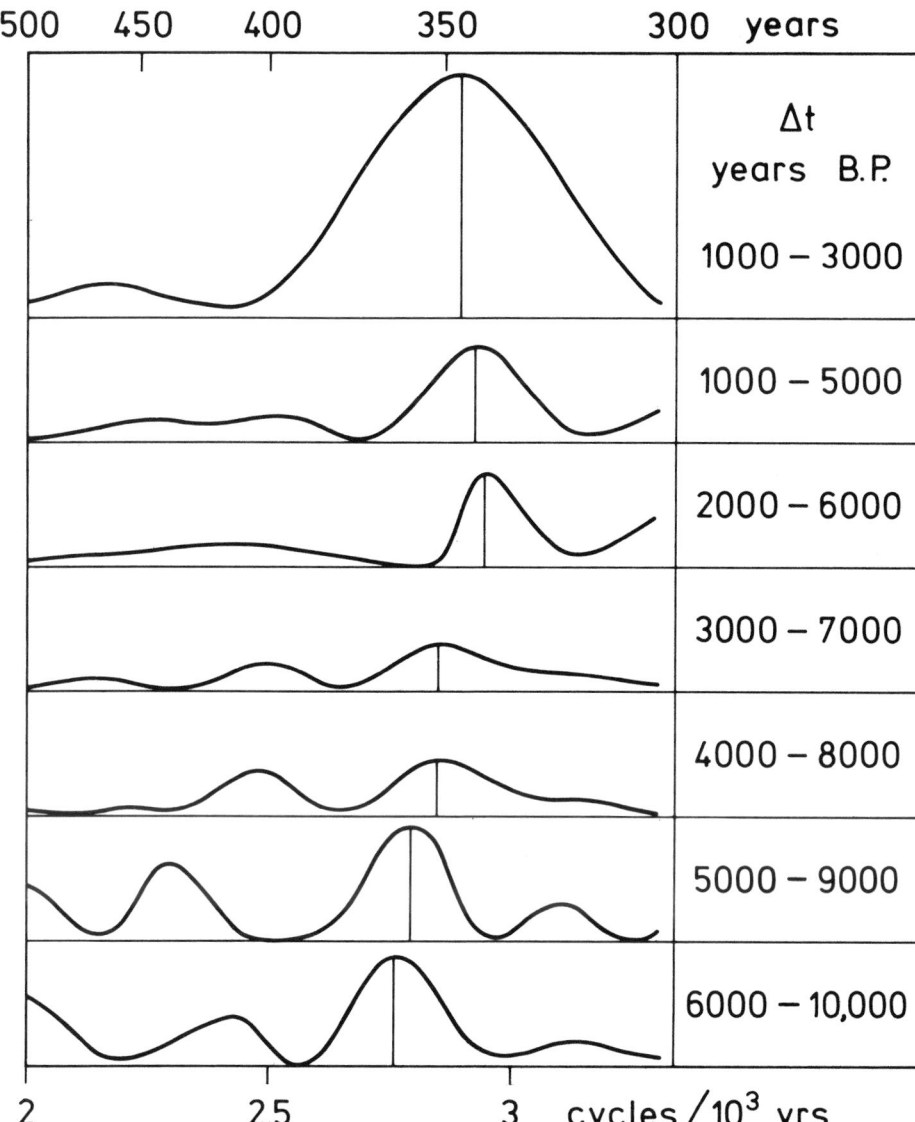

Fig. 5. Fourier power spectra of δO^{18} data on increments of the ice core, each representing ca. 50 years in the Holocene. The only persistent peak corresponds to a period of ca. 350 years in the preliminary time scale.

At the end of the glaciation we recognize the Bølling-Older Dryas-bipartite Allerød-Younger Dryas sequence known from the pollen record (Iversen 1954). In Table 1 our corrected ages for these periods are compared with both C14 dates and varve dates according to the Swedish varve chronology for the same periods

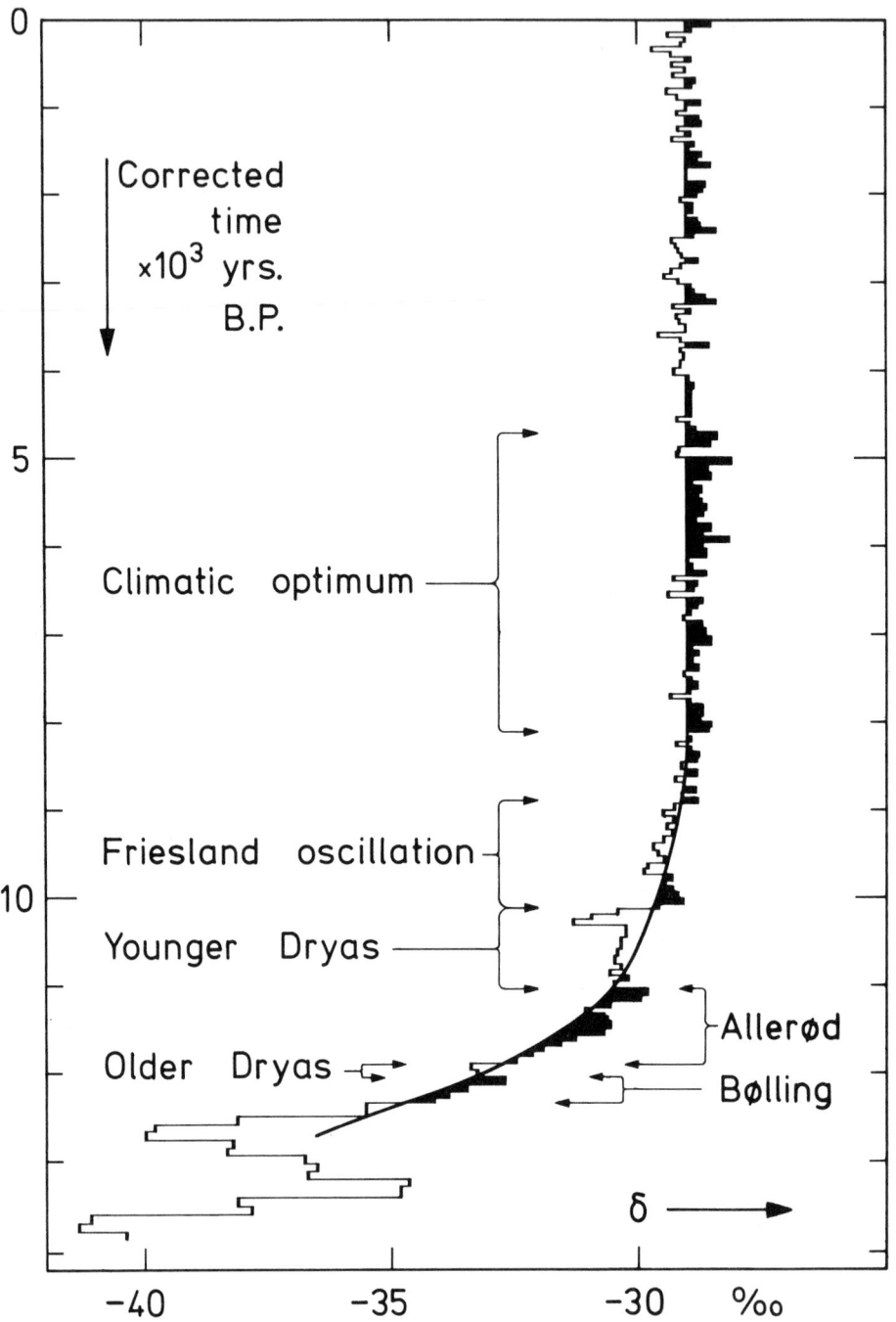

Fig. 6. Late glacial and Holocene δ variations plotted on a corrected time scale. All δ's exceeding a smoothed curve are set off in black. Climatic oscillations during the transition period fit the European sequence; they are superimposed by a general increase in δ (14,000–10,000 B.P.), part of which might be due to the gradual opening of Davis Strait and Baffin Bay, or to a changing meteorological circulation pattern connected with the extinction of the large ice sheets.

Table 1. Late Glacial Climatic Variations

	Corrected t acc. to Fig. 7 (*yrs B.P.*)	*C14 age* (*yrs B.P.*)	*Varve chronology* (*yrs B.P.*)
Bølling	12,400–12,100	12,400–12,000	12,400–12,100
Older Dryas	12,100–11,850	12,000–11,800	12,100–11,900
Allerød	11,850–10,900	11,800–11,000	11,900–11,100
Younger Dryas	10,900–10,100	11,000–10,200	11,100–10,300

(Tauber 1970). The agreement within less than $\pm$ 1 percent is remarkable, but it should not be considered typical of the accuracy of the time scale in the entire Holocene; $\pm$ 10 percent might be more realistic.

Core Depths 0 to 1373 Meters ($17 \leqslant y \leqslant 1370$ m)

In this section we shall consider the entire core, disregarding the lower 17 meters of silty ice, the creation of which is still open to discussion (Hansen et al. 1966; Weertman 1968a). In the preliminary time scale (equation 1 with $\lambda_H = 0.35$ m), $y = 17$ m corresponds to $t = 157,000$ B.P.

A continuous series of core increments, representing 200 years each, was analyzed. Figure 7 shows the δ curve in the preliminary time scale (to the left). The curve has been smoothed to a "resolution power" of 500 years.

In order to correct the time scale, Fourier power spectra were calculated on the 200-year samples in sliding 10,000-year intervals. The considered frequencies ranged from $1.4 \cdot 10^{-4}$ to $8.3 \cdot 10^{-4}$ year^{-1}, corresponding to the periods $1200 \leqslant T \leqslant 7000$ years; 62 spectra were calculated, covering 157,000 years in the preliminary time scale.

In Figure 8 a given spectrum representing $t_1 \leqslant t \leqslant t_2$ is described at $t = (t_1 + t_2)/2$ by a number of dots. Each dot represents a peak in the spectrum and is plotted at the frequency of the peak. Filled dots stand for relatively well-defined peaks, and open dots represent more or less diffuse peaks. Most of the open dots reflect noise in the δ curve, whereas the long series of generally filled dots represents a persistent δ oscillation. In most cases, shown by solid lines, the connection between the main peaks in two neighboring spectra can be determined by phase considerations.[5]

5. If two adjoining intervals, $t_1 \leqslant t \leqslant t_2$ and $t_2 \leqslant t \leqslant t_3$, contain a harmonic oscillation going through their common boundary, its phase at the end of the former interval should, of course, equal its phase at the beginning of the latter interval. In case the time scale is not linear in relation to absolute chronology, not only will the corresponding peak show up at two different frequencies in the spectra, but the phase at $t = t_2$ also will appear different in the two intervals. This involves the risk that the real peak in the first interval is identified with a false peak (due to noise) in the second interval. Spectra of three overlapping intervals between the two mentioned help to determine the apparent shift in both frequency and phase, because these shifts are less in case of overlapping intervals.

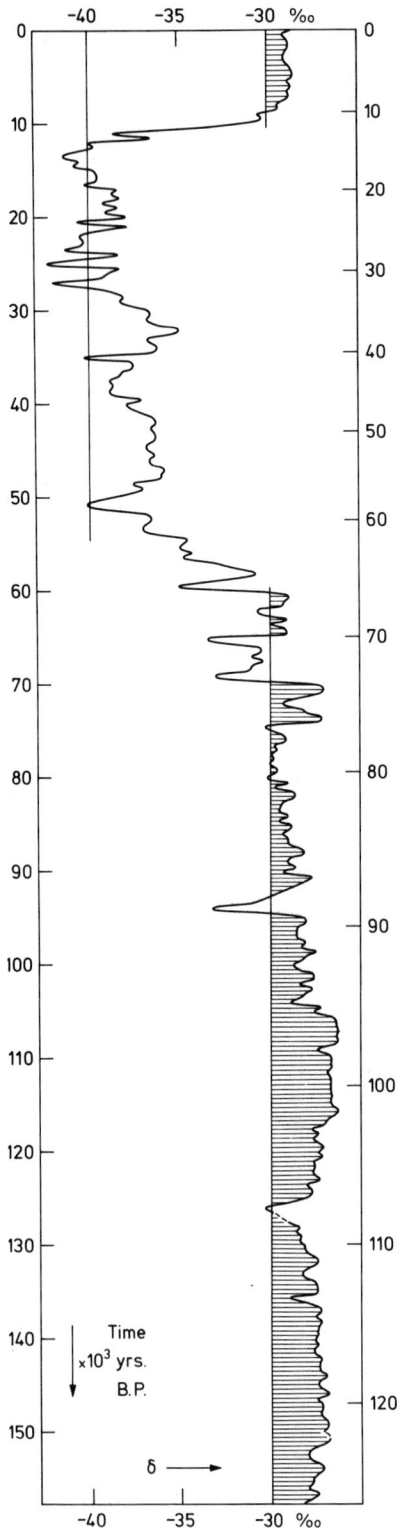

Fig. 7. Continuous δ profile along the entire ice core except for the deepest 17 meters, plotted on the preliminary time scale (left) and on the corrected time scale (right).

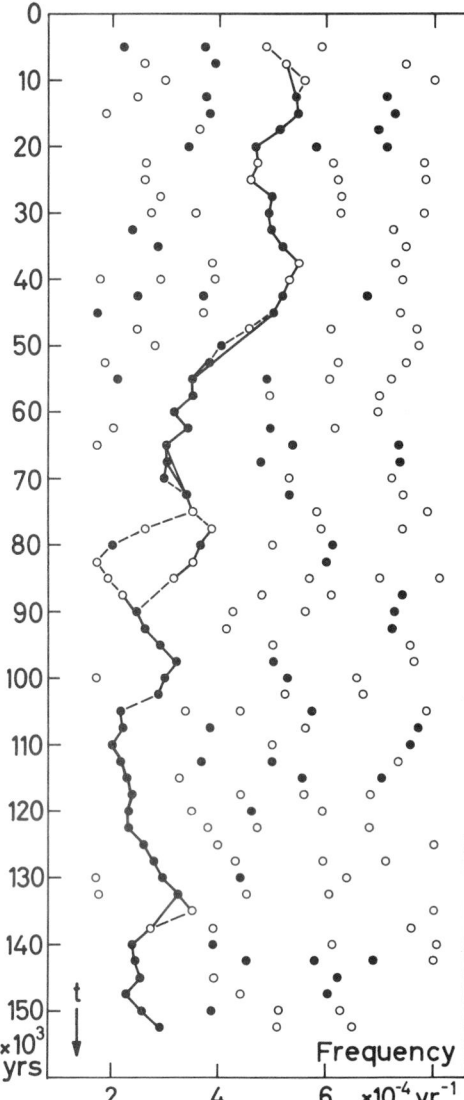

Fig. 8. Dominating (filled circles) and secondary peaks (open circles) in Fourier power spectra covering sliding intervals of 10,000 years in the preliminary time scale; e.g. the spectrum at $t = 150,000$ B.P., covering the interval $145,000 \leqslant t \leqslant 155,000$, contains two dominating peaks (at the frequencies $2.6 \cdot 10^{-4}$ and $3.9 \cdot 10^{-4}$ yr^{-1}) and two secondary peaks at higher frequencies. Solid connecting lines between two circles indicate that the correspondence is supported by phase considerations. The long curve is used for correction of the preliminary time scale, assumed that the period of the oscillation has remained constant in an absolute time scale.

Apparently, the frequency of the persistent oscillation was close to $5 \cdot 10^{-4}$yr^{-1} ($\sim$ 2000-year period) back to 45,000 B.P., then decreasing to about $2.5 \cdot 10^{-4}$yr^{-1} ($\sim$ 4,000-year period) previous to 100,000 B.P. However, such a shift of frequency is not very likely if the cause for the climatic oscillation is to be ascribed to varying solar activity, as suggested by an antiphase correlation between the Holocene δ's and a 2400-year period in the C^{14} concentration in tree rings (Suess, unpublished). Therefore the best we can do at present is to assume that the period of the oscillation observed has remained constant during the entire time range spanned, and that the apparent change of frequency is due to the deviation of the preliminary

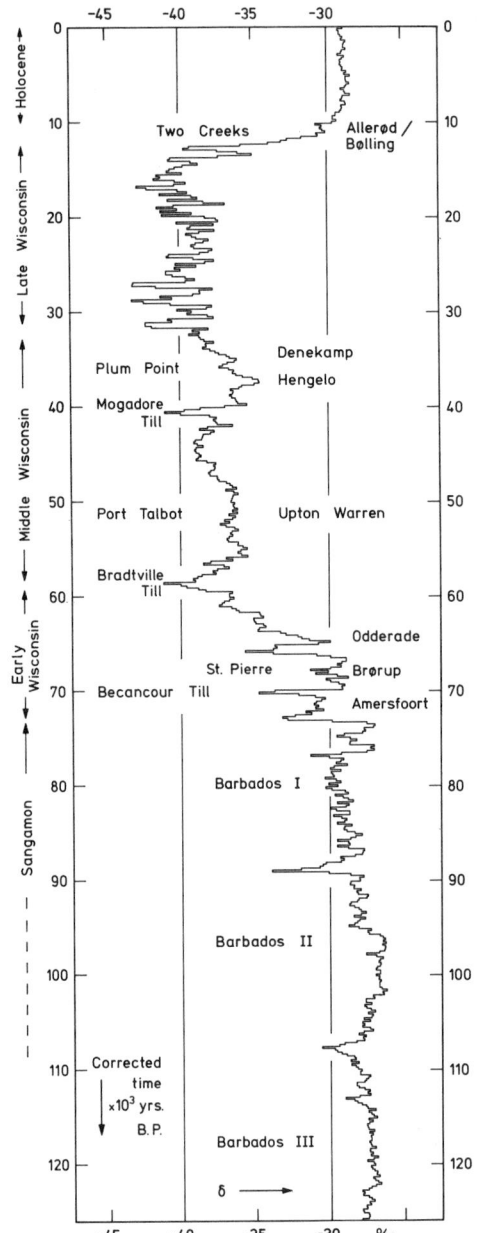

Fig. 9. δ in 200-year intervals plotted on the corrected time scale. Tentative interpretations in European and American terminology are shown to the right and left respectively. At the far left is a proposed division of the Wisconsin glaciation in accordance with the characteristic features of the δ curve.

time scale more and more from absolute chronology as we approach the bottom of the ice sheet.

Accordingly, the frequency curve shown in Figure 8 may serve as basis for a stepwise correction of the time scale: an interval of 10,000 years in the preliminary time scale corresponds to 2400 • f years in the corrected scale, f being the number of periods in the interval. The corrected time scale has been applied to the right in Figure 7. It is independent of the C^{14} scale, of all ice flow parameters, and of their possible temporal changes.

As a last step, a detailed δ curve ("resolution power" 200 years) is presented in Figure 9, plotted on a linear, corrected time scale, and provided with a tentative interpretation, both in American and European terminology.

The main features of the curve suggest a division of the Wisconsin glaciation into three intervals, as shown to the left in Figure 9: (1) early Wisconsin (73,000–59,000 B.P.) comprising the main cooling period from the Sangamon interglacial; (2) middle Wisconsin (59,000–32,000) with generally more stable conditions; and (3) late Wisconsin (32,000–13,000) comprising the coldest part of the glaciation and further characterized by greatly varying conditions. In broad outline, the abrupt transition to the Holocene supports the conception advanced by other authors (Weertman 1964; Broecker and Van Donk 1969) that the disappearance of an ice sheet is a much faster process than its creation. The transitional interstadial Bølling, which appeared in Figure 6, is smoothed out in this coarser presentation.

The most obvious basis for comparison between the δ record and American glacial chronology is the stratigraphic study of Pleistocene deposits in the Ontario and Erie basins (Goldthwait et al. 1965). In the period 18,000 to 12,500 B.P. it shows a slow general retreat interrupted by numerous minor readvances, which fits pretty well with the δ record. Before 18,000 B.P. the geological record is less detailed, but the major climatic events agree satisfactorily with the main course of the δ curve: (1) the tripartite Plum Point interstadial appears at 25,000 to 39,000 B.P. (δ curve 32,000–40,000); (2) the Mogadore Till dates at 41,000 (δ curve 41,000); (3) the bipartite Port Talbot interstadial at 44,000–57,000 (δ curve 47,000–57,000); (4) Bradtville Till at 60,000 (δ curve 58,000); (5) the St. Pierre interstadial at 65,000–69,000 (δ curve 66,000–70,000); and (6) the Bécancour Till at 70,000 (δ curve 70,000). Of course, other interpretations are possible, but the above seem most obvious.

The identification of the Brørup interstadial in the European sequence is considered fairly certain, because Brørup is known (Andersen 1961) to have been a really warm, tripartite interstadial; moreover, it was preceded by a cooler interstadial (Amersfoort) and succeeded by a short-lived interstadial (Odderade).

Prior to the Wisconsin glaciation, the Barbados I, II, and III high sea levels are dated by Pa^{231} and Th^{230} at 82,000, 103,000, and 122,000 B.P. (Veeh 1961; Broecker et al. 1968). In view of the uncertainties involved, they coincide with the three broad high δ levels separated by the δ minima at 89,000 and 108,000 years B.P. It should be noticed that, unlike Emiliani's paleotemperature curve

(1966), or paleoglaciation curve (Dansgaard and Tauber 1969), the δ record renders no support for a major glaciation between 120,000 and 100,000 B.P. Furthermore, the apparent duration of more than 50,000 years of what is called the Sangamon in Figure 9 does not tempt one to interpret it as an early interstadial in the Wisconsin (Ericson and Wollin 1970). However, one should bear in mind that the dating of the deepest strata in the ice core is fraught with considerable uncertainty, in spite of the agreement with the Barbados high sea levels.

It should be stressed that the fit between the δ curve and other records allows no conclusion as to temporal variation of the rate of accumulation, because the present thickness of an old annual layer depends on several parameters not considered here, e.g. the temperatures to which the layer has been exposed since the time of deposition.

Neither is it justified to apply the present geographical correlation between δ and air temperature in order to convert the temporal δ variations into a temperature curve, because (1) the deeper strata originated farther inland, where perhaps slightly different climatic conditions existed; (2) the isotopic composition of sea water, which provides the moisture for the precipitation, changed; (3) the ratio of summer to winter precipitation possibly changed; (4) the main meteorological wind patterns possibly changed; (5) the flow pattern of the ice in the accumulation area possibly changed; and (6) the thickness of the ice sheet changed. As to the last point, a major increase (or decrease) of the surface altitude could hardly occur without a general cooling (or warming) of the climate. Both of these effects influence the δ's in the same direction, but it is not yet possible to distinguish among the individual contributions to an observed change in δ. Thus, a given increase (decrease) may be interpreted as a consequence of a warming (cooling), but it should not be specified on a temperature scale.

Conclusion

Stable-isotope investigation of deep ice cores from high polar regions constitutes a new and powerful tool in paleoclimatology. Ice core data provide far greater climatological detail than any hitherto known method; furthermore, unlike other terrestrial deposits, ice cores from some dry-snow zones can provide continuous sedimentary records spanning more than 100,000 years.

An ice core from the southern dome of the Greenland ice sheet is desirable for comparison with the Camp Century ice core. Moreover, like Europe and the eastern part of North America, southern Greenland has a climate directly influenced by the Atlantic Ocean. Therefore, a South Greenland ice core may contain more direct information about the conditions that led to buildup and extinction of the large Scandinavian and Laurentide ice sheets.

Acknowledgments. Financial support for the core drilling program was contributed by the U.S. National Science Foundation; the Carlsberg Foundation, Copenhagen, supported the cutting and measuring programs.

The authors are deeply indebted to Prof. Børge Jessen and Dr. Henrik Tauber for valuable discussions and constructive criticism.

REFERENCES

Aegerter, S., H. Oeschger, A. Renaud, and E. Schumacher, Studies on the tritium content of ice samples, *Medd. Groenland, 177*, no. 2, 76, 1969.

Ambach, W., H. Eisner, and G. Sauzay, Tritium profiles in two firn cores from alpine glaciers and tritium content in precipitation in the alpine area, *Arch. Meteorol. Geophys. Bioklimatol., Ser. B, 17*, 93, 1969.

Andersen, S. T., Vegetation and its environments in Denmark in the early Weichselian glacial (last glacial), *Geol. Surv. Denmark, Ser. II*, no. 75, 175 pp., 1961.

Arnason, B., The exchange of hydrogen isotopes between ice and water in temperate glaciers, *Earth Planetary Sci. Letters, 6*, 423, 1969.

Broecker, W. S., and J. v. Donk, Insolation changes, ice volumes and the O^{18} record in deep sea cores, 28 pp., 1969, preprint manuscript.

Broecker, W. S., D. L. Thurber, J. Goddard, T. L. Ku, R. K. Matthews, and K. J. Mesolella, Milankovitch hypothesis supported by precise dating of coral reefs and deep-sea sediments, *Science, 159*, 297, 1968.

Clausen, H. B., B. Buchmann, and W. Ambach, Si-32 dating of an Alpine glacier, I. A. S. H. Publ. no. 79, 135, 1967.

Craig, H., Standard for reporting concentrations of deuterium and oxygen-18 in natural waters, *Science, 133*, 1833, 1961.

Crozaz, G., and C. C. Langway, Jr., Dating Greenland firn-ice cores with Pb^{210}, *Earth Planetary Sci. Letters, 1*, 194, 1966.

Dansgaard, W., The O^{18}-abundance in fresh water, *Geochim. Cosmochim. Acta, 6*, 241, 1954.

Dansgaard, W., Stable isotopes in precipitation, *Tellus, 16*, 436, 1964.

Dansgaard, W., and S. J. Johnsen, A flow model and a time scale for the ice core from Camp Century, Greenland, *J. Glaciology, 8*, 215, 1969.

Dansgaard, W., and H. Tauber, Glacier oxygen-18 content and Pleistocene ocean temperatures, *Science, 166*, 499, 1969.

Dansgaard, W., E. Roth, and G. Nief, Isotopic distribution in a Greenland iceberg, *Nature, 185*, 232, 1960.

Dansgaard, W., H. B. Clausen, and A. Aarkrog, The Si^{32} fallout in Scandinavia: A new method for ice dating, *Tellus, 18*, 187, 1966.

Dansgaard, W., S. J. Johnsen, H. B. Clausen, and C. C. Langway, Jr., Ice cores and paleoclimatology, in I. U. Olsson, ed., Twelfth Nobel Symp., *Radiocarbon Variations and Absolute Chronology*, Uppsala, 1969, Almquist & Wiksell, Stockholm, and Wiley, New York, 1970.

Dansgaard, W., S. J. Johnsen, J. Møller, and C. C. Langway, Jr., One thousand centuries of climatic record from Camp Century on the Greenland ice sheet, *Science, 166*, 377, 1969.

Emiliani, C., Isotopic paleo temperatures, *Science, 154*, 851, 1966.

Epstein, S., and R. P. Sharp, Oxygen isotope studies, *Trans. Am. Geophys. Union, 40*, 81, 1959.

Ericson, D. B., and G. Wollin, Pleistocene climates in the Atlantic and Pacific oceans: A comparison based on deep-sea sediments, *Science, 167*, 1483, 1970.

Gleissberg, W., A table of secular variations of the solar cycle, *Terr. Magn. Stru. Telectr.*, *49*, 243, 1944.

Goldberg, E. D., Geochronology with lead-210, in *Proc. Symp. Radioact. Dating, I. A. E. A., Athens*, p. 121, 1963.

Goldthwait, R. P., A. Dreimanis, J. L. Forsyth, P. F. Karrow, and G. W. White, in H. E. Wright, Jr., and D. G. Frey, eds., *The Quaternary of the United States*, p. 737, Princeton Univ. Press, Princeton, N.J., 1965.

Gonfiantini, R., V. Togliatti, E. Tongiorgi, W. de Breuck, and E. Picciotto, Snow stratigraphy and oxygen isotope variations in the glaciological pit of King Baudouin Station, Queen Maud Land, Antarctica, *J. Geophys. Res.*, *68*, 3791, 1963.

Hansen, B. L., and C. C. Langway, Jr., Deep core drilling in ice and core analysis at Camp Century, Greenland, 1961–1966, *Antarctic J. U.S.*, *1*, 207, 1966.

Iversen, J., The late-glacial flora of Denmark and its relation to climate and soil, *Geol. Surv. Denmark, Ser. II*, no. 80, 87, 1954.

Johnsen, S. J., Diffusion of the water molecule in ice sheet deposits, manuscript in preparation.

Johnsen, S. J., W. Dansgaard, H. B. Clausen, and C. C. Langway, Jr., Climatic oscillations 1200–2000 AD, *Nature*, *227*, 482, 1970.

Mock, S. J., Snow accumulation studies on the Thule Peninsula, Greenland, *J. Glaciology*, *7*, 59, 1968.

Nye, J. F., The flow of glaciers and ice sheets as a problem in plasticity, *Proc. Roy. Soc. (London), Ser. A.*, *207*, 554, 1951.

Nye, J. F., The distribution of stress and velocity in glaciers and ice sheets, *Proc. Roy. Soc. (London), Ser. A*, *239*, 113, 1957.

Nye, J. F., The motion of ice sheets and glaciers, *J. Glaciology*, *3*, 493, 1959.

Oeschger, H., B. Alder, and C. C. Langway, Jr., An in situ gas extraction system for radiocarbon dating glacier ice, *J. Glaciology*, *6*, 939, 1967.

Picciotto, E., X. de Maere, and I. Friedman, Isotopic composition and temperature of formation of Antarctic snows, *Nature*, *187*, 857, 1960.

Scholander, P. F., W. Dansgaard, D. C. Nutt, H. de Vries, L. K. Coachman, and E. Hemmingsen, Radio-carbon age and oxygen-18 content of Greenland icebergs, *Medd. Groenland*, *165*, 1, 1962.

Schove, D. J., The sunspot cycle, 649 B.C. to A.D. 2000, *F. Geophys. Res.*, *60*, 127, 1955.

Suess, H., The three causes of the secular carbon-14 fluctuations, their amplitudes and time constants, in I. U. Olsson, ed., Twelfth Nobel Symp., *Radiocarbon Variations and Absolute Chronology*, Uppsala, 1969, Almquist & Wiksell, Stockholm, and Wiley, New York, 1970.

Tauber, H., The Scandinavian varve chronology and C-14 dating, ibid.

Veeh, H. H., Th^{230}/U^{238} and U^{234}/U^{238} ages of Pleistocene high sea level stand, *J. Geophys. Res.*, *71*, 3379, 1966.

Vibe, C., Arctic animals in relation to climatic fluctuations, *Medd. Groenland*, *170*, no. 5, 227 pp., 1967.

Weertman, J., Rate of growth or shrinkage of nonequilibrium ice sheets, *J. Glaciology*, *5*, 145, 1964.

Weertman, J., Bubble coalescence in ice as a tool for the study of its deformation history, *J. Glaciology*, *7*, 155, 1968a.

Weertman, J., Comparison between measured and theoretical temperature profiles of the Camp Century, Greenland, borehole, *J. Geophys. Res.*, *73*, 2691, 1968b.

Minze Stuiver

4. EVIDENCE FOR THE VARIATION OF ATMOSPHERIC C¹⁴ CONTENT IN THE LATE QUATERNARY

The usefulness of the carbon-14 dating technique strongly depends on the limits that can be placed on the variability of specific atmospheric C^{14} content during the past. In the early stages of the development of C^{14} dating, these variations were indistinguishable from those associated with technical imperfections (Libby 1955), but upon further improvement of the technique an assortment of C^{14} variations was found. The variations destroy the simple "one C^{14} year equals one calendar year" relationship and complicate things for archaeologists (Stuiver and Suess 1966). To the geophysicist the variations are very pleasing because extensive information on time-dependent changes in earth magnetic field, ocean mixing, and solar sunspot cycle can be obtained. The geophysical mechanisms for atmospheric C^{14} fluctuations are given in a cursory manner only, since this paper is mainly concerned with the evidence for them.

Variations in C^{14} content of the oceanic reservoirs are not necessarily in phase with atmospheric C^{14} variations, nor equal in amplitude. Buddemeyer (1969) analyzed a 2000-year varve series from Saanich Inlet, Canada, and discussed the local oceanic circulation changes possibly associated with the changes in C^{14} content found in the varves. All other investigations give data on atmospheric C^{14} only.

TREE RINGS

The most suitable samples for precise determinations of atmospheric C^{14} variations during the past are dendrochronologically dated tree ring samples (Fergusson 1968). Each ring is formed during one year only and afterward ceases to accumulate or exchange carbon. By measuring the present C^{14} concentration in wood samples whose ages are dendrochronologically determined it is possible to calculate the initial concentration in the sample at the time of formation. The initial C^{14} activity in atmospheric CO_2 is calculated from the initial C^{14} activity of the wood sample by applying a correction for the isotope fractionation between wood and atmospheric CO_2.

The Laboratory of Tree Ring Research of the University of Arizona has been the main supplier of dendrochronologically dated *Sequoia gigantea* and *Pinus aristata* tree ring samples that cover a time span of 7400 years. This age range can perhaps be extended to about 10,000 years when older fossil wood becomes available.

The total number of tree ring samples analyzed by several laboratories is rapidly approaching one thousand. Within statistics the results generally agree with each other, although occasional systematic errors like a 5 per mil difference between the La Jolla and University of Arizona Laboratories are troublesome (Damon 1969). Both long- and short-term variations in the C^{14} concentration of wood samples have been proven. Many data are available for the short-term oscillations lasting about 100–200 years (DeVries 1958; Willis et al. 1960; Stuiver 1965; Suess 1965, 1969; Damon et al. 1966). These short-term oscillations have been shown to correlate significantly with solar cycle modulation of the cosmic ray flux (Stuiver 1961, 1965; Lingenfelter 1963; Suess 1965; Grey 1969). In addition, a 400-year cycle has been suggested (Suess 1969).

The long-term variation in atmospheric C^{14} concentration, as derived from tree ring data, is given in Figure 1. This latest curve for the average long-term trend was suggested by Suess (1969) and agrees in its basic features with the results obtained by Damon (1969) and by Michael and Ralph (1969).

Fossil tree sections, too old for dendrochronological connection with calendar years, can still be employed for the investigation of trends in atmospheric C^{14} content; such an approach suggests that the content was not changing between 7050 and 7350 C^{14} years ago (Vogel et al. 1969).

HISTORICALLY DATED SAMPLES

In the early stages of the investigation of C^{14} variations, historically dated samples were of fundamental importance in corroborating the long-term deviation of atmospheric C^{14} content between 3000 and 5000 years ago. However, archaeological samples are generally associated with a certain period and often not well defined historically. Sample materials also may have formed over considerable time intervals. When these factors are taken into account, good agreement is obtained between the historical ages of samples up to 5000 years old and their C^{14} ages, corrected for the deviations in atmospheric C^{14} as derived from the tree ring chronology (Michael and Ralph 1969).

VARVES

Most investigations of atmospheric C^{14} variations have utilized tree ring material and samples of known historical age. The time span of the tree ring chronology, at present 7400 years, is unlikely to be extended more than a few thousand years

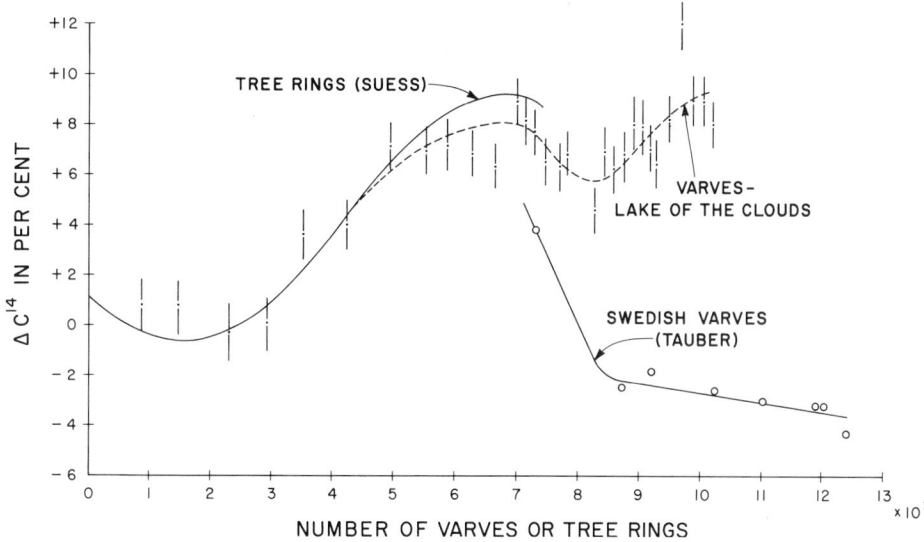

Fig. 1. ΔC^{14}, in percentage of atmospheric content, plotted against Lake of the Clouds varve years (dashed line and experimental points), varve years of the Swedish varve chronology (solid line at bottom), and number of tree rings (solid line at left). Lines drawn are approximate only.

because of the lack of suitable fossil tree specimens. Varve series covering ca. 12,000 years are used to extend the study of C^{14} variations beyond the reach of dendrochronology.

Varves are produced in sedimentary environments by annual deposition of layers of different composition or texture. The most extensively studied varve series in the late Quaternary consist of silt and clay laminas deposited as couplets in proglacial lakes. Prominent in these investigations is the Swedish varve chronology of De Geer. The portion of the chronology essential for the C^{14} investigations consists of two parts: (1) the varve count at the Ångerman Älven ranging from a stage close to historic time back to the time the region was deglaciated; (2) the older, Finniglacial part corresponding to the ice recession from Stockholm to Indals Älven and Angermanland.

The postglacial varves in the Ångerman Älven Valley reflect a pattern of annual deposition due to seasonal variations in flooding of the Ångerman Älven. Because of the high rate of land upheaval in this region the varved sediments were continuously displaced eastward into the Baltic. A series of 7500 varves is preserved along the riverbanks (Lidén 1938). The youngest varve forms the surface of a delta terrace which in A.D. 1900 was situated 10.2 meters above sea level. Lidén, on the basis of rates of uplift, estimated that the last varve was deposited around A.D. 920. A correction for the rates of uplift employed by Lidén makes the youngest varve 200 to 500 years older (Wenner 1968; Tauber 1969).

The Swedish varves contain insufficient authigenic organic material for precise C^{14} measurements. However, C^{14}-dated climatic episodes of northern Europe can be correlated with varve-dated rates of recession and geomorphological features in southern Sweden. In addition, pollen zone boundaries, varve dated in the Ångerman series by Fromm (1938), can be correlated with C^{14}-dated pollen zone boundaries in nearby peat bogs (Wenner 1968). Employing both techniques, Tauber (1969) calculated the deviations resulting from the differences in C^{14} age and the corrected Swedish varve age (Fig. 1). The corrected varve ages were obtained by adding 200 years to Lidén's calculated age of the youngest varve of the Ångerman Älven series.

Both C^{14} ages and ages of the Swedish varve chronology agree closely for the interval of 8500–12,500 varve years B.P., although the close agreement depends somewhat on Tauber's interpretation of the Bølling and Allerød climatic oscillations. If Nilsson's (1968) varve ages are accepted for these episodes, the C^{14} level would be about 5 percent higher around 12,500 varve years ago. The classic proglacial varves found throughout Sweden were used for the varve chronology of the interval of 8500–12,500 varve years B.P. The chronology has been derived from closely spaced overlapping varve sequences that have been studied extensively. This interval is probably the most reliable part of the Swedish varve chronology. Thus it seems likely that the atmospheric C^{14} level underwent little change during these 8500–12,500 varve years B.P.

The exact position of this more or less constant C^{14} level depends entirely on the varve count of the Ångerman Älven River varve series. The C^{14} control of this series is very scanty, and the reliability has not been satisfactorily proven for the age range covered by the tree ring chronology. It is therefore possible that the good agreement between C^{14} and Swedish varve ages for the interval of 8500–12,500 varve years is fortuitous. It may result from a single error in the Ångerman Älven varve series that adds an identical anomaly to all the older proglacial varve dates.

A few lakes in temperate zones also contain rhythmites. Welten (1944) discussed a series of rhythmites deposited in a lake near Interlaken in Switzerland. Each couplet consists of a light-colored layer rich in calcium carbonate and a dark layer rich in organic material. From the stratification of pollen grains of various genera according to the season of blooming, from pollen content per unit of volume, and from diatom abundance it was concluded that the light layers were deposited in the summer and the dark layers during the other seasons. Vogel (1969) attempted to correlate C^{14} ages and varve counts by comparing the C^{14} ages of pollen zone boundaries with similar boundaries determined by Welten in the varved core. Little variation was found for the C^{14} activity between 6500 and 8500 varve years B.P., but a steep decline in activity of about 25 percent was found for the 8500- to 10,500-year interval. Such a change is not encountered in the Swedish varve chronology. It is likely that the varve counts are not sufficiently precise over the 8500- to 10,500-year interval. Perhaps they are more accurate over the 6500–8500-year interval where the calculated variation in atmospheric C^{14} content is in agreement with the North American C^{14}-dated varve chronology discussed below.

Extensive laminations of sediments flooring Lake of the Clouds in northern Minnesota were discovered a few years ago by H. E. Wright of the University of Minnesota. The laminations may be the result of seasonal incursions of oxygen into the iron monimolimnion of the lake, resulting in light layers rich in iron carbonate (H. E. Wright, personal communication). Parts of the C^{14} analyses for this lake have been reported previously (Stuiver 1969). The complete results are now available and are given in Figure 1 and Table 1.

A continuous 5-meter core was obtained by H. E. Wright and E. J. Cushing in 1968 from the floor of Lake of the Clouds and made available for a detailed C^{14} study. The core was laminated over its entire length. The sediment consisted chiefly of algal microfragments, and core sections 2 to 3 cm thick contained sufficient organic carbon for C^{14} analysis.

The laminated sediments are found only in the deepest parts of the lake. Because of the deep water at the coring location, ca. 30 meters, the mud–water interface is not precisely located. The top 19.5 cm of the core was disturbed and contained, by estimate, about 200 varves. A separate short core that contained a sequence through the mud–water interface was used for the calibration of the 5-meter core. The C^{14} age of a section of the short core between 127 and 310 varves below the mud–water interface is 440 ± 60 years (after correction for isotopic fractionation). The actual C^{14} age for a sample averaging 220 calendar years is about 100 years (Stuiver and Suess 1966). Thus the C^{14} deficiency of the lake amounts to ca. 440 − 100, or 340 years. The age of the youngest sample of the 5-meter core is 1120 ± 90 C^{14} years. This would indicate that this sample is about 1120 − 340 or 780 varves below the mud–water interface. The actual count by Craig in the 5-meter core is 242 varves. Evidently ca. 780 − 242 or 538 varves are missing or in the disturbed portion at the top of the 5-meter core. A similar calculation for sample Y-2154 with a net C^{14} age of 2930 ± 80 years (ca. 3010 calendar years) yields 668 missing varves. A rounded figure of 600 missing varves has been used in this paper for the 5-meter core. This number is only approximate and may be in error by about 100 varves.

Alan Craig, at the University of Minnesota, counted about 9500 varves through the entire core, with the exception of the lowermost 40 cm where a continuous sequence was lacking. His counts are used in this paper. All C^{14} ages are based on the 5730-year half-life and have been corrected for isotopic fractionation. A deduction of 340 years was made to correct for the C^{14} deficiency of the lake water.

A section containing 4850 varves was counted independently by the writer. Craig counted 5000 varves for the same section (4800 to 9800 in Fig. 1). The writer's count was consistently 3 percent lower over the entire interval. Most likely Craig's count is more accurate as he spent considerably more time counting the

Table 1. Varve Counts and C¹⁴ Ages for Lake of the Clouds, Minnesota

	Craig varve count + 600	C¹⁴ Age 5730 yr. half-life C¹³ corrected	C¹⁴ Age corr. for Lake C¹⁴ def.	$\triangle$C¹⁴ (%)
Y-2151	842	1120 ± 90	780 ± 90	+ 0.8 ± 1.1
Y-2152	1479	1750 ± 70	1410 ± 70	+ 0.8 ± 0.9
Y-2153	2298	2660 ± 100	2320 ± 100	− 0.3 ± 1.2
Y-2154	2942	3270 ± 80	2930 ± 80	+ 0.1 ± 1.0
Y-2155	3525	3570 ± 70	3230 ± 70	+ 3.6 ± 0.9
Y-2156	4232	4250 ± 80	3910 ± 80	+ 4.0 ± 1.0
Y-2157	4960	4730 ± 70	4390 ± 70	+ 7.1 ± 0.9
Y-2158	5515	5300 ± 70	4960 ± 70	+ 6.9 ± 0.9
Y-2159	5877	5650 ± 100	5310 ± 100	+ 7.1 ± 1.2
Y-2160	6283	6080 ± 70	5740 ± 70	+ 6.8 ± 0.9
Y-2161	6658	6490 ± 70	6150 ± 70	+ 6.3 ± 0.9
Y-2162	7001	6640 ± 80	6300 ± 80	+ 8.9 ± 1.0
Y-2163	7150	6850 ± 90	6510 ± 90	+ 8.1 ± 1.1
Y-2165	7290	7020 ± 80	6680 ± 80	+ 7.7 ± 1.0
Y-2166	7464	7280 ± 80	6940 ± 80	+ 6.5 ± 1.0
Y-2167	7696	7530 ± 80	7190 ± 80	+ 6.3 ± 1.0
Y-2168	7924	7720 ± 80	7380 ± 80	+ 6.8 ± 1.0
Y-2171	8269	8240 ± 80	7900 ± 80	+ 4.6 ± 1.0
Y-2172	8430	8260 ± 80	7880 ± 80	+ 6.9 ± 1.0
Y-2173	8577	8420 ± 70	8080 ± 70	+ 6.2 ± 0.9
Y-2174	8735	8530 ± 100	8190 ± 100	+ 6.8 ± 1.2
Y-2175	8894	8600 ± 80	8260 ± 80	+ 8.0 ± 1.0
Y-2176	9040	8750 ± 100	8410 ± 100	+ 7.9 ± 1.2
Y-2177	9180	8960 ± 80	8620 ± 80	+ 7.0 ± 1.0
Y-2178	9270	9100 ± 100	8760 ± 100	+ 6.4 ± 1.2
Y-2179	9471	9080 ± 100	8740 ± 100	+ 9.2 ± 1.2
Y-2180	9670	9070 ± 100	8730 ± 100	+ 12.0 ± 1.2
Y-2181	9867	9430 ± 110	9090 ± 110	+ 9.9 ± 1.3
Y-2182	10034	9600 ± 100	9260 ± 100	+ 9.9 ± 1.2
Y-2183	10200	9830 ± 110	9490 ± 110	+ 9.0 ± 1.3

varves, which are thin so that a slight difference in eye resolution can explain the small difference between the counts.

The good agreement between both counts indicates an error of only a few percent in the varve counting. The varve C¹⁴ ages generally agree within about 100 years with the varve counts when corrected for atmospheric C¹⁴ deviations known from the tree ring chronology. Evidently the varves are annual over at least a 7400-year interval. Appreciable variations in C¹⁴ deficiency of the lake water are absent as they would result in large deviations of corrected C^{14} varve ages. In addition, C^{13}/C^{12} ratios of the organic fraction of thirty samples, distributed over the whole core, fall within a narrow range of -27.0 ± 1.8 per mil on the PDB scale. The relative stability of the C^{13}/C^{12} ratios make it probable that the C^{14} deficiency of the lake water remained unchanged over the entire interval of 10,000 varve years. These properties seem to make this lake an ideal tool for the investigation of long-term C^{14} variations.

The Lake of the Clouds varve ages, when compared with C^{14} ages, result in atmospheric C^{14} contents of $+7$ to $+10$ percent for the interval of 5000–10,000 varve years B.P. The Swedish varve chronology, however, gives a C^{14} level of -2 to -4 percent over the 8500–12,500-year interval. The tree ring chronology indicates a level of ca. $+9$ percent around 7500 years B.P. The -2 percent in atmospheric C^{14} content around 8500 years B.P. for the Swedish varve chronology can only be attained if atmospheric C^{14} activity rose about 11 percent between 8500 and 7500 years B.P. (from -2 percent at 8500 years to $+9$ percent around 7500 years, as indicated by the tree ring chronology). At 9300 varves for the Lake of the Clouds chronology the C^{14} age is about 8700 years. Agreement between C^{14} age and varve age (like the Swedish varve chronology) can be obtained if the Lake of the Clouds varve count of 9300 is actually only 8500. This implies that the 7500–9300 Lake of the Clouds varve interval is wrong and that it actually should be 7500–8500 varves. The possibility of counting 800 extra varves over a 1000-year interval cannot be excluded, but it is unlikely because the proven error in the number of varves over a 7400-year interval is only 2 percent. A large change in atmospheric C^{14} content between 7500 and 8500 calendar years B.P. also conflicts with the constancy in atmospheric C^{14} content indicated by a floating tree ring chronology between 7050 and 7350 C^{14} years ago (ca. 7700–8000 calendar years) (Vogel et al. 1969).

The discrepancy between both varve chronologies may be due to the absence of about 800 varves from the Swedish chronology. Little C^{14} control is available for this chronology for the last 7500 years, and an error of less than 2 percent has not yet been proven. In addition, the possibility of missing varves around 8000 years ago (for instance at the transition from proglacial varves to nonglacial varves) cannot be *a priori* excluded. The difference between both varve chronologies can be solved with the extension of the tree ring chronology by an additional 1000 years. If the Lake of the Clouds chronology is proved to be correct, the following overall picture seems indicated:

1. Atmospheric C^{14} concentrations over the last 2500 years were within a few percent of the A.D. 1850 level

2. Between 5500 and 2500 years B.P. atmospheric C^{14} content was reduced from ca. $+9$ percent to the present base line

3. For the entire interval of 5500 to 13,000 years B.P., atmospheric C^{14} level was ca. 7 to 12 percent higher than at present, and may have resulted from:

a. Changes in earth magnetic dipole moment
b. Changes in exchange rate between atmosphere and the oceans, caused by a lower mean ocean temperature, lower sea level, and increasing ice cover

For the Lake of the Clouds varve chronology the change in exchange rate may explain the gradual reduction in atmospheric C^{14} content from ca. $+10$ percent 10,000 years ago to the present level. Changes in earth magnetic field intensity would add a C^{14} oscillation with an amplitude of ca. 4 percent and with a period of ca. 7000 years on this general trend.

LAKE SEDIMENTATION RATES

The long-term variations in atmospheric C^{14} content result in a contraction or expansion of the C^{14} time scale. Such time scale changes should be noticeable in C^{14}-dated lake sedimentation rates. A portion of these sedimentation rate variations is local and cannot be correlated among widely scattered lakes. Widespread synchronous changes in sedimentation rates, when they do occur, are either real changes due to worldwide climatic events or apparent changes due to the influence of long-term changes in atmospheric C^{14} concentration on the controlling dates. These "fictitious" changes in sedimentation rates, calculated from C^{14} dates, provide information on the absolute magnitude of long-term variations in atmospheric C^{14} content (Stuiver 1967, 1969). The lakes investigated are from widely different latitudes to avoid synchronous variations in sedimentation rate induced by local climatic changes. The sedimentation rates of noncombustible material, based on C^{14} years, are given in Figure 2 for five lakes; they have been normalized by equating the sedimentation rates for the last few thousand years to one. Four of the lakes (Lake Victoria, Rogers Lake, Lake Yueh Tan, and Jacobson Lake) show increased sedimentation "rates" around 3000 years ago, in agreement with the time scale change derived from the tree ring chronology. Important sedimentation rate changes, common to all lakes, are not found for the 3000–9000 C^{14} interval. The average sedimentation rate between 3800 and 9000 C^{14} years B.P. for Lake Victoria, Jih Tan, Yueh Tan and Jacobson Lake (with the exception of the 8000–9000-year interval B.P. for Jacobson Lake) is given in the upper curve of Figure 2. The maximum change in average sedimentation rate, expressed in C^{14} years, is about 20 percent. This suggests that a change exceeding 20 percent in C^{14} time scale for long periods during the 3800–9000 C^{14}-year interval is unlikely. A larger change in C^{14} time scale is possible only when the effect of long-term atmospheric C^{14} changes on apparent sedimentation rates has been masked by synchronous changes in real sedimentation rate.

An evaluation of the above restriction of about 20 percent on changes in time scale leaves the following alternatives (Stuiver 1969): (1) a continuation, or perhaps slight acceleration of the rising C^{14} activity trend between 2500 and 5500 calendar years B.P. back to 9000 C^{14} years B.P.; (2) a general leveling rather than rise of the trend of C^{14} activity during this time interval.

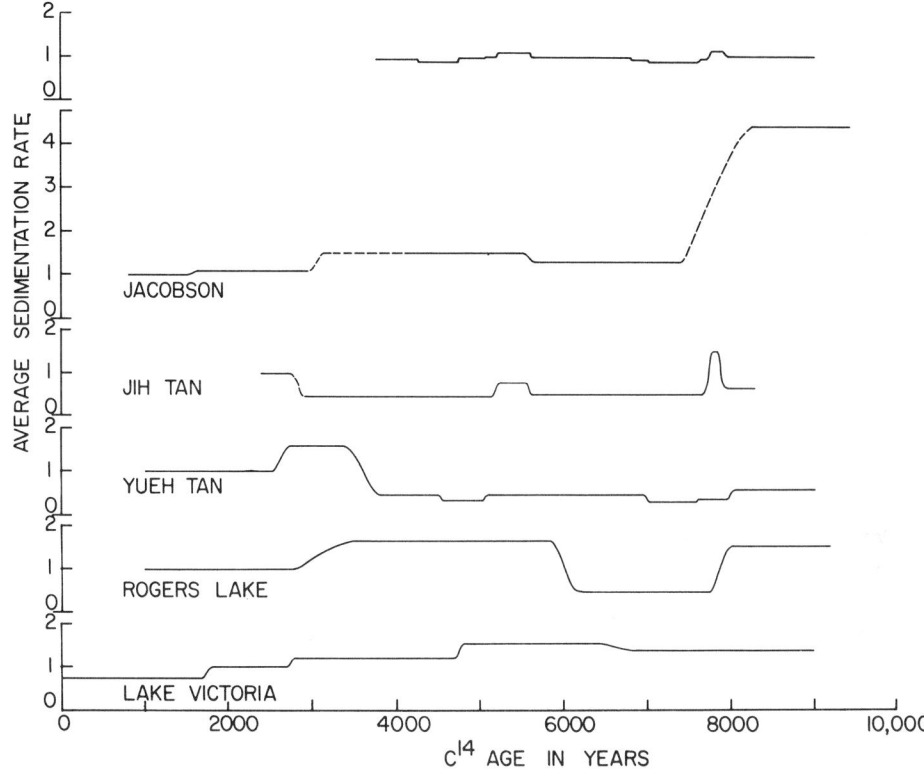

Fig. 2. C^{14} sedimentation rates for five lakes as a function of C^{14} age. The sedimentation rates have been normalized by equating sedimentation rates for the last few thousand years to one. The upper curve gives the average sedimentation rate for four lakes (see text).

The restriction on time scale changes seems to exclude the return to "normal" atmospheric C^{14} activities, as suggested by the Swedish varve chronology.

CHRONOLOGICAL DISTRIBUTIONS

Short-term atmospheric C^{14} oscillations appear indirectly in frequency distributions based on C^{14} dates. An example is the cycle of about 400 years with amplitude of about 3 percent, suggested by Suess (1969). The idealized oscillation given in Figure 3 has an amplitude of 2.5 percent and a period of 400 years. The ages of a large number of samples, formed uniformly over the interval of 800 calendar years, will cluster for this example around 200 and 600 C^{14} years when plotted as a function of C^{14} age. Samples formed during the entire 100–300 calendar years all have C^{14} ages close to 200 years because the decay (1 percent per 80 years) approximately equals the reduction in atmospheric C^{14} content. The resulting

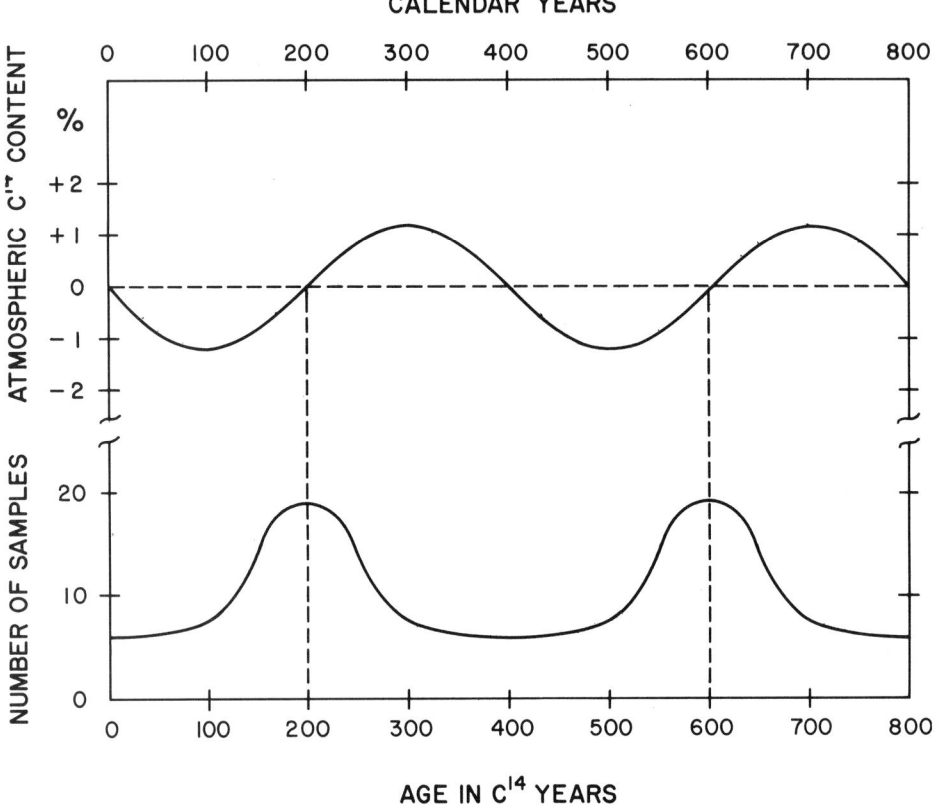

Fig. 3. The number of sample ages, in C¹⁴ years, for each decade plotted against C¹⁴ age (bottom curve). The C¹⁴ age distribution is caused by the oscillation in atmospheric C¹⁴ content given in the top curve. The number of sample ages, in calendar years, is 10 for each decade over the entire 800-year interval.

distribution depends somewhat on the accuracy with which the C¹⁴ age can be determined. Greater accuracy increases the difference between maxima and minima. The frequency distribution in Figure 3 is calculated for C¹⁴ dates with a standard deviation of about sixty years and a uniform distribution of ten samples per decade (calendar years).

For the 3500 to 6000 year interval B.P., the tree ring curve (Suess 1969) shows minima in atmospheric C¹⁴ activity (after removal of the long-term trend) at 3700, 4150, 4550, 4950, 5350, and 5700 years. The maxima in a frequency distribution of C¹⁴ dates of events randomly distributed in time are expected to be about 100 years older for a 400-year cycle. After conversion of calendar years to C¹⁴ ages, the maxima in the frequency distribution curve are predicted at 3450, 3800, 4100, 4350, 4700, and 5100 C¹⁴ years B.P.

Events dated by carbon-14 are normally not randomly distributed in time but are associated with specific climatic events, cultural layers, etc. Possibly excepted are dates obtained from sea level investigations. Sample depth is the only essential variable and stratigraphical considerations are generally absent. All available C^{14} dates of peat from North Sea marshes were recently compiled by Geyh (1969). The frequency distribution of the dates of a group of 330 samples showed distinct maxima and minima. Between 3000 and 5500 C^{14} years B.P. the following maxima for the North Sea coast are listed: 3100, 3700, 4050, 4350, 4750, and 5200 C^{14} years. The predicted ages for these maxima were 3450, 3800, 4100, 4350, 4700, and 5100 C^{14} years B.P. With the exception of the first maximum around 3100 years, the agreement is excellent between both series.

Various causes may be responsible for the shape of frequency distribution of C^{14} dates connected with sea level changes. Some maxima and minima are perhaps correlative with regression and transgression of sea level (Geyh 1969). However, the good agreement between observed and calculated maxima makes it likely that the C^{14} variations cause the main part of the frequency distribution variations. Conversely, the frequency distribution found for the C^{14} ages of the North Sea samples can be considered as a confirmation of the 400-year cycle in the tree ring chronology.

REVERSALS

Earth geomagnetic dipole moment, according to present-day views, was reduced about 60 to 80 percent during field reversals. The time required to complete a transition between polarity states is estimated to be from 1000 to 10,000 years (Cox 1969). The average intensity of the field in both polarity states appears to be about the same (Smith 1967).

The cosmic ray flux reaching the atmosphere will be nearly doubled when the earth geomagnetic dipole moment is reduced by 80 percent. The resulting increase in global C^{14} production causes the C^{14} level in the atmospheric ocean system to increase with a characteristic time of about 8000 years. If the duration of the transition between polarity states is 8000 years or more, and if the geomagnetic field is reduced by 80 percent over the entire interval, almost a twofold increase in atmospheric C^{14} content is expectable. For shorter transitions between polarity states, and with a reduction of 80 percent in geomagnetic field over only part of the transition period, the increase in atmospheric C^{14} is appreciably smaller. For a 3000-year transition period, and a harmonic magnetic field change with an 80 percent maximum, the expected increase is around 15 percent.

A reversal in earth magnetic field has been demonstrated for the late Quaternary in two formations of the Chain des Puys (Auvergne, France), one of those the Puy de Laschamp (Bonhommet and Babkine 1967). Potassium-argon determinations show an upper limit of about 20,000 years for the end of the Laschamp polarity event (Bonhommet and Zähringer 1969). A lower limit is given by a C^{14} age of 8700 years for overlying material.

This reversal is perhaps reflected in the C^{14} age anomaly found for Lake Jih

Tan sediment (Fig. 4). Samples from this lake were kindly provided by M. Tsu-kada. The fluctuation amounts to a 20 percent change in atmospheric C^{14} content and is independent of sedimentation rate assumptions because it involves an inversion in the time scale. The "oscillation" is perhaps the result of sample contamination and has to be confirmed in other cores before it can be associated unambiguously with the Laschamp polarity event.

If constant sedimentation rate is assumed, the polarity transition lasted approximately from 24,000 to 20,000 years B.P., with an increase in atmospheric C^{14} content of about 20 percent and with a return to normal activity in about 1000 years. The return to normal is somewhat too fast, but the order of magnitude of the C^{14} anomaly agrees with the calculated values.

The only other time reversal known to the writer is for Searles Lake, where an age of 24,700 years is found for a sample that, based on a series of dates in stratigraphical order should have been about 26,300 years (Stuiver 1964, and unpublished 1968 results). Correction for the C^{14} deficiency of the lake water reduces the ages to about 22,500 and 24,100 years. This anomaly seems about 4000

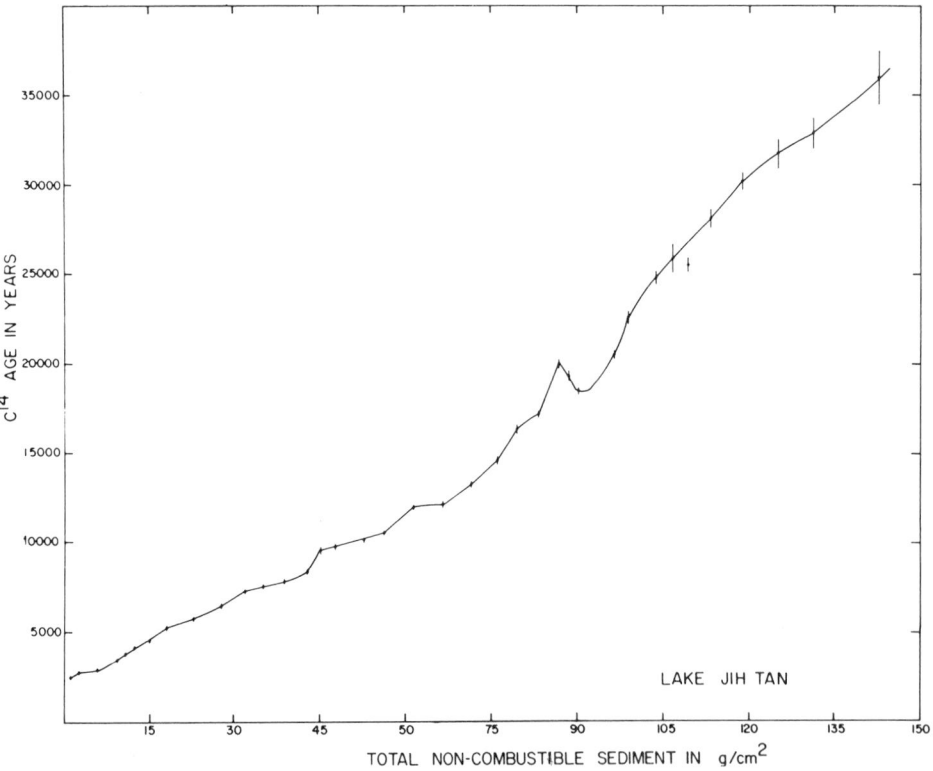

Fig. 4. C^{14} age of Lake Jih Tan sediment as a function of total accumulated noncombustible sediment.

years older than the one discussed for Lake Jih Tan. Perhaps the anomaly is associated with the beginning of the polarity event. If so, the total duration of the Laschamp polarity event would have been restricted to about 4000 years.

It should be realized that additional evidence for the C^{14} anomalies is needed before the above picture of the Laschamp polarity event can be completely accepted.

Summary. The evidence is reviewed for late Quaternary C^{14} variations obtained from tree rings, historically dated samples, varves, lake sedimentation rates, chronological distribution of C^{14} dates, and time reversals.

Excellent agreement is found for the various methods over the past 7500 calendar years. New data for the Lake of the Clouds varve chronology disagree with the Swedish varve chronology for the interval of 7500–10,000 years B.P. The end of the Laschamp magnetic polarity event is tentatively dated at 20,000 years B.P.

Acknowledgments. The C^{14} research reported was supported by NSF grants GA 1157 and GA 17910. The stable isotope measurements were made under AEC contract AT (30-1) 3204.

REFERENCES

Bonhommet, N., and J. Babkine, Magnétisme terrestre, *Compt. Rend. Ser. B, 264*, 92, 1967.

Bonhommet, N., and J. Zähringer, Paleomagnetism and potassium argon determinations of the Laschamp geomagnetic polarity event, *Earth Planetary Sci. Letters, 6*, 43, 1969.

Buddemeyer, R. W., A radiocarbon study of the varved marine sediments of Saanich Inlet, British Columbia, Ph.D. thesis, University of Washington, Seattle, 1969.

Cox, A., Geomagnetic reversals, *Science, 163*, 237, 1969.

Damon, P. E., Climatic vs. magnetic perturbation of the atmospheric carbon-14 reservoir, In *Radiocarbon Variations and Absolute Chronology*, edited by I. U. Olsson, Twelfth Nobel Symposium, Almquist & Wiksell, Stockholm, and Wiley, New York, 1970.

Damon, P. E., A. Long, and D. C. Grey, Fluctuations of atmospheric C^{14} during the last six millennia, *J. Geophys. Res., 71*, 1055, 1966.

DeVries, H., Variation in concentration of radiocarbon with time and location in earth, *Koninkl. Ned. Akad. Wetenschap., Proc., Ser. B, 61*, 94, 1958.

Fergusson, C. W., Bristlecone pine: Science and esthetics, *Science, 159*, 839, 1968.

Fromm, E., Geochronologisch datierte Pollen-diagramme und Diatoméen-analysen aus Ångermanland, *Geol. Foren. Stockholm Forh., 60*, 365, 1938.

Geyh, M. A., Versuch einer chronologischen Gliederung des marinen Holozäns an der Nordsee Küste mit Hilfe der statistischen Auswertung von ^{14}C Daten, *Z. Deut. Geol. Ges., 118*, 351, 1969.

Grey, D. C. Geophysical mechanisms for C^{14} variations, *J. Geophys. Res., 74*, 6333, 1969.

Libby, W. F., *Radiocarbon Dating*, 174 pp., University of Chicago Press, Chicago, 1955.

Lidén, R., Den senkvartära strandförskjutningens förlopp och Kronologi i Ångermanland, *Geol. Foren. Stockholm Forh., 60*, 397, 1938.

Lingenfelter, R. E., Production of carbon 14 by cosmic-ray neutrons, *Rev. Geophys., 1*, 35, 1963.

Michael, H. N., and E. K. Ralph, Correction factors applied to Egyptian radiocarbon dates from the B.C. era, in *Radiocarbon Variations and Absolute Chronology*, edited by Olsson, Twelfth Nobel Symposium (see Damon, above), 1970.

Nilsson, E., Södra Sveriges senkvartära historia, *Kgl. Svenska Vetensk. Akad. Handl.*, ser. 4, 12(1), 1, 1968.

Smith, P. J., The intensity of the ancient geomagnetic field: A review and analysis, *Geophys. J.*, 12, 321, 1967.

Stuiver, M., Variations in radiocarbon concentration and sunspot activity, *J. Geophys. Res.*, 66, 273, 1961.

Stuiver, M., Carbon isotopic distribution and correlated chronology of Searles Lake sediments, *Am. J. Sci.*, 262, 377, 1964.

Stuiver, M., Carbon-14 content of 18th and 19th century wood; variations correlated with sunspot activity, *Science*, 149, 533, 1965.

Stuiver, M., Origin and extent of atmospheric ^{14}C variations during the past 10,000 years, Proc. Radioactive dating and methods of low-level counting, Intern. Atomic Energy Agency, STI/Pub/152, Vienna, 27, 1967.

Stuiver, M., Long-term C^{14} variations, in *Radiocarbon Variations and Absolute Chronology*, edited by Olsson, p. 197, Twelfth Nobel Symposium (see Damon, above), 1970.

Stuiver, M., and H. E. Suess, On the relationship between radiocarbon dates and true sample ages, *Radiocarbon*, 8, 534, 1966.

Suess, H. E., Secular variations of the cosmic-ray produced carbon-14 in the atmosphere and their interpretations, *J. Geophys. Res.*, 70, 5937, 1965.

Suess, H. E., The three causes of the secular carbon-14 fluctuations, their amplitudes and time constants, in *Radiocarbon Variations and Absolute Chronology*, edited by Olsson, Twelfth Nobel Symposium (see Damon, above), 1970.

Tauber, H., The Scandinavian varve chronology and C^{14} dating, in *Radiocarbon Variations and Absolute Chronology*, edited by Olsson, Twelfth Nobel Symposium (see Damon, above), 1970.

Vogel, J. C., C^{14} trends before 6000 B.P., in *Radiocarbon Variations and Absolute Chronology*, edited by Olsson, Twelfth Nobel Symposium (see Damon, above), 1970.

Vogel, J. C., W. A. Casparie, and A. V. Munaut, Carbon-14 trends in subfossil pine stubs, *Science*, 166, 1143, 1969.

Welten, M. Pollenanalytische, stratigraphische und geochronologische Untersuchungen aus dem Faulenseemoss bei Spiez, *Veroeffentl. Geobot. Inst. Rubel Zurich*, 21, 201 pp., 1944.

Wenner, C. G., Comparison of varve chronology, pollen analysis and radiocarbon dating, *Stockholm Contrib. Geol.*, 18(3), 75, 1968.

Willis, E. H., H. Tauber, and K. O. Munnich, Variations in the atmospheric radiocarbon concentration over the past 1300 years, *Radiocarbon*, 2, 1, 1960.

John Imbrie and Nilva G. Kipp

5. A NEW MICROPALEONTOLOGICAL METHOD FOR QUANTITATIVE PALEO-CLIMATOLOGY: APPLICATION TO A LATE PLEISTOCENE CARIBBEAN CORE

Since the pioneer work of Schott (1935), micropaleontologists have used fossil plankton in deep-sea cores to infer Pleistocene marine climates. The general approach has been to associate key species or groups of species with broad latitudinal ranges and to make semiquantitative interpretations in terms of climatic zones. Phleger et al. (1953), for example, drew curves recording the proportions of high-, mid-, and low-latitude forms as a function of core depth. Ericson and his colleagues (e.g. Ericson et al. 1964) have used abundance curves of selected species and coiling varieties of planktonic foraminifera as a basis for drawing paleoclimatic curves ranging from "warm" to "cold." Other workers, including McIntyre (1967), have made quantitative interpretations using ecological range data on Recent and sub-Recent forms to infer the geographic position of Pleistocene isotherms.

All paleoecological work on Pleistocene plankton, including the present study, is based on the fundamental assumption that the pelagic ecosystem being sampled today has remained essentially unchanged during the Pleistocene. In particular, for a given species or species assemblage it is assumed that its ecological responses to physical and chemical parameters of the ocean are unchanged. Our study carries this assumption to its logical limit: writing equations relating portions of the biological side of this ecosystem to selected physical parameters of the oceans—and then using those equations on samples from cores to make fully quantitative estimates of past marine climates.

So stated, our rationale is clear and our scientific objective seems simple. This simplicity is deceptive. It has taken several years to work out optimum mathematical, taxonomic, and laboratory procedures. As a result the reader may be disappointed to learn that we report here only one actual application of the method. However, the results, reported below, are internally consistent, reasonable in magnitude, and in fair agreement with the limited amount of independent external evidence available. Thus encouraged, we are proceeding to analyze a group of cores distributed throughout the Atlantic. When the work on this set of cores is complete we will have, for the first time, a really firm basis for evaluating the method. Meanwhile, we hope that our efforts will encourage others. Although our work to date has been exclusively on planktonic foraminifera, in principle the method should work with a wide variety of materials including fossil pollen, benthic foraminifera, coccoliths, and radiolaria. The value of studying several groups of plankton simultaneously is clearly shown by Bandy (1961).

In the final section of this paper we have used our results on one core to test a major climatic theory. One objective here is to demonstrate that fully quantified paleoclimatic curves are amenable to more rigorous treatment than other types. In this case we use the classic techniques of time series analysis to identify significant component frequencies in the quasiperiodic paleoclimate curves, and compare the calculated spectra with frequencies predicted by the Milankovitch theory.

Considerable experience in explaining the method has shown that geologists and paleontologists find no difficulty if they acquire first a clear understanding of the nature of the raw data. For this reason we present in the next section a brief description of the data. Following that the method itself is developed in the abstract, and then applied to the actual data for the core tops and core V12-122. Finally, the quantitative results for the core are used as a test of the Milankovitch theory.

DATA

Core-Top Samples

Initially, 71 deep-sea cores from the Lamont-Doherty collection were selected for study with the advice and help of David B. Ericson and Goesta Wollin. Choice of these cores was governed by the following objectives:

> 1. To study core tops with a reasonably even geographic distribution over the Atlantic and adjacent portions of the Indian Oceans.
> 2. To sample bottom sites underlying surface waters with a wide range of salinities and temperatures, and with different combinations of salinity and temperature values.
> 3. To avoid core tops with evidence of carbonate dissolution.
> 4. To avoid core tops with a significant fraction of nonpelagic components.
> 5. To study core tops containing only Recent and near-Recent foraminifera.

Ten of the core tops originally selected were eliminated from consideration because they contained high ratios of benthic to planktonic foraminifera, or showed other evidence of solution, or had unusual quantities of detrital quartz or

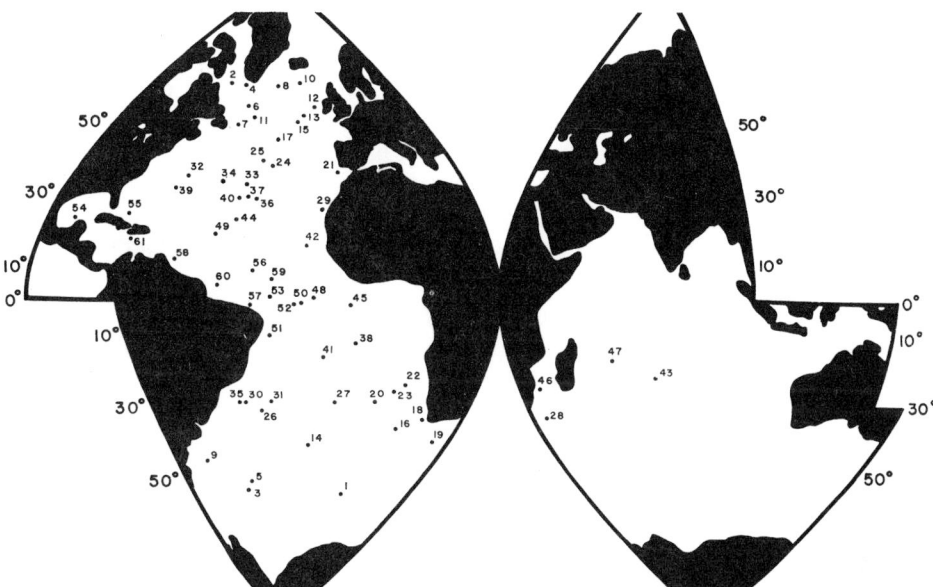

Fig. 1. Location of 61 core top samples. For description of the cores see Table 1.

other nonpelagic debris. Location of the remaining 61 core top samples is shown in Figure 1. With one exception (item 5), the objectives listed above have been reasonably well achieved. The geographic distribution is wide, ranging from latitude 60°N to 54°S. The Gulf of Mexico and Caribbean are each represented by one sample, the west-central Indian Ocean by four. The most significant geographic gaps are probably the uneven coverage of the western part of the North and South Atlantic. Samples without dissolution are difficult to come by in these areas. Sampled depths range from 1018 to 4610 meters; 82 percent of the cores were raised from depths less than 4000 meters. Average winter temperatures at the surface of the ocean at the 61 localities range from about −1° to 26°C, and the corresponding surface salinities from about 33 to 37 per mil. Exact data on location and depth are given in Table 1; data on temperature and salinity in Table 13.[*]

Ideally, each core top sample should represent the accumulation of tests of plankton being deposited at a particular site over the past century or some similarly short period of time. Given a core top 1 cm in thickness, a theoretically perfect sample would contain the record of the past 500 years in a typical Atlantic core with a deposition rate of 2 cm per 1000 years. But the action of burrowing animals must result in an admixture of older materials, the magnitude of the admixture decreasing exponentially with distance from the sediment surface. The intensity of this effect is not easy to estimate and can be expected to vary from place to place with variations in deposition rate and other factors. As noted by

[*] Tables are at end of text.

Berger and Heath (1968), published estimates of normal mixing depths range from 2 to 5 cm. Most workers seem to have followed the conclusion of Arrhenius (1963, p. 659) that "the mean mixing depth, above which 50% of the extraneous material is located, is of the order of 4–5 cm." In some cases this estimate is clearly too high. Alpha track data given by Arrhenius (1963, Fig. 3) for mixing below an unconformity suggests, for example, that intense burrowing is limited in some places to a 2- or 3-cm layer. Unpublished work by Madeleine Briskin in our laboratory bears on this point. Sampling a South Atlantic core (V12-18) with an average sedimentation rate of 0.4 cm per 1000 years, she has noted rapid fluctuations of constituent particles between samples located 5 cm apart. We conclude that in a typical Atlantic core top affected only by burrowing we will find fossils deposited mostly during the past 2500 years, with a proportionately greater representation of younger materials.

For two reasons even this limited ideal is difficult to achieve. In the first place there is always the possibility that the surface layer may be lost in the process of taking a core—a danger that is particularly real in dealing with a set of cores like ours in which both gravity and piston techniques were used. Second, the core may penetrate a stratigraphic hiatus in which the Recent sediments have been removed. These difficulties cannot be entirely avoided, but they can be minimized by making a biostratigraphic analysis of fossils in the upper portion of the cores. Wherever David Ericson or the authors have been able to check the sequence of faunas toward the top of a core, they give evidence that the top sample is postglacial—i.e. contains fossils appropriate to the Recent Z zone overlying the glacial Y zone fauna. Typically, three or four equally spaced samples constitute the record of the Z zone. Taking the duration of this interval as 11,000 years we may assume that our typical sample should represent no more than the past 4000 years.

From the biostratigraphic arguments just given, and the scant data on burrowing, we conclude that all the core top samples represent postglacial deposition, that most represent the last 2000 to 4000 years, and that some may contain materials deposited in the age range 4000–8000 B.P. An independent argument for this conclusion can be made from the results of our ecological analysis reported below, where we demonstrate that a clear quantitative relationship exists between core top faunas and surface ocean temperatures; and that objective, quantitative measures of faunal composition display systematic geographic gradients. We do not, however, claim that all of our core top samples represent the past 2500 years or even the same fraction of the past 11,000. In fact, we surmise that our largest single source of error is the chronological heterogeneity of our core top samples. Other possible sources of error include sampling, identification, and inadequacies in the mathematical model.

One advantage of the mathematical model developed below should be pointed out here: because a least-squares fit is employed in writing the paleoecological equation, the effect of nonsystematic errors—which include chronologic heterogeneity, sampling error, and identification error—should not cause serious discrepancies in the paleoenvironmental estimate.

Core V12-122

In a later section of this paper equations are derived relating quantitative data from 61 core top samples to physical oceanographic parameters of the overlying surface waters. One object of this procedure—although by no means the only one —is to provide a means by which paleontological samples from deeper layers in any Atlantic core may be made to yield quantitative paleoclimatic estimates. For the initial test of this paleoecological method we selected one core: Caribbean core V12-122 (code number 61 on Fig. 1). This core was selected for a number of reasons.

1. Work by Ericson and Wollin (1968) has shown that a continuous biostratigraphic record of zones U through Z are represented without evident hiatus. According to their chronology this interval spans approximately the past 450,000 years of Cenozoic time.

2. Its location in the Caribbean makes it likely that carbonate dissolution effects are minimal or absent.

3. Measurements of Pa^{231} and Th^{230} and calculations by Broecker and Ku (1969) and Broecker and Van Donk (1970) suggest that the accumulation rate in V12-122 is approximately constant, averaging about 2.35 cm per 1000 years. This radiometrically determined rate places the paleontological U–V zonal boundary at 890 cm depth about 380,000 years B.P., a date consistent with independent estimates for the boundary (Ericson and Wollin 1968) made by interpolation in the magnetic time scale. Rona and Emiliani (1969), using similar radiometric methods on other cores whose correlation with V12-122 is clear, argue for a time scale shorter by 25 percent. For explanations of the analytical differences reflected in these conflicting estimates the reader is referred to Rona and Emiliani (1969), Emiliani and Rona (1969), and Broecker and Ku (1969). As stressed by Broecker and Van Donk (1970), the independent dating of the U–V boundary provided by magnetic stratigraphy strongly supports the longer time scale, and we have therefore adopted it in this paper.

4. Extensive work by Emiliani (e.g. 1955, 1966) provides a number of O^{18}/O^{16} curves for Pleistocene cores in the Caribbean and equatorial Atlantic showing remarkably consistent fluctuation patterns. Although their exact interpretation is uncertain, there can be no doubt that these fluctuations contain important paleooceanographic information: on ice volume, water composition, water temperature, or some combination of these effects. One purpose in choosing a Caribbean core is to confront isotopic measurements with independent, quantitative paleoclimatic interpretations based on fossils. Van Donk (Broecker and Van Donk 1970) has measured oxygen isotopes in samples of *Globigerinoides ruber* and *G. sacculifer* picked by us from V12-122 (Fig. 32). His results correlate well with curves published by Emiliani.

5. There is no evidence of turbidite deposition or significant influx of coarse terrigenous debris.

6. The length of the core (10.9 meters) is sufficient to permit the construc-

tion of a reasonably detailed paleoclimatic curve. Samples spaced 10 cm apart yield 110 points.

Detailed work on this core has confirmed our initial impressions of depositional continuity and lack of dissolution. For evidence on the former we cite the O^{18}/O^{16} curves calculated by Van Donk. These show fluctuation patterns strikingly similar to those published by Emiliani for nearby sites in the Caribbean and Atlantic (Broecker and Van Donk 1970). Several observations confirm the lack of calcite dissolution in the core. Of particular importance is the fact that benthic foraminifera average 0.52 percent and never exceed 2 percent of the total foraminiferal content of the > 149-μ fraction. Forty-seven of the samples have no benthonics at all. As shown by Arrhenius (1952), this ratio increases markedly when carbonate is dissolved. Parker (1970) regards "with suspicion" any sample with more than 1 percent benthonics. Another evidence is the relatively high content throughout the core of *Globigerinoides ruber*, cited by Berger (1968) as the planktonic species most sensitive to dissolution.

All in all we are satisfied that V12-122 provides a fair first test of our method. The one disadvantage of using a Caribbean core is that the rather limited range of paleoclimatic fluctuation to be expected—at least compared to that at high latitudes—will almost certainly give rise to a relatively low signal/noise ratio.

Paleontological Data

Each of the 61 core top samples and the 110 core samples were processed in essentially the same way, as described below. For all V12-122 samples, and some of the core top samples, steps 1–4 were done in the Lamont-Doherty Observatory under the direction of David B. Ericson. The remainder of the samples were processed by us. The procedure is as follows:

Step 1. A 10- to 20-gm sample of a 1-cm core slice is dried and weighed.

Step 2. The sample is disaggregated by soaking in water. For samples processed by us, H_2O_2 is added and the sample cleaned ultrasonically.

Step 3. The sample is wet-sieved: a 74-μ sieve for the Lamont-Doherty laboratory, a 63-μ sieve for ours.

Step 4. The coarse fraction (> 74 μ; or > 63 μ) is oven-dried and weighed. The weight percent of the coarse fraction is calculated.

Step 5. The coarse fraction is randomly split as many times as needed to yield a study sample of approximately 300 whole foraminifera > 149 μ.

For our core top samples more than 300 foraminifera were studied; the average number counted was 656 specimens. The smallest sample had 250 and the largest 1550 specimens. For samples in V12-122, an average of only 207 specimens per sample was counted. The smallest sample (at 720 cm) is 84; the largest (at 0 cm) is 614 specimens. The smaller sample size in V12-122 is the result of an investigation to determine the optimum mesh size. In future work we will count approximately 300 specimens.

Quantitative paleontological data on foraminifera can only be obtained by taking a census of a defined size fraction of the sediment. It is therefore of the utmost importance for the future development of such work that different laboratories scan the same fraction. With this in mind, we devoted many months to experimenting with different mesh sizes, including 105 μ, 125 μ, 149 μ, and 177 μ. Each set of data was subjected to quantitative manipulation, including factor analysis. Based on this work we use and recommend the $>$ 149-μ fraction. Smaller fractions give rise to too many uncertainties in the identification of small specimens and require too long to process. Larger mesh sizes yield undesirable loss of small species, and small specimens of larger species. The 149-μ mesh yields the optimum results in terms of a ratio of ecological information obtained to effort expended. A skilled worker, having gained familiarity with the fauna, can sort and count twenty samples per week.

Step 6. The random split is sorted through a 149-μ sieve and the finer residue saved for reference.

Step 7. The study sample (approximately 300 specimens of the $>$ 149-μ fraction) is strewn on a standard 60-square micropaleontological slide, sorted into taxa, and counted. All whole or essentially whole specimens are identified. Each taxonomic category is segregated on a portion of the slide and glued into position. We recognize 27 species and coiling varieties of planktonic foraminifera that occur in at least one of our samples with an abundance of 1 percent. Benthic foraminifera are grouped together and counted. Inevitably, specimens are encountered which defy certain identification. These are segregated as "unidentified" and included in the census.

Step 8. Raw census data for each sample are punched for computer processing across a set of cards. A utility routine in our Q-mode factor analysis program (CABFAC) recalculates the raw census data for the planktonic taxa into percent form and punches the result on cards, including that for "unidentified." For all subsequent data processing the unidentified percentage is ignored.

It hardly needs to be stressed that the taxonomic portion of this study is the foundation upon which all else rests. We recognize some 27 species and coiling varieties of planktonic foraminifera and base our species definitions and nomenclature on that of Allan Bé (Bé and Hamlin 1967) and Parker (1962, 1967). Where there is a nomenclatural controversy for the same form we have followed Bé. Two points of difference may be noted between our work and that of some previously published works on the Atlantic. (1) We have considered *Globorotalia menardii* as a single indivisible complex because we have not been able to separate *G. tumida* and *G. menardii s. s.* (strict sense) both in core top samples and in all levels of V12-122 with sufficient certainty to take a census of all specimens in the $>$ 149-μ fraction. We did, however, perform one factor analysis on core top samples in which the two forms were split. The result added nothing to a factor analysis of the same data with *G. menardii* considered as a single complex. (2) We recognize and find ecologically useful *Globigerina falconensis* Blow and *G. calida* Parker.

Tables 2 and 3 give a complete listing of taxonomic categories used in this report.

Oceanographic Data

At each of the 61 core sites an estimate of three oceanographic parameters of the surface water was obtained: average winter temperature, average summer temperature, and average salinity. These estimates were made by interpolation from charts compiled by Defant (1961, Plates 3A, 3B, 5). They are given in Table 13 as "observed" values to facilitate comparison with estimated values obtained later from the paleoecological equation.

A general picture of the relation between salinity and temperature may be had by inspecting the pattern of 61 core locations plotted on Figure 31. The general and well-known tendency for cooler waters to have lower salinities is shown. From the point of view of ecological analysis this phenomenon poses a problem: it makes difficult the aim of studying separately the role of salinity and temperature. Only in low latitudes, where the relatively low salinity of tropical water runs counter to the worldwide pattern, is it possible to consider the variables as statistically independent.

THE METHOD

The aim of the method described below is to extract from paleontological data on a given deep-sea core an objective and quantitative estimate of the physical state of the surface waters during its deposition. Such an estimate can, in principle, be based either on a knowledge of the distribution of living plankton or on paleontological data from core tops. Because core top faunas are to some extent chronologically heterogeneous, a method based on plankton might be preferred. However, it is widely recognized that plankton data are not directly comparable to fossil data even if comparison is made to a set of core top samples known to represent the last centimeter of accumulation. At least four reasons can be cited. (1) It is difficult to compare abundance data obtained by sieving dead, spineless, tissue-free skeletons with those obtained from live material in plankton tows. (2) Differential preservation of carbonate tests before and after accumulation on the sea floor distorts the fossil record. (3) Burrowing activities of the benthos mix older fossil material with younger. (4) Seasonal succession in plankton makes it necessary to compare fossil faunas not with spot plankton tow data but with calculated average yearly plankton production. (5) Although the authors know of no data to support the hypothesis, the possibility must be kept in mind that transportation effects intervening between death and deposition commonly and significantly distort the fossil record.

With these factors in mind, the virtue of core top data, on which this study is based, is apparent: they are directly comparable to older samples in the stratigraphic record. We freely acknowledge, however, that the ultimate objective of Pleistocene paleoecological work will not be achieved until the complex links between plankton and core top faunas have been thoroughly and quantitatively analyzed. Work of this kind is being carried on and gives promise of important results (e.g. Bé 1959, 1960, 1969; Belyaeva 1964; Cifelli 1967; Berger 1968; Jones 1968; Ruddiman 1968; Tolderlund 1969; Boltovskoy 1969; Kennett 1969; Parker 1970; Bé and Tolderlund 1970).

Our method is based on seven assumptions:

1. That core top faunas are systematically related to the physical nature of the overlying surface waters.

2. That yearly average salinity and temperature data are, or are linearly related to, ecologically significant aspects of the surface waters.

3. That geographic variations in the composition of foraminiferal faunas can be meaningfully represented by a linear mixing model involving a small number (m) of end-member assemblages, as expressed in Q-mode factor analysis.

4. That the m assemblages respond to no more than $m - 1$ statistically independent sets of physical oceanographic parameters. We do not require that $m - 1$ independent physical factors be identified or measured—only that no more than $m - 1$ significant factors operate; and, as stated in assumption 2, that our temperature and salinity data are linearly related to one or more of them.

5. That the response of each assemblage can be approximately represented as an m-dimensional paraboloid. Algebraically, this is equivalent to considering quantitative data on each assemblage as the dependent variable of an equation of the second degree in the $m - 1$ independent variables representing the physical parameters of the ecosystem.

6. That the ecosystem under study (Atlantic plankton) has not changed significantly during the last 450,000 years. Specifically we require that Recent species of planktonic foraminifera are essentially the same biological entities as they have been during the study interval, and that they have not changed significantly their ecological responses.

7. That differential solution of carbonate at the site of Caribbean core V12-122 has either not been significant or has not changed during the deposition of the core.

Our method can best be described in terms of five sequential procedures. (A) The raw paleontological data on core tops and cores are collated and suitably transformed. (B) Core top data are factor-analyzed into varimax assemblages. (C) A least-squares technique is used to write a set of paleoecological equations relating the varimax assemblages to observed oceanographic parameters. (D) The fossil data from a core are described in terms of the core top varimax assemblages. (E) The paleoecological equations are used to estimate paleoenvironments. For clarity of explanation it is useful to subdivide these procedures into fifteen computational steps. In the outline below, these steps are numbered continuously from A-1 through E-15.[1]

---

1. With one exception, programs used to make necessary calculations are standard and available at any computer center with a 210 K byte core memory. The program CABFAC, which performs factor analysis with several unique options necessary for the method, will be available sometime during 1971 (Klovan and Imbrie 1971).

Bold-face capital letters stand for matrices.

A. Collation and Transformation of Paleontological Data

Step A-1: Assemble percent data for core tops ($_p\mathbf{X}_{ct}$) *and core* ($_p\mathbf{X}_c$). Percent data for the core top samples are designated as the matrix $_p\mathbf{X}_{ct}$; for the core samples as $_p\mathbf{X}_c$. Each matrix has n columns corresponding to n observed species.

Step A-2: Eliminate rare species. In order to minimize sampling error, species are eliminated from consideration which do not exceed in at least one core top and one core sample a given percentage figure. In the actual data of this study a cutoff figure of 2 percent is adopted, based on considerable experimentation with higher and lower figures. With each trial the entire procedure outlined below was carried at least through step B-8. Runs with as few as 12 and as many as 26 species were made. Although these experiments clearly demonstrated the stability of the system and its ability to perceive similar ecological relationships in data matrices of varying width, an optimum signal/noise ratio was achieved when 22 species were employed, corresponding to a cutoff value of 2 percent.

Step A-3: Calculate percent range data: $\mathbf{X}_{ct}$ *and* $\mathbf{X}_c$. The investigator must now make an important decision: whether to treat the data "as is" in percent form —assuming that the relative ecological importance of each species is accurately represented by its relative abundance in the > 149-μ fraction—or to transform the data in some way to give each species equal weight in the computations which follow, regardless of its absolute representation in the >149-μ fraction. The second procedure has the advantage that it extracts the maximum ecological information from the data, since there is no reason to suppose that the ecological importance of a species is correlated with its average dimensions. It has, however, the disadvantage of magnifying counting error. For species represented by relatively few specimens the noise thus generated can be troublesome. We are conducting extensive tests to determine which procedure yields the best results. In the present study we employ a transformation that gives each species equal weight in the calculations.[2]

The method of transformation we use expresses the abundance of each species as percentage of its total observed range. Abundance data for each species are scanned over the entire core and core top matrices; the maximum and minimum observed values are noted; and each original figure for a sample is expressed as a percentage of the total range observed. In each of the species recorded here the minimum observed abundance is zero. The observed maximum abundances are given in Table 8. After transformation the percent range data matrices are defined as $\mathbf{X}_{ct}$ and $\mathbf{X}_c$ for the tops and core respectively.

Other transformations could be used. Statisticians sometimes advocate a procedure in which a variable is linearly transformed to have zero mean and unit

---

2. Note added in proof: Experience gained since completing this paper has convinced us that more reliable paleoecological equations are obtained by eliminating Step A-3, i.e. by factoring the untransformed percent data. Not only are the resulting equations characterized by slightly higher multiple correlation coefficients, but the plots of temperature estimates as a function of core depth have significantly less high-frequency noise.

standard deviation. Such a procedure is not, however, appropriate for data of the sort represented by our core tops: the value of the standard deviation reflects too strongly the frequency distribution of samples in various environments.

The percent range transformation is calculated as an option of the Fortran IV program CABFAC.

B. Factor Analysis of Core Top Data: $\mathbf{U}_{ct} = \mathbf{B}_{ct}\mathbf{F} + \mathbf{E}$

Our chief aim in studying core tops is to write equations relating surface temperatures and salinity to the core top faunas. If we were dealing with only one species we could proceed immediately. However, the fact that we are dealing with a number of species ($n = 22$) that are generally not statistically independent requires that we first process the data in some way to extract nonredundant taxonomic information, expressed as m statistically independent parameters. We are entitled to hope that $m << n$.

This objective can be achieved simply, quickly, and with a high degree of objectivity by means of a mathematical procedure known as Q-mode factor analysis. Details of the mathematical theory are available elsewhere (Imbrie and Van Andel 1964; Manson and Imbrie 1964; Krumbein and Graybill 1965) and we will here confine ourselves to an outline of the procedure. All the necessary computations (steps A-3 through B-8) are performed by means of a Fortran IV program CABFAC in a minute or so of high-speed digital computer time.

Consider the hypothetical data of Figure 2, a 15-station traverse with abundance data on seven species. The postulated situation represents the maximum possible complexity: each of the seven species varies independently of others (correlations of zero), and it would therefore require seven parameters ($n = m = 7$) to describe the geographic variations in the fauna. If we consider each param-

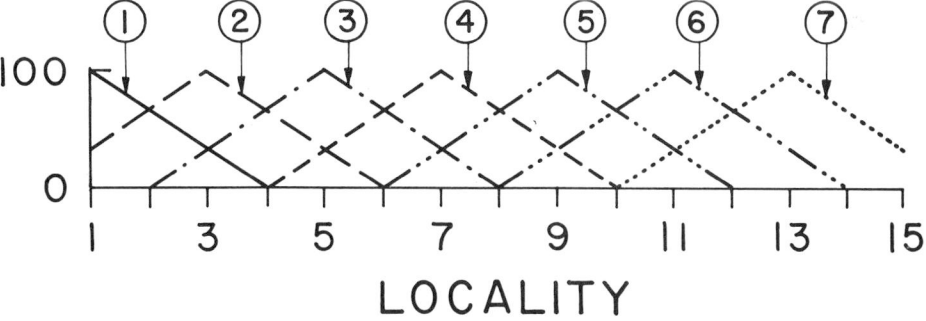

Fig. 2. Plot of hypothetical data on the abundance of seven species varying in response to an ecological gradient sampled at fifteen localities along a traverse. Abundance scale arbitrary. The model is designed to illustrate the maximum synecological complexity possible, with all seven species having different optima and exhibiting independent responses. A factor analysis of these data would yield exactly seven monospecific assemblages.

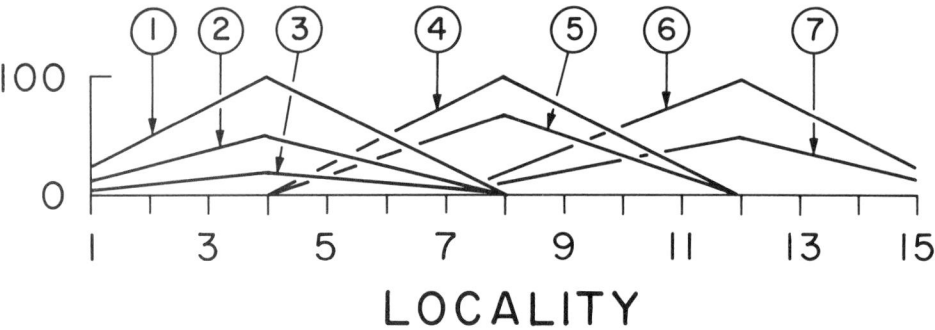

Fig. 3. Plot of hypothetical data on the abundance of seven species varying in response to an ecological gradient sampled at fifteen localities along a traverse. Abundance scale arbitrary. In contrast to the model illustrated on Figure 2, three species
assemblages react independently; within each assemblage constituent species retain
constant proportions as its contribution to the total fauna varies. A factor analysis of these
data would yield exactly three varimax assemblages, each represented approximately
by observed samples at localities 4, 8, and 12.

eter as an assemblage, then each assemblage consists of a single species. Figure 3
shows another possible set of data for the same traverse. Here we postulate ecological interrelationships yielding simple covariation patterns. Three assemblages can be defined. Within each, the proportion of constituent species remains
constant. A Q-mode factor analysis of these data would reveal this covariation
pattern by expressing the observed fauna at every locality as proportions of three
assemblages (also called factors or end members) within which the proportions
of constituent species remain constant. If we ask, "What are the assemblages?"
we could describe them by noting the composition of samples at localities 4, 8,
and 12 where they exhibit their maximum abundance. If we wished to study geographic variation we could prepare three maps, each portraying the distribution
of an assemblage, rather than seven maps showing the individual species patterns.
The factor analysis would not only eliminate ecological redundancy and simplify
problems of data display but offer insights into the underlying structure and complexity of the ecosystem as well. Blackman and Somayajulu (1966) obtained significant results on Pacific foraminifera by means of this method.

The form of Q-mode factor analysis just described is known as an *oblique
solution* (see Imbrie and Van Andel 1964). This designation refers to the fact
that the assemblages are defined as the *m* samples exhibiting the maximum observed compositional dissimilarity—in this case, samples at localities 4, 8, and 12.
For samples 4 and 8 the two samples have, in fact, nothing in common, hence the
7-dimensional vectors representing them are *orthogonal*. For samples 8 and 12,
however, the presence of species 6 and 7 at both localities is reflected geo-
metrically by an *oblique* angle between the corresponding vectors.

For the present problem, we require a form of factor analysis that will resolve the data into m theoretical assemblages (reference vectors) which are completely distinct taxonomically—i.e. are represented by mutually orthogonal n-dimensional vectors. For simplicity in interpretation we also require that each of these theoretical assemblages (reference vectors) be as close as possible to the m reference samples of the oblique solution. These objectives are achieved by a *varimax solution* of a Q-mode factor analysis. The oblique solution is conceptually the simpler. Unfortunately, the varimax solution is required if we are to attain our objectives.

Before discussing the algebraic formulation of factor analysis it is well to consider in the abstract possible *ecological* interpretations that could be given to varimax assemblages derived by computation from data of the sort modeled in Figure 3. There are at least five. (1) That in areas of overlap the three assemblages occupy the same volume of water at the same time, and are distinct there only in terms of their independent reactions to a physical gradient. In this case the varimax assemblages are only a means of studying different patterns of ecological response. (2) That the three assemblages occupy the same volume of water in areas of overlap, but do so at different times of the year. The work of Bé (1960) and Tolderlund (1969) has demonstrated clearly this phenomenon of seasonal succession in planktonic foraminifera. (3) That the three assemblages occupy the same volume of water in areas of overlap but do so at different intervals of geologic time relatively short compared to the time represented in a typical sediment sample. (4) That the three assemblages occupy different, vertically stacked water masses in areas of apparent overlap. (5) That combinations of the four explanations given above apply.

Thus a successful varimax solution to a paleontologic problem does not constitute an ecological solution. The proper role of the machinery of matrix algebra used in factor analysis is to summarize parsimoniously, objectively, and quantitatively general tendencies in a given body of faunal data; to make tendencies clear which are obscured by random error or other, quantitatively dominant, effects; and to focus attention on ecological or diagenetic questions worthy of further inquiry.

We are now in a position to outline in matrix terms the procedures of factor analysis.

Step B-4: Calculate row-normal data matrix U_{ct}. Data input to the factor analysis is the 22-column matrix of percent range data X_{ct}. In principle, this matrix can be factored directly. In practice, if the raw data are in percentage form it is convenient to multiply the values of each row of X_{ct} by a constant so that the transformed row vector has unit length.[3] The resulting *row-normalized matrix* U_{ct} is much simpler to deal with computationally; and (for present purposes) its essential features are undistorted. Multiplication of a row by a constant does not, after all, disturb the proportional significance of the constituent species. Furthermore, the fact that the length of each row vector is unity provides a simple method of recording how effective a given number of assemblages is in account-

---

3. The constant is so chosen that the sum of the squares of the elements in the normalized vector equals unity.

ing for the compositional information in the system. For example, if there are N samples in $\mathbf{U}_{ct}$, and if we define the *total information in each sample* as the corresponding vector length, then for the matrix $\mathbf{U}_{ct}$ there are N units of compositional information to account for. For the model data displayed in Figure 3, for example, there are 15 (N) units of information. Because the data conform to the linear mixing model, all 15 units will be accounted for by the three (m) varimax assemblages derived by factor analysis. In real data, of course, random sampling errors —as well as inadequacies of the linear mixing model—would be reflected in the failure of three assemblages to account for all N units of information input. A given varimax solution of complexity m can therefore be usefully evaluated by calculating the total information explained, and expressing it as a fraction of N. As outlined in Table 4, bookkeeping on "total information" involves both the square of the vector lengths (communalities) and the vector lengths themselves (fraction of the original observational data explained).

Step B-5: Estimate the number of assemblages (m). We now have ready for factor analysis an N by n matrix of normalized data ($\mathbf{U}_{ct}$), representing information on N samples and n species. Before proceeding it is necessary to estimate m, the number of assemblages to be considered in the calculation. In spite of a large literature to the contrary, there is no ecologically valid *a priori* criterion. Through an iterative trial and error approach, however, an objective and meaningful determination of m can be made. The first step is to assume a value of m somewhat larger than the number of independent dimensions of compositional variation anticipated. If there is any observed change in the proportional composition of the samples at all, we know that m must be at least 2; and it cannot exceed n. In the analysis of core top data below, for example, we took advantage of work by Bé (1969), showing five world distributional zones of planktonic foraminifera, and estimated m initially as 7. Evaluation of these assemblages showed that two did not exhibit a systematic geographic pattern. These were eliminated and a final calculation made with $m = 5$.

Step B-6: Calculation of the varimax matrix ($\mathbf{B}_{ct}$) *and the assemblage description matrix* ($\mathbf{F}$). Although the data matrix $\mathbf{X}_{ct}$ (and a value of m) is the actual input into a factor analysis program such as CABFAC, the results of the analysis are best understood by relating them to the row-normalized data matrix $\mathbf{U}_{ct}$. The program output is an m by n matrix $\mathbf{F}$ which describes the composition of the m varimax assemblages in terms of the n species; and an N by m matrix $\mathbf{B}_{ct}$ which displays down each of the m colunms the varying contribution of a varimax assemblage to the N samples. If the value of m were taken as n, then the entire analysis can be exactly expressed as the matrix equation

$$\mathbf{U}_{ct} = \mathbf{B}_{ct}\,\mathbf{F}$$

If m is taken as less than n, as is usually the case, then the results of factor analysis can be expressed as the matrix equation

$$\mathbf{U}_{ct} = \mathbf{B}_{ct}\,\mathbf{F} + \mathbf{E}$$

where $\mathbf{E}$ is an N by n matrix of errors. As m approaches n, $\mathbf{E}$ approaches $\mathbf{O}$. The total information explained (expressed as sum of squares) is $N - \text{tr } \mathbf{E}'\mathbf{E}$. Divided by N, it yields the total communality.

The varimax matrix is derived by calculating the eigenvalues and eigenvectors of $\mathbf{U}_{ct}\,\mathbf{U}_{ct}'$. The sum of the eigenvalues of this matrix equals the total sum of squares in $\mathbf{U}_{ct}$; and the first m eigenvectors, rotated to a "best fit" position according to the varimax criterion, become the row vectors of $\mathbf{F}$. The elements in each row of $\mathbf{B}_{ct}$ may be regarded as projections of a sample vector on the row vectors of $\mathbf{F}$.

Step B-7: Evaluation of $\mathbf{B}$ *and* m. The varimax matrix is the basis for all further computations, and it is therefore important, before proceeding, to evaluate both the number and the ecological validity of the assemblages represented. This evaluation is accomplished by preparing m maps; and by scrutinizing the row sum of squares of $\mathbf{B}$.

If the numbers in a column of $\mathbf{B}$ display simple regional gradients when plotted on a map, or vary systematically with respect to known oceanographic parameters, then the corresponding assemblage is judged to be both statistically and ecologically significant, and is retained. If these criteria are not met, the assemblage is discarded. The total number of columns remaining after this evaluation is taken as the final value of m.

The m elements in any row of $\mathbf{B}$ represent proportional contributions of the m varimax assemblages to that sample. How much of the compositional information in the original sample (i.e. in $\mathbf{U}_{ct}$) is accounted for by the varimax model? To answer this question we recall that the sum of the squares in every sample of $\mathbf{U}_{ct}$ is unity. The sum of squares of any row in $\mathbf{B}_{ct}$ is then defined as the *communality of the sample*. This sample communality will be unity if all the original information is retained in the m-dimensional varimax model, and will decrease to zero as the model fails to account for the original information. The CABFAC program calculates the communality for each sample in $\mathbf{B}_{ct}$. By scanning the communality column, therefore, the investigator can determine at a glance how well each sample fits the general varimax model. Low values may be associated with compositional anomalies due to differential dissolution, to stratigraphic mixing, or to real but local ecological phenomena.

The sample communalities averaged over the N samples equal the total communality, i.e. the total compositional information retained in the m-dimensional varimax model.

Negative numbers may arise owing to constraints of the orthogonal varimax reference system. Although this may appear odd, these values are meaningful both mathematically and ecologically. If we ask, for example, how many polar bears live in New Haven, one answer is "none." But if we plot on a map gradients of polar bear distribution, the zero isopleth would lie far to the north, and New Haven would be characterized by negative values. The foraminiferal analog of the polar bears is not far to seek. If the reader will reflect on the long-standing controversy over *Globorotalia menardii* and δO^{18} data, he will recognize that part of the problem was the inability to record negative *G. menardii* during severe portions of a glacial epoch.

Step B-8: Calculate the factor score matrix F_s. The matrix F describes the species composition of the m varimax assemblages. Magnitudes of the n numbers in any row of F reflect the relative importance of the n species in the corresponding assemblage. Their absolute magnitudes have no special significance and in fact derive from a convention that each row vector in F have unit length. This convention makes it awkward to compare the results of two analyses involving different numbers of species. To simplify such comparisons, and provide a more readily understandable scale, each row of F is multiplied by a constant so that the sum of squares of the transformed elements equals n. Thus, if all species were equally important in an assemblage, all elements in the corresponding row of the transformed matrix equal unity. The F matrix so transformed is called the factor score matrix, F_s.

As a species in nature may play a role in more than one community, so in F_s a given species may figure significantly in more than one assemblage.

After examining B, the investigator will normally wish to study the species composition of the varimax assemblages. This objective is best accomplished by examining the rows of F_s. High absolute values (which cannot exceed $\sqrt{n}$) indicate that a species is an important component. Absolute values over one indicate that a species is more important than would be the case if all species were equally represented. Values approaching zero indicate that a species is of no importance in an assemblage. Positive signs indicate that a species is directly related to high positive values in B. Negative values indicate that a species exhibits inverse correlation with values in B and has a distribution pattern inverse to those elements in the assemblage having the opposite sign.

Summary. Nonredundant ecological information in an n-column matrix of raw data is expressed in terms of an m-column varimax factor matrix B_{ct}. Each column in B_{ct} may be mapped to display regional variation patterns of the m assemblages described in F and F_s. The m columns of B_{ct} are statistically independent biological variates useful in writing ecological and paleoecological equations. An objective estimate of the complexity of the ecosystem under study is expressed as m. The effects of random error in the data are suppressed because the columns of B_{ct} are linear combinations of the n observational variables. By the central limit theorem of statistics, such linear combinations are more reliable than the original variables.

C. Derivation of Paleoecological Equations

Physical or chemical observations of the surface waters of the ocean over the N core top sites are obtained and entered into a column vector Y_{ct}. Our objective now is to relate Y_{ct} to the biological parameters in B_{ct}. Specifically, we wish to write an empirical paleoecological equation of the form

$$Y_{ct} = \phi(B_{ct})$$

but in order to do so we must postulate the form of this function. The simplest form of the model chosen is shown on Figure 4, where two species are each assumed to have a parabolic response to a single ecological parameter t. A considerable body of experimental and field data supports this generalization, typified by Bradshaw's (1957) experimental work on benthic foraminifera, Tolderlund's (1969) field studies on planktonic foraminifera, and Braarud's (1961) for other plankton. These studies confirm the intuitive notion that if the response of a species to a single parameter such as temperature or salinity is considered, the species will show an optimum for some value of the parameter and decrease in abundance more or less smoothly to zero as conditions increase or decrease from the optimum. In two respects the parabolic model pictured in Figure 4 is inaccurate: the response may be asymmetric, and negative abundances are not observed for real species. The first inaccuracy is accepted as of minor significance compared to

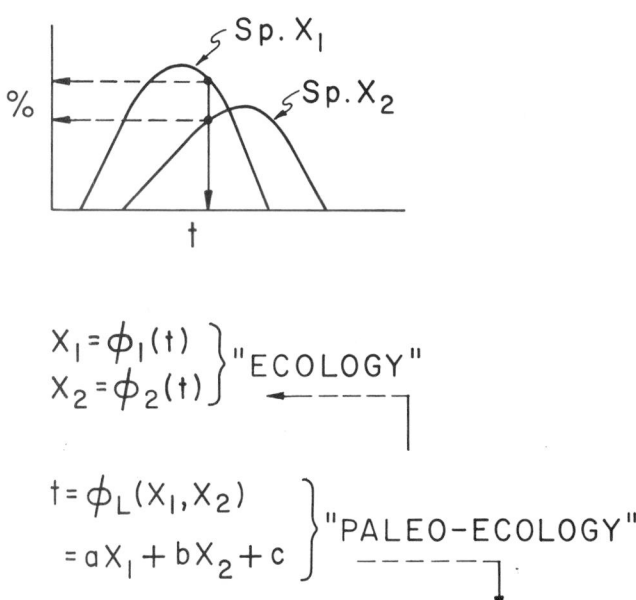

Fig. 4. Model I. A simple model for ecological response and paleoecological estimation involving one ecologically significant parameter t and two species, x_1 *and* x_2. Abundance of the species is recorded in percent. In ecology, each species is viewed as a function of t; the form of these functions is here assumed to be parabolic. In paleoecology, the parameter t is considered as an empirical function of x_1 and x_2. If ϕ_1 and ϕ_2 are parabolic, then ϕ_L is linear.

other sources of error. Furthermore, the effect of this inaccuracy on the *paleo-ecological* as opposed to the *ecological* function is much less severe than one would anticipate. The second inaccuracy can be minimized by employing vari-max assemblages rather than species. As we have seen, values in $\mathbf{B}_{ct}$ do assume negative values. Furthermore, our empirical results have shown that the ecologi-cal range of varimax assemblages is rather broad, and by ignoring one assemblage (the polar assemblage) it is possible to deal mainly with broadly overlapping assemblages.

Returning to Figure 4, we note that our ecological model is a parabolic re-sponse for each species:

$$x_1 = d\,t^2 + e\,t + f$$
$$x_2 = g\,t^2 + h\,t + i$$

Our objective, however, is a paleoecological equation with t specified as a func-tion of x_1 and x_2. To find this, we solve the two ecological equations si-multaneously. Except in the special case of two species with the same optimum, we can eliminate the square term and derive a linear equation of the form

$$t = a\,x_1 + b\,x_2 + c$$

If two physical parameters t_1 and t_2 are controlling the distribution of three or more species with overlapping ranges, then we may picture the ecological functions as paraboloids, as shown in Figure 5. In this case it can be shown that in general the paleoecological solution is two linear equations of the form

$$t_1 = a\,x_1 + b\,x_2 + c\,x_3 + d$$
$$t_2 = e\,x_1 + f\,x_2 + g\,x_3 + h$$

The implications of this simple result are far-reaching. Suppose, for example, that t_1 represents surface water temperature and that data are available; and t_2 represents some parameter (say the abundance of PO_4 in March) on which there are no data. Then, given only data on temperature and on the abundances of the three species x_1, x_2, and x_3, we can still write the empirical equation for t_1 and compute it *exactly* from the species data, even though we have no data on param-eter t_2 which *is* ecologically significant. The model in Figure 5 will make this clear. Given information that species x_1 is 30 percent of the fauna, x_2 10 percent, and x_3 40 percent, the values of t_1 and t_2 are both fixed.

Figure 5 presents the model for three species and two parameters. The model can in fact be generalized to an ecosystem involving m species (or assem-blages) controlled by $m - 1$ or fewer independent influences. In the general case, the ecological response model is a hyperparaboloid; but the paleoecological equations remain linear.

One final matter needs to be explained before we outline the computational steps in writing a paleoecological equation. In the algebraic analysis of the model given in Figure 5 it was assumed that the response surfaces continued their para-

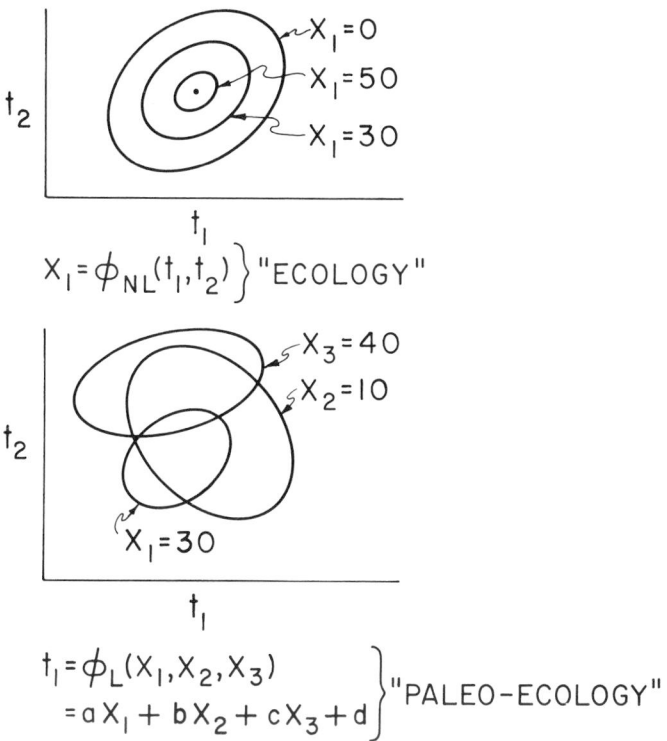

Fig. 5. Model II. A more complex model for ecological response and paleoecological estimation involving two ecologically significant, independent parameters t_1 and t_2 and three species x_1, x_2, and x_3. Abundance data on species are contoured as percent. In ecology, each species (e.g. x_1 illustrated at the top of the figure) is here assumed to be a paraboloidal function of t_1 and t_2, with abundance decreasing to zero away from a joint optimum in the t_1–t_2 field. In paleoecology, t_1 and t_2 will each be a linear function of x_1, x_2, and x_3.

bolic form below the $t_1 - t_2$ plane—i.e. that negative abundances were recorded—so that all three species distributions were everywhere overlapping. Whenever, in actual data, an assemblage abundance remains zero over a part of the total range, a degree of indeterminacy is introduced into the system. This problem can be minimized in two ways: first, by treating varimax assemblages rather than individual species—our experience has shown that the varimax model yields assemblages with broad, overlapping regional gradients, whereas many individual species tend to have restricted ranges; second, in writing the paleoecological equation it is advantageous to relax the restriction of linearity and allow the plane (or hyperplane) to warp into hyperparabolic form. Although the departure of our empirically derived equations is never far from linear, allowing the surface to flex does improve the overall fit of the model to the data.

Step C–9: Assemble oceanographic data ($\mathbf{Y}_{ct}$). At each of the N localities where core tops have been taken, obtain measurements or estimates of physical oceanographic parameters judged to be ecologically significant. Define each set of measurements as a column vector $\mathbf{Y}_{ct}$.

Step C–10: Form cross-product matrix $\mathbf{B}^2_{ct}$. In order to write a paleoecological equation of the second degree, and thereby permit the paleoecologic prediction surface to warp, it is necessary to form the squares and cross-products of the m vectors in $\mathbf{B}_{ct}$ and to arrange them along with columns of $\mathbf{B}_{ct}$ as a new matrix. The resulting matrix ($\mathbf{B}^2_{ct}$) has $(m^2 + 3m)/2$ columns. If $m = 2$, for example, each row in $\mathbf{B}^2_{ct}$ has the following five elements:

$$b_1^2 \; b_2^2 \; b_1 b_2 \; b_1 \; b_2$$

This operation is carried out with a program called POWER and the result punched on cards for the next operation.

Step C-11: Write the paleoecological equation, $\hat{\mathbf{Y}}_{ct} = \mathbf{B}^2_{ct}\mathbf{K} + \mathbf{k}_o$. Using standard techniques of linear regression, find $(m^2 + 3m)/2$ coefficients corresponding to the columns of $\mathbf{B}^2_{ct}$ and array them in a column vector $\mathbf{K}$. Call the constant term $\mathbf{k}_o$. The N estimates of $\mathbf{Y}_{ct}$ derived by the regression equation may be placed in a column vector $\hat{\mathbf{Y}}_{ct}$. The techniques of regression select the coefficients in such a way as to minimize the sum of the squares of the errors of the N estimates. The entire result can be expressed in the equation

$$\hat{\mathbf{Y}}_{ct} = \mathbf{B}^2_{ct}\mathbf{K} + \mathbf{k}_o$$

The statistical significance of the result can be evaluated in a number of ways. The simplest is to calculate a correlation coefficient between $\mathbf{Y}_{ct}$ and $\mathbf{Y}_{ct}$. This measure, known as the multiple correlation coefficient, can be used to calculate the probability that the observed result could have been achieved by chance alone. The square of this coefficient represents the fraction of the variance of $\mathbf{Y}_{ct}$ which is accounted for by the equation. The remaining fraction of the variance of $\mathbf{Y}_{ct}$ is unexplained by the equation, i.e. must be accounted for by random error or in other ways.

D. Description of core faunas in terms of varimax assemblages ($\mathbf{B}_c$)

By using factor analysis, data on core top samples have been analyzed in terms of m varimax assemblages:

$$\mathbf{U}_{ct} = \mathbf{B}_{ct}\mathbf{F}$$

Because $\mathbf{F}$ is a row-wise orthonormal matrix, we may solve the equation for $\mathbf{B}$:

$$\mathbf{B}_{ct} = \mathbf{U}_{ct}\mathbf{F}'$$

and view $\mathbf{F}'$ ($\mathbf{F}$ transpose) as an operator yielding a varimax solution from a matrix of normalized data by postmultiplication. We may use this result to describe *any* set of core faunas in terms of the *m* varimax assemblages obtained from the *core top* study. The procedure is simple in that it involves only matrix multiplication; it does *not* involve a factor analysis of core data. By contrast, Oba (1969) factor-analyzes cores directly.

Step D-12: Calculate normalized data matrix for core: U_c. Starting with the percent range data matrix for the core samples X_c, the row-normalized data matrix U_c is calculated.

Step D-13: Resolve core faunas into varimax assemblages: $B_c = U_c F'$. Post-multiplication of U_c by F' yields a matrix B_c in which each core sample is resolved into contributions of the *m* varimax assemblages. The degree to which the core top assemblage model fits the faunal data from the core can be measured by calculating the communality for each row of B_c. It is also possible to examine that fraction of the data in U_c which is not accounted for by calculating the error matrix

$$E = U_c - B_c F$$

If communalities for a given sample are low, the values in the corresponding row of E will be large. Possible interpretations for such samples would include differential dissolution, ecological phenomena not represented in the core top data, and stratigraphic mixing. As older levels in cores are analyzed—in which Recent species are replaced by now-extinct counterparts—values in the E matrix will record not error but evolution.

E. Paleoenvironmental Estimation

In section C we derived constants K and k_o in the equation

$$\widehat{Y}_{ct} = B^2_{ct}K + k_o$$

This equation gives estimates of a modern physical oceanographic parameter as a function of core top assemblages B_{ct}. In section D we calculated core assemblages B_c. We can now combine these results and make quantitative paleoenvironmental estimates.

Step E-14: Form cross-product matrix B^2_c. As with the core top data, it is necessary to form cross-product terms B^2_c from the varimax matrix B_c.

Step E-15: Calculate paleoenvironmental estimates: $\widehat{Y}_c = B^2_c K + k_o$. Given the coefficient vector K and the constant k_o derived in section C above, and the cross-product matrix B^2_c, calculate Y_c, estimates of paleoclimatic conditions during deposition of the core samples:

$$\widehat{Y}_c = B^2_c K + k_o$$

APPLICATION OF THE METHOD

Paleontological Data

Percentage data for planktonic foraminifera identified in 61 core top samples and in 110 core samples from V12-122 are given in Tables 2 and 3. Only species having abundances exceeding 1 percent in at least one sample are given: 26 species and coiling varieties for the tops; 27 for the core samples. As discussed above, species were eliminated from subsequent computations which did not exhibit a 2 percent abundance in at least one core and one core top sample; 22 species meet this criterion and are listed in Table 8.

Before factoring, percent data were transformed to percent range, as discussed in A-3 above. Maxima used in these calculations are given in Table 8. Minima for all species are zero.

For each sample a census was taken of the total benthic foraminifera. Values for the core tops are given in Table 16. For V12-122, the average percentage is only 0.52. Only two samples exceed 2 percent: 2.08 percent for the sample at 110 cm, and 2.17 percent for the sample at 450 cm. Forty-seven of the samples contained no benthonic foraminifera.

Specimens of pteropods and other aragonitic micromollusks were noted in fair abundance only in the upper part of the Y zone. A few specimens occur in the Z zone.

Factor Analysis of Core Top Data

Results of the factor analysis are summarized in the varimax matrix (Table 4); the factor score matrix (Table 5); the F matrix (Table 6); five varimax assemblage maps (Figs. 6–10); and a dominant assemblage map (Fig. 14). The following conclusions seem justified:

1. Five varimax assemblages are ecologically meaningful. This conclusion is supported most convincingly by the systematic variations exhibited on Figures 6–10.

2. Four are clearly related to climatic zones, and are named the tropical, subtropical, subpolar, and polar assemblages. An indication that water temperatures are a significant factor in their distribution derives from the varimax matrix (Table 4). The rows of this table have been arranged in order of increasing average surface ocean temperature. Values representing the contribution of factor 1 (tropical assemblage) show a clear preference for the warmest water. Those for factor 3 (polar assemblage) show the opposite tendency. The subtropical assemblage (factor 2) and the subpolar assemblage (factor 4) peak at intermediate temperatures in the vicinity of samples 25 and 11 respectively. Other evidence of temperature control is discussed below (see Figs. 15–19).

3. Varimax assemblage 5 is named the *gyre margin* assemblage to epitomize the distribution pattern exhibited on Figure 10. In general, this assemblage favors the periphery and avoids the central areas of the North and South Atlantic gyres. Lower values are found in the western part of the South Atlantic and in the eastern part of the North Atlantic. Temperature alone seems not to be an important control, to judge from the wide latitudinal and temperature range encompassed by the maximum band. In the North and South Atlantic the pattern is strongly correlated with standing crop of planktonic foraminifera as measured by Bé and Tolderlund (1970). In the South Atlantic it is strongly correlated with the

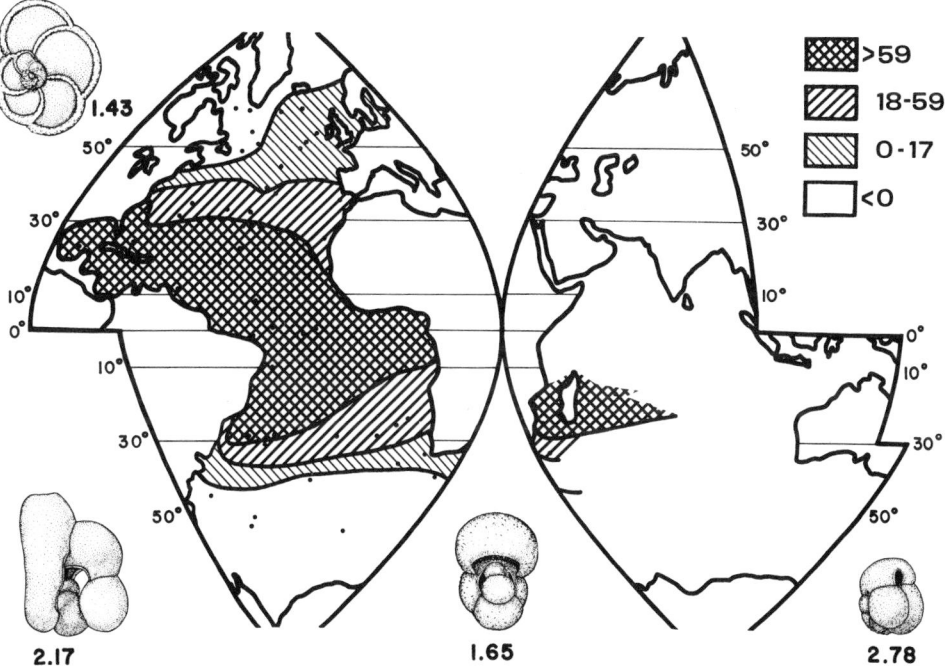

Fig. 6. Geographic variation of the contribution of the tropical assemblage to the 61 samples of Figure 1. The values plotted are taken from the first column of the varimax matrix B_{rt} given in Table 4, and represent proportions. The four numerically dominant species in the assemblage are illustrated. By each illustration the appropriate value from the factor score matrix (Table 5) indicates the relative importance of that species in the assemblage. In decreasing order they are: *Globigerinoides ruber, G. sacculifer, Globigerinella aequilateralis,* and *Globorotalia menardii s.l.* Here, and in Figures 7–10, the illustrations are from Parker (1962).

standing crop of total plankton plotted by Sverdrup et al. (1942, p. 786). We surmise that factors influencing total foraminiferal productivity are chiefly responsible for fluctuations in the gyre margin assemblage. The main argument for this conclusion is based on a comparison between Figures 10 and 11.

At this point it is well to scrutinize the possible role of dissolution in increasing the abundance of the gyre margin assemblage. The dominant species given in Table 5 show a pattern that would, in fact, be affected by dissolution. We therefore plotted the percentage of benthic foraminifera against the abundance of the gyre margin assemblage in 61 core tops (Fig. 34). We judge the pattern to show some positive correlation, but the effect is not marked for samples having less than 4 percent benthics, and we conclude that the relationship of the fauna to productivity is the dominant effect. Parker and Berger (in press) point out that in areas of high productivity dissolution takes place even in rather shallow depths, owing to "carbon dioxide production and low pH within these carbon-rich sediments." Thus high productivity may increase the gyre margin abundance both

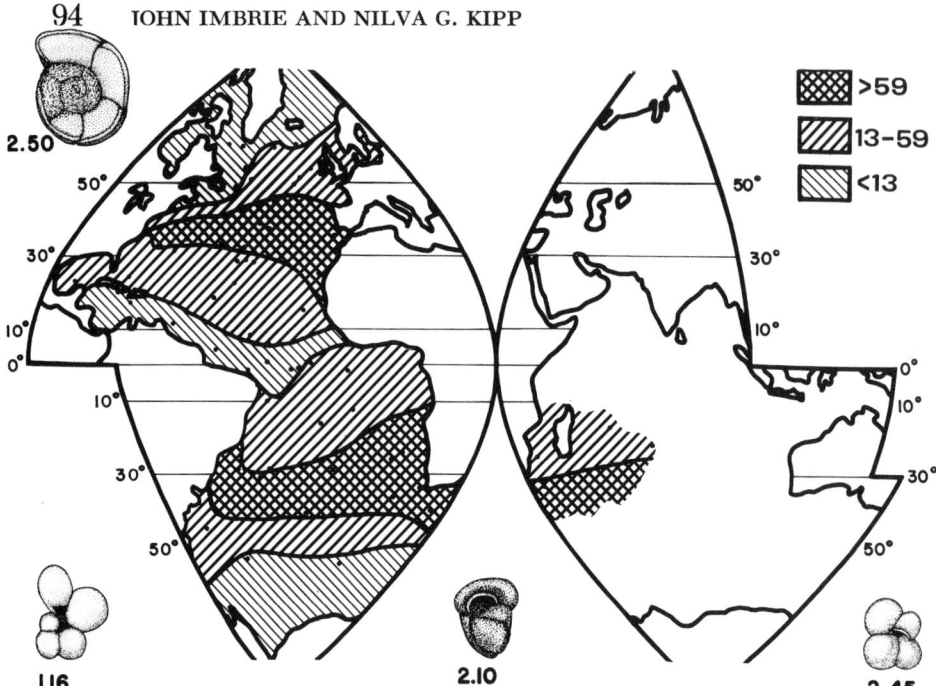

Fig. 7. Geographic variation of the contribution of the subtropical assemblage to the 61 samples of Figure 1. The values plotted are taken from the second column of the varimax matrix $\mathbf{B}_{\mathrm{ct}}$ given in Table·4, and represent proportions. The four numerically dominant species in the assemblage are illustrated. By each illustration the appropriate value from the factor score matrix (Table 5) indicates the relative importance of that species in the assemblage. In decreasing order they are: *Globorotalia truncatulinoides* L, *Globigerina falconensis*, *Globorotalia inflata*, and *Globigerina calida*.

directly, by encouraging the growth of key species in surface waters, and indirectly through diagenetic effects in the sediment.

4. The five-assemblage factor model has a total communality of 0.833 and accounts for 91 percent of the original data (Table 4). Considering the noise-magnification properties of the percent range transformation, the possibilities of sampling and identification error, and the limitations of a linear model, this result is judged to be surprisingly good. The contribution of each assemblage to the total communality is, in order: 33.4, 22.5, 10.1, 12.7, and 4.6 percent. Inspection of the communality column (Table 4) indicates how well any particular sample is explained by the model. These values range from a low of 50.6 percent for sample 39 to a high of 99.4 percent for sample 3.

5. The species composition of the five varimax assemblages are portrayed by the factor score matrix (Table 5). For the tropical assemblage the two most important species are *Globigerinoides ruber* and *G. sacculifer*, followed in order by *Globigerinella aequilateralis*, *Globorotalia menardii*, and *Pulleniatina obliquiloculata*. The important point to note here is that *G. menardii* and *P. obliquiloculata*,

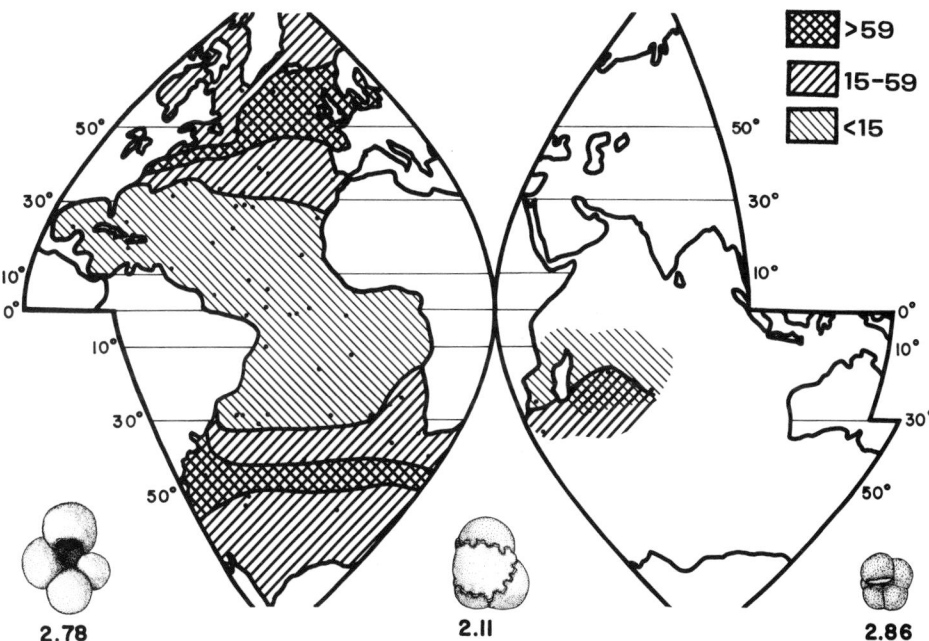

Fig. 8. Geographic variation of the contribution of the subpolar assemblage to the 61 core top samples of Figure 1. The values plotted are taken from the fourth column of the varimax matrix $\mathbf{B}_{ct}$ given in Table 4, and represent proportions. The three numerically dominant species in the assemblage are illustrated. By each illustration the appropriate value from the factor score matrix (Table 5) indicates the relative importance of that species in the assemblage. In decreasing order they are: right-coiling *Globigerina pachyderma*, *G. bulloides*, and *Globigerinita glutinata*.

which are important although secondary elements in the tropical assemblage, are the most characteristic species in the gyre margin assemblage, followed by *Globorotalia inflata* and (in a negative or inverse sense) by *Globigerinoides tenellus* and *G. ruber*. Thus, with regard to transitions from tropical to nontropical sites, *G. ruber* and *Globorotalia menardii* tend to fluctuate together; but with regard to transitions from gyre center to gyre margin positions, they fluctuate inversely. Maps of these two species (Figs. 12, 13) clearly demonstrate this tendency: although both are, in a sense, "tropical species," they react differently to some influence independent of temperature. What this influence may be is not clear. Possibilities for consideration include water mass productivity, salinity, and differental solution.

Dominant species in the subtropical assemblage are, in order, *Globorotalia truncatulinoides* L, *Globigerina falconensis*, *Globorotalia inflata*, and *Globigerina calida*. The importance of *G. falconensis* and *G. calida* as ecologic indicators has also been stressed by Parker (1970). Superficially these species resemble *G. bulloides* and *Globigerinella aequilateralis*.

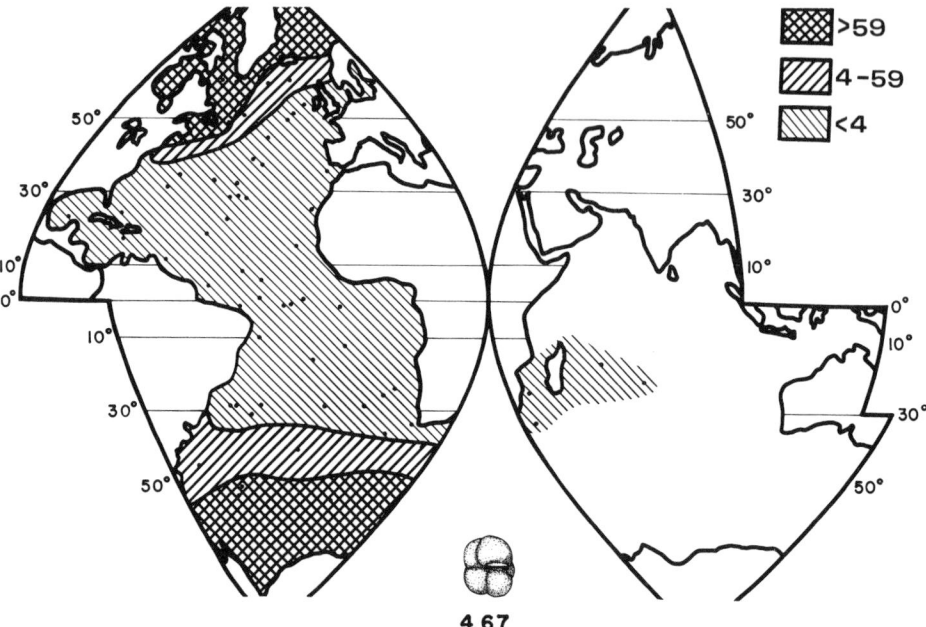

Fig. 9. Geographic variation of the contribution of the polar assemblage to the 61 core top samples of Figure 1. The values plotted are taken from the third column of the varimax matrix $\mathbf{B}_{ct}$ given in Table 4, and represent proportions. One species is dominant numerically: left-coiling *Globigerina pachyderma*, which has a value of 4.67 in the factor score matrix of Table 5.

The polar assemblage consists of one species: *Globigerina pachyderma* L. Three species dominate the subpolar assemblage: *G. pachyderma* R, *G. bulloides*, and *Globigerinita glutinata*.

6. A map showing areas dominated by one assemblage (Fig. 14) strongly suggests that temperature is the main ecological control for all but the gyre margin assemblage. In preparing this map, all samples numerically dominated by a given assemblage (as read from $\mathbf{B}_{ct}$) are classified together. Inspection of the map, and of the varimax matrix, reveals that relatively sharp boundaries separate areas of dominance, and that these ecotones correspond in *both hemispheres* very closely to the 2°, 12°, and 20° winter surface isotherms. In order to emphasize the close relationship between the four objectively defined faunal regions and the winter isotherms, we have actually used these isotherms as boundary lines in preparing Figure 14. Only 2 of the 61 core tops are incorrectly classified by this procedure: samples 30 and 31 are actually dominated by the tropical assemblage, rather than the subtropical as indicated on Figure 14. A glance at Figure 1 will show that the geographic magnitude of this discrepancy is trivial.

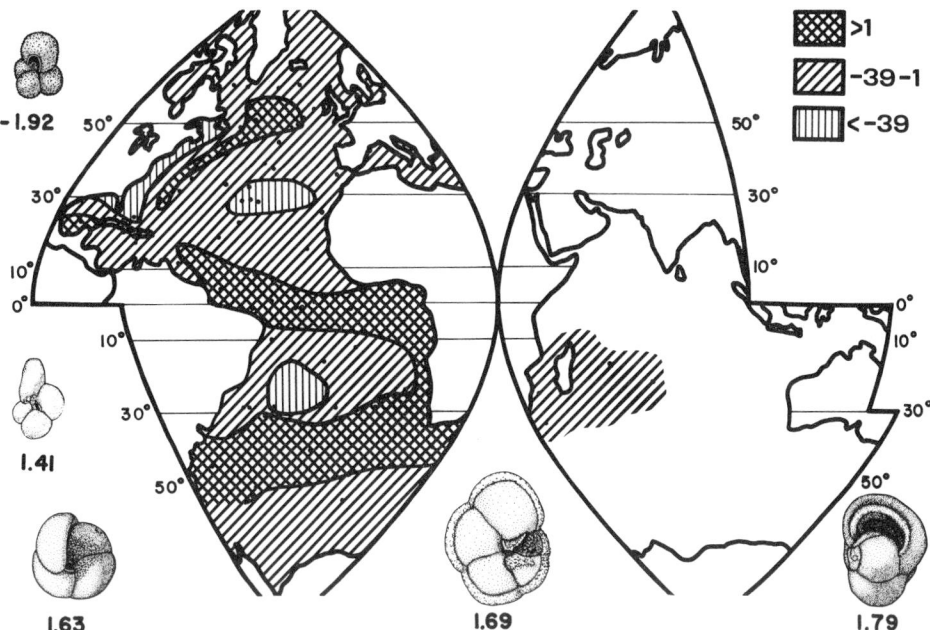

Fig. 10. Geographic variation of the contribution of the gyre margin assemblage to the 61 samples of Figure 1. Control for the Gulf of Mexico, Caribbean, and western North Atlantic is poor. The values plotted are taken from the last column of the varimax matrix $\mathbf{B}_{ct}$ given in Table 4, and represent proportions. The five numerically dominant species in the assemblage are illustrated. By each illustration the appropriate value from the factor score matrix (Table 5) indicates the relative importance of that species in the assemblage. In algebraically decreasing order of importance they are: *Pulleniatina obliquiloculata*, *Globorotalia menardii s.l.*, *G. inflata*, *Globigerinoides sacculifer*, and *G. tenellus*. Ruddiman (1968) recognizes an "equatorial" ecological water mass in the North Atlantic similar to this assemblage which is believed by him to reflect "warm peripheral water of the subtropical gyre."

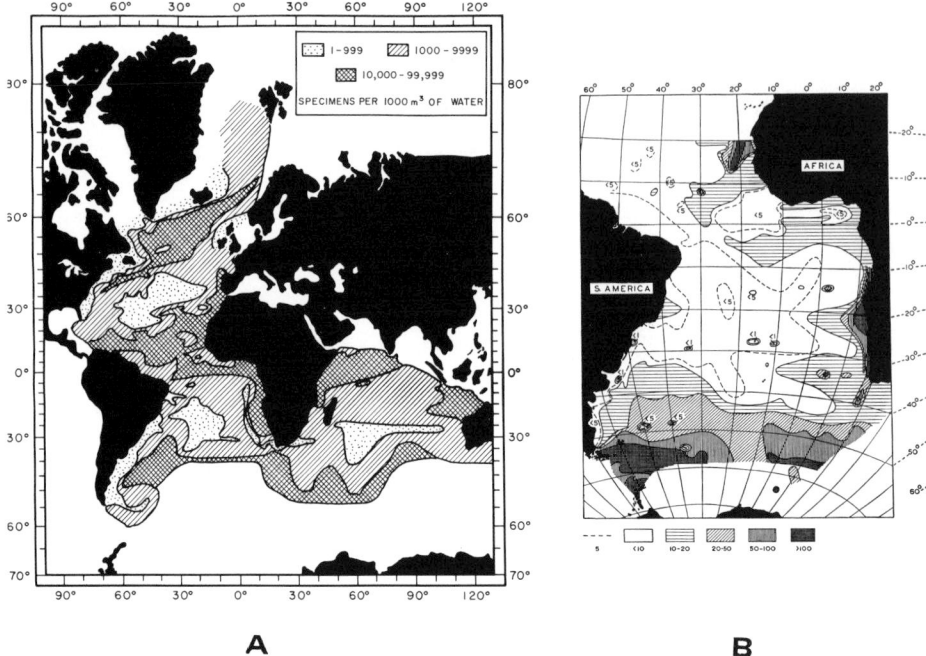

Fig. 11. Absolute concentrations of Atlantic plankton. A. Specimens per 1000 cubic meters of planktonic foraminifera, 0–10 meters (from Bé and Tolderlund 1970). B. Thousands of specimens per liter of total plankton, 0–50 meters, in the South Atlantic (from Sverdrup et al. 1942, Fig. 216).

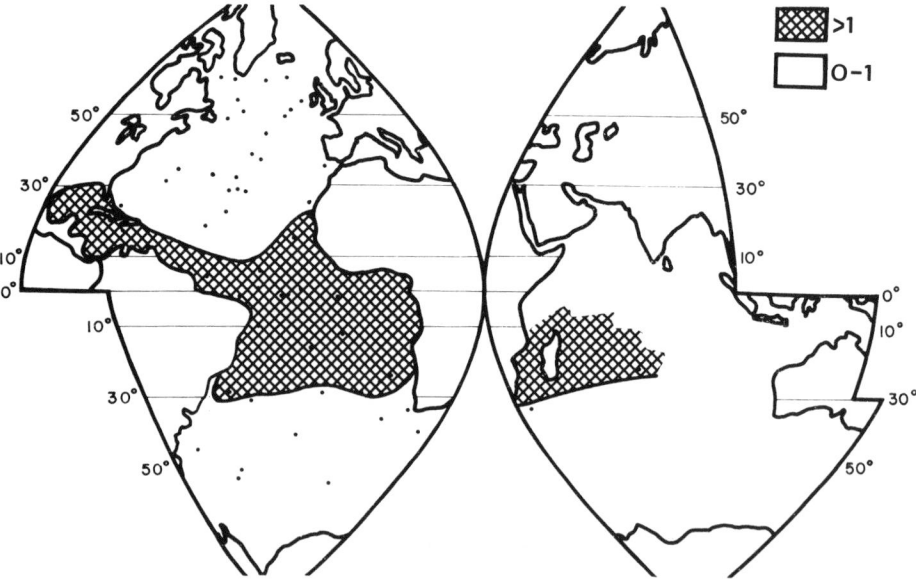

Fig. 12. Abundance of the *Globorotalia menardii* & *G. tumida* complex in 61 core top samples on Figure 1, expressed in percent of specimens > 149 μ.

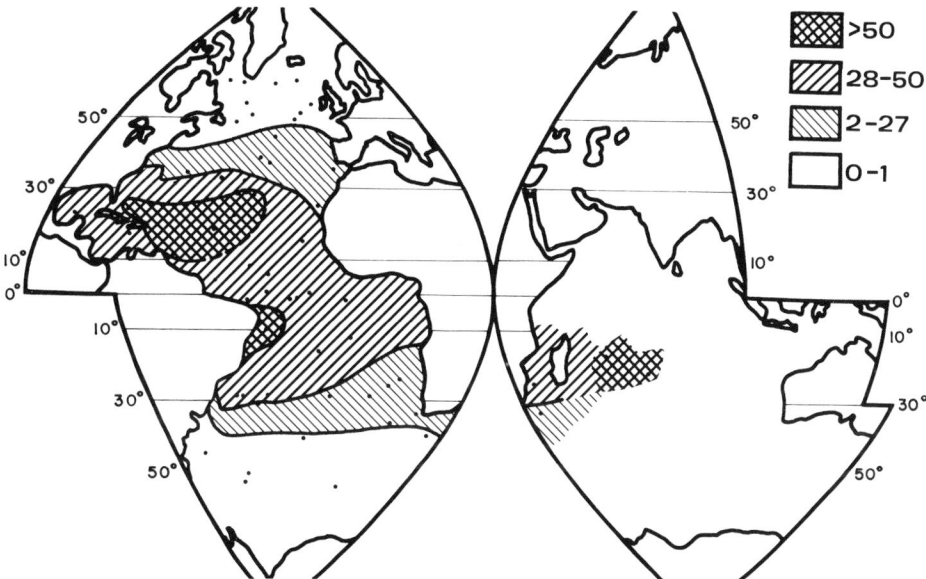

Fig. 13. Abundance of *Globigerinoides ruber* in 61 core top samples shown on Figure 1, expressed in percent of specimens > 149 μ.

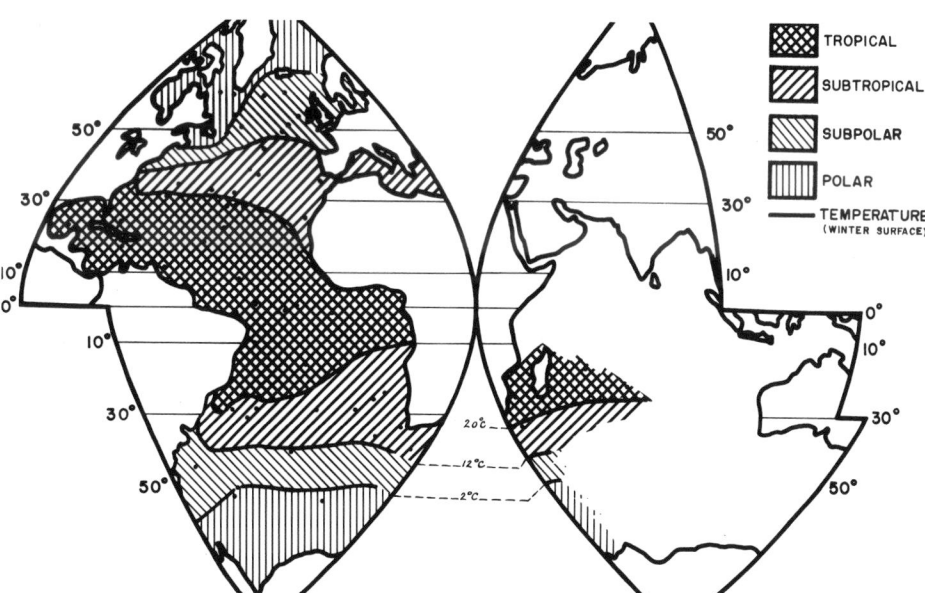

Fig. 14. Distribution of core top samples dominated by an assemblage defined in factor analysis, based on the varimax factor matrix given in Table 4. Regions of dominance are delineated by 2°, 12° and 20°C winter surface isotherms (Defant 1961) in both hemispheres. Subtropical dominance in the Mediterranean region inferred from data in Parker (1958). Location of 61 control points indicated on Figure 1.

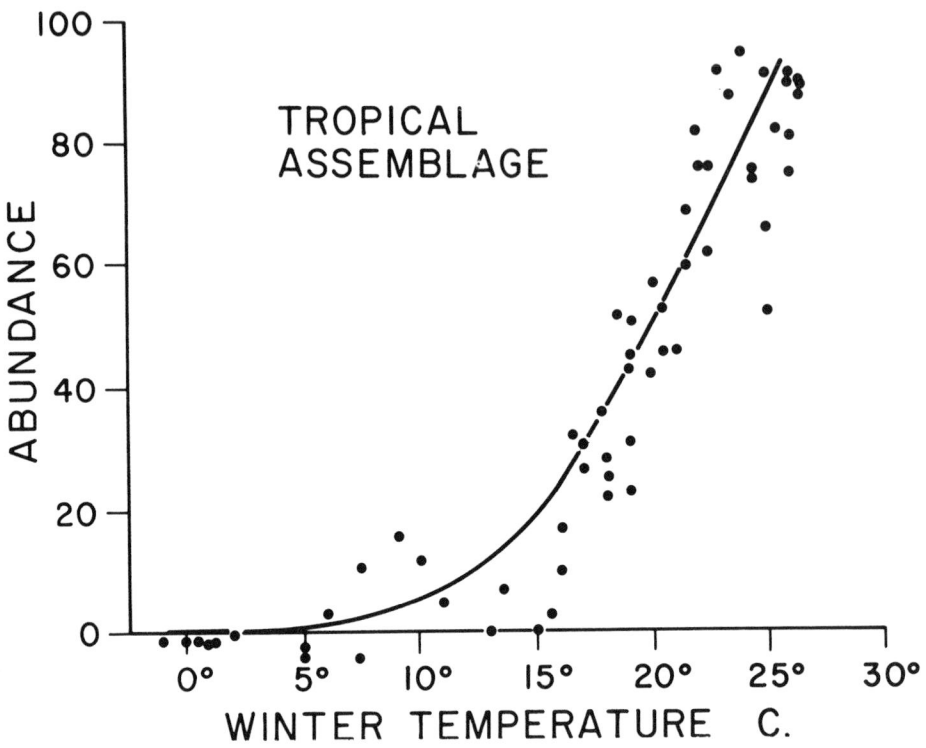

Fig. 15. Abundance of the tropical assemblage versus winter surface temperature for 61 core top samples. Data from Tables 4 and 13. Curve fitted by eye.

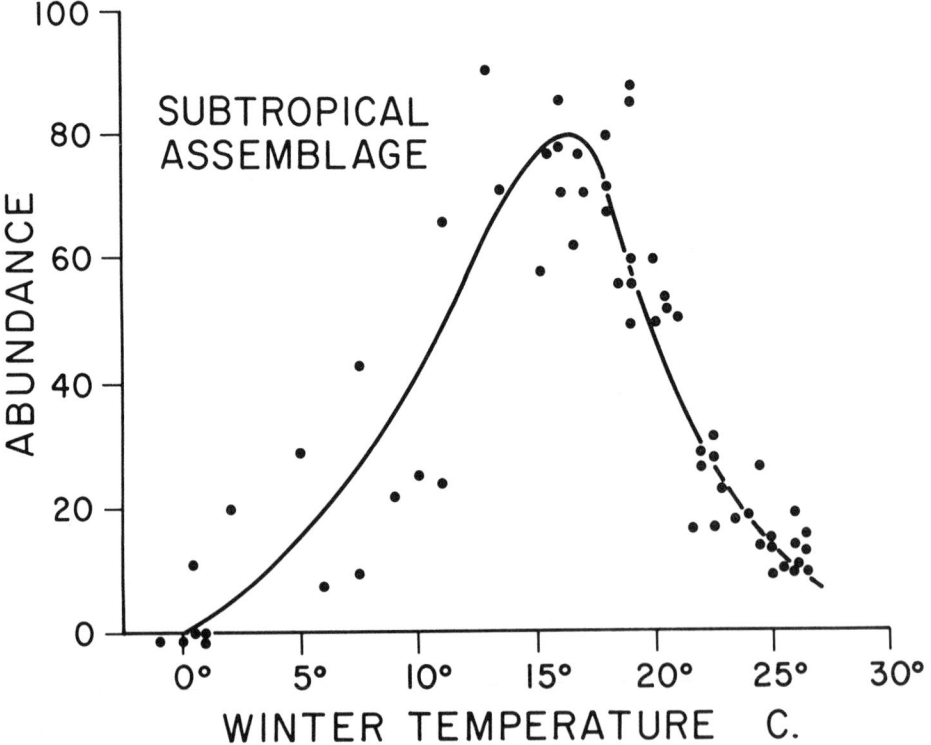

Fig. 16. Abundance of the subtropical assemblage versus winter surface temperature for 61 core top samples. Data from Tables 4 and 13. Curve fitted by eye.

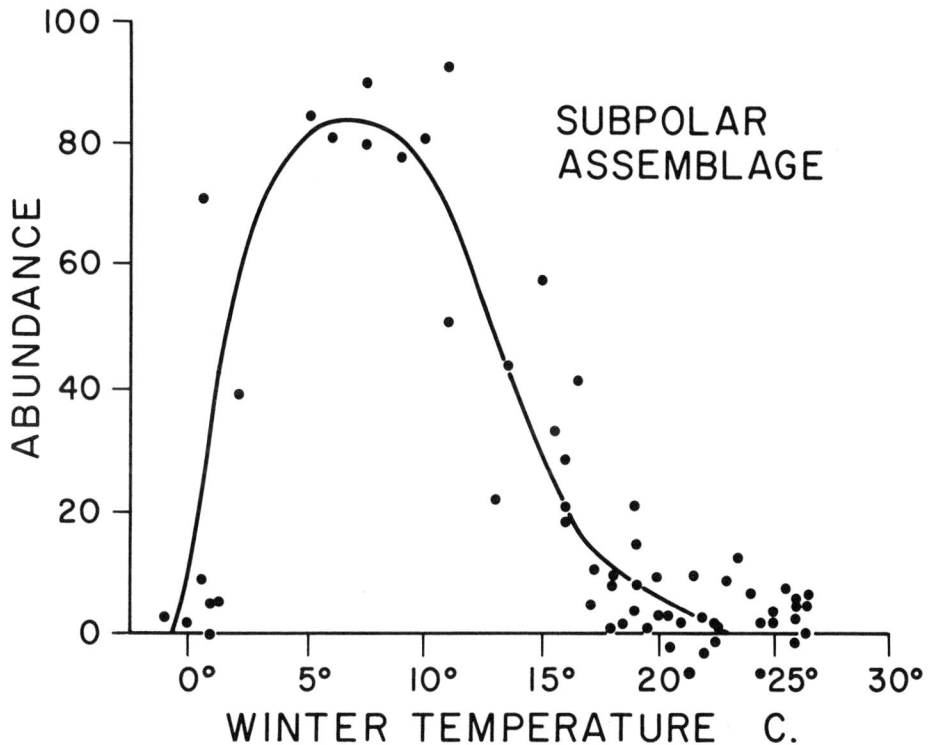

Fig. 17. Abundance of the subpolar assemblage versus winter surface temperature for 61 core top samples. Data from Tables 4 and 13. Curve fitted by eye.

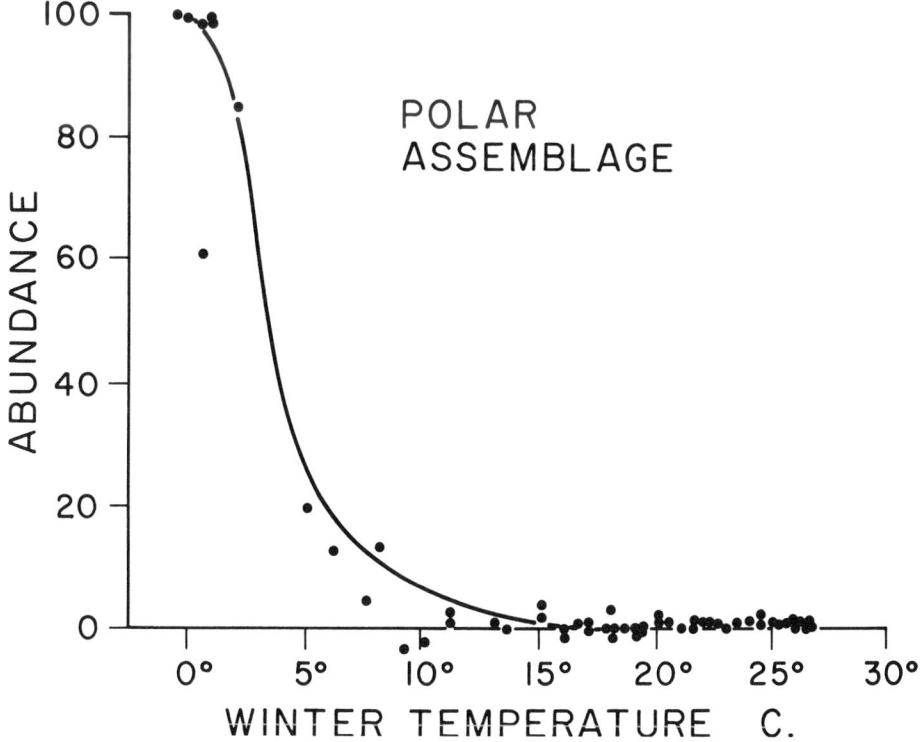

Fig. 18. Abundance of the polar assemblage versus winter surface temperature for 61 core top samples. Data from Tables 4 and 13. Curve fitted by eye.

Derivation of Paleoecological Equations

Before writing the desired equations it is instructive to test the ecological model described in Figures 4 and 5 against real data. Figures 15–18 demonstrate that the first four varimax assemblages, when plotted against winter surface temperatures, do conform approximately to the parabolic response model. The scatter around the hand-drawn curves could reflect a number of influences: chronologic heterogeneity of the samples, the operation of ecologic influences not correlated with temperature, sampling error, or imperfections in the model. Later research, it is hoped, will improve our understanding. For the present it is sufficient to conclude that the parabolic model does approximate the real world, and that the ranges of all but the polar assemblage are broad.

Figure 19, which summarizes the observed parabolic tendencies, dramatically illustrates the success of the varimax criterion in extracting from highly inter-correlated data a set of statistically independent parameters. Considered as "waves" traversing a temperature field, each wave is exactly out of phase with its neighbor.

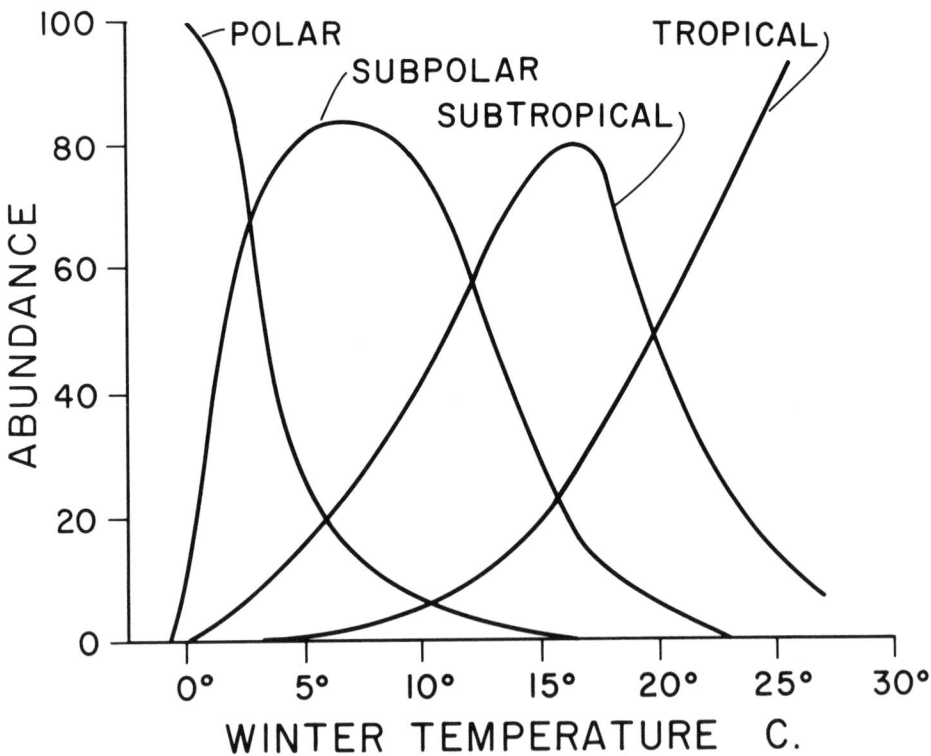

Fig. 19. General abundance trends for four of the varimax assemblages related to winter surface temperature. Curves taken from Figures 15–18.

Plots of the gyre margin assemblage versus surface temperatures and against surface salinity show a nearly random relationship. Considered as a joint function of winter temperature *and* salinity, however, a crudely paraboloidal tendency can be discerned (Fig. 20), in which the maximum observed values of the assemblage tend to lie along a line in the temperature–salinity field, and smaller values are distributed at distances from the assumed optimum. Here the observational limitations imposed by the fact that temperatures and salinities tend to be highly correlated are severe. In any event the actual control may well be related to the productivity or other characteristics of the water mass (or sea bottom) associated with the gyre margin position. All that can be concluded here is that the varimax

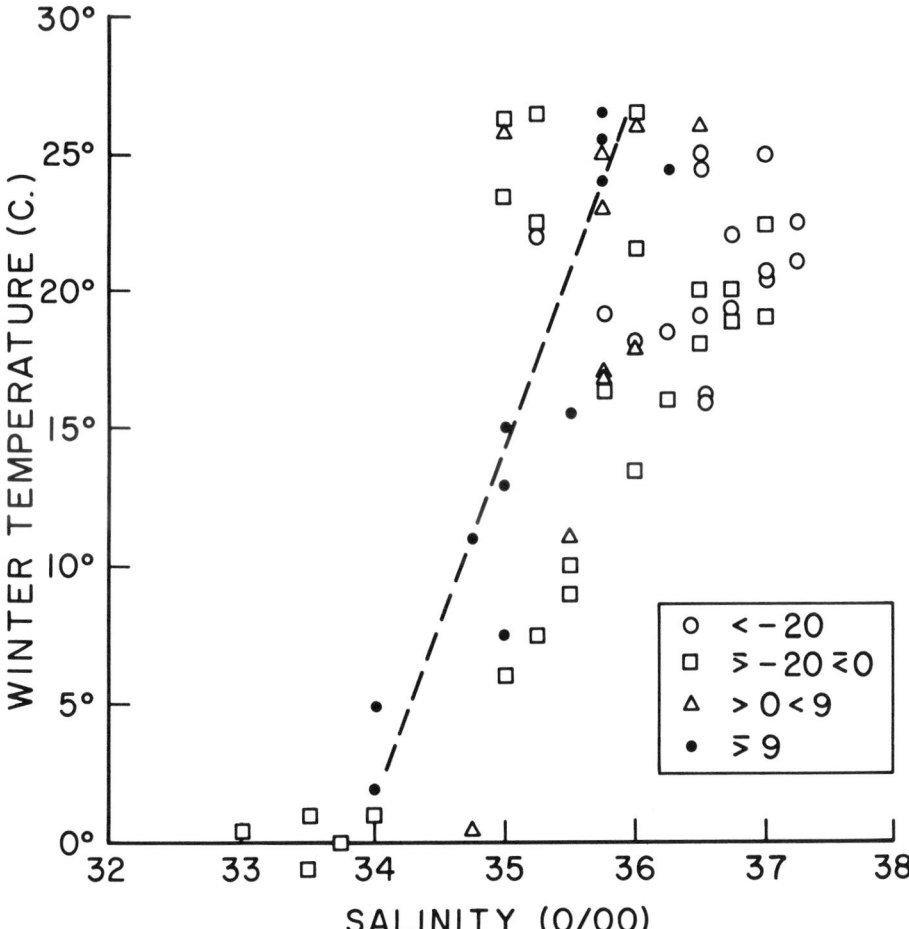

Fig. 20. Abundance of gyre margin assemblage considered as a joint function of the salinity and winter temperature of surface waters overlying 61 core tops shown on Figure 1. The dashed line indicates a general tendency for maximum values.

assemblage identified as gyre margin in distribution has observed optima over a very wide range of temperature and a fair range of salinity.

The desired paleoecological equations are derived by standard multivariate linear regression techniques, fitting least-squares surfaces to empirical data on surface temperatures and salinities (Table 13) and to the matrix $\mathbf{B}^2_{ct}$. Initially, we used all five varimax assemblages (i.e. the 20 cross-product variables derived from them) in the regression. The results are not significantly better than those reported below, in which the polar assemblage is eliminated because of its relatively restricted ecologic range. This assemblage is virtually absent from Caribbean core V12-122 in any event, so that nothing would be gained by including it in the equation. Recalling that five assemblages yield a total of 20 cross-product variables and that four assemblages yield only 14, there is from a statistical point of view a definite advantage in using the smaller number of terms.

In writing the equations a standard stepwise procedure is employed. At each step a single term is added to the equation and a statistical evaluation is made of the success of the equation in predicting the dependent variable. The program is so designed that, at each step, the variable is selected which will contribute most to the success of the prediction. In this way one can discover how useful a given variable is. In addition, we impose a special restriction on the first four steps: that for these calculations the choice of variable be limited to the "linear" variables—i.e. to the actual tropical, subtropical, subpolar, and gyre margin variables. Starting at step 5, and continuing through the final or 14th step, a free choice is made among the squared and cross-product variables derived from the linear variables. This restriction results in the same final equation that would otherwise be achieved but ensures that the results of the linear equation written at step 4 can be compared directly with the results of the nonlinear equation written at step 14.

Coefficients of the equations derived from the stepwise regression procedure are given in Tables 11 and 12. These tables also contain values of the coefficient of multiple correlation, and other information useful in evaluating the equations. In Table 13 the observed values of temperature and salinity are compared with those estimated by the equations. Figures 21–23 give plots of the observed versus the estimated values.

The following points deserve emphasis or clarification.

1. The tables are organized to permit evaluation of the final linear and nonlinear equations, as well as those obtained at intermediate steps. As an example, consider the parameters of linear regression given in part of Table 11. Coefficients of the final linear equation are given in column 2. In this case, the equation for estimating the summer surface temperature (T_s) is:

$$T_s = 19.7\,A + 11.6\,B + 2.7\,C + 0.3\,D + 7.6$$

where the letters A–D stand for observed values of the tropical, subtropical, subpolar, and gyre margin assemblages and T_s is the summer temperature in cen-

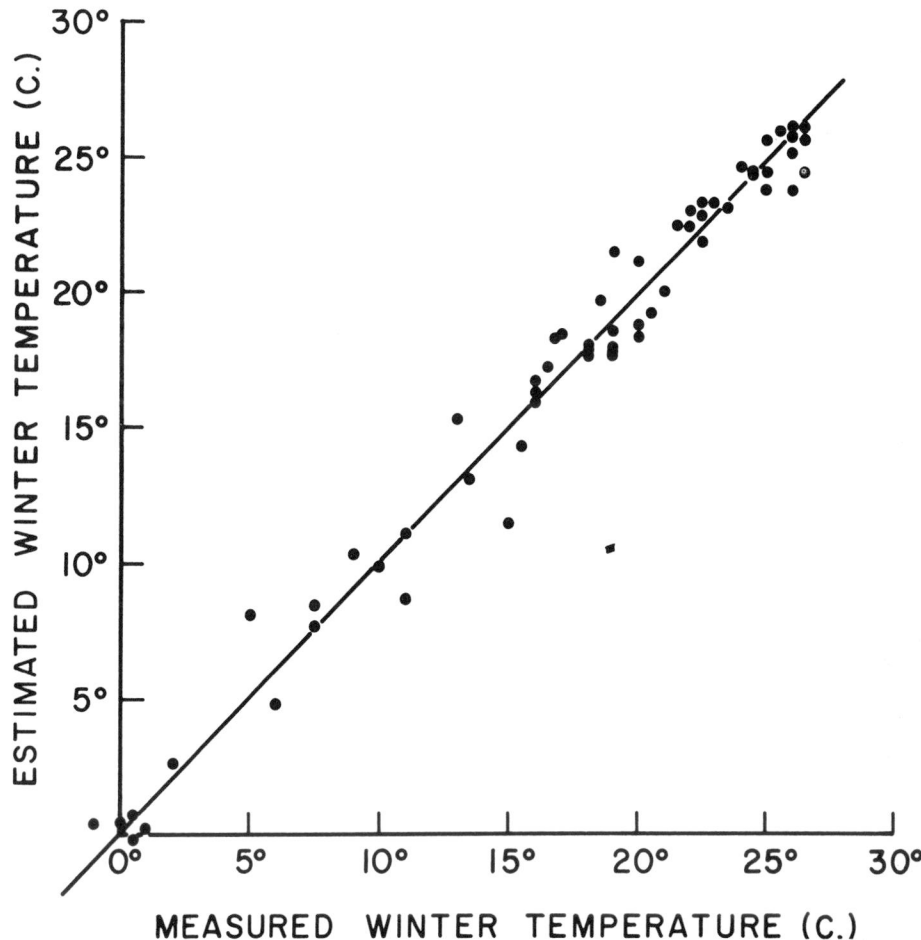

Fig. 21. Winter surface temperatures "measured" by Defant (1961) versus those estimated from the fauna in 61 core top samples (Table 13) by means of a paleoecological equation given in Table 12.

tigrade degrees. The constant term has no special significance beyond reflecting the scale of measurement.

The order of terms in columns one and two reflects the order of stepwise selection. For example, the tropical assemblage (A) provides the best single criterion of summer surface temperature. Although the equation itself (two terms, including a constant) is *not* given in Table 11, the success of the equation can be evaluated by examining columns 3–5 in row one. Column 3 records the proportion of the total variance explained: 68.6 percent of the variance of the observations on T_s.

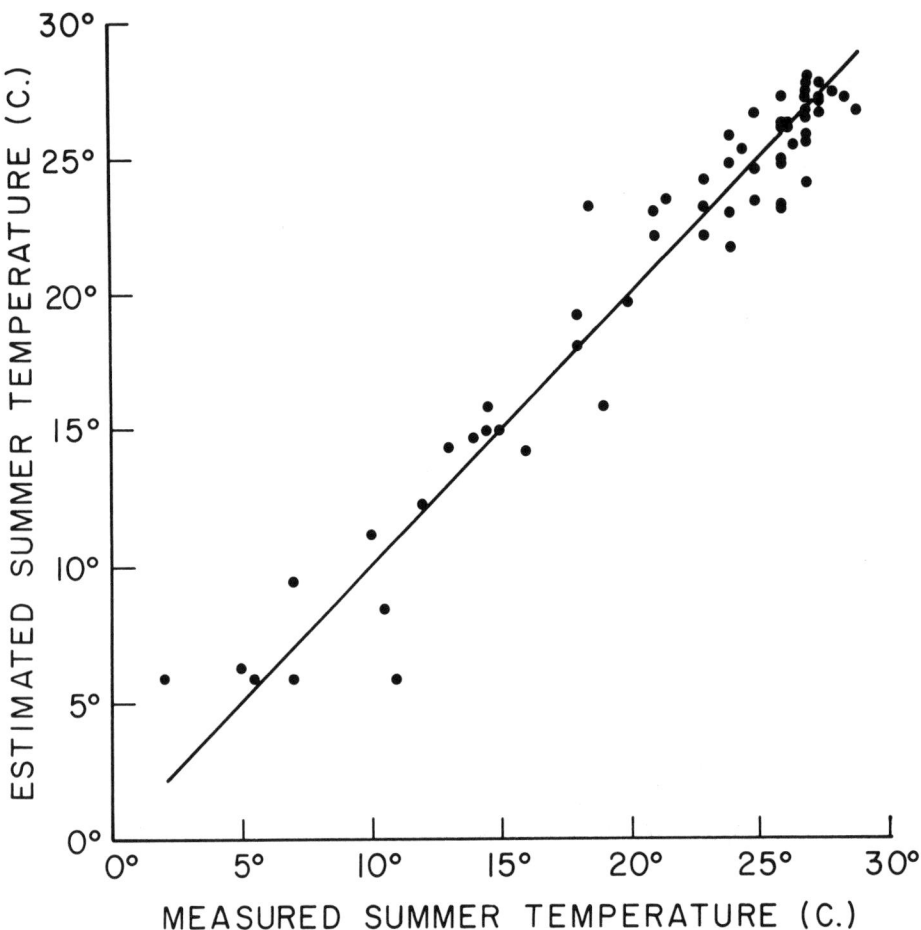

Fig. 22. Summer surface temperatures "measured" by Defant (1961) versus those estimated from the fauna in 61 core top samples (Table 13) by means of a paleoecological equation given in Table 12.

Column 4 is the multiple correlation coefficient. This coefficient is zero if there
is random relationship between observed and predicted values, and unity for a
perfect linear relationship. The multiple correlations reported in Tables 11 and
12 have been adjusted to take into account the degrees of freedom lost when the
number of terms in the equation exceeds one. This adjustment is convenient, for
it allows the investigator to judge the statistical significance of the prediction
quite simply, in the same manner that the significance of an ordinary correlation
coefficient is tested. For the problem in hand, with $N = 61$, any correlation exceeding 0.328 is significant at the 0.01 level. Any exceeding 0.411 is significant at
the 0.001 level—i.e. there is less than one chance in a thousand that the observed

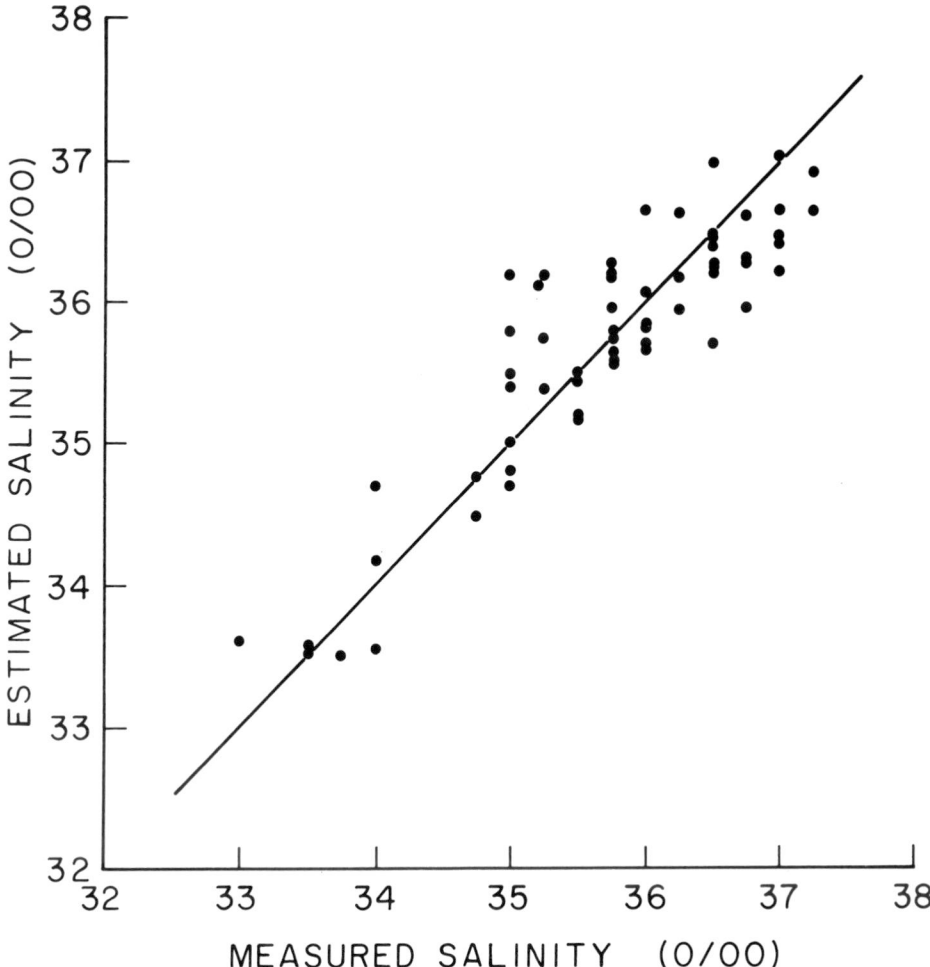

Fig. 23. Average surface salinities "measured" by Defant (1961) versus those estimated from the fauna in 61 core top samples (Table 13) by means of a paleoecological equation given in Table 12.

correlation would occur randomly. Clearly, all correlations tabled are massively significant, with significance levels orders of magnitude below 0.001.

Column 5 gives the standard error of estimate, i.e. the standard deviation of the discrepancies between known and estimated values of T_s. This value also has been adjusted for the number of degrees of freedom. For the equation considered in the first step of Table 11, approximately 68 percent of all estimates made should fall within 4.0°C of the actual value, if errors are normally distributed and the statistical universe being sampled remains the same.

Values in column 3 will always either increase or remain constant as the number of terms in the equation is increased. In general this is true also of columns 4 and 5; but here the adjustment for degrees of freedom may cause a lowering of the reliability as new terms are added to the equation. Values in column 3 record the absolute accuracy of the prediction for the data in hand; those in columns 4 and 5 estimate the reliability of the equation as a prediction tool for new data.

2. For all three oceanographic parameters—summer surface temperature T_s; winter surface temperature T_w; and average surface salinity 0/00—both the linear and the nonlinear paleoecological equations provide highly reliable estimates. Taking the values of the (adjusted) multiple correlation coefficient as the best measure of accuracy, we have the following:

	Linear	Nonlinear
T_s	0.938	0.964
T_w	0.952	0.986
0/00	0.833	0.866

In each case, the nonlinear model yields a somewhat better prediction, and we therefore use the nonlinear equation in making paleoenvironmental estimates for V12-122. It is worth noting, however, that the linear model itself provides a good fit to the data, a fact which is taken as a justification of the arguments presented earlier for the essentially linear nature of paleoecological equations.

3. The most reliable prediction is for T_w; that for T_s is slightly inferior. The salinity prediction is significantly poorer. We interpret this difference to indicate the relative importance of the three parameters as ecologic controls on planktonic foraminifera.

4. Discrepancies between the predicted and the measured parameters seem to be randomly distributed with respect to geographic position, temperature, and salinity. In Table 13 the nonlinear estimates are compared with original observations and the samples ordered by average surface temperature. No pattern is discerned. Note that for winter temperature only two errors exceed 3.0°C over a total observed range of 27.5°. For salinity, the maximum error is 0.9 per mil, with an observed range of 4.2 per mil. In Figures 21–23 the errors appear to be randomly distributed.

5. For the two temperature estimates the most useful predictors are the tropical and subtropical faunas. For salinity, however, the gyre margin assemblage is the most useful predictor. This fact reinforces the suggestion based on data presented in Figures 10 and 20 that the gyre margin assemblage is much more sensitive to changes in salinity than temperature. We consider it possible, however, that the correlation with salinity is fortuitous, and that the more important factor controlling the gyre margin assemblage is productivity. If this should prove to be the case, then the reliability of our estimates of paleosalinity will be adversely affected.

Description of Core Faunas in Terms of Core Top Assemblages

Factor analysis of the core top data described above has identified five varimax assemblages. We now describe each sample of V12-122 in terms of these assemblages, arranging the results in $\mathbf{B}_c$. This matrix, calculated from the relation $\mathbf{B}_c = \mathbf{U}_c \, \mathbf{F}'$, constitutes one of the main contributions of this paper. It is given in Table 9, and the matrices used to obtain it are given in Tables 6 and 7.

1. Numbers in each row of $\mathbf{B}_c$ are interpreted as proportions of constituent assemblages. If the raw data on a sample are completely accounted for by the five-assemblage model, the corresponding row sum of squares or communality (h^2) equals unity. Communalities calculated for $\mathbf{B}_c$ are summarized in Table 17, and compared to core top communalities. Although the average for the core (0.69) is significantly less than that for the core tops (0.83), the fraction of information explained is satisfactory, particularly when we remember that h^2 (like a variance) is calculated from a sum of squares. A more satisfactory general measure of the fraction of *data* explained is h, which is analogous to a standard deviation. For the core this is 83 percent.

2. Consideration of the average h^2 for each zone in the core (Table 17), reveals an interesting pattern of downward decrease. The change is fairly rapid to the middle of the Y zone, and then generally constant back through the U. One interpretation is that evolution, occurring on a community level, has been marked by shifts in the relative abundance of various species.

3. The polar assemblage is virtually absent during the deposition of the core. Only two samples contain nontrivial proportions: 0.06 and 0.05 for samples at 670 and 680 cm. Other evidence developed below indicates that these levels do not reflect cold periods; and we therefore regard the percentages of *Globigerina pachyderma* L observed in these samples as climatically insignificant.

4. The tropical assemblage dominates every sample in V12-122 (see Fig. 24). The lowest proportion observed is 0.53 at 720 cm. With one exception, all other samples contain more than 0.63. One conclusion we draw is suggested by the dominant assemblage map (Fig. 14): winter surface temperatures in the Caribbean never fell below 20°C.

5. The pattern in Figure 24 can presumably be interpreted as a very rough indication of temperature fluctuation because the tropical assemblage, unlike the subtropical and subpolar, has a temperature optimum at or beyond the extreme temperature likely to be observed (Fig. 19). On this basis the general variation pattern observed for the upper half of the core is approximately that expected from other work: a declining temperature back into the Y zone; a warm X; a cold W; and periods of warm and cold in the V.

6. What fraction of the variation patterns plotted on Figures 24–27 represents random effects? To aid in answering this question we have plotted 110 random two-digit numbers in Figure 37. The characteristic feature of such a plot is that between any two samples there is an equal chance of any given scale of

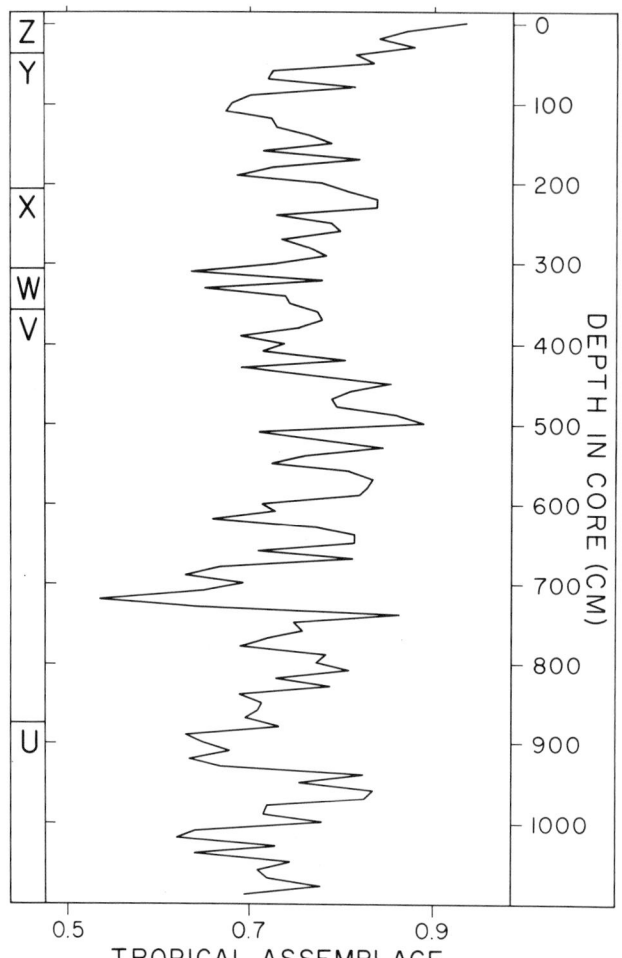

Fig. 24. Abundance of tropical assemblage versus depth in Caribbean core V12–122. Data given in Table 9. Ericson faunal zones U through Z represent intervals of presence or absence of *Globorotalia menardii* (see Table 3).

fluctuation. Visually we see this as a complex zigzag pattern lacking long runs at any given signal level, and without general trends. A comparison with Figures 24–27 shows that significant nonrandom patterns are present in B_c. In Figure 24, for example, there is a relatively simple trend from 0 to 100 cm. In Figure 25 high values are concentrated significantly in the U, parts of the V, and the Y zones. In Figure 26 the isolated groups of samples with high values near 100, 580, 900, and 1000 cm are clear, nonrandom effects. In general, the impression is given of quasiperiodic curves with fairly simple features partially masked by high-frequency noise.

7. Another method of evaluation is to inspect the four curves *simultaneously* in that portion of the core for which there is abundant, external, paleoclimatic evidence: the upper 400 cm.

Before examining our curves in this way it will be helpful to consider what

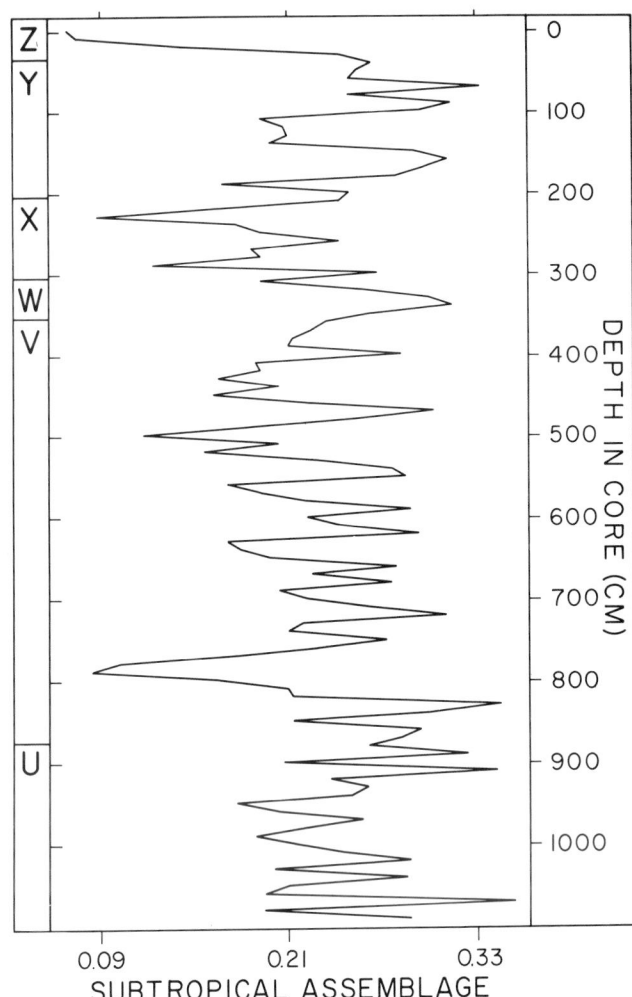

Fig. 25. Abundance of subtropical assemblage versus depth in Caribbean core V12–122. Data given in Table 9. Ericson faunal zones U through Z as in Figure 24.

SUBTROPICAL ASSEMBLAGE

0.09   0.21   0.33

responses are to be expected from a species (or an assemblage) as some controlling factor in the environment undergoes a cyclic change. The usual intellectual model is both implicit and unabashedly linear: when it is warm a "warm" species will increase; when it is cold it will decrease. Such a model may be satisfactory for the case of a species such as *Globogerinoides ruber* whose optimum abundance is at or near the maximum fluctuation of the system under study, or for *Globigerina pachyderma* which delights in the coldest water; but it is not adequate for species or assemblages with optima within the range of observed extremes. The non-monotonic model we assume is shown in Figure 28. Every species is assumed to have a single temperature optimum (using temperature to represent any ecologic parameter judged to be significant) and a maximum and a minimum temperature beyond which the species will not survive. Consider two

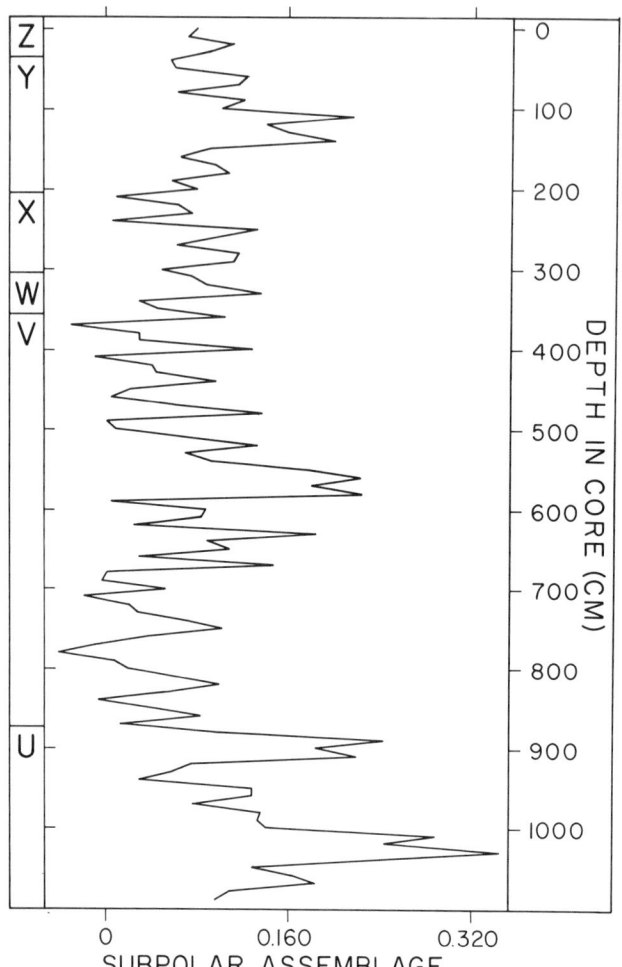

Fig. 26. Abundance of subpolar assemblage versus depth in Caribbean core V12–122. Data given in Table 9. Ericson faunal zones U through Z as in Figure 24.

species: Y, with a high optimum; and X, with a lower optimum. Suppose that a temperature fluctuation occurs as shown in the lower left. The optimum of X is taken to be within the observed range, and that of Y at the top of the range. The resulting paleontologic abundance curves are shown in the lower right: Y will have a simple abundance peak corresponding to the time of maximum temperature. Here an implicit linear response model works well. For species X, however, *two* abundance peaks occur as the ambient temperature attains the optimum value: once during an episode of temperature increase and once during an episode of temperature decrease.

This phenomenon, which must occur for all species except those whose temperature optima are at or beyond the extremes of the system, has both stratigraphic and ecologic consequences which do not seem to be widely recognized.

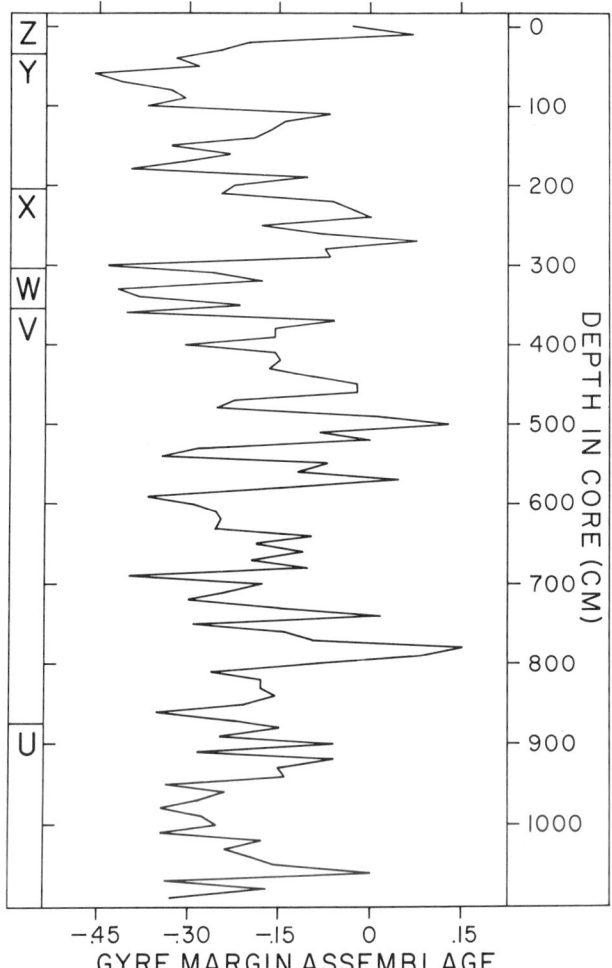

Fig. 27. Abundance of gyre margin assemblage versus depth in Caribbean core V12–122. Data given in Table 9. Ericson faunal zones U through Z as in Figure 24.

Suppose, for example, that two paleontologists regard (correctly) a pair of species as "warm-water forms." Then if separate interpretations are made, each based on a single species and a linear response model, a paleoecological conflict can occur—but the problem is one of concept, not of data. The stratigraphic consequences are more subtle: the number, sense, and absolute chronology of the fluctuations of a single species will vary from place to place depending on the range of temperature occurring at a site relative to the species optimum.

Returning to the upper 400 cm of V12-122, we recall first that the tropical assemblage (Fig. 24) shows a pattern generally consonant with Pleistocene theory: peaks in the Z and X zones, and lows in the W and Y. The subpolar assemblage (Fig. 26) peaks once in the middle of the Y zone; the subtropical assemblage (Fig. 25), twice. The pattern is displayed more clearly in Figure 29

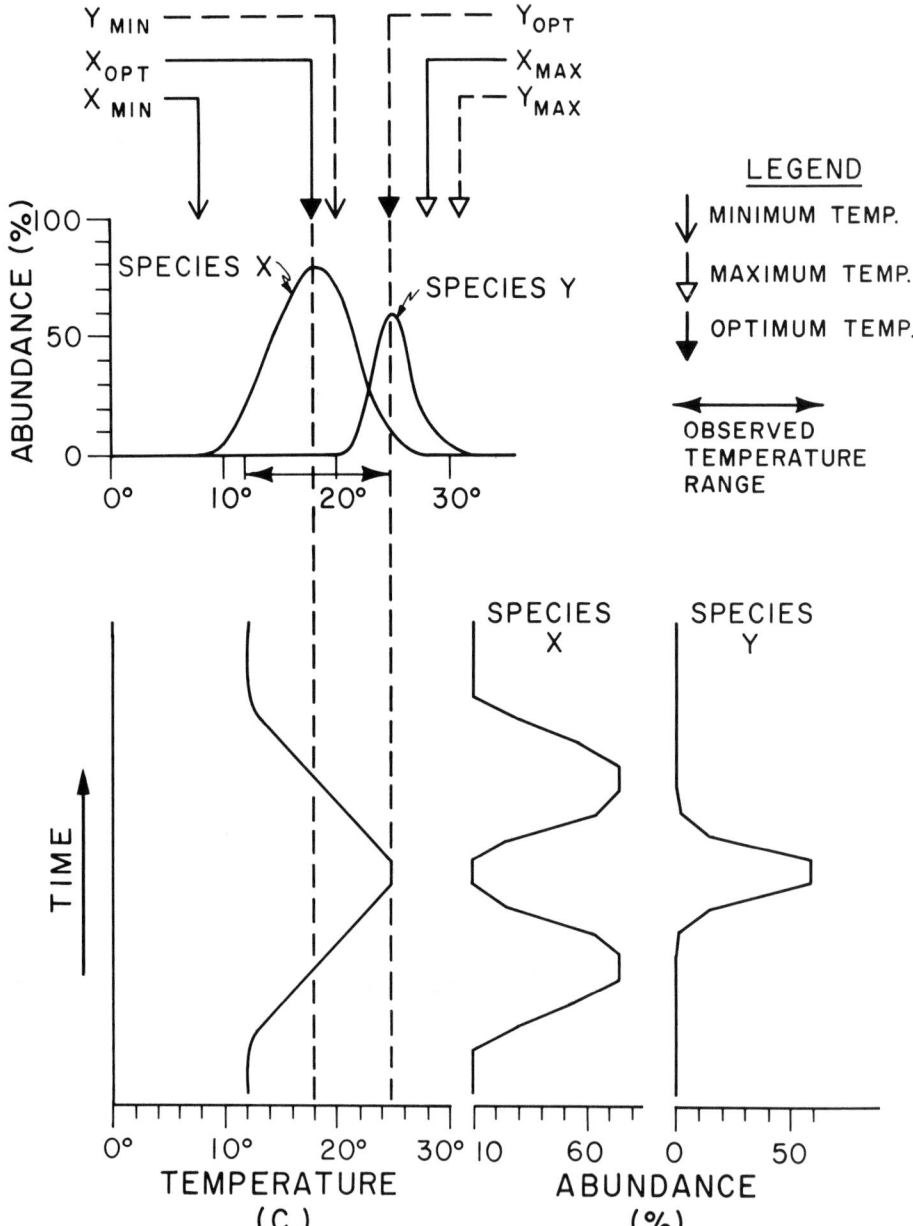

Fig. 28. Hypothetical model showing the abundance–depth curves resulting from the response of two species (X and Y) having different temperature optima to an interglacial pulse of temperature increase. The contrasting stratigraphic records (lower right) of the postulated pulse (lower left) are caused by the different location of species optima in the observed temperature range (upper left). Minimum, maximum, and optimum temperatures for each species indicated by appropriate subscripts (upper left).

where three-sample running averages are plotted, and the letters A, B, and C are used to designate, in order, the tropical, subtropical, and subpolar assemblages. Ignoring the gyre margin assemblage (D), the three assemblages known to be related primarily to water temperature show an arrangement of abundance peaks that is both cyclic and symmetric for the upper 400 cm. Reading from the top down: A–B–C–B–A–B–C–B–A–B. We consider that this cyclothemic arrangement argues strongly for the reality of the varimax faunal analysis, particularly the double abundance peaks (in the Y and near the W zones) of the subtropical fauna.

V-zone faunas (still ignoring the gyre margin assemblage) show a warmer aspect, characterized by alternating episodes of dominance by tropical and subtropical assemblages, with only two subpolar peaks near 600 cm. U-zone faunas are colder in aspect, with two subpolar peaks and a roughly cyclical distribution of warmer assemblages about them.

The gyre margin assemblage fluctuates independently of the other four assemblages, as would be expected from our analysis of core top ecology.

8. The four abundance peaks of the subpolar assemblage (Fig. 26) represent a significant quantitative ecologic anomaly. The argument runs as follows. Quantitative paleotemperature estimates made in the next section confirm the conclusion already reached by inspection of B_c and Figure 14 that Caribbean surface water near V12-122 did not fall below 20°C during its deposition.

The staggered abundance peaks well shown in the upper portion of Figure 29 imply that during a glacial pulse as the tropical assemblage decreases in abundance, the values first for subtropical and then for subpolar assemblages rise to subdued maxima. The first step in this succession of peaks (tropical to subtropical) is quantitatively consistent with the data summarized in Figure 19 in that the observed relative abundances of the two assemblages (Table 9) can be matched in the core top data without going below 20°C. The second step—from subtropical peak to subpolar peak—is not consistent. During the four intervals with exceptional abundances of the subpolar assemblage (near 110, 560, 890, and 1030 cm), both the absolute magnitude of the subpolar assemblage and its abundance relative to that of the other assemblages are inconsistent with data summarized in Figure 19. The sample at 1030 cm, for example, has 0.73 for tropical, 0.34 for subpolar, and 0.20 for subtropical proportions. But nowhere in the 61 core tops is such a combination found; and nowhere in waters as warm as 20°C is such a high level of subpolar asssmblage abundance observed.

How is this anomaly to be explained? One possibility is that the varimax model fails to describe the faunas adequately during a glacial pulse. An inspection of the communalities in Table 17 indicates that this is not the case: the average communality for the ten most anomalous samples is no lower than others. We are therefore driven to conclude that during glacial pulses a *combination* of our defined varimax assemblages lives in the Caribbean, which is not found there or in the Atlantic today. In fact, just this combination of tropical and subpolar forms is found now in the northern part of the Gulf of Mexico. Phleger (1960) interpreted the high values of *Globigerina bulloides* there as a relict glacial

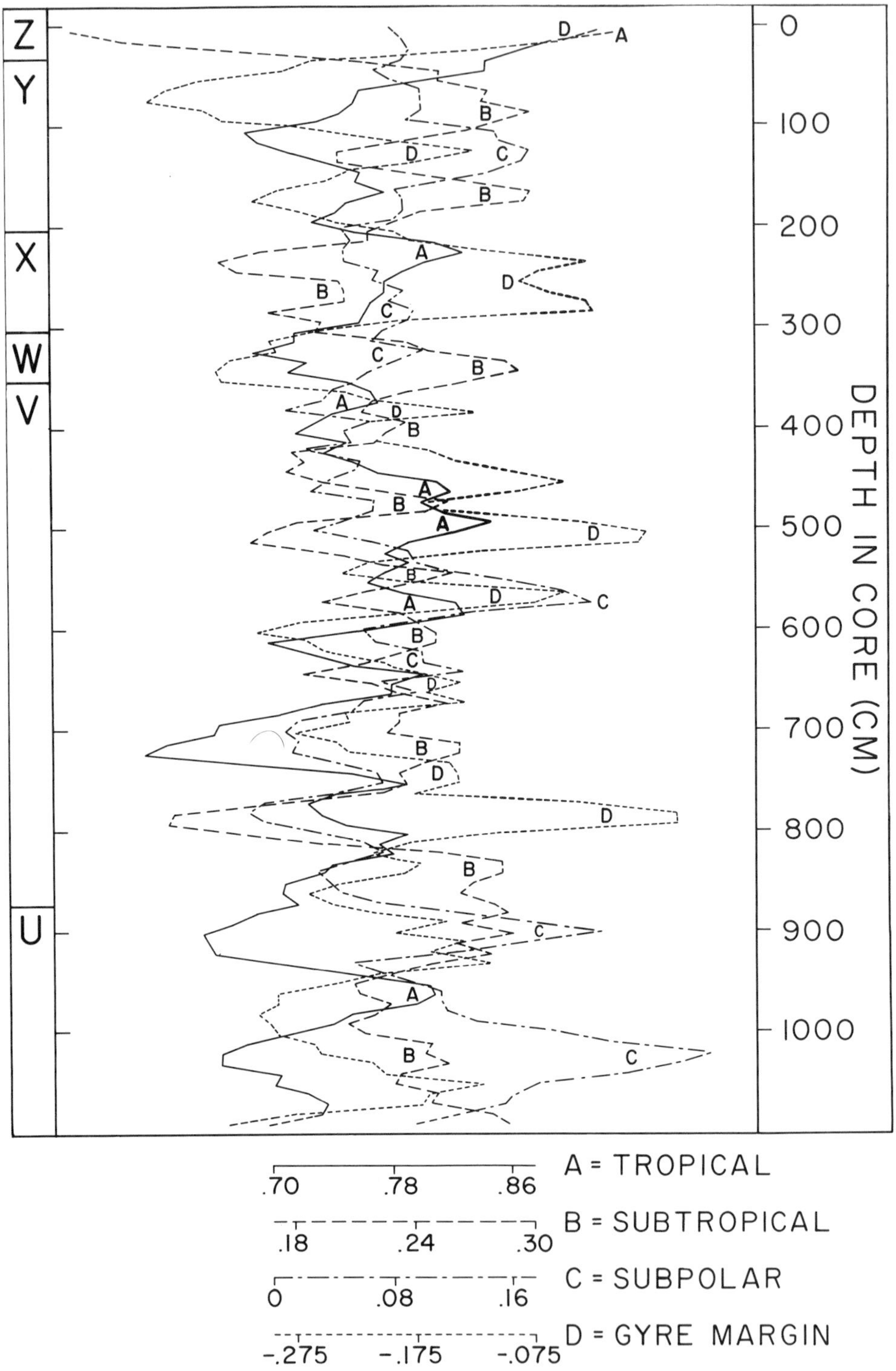

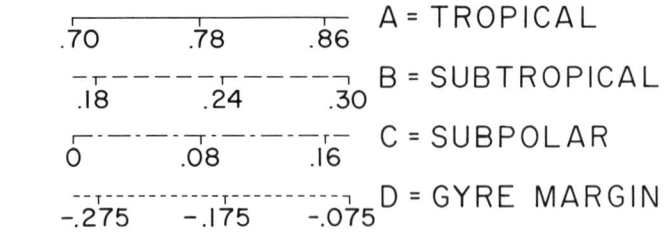

Z
Y
X
W
V
U
0
100
200
300
400
500
600
700
800
900
1000
DEPTH IN CORE (CM)
A
B
C
D
A = TROPICAL
.70 .78 .86
B = SUBTROPICAL
.18 .24 .30
C = SUBPOLAR
0 .08 .16
D = GYRE MARGIN
-.275 -.175 -.075

CARIBBEAN CYCLOTHEMS

fauna; and Parker (1970) reports right-coiling G. *pachyderma* in the northeast Gulf. These two species are the dominants in our subpolar assemblage. Unfortunately, our sampling net does not include a station in the northern Gulf.

The combined evidence makes it quite clear that although glacial Caribbean faunas continued to be dominated in shallow depths by the tropical assemblage, at deeper levels an ecologic water mass originating at high latitudes brought in significant contributions of the subpolar assemblage. Y-zone and other glacial faunas in V12-122 contain the historical record of the entry into the American Mediterranean region of the fauna now relict in the Gulf. The supposed anomaly is actually evidence of tropical submergence.

Paleoenvironmental Estimation

After processing the matrix $\mathbf{B}_c$ to yield $\mathbf{B}^2_c$ (Table 10), the paleoecological equations summarized in Table 12 are used to make quantitative estimates of surface temperatures and salinities for the 110 samples of V12-122. The results are given in Table 14 and plotted on Figure 30.

1. Estimated winter temperatures range through 5.8°C, from 19.9° just below the U–V boundary, to 25.7° at the core top. Summer temperatures are estimated to range through 4.6°C, with a minimum of 23.1° in the U zone through a maximum of 27.7° at 500 cm. The estimated extreme fluctuation of average temperatures of Caribbean surface water during this interval is therefore about 5°C. The estimated range of salinity fluctuations is about 1.0 per mil. All calculated ranges seem generally consistent with previous estimates and fall within the ranges of values observed today in low Atlantic latitudes (Fig. 31).

2. The overall plots of temperature versus depth reveal a small but significant secular trend—or at least differences in the climatic regimen above and below the U–V boundary. The U zone displays a definitely cooler average temperature and two episodes during which temperatures were lower than they have been at any time since. By contrast, post U-zone temperatures have been on the average

---

Fig. 29. Abundance of four assemblages versus depth in Caribbean core V12–122. Abundance curves have been smoothed by taking a three-sample running average of data given in Table 9 and plotted raw in Figures 24–27; the result represents an 8,000-year moving average and damps high frequency fluctuations. Ericson faunal zones U through Z as in Figure 24. Note that the tropical fauna quantitatively dominates the entire core and that the scales have been shifted to portray clearly the stratigraphic position of the various abundance peaks identified by letter code. Ignoring the gyre margin assemblage which fluctuates independently, peaks of the other assemblages show a clear cyclic arrangement in the top 400 cm. The middle section, from 400 to 700 cm, is marked by considerably damped fluctuations of the three temperature-dependent assemblages and, except for a subpolar peak about 570 cm, records simple alternations between the tropical and subtropical peaks. The lower portion of the core has wide fluctuation ranges and, with the exception of the subpolar peak at 1030 cm, displays a rough cyclic arrangement of peaks.

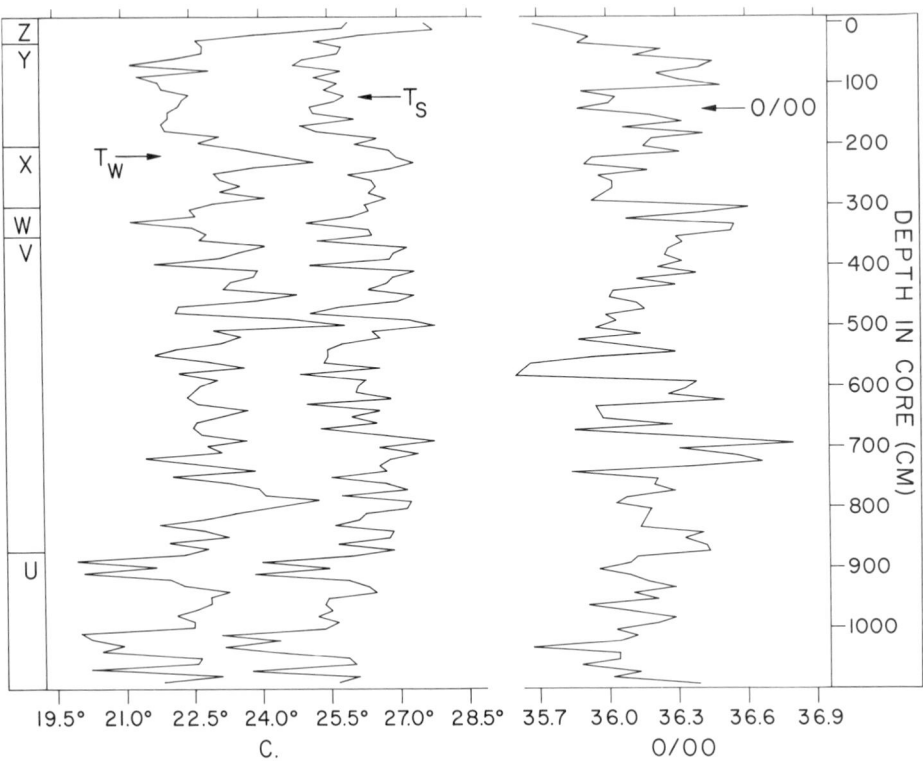

Fig. 30. Paleoclimatic estimates for 110 samples of Caribbean core V12–122, based on paleoecological equations (Table 12) derived from 61 core tops. T_w = winter surface temperature; T_s = summer surface temperature; o/oo = average surface salinity. Ericson zones U through Z as in Figure 24. Data in Table 14.

distinctly higher and are marked by at least four periods at or near current levels. These trends are made clearer by eliminating some of the higher-frequency fluctuations. In Figure 32 we plot a three-sample (8000-year) moving average for estimates of the summer temperature.

3. The secular trend just discussed contrasts rather sharply with the lack of such a trend exhibited by oxygen isotope ratios. On Figure 32 the δO^{18} curve calculated for *Globigerinoides ruber* specimens in V12-122 is displayed (Broecker and Van Donk 1970). Here, as in other isotope work, a long-term trend is not found: relatively recent highs and lows on the curve are at approximately the same levels as older fluctuations. One way of explaining this discrepancy is to assume that the oxygen isotopes reflect primarily the regimen of ice volume which, as indicated by the termini of many morainal systems, presumably fluctuates between two fairly constant extreme values.

4. Another difference between the isotope and the paleontologically derived temperature curves is the simple, lower-frequency fluctuation pattern of the

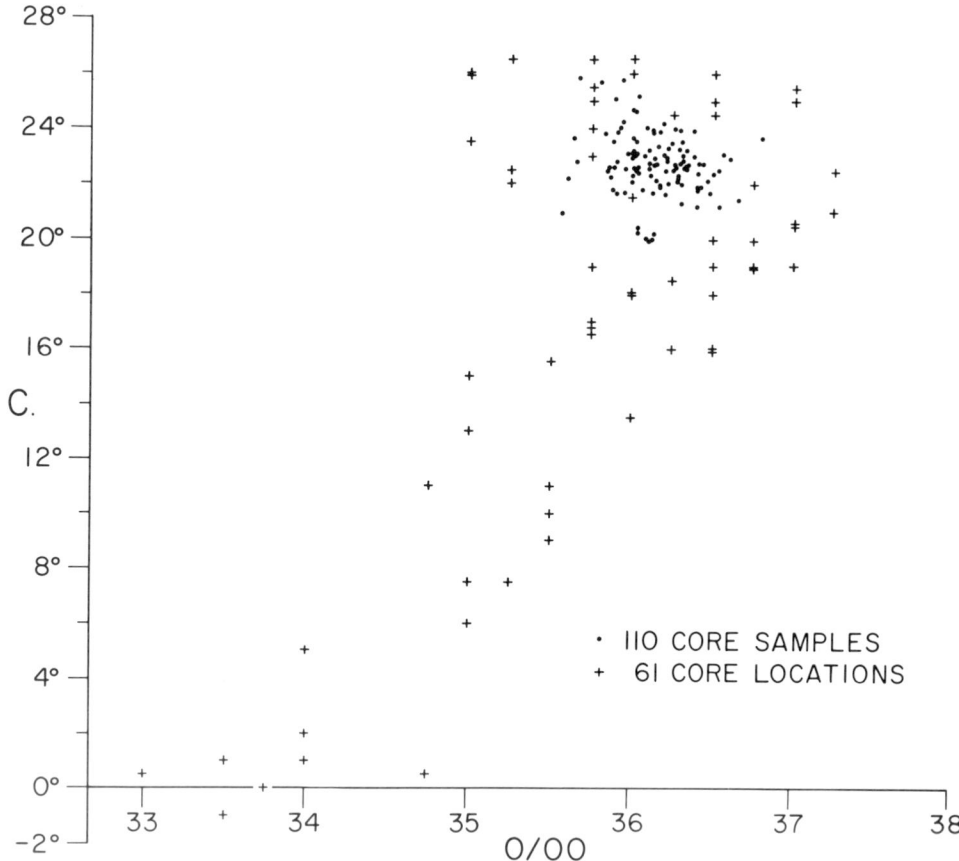

Fig. 31. Plot of salinity versus winter temperatures known for surface waters overlying 61 core top samples; and estimated for 110 samples of Caribbean core V12–122. Data given in Tables 13 and 14.

former. In this respect the 8000-year moving average T_s curve more nearly resembles the isotope curve than the raw T_s curve. One explanation for this could be that the isotope curve, reflecting ice volume more than local temperature, sees an integrated record of worldwide ice volume.

5. Broecker and Van Donk (1970) have emphasized that most of the isotope fluctuations show abrupt terminations of presumed glacial events, and slow growth of glacial conditions. This asymmetrical, sawtooth character of the isotope curves, well shown on Figure 32, is generally not characteristic of our estimated temperature curves. Again, if we assume that the isotope fluctuations reflect primarily ice volume, and that glaciers melt much faster than they grow, we are at liberty to interpret our curves as reflecting a linear response by the oceans to a simple, symmetrical driving function.

6. In general, there is an impressive agreement for zones W–Z between the

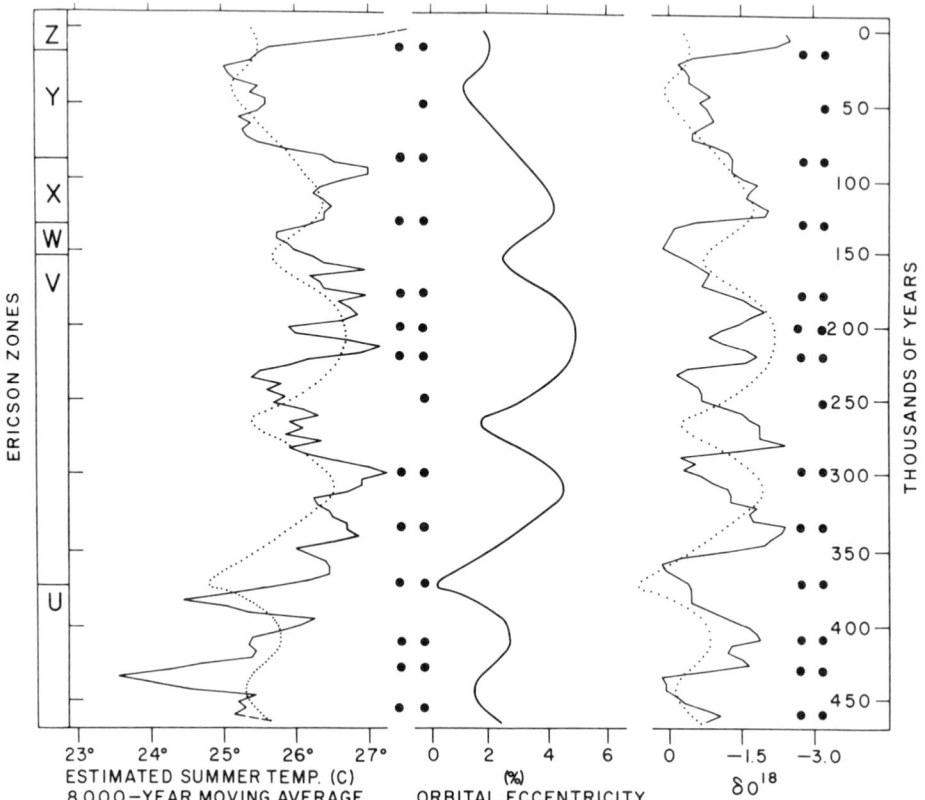

Fig. 32. Calculated perturbations in the earth's orbit (Vernekar 1968) compared with estimated surface summer temperatures and measured δO^{18} in Caribbean core V12–122. The paleotemperature curve is a three-sample running average of data given in Table 14 and plotted raw in Figure 30. Curve for orbital eccentricity duplicated twice as a dotted line to facilitate comparisons. A single heavy dot is placed to correspond with significant peaks in the total Milankovitch insolation curve calculated by Vernekar for 65°N latitude; a second dot is placed beside the first when a peak in the total Milankovitch curve for 15°N latitude coincides with that for 65°N. δO^{18} values for *Globigerinoides ruber* are given in per mil compared to Pee Dee Belemnite standard (Van Donk 1970). Ericson faunal zones U through Z represent presence and absence of *Globorotalia menardii*. Time scale based on calculated average accumulation rate of 2.35 cm per 1000 years calculated for this core by Broecker and Van Donk (1970).

isotope and the paleontological temperature curves: a short cool W zone, a warm X zone, a cool Y zone with a small episode of higher temperature contained within it, and a rapid transition from glacial Y to Recent Z zone. This pattern closely matches the isotope fluctuation patterns originally documented by Emiliani (1955, 1966).

7. Below the W zone the two curves diverge quite markedly. In the V zone,

our temperature curve shows two broad warm periods, separated by a distinct cooler period. The isotope curve during the same interval shows three distinct peaks, the upper one split into two subpeaks. In the lower portion of the core, the paleontological temperature curve shows two distinct cold intervals, the upper one just below the U–V boundary. The isotope curve has two lows here. The lower one coincides with ours, but the upper occurs distinctly later, above the U–V boundary.

8. Figure 31 shows the position on a T–S diagram of surface waters overlying the 61 core top samples, and the position calculated for 110 samples in V12-122. The core sample estimates fall about in the middle of an elliptical swarm of points representing modern localities warmer than 20°C. In general, the core samples show a tendency to be more saline during colder periods, a relation expected both in magnitude and direction as a response to glaciation.

9. It is instructive to compare our paleotemperature curves in Figures 30 and 32 with other paleontologically based estimates. McIntyre (1967), for example, concludes from a study of Atlantic coccoliths that Wisconsin surface water temperatures in the Caribbean were somewhat warmer than 20°C. This conclusion matches ours for the coldest part of the Y zone. In a general way our curves also correspond to climatic inferences made by Ericson and Wollin (1968) for V12-122, based on fluctuations in the abundance of *Globorotalia menardii*. As with the isotopes the correspondence is good only for post V-zone deposition; but we also agree that the V zone was generally warmer than conditions represented in the W and U zones. But a number of significant discrepancies occur. For example, our temperature peak at about 950 cm occurs during a period when *G. menardii* is completely absent, whereas the smaller peak at the bottom coincides with modest *G. menardii* abundance. Furthermore, the timing of *G. menardii* peaks within the V zone does not correspond well with our curve. We conclude that *G. menardii* (which is part of our tropical and gyre margin assemblages) responds both to water temperature and to water mass productivity. Other factors, as yet unspecified, must also play a role.

10. One way of studying the relationships between our paleotemperature estimates and data on *G. menardii* and oxygen isotopes is to focus attention on the coincidence of major peaks in depth–amplitude curves representing these variables (Table 19). Eight peaks are represented. Only three of them are shared by all three curves: the X-zone peak in the range 230–300 cm; and two V-zone peaks at 500 and 790 cm. For any pair of curves about half the peaks represented in both curves coincide.

11. We are much more confident of the temperature estimates than those for salinity. Not only are the latter weaker statistically, but the arguments presented for a relation of the gyre margin fauna to productivity suggest that the equation based on our data might give erroneous estimates. Ryther et al. (1967), for example, show that the Amazon outflow is reflected in Caribbean salinities. Final judgment must, in any event, be suspended until a set of cores has been analyzed by our method.

A Model for the Glacial Atlantic

In the preceding section it was concluded that estimates of temperature made by our method are reasonable in range and exhibit stratigraphic patterns that can be interpreted as consistent with isotopic and other paleontologic data—provided that an appropriate response model is used. The model we suggest considers the isotope curve as a function of ice volume and temperature, the *Globorotalia menardii* curve as a function of productivity and temperature, and our T_w and T_s curves as functions of temperature alone. Observed correlations among the curves therefore represent the common factor of temperature, and the discrepancies reflect intervals during which ice volume or productivity fluctuated independently of local Caribbean temperature.

If we assume this model is true we are led to ask what conditions in the Atlantic might cause *G. menardii*, water temperature, and isotope values to fluctuate independently. Useful insights into this question can be achieved by examining two paleontological curves which correlate well with the O^{18} curve throughout the deposition of V12-122 (Fig. 33). One of these curves is a smoothed version of the gyre margin assemblage abundance. The other is the coarse-fraction curve, representing that portion of the sediment >74 μ. The correlation between coarse fraction and isotope curves in equatorial Atlantic cores has been known and stressed for many years (Emiliani 1955). Our contribution is to point out the correlation of both to the gyre margin assemblage.

How are the correlations presented in Figure 33 to be explained? We first note that the coarse-fraction curve represents essentially a plot of the density of foraminifera in the sediment, and that both curves have high values in known interglacial intervals when the $CaCO_3$ content is also higher than the average (Table 18). We have already stressed our evidence that the gyre margin abundance curve is controlled primarily by water mass productivity (Figs. 10, 11, 34). The same cause can now be invoked to explain parallel fluctuations in the coarse-fraction curve (Beers et al. 1968).

In order to test this idea we examined coarse-fraction data for 32 of the 61 core tops (those that had been processed by our laboratory). The ten highest values are located in the South Atlantic and Indian Ocean. Of the next highest seven values, one lies in the South Atlantic off the mid-Atlantic ridge, and six lie on or near the ridge in the North Atlantic. We conclude that the coarse fraction represents primarily surface productivity of the water, and secondarily a preservational effect of shallow topography.

Waters of the South Atlantic, particularly surface waters along the African coast, are well known for high nutrient values. These waters feed into the southern portion of the equatorial Atlantic and are in part responsible for the relatively high productivity values there. During summers of the present epoch the belt of equatorial doldrums and flanking trades moves north (Knauss 1963). As indicated by Defant (1961, Plates 8, 8a) the surface waters of the entire equatorial system in summer contain a higher percentage of South Atlantic water. Because these

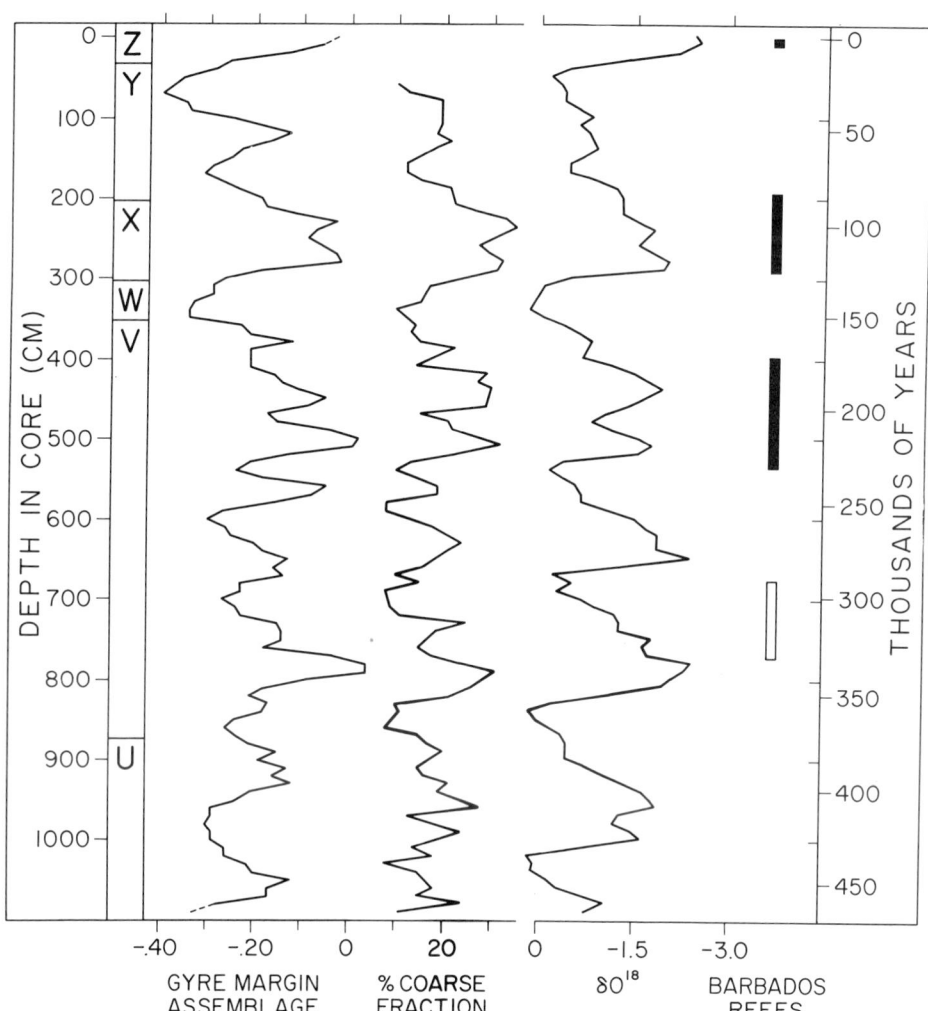

Fig. 33. Plot of gyre margin assemblage abundance, coarse-fraction percentage, and δO^{18} values in Caribbean core V12–122. Occurrences of reef terraces on Barbados, W.I., well controlled by radiometric dates (Mesolella et al. 1969), are indicated by three solid bars; a terrace complex whose chronology is only approximately established is shown by an open bar. Chronology for V12–122 curves is established as explained in Figure 32. The curves show a strong correlation. With the exception of the lowest terrace, Barbados reefs can be correlated with concurrent peaks on the three curves. The gyre margin curve is a three-sample running average of data given in Table 9 and plotted raw on Figure 27. Coarse-fraction curve (calculated by D. B. Ericson) represents the percent of sediment coarser than 74 μ.

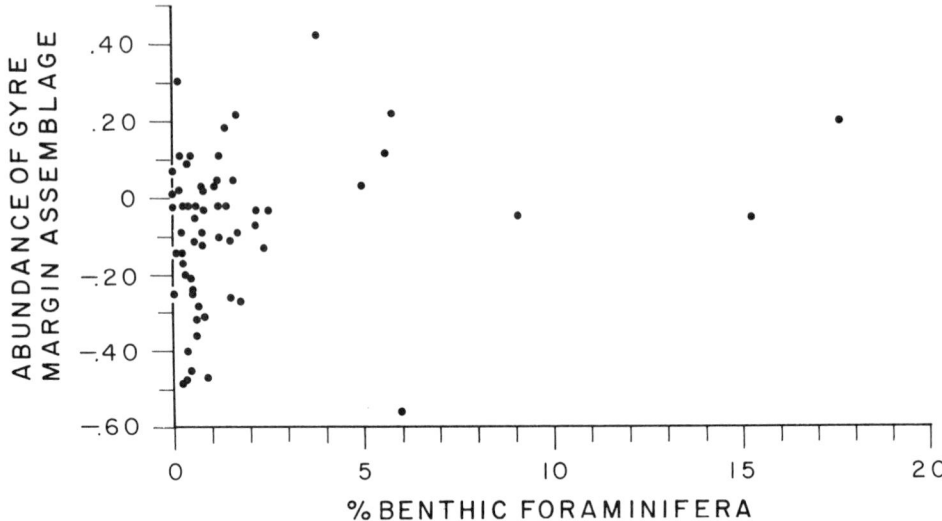

Fig. 34. Percent benthic foraminifera versus abundance of gyre margin assemblage in 61 core top samples. Benthic foraminifera are expressed as percent of total foraminifera.

waters flow into the Caribbean it is our suggestion that the observed fluctuations of coarse-fraction and gyre margin faunal values in V12-122 reflect changing proportions of productive South Atlantic water (Fig. 10).

It is noteworthy that Mesolella et al. (1969) documented periods of reef growth on Barbados which correspond to our inferred water temperature maxima. As indicated on Figure 33, these correspond also to our presumed productivity maxima. Conceivably, the early reef terrace (the dating of which is subject to considerable uncertainty) corresponds to a conjunction of peaks near 785 cm on the winter temperature curve (Fig. 30), the gyre margin curve, the coarse-fraction curve, and the isotope curve. All similar conjunctions in the core do, in fact, correlate with times of Barbados reef development that are well dated by radiometric methods. If the T_w curve does represent winter temperatures, the gyre margin and coarse-fraction curves reflect productivity, and the isotope curve correlates with ice volume and sea level, then the construction of reef terraces would be logically related to high sea stands occurring when West Indian waters were warm and highly productive.

Accordingly, summers of the present epoch are a model for interglacial, and winters are a model for glacial times. As pointed out by Lamb (1961, p. 31) "A slight shift southward of the climatic zones of the North and South Atlantic . . . might well shift and weaken the ocean surface currents including the Equatorial Current, more of which would be deflected south along the Brazilian coast from 5°S. Both aspects should weaken the supply of equatorial warm water to the Gulf Stream."

Van Donk (1970) has pointed out that of all the oxygen isotope fluctuation curves so far constructed for deep-sea cores on planktonic foraminifera, only those in the Caribbean, equatorial Atlantic, and Indian Ocean show wide fluctuations in δO^{18}. It would be interesting to learn if the isotope pattern of the South Atlantic, which is now unknown, resembles that of the equatorial Atlantic. When south Atlantic isotope data are in hand it may be possible to explain why portions of the Atlantic exhibit isotopic fluctuation patterns such as those recorded in V12-122. In a general way it is clear that the pattern reflects the volume of glacial ice, but the mechanism of this relationship remains obscure.

TEST OF THE MILANKOVITCH THEORY

In the preceding section we argue for the general validity of our estimates of Caribbean paleoclimate. Here we assume that they are valid and use them to test a theory that makes quantitative predictions of long-term fluctuations in the global temperature pattern.

The Milankovitch theory, named for the worker whose calculations and publications during the 1930s gave it a solid form, assumes that perturbations in the earth's orbit and its axis of rotation are responsible for climatic fluctuation. These perturbations result from changes in the gravitational field accompanying motions of the planets, and are calculated from astronomical data on their masses and orbits. The general nature of this theory is well known, and has been well outlined elsewhere (e.g. Broecker and Van Donk 1970). Here we briefly summarize those elements in the theory which give rise to our strategy of testing.

Three independent parameters are used to describe changes in the earth's motion as a function of time (Fig. 35): eccentricity of the orbit; tilt, or obliquity of the ecliptic; and precession. These parameters are quasiperiodic functions of time, with periods averaging 92,000 years for eccentricity, 40,000 years for tilt, and 21,000 years for precession (Fig. 36). From calculations of Vernekar (1968) it is clear that the periods for tilt and precession do not vary much from the average, but that for eccentricity does. For the purposes of our test, therefore, it is important to consider the periodicities calculated for the interval of time represented in V12-122, i.e. the last 465,000 years. These periods are 98,000, 40,000, and 22,000 years (Vernekar 1968).

From the quasiperiodic changes in the earth's motion just described, calculations are made of the influx of solar energy received by the upper atmosphere at a given latitude and for a specified season. Results, expressed in langleys per day, are plotted as a function of time for a given latitude. Such curves—here called total Milankovitch curves—are commonly used in discussions of Milankovitch theory. It is important to realize that in this model the total influx of heat received by the earth over a year is constant; what changes is the time–space distribution pattern.

The usual approach to Milankovitch testing is to select a high northern latitude and examine the corresponding total summer insolation curve. Such a

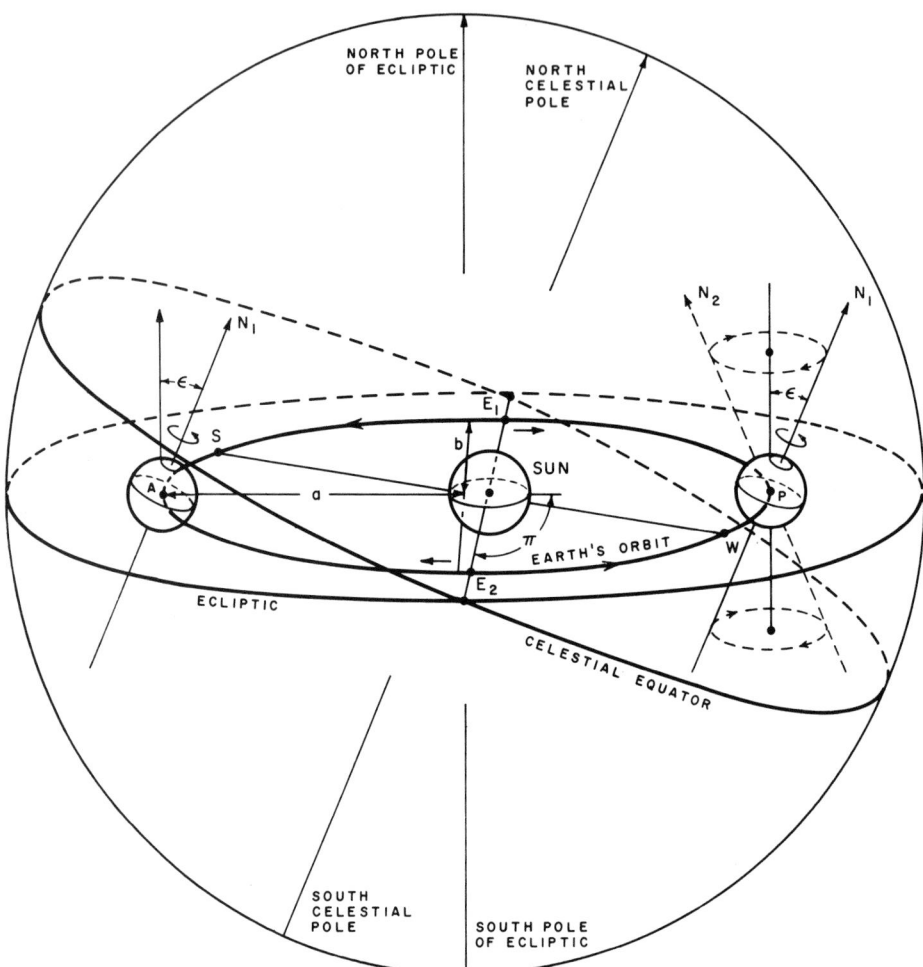

Fig. 35. Sketch of the earth's orbit around the sun in relation to the celestial sphere (from Vernekar 1968). A = aphelion; P = perihelion; S = summer solstice; W = winter solstice; a = the major axis of the earth's elliptical orbit; b = the minor axis. Eccentricity is expressed as $e = \sqrt{a^2 - b^2}\,/a$. E_1 = the vernal equinox; E_2 = autumnal equinox. Arrows near the equinoxes indicate their precession. N_1 is the present position after half-orbit of precession.

choice, made necessary by the fact that every latitude has a different time–amplitude pattern, is commonly justified by the assumption that heat received in high latitudes during summers is critical for the growth and decay of the northern hemisphere ice sheet. We have no quarrel with this assumption. Our concern is to point out that total Milankovitch curves for high latitudes reflect predominantly the effect of tilt, whereas low-latitude curves are dominated by precession. But if

MILANKOVITCH MODEL				CORE V12-122			
CAUSE	AVERAGE PERIOD	AVERAGE PERIOD FOR PAST 465,000 YEARS	AVERAGE FREQUENCY (c/cm) IN V12-122*	PEAK FREQUENCIES* IN POWER SPECTRUM OF DEPTH-T AND DEPTH-0/00 CURVES			AVERAGE PERIOD FOR DEPTH-T CURVES
				T_W	T_S	0/00	
ECCENTRICITY	92,000 YEARS	98,000 YEARS	.0043	.0035-.0045	.001-.005	.0015-.0045	100,000 YEARS
TILT	40,000 YEARS	40,000 YEARS	.011		.008 / .010†	.008 / .010†	—
				.014	.0145	.014	
PRECESSION	21,000 YEARS	22,000 YEARS	.019	.018† / .027	.019† / .0265		—

* USING BROECKER'S ACCUMULATION RATE OF 2.35 cm/1000 YEARS.
† HARMONIC MEAN OF TWO OBSERVED FREQUENCIES.

Fig. 36. Frequency peaks in the climatic spectra of the Milankovitch model compared with frequency peaks observed in the spectra calculated from empirical paleoclimate curves of Caribbean core V12–122.

the total reaction of the ocean involves global temperature variations we must take both high- and low-latitude effects into account in any test. The two independent effects might be weighed differently, depending on the nature and location of the geologic phenomena under observation. For effects related to ice volume, the high-latitude insolation might dominate over the entire ocean. For effects related to local surface water temperatures, however, the observed result might reflect both high- and low-latitude insolation in varying proportion depending on the site.

Our version of a Caribbean test for the Milankovitch theory assumes that local water temperatures will be affected in three ways. (1) The dominant effect will be the general low-latitude insolation budget as expressed in the time–amplitude plot of orbital eccentricity. This use of the eccentricity curve is suggested by the work of Mesolella et al. (1969), who document a suggestive correlation between times of eccentricity maxima and reef growth on Barbados. The underlying assumption here is that the main influence eccentricity has on the earth's insolation budget is via the low-latitude precessional effect. By using the eccentricity curve, therefore, we are in effect integrating over the low latitudes. As pointed out by Mesolella et al. (1969), times of high eccentricity will be accompanied in low latitudes by sharp seasonality, with hot summers and cold winters. (2) A second-order effect for a Caribbean core will be that of high-latitude insolation. (3) A third-order effect will be the insolation at the latitude of core V12-122.

Time Series Analysis of V12-122 Paleoclimate Curves

The paleoclimatic curves given in Figures 30 and 32 can be analyzed by techniques of time-series analysis in two ways. We can, in the first place, examine each signal in the *time domain*, and ask if the observed peaks and valleys correspond in time to those predicted by the Milankovitch model. Second, we can study the *spectrum* of the signal and ask if there are periodic components in it, and if so whether or not the frequencies of these components correspond to those predicted. In order to apply these tests the first requisite is a time scale for the core. As discussed previously we base our chronology on Broecker and Van Donk's (1970) calculated average sedimentation rate of 2.35 cm per 1000 years. It cannot be emphasized too strongly that most of the conclusions we draw in the remaining portion of this paper are directly influenced by this figure—and the important fact that the sedimentation rate calculated is an average.

Time Domain Test

Data plotted in Figure 32 constitute our time domain test. The time scale follows from the sedimentation rate. Here we pick the bottom sample (1090 cm) as corresponding to 465,000 years and assume that accumulation has been constant during the deposition of the core. The 8000-year moving average curve is a simple form of filter analysis, in which the higher frequencies (seen in the plot of the unfiltered data in Fig. 30) are eliminated by a 3-point moving average. We have experimented with a number of more sophisticated filters, including one of the form $f_n = 0.25 \, x_{n-1} + 0.50 \, x_n + 0.25 \, x_{n+1}$, and find that for the spectrum represented by T_s there is little difference. For simplicity, therefore, we present the simple moving averages.

The orbital eccentricity curve calculated by Vernekar (1968) is plotted in triplicate on Figure 32. This curve represents the first-order cause postulated in our version of the theory. The second- and third-order effects are represented by the rows of dots on the figure. A single dot has been placed at times when the total Milankovitch curve for 65°N latitude shows a conspicuous peak (from data given in Vernekar 1968). The spacing of these peaks reflects primarily the tilt frequency. The third-order effect postulated is represented by the second row of dots. A dot has been placed beside the first whenever the total Milankovitch curve for 15°N Latitude (approximately the latitude of V12-122) coincides with that for 65°N.

We conclude the following. (1) There is a surprisingly good correlation between the eccentricity curve and the filtered T_s curve for V12-122, particularly when the arbitrariness of the assumption of uniform accumulation rate is kept in mind. The fit is reasonably good except for the upper V zone. (2) Considering the pattern of dots reflecting total Milankovitch peaks for 65°N and 15°N, one can explain many of the second-order fluctuations in T_s. The coincidence of peaks and dots for all but those in the upper V zone strikes us as being significant. Others may have contrary opinions. In any event, final judgment must be suspended until many other cores have been analyzed.

The time–domain test of oxygen isotope data gives positive results in the top portion of the core, generally negative below.

Frequency Domain Test

Using standard techniques of power spectrum analysis developed by Blackman and Tukey (1958), we will estimate the spectra of our paleoclimatic curves and compare them with theoretical Milankovitch periodicities. Both objectivity and convenience are served by converting periods to frequencies, and measuring all frequencies as cycles per centimeter of accumulation in V12-122. On this scale, the Milankovitch eccentricity frequency is 0.0043, that for tilt is 0.011, and that for precession 0.019 (Fig. 36). We have calculated spectra for our empirical curves by a standard procedure adapted from Dixon (1965, p. 459): hamming the Fourier transform of an autocorrelation function computed for a specified number of lags on detrended data.

Before examining the empirical paleoclimatic spectra it will be useful to consider the spectrum calculated from a random signal of 110 points representing a hypothetical curve for a deep-sea core (Fig. 37). Power spectrum analysis considers time series data as a signal varying around a mean position. For any given length of record a certain amount of variation (expressed in statistical terms as variance and in engineering parlance as power) is present, and the object of the calculation is to estimate how much of the total variation observed is characteristic of various frequency bands. In a random signal of infinite length all frequencies are present in equal amounts (white noise). For a small-record length, such as the data for our hypothetical and real cores, the calculated spectrum will only approach this ideal. Figure 38 shows the spectrum calculated from data graphed in Figure 37. Note that the time scale in this model spectrum is taken as 10 cm of "time" between sampling points. The highest frequency that can be meaningfully resolved in a given signal is the reciprocal of twice the sampling interval. This frequency, called the Nyquist frequency, is 0.050 cycle per centimeter for our data. The lowest frequency is zero, i.e. a straight line. The spectrum in Figure 38 shows a distribution of frequency peaks of approximately equal amplitude ranging across the whole statistically visible spectrum, and it is not possible to identify any peak as rising clearly above a background noise level. Such a spectrum characterizes a fairly short random record.

Results of spectrum analysis for winter temperature, summer temperature, salinity, and 0^{18} curves are given in Figures 39–42 and partially summarized on Fig. 36. The following conclusions seem justified.

1. None of the spectra could have been derived from random signals; all show statistically significant, nonrandom periodicities.

2. The three paleontologically derived curves all show a dominant frequency near the postulated 0.0043 eccentricity value. Because the core length is too short to permit accurate spectral estimates in this low range, it is actually preferable to estimate periods corresponding to the low-frequency peaks by examining original and smoothed signals and calculating the *average* period directly. The

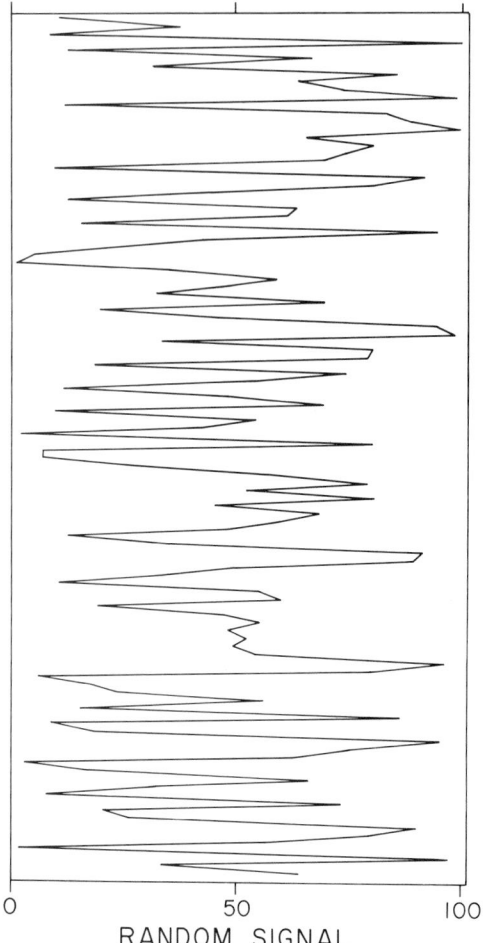

0   50   100

RANDOM SIGNAL

Fig. 37. Random signal generated by plotting 110 random two-digit numbers (Dixon and Massey 1951, Table 1).

result of this computation is exactly 100,000 years for T_s and T_w curves. Thus, our data clearly show a dominant periodicity corresponding closely to that predicted for the Milankovitch eccentricity effect.

3. The isotope spectrum (Fig. 42) shows a clear, dominant frequency peak near 0.005, corresponding to a period of 80,000 years. A glance at the original data (Fig. 32) confirms this estimate: five major cycles occur during the past 400,000 years. This basic isotope cycle is quite distinct from the 100,000-year cycle for the T_s and T_w curves, where four cycles span the same 400,000-year interval, as predicted by the Milankovitch model. Although the statistical significance of this difference is clear, its geological meaning is not.

4. The expected frequency components for tilt (0.011) and precession (0.019) are not found in any of the four spectra.

5. One frequency peak at about 0.014 is found in all four spectra. One peak

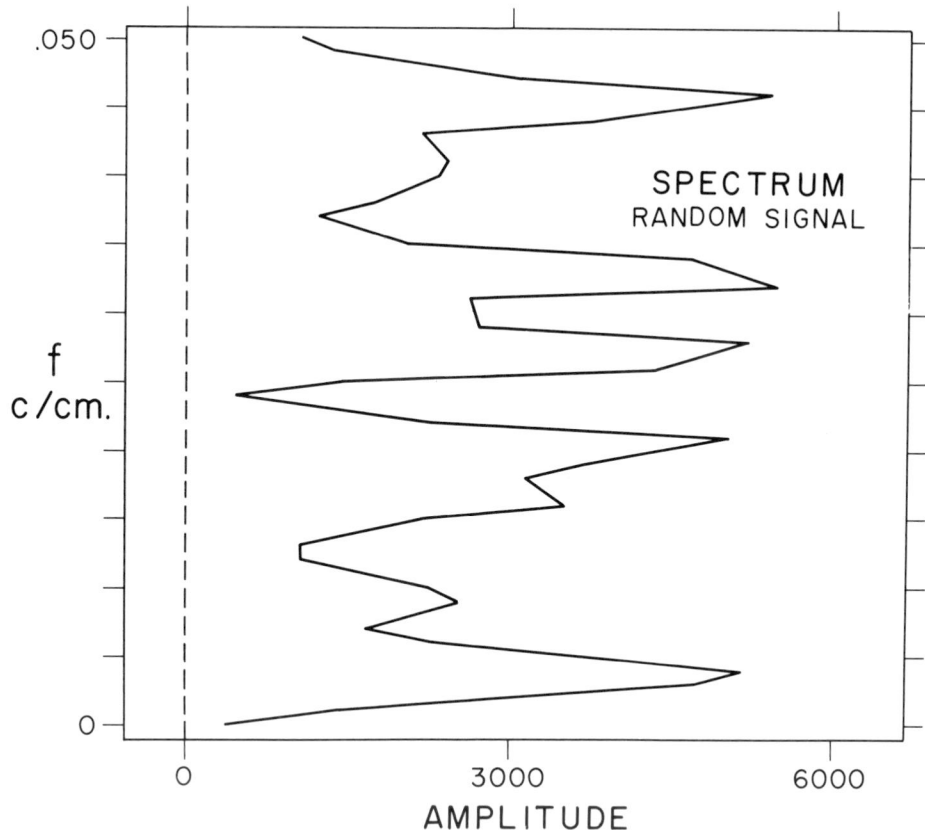

.050

f
c/cm.

0

0   3000   6000

SPECTRUM
RANDOM SIGNAL

AMPLITUDE

Fig. 38. Spectrum calculated from random signal plotted in Figure 37, based on an autocovariance function of 34 lags. Frequencies expressed in cycles per centimeter, based on a 10-cm interval assumed between the 110 sample points. Statistically visible spectrum runs from a frequency of 0 to the Nyquist frequency of 0.050.

at 0.027 is found both in the T_w and T_s spectra. One at 0.008 is found in the T_s and 0/00 spectra. Although some of these peaks do not rise far above the background noise level, taken as a whole we judge them to be statistically significant. As further cores are studied the question of statistical significance of these higher-frequency peaks will be reassessed.

6. If we assume that the high-frequency peaks in the spectra are statistically significant, we must ask why *these* frequencies rather than the model Milankovitch frequencies are found. One explanation, and surely the simplest, is that the Milankovitch model is wrong—for to deny the tilt and precessional effects, while accepting the eccentricity effect, would violate the model. Without tilt and precession the yearly insolation at any latitude is constant.

7. Another possibility is to deny the validity of the paleontological tempera-

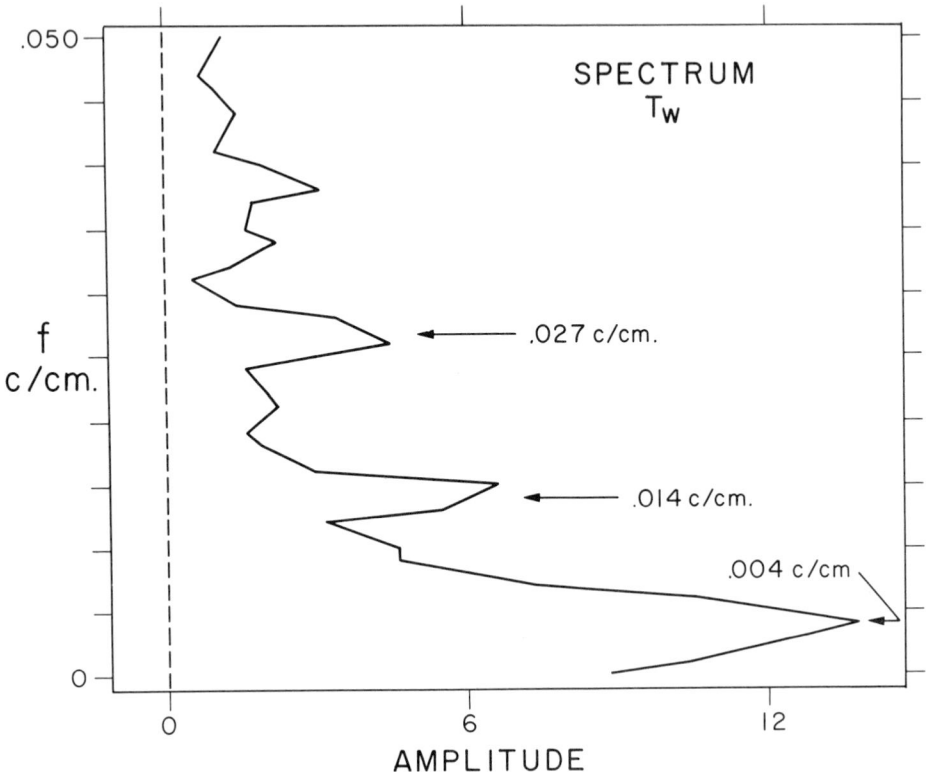

Fig. 39. Spectrum calculated from estimated winter temperature data for V12–122 given in Table 14, and based on an autocovariance function of 34 lags. Scales as shown for Figure 38.

ture estimates, accept the isotope curve as a paleotemperature curve, and decrease the accumulation rate in the core by 25 percent. The last assumption would change the interpretation of the observed frequencies in the isotope spectrum —peaks at 0.005, 0.014, and 0.020 (?) corresponding to periods of 80,000, 30,000, and 19,000 years, assuming 2.35 cm per 1000 years—to frequencies corresponding to 100,000 years, 38,000 years, and 24,000 years. But to decrease the sedimentation rate is to *lengthen* the time span of the core, and none of the investigators involved with dating this core (as discussed above) considers this a real possibility. Rona and Emiliani (1969) in fact argue for a time scale 25 percent *shorter* than the one assumed in this paper. To accept the isotope curve as a paleotemperature curve seems to argue against the Milankovitch model.

8. Another possibility, suggested in discussions with Wallace Broecker and R. K. Matthews, is to postulate a sedimentation rate fluctuating around the average of 2.35 cm per 1000 years. C^{14} dates on two samples from V12-122 bear on this point (Ku and Broecker 1966). The interval 25–30 cm gives a date of 8950 $\pm$

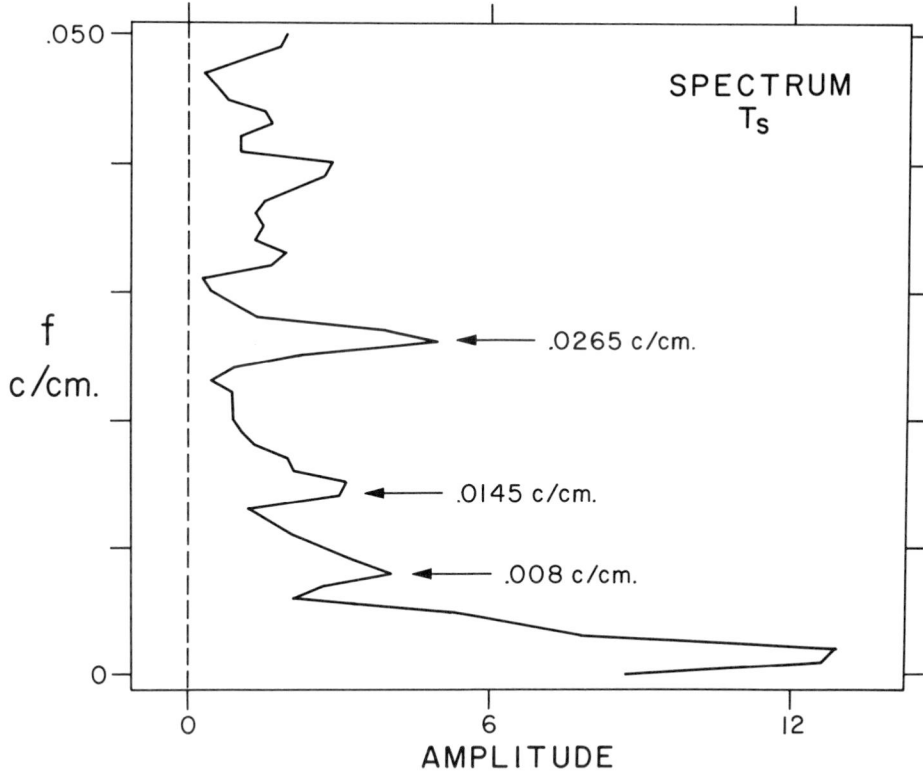

Fig. 40. Spectrum calculated from estimated summer temperature data for V12–122 given in Table 14, and based on an autocovariance function of 44 lags. Scales as shown for Figure 38.

550 years; the interval 55–60 cm gives 18,300 ± 1200 years. If we assume the core top has a zero date, the three points fall on a straight line corresponding to a sedimentation rate of 3.1 cm per 1000 years. Compared with the average accumulation rate calculated by Broecker and Van Donk (1970) for the entire core (2.35 cm per 1000 years), we find that the rate for the Z zone is 32 percent above the average. We therefore conclude that accumulation rates in V12-122 must fluctuate at least by ± 32 percent, i.e. in the range 1.60–3.1 cm per 1000 years. The stratigraphic distribution of these fluctuations is, of course, not known. One possibility is that the sedimentation rate is itself related systematically to a 100,000-year climatic cycle, and we have constructed a digital model to demonstrate how such a situation could distort the higher frequency portion of the spectrum while leaving the lower portion undistorted.

Data for the model are given in Table 15, presented graphically in Figure 43, and subjected to power spectral analysis in Figures 44–46. Two regular, cyclic components are assumed (signals 1 and 2) and combined to yield signal 3. The

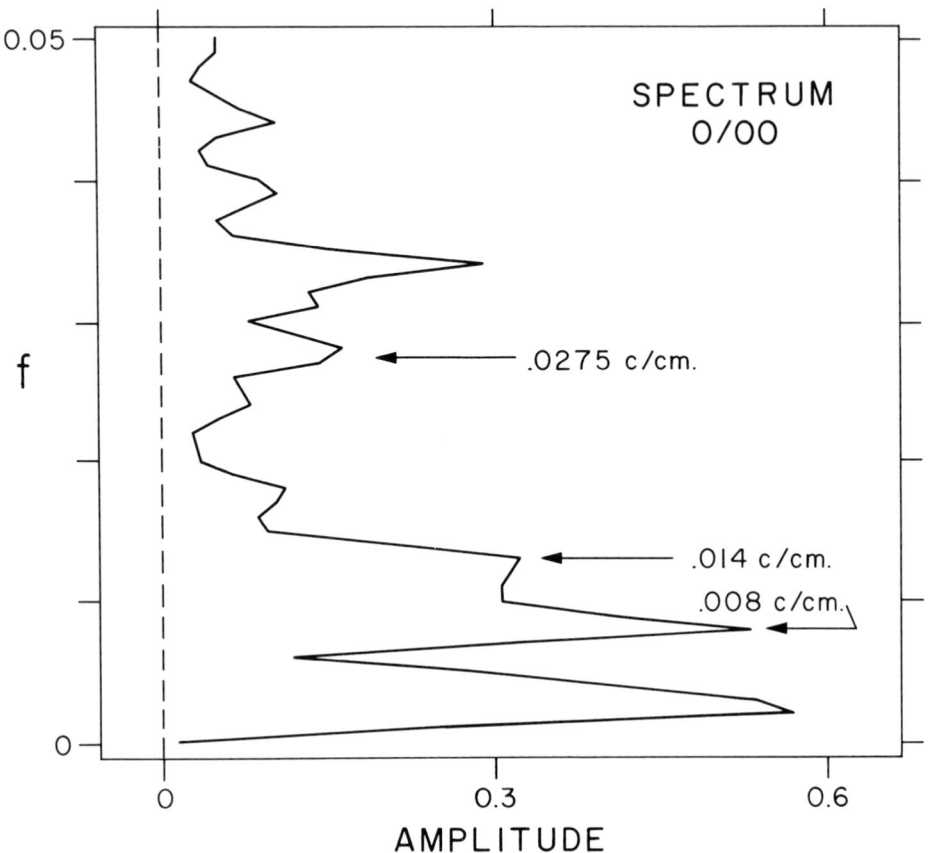

Fig. 41. Spectrum calculated from estimated salinity data for V12–122 given in Table 14, and based on an autocovariance function of 44 lags. Scales as shown for Figure 38.

calculated spectrum is given in Figure 44, where it can be seen that the method correctly resolves the input frequencies of 0.020 and 0.004. Next, a systematic distortion of the depositional rate is assumed in which, during a "glacial" period, a 33 percent increase in accumulation occurs; and during an "interglacial" a decrease of the same magnitude occurs. This effect was achieved by considering all points below the 100-unit midline of signal 3 as glacial, the remainder as interglacial, and spacing the points accordingly. If we make the opposite assumption, that interglacials have higher rates, the calculated spectrum will be the same. The record of signal 3 as distorted by differential sedimentation is plotted as signal 4, in which the accordion effect is easily seen. A spectrum calculated for signal 4 is given on Figure 45. Note that the distortion, as anticipated, has split the input peak at 0.020 into two apparent frequencies of 0.014 and 0.026 while leaving the lower peak unchanged. Theoretical apparent frequencies (i.e. those calculated from the known altered sedimentation rate) are

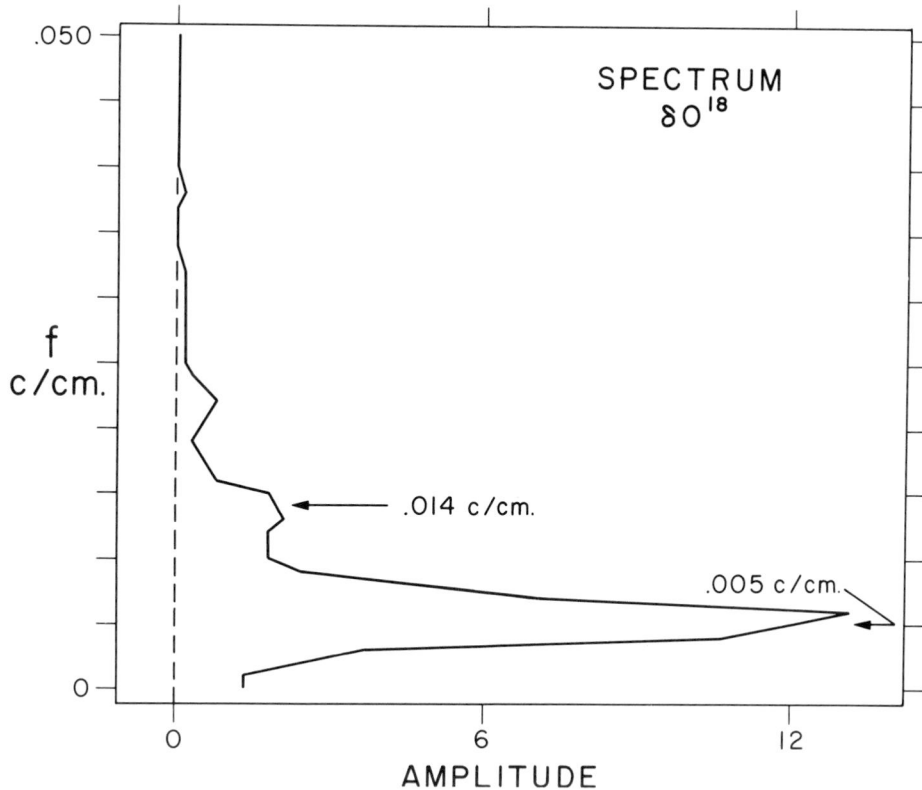

Fig. 42. Spectrum calculated from δO^{18} data on *Globigerinoides ruber* in Caribbean core V12–122, and based on an autocovariance spectrum of 34 lags. Scales as shown on Figure 38.

indicated on the figure. The harmonic mean of the apparent frequencies (0.019) is close to the actual input frequency 0.020.

To approach the real situation more closely, a random normal noise (signal 5 on Fig. 43) equal in magnitude to the strength of signal 4 has been added algebraically to the latter. The resulting signal, distorted both by a substantial random error and by differential sedimentation, constitutes signal 6. Perhaps the reader, like the authors, will find the contemplation of this model signal rewarding. In general appearance it resembles somewhat the calculated T_w curve of Figure 30, for example. We have found it instructive to present signal 6 to an unsuspecting colleague and ask him to discern by inspection and interpret significant frequencies (if any) contained in the curve. To date no one has succeeded in unraveling the *apparent* spectrum correctly, much less the distortion pattern built into it. Yet the calculated spectrum (Fig. 46) clearly reveals the apparent spectrum. The only effect the 1/1 signal/noise ratio has is to broaden the peaks and shift them slightly off their undistorted positions. Whatever may be the merits of the differ-

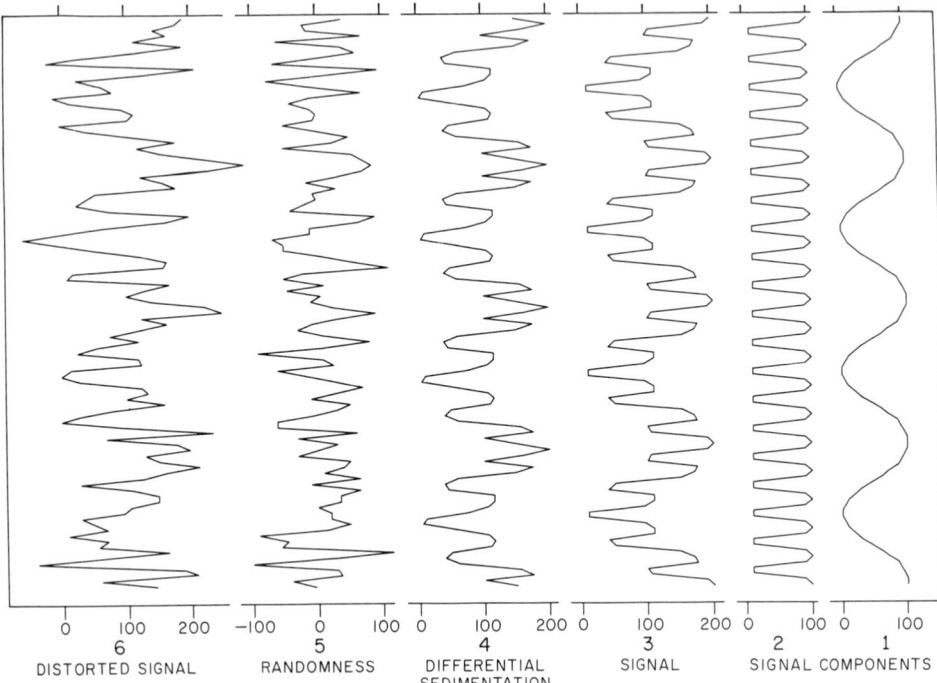

0 100 200	−100 0 100	0 100 200	0 100 200	0 100	0 100
6	5	4	3	2	1
DISTORTED SIGNAL	RANDOMNESS	DIFFERENTIAL SEDIMENTATION	SIGNAL	SIGNAL COMPONENTS	

Fig. 43. Digital model of deep-sea core data (101 sample points) designed to study distortion of a cyclical signal by differential sedimentation and random normal noise. Signal 1 is a low-frequency component. Signal 2 is a higher-frequency component. Signal 1 + signal 2 = signal 3. Signal 4 is derived from signal 3 by assuming an increase in accumulation rate 33 percent above the average during "glacial" and a decrease of the same magnitude during "interglacial" periods. For details, see text. Signal 5 is a random normal noise (Dixon and Massey 1951) which, added to signal 4, yields a pattern (signal 6) distorted both by differential accumulation and randomness. Data given in Table 15.

ential sedimentation model that is the occasion for this effort, we have at least been convinced of the usefulness of spectrum analysis in examining deep-sea cores for periodicities.

Returning to the real spectra and their apparent frequencies listed in Figure 36 and graphed in Figures 39–41, we note that for T_w data, if we consider the peaks at 0.014 and 0.027 as apparent frequencies derived by distortion from a single input frequency, the harmonic mean estimate of the latter is 0.018, which would correspond closely with the predicted precession frequency 0.019. For the salinity spectrum the same calculation on the apparent 0.008 and 0.014 peaks yields an estimate of 0.010 for the predicted tilt frequency of 0.011. For the T_s spectrum we must consider that the observed apparent frequency 0.0145 repre-

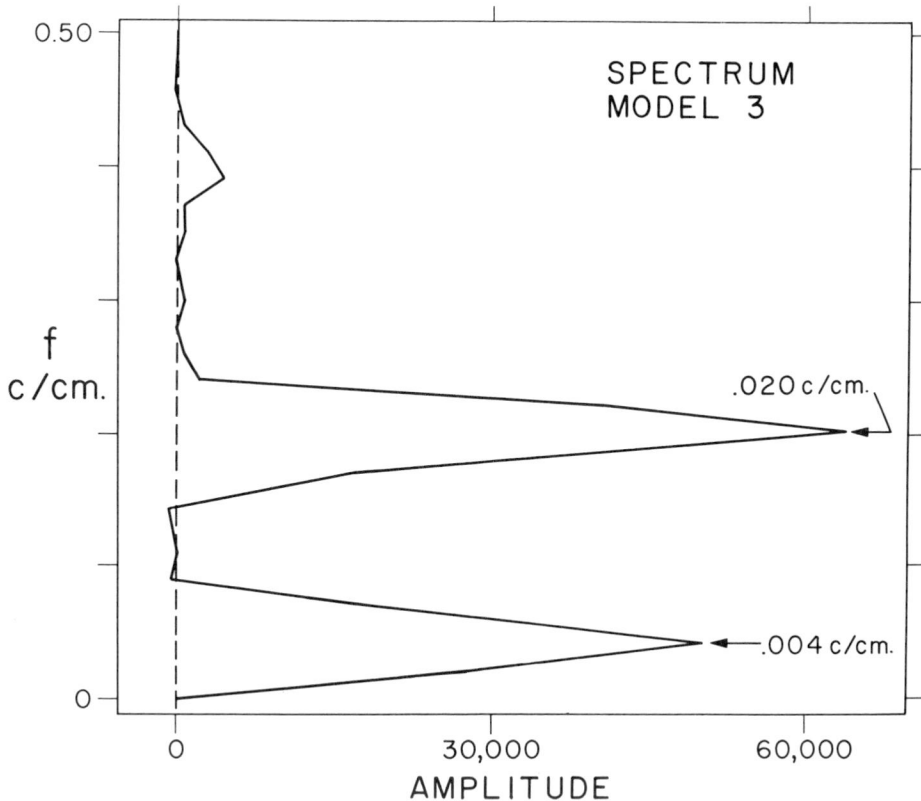

Fig. 44. Spectrum calculated from signal 3 of the distortion model plotted in Figure 43, and based on an autocovariance function of 23 lags. Scales as in Figure 38.

sents distorted contributions from both higher and lower input frequencies. The corresponding harmonic mean estimates for tilt and precession input frequencies are 0.010 and 0.019 respectively.

That the presumed tilt frequencies should dominate the salinity spectrum and be weakly represented on the temperature spectra conforms to the concept that the high-latitude tilt effect would control ice volume and hence salinity. All things considered, however, we are unconvinced that the distortion model for explaining the apparent high frequencies is a valid explanation, and we put it forward here primarily to show that it is at least possible to retain confidence in the estimate of 2.35 cm per 1000 years in V12-122, and in our paleoclimatic method, without discarding our Milankovitch model for the Caribbean. Certainly the spectra alone are unconvincing for the higher expected frequencies. However, the dominant observed low-frequency component of our paleoclimatic curves does match the Milankovitch prediction. This coincidence, combined with

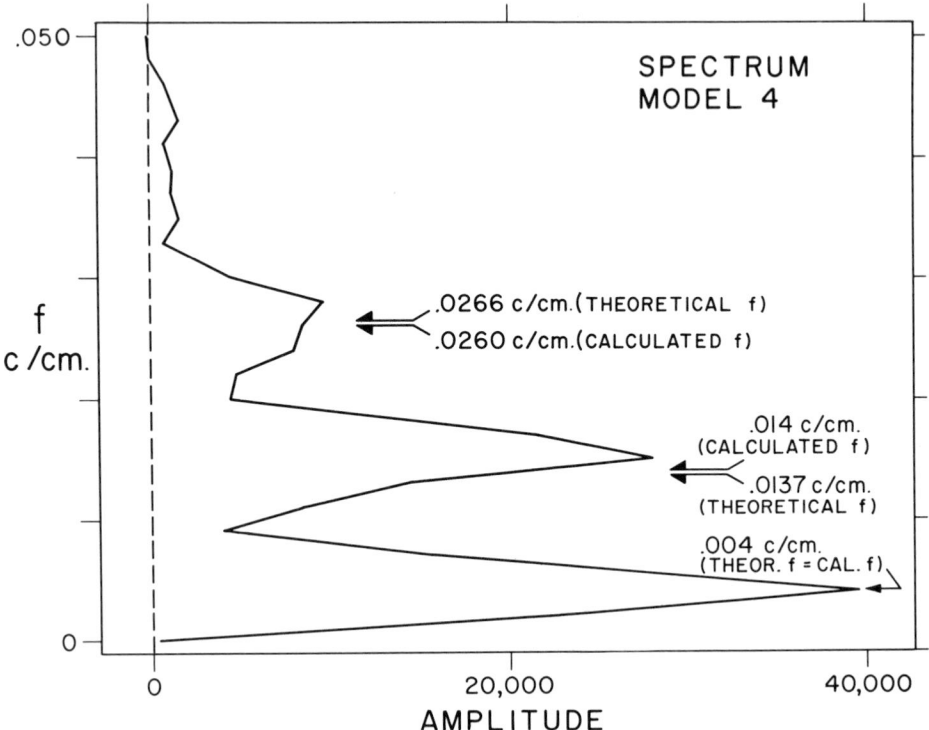

Fig. 45. Spectrum calculated from signal 4 of the distortion model plotted in Figure 43, and based on an autocovariance function of 23 lags. Scales as shown in Figure 38.

the time–domain study summarized in Figure 32, suggests that the Milankovitch theory cannot be lightly dismissed.

9. Another possibility, and one which at this writing seems to us the most likely, is that the Milankovitch effect is real, but that the Caribbean is poorly situated for a test. Its surface and deep waters originate in both the North and South Atlantic and, as discussed previously, it is likely that the proportions of North and South Atlantic contributions to Caribbean surface waters have fluctuated considerably during the Pleistocene. Thus neither Caribbean surface water temperatures nor the isotopic composition of its planktonic foraminifera can be expected to respond in a simple way either to high- or low-latitude fluctuations in insolation.

10. It is instructive to compare our spectral analysis with the results of Kemp and Eger (1967), who pioneered the application of time series techniques to isotopic and paleontologic data in deep-sea cores. Their δO^{18} spectrum of Caribbean core P6304–9 shows two significant periods of 180 and 240 cm. According to Broecker and Van Donk's (1970) sedimentation rate of 2.70 cm per 1000 years for this core, these periods represent 67,000 and 89,000 years. An inspection of

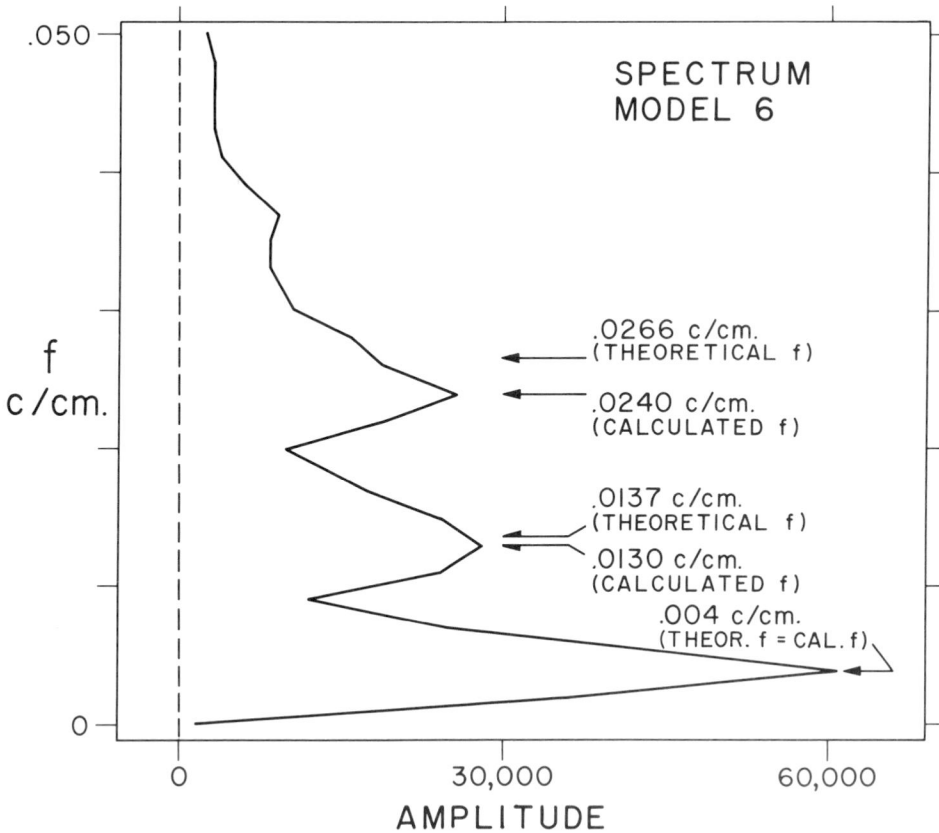

Fig. 46. Spectrum calculated from signal 6 of the distortion model plotted in Figure 43, and based on an autocovariance function of 23 lags. Scales as in Figure 38.

the raw data shows that the longer period reflects a periodicity in the upper portion of the core and the shorter period reflects a periodicity in the lower portion of the core. We conclude that the accumulation rate was higher in the upper portion, and that the average of the two apparent periods (78,000 years) is a fair match to our 80,000-year δO^{18} period.

Using paleontologic data of Lidz (1966), Kemp and Eger also calculate a *Globigerinoides ruber* spectrum for Caribbean core P6304–8. This species best represents our tropical assemblage and should give temperature interpretations resembling those calculated by our method. Again, Kemp and Eger's calculated spectrum shows two peaks, closely situated around an average period of 277 cm. Using Broecker and Van Donk's sedimentation rate for this core (2.90 cm per 1000 years), we obtain a 96,000-year period, which corresponds closely with the 100,000-year period calculated by us for V12-122 temperature curves.

SUMMARY AND CONCLUSIONS

1. Quantitative data on the relative abundance in the > 149-μ fraction of some 26 species and coiling varieties of planktonic foraminifera are given for 61 core top samples distributed widely about the Atlantic; and for 110 samples taken at 10-cm intervals down Caribbean core V12-122.

2. Factor analysis of core top faunas yields five statistically independent and ecologically significant assemblages designated tropical, subtropical, subpolar, polar, and gyre margin. Each assemblage displays systematic geographic gradients of abundance with an area of maximum development epitomized in its name. In the aggregate, the five-assemblage model accounts for 90 percent of the compositional information contained in the raw data.

3. Distributions of the tropical, subtropical, subpolar, and polar assemblages are clearly related to surface water temperature. Areas of assemblage dominance are exactly delineated in both hemispheres by $2°$, $12°$, and $20°$ winter surface isotherms. In the south, the $2°$ and $12°$ limits coincide with the polar front and the poleward limit of the subtropical convergence; in the north, they do not. Plotted as a function of winter temperature, each of the four assemblages shows a broad, non-monotonic pattern of roughly parabolic form with temperature optima at $25°$, $17°$, $6°$, and $0°C$, and abundance levels decreasing slowly as ambient temperatures depart from the optima.

4. Although factors responsible for the gyre margin assemblage are not so clear, several lines of evidence favor the interpretation that the level of biological productivity is the most important control. In the core tops the assemblage favors gyre margin positions. A wide range of temperature is tolerated. A rough correlation with positions in the salinity–temperature field is displayed with abundance maxima in warmer waters coinciding on the diagram and in the ocean with salinities between 35.5 and 36.0 per mil.

5. The effects of carbonate dissolution have been moderately well suppressed in the core top data by eliminating samples with obvious solution features such as high benthic/planktonic ratios. The dissolution tolerances of dominant species in the gyre margin assemblage, however, make it especially sensitive, and this diagenetic effect is possibly compounded with the presumed primary ecological relationship to water mass productivity.

6. In V12-122 a clear correlation is displayed between the coarse-fraction percentage of the sediment and the abundance of the gyre margin assemblage. Both curves may reflect variations in the productivity of Caribbean surface waters. Seasonal differences in circulation patterns now observed in the Atlantic are invoked as a Pleistocene model: during northern hemisphere summers (and during interglacials) a northward shift of the equatorial wind and surface current systems increases the proportion of warm, productive South Atlantic water entering the North Atlantic and flowing into the Caribbean.

7. The success of the five-assemblage model in identifying ecologically significant variation patterns, and in accounting for high percentages of the original

compositional data (90 percent in the tops, 80 percent in the core) indicates that the linear mixing model implicit in factor analysis conforms to actual tendencies in foraminiferal community structure.

8. An ecological response model is developed which assumes that each assemblage responds parabolically to p independent sets of physical parameters. Based on this assumption, which is confirmed at least for temperature control by evidence cited in paragraph 3 above, a paleoecological estimation model is constructed which enables the calculation of any studied physical parameter linearly related to the p sets of controls, provided that data are available on $p + 1$ independent biological parameters and the physical parameter of interest. The corresponding paleoecological equation is linear, or nearly so; small parabolic deviations from linearity are to be expected as a consequence of incomplete overlap of the ranges of the biological parameters, or of deviations from the parabolic form of the ecological function.

9. Data on 61 Atlantic core tops confirm the reality of the paleoecological model. The best result is a nearly linear equation estimating winter surface temperatures from information on the abundance of the five assemblages with a multiple correlation of 0.986; the poorest is that for salinity, with a correlation of 0.866. Statistically, both are massively significant.

10. By means of an operator matrix (F') derived from factor analysis of core top data, faunal information on 110 samples of Caribbean core V12-122 is analyzed into proportions of the assemblages identified in the core tops. An average of about 80 percent of the original compositional information in the core data is explained by the core top model (compared with 90 percent for the latter). Apparently, the ecosytem being studied has not varied significantly during the cored interval, with the exception of a small change in the composition of assemblages during the Y-zone glacial.

11. The tropical assemblage dominates the fauna throughout the deposition of the core. Apparently, Caribbean winter surface temperatures never dropped below 20°C during this interval. The polar assemblage is completely absent.

12. Although the tropical fauna numerically dominates the entire core, fluctuations in abundance of the tropical, subtropical, and subpolar assemblages in V12-122 are marked by cyclically arranged peaks of relative abundance, especially during post V-zone times. Peaks of subpolar assemblage abundance (for example in the Y zone) are presumed to represent glacial maxima. Such peaks are flanked stratigraphically by subtropical peaks recording local optimum conditions for that assemblage during warming or cooling trends.

13. During the four intervals of maximum subpolar assemblage abundance, presumed to represent glacial episodes, numerical levels of that assemblage are much higher, relative to the tropical assemblage, than any found in the Caribbean or Atlantic today. These stratigraphic horizons apparently contain the historical record of entry into the American Mediterranean region of a fauna characterized by *Globigerina bulloides* and right-coiling *G. pachyderma,* now relict in the northern part of the Gulf of Mexico. These episodes of subpolar influx

can reasonably be interpreted as tropical submergence—i.e. they occurred below the surface waters where the tropical assemblage still dominated, in an ecological water mass originating at high latitudes.

14. Quantitative estimates of temperature and salinity are calculated for surface waters of the Caribbean overlying the site of V12-122, by means of paleoecological equations developed and tested for the core top data. Apparently, surface Caribbean waters did not fluctuate through more than 5°C and 1.0 per mil.

15. Unlike isotopically derived paleoclimatic curves, ours show a secular trend in which temperatures during the U zone were colder than any since; and those in mid-V zone and later times include warmer peaks than any noted earlier.

16. A comparison of *Globorotalia menardii*, O^{18}, and our paleotemperature curves shows a concordance of three out of a total of eight major peaks, and significant but imperfect linear intercorrelations. This pattern of partial independence is explained by assuming that *G. menardii* is controlled by productivity and temperature, that the isotope curve reflects ice volume and temperature, and that our paleotemperature estimates reflect temperature alone.

17. Given the sedimentation rate of 2.35 cm per 1000 years calculated by Broecker and Van Donk, fluctuations in the various paleoclimatic curves are compared with patterns predicted by the Milankovitch theory. The patterns are compared in the time and frequency domains.

18. In the time domain, the position of observed peaks and troughs in the paleontological temperature curves matches fairly well that predicted by a version of the Milankovitch model which emphasizes, in decreasing order of importance, the eccentricity, tilt, and local precessional effects. The match for the isotopic curve is poor, except for the post V-zone portions of the core.

19. With power spectrum analysis to examine the frequency domain, the paleontologically derived paleoclimatic and isotopic curves are examined for periodicities corresponding to those predicted by the Milankovitch model for the interval covered by the core: 98,000 years for eccentricity, 40,000 for tilt, and 22,000 for precession. The expected eccentricity frequency does in fact dominate the observed spectrum in the paleontological curves.

20. The expected tilt and precessional frequencies are not found either in the paleontological or the isotopic spectra. Instead, in the paleontological curves, three small marginally significant frequency peaks occur whose location straddles but does not match those expected. A digital model is constructed to show that a fluctuating accumulation rate could shift peaks in an input Milankovitch spectrum to yield an apparent spectrum like those measured in V12-122. Other possible interpretations are that the Milankovitch model is wrong, or that higher frequency peaks observed in our spectra are statistically insignificant.

21. Another possibility, and one which at this writing seems to us most likely, is that the Milankovitch effect is real, but that the Caribbean is poorly situated for a simple test.

22. Spectrum analysis of the oxygen isotope data shows a dominant period of 80,000 years, and two smaller, marginally significant peaks. If the accumulation rate of 2.35 cm per 1000 years is assumed to be too large by 25 percent, then the observed isotope spectrum would approximate the Milankovitch model. However, such a correction in rate is in the opposite sense to that suggested by Emiliani after his detailed consideration of depositional rates in Caribbean and other cores for which isotope curves have been calculated.

23. Time series analysis, including spectrum and filter techniques, provide useful and objective tools for the study of periodicities in deep-sea cores. One virtue of the paleoecological method introduced here is that it enables the application of such techniques in paleontologically based paleo-oceanographic research.

24. Final evaluation of the method proposed here, and of the conclusions regarding the Milankovitch theory, must await the study of a suite of cores by the same method.

25. In principle, it seems possible that data on pollen, coccoliths, radiolaria, and other microfossils should be amenable to the method used here.

Summary. A method is described which extracts from data on planktonic microfossils an objective and quantitative estimate of the physical state of surface waters overlying a site during deposition of a core. It is based on an ecological response model in which m species (or species groups) respond parabolically to $m - 1$ or fewer independent physical parameters t. The simple case of one physical parameter (Fig. 4) generalizes to higher dimensions. Conveniently, the inverse or paleoecological function is linear: $t = ax_1 + bx_2 + c$, for the simple case. Because the method requires that the biological variates be statistically independent, raw data on the abundance of n species are resolved by factor analysis into m end-member, varimax assemblages. In Figure 3, $n = 7$ and $m = 3$.

Quantitative data on planktonic foraminifera in 61 core tops widely distributed about the Atlantic Ocean conform to the model. Four assemblages are clearly related to surface temperature as shown by plots of observed temperature versus the abundance of the assemblages (Fig. 19); and by maps of assemblage distribution (Figs. 6–9). A fifth assemblage is related to surface water productivity (Fig. 10). Paleoecological equations specifying various physical parameters of the ocean as a function of observed assemblage abundance fit the data (Figs. 21–23).

By means of an operator matrix derived from the core top factor analysis, data on 110 samples spaced at 10-cm intervals down Caribbean core V12–122 are expressed in terms of the five assemblages, and the paleoecological equations are applied to yield paleoenvironmental estimates. A smoothed curve representing estimates of summer Caribbean surface temperature for the past 450,000 years is compared to δO^{18} data (Fig. 32). A model is developed in which abundance data for *Globorotalia menardii* reflects productivity and temperature; δO^{18} reflects ice volume and temperature; and the foraminifer-based paleotemperature curve reflects temperature alone.

Spectra calculated from the paleotemperature curves yield a dominant frequency matching that of orbital eccentricity. However, weakly significant higher frequency peaks do not match Milankovitch expectations for tilt and precession. A causal relationship between orbital eccentricity and Caribbean climate is inferred. The mechanism of this relationship is obscure.

Acknowledgments. It is a pleasure to acknowledge the help of many persons who have contributed to our work. David B. Ericson and Goesta Wollin made the study possible by helping us select appropriate cores from the Lamont-Doherty Core Collection, and by giving continued counsel and encouragement. Through publications and personal guidance, Frances L. Parker, Allan W. H. Bé, S. Stephen Streeter, and Tsunemasa Saito made it possible for us to formulate a workable taxonomic scheme. We also record our debt to four people who cared enough about our work to share data and ideas about Pleistocene climate with us: Wallace S. Broecker, Jan Van Donk, Cesare Emiliani, and R. K. Matthews. Our ecological concepts about planktonic foraminifera have been significantly influenced by discussions with Allan Bé, Fred R. Phleger, William Ruddiman, and Douglas Tolderlund. Several key mathematical steps used in the text originated in discussions with Vincent Manson and Edward Klovan; the latter also wrote the initial version of our factor analysis program CABFAC. Harvey Sachs suggested the use of time series analysis. Madeleine Briskin drafted several figures, shared the computational work, and stimulated us by applying the method to a South Atlantic core. Margaret Cummings helped us in many ways to prepare the manuscript for publication.

The senior author wishes to acknowledge his debt to Norman D. Newell, whose classic paper (Newell 1948) first made clear to him the scientific potential of statistical work in invertebrate paleontology; and to G. G. Simpson and Anne Roe, whose textbook of quantitative zoology (Simpson and Roe 1939) provided a lucid introduction to the methodology of the field.

Financial support has been provided by National Science Foundation grants NSF–GP–4994, NSF–GA–1303, and NSF–GA–14853. All samples used in this study are from Lamont-Doherty cores.

REFERENCES

Arrhenius, Gustaf, Sediment cores from the East Pacific, *Repts. Swedish Deep-Sea Exped., 1947–1948, 5,* fasc. 1, 6, 1952.

Arrhenius, Gustaf, Pelagic sediments, in *The Sea,* vol. 3, 963 pp., Interscience, New York, 1963.

Bandy, O. L., Distribution of foraminifera, radiolaria, and diatoms in sediments of the Gulf of California, *Micropaleontology, 7,* no. 1, 1, 1961.

Bé, Allan W. H., Ecology of Recent planktonic foraminifera. Part 1. Areal distribution in the western North Atlantic, *Micropaleontology, 5,* no. 1, 77, 1959.

Bé, Allan W. H., Some observations on Arctic planktonic foraminifera, *Cushman Found. Foram. Res., Contr., 11,* Pt. 2, 64, 1960.

Bé, Allan W. H., Planktonic foraminifera, in Distribution of selected groups of marine invertebrates in waters south of 35° S Latitude, Folio 11, Antarctic Map Folio Series, American Geographical Society, p. 9, 1969.

Bé, Allan W. H., and William H. Hamlin, Ecology of Recent planktonic foraminifera, *Micropaleontology, 13,* 87, 1967.

Bé, Allan W. H., and Douglas S. Tolderlund, Distribution and ecology of living planktonic foraminifera in surface waters of the Atlantic and Indian Oceans, in *Micro-*

paleontology of Marine Bottom Sediments, symposium sponsored by Scientific Committee on Oceanic Research, Cambridge, England, 1970, in press.

Beers, John R., D. M. Steven, and J. B. Lewis, Primary productivity in the Caribbean Sea off Jamaica and the tropical North Atlantic off Barbados, *Bull. Marine Sci., 18,* no. 1, 86, 1968.

Belyaeva, N. V., Distribution of planktonic foraminifera in the water and on the floor in the Indian Ocean, *Trudy Inst. Okeanol., Akad. Nauk SSSR, 68,* 12, 1964.

Berger, W. H., Planktonic foraminifera: Shell production and preservation, 237 pp., Ph.D. dissertation, University of California at San Diego, 1968.

Berger, W. H., and G. R. Heath, Vertical mixing in pelagic sediments, *J. Marine Res., 142,* 134, 1968.

Blackman, R. B., and J. W. Tukey, *The measurement of Power Spectra from the Point of View of Communication Engineering,* 190 pp., Dover Publications, New York, 1958.

Blackman, Abner, and B. L. K. Somayajulu, Pacific Pleistocène cores: Faunal analyses and geochronology, *Science, 154,* 886, 1966.

Boltovskoy, Esteban, Living planktonic foraminifera at the 90°E meridian from the equator to the Antarctic, *Micropaleontology, 15,* no. 2, 237, 1969.

Braarud, Trygve, Cultivation of marine organisms as a means of understanding environmental influences on populations, in *Oceanography,* edited by Mary Sears, p. 271, Washington, D.C., A. A. A. S. Pub. 67, 1961.

Bradshaw, J. S., Laboratory studies on the rate of growth of the foraminifera *Streblus beccarii* (Linné) var. *tepida Cushman, J. Paleontol., 31,* 1138, 1957.

Broecker, W. S., and T. L. Ku, Absolute dating of Caribbean cores P6304–8 and P6304–9, *Science, 166,* 404, 1969.

Broecker, W. S., and Jan Van Donk, Insolation changes, ice volumes, and the O^{18} record in deep-sea cores, *Rev. Geoph. Space Phys.,* 8, 1, 169, 1970.

Cifelli, Richard, Distribution analysis of North Atlantic foraminifera collected in 1961 during cruises 17 and 21 of R/V Chain, *Cushman Found. Foram. Res., Contr., 18,* 118, 1967.

Defant, Albert, *Physical Oceanography,* vol. 1, 729 pp., Pergamon, New York, 1961.

Dixon, W. J., ed., *BMD Biomedical Computer Programs,* 620 pp., Los Angeles School of Medicine, University of California, 1965.

Dixon, W. J., and F. J. Massey, Jr., *Introduction to Statistical Analysis,* 370 pp., McGraw-Hill, New York, 1951.

Emiliani, Cesare, Pleistocene temperatures, *J. Geol., 63,* 538, 1955.

Emiliani, Cesare, Paleotemperature analysis of Caribbean cores P6304–8 and P6304–9 and a generalized temperature curve for the past 425,000 years, *J. Geol., 74,* no. 2, 109, 1966.

Emiliani, Cesare, and E. Rona, Caribbean cores P6304–8 and P6304–9: New analysis of absolute chronology, a reply, *Science, 166,* 1551, 1969.

Ericson, David B., and Goesta Wollin, Pleistocene climates and chronology in deep-sea sediments, *Science, 162,* 1227, 1968.

Ericson, David B., Maurice Ewing, and Goesta Wollin, The Pleistocene Epoch in deep-sea sediments, *Science, 146,* 723, 1964.

Imbrie, John, and T. H. Van Andel, Vector analysis of heavy-mineral data, *Bull. Geol. Soc. Am., 75,* 1131, 1964.

Jones, J. I., The relationship of planktonic foraminiferal populations to water masses in the western Caribbean and Lower Gulf of Mexico, *Bull. Marine Sci.*, 18, 946, 1968.

Kemp, W. C., and D. T. Eger, The relationships among sequences with applications to geologic data, *J. Geophys. Res.*, 72, 739, 1967.

Kennett, J. P., Distribution of planktonic foraminifera in surface sediments southeast of New Zealand, *Proc. Intern. Conf. Plank. Microfoss.*, 1st, 2, 307, 1969.

Klovan, J. E., and John Imbrie, An algorithm and Fortran IV program for large scale Q-mode factor analysis, *J. Intern. Assoc. Mathematical Geol.*, 3 (1), 1971.

Knauss, J. A., Equatorial current systems, in *The Sea*, vol. 2, p. 235, Interscience, New York, 1963.

Krumbein, W. C., and F. A. Graybill, *An Introduction to Statistical Models in Geology*, 475 pp., McGraw-Hill, New York, 1965.

Ku, T. L., and W. S. Broecker, Atlantic deep-sea stratigraphy: Extension of absolute chronology to 320,000 years, *Science, 151*, 448, 1966.

Lamb, H. H., Fundamentals of climate, in *Descriptive Paleoclimatology*, edited by A. E. M. Nairn, p. 8, Interscience, New York, 1961.

Lidz, Louis, Deep-sea Pleistocene biostratigraphy, *Science, 154*, 1448, 1966.

McIntyre, Andrew, Coccoliths as paleoclimatic indicators of Pleistocene glaciation, *Science, 158*, 1314, 1967.

Manson, Vincent, and John Imbrie, Fortran program for factor and vector analysis of geologic data using an IBM 7090 or 7094 computer system, *Kansas Geol. Surv. Computer Contrib., Spec. Publ.*, no. 13, 1, 1964.

Mesolella, K. J., R. K. Matthews, W. S. Broecker, and D. L. Thurber, The astronomical theory of climatic change: Barbados data, *J. Geol.*, 77, no. 3, 250, 1969.

Newell, N. D., Infraspecific categories in invertebrate paleontology, *J. Paleontol.*, 22, 225, 1948.

Oba, Tadamichi, Biostratigraphy and isotopic paleotemperature of some deep-sea cores from the Indian Ocean, *Sci. Rept. Tohoku Univ., Sendai, Second Ser.* (Geology), 41, no. 2, 129, 1969.

Parker, F. L., Eastern Mediterranean foraminifera, *Repts. Swedish Deep-Sea Exped.*, 8, no. 4, 219, 1958.

Parker, F. L., Planktonic foraminiferal species in Pacific sediments, *Micropaleontology*, 8, no. 2, 219, 1962.

Parker, F. L., Late Tertiary biostratigraphy (planktonic foraminifera) of tropical Indo-Pacific deep-sea cores, *Bull. Am. Paleontol.*, 52, no. 235, 115, 1967.

Parker, F. L., Distribution of planktonic foraminifera in recent deep-sea sediments, SCOR Symposium on Micropaleontology of Marine Bottom Sediments, 1970, in press.

Parker, F. L., and W. H. Berger, Faunal and solution patterns of planktonic foraminifera in surface sediments of the South Pacific, *Deep-Sea Res.*, in press.

Phleger, F. B., *Ecology and Distribution of Recent Foraminifera*, 297 pp., Johns Hopkins Press, Baltimore, 1960.

Phleger, F. B., F. L. Parker, and J. F. Peirson, North Atlantic core foraminifera, *Repts. Swedish Deep-Sea Exped.*, 7, no. 1, 1, 1953.

Rona, E., and Cesare Emiliani, Absolute dating of Caribbean cores P6304–8 and P6304–9, *Science, 163*, 66, 1969.

Ruddiman, W. F., Planktonic foraminifera of the subtropical North Atlantic gyre, Ph.D. dissertation, Columbia University, 1968.

Ryther, J. H., D. W. Menzel, and Nathaniel Corwin, Influence of the Amazon River outflow on the ecology of the Western Tropical Atlantic. I. Hydrography and Nutrient Chemistry, *J. Marine Res.*, 25, 69, 1967.

Schott, Wolfgang, Die Foraminiferen in den Aquatorialen Teil des Atlantischen Ozeans, *Deutsche Atlantische Exped.*, 11, heft 6, 411, 1935.

Simpson, G. G., and Ann Roe, *Quantitative Zoology*, 414 pp., McGraw-Hill, New York, 1939.

Sverdrup, H. U., M. V. Johnson, and R. H. Fleming, *The Oceans, Their Physics, Chemistry, and General Biology*, 1987 pp., Prentice Hall, Englewood Cliffs, 1942.

Tolderlund, D. S., Seasonal distributional patterns of planktonic foraminifera at five ocean stations in the western North Atlantic, Ph.D. thesis, Columbia University, 1969.

Van Donk, Jan, The oxygen isotope records in deep-sea sediments, Ph.D. dissertation, Columbia University, 1970.

Vernekar, A. D., Long-period global variations of incoming solar radiation, in *Research on the Theory of Climate*, vol. 2, 289 pp., Report of the Travelers Research Center, Hartford, Conn., 1968.

Table 1. Locations of Core Tops

No.	Core	Lat.	Long.	Depth (m)
1	V14–61	54° 28′ S	2° 36′ W	1835
2	V17–196	60° 44′ N	57° 50′ W	2818
3	V18–110	53° 35′ S	44° 42′ W	2610
4	V16–227	60° 03′ N	50° 50′ W	3305
5	V14–47	50° 47′ S	42° 09′ W	1690
6	V23–22	54° 12′ N	45° 58′ W	3669
7	V2–12	49° 08′ N	49° 20′ W	1267
8	V23–29	59° 57′ N	32° 51′ W	2186
9	V12–43	45° 19′ S	57° 59′ W	3880
10	R9–7	59° 39′ N	22° 46′ W	2770
11	A157–3	50° 56′ N	41° 45′ W	4025
12	V23–81	54° 15′ N	16° 50′ W	2393
13	V23–82	52° 35′ N	21° 56′ W	3974
14	V12–53	40° 54′ S	20° 23′ W	3797
15	V23–83	49° 52′ N	24° 15′ W	3871
16	V12–56	36° 30′ S	8° 06′ E	3222
17	A152–84	44° 21′ N	30° 16′ W	2750
18	V16–50	33° 21′ S	16° 26′ E	2376

(*cont.*)

Table 1 (*cont.*)

No.	Core	Lat.	Long.	Depth (m)
19	V22–122	39° 35′ S	24° 35′ E	3272
20	V16–41	27° 52′ S	1° 06′ W	4462
21	V4–32	35° 03′ N	11° 37′ W	2230
22	V12–66	22° 59′ S	7° 01′ E	2760
23	V19–245	26° 12′ S	4° 41′ E	2725
24	V4–8	37° 14′ N	33° 08′ W	1655
25	A180–15	39° 16′ N	36° 42′ W	4610
26	V18–34	31° 21′ S	36° 49′ W	3252
27	V20–213	28° 20′ S	13° 09′ W	2175
28	V19–222	33° 22′ S	34° 24′ E	2005
29	A180–39	25° 50′ N	19° 18′ W	3470
30	V16–189	28° 50′ S	41° 02′ W	3781
31	V12–18	28° 42′ S	34° 30′ W	2935
32	V7–67	34° 40′ N	61° 28′ W	4308
33	V17–165	32° 45′ N	41° 54′ W	3924
34	V19–310	33° 18′ N	48° 16′ W	4607
35	V16–190	27° 57′ S	42° 27′ W	2919
36	A153–154	28° 00′ N	38° 47′ W	4020
37	V19–308	29° 01′ N	41° 24′ W	3197
38	V22–172	12° 40′ S	9° 49′ W	4127
39	V10–98	31° 26′ N	64° 11′ W	4299
40	V22–219	27° 55′ N	43° 38′ W	2582
41	V16–33	15° 20′ S	19° 43′ W	4360
42	V22–204	15° 01′ N	23° 14′ W	1723
43	V20–167	21° 03′ S	72° 30′ E	3634
44	V10–89	23° 02′ N	43° 48′ W	3523
45	V12–79	1° 31′ S	11° 47′ W	3823
46	V19–216	25° 20′ S	36° 47′ E	2206
47	V14–90	16° 23′ S	61° 09′ E	3314
48	A180–72	0° 36′ N	21° 47′ W	3841
49	V16–21	17° 17′ N	48° 25′ W	3975
50	A180–76	0° 46′ S	26° 02′ W	3512
51	V15–164	9° 45′ S	34° 24′ W	3588
52	A180–78	1° 30′ S	27° 01′ W	4261
53	V14–5	0° 51′ N	32° 51′ W	3255
54	V3–128	23° 46′ N	92° 29′ W	3495
55	A179–13	23° 56′ N	75° 45′ W	1847
56	V9–31	8° 14′ N	37° 52′ W	4204
57	V20–230	1° 57′ S	39° 02′ W	3294
58	V20–7	11° 33′ N	60° 31′ W	1018
59	V20–234	5° 19′ N	33° 02′ W	3133
60	V18–21	4° 14′ N	47° 45′ W	2374
61	V12–122	17° 00′ N	74° 24′ W	2800

Table 2. Percentage of Taxonomic Categories in Sixty-One Core Tops

No.: Locality code number. Sample name: Core identification and sieve size (μ). Column 1: *Orbulina universa*. 2: *Globigerinoides conglobatus*: 3: *G. ruber*. 4: *G. tenellus*. 5: *G. sacculifer*. 6: *Globigerina rubescens*. 7: *G. quinqueloba*. 8: *G. pachyderma* (left coiling). 9: *G. pachyderma* (right coiling). 10: *G. bulloides*. 11: *G. falconensis*. 12: *G. calida*. 13: *Globigerinella aequilateralis*. 14: *Globigerinita glutinata*. 15: *Globoquadrina dutertrei*. 16: *Globorotalia inflata*. 17: *G. truncatulinoides* (left coiling). 18: *G. truncatulinoides* (right coiling). 19: *G. crassaformis*. 20: *G. hirsuta*. 21: *G. scitula*. 22: *G. menardii* and *G. tumida*. 23: *Pulleniatina obliquiloculata*. 24: *Candeina nitida*. 25: *Sphaeroidinella dehiscens*. 26: *Globigerina digitata*. 27: Other species and unidentified.

NO.	SAMPLE NAME	1 O.UNIV	2 G.CGLOB	3 G.RUBER	4 G.TEN	5 G.SAC	6 G.RUBES	7 G.QUIN	8 G.PAC L	9 G.PAC R	10 G.BULL
1	V14-61 >149	0.0	0.0	0.0	0.0	0.0	0.0	0.0	98.972	0.900	0.0
2	V17-196 >149	0.0	0.0	0.0	0.0	0.0	0.0	0.0	98.131	0.935	0.467
3	V18-110 >149	0.0	0.0	0.0	0.0	0.0	0.0	0.0	96.286	1.714	1.000
4	V16-227 >149	0.0	0.0	0.0	0.0	0.0	0.0	0.0	94.334	4.816	0.850
5	V14-47 >149	0.0	0.0	0.107	0.0	0.107	0.107	0.322	67.811	2.682	10.837
6	V23-22 >149	0.0	0.0	0.0	0.0	0.0	0.0	3.141	55.694	16.623	19.895
7	V2-12 >149	0.0	0.0	0.0	0.0	0.0	0.0	0.0	96.311	3.279	0.0
8	V23-29 >149	0.0	0.0	0.0	0.0	0.0	0.0	17.634	17.963	20.372	30.340
9	V12-43 >149	0.0	0.0	0.0	0.0	0.0	0.0	6.230	18.850	31.949	14.696
10	R9-7 >149	0.0	0.0	0.0	0.0	0.0	0.0	7.843	9.412	29.020	21.569
11	A157-3 >149	0.0	0.0	0.569	0.0	0.142	0.0	1.471	13.371	33.144	21.337
12	V23-81 >149	0.0	0.0	0.0	0.0	0.0	0.0	0.427	0.0	34.706	15.882
13	V23-82 >149	2.353	0.0	0.0	0.0	0.0	0.0	0.376	2.256	40.602	8.647
14	V12-53 >149	0.752	0.0	0.406	0.0	0.0	0.0	2.300	0.947	16.644	25.846
15	V23-83 >149	0.135	0.0	0.0	0.0	0.0	0.0	1.186	1.383	40.443	19.960
16	V12-56 >149	0.395	0.0	2.983	0.0	0.239	0.0	0.835	0.358	11.695	13.604
17	A152-84 >149	0.597	0.0	6.699	0.0	0.490	0.0	1.634	0.817	17.484	19.281
18	V16-50 >149	0.0	0.0	4.259	0.0	0.185	0.0	0.185	2.037	14.259	5.000
19	V22-122 >149	0.185	0.0	0.894	0.0	0.745	0.0	0.149	0.596	11.773	26.528
20	V16-41 >149	0.745	0.0	15.155	0.149	5.963	0.0	0.0	0.373	3.727	1.615
21	V4-32 >149	1.118	0.142	15.121	0.124	6.562	0.571	1.427	0.571	10.556	9.130
22	V12-66 >149	1.997	0.149	9.469	1.854	3.285	0.386	0.483	1.256	24.831	6.570
23	V19-245 >149	1.546	0.248	10.591	1.932	7.703	0.189	1.230	0.275	10.041	2.338
24	V4-8 >149	1.513	0.428	15.705	0.825	2.176	0.0	0.205	0.284	4.257	17.975
25	A180-15 >149	0.095	0.550	17.008	1.419	2.049	0.786	0.0	0.410	9.631	16.393
26	V18-34 >149	0.589	0.757	27.308	0.410	4.519	0.0	0.0	0.196	8.055	4.912
27	V20-213 >149	1.342	0.205	26.510	2.685	2.013	1.119	0.103	0.224	8.166	3.244
28	V19-222 >149	1.436	0.736	18.872	1.333	5.949	0.410	0.0	0.205	4.718	9.231
29	A180-39 >149	1.378	0.336	26.406	2.985	14.122	0.459	0.0	0.0	8.496	0.918
30	V16-189 >149	2.490	4.149	43.568	0.415	8.506	0.0	0.0	0.0	0.622	0.0

(cont.)

Table 2 (cont.)

	11	12	13	14	15	16	17	18	19	20
NO. SAMPLE NAME	O.UNIV	G.CGLUB	G.RUBER	G.TEN	G.SAC	G.RUBES	G.QUIN	G.PAC L	G.PAC R	G.BULL
31 V12-18 >149	1.267	1.774	49.937	1.774	4.056	0.380	0.0	0.0	1.774	1.774
32 V7-67 >149	0.0	0.602	22.892	0.402	3.815	0.0	0.0	0.0	2.209	8.032
33 V17-165 >149	0.323	0.0	30.860	1.398	1.828	0.215	0.316	0.108	1.935	7.742
34 V19-310 >149	0.633	0.475	29.747	0.316	1.741	0.0	0.0	0.0	3.797	6.329
35 V16-190 >149	1.300	2.080	40.295	1.386	5.026	0.693	0.0	0.087	5.633	3.380
36 A153-154>149	1.183	0.789	51.282	2.564	3.748	0.789	0.0	0.0	0.0	3.156
37 V19-308 >149	2.450	0.891	51.782	4.343	3.341	0.780	0.0	0.0	0.759	0.891
38 V22-172 >149	0.506	0.253	49.873	1.519	8.861	0.0	0.0	0.0	3.954	0.759
39 V10-98 >149	1.145	1.873	32.778	0.0	3.226	1.873	0.0	0.312	0.372	4.579
40 V22-219 >149	2.793	1.304	51.583	3.166	3.166	0.931	0.0	0.186	0.0	2.235
41 V16-33 >149	2.433	6.813	28.954	1.622	35.036	0.0	0.0	0.0	2.162	0.243
42 V22-204 >149	0.541	0.0	36.757	0.345	12.703	4.595	0.0	0.0	0.164	3.243
43 V20-167 >149	1.209	1.900	53.195	3.618	13.472	0.518	0.0	0.0	2.170	0.0
44 V10-89 >149	0.329	2.796	55.921	0.723	9.211	2.303	0.0	0.0	1.348	0.658
45 V12-79 >149	0.904	1.343	41.049	0.674	14.828	0.904	0.0	0.173	2.418	1.447
46 V19-216 >149	1.482	2.245	38.410	1.727	11.456	0.809	0.0	0.221	1.438	2.022
47 V14-90 >149	0.345	0.111	39.033	0.774	12.435	1.036	0.0	0.0	0.0	2.763
48 A180-72 >149	0.111	1.537	61.581	1.317	21.792	0.664	0.0	0.0	1.119	0.664
49 V16-21 >149	1.207	0.224	47.875	1.790	14.270	1.647	0.0	0.0	7.595	0.0
50 A180-76 >149	0.0	0.337	59.256	0.837	21.924	1.119	0.0	0.253	0.0	0.224
51 V15-164 >149	0.744	0.506	22.785	0.431	12.744	0.744	0.0	0.0	0.157	0.093
52 A180-78 >149	0.253	0.431	45.043	0.157	18.734	0.127	0.0	0.0	7.595	0.633
53 V14-5 >149	1.293	0.628	49.765	0.431	25.431	0.431	0.0	0.0	0.157	0.216
54 V3-128 >149	1.413	3.431	55.882	6.373	11.460	0.157	0.0	0.395	4.348	0.314
55 A179-13 >149	1.471	0.395	39.526	0.395	2.941	1.961	0.0	0.0	1.645	1.186
56 V9-31 >149	0.0	0.164	44.901	0.0	18.972	1.581	0.0	0.0	0.0	0.0
57 V20-230 >149	0.164	0.0	60.714	0.0	18.914	1.974	0.0	0.0	0.363	0.0
58 V20-7 >149	3.061	0.0	46.279	1.089	20.918	0.0	0.0	0.0	0.0	0.181
59 V20-234 >149	1.089	0.0	47.468	0.211	29.038	0.544	0.0	0.0	0.157	0.0
60 V18-21 >149	0.211	0.422	43.485	0.211	18.987	0.0	0.0	0.0	0.163	0.0
61 V12-122 >149	1.792	0.489		0.814	25.570	0.651	0.0	0.392		

	11	12	13	14	15	16	17	18	19	20
NO. SAMPLE NAME	G.FALC	G.CALID	G.AEQUI	G.GLUT	G.DUTER	G.INFLA	G.TRN L	G.TRN R	G.CRASF	G.HIRSU
1 V14-61 >149	0.0	0.0	0.0	0.0	0.0	0.0	0.0	0.0	0.0	0.0
2 V17-196 >149	0.0	0.0	0.0	0.467	0.0	0.0	0.0	0.0	0.0	0.0
3 V18-110 >149	0.0	0.0	0.0	0.0	0.0	0.571	0.0	0.0	0.0	0.0
4 V16-227 >149	0.0	0.0	0.0	0.0	0.0	0.0	0.0	0.0	0.429	0.0
5 V14-47 >149	0.0	0.0	0.0	1.931	0.0	14.056	0.107	0.0	0.0	0.131
6 V23-22 >149	0.0	0.0	0.0	2.291	0.0	1.571	0.0	0.0	0.0	0.0
7 V2-12 >149	0.0	0.0	0.0	0.0	0.0	0.0	0.0	0.0	0.0	0.0
8 V23-29 >149	0.160	0.0	0.0	10.515	0.0	0.657	0.0	0.0	0.0	0.0
9 V12-43 >149	0.160	0.0	0.0	2.077	0.0	21.565	1.917	0.0	0.0	0.0
10 R9-7 >149	0.392	0.0	0.0	23.922	0.0	2.353	0.0	0.392	0.0	0.0

	Core									
11	A157-3 >149	0.427	0.0	1.991	0.0	23.044	1.138	0.711	0.0	1.991
12	V23-81 >149	0.588	0.588	26.765	0.0	14.118	0.0	0.588	0.0	1.765
13	V23-82 >149	0.0	0.752	23.684	0.0	18.421	0.376	0.0	0.135	1.128
14	V23-53 >149	0.812	1.083	1.488	0.0	34.100	11.096	0.135	0.0	1.488
15	V23-83 >149	0.0	0.0	11.858	0.0	14.822	0.198	0.593	0.0	0.593
16	V12-56 >149	18.377	0.335	1.790	0.0	30.549	8.353	1.193	0.119	2.983
17	A152-84 >149	11.275	1.634	1.353	0.0	17.647	1.307	1.634	0.556	2.941
18	V16-50 >149	5.741	0.926	2.037	0.741	52.407	2.778	0.0	0.596	2.963
19	V22-122 >149	0.596	0.0	0.298	1.341	47.839	4.620	0.149	0.124	1.043
20	V16-41 >149	6.460	1.118	5.342	0.124	34.037	2.857	5.963	0.713	4.969
21	V4-32 >149	14.693	2.568	5.991	0.0	10.128	2.425	2.995	0.097	1.427
22	V12-66 >149	3.768	2.029	4.251	0.870	22.222	2.222	3.865	0.413	1.159
23	V19-245 >149	4.677	2.201	2.889	0.963	30.124	3.989	4.127	0.0	4.264
24	V4-8 >149	14.759	2.649	7.096	0.189	9.366	3.690	3.974	6.680	2.933
25	A180-15 >149	13.556	0.615	6.557	0.0	15.984	2.554	2.254	0.559	5.123
26	V18-34 >149	12.304	1.965	0.982	0.0	15.914	1.119	7.269	0.308	0.196
27	V20-213 >149	5.949	2.349	6.600	0.224	14.094	0.205	7.718	0.918	1.119
28	V19-222 >149	6.774	4.821	5.769	2.256	24.615	7.118	1.026	0.0	0.718
29	A180-39 >149	2.697	1.493	4.363	0.689	12.400	0.415	7.118	0.201	0.344
30	V16-189 >149	3.422	2.075	7.469	0.0	7.469	2.155	0.432	0.186	5.394
31	V12-18 >149	6.627	2.231	9.438	2.811	4.816	10.643	5.070	0.316	2.662
32	V7-67 >149	15.806	1.506	7.097	0.215	14.257	6.452	1.406	2.340	6.426
33	V17-165 >149	7.595	2.043	7.595	1.108	9.570	6.329	2.365	0.197	4.946
34	V19-310 >149	3.640	2.057	1.424	0.693	22.468	3.813	1.424	0.111	3.323
35	V16-190 >149	6.903	0.780	4.593	0.394	11.179	3.945	2.686	2.532	1.473
36	A153-154 >149	5.568	0.986	8.481	0.111	6.312	5.345	4.339	8.012	0.592
37	V19-308 >149	1.519	2.395	4.900	3.038	3.898	0.0	1.559	0.487	1.893
38	V22-172 >149	8.845	2.025	9.620	2.810	2.025	1.561	1.519	0.270	0.0
39	V10-98 >149	4.842	1.561	3.434	0.0	15.088	3.911	3.434	0.345	1.249
40	V22-219 >149	0.730	2.421	5.587	1.217	6.518	0.730	4.097	0.658	1.117
41	V16-33 >149	1.622	2.607	3.650	2.973	7.786	0.0	3.650	1.266	0.0
42	V22-204 >149	0.345	0.487	8.649	1.209	2.703	1.209	0.270	1.213	0.0
43	V20-167 >149	1.645	1.351	6.736	0.0	0.345	0.0	2.703	0.518	0.0
44	V10-89 >149	4.340	3.232	8.717	3.797	0.822	0.493	2.303	0.221	0.658
45	V12-79 >149	1.213	2.961	3.765	6.739	0.904	0.0	0.0	0.220	0.0
46	V19-216 >149	2.418	0.723	3.100	5.181	9.030	0.135	3.404	0.279	0.539
47	V14-90 >149	3.650	3.504	12.435	2.655	0.173	0.0	4.209	1.392	0.0
48	A180-72 >149	0.329	0.864	10.730	0.549	0.221	0.0	0.332	0.727	0.111
49	V16-21 >149	1.119	0.774	7.355	1.342	0.0	0.220	0.559	0.493	0.0
50	A180-76 >149	0.0	0.988	8.277	0.837	0.0	0.0	0.447	0.220	0.0
51	V15-164 >149	0.506	0.224	6.140	3.418	0.127	0.0	0.651	0.279	0.0
52	A180-78 >149	0.216	2.233	5.949	1.078	0.314	0.0	1.772	1.392	0.0
53	V14-5 >149	0.314	0.127	5.328	0.980	0.0	0.314	0.0	0.727	0.0
54	V3-128 >149	0.490	0.216	2.512	0.216	0.314	0.980	8.320	1.727	0.0
55	A179-13 >149	0.395	1.099	9.181	1.099	0.0	0.0	2.451	0.0	0.0
56	V9-31 >149	0.164	1.471	7.115	8.300	0.0	0.0	3.953	0.0	0.0
57	V20-230 >149		1.581	9.539	3.289	0.0	0.0	2.632	0.493	0.0

(cont.)

SAMPLE	NAME	11 G.FALC	12 G.CALID	13 G.AEQUI	14 G.GLUT	15 G.DUTER	16 G.INFLA	17 G.TRN L	18 G.TRN R	19 G.CRASF	20 G.HIRSU
58	V20-7 >149	0.0	0.0	3.061	0.0	1.653	0.0	0.0	0.0	0.0	0.0
59	V20-234 >149	0.0	1.089	3.267	7.985	1.996	0.0	0.0	0.544	0.0	0.0
60	V18-21 >149	0.0	1.055	4.430	6.540	2.532	0.0	0.0	0.0	0.0	0.0
61	V12-122 >149	0.163	0.326	3.257	8.958	4.560	0.163	0.163	0.0	0.0	0.0

NO.	SAMPLE NAME	21 G.SCITU	22 G.MENTU	23 P.OBLIQ	24 C.NITID	25 S.DEHIS	26 G.DIGIT	27 OTHER
1	V14-61 >149	0.0	0.0	0.0	0.0	0.0	0.0	0.129
2	V17-196 >149	0.0	0.0	0.0	0.0	0.0	0.0	0.0
3	V18-110 >149	0.0	0.0	0.0	0.0	0.0	0.0	0.429
4	V16-227 >149	0.107	0.107	0.107	0.0	0.0	0.0	0.0
5	V14-47 >149	0.0	0.0	0.0	0.0	0.0	0.0	1.180
6	V23-22 >149	0.0	0.0	0.0	0.0	0.0	0.0	0.654
7	V2-12 >149	0.438	0.0	0.0	0.0	0.0	0.160	0.410
8	V23-29 >149	0.0	0.0	0.0	0.0	0.0	0.0	1.972
9	V12-43 >149	1.176	0.0	0.0	0.0	0.0	0.0	2.396
10	R9-7 >149	0.0	0.0	0.0	0.0	0.0	0.0	3.922
11	A157-3 >149	0.0	0.142	0.0	0.0	0.0	0.0	1.422
12	V23-81 >149	1.128	0.0	0.0	0.0	0.0	0.0	1.176
13	V23-82 >149	0.271	0.0	0.0	0.0	0.0	0.135	1.880
14	V12-53 >149	1.186	0.0	0.0	0.0	0.0	0.0	2.977
15	V23-83 >149	0.716	0.119	0.0	0.0	0.0	0.0	1.186
16	V12-56 >149	5.065	0.185	0.0	0.0	0.0	0.163	2.506
17	A152-84 >149	1.607	0.298	0.0	0.0	0.0	0.370	3.431
18	V16-50 >149	0.298	3.602	0.0	0.0	0.745	0.0	2.778
19	V22-122 >149	1.118	3.382	0.0	0.0	0.285	0.124	1.192
20	V16-41 >149	4.137	4.402	0.285	0.0	0.413	0.143	1.366
21	V4-32 >149	0.966	0.205	0.0	0.095	0.0	0.193	4.280
22	V12-66 >149	1.376	0.196	0.275	0.0	0.0	0.0	3.188
23	V19-245 >149	4.920	0.671	0.0	0.0	0.0	0.189	1.926
24	V4-8 >149	1.639	0.513	0.205	0.0	0.112	0.0	3.311
25	A180-15 >149	0.0	0.804	0.0	0.0	0.0	0.0	3.484
26	V18-34 >149	3.020	3.527	0.0	0.0	0.0	0.0	3.536
27	V20-213 >149	1.949	1.521	0.112	0.336	0.0	1.230	1.566
28	V19-222 >149	0.459	1.205	0.923	0.0	0.0	0.615	1.744
29	A180-39 >149	0.622	0.108	1.263	0.415	0.127	0.344	2.641
30	V16-139 >149	2.281	0.475	0.207	0.507	0.0	0.415	1.037
31	V12-18 >149	0.201	1.993	0.0	0.0	0.0	0.507	1.141
32	V7-67 >149	1.613	0.0	0.602	0.0	0.0	0.0	3.012
33	V17-165 >149	0.158	0.0	0.475	0.0	0.0	0.0	2.473
34	V19-310 >149	0.867	0.0	0.0	0.0	0.0	0.0	1.899
35	V16-190 >149	0.394	0.0	1.040	0.173	0.0	0.0	2.600
36	A153-154 >149	0.0	0.0	0.0	0.0	0.0	0.0	1.183
37	V19-308 >149	0.780	0.0	0.0	0.0	0.0	0.557	2.339

#	Label							
38	V22-172 >149	0.506	4.557	1.772	0.253	0.0	0.253	1.519
39	V10-98 >149	0.520	0.520	0.208	0.0	0.104	0.520	1.665
40	V22-219 >149	1.117	0.186	0.0	0.186	0.0	0.186	1.304
41	V16-33 >149	0.0	1.946	0.0	0.487	0.0	0.243	1.460
42	V22-204 >149	1.351	9.189	0.541		0.173	1.081	2.162
43	V20-167 >149	0.0	3.109	1.382		0.0	0.345	1.036
44	V10-89 >149	0.542	1.151	0.0	0.0	0.181	0.164	1.645
45	V12-79 >149	0.674	5.425	4.702	0.0	0.0	0.362	2.532
46	V19-216 >149	0.173	2.830	2.561	0.0	0.0	0.0	3.774
47	V14-90 >149	0.442	4.663	4.663	0.0	0.111	0.0	3.282
48	A180-72 >149	0.110	6.858	4.535		0.110	0.111	0.996
49	V16-21 >149	0.224	0.878	0.220	0.988	0.093	0.439	1.537
50	A180-76 >149	0.279	5.369	4.474		0.127	0.651	1.566
51	V15-164 >149	0.506	3.070	0.279	2.326	0.0	0.633	1.767
52	A180-78 >149	0.647	16.335	11.399	0.0	0.157	0.431	1.772
53	V14-5 >149	0.0	7.759	3.664	0.216	0.0	0.0	0.862
54	V3-128 >149	1.186	7.692	7.849		0.0	1.471	1.256
55	A179-13 >149	0.164	0.0	0.0	0.490	0.164	0.0	1.471
56	V9-31 >149	0.0	0.395	5.929	0.0	0.181	0.0	3.162
57	V20-230 >149	1.270	1.316	6.743	0.0	0.422	0.0	1.151
58	V20-7 >149	0.633	3.061	0.0		0.163	0.0	1.531
59	V20-234 >149	0.163	3.630	0.907			0.0	0.544
60	V18-21 >149		12.447	3.376	0.211		0.0	1.055
61	V12-122 >149		7.492	0.977	0.651		0.0	0.0

Table 3. Percentage of Taxonomic Catagories in Core V12–122

No.: Sample number: Sample name: Core identification and depth in core (c). Column 1: *Orbulina universa*. 2: *Globigerinoides conglobatus*. 3: *G. ruber*. 4: *G. tenellus*. 5: *G. sacculifer*. 6: *Globigerina rubescens*. 7: *G. pachyderma* (left coiling). 8: *G. pachyderma* (right coiling). 9: *G. bulloides*. 10: *G. falcomensis*. 11: *G. calida*. 12: *Globigerinella aequilateralis*. 13: *Globigerinita glutinata*. 14: *Globoquadrina dutertrei*. 15: *Globorotalia inflata*. 16: *G. truncatulinoides* (left coiling). 17: *G. truncatulinoides* (right coiling). 18: *G. crassaformis*. 19: *G. scitula*. 20: *G. menardii* and *G. Tumida*. 21: *Pulleniatina obliquiloculata*. 22: *Candeina nitida*. 23: *Sphaeroidinella dehiscens*. 24: *Globigerina digitata*. 25: *Globoquadrina hexagona*. 26: *G. conglomerata*. 27: *Hastigerina* (?) sp. 28: Other species and unidentified.

NO.	SAMPLE NAME		1 O.UNIVE	2 G.CGLOB	3 G.RUBER	4 G.TENEL	5 G.SACCU	6 G.RUBES	7 G.PAC L	8 G.PAC R	9 G.BULLO	10 G.FALCO
1	V12-122	0	1.792	0.489	43.485	0.814	25.570	0.651	0.0	0.163	0.0	0.163
2	V12-122	10	3.203	0.712	37.722	0.356	30.961	0.712	0.0	0.356	0.0	0.0
3	V12-122	20	2.564	1.709	47.009	0.855	20.513	1.709	0.0	1.282	0.427	0.0
4	V12-122	30	1.124	0.562	47.191	1.124	12.360	2.247	0.0	3.933	0.562	0.562
5	V12-122	40	0.671	1.007	43.624	3.020	15.436	1.007	0.0	0.336	0.671	0.336
6	V12-122	50	1.149	0.756	52.874	0.766	12.261	0.0	0.0	0.383	2.299	0.0
7	V12-122	60	1.990	0.498	53.234	3.980	6.965	0.0	0.0	0.498	0.995	0.0
8	V12-122	70	2.222	2.222	45.926	2.222	13.333	2.963	0.0	1.481	1.481	1.481
9	V12-122	80	1.786	1.190	49.405	1.786	10.714	1.786	0.0	0.595	0.595	0.0
10	V12-122	90	0.621	0.621	36.025	2.484	10.559	0.621	0.0	1.242	1.863	0.0
11	V12-122	100	1.418	0.0	46.099	2.837	9.220	4.255	0.0	0.709	2.837	0.0
12	V12-122	110	0.0	0.0	38.298	0.709	11.348	2.837	0.0	1.418	5.674	0.0
13	V12-122	120	0.498	0.498	48.756	0.0	5.970	1.990	0.498	0.498	2.985	0.690
14	V12-122	130	1.379	1.034	42.069	0.690	8.621	2.069	0.0	2.759	1.724	0.690
15	V12-122	140	0.652	0.0	46.358	0.0	11.921	0.0	0.0	1.987	3.311	0.0
16	V12-122	150	3.429	1.143	45.714	1.143	14.286	1.714	0.0	0.571	3.429	0.571
17	V12-122	160	2.899	2.899	42.995	0.0	14.010	1.449	0.0	2.415	2.415	0.483
18	V12-122	170	1.198	1.796	50.299	1.198	8.383	2.994	0.0	0.599	0.599	0.599
19	V12-122	180	1.887	2.516	38.994	3.145	7.547	2.516	0.0	1.258	1.258	0.0
20	V12-122	190	5.143	2.357	38.286	0.0	13.714	1.143	0.0	1.143	1.143	0.0
21	V12-122	200	3.067	0.613	37.423	1.227	13.497	2.761	0.0	1.227	0.0	0.307
22	V12-122	210	1.961	2.614	41.830	3.268	11.765	1.307	0.654	1.307	0.654	0.0
23	V12-122	220	1.515	2.020	37.374	1.010	12.626	2.020	0.0	0.0	0.505	0.0
24	V12-122	230	1.422	2.344	38.389	1.422	16.114	0.948	0.0	0.0	0.474	0.0
25	V12-122	240	1.630	1.630	36.957	2.174	10.870	2.174	0.0	0.0	0.0	0.0
26	V12-122	250	1.571	1.571	37.696	1.571	10.995	4.188	0.0	2.094	2.618	1.047
27	V12-122	260	1.326	3.196	36.073	0.913	12.329	2.283	0.0	0.457	0.913	0.457
28	V12-122	270	0.926	3.241	28.241	0.463	12.037	0.926	0.0	0.463	1.852	0.463
29	V12-122	280	1.379	2.414	35.517	0.345	11.379	0.345	0.0	0.0	4.828	0.0
30	V12-122	290	1.036	6.218	34.197	1.036	14.508	0.518	0.0	0.0	1.554	0.518
31	V12-122	300	0.649	3.896	39.610	3.896	13.636	1.299	0.0	0.543	0.649	0.0

32	V12-122	310	1.485	7.426	29.208	2.475	15.842	1.485	0.0	2.970	1.485	0.0
33	V12-122	320	1.087	0.0	42.391	1.630	15.761	1.630	0.0	2.174	1.087	0.0
34	V12-122	330	3.404	0.426	32.766	4.255	13.191	2.128	0.0	3.830	0.851	1.702
35	V12-122	340	1.429	0.476	42.381	2.857	10.952	1.905	0.0	0.476	0.952	1.905
36	V12-122	350	1.449	3.623	36.957	0.0	15.942	3.623	0.0	0.725	1.449	0.725
37	V12-122	360	1.685	1.685	48.315	2.809	10.674	1.124	0.0	1.124	1.124	0.0
38	V12-122	370	0.772	0.386	40.927	0.772	15.444	2.703	0.0	0.0	0.772	0.386
39	V12-122	380	1.266	1.266	37.975	2.554	18.143	3.376	0.0	2.110	0.422	0.0
40	V12-122	390	3.627	0.518	41.451	1.554	16.580	0.518	0.0	2.591	1.554	0.0
41	V12-122	400	1.869	1.402	37.850	2.804	12.617	2.336	0.0	9.813	0.467	0.935
42	V12-122	410	3.509	2.456	42.105	2.105	14.759	1.053	0.351	1.506	0.0	0.0
43	V12-122	420	1.449	3.904	44.578	1.205	12.560	0.602	0.301	2.899	0.602	0.0
44	V12-122	430	3.865	0.483	43.961	3.865	17.208	1.948	0.0	4.545	1.449	0.0
45	V12-122	440	1.299	0.649	38.961	0.325	22.222	2.222	0.0	0.741	1.948	0.0
46	V12-122	450	2.513	0.741	35.333	1.005	20.603	0.0	0.0	0.0	1.299	0.0
47	V12-122	460	1.026	4.523	35.176	2.051	16.410	2.051	0.0	0.513	2.513	0.0
48	V12-122	470	0.565	0.513	42.051	3.955	10.169	1.695	0.0	9.605	1.026	0.0
49	V12-122	480	1.523	0.565	44.068	2.030	20.305	2.030	0.0	1.523	1.523	0.0
50	V12-122	490	0.508	0.0	34.518	0.508	21.827	0.508	0.0	3.046	0.508	0.0
51	V12-122	500	0.0	0.0	40.609	1.622	24.324	3.784	0.0	2.162	0.0	0.0
52	V12-122	510	0.629	2.703	28.649	0.629	10.063	3.145	0.0	2.660	0.629	0.0
53	V12-122	520	2.400	4.403	39.623	1.600	11.200	2.400	0.0	4.800	0.800	0.0
54	V12-122	530	1.630	2.490	50.400	2.174	7.609	3.804	0.0	1.630	1.630	0.0
55	V12-122	540	0.0	0.543	54.348	1.087	11.413	4.891	0.0	3.804	0.0	0.0
56	V12-122	550	1.622	1.081	32.609	2.162	16.892	3.784	0.0	9.730	1.622	0.709
57	V12-122	560	1.762	2.162	32.973	0.709	11.348	2.679	0.0	9.251	2.717	0.893
58	V12-122	570	1.418	0.0	33.921	2.273	11.932	4.255	0.0	0.568	2.717	0.0
59	V12-122	580	1.136	2.273	36.879	5.357	13.393	2.273	0.0	4.464	2.643	0.0
60	V12-122	590	0.893	5.357	49.432	2.424	6.061	2.679	0.0	0.568	4.965	0.893
61	V12-122	600	3.636	2.424	33.036	1.478	14.778	6.061	0.0	3.030	0.893	0.0
62	V12-122	610	3.448	1.478	35.758	2.685	12.081	4.433	0.0	2.955	0.0	0.0
63	V12-122	620	1.342	2.685	29.064	3.356	17.742	2.685	0.0	2.685	4.027	0.0
64	V12-122	630	4.435	2.158	34.228	0.806	12.081	3.226	0.0	0.719	4.032	0.0
65	V12-122	640	2.158	4.545	33.468	2.158	15.827	5.036	0.0	2.273	2.158	0.0
66	V12-122	650	0.0	0.0	34.532	2.158	15.152	1.515	0.0	0.617	2.273	0.758
67	V12-122	660	1.235	0.0	38.636	4.545	12.346	1.852	7.407	2.273	2.469	0.0
68	V12-122	670	1.508	1.508	41.975	0.0	13.518	1.508	4.523	0.617	2.010	0.503
69	V12-122	680	3.550	2.367	38.191	0.503	5.917	10.059	0.0	1.508	0.592	0.0
70	V12-122	690	5.344	1.527	47.337	2.367	13.740	6.870	0.0	0.763	0.0	0.0
71	V12-122	700	5.455	0.0	39.695	1.527	10.303	7.273	1.212	0.606	0.0	1.190
72	V12-122	710	0.0	0.606	43.636	1.818	4.762	9.524	0.0	0.606	0.435	0.0
73	V12-122	720	2.609	1.304	38.095	1.739	9.130	3.913	0.870	3.571	3.209	0.535
74	V12-122	730	1.604	1.604	33.043	1.739	19.251	2.139	0.633	3.209	1.266	0.408
75	V12-122	740	1.899	0.0	33.690	2.532	12.025	4.430	0.0	2.532	1.224	0.0
76	V12-122	750	2.041	0.316	34.177	2.041	20.000	2.449	0.0	2.449	0.595	0.0
77	V12-122	760	0.595	2.976	36.327	0.0	7.738	6.548	0.0	2.381	0.383	0.408
78	V12-122	770	0.0	6.130	50.000	0.316	10.728	0.0	0.0	0.383	0.372	0.0
79	V12-122	780	0.595	6.130	35.249	0.0	10.728	6.548	0.633	2.381	0.383	0.0
80	V12-122	790	0.372	5.576	37.918	0.372	15.613	0.743	0.0	0.0	0.372	0.0

(cont.)

NO.	SAMPLE NAME	1 O.UNIVE	2 G.CGLOB	3 G.RUBER	4 G.TENEL	5 G.SACCU	6 G.RUBES	7 G.PAC L	8 G.PAC	9 G.BULLO	10 G.FALCO
81	V12-122 800	3.582	5.373	38.209	0.896	17.015	0.896	0.0	0.0	0.896	0.299
82	V12-122 810	2.362	2.362	36.220	3.150	14.173	1.969	0.0	0.787	1.575	0.0
83	V12-122 820	2.105	4.211	26.842	1.053	13.684	4.737	0.526	5.263	2.105	0.0
84	V12-122 830	2.381	3.175	32.143	1.190	17.460	1.587	0.0	0.397	1.190	0.0
85	V12-122 840	0.455	0.909	37.273	0.455	24.091	3.182	0.0	0.455	0.455	0.909
86	V12-122 850	0.858	3.863	31.760	1.717	21.888	7.296	0.0	4.721	0.858	0.0
87	V12-122 860	2.769	1.231	43.385	1.231	2.769	4.000	0.0	6.462	3.077	0.0
88	V12-122 870	0.658	1.316	52.632	0.0	3.289	1.974	0.0	3.947	0.658	0.0
89	V12-122 880	3.448	0.575	35.632	1.149	14.368	0.0	0.0	4.598	0.575	0.0
90	V12-122 890	1.689	0.076	26.689	2.027	8.108	4.392	0.338	13.176	2.027	1.689
91	V12-122 900	1.533	0.0	35.249	0.383	9.195	2.682	1.533	13.793	1.533	0.0
92	V12-122 910	1.064	0.0	40.957	1.596	6.915	2.660	0.0	3.723	2.660	0.0
93	V12-122 920	1.394	0.348	36.585	1.045	8.014	3.833	0.0	6.969	1.394	0.0
94	V12-122 930	1.970	0.0	35.533	1.015	13.706	7.614	0.0	3.553	0.493	0.493
95	V12-122 940	1.471	2.463	39.901	0.493	15.764	3.941	0.0	0.985	0.735	0.0
96	V12-122 950	1.513	2.206	34.559	2.941	15.441	1.471	0.0	0.0	0.403	0.0
97	V12-122 960	0.0	0.403	42.742	1.210	16.129	2.823	0.0	2.823	0.995	0.498
98	V12-122 970	0.448	0.493	44.776	2.488	19.900	2.242	0.0	1.990	0.897	0.0
99	V12-122 980	2.717	0.448	40.359	4.484	12.556	1.087	0.0	6.278	1.087	0.0
100	V12-122 990	1.387	0.0	32.065	3.261	15.761	1.415	0.0	6.522	1.415	0.472
101	V12-122 1000	1.342	1.397	34.906	1.415	12.264	1.342	0.0	3.302	0.0	0.671
102	V12-122 1010	1.533	2.013	24.161	3.356	11.409	0.408	0.0	9.396	2.041	0.0
103	V12-122 1020	1.548	0.316	24.898	2.449	6.531	0.0	0.0	12.245	4.644	0.0
104	V12-122 1030	1.093	0.310	31.269	1.548	9.288	0.0	0.0	9.288	4.372	1.747
105	V12-122 1040	1.878	0.546	31.694	1.639	14.208	0.437	0.0	19.072	1.747	0.0
106	V12-122 1050	2.286	1.747	33.188	0.437	13.974	1.408	0.0	9.367	0.939	0.0
107	V12-122 1060	1.747	0.469	24.883	1.878	14.085	1.714	0.0	9.390	4.571	0.0
108	V12-122 1070	0.469	2.286	37.143	1.714	8.000	1.117	0.0	8.000	0.559	0.0
109	V12-122 1080	3.911	2.793	32.961	1.117	14.525	1.316	0.0	2.793	1.974	0.0
110	V12-122 1090	0.658	0.658	34.868	4.605	15.789	1.316	0.0	3.947	1.316	0.0

NO.	SAMPLE NAME	11 G.CALID	12 G.AEQUI	13 G.GLUTI	14 G.DUTEK	15 G.INFLA	16 G.TRN L	17 G.TRN R	18 G.CRASF	19 G.SCITU	20 G.MENTU
1	V12-122 0	0.326	3.257	8.958	4.560	0.163	0.163	0.0	0.0	0.163	7.492
2	V12-122 10	0.0	2.491	8.185	5.694	0.0	0.712	0.356	0.0	0.0	5.694
3	V12-122 20	0.855	0.855	9.402	5.556	0.0	0.427	0.855	0.562	0.855	2.991
4	V12-122 30	2.247	5.056	7.865	6.742	1.124	0.0	1.685	0.562	0.562	1.124
5	V12-122 40	1.678	8.054	9.396	3.691	4.698	0.336	2.349	1.342	0.336	1.007
6	V12-122 50	1.916	6.897	7.663	4.981	4.215	0.0	2.682	0.766	0.0	0.0
7	V12-122 60	0.497	4.478	12.935	5.185	5.473	0.497	1.990	0.497	0.497	0.0
8	V12-122 70	2.222	1.481	9.630	5.185	2.222	0.0	2.222	1.481	2.222	0.0
9	V12-122 80	0.595	5.952	7.738	7.738	4.167	0.0	3.571	0.595	0.595	0.0
10	V12-122 90	1.863	5.590	9.938	13.043	7.453	0.621	3.106	1.242	1.863	0.0

No.	Sample	Depth	1	2	3	4	5	6	7	8	9	10
11	V12-122	100	2.128	2.837	6.383	11.347	3.546	0.0	2.837	0.0	2.128	0.0
12	V12-122	110	0.709	1.418	13.475	14.894	2.128	2.128	2.128	0.0	0.0	0.0
13	V12-122	120	0.497	2.488	10.448	12.935	4.975	0.0	4.138	0.497	0.497	0.0
14	V12-122	130	0.0	2.759	10.690	13.793	5.172	0.0	4.975	0.0	0.690	0.0
15	V12-122	140	0.662	3.311	13.907	9.934	1.987	0.0	5.298	1.143	0.662	0.0
16	V12-122	150	2.286	4.571	8.571	3.382	2.857	1.449	4.000	0.483	0.0	0.0
17	V12-122	160	0.966	3.865	7.246	3.593	2.415	1.796	8.696	0.599	0.599	0.0
18	V12-122	170	0.599	4.192	11.976	8.805	1.796	2.516	4.192	1.258	0.0	0.0
19	V12-122	180	1.258	7.547	15.094	16.000	0.0	0.0	3.145	2.286	0.0	0.0
20	V12-122	190	1.714	3.429	10.736	12.270	1.143	0.307	0.571	2.761	0.920	0.920
21	V12-122	200	1.840	4.601	5.229	9.804	0.0	0.0	3.067	1.961	0.0	6.536
22	V12-122	210	2.614	3.922	8.586	14.141	0.0	0.505	3.268	1.010	0.920	6.566
23	V12-122	220	0.505	6.566	9.005	10.900	0.0	0.0	1.010	0.0	0.0	10.900
24	V12-122	230	0.0	1.896	5.978	7.609	0.0	0.543	3.317	1.630	1.010	16.848
25	V12-122	240	1.087	0.543	10.471	7.330	0.0	0.913	7.065	1.047	0.0	7.853
26	V12-122	250	0.524	2.094	10.502	6.849	0.0	0.0	5.236	3.196	0.457	13.959
27	V12-122	260	1.370	2.740	10.185	10.648	0.0	1.036	3.653	0.463	0.463	13.426
28	V12-122	270	1.852	4.167	8.621	11.034	0.345	0.0	1.389	3.793	1.034	8.965
29	V12-122	280	1.724	4.138	12.953	7.772	3.247	1.087	0.0	0.0	0.0	10.363
30	V12-122	290	0.0	4.145	7.792	5.844	4.455	0.0	2.072	0.0	1.299	10.649
31	V12-122	300	2.597	8.442	5.844	11.386	4.891	0.725	3.896	2.475	0.0	0.0
32	V12-122	310	0.495	3.960	8.416	11.413	5.106	0.0	4.455	0.543	0.0	0.0
33	V12-122	320	1.087	2.717	8.152	5.957	5.957	0.422	3.804	0.425	0.0	0.0
34	V12-122	330	2.128	2.128	13.191	6.190	5.190	0.0	5.957	0.0	0.0	0.0
35	V12-122	340	2.857	9.048	7.143	6.522	1.449	0.0	3.809	1.449	0.725	1.124
36	V12-122	350	0.725	5.797	7.246	8.427	0.562	1.449	7.246	1.158	0.0	3.861
37	V12-122	360	2.247	2.309	11.798	13.513	3.861	0.0	4.494	0.844	1.158	4.641
38	V12-122	370	0.772	6.564	1.930	13.924	0.0	0.0	6.178	1.554	1.266	2.072
39	V12-122	380	1.266	1.688	3.375	12.953	3.627	1.005	3.797	0.0	0.0	3.271
40	V12-122	390	1.036	1.036	2.591	18.246	0.467	0.565	5.181	1.403	0.935	3.158
41	V12-122	400	1.402	3.869	5.607	14.157	0.0	0.0	6.075	0.301	0.602	3.012
42	V12-122	410	1.053	3.860	1.754	15.942	0.0	0.0	3.509	0.602	0.0	3.865
43	V12-122	420	0.904	3.014	5.723	3.704	0.0	0.0	5.120	0.0	1.299	2.273
44	V12-122	430	0.0	0.966	6.280	0.0	0.0	0.0	2.899	1.299	0.741	13.333
45	V12-122	440	1.948	2.273	4.221	0.513	0.0	0.0	2.922	0.741	0.0	9.543
46	V12-122	450	1.481	4.444	5.185	0.565	0.0	0.0	4.444	0.502	1.026	3.590
47	V12-122	460	3.015	2.513	6.533	7.107	2.564	0.0	6.030	0.513	1.130	6.215
48	V12-122	470	4.615	2.564	7.179	9.645	0.0	0.0	4.103	1.130	0.0	13.198
49	V12-122	480	1.130	5.650	4.520	11.892	0.0	0.0	1.695	0.508	1.081	10.660
50	V12-122	490	2.538	4.568	2.538	8.176	0.0	0.0	3.553	0.0	0.629	2.703
51	V12-122	500	0.508	4.568	4.324	4.000	1.081	0.0	4.061	0.540	0.0	9.434
52	V12-122	510	0.540	3.243	1.887	1.087	1.258	0.0	7.568	0.629	0.540	1.600
53	V12-122	520	0.0	1.887	3.145	0.543	7.200	0.0	3.774	0.0	0.629	0.0
54	V12-122	530	0.800	4.000	4.000	2.703	10.870	0.0	1.600	0.0	0.0	0.0
55	V12-122	540	0.543	3.804	5.435		10.870	0.0	2.717	0.0	0.0	0.0
56	V12-122	550	0.543	3.261	10.870		0.0	0.0	7.065	2.174	0.543	2.174
57	V12-122	560	0.540	2.162	14.054			0.540	4.865	1.081	0.0	4.865

(cont.)

Table 3 (cont.)

			11	12	13	14	15	16	17	18	19	20
			G.CALID	G.AEQUI	G.GLUTI	G.DUTER	G.INFLA	G.TRN L	G.TRN R	G.CRASF	G.SCITU	G.MENTU
58	V12-122	570	0.440	3.524	7.048	3.084	1.372	0.0	7.048	0.0	0.440	4.846
59	V12-122	580	2.128	4.964	12.057	1.418	1.418	0.0	1.418	1.418	0.709	2.837
60	V12-122	590	2.841	5.114	5.114	3.977	2.841	0.0	4.545	0.0	0.568	2.273
61	V12-122	600	0.893	4.464	7.143	6.250	0.0	1.818	3.571	0.606	1.786	3.929
62	V12-122	610	1.818	5.454	10.909	10.303	0.606	0.493	9.852	0.493	0.0	6.667
63	V12-122	620	1.970	3.448	4.433	7.882	3.448	0.493	4.698	0.493	0.493	6.404
64	V12-122	630	0.671	2.685	12.752	8.054	0.0	0.0	2.016	0.0	0.0	6.040
65	V12-122	640	1.613	4.839	6.855	7.258	0.0	0.0	1.439	0.806	0.0	8.468
66	V12-122	650	2.158	3.597	9.352	5.755	0.719	0.0	9.091	0.719	0.719	10.072
67	V12-122	660	2.273	3.030	3.788	5.303	3.030	0.0	9.091	0.0	0.758	6.818
68	V12-122	670	1.852	3.086	9.876	9.259	0.617	0.0	2.469	0.0	1.235	6.533
69	V12-122	680	1.005	3.518	3.518	13.568	0.502	1.005	9.045	1.507	0.0	0.592
70	V12-122	690	0.592	2.290	4.142	3.817	1.183	0.0	6.509	0.592	0.0	8.397
71	V12-122	700	2.290	2.424	7.634	6.061	3.817	0.763	2.290	0.763	0.0	3.571
72	V12-122	710	2.424	2.381	3.030	13.913	0.0	0.606	6.061	1.818	1.190	5.652
73	V12-122	720	3.571	3.913	5.952	6.952	2.381	0.0	10.714	1.190	0.0	9.091
74	V12-122	730	0.435	4.813	5.217	6.962	0.435	0.526	10.870	0.870	0.0	6.962
75	V12-122	740	1.604	4.430	5.348	4.082	0.0	1.587	3.743	2.139	1.266	7.347
76	V12-122	750	4.430	2.041	9.494	5.952	0.408	0.0	1.899	0.633	0.0	8.929
77	V12-122	760	1.224	1.786	4.898	11.494	0.0	0.0	8.980	0.0	0.0	14.559
78	V12-122	770	1.190	3.448	2.381	8.922	0.0	0.308	6.548	0.0	0.0	14.870
79	V12-122	780	0.0	2.974	1.533	6.693	0.0	0.0	1.533	4.215	0.575	11.045
80	V12-122	790	1.115	2.985	4.833	3.158	0.0	1.149	0.372	1.115	1.351	7.087
81	V12-122	800	1.791	5.512	3.881	5.556	1.053	0.0	0.0	0.597	1.064	3.684
82	V12-122	810	1.968	8.947	7.874	4.091	7.936	1.064	0.526	2.756	0.394	2.381
83	V12-122	820	1.579	4.365	9.474	7.296	8.182	0.348	0.794	4.210	0.397	1.364
84	V12-122	830	2.778	2.727	7.143	6.461	1.288	0.508	2.273	2.381	0.0	3.863
85	V12-122	840	5.909	3.863	3.182	7.237	0.308	0.493	2.146	0.0	0.429	4.000
86	V12-122	850	3.863	8.000	3.004	14.368	2.632	0.735	4.000	0.0	0.0	7.237
87	V12-122	860	4.000	3.947	5.846	12.162	5.117	0.403	2.632	0.0	0.575	0.0
88	V12-122	870	5.921	4.598	3.947	15.709	9.459	0.497	0.575	0.0	1.351	0.0
89	V12-122	880	2.874	4.392	7.471	9.043	6.513	0.448	1.013	1.351	1.064	0.0
90	V12-122	890	1.351	2.682	7.770	18.467	11.170	0.543	0.0	1.533	2.030	0.0
91	V12-122	900	1.916	1.596	4.598	12.690	9.408	1.064	2.439	0.532	0.493	0.0
92	V12-122	910	2.660	2.787	11.702	9.852	2.538	0.348	4.061	0.348	0.0	0.0
93	V12-122	920	1.742	4.568	2.091	11.765	3.448	0.508	2.463	1.523	0.403	0.0
94	V12-122	930	2.030	4.433	5.584	7.661	2.016	0.493	1.471	1.478	1.492	0.0
95	V12-122	940	1.478	3.676	5.419	3.980	3.483	0.735	0.0	0.0	0.0	0.0
96	V12-122	950	2.206	2.823	14.706	9.865	0.897	0.403	2.488	1.613	0.543	0.0
97	V12-122	960	2.016	4.478	10.887	13.043	0.0	0.497	1.345	0.0	0.472	0.0
98	V12-122	970	0.995	3.587	5.970	11.792	2.358	0.448	1.087	1.087	2.013	0.0
99	V12-122	980	2.242	8.696	9.865	11.409	2.013	0.543	1.415	1.415	0.543	0.0
100	V12-122	990	0.543	5.189	10.326	13.043	12.653	0.472	1.415	1.342	0.472	0.0
101	V12-122	1000	1.415	3.356	11.321	11.792	2.358	0.671	2.013	0.816	2.013	0.0
102	V12-122	1010	1.342	2.857	20.134	11.409	2.013	0.0	2.013	1.342	0.0	0.0
103	V12-122	1020	2.449	2.857	8.980	14.694	12.653	0.0	2.041	0.816	0.0	2.857

(cont.)

NO.										
104	V12-122 1030	0.619	4.334	15.170	10.836	4.334	0.0	2.786	0.0	0.310	1.238
105	V12-122 1040	4.918	1.639	2.186	7.650	0.0	0.0	2.186	0.0	0.546	4.372
106	V12-122 1050	0.873	5.240	7.424	13.537	5.677	0.0	1.747	0.0	0.437	2.183
107	V12-122 1060	1.878	2.347	6.103	15.493	7.512	0.0	1.408	0.0	0.469	9.390
108	V12-122 1070	2.857	3.429	8.000	5.714	5.143	0.571	5.714	0.571	0.0	2.286
109	V12-122 1080	1.676	5.028	10.056	11.732	3.911	0.0	1.117	1.117	0.0	4.469
110	V12-122 1090	3.289	2.632	9.210	5.921	3.289	0.658	3.289	1.974	0.0	3.289

NO.	SAMPLE NAME	21 P.OBLIQ	22 C.NITID	23 S.DEHIS	24 G.DIGIT	25 G.HEXAG	26 G.CGLOM	27 CFH.PEL	28 UNIDEN
1	V12-122 0	0.977	0.651	0.163	0.0	0.0	0.0	0.0	0.0
2	V12-122 10	1.423	0.0	0.356	0.0	0.0	0.0	0.0	1.068
3	V12-122 20	0.855	0.855	0.0	0.0	0.0	0.0	0.0	0.427
4	V12-122 30	2.247	0.0	0.0	0.0	0.0	0.0	0.0	1.124
5	V12-122 40	0.671	0.0	0.0	0.0	0.0	0.0	0.0	0.336
6	V12-122 50	0.0	0.0	0.0	0.0	0.0	0.0	0.0	0.383
7	V12-122 60	0.0	0.0	0.0	0.0	0.0	0.0	0.0	0.498
8	V12-122 70	0.0	0.0	0.0	0.0	0.0	0.0	0.0	0.0
9	V12-122 80	0.0	0.621	0.0	0.0	0.0	0.0	0.0	1.190
10	V12-122 90	0.0	0.0	0.0	0.0	0.0	0.0	0.0	0.621
11	V12-122 100	1.418	0.0	0.0	0.0	0.0	0.0	0.0	1.418
12	V12-122 110	0.995	0.0	0.0	0.0	0.0	0.0	0.0	1.418
13	V12-122 120	1.034	0.0	0.0	0.0	0.0	0.0	0.0	0.498
14	V12-122 130	0.0	0.0	0.0	0.0	0.0	0.0	0.0	0.690
15	V12-122 140	0.483	0.0	0.0	0.0	0.0	0.0	0.0	0.0
16	V12-122 150	2.395	0.0	0.0	0.0	0.0	0.0	0.0	1.449
17	V12-122 160	0.629	0.0	0.0	0.0	0.0	0.0	0.0	0.599
18	V12-122 170	2.857	0.0	0.0	0.0	0.0	0.0	0.0	0.629
19	V12-122 180	1.534	0.307	0.0	0.0	0.0	0.0	0.0	1.143
20	V12-122 190	1.307	0.0	0.0	0.0	0.0	0.0	0.0	0.513
21	V12-122 200	2.525	0.0	0.0	0.0	0.0	0.0	0.0	0.0
22	V12-122 210	0.948	0.474	0.0	0.0	0.0	0.0	0.505	0.0
23	V12-122 220	1.630	0.543	0.0	0.0	0.0	0.0	0.474	0.474
24	V12-122 230	0.0	0.524	0.0	0.0	0.524	0.0	0.0	1.087
25	V12-122 240	0.457	0.0	0.0	0.0	0.0	0.0	0.0	1.047
26	V12-122 250	1.389	1.034	0.0	0.0	0.463	0.0	1.389	0.457
27	V12-122 260	0.0	0.0	0.0	0.0	0.345	0.0	1.379	1.379
28	V12-122 270	1.554	0.0	0.0	0.0	0.518	0.0	0.0	0.0
29	V12-122 280	0.0	0.649	0.0	0.0	0.0	0.0	0.0	1.379
30	V12-122 290	0.649	0.495	0.0	0.0	0.0	0.0	0.495	0.0
31	V12-122 300	0.0	0.0	0.0	0.0	0.0	0.0	0.0	0.549
32	V12-122 310	0.0	0.0	0.0	0.0	0.0	0.0	0.0	0.990
33	V12-122 320	0.0	0.0	0.0	0.0	0.0	0.0	0.0	0.543
34	V12-122 330	0.0	0.0	0.0	0.0	0.0	0.0	0.0	1.277
35	V12-122 340	0.0	0.0	0.0	0.0	0.0	0.0	0.0	1.905

Table 3 (*cont.*)

			21 P.OBLIQ	22 C.NITID	23 S.DEHIS	24 G.DIGIT	25 G.HEXAG	26 G.CGLOM	27 CFH.PEL	28 UNIDEN
36	V12-122	350	0.725	0.725	0.0	0.0	0.0	0.0	0.0	2.174
37	V12-122	360	0.0	0.0	0.0	0.0	0.0	0.0	0.0	0.0
38	V12-122	370	0.0	0.0	0.0	0.0	0.0	0.0	0.0	0.0
39	V12-122	380	0.422	0.422	0.0	0.0	0.0	0.0	0.422	0.422
40	V12-122	390	0.0	1.036	0.0	0.0	0.0	0.0	0.0	0.518
41	V12-122	400	0.467	0.0	0.0	0.0	0.0	0.0	0.467	2.336
42	V12-122	410	0.702	0.0	0.0	0.0	0.0	0.0	0.0	0.0
43	V12-122	420	0.301	0.301	0.0	0.0	0.0	0.0	0.0	0.602
44	V12-122	430	0.966	0.0	0.0	0.0	0.966	0.0	0.0	0.0
45	V12-122	440	0.649	0.325	0.325	0.0	0.649	0.0	0.0	1.948
46	V12-122	450	1.481	0.741	0.0	0.0	0.741	0.0	0.0	1.481
47	V12-122	460	6.533	0.0	0.513	0.503	0.0	0.0	0.0	1.538
48	V12-122	470	4.615	0.0	0.0	0.0	0.0	0.0	0.0	0.0
49	V12-122	480	3.390	0.0	0.0	0.0	0.0	0.0	0.0	1.015
50	V12-122	490	2.030	0.0	0.0	0.0	0.0	0.0	0.0	0.0
51	V12-122	500	1.015	0.0	0.0	0.541	0.0	0.0	0.0	0.629
52	V12-122	510	0.0	0.0	0.0	0.800	0.0	0.0	0.0	0.0
53	V12-122	520	1.887	0.0	0.800	0.0	0.0	0.0	0.800	1.087
54	V12-122	530	0.0	0.0	0.0	0.0	0.0	0.0	0.543	0.0
55	V12-122	540	0.543	0.0	0.0	0.0	0.0	0.0	0.0	0.441
56	V12-122	550	3.804	0.0	0.0	0.0	0.0	0.0	0.0	1.136
57	V12-122	560	5.405	0.0	0.0	0.0	0.0	0.0	0.0	0.0
58	V12-122	570	4.846	0.0	0.0	0.0	0.0	0.0	0.0	0.0
59	V12-122	580	5.674	0.709	0.0	0.0	0.0	0.0	0.0	0.0
60	V12-122	590	1.136	0.0	0.0	0.0	0.0	0.0	0.0	0.0
61	V12-122	600	1.786	0.0	0.0	0.0	0.0	0.985	0.0	0.0
62	V12-122	610	1.212	0.0	0.0	0.0	0.493	0.0	0.0	0.0
63	V12-122	620	0.0	0.0	0.0	0.0	0.671	0.0	0.0	0.0
64	V12-122	630	1.342	0.0	0.403	0.0	0.719	0.0	0.0	0.403
65	V12-122	640	1.210	0.0	0.0	0.0	0.0	0.0	0.0	0.0
66	V12-122	650	0.0	0.0	0.0	0.0	0.0	0.617	0.0	0.758
67	V12-122	660	0.0	0.0	0.0	0.0	0.0	3.015	0.0	0.617
68	V12-122	670	0.503	0.503	0.0	0.0	0.0	0.0	0.0	0.503
69	V12-122	680	1.183	1.183	0.0	1.183	0.0	0.0	0.0	1.183
70	V12-122	690	0.0	0.0	0.0	0.0	0.0	0.0	0.0	0.0
71	V12-122	700	0.0	0.0	0.0	0.0	0.0	0.0	0.0	0.606
72	V12-122	710	1.190	0.0	0.0	0.0	0.0	0.0	0.0	0.0
73	V12-122	720	0.0	0.0	0.0	0.0	0.0	0.0	0.0	0.0
74	V12-122	730	0.0	0.0	0.0	0.0	0.0	2.174	0.0	0.0
75	V12-122	740	0.0	0.0	0.0	0.0	0.0	0.535	0.0	0.535
76	V12-122	750	0.633	0.0	0.633	0.0	0.633	1.899	0.0	0.633
77	V12-122	760	0.0	0.0	0.408	0.0	0.816	1.224	0.0	0.816
78	V12-122	770	0.0	0.0	0.595	0.0	0.595	0.595	0.0	0.595
79	V12-122	780	1.149	0.0	2.682	0.0	3.831	1.533	0.0	1.149
80	V12-122	790	1.115	0.0	0.743	0.0	1.859	0.743	0.0	0.372

81	V12-122 800	0.299	0.0	2.090	0.0	0.896	1.194	0.0	0.597
82	V12-122 810	0.394	0.394	0.394	0.0	0.787	1.575	0.394	1.181
83	V12-122 820	1.579	0.0	0.0	0.0	1.053	1.053	0.0	2.632
84	V12-122 830	0.794	0.0	0.0	0.0	0.794	1.984	0.794	0.794
85	V12-122 840	0.0	0.0	0.0	0.0	0.0	0.909	0.0	3.162
86	V12-122 850	0.0	0.0	0.0	0.0	0.0	0.858	0.0	0.429
87	V12-122 860	0.0	0.0	0.0	0.0	0.0	0.308	0.0	1.846
88	V12-122 870	2.874	0.0	0.0	0.0	0.0	0.658	0.0	1.316
89	V12-122 880	0.0	0.0	0.0	0.0	0.0	0.0	0.0	0.0
90	V12-122 890	0.765	0.0	0.0	0.0	0.0	0.338	0.0	0.0
91	V12-122 900	0.532	0.0	0.0	0.0	0.0	0.0	0.0	0.383
92	V12-122 910	1.742	0.0	0.0	0.508	0.0	0.348	0.0	1.064
93	V12-122 920	1.523	0.0	0.0	0.0	0.0	0.508	0.0	0.697
94	V12-122 930	2.463	0.735	1.471	0.0	0.985	0.0	0.0	0.508
95	V12-122 940	1.471	0.403	1.210	0.0	0.735	0.0	0.0	0.493
96	V12-122 950	1.210	0.498	0.498	0.0	1.210	0.0	0.0	1.471
97	V12-122 960	0.995	0.0	0.448	0.0	1.493	0.0	0.0	0.0
98	V12-122 970	2.691	0.0	0.0	0.0	0.0	0.0	0.0	0.995
99	V12-122 980	0.0	0.0	0.0	0.0	0.0	0.0	0.0	0.0
100	V12-122 990	1.887	0.943	1.415	0.0	0.0	0.0	0.0	1.530
101	V12-122 1000	1.342	0.0	0.671	0.0	0.0	0.0	0.0	0.343
102	V12-122 1010	0.0	0.0	0.0	0.0	0.0	0.0	0.0	0.0
103	V12-122 1020	0.0	0.0	0.0	0.0	0.0	0.0	0.0	1.633
104	V12-122 1030	0.0	0.0	0.0	0.0	0.0	0.0	0.0	2.477
105	V12-122 1040	0.0	0.0	0.0	0.0	0.0	0.0	0.0	3.279
106	V12-122 1050	0.0	1.310	0.0	0.0	0.0	0.0	0.0	1.747
107	V12-122 1060	0.0	0.0	0.0	0.0	0.0	0.0	0.0	0.469
108	V12-122 1070	0.559	0.0	0.0	0.0	0.559	0.0	0.0	0.0
109	V12-122 1080	0.0	0.0	0.0	0.0	0.0	0.0	0.0	0.0
110	V12-122 1090	0.0	0.0	0.0	0.0	0.0	0.0	0.0	2.632

Table 4. Varimax Factor Matrix (B_{ct}) from Factor Analysis of Core Top Data

Identification column: Locality code number, core and sieve size (μ). Comm. = communality. Column 1: Tropical assemblage. 2: Subtropical assemblage. 3: Polar assemblage. 4: Subpolar assemblage. 5: Gyre margin assemblage. Each row in the table corresponds to a core top sample. Values in the five numbered columns represent proportional contributions of the five defined assemblages to a given sample. The sum of squares of the i'th row is defined as the *communality of the i'th sample*, h^2_i. These sample communalities represent that fraction of the sum of squares of the i'th row of the normalized data matrix U_{ct} accounted for by the five-assemblage model, and will equal 1.000 only if the model represents the original without error. The squared value of the elements in each of the five columns, averaged over the N samples, is the fraction of the total original sum of squares accounted for by the corresponding assemblage. Expressed as percent, these figures are entered in the row labeled variance. The cumulative sum of these variances, entered in the last row, expresses in percent how much of the original sum of squares in U is accounted for by a factor model having 1 through 5 assemblages. The five-assemblage model therefore has a *total communality* of 0.833; i.e. it accounts for 83.3 percent of the original sum of squares. The square root of this figure (0.91), not shown, measures the *fraction of the original observational data* (as opposed to the squares of the observations) *explained* by the five-assemblage model: 91 percent.

			COMM.	1	2	3	4	5
1	V14-61	>149	C.991	-0.013	-0.014	0.995	0.014	-0.022
2	V17-156	>149	C.991	-0.010	-0.012	0.994	0.031	-0.025
3	V18-110	>149	0.994	-0.016	-0.003	C.996	0.047	-0.015
4	V16-227	>149	0.988	-0.016	-0.006	C.990	0.085	-0.017
5	V14-47	>149	C.927	-0.025	0.219	0.849	0.379	0.117
6	V23-22	>149	0.891	-C.040	0.122	C.628	0.692	0.034
7	V2-12	>149	0.987	-0.014	-0.012	C.992	0.046	-0.020
8	V23-29	>149	C.843	0.009	0.146	C.177	0.889	-0.021
9	V12-43	>149	C.869	-0.063	0.380	C.220	0.796	0.197
10	R9-7	>149	0.903	0.101	C.124	C.054	0.927	-0.123
11	A157-3	>149	C.885	-C.050	0.362	C.154	0.833	0.184
12	V23-81	>149	0.817	0.161	0.164	-C.021	0.867	-0.109
13	V23-82	>149	C.804	0.122	0.211	-C.015	0.859	-0.085
14	V12-53	>149	C.750	-0.085	0.691	C.040	0.465	0.216
15	V23-83	>149	C.949	C.032	0.236	C.012	0.944	0.014
16	V12-56	>149	C.889	-0.003	C.914	0.016	0.205	0.105
17	A152-84	>149	0.693	0.073	0.650	-C.002	0.503	-0.112
18	V16-50	>149	C.741	0.016	0.762	0.023	0.333	0.222
19	V22-122	>149	0.763	-0.055	C.609	0.047	0.547	0.297
20	V16-41	>149	C.763	0.425	0.751	C.006	0.127	0.046
21	V4-32	>149	C.762	C.277	0.747	-C.006	0.233	-0.271
22	V12-66	>149	C.780	0.343	0.658	C.017	0.467	-0.100
23	V19-245	>149	C.816	C.447	0.766	C.008	0.171	0.026
24	V4-8	>149	0.741	C.206	C.747	-C.002	0.270	-0.259
25	A180-15	>149	C.845	0.165	C.810	C.015	0.358	-0.050
26	V18-34	>149	0.694	0.228	0.800	0.028	-0.021	0.047
27	V20-213	>149	0.877	0.292	0.837	-C.006	0.108	-0.282
28	V19-222	>149	0.718	C.461	0.633	C.007	0.245	-0.212
29	A180-39	>149	C.823	C.562	0.697	C.011	0.047	-0.136
30	V16-189	>149	0.722	0.695	0.373	C.008	0.040	-0.314
31	V12-18	>149	0.919	0.672	C.508	C.005	0.063	-0.454
32	V7-67	>149	C.735	C.357	C.768	0.011	0.130	-0.025
33	V17-165	>149	0.892	C.329	0.843	C.004	0.080	-0.240
34	V19-310	>149	C.875	0.382	0.835	C.014	0.167	-0.071
35	V16-190	>149	0.877	0.664	0.621	0.024	0.100	-0.200
36	A153-154	>149	C.888	C.623	0.585	C.013	0.040	-0.355
37	V19-308	>149	C.903	0.598	0.559	0.012	-0.041	-0.481
38	V22-172	>149	0.884	C.868	0.276	0.014	0.056	-0.228

Table 4 (*cont.*)   163

39	V10-98	>149	0.506	0.391	0.593	C.030	0.020	-0.033
40	V22-219	>149	0.922	C.603	0.577	0.012	0.011	-0 474
41	V16-33	>149	C.524	0.699	0.180	0.018	0.006	-0.049
42	V22-204	>149	C.771	C.824	C.232	C.007	0.143	-0.123
43	V20-167	>149	C.875	0.855	0.282	C.017	-0.009	-0.253
44	V10-89	>149	C.881	C.751	0.295	C.017	0.022	-0.478
45	V12-79	>149	0.919	C.922	0.232	C.007	0.118	0.025
46	V19-216	>149	C.789	0.806	0.345	C.020	0.031	-0.136
47	V14-90	>149	0.867	0.891	C.180	0.008	0.183	-0.092
48	A180-72	>149	C.921	C.936	0.158	C.005	0.113	0.085
49	V16-21	>149	C.972	C.849	0.159	0.018	0.029	-0.356
50	A180-76	>149	0.904	C.944	0.091	C.012	0.074	0.010
51	V15-164	>149	0.675	0.744	0.144	C.017	0.013	-0.317
52	A180-78	>149	C.793	C.780	0.045	C.009	0.098	0.420
53	V14-5	>149	C.941	0.963	0.080	C.010	0.063	0.066
54	V3-128	>149	C.698	0.759	0.209	C.022	-0.044	0.112
55	A179-13	>149	0.817	0.665	0.241	0.013	0.018	-0.562
56	V9-31	>149	C.722	0.822	0.178	C.009	0.118	0.024
57	V20-230	>149	C.858	0.916	C.115	C.011	0.074	0.020
58	V20-7	>149	C.710	0.834	0.103	0.025	-0.043	-0.047
59	V20-234	>149	C.860	C.909	0.140	C.006	0.079	-0.089
60	V18-21	>149	C.900	0.938	0.089	0.012	0.039	0.105
61	V12-122	>149	0.879	C.931	0.072	0.009	0.081	-0.018
		VARIANCE	33.365	22.465	10.114	12.707	4.612	
		CUM. VAR	33.365	55.831	65.945	78.652	83.263	

Table 5. Factor Score Matrix from Factor Analysis of Core Top Data (F's)

Species[*]	Variable no.	Factor 1 Tropical	2 Subtrop.	3 Polar	4 Subpolar	5 Gyre margin
O. univ.	1	0.587	0.435	-0.032	0.121	-0.735
G. cglob.	2	0.444	0.117	0.044	-0.073	-0.778
G. ruber	3	2.780	0.482	0.139	-0.322	-1.133
G. tenel.	4	0.385	0.403	0.019	-0.035	-1.924
G. saccu.	5	2.168	-0.271	0.041	0.004	1.411
G. rubes.	6	0.335	-0.006	0.017	0.020	-0.266
G. pac. L	7	-0.058	-0.069	4.666	0.010	-0.105
G. pac. R	8	-0.078	0.141	-0.020	2.861	0.146
G. bullo.	9	-0.293	0.740	0.256	2.779	0.301
G. falco.	10	-0.272	2.448	-0.034	-0.583	-0.131
G. calid.	11	0.469	1.156	-0.004	-0.196	-1.111
G. aequi.	12	1.646	0.464	0.039	-0.111	-0.639
G. gluti.	13	0.996	-0.292	-0.268	2.113	-1.060
G. duter.	14	0.722	-0.034	0.009	0.003	0.693
G. infla.	15	-0.237	2.099	0.054	0.627	1.627
G. trn. L	16	-0.347	2.498	0.016	-0.549	0.507
G. trn. R	17	0.436	0.916	0.012	-0.208	-0.346
G. crasf.	18	0.233	0.657	0.106	-0.387	0.591
G. scitu.	19	0.008	1.063	-0.159	0.460	-0.728
G. mentu.	20	1.427	-0.136	0.009	0.010	1.689
P. obliq.	21	1.273	-0.217	0.012	0.004	1.785
C. nitid.	22	0.305	-0.153	0.019	0.060	-0.700

[*]See Table 3 for abbreviations of species.

Table 6. **F′** Matrix from Factor Analysis of Core Top Data

Species	Variable no.	Factor 1 Tropical	2 Subtrop.	3 Polar	4 Subpolar	5 Gyre margin
O. univ.	1	0.125	0.093	−0.007	0.026	−0.157
G. cglob.	2	0.095	0.025	0.009	−0.016	−0.166
G. ruber	3	0.593	0.103	0.030	−0.069	−0.242
G. tenel.	4	0.082	0.086	0.004	−0.007	−0.410
G. saccu.	5	0.462	−0.058	0.009	0.001	0.301
G. rubes.	6	0.071	−0.001	0.004	0.004	−0.057
G. pac. L	7	−0.012	−0.015	0.995	0.002	−0.022
G. pac. R	8	−0.017	0.030	−0.004	0.610	0.031
G. bullo.	9	−0.063	0.158	0.055	0.593	0.064
G. falco.	10	−0.058	0.522	−0.007	−0.124	−0.028
G. calid.	11	0.100	0.246	−0.001	−0.042	−0.237
G. aequi.	12	0.351	0.099	0.008	−0.024	−0.136
G. gluti.	13	0.212	−0.062	−0.057	0.451	−0.226
G. duter.	14	0.154	−0.007	0.002	0.001	0.148
G. infla.	15	−0.051	0.447	0.012	0.134	0.347
G. trn. L	16	−0.074	0.533	0.003	−0.117	0.108
G. trn. R	17	0.093	0.195	0.003	−0.044	−0.074
G. crasf.	18	0.050	0.140	0.023	−0.083	0.126
G. scitu.	19	0.002	0.227	−0.034	0.098	−0.155
G. mentu.	20	0.304	−0.029	0.002	0.002	0.360
P. obliq.	21	0.271	−0.046	0.003	0.001	0.381
C. nitid.	22	0.065	−0.033	0.004	0.013	−0.149

Table 7. Extract from Normalized Data Matrix ($\mathbf{U}_c$) for Ten Samples of Core V12–122

No: Sample number. Sample name: Core identification and sieve size(μ). See Table 3 for abbreviations of species.

NO.	SAMPLE NAME	1 O.UNIVE	2 G.CGLOB	3 G.RUBER	4 G.TENEL	5 G.SACCU	6 G.RUBES	7 G.PAC L	8 G.PAC R	9 G.BULLO	10 G.FALCO
1	V12-122 0	0.248	0.050	0.532	0.115	0.550	0.049	0.0	0.003	0.0	0.007
2	V12-122 10	0.424	0.069	0.442	0.048	0.638	0.051	0.0	0.036	0.0	0.0
3	V12-122 20	0.360	0.176	0.584	0.122	0.448	0.130	0.0	0.021	0.011	0.0
4	V12-122 30	0.162	0.059	0.602	0.165	0.277	0.176	0.0	0.067	0.015	0.024
5	V12-122 40	0.083	0.092	0.479	0.381	0.298	0.068	0.0	0.005	0.015	0.012
6	V12-122 50	0.155	0.076	0.630	0.105	0.257	0.0	0.0	0.006	0.056	0.0
7	V12-122 60	0.253	0.046	0.599	0.515	0.138	0.0	0.0	0.007	0.023	0.0
8	V12-122 70	0.298	0.219	0.545	0.303	0.278	0.215	0.0	0.023	0.036	0.059
9	V12-122 80	0.241	0.118	0.590	0.245	0.225	0.131	0.0	0.009	0.014	0.0
10	V12-122 90	0.078	0.057	0.401	0.318	0.207	0.042	0.0	0.018	0.042	0.0

NO.	SAMPLE NAME	11 G.CALID	12 G.AEQUI	13 G.GLUTI	14 G.DUTER	15 G.INFLA	16 G.TRN L	17 G.TRN R	18 G.CRASF	19 G.SCITU	20 G.MENTU
1	V12-122 0	0.041	0.271	0.252	0.186	0.002	0.011	0.0	0.0	0.024	0.335
2	V12-122 10	0.0	0.199	0.221	0.223	0.0	0.046	0.024	0.0	0.0	0.244
3	V12-122 20	0.111	0.072	0.269	0.230	0.0	0.029	0.060	0.0	0.129	0.136
4	V12-122 30	0.298	0.439	0.231	0.287	0.017	0.0	0.122	0.055	0.087	0.052
5	V12-122 40	0.192	0.602	0.237	0.135	0.061	0.020	0.146	0.113	0.045	0.040
6	V12-122 50	0.237	0.559	0.210	0.198	0.059	0.0	0.181	0.070	0.0	0.0
7	V12-122 60	0.058	0.343	0.335	0.168	0.072	0.031	0.127	0.043	0.068	0.0
8	V12-122 70	0.274	0.120	0.263	0.205	0.031	0.0	0.149	0.135	0.321	0.0
9	V12-122 80	0.074	0.484	0.213	0.308	0.058	0.0	0.242	0.055	0.086	0.0
10	V12-122 90	0.216	0.423	0.254	0.484	0.097	0.038	0.196	0.106	0.252	0.0

NO.	SAMPLE NAME	21 P.OBLIQ	22 C.NITID
1	V12-122 0	0.062	0.211
2	V12-122 10	0.086	0.0
3	V12-122 20	0.055	0.281
4	V12-122 30	0.148	0.0
5	V12-122 40	0.038	0.0
6	V12-122 50	0.0	0.0
7	V12-122 60	0.0	0.0
8	V12-122 70	0.0	0.0
9	V12-122 80	0.0	0.0
10	V12-122 90	0.0	0.183

Table 8. Maximum Percentage Observed for Twenty-two Species
in V12–122 and Sixty-one Core Tops

Species	Maximum percentage
Orbulina universa	5.455
Globigerinoides conglobatus	7.426
G. ruber	61.581
G. tenellus	6.373
G. sacculifer	35.036
Globigerina rubescens	10.059
G. pachyderma (left coiling)	98.972
G. pachyderma (right coiling)	46.443
G. bulloides	30.340
G. falconensis	18.377
G. calida	5.921
Globigerinella aequilateralis	9.048
Globigerinita glutinata	26.765
Globoquadrina dutertrei	18.467
Globorotalia inflata	52.407
G. truncatulinoides (left coiling)	11.096
G. truncatulinoides (right coiling)	10.870
G. crassaformis	8.012
G. scitula	5.065
G. menardii and G. tumida	16.848
Pulleniatina obliquiloculata	11.899
Candeina nitida	2.326

Table 9. Calculated Varimax Factor Matrix (**B**$_c$) for Core V12–122

For explanation of the five columns, see caption to Table 4. Although the communalities are not entered in Table 9, they are summarized in Table 17.

Identification column: Depth in core (cm). Column 1: Calculated tropical assemblage. 2: Calculated subtropical assemblage. 3: Calculated polar assemblage. 4: Calculated subpolar assemblage. 5: Calculated gyre margin assemblage.

	1	2	3	4	5
0	0.9309	0.0735	0.0090	0.0801	-0.0253
10	0.8717	0.0764	0.0072	0.0723	0.0647
20	0.8407	0.1417	0.0049	0.1112	-0.2042
30	0.8802	0.2439	0.0114	0.0904	-0.2471
40	0.8137	0.2641	0.0135	0.0551	-0.3258
50	0.8359	0.2550	0.0196	0.0614	-0.2856
60	0.7269	0.2476	0.0049	0.1230	-0.4575
70	0.7204	0.3321	0.0011	0.1160	-0.4095
80	0.8154	0.2475	0.0132	0.0596	-0.3288
90	0.7011	0.3164	0.0035	0.1218	-0.3102
100	0.6789	0.2933	0.0059	0.0987	-0.3671
110	0.6752	0.1921	0.0068	0.2177	-0.0689
120	0.7246	0.2073	0.0138	0.1390	-0.1445
130	0.7300	0.2085	0.0036	0.1593	-0.1680
140	0.7635	0.1985	0.0014	0.2010	-0.1913
150	0.7895	0.2914	0.0150	0.0921	-0.3274
160	0.7143	0.3129	0.0149	0.0635	-0.2317
170	0.8199	0.2957	0.0066	0.0943	-0.3083
180	0.7232	0.2784	0.0048	0.1077	-0.3938
190	0.6865	0.1671	0.0106	0.0573	-0.1024
200	0.7775	0.2482	0.0039	0.0790	-0.2232
210	0.8036	0.2417	0.0234	0.0070	-0.2492
220	0.8423	0.1641	0.0074	0.0645	-0.0597
230	0.3390	0.0874	0.0095	0.0741	-0.0263
240	0.7314	0.1763	0.0130	0.0043	0.0034
250	0.7912	0.1922	0.0105	0.1301	-0.1801
260	0.8088	0.2400	0.0100	0.0676	-0.0718
270	0.7373	0.1872	0.0138	0.0598	0.0754
280	0.7632	0.1920	0.0159	0.1144	-0.0776
290	0.7874	0.1222	0.0062	0.1111	-0.0658
300	0.7304	0.2670	0.0098	0.0466	-0.4379
310	0.6343	0.1911	0.0164	0.0744	-0.2614
320	0.7806	0.2540	0.0131	0.0868	-0.1817
330	0.6502	0.2986	-0.0007	0.1358	-0.4216
340	0.7390	0.3142	0.0133	0.0272	-0.3794
350	0.7464	0.2652	0.0143	0.0431	-0.2200
360	0.7732	0.2327	0.0069	0.1021	-0.4063
370	0.7820	0.2245	0.0234	-0.0310	-0.0571
380	0.7569	0.2135	0.0129	0.0267	-0.1561
390	0.6900	0.2111	0.0195	0.0299	-0.1583
400	0.7378	0.2826	0.0072	0.1283	-0.3093
410	0.7146	0.1887	0.0210	-0.0129	-0.1608
420	0.8057	0.1911	0.0157	0.0391	-0.1501
430	0.6876	0.1635	0.0115	0.0454	-0.1687
440	0.7725	0.2043	0.0103	0.0966	-0.0989
450	0.8540	0.1607	0.0129	0.0195	-0.0235
460	0.8084	0.2222	0.0118	0.0035	-0.0216
470	0.7918	0.3030	0.0095	0.0619	-0.2216
480	0.7948	0.2520	0.0181	0.1354	-0.2529
490	0.8610	0.1845	0.0180	0.0015	0.0183
500	0.8887	0.1163	0.0192	0.0068	0.1242
510	0.7120	0.2030	0.0154	0.0689	-0.0831
520	0.7688	0.1573	0.0233	0.1307	-0.0006
530	0.8437	0.2263	0.0225	0.0690	-0.2882
540	0.7617	0.2772	0.0236	0.0907	-0.3450
550	0.7238	0.2851	0.0106	0.1745	-0.0722
560	0.8078	0.1710	-0.0004	0.2259	-0.1232
570	0.8328	0.1912	0.0094	0.1816	0.0443
580	0.8296	0.2210	0.0077	0.2228	-0.1531
590	0.8205	0.2890	0.0166	0.0020	-0.3651
600	0.7133	0.2210	0.0049	0.0889	-0.2923
610	0.7305	0.2433	0.0037	0.0857	-0.2519

(*cont.*)

	1	2	3	4	5
620	0.6575	0.2937	0.0091	0.0221	-0.2471
630	0.7725	0.1708	0.0070	0.1849	-0.2586
640	0.8143	0.1800	0.0140	0.0877	-0.0980
650	0.3151	0.1976	0.0072	0.1073	-0.1882
660	0.7090	0.2791	0.0172	0.0292	-0.1146
670	0.8174	0.2244	0.0642	0.1493	-0.1914
680	0.6684	0.2745	0.0499	0.0016	-0.1042
690	0.6302	0.2050	0.0172	-0.0044	-0.3988
700	0.6934	0.2234	0.0061	0.0511	-0.1803
710	0.6521	0.2597	0.0212	-0.0208	-0.2393
720	0.5327	0.3115	0.0073	0.0205	-0.3035
730	0.6476	0.2201	0.0168	0.0299	-0.1511
740	0.8644	0.2108	0.0211	0.0739	0.0176
750	0.7492	0.2718	0.0050	0.1027	-0.2901
760	0.7596	0.2283	0.0138	0.0370	-0.1410
770	0.7204	0.1763	0.0217	-0.0079	-0.1010
780	0.6875	0.1007	0.0264	-0.0426	0.1517
790	0.7863	0.0849	0.0184	0.0067	0.0831
800	0.7739	0.1635	0.0138	0.0194	-0.1006
810	0.8122	0.2095	0.0150	0.0609	-0.2646
820	0.7310	0.2141	0.0193	0.0983	-0.1786
830	0.7898	0.3447	0.0154	0.0528	-0.1822
840	0.6919	0.2995	0.0164	-0.0099	-0.1547
850	0.7160	0.2125	0.0161	0.0366	-0.2109
860	0.7117	0.2952	0.0145	0.0854	-0.3540
870	0.6964	0.2806	0.0165	0.0121	-0.2246
880	0.7339	0.2620	0.0043	0.1002	-0.1528
890	0.6282	0.3246	0.0057	0.2439	-0.2448
900	0.6525	0.2057	0.0276	0.1858	-0.0578
910	0.6779	0.3434	0.0009	0.2202	-0.2818
920	0.6343	0.2360	0.0187	0.0764	-0.0624
930	0.6701	0.2620	0.0063	0.0559	-0.1507
940	0.8227	0.2507	0.0167	0.0283	-0.1392
950	0.7567	0.1781	0.0005	0.1299	-0.3402
960	0.8340	0.2046	0.0070	0.1284	-0.2411
970	0.3272	0.2592	0.0117	0.0745	-0.2877
980	0.7196	0.2210	0.0041	0.1375	-0.3481
990	0.7130	0.1890	0.0044	0.1310	-0.2744
1000	0.7805	0.2149	0.0065	0.1400	-0.2561
1010	0.6394	0.2450	-0.0201	0.2861	-0.3448
1020	0.6177	0.2883	0.0059	0.2420	-0.1818
1030	0.7282	0.2024	-0.0025	0.3447	-0.2425
1040	0.6423	0.2859	0.0148	0.2324	-0.1988
1050	0.7345	0.2043	0.0091	0.1246	-0.1658
1060	0.7106	0.1957	0.0050	0.1647	0.0028
1070	0.7179	0.3538	0.0153	0.1829	-0.3397
1080	0.7820	0.1944	0.0062	0.1089	-0.1697
1090	0.6955	0.2891	0.0126	0.0967	-0.3308

Table 10. Extract from Cross-Product Matrix B^2_c

For the top ten samples in V12–122 the square, cross-product, and linear terms derived from B_c are given. A: Tropical assemblage. B: Subtropical assemblage. C: Subpolar assemblage. D: Gyre margin assemblage.

No.	Sample name Depth in core (cm)	1 A square	2 B square	3 C square	4 D square	5 AB	6 AC	7 AD
1	00	0.867	0.005	0.006	0.001	0.068	0.075	−0.024
2	10	0.760	0.006	0.005	0.004	0.067	0.063	0.056
3	20	0.707	0.020	0.012	0.042	0.119	0.093	−0.172
4	30	0.775	0.059	0.008	0.061	0.215	0.080	−0.217
5	40	0.662	0.070	0.003	0.106	0.215	0.045	−0.265
6	50	0.699	0.065	0.004	0.082	0.213	0.051	−0.239
7	60	0.528	0.061	0.015	0.209	0.180	0.089	−0.333
8	70	0.519	0.110	0.013	0.168	0.239	0.084	−0.295
9	80	0.665	0.061	0.004	0.108	0.202	0.049	−0.268
10	90	0.492	0.100	0.015	0.096	0.222	0.085	−0.217

No.	Sample name Depth in core (cm)	8 BC	9 BD	10 CD	11 A	12 B	13 C	14 D
1	00	0.006	−0.002	−0.002	0.931	0.073	0.080	−0.025
2	10	0.006	0.005	0.005	0.872	0.076	0.072	0.065
3	20	0.016	−0.029	−0.023	0.841	0.142	0.111	−0.204
4	30	0.022	−0.060	−0.022	0.880	0.244	0.090	−0.247
5	40	0.015	−0.086	−0.018	0.814	0.264	0.055	−0.326
6	50	0.016	−0.073	−0.018	0.836	0.255	0.061	−0.286
7	60	0.030	−0.113	−0.056	0.727	0.248	0.123	−0.457
8	70	0.039	−0.136	−0.048	0.720	0.332	0.116	−0.409
9	80	0.015	−0.081	−0.020	0.815	0.247	0.060	−0.329
10	90	0.039	−0.098	−0.038	0.701	0.316	0.122	−0.310

Table 11. Parameters of Linear Regression

Columns 1: Independent variable. A: Tropical assemblage. B: Subtropical assemblage. C: Subpolar assemblage. D: Gyre margin assemblage. K = constant. Columns 2: Regression coefficient. Columns 3: Cumulative proportion of variance reduced. Columns 4: Multiple correlation coefficient (adjusted for degrees of freedom). Columns 5: Standard error of estimate (adjusted for degrees of freedom).

Summer Surface Temperature				
1	2	3	4	5
A	19.716	0.686	0.828	3.998
B	11.567	0.877	0.936	2.542
C	2.752	0.885	0.939	2.506
D	0.290	0.885	0.938	2.549
K	7.549			

Winter Surface Temperature				
1	2	3	4	5
A	23.607	0.790	0.889	3.730
B	10.395	0.896	0.945	2.673
D	3.660	0.905	0.949	2.599
C	2.728	0.910	0.952	2.568
K	2.046			

Surface Salinity				
1	2	3	4	5
D	−1.565	0.277	0.526	0.857
A	1.956	0.425	0.644	0.777
B	1.947	0.676	0.815	0.594
C	0.831	0.709	0.833	0.573
K	33.781			

Table 12. Parameters of Curvilinear Regression
For explanation see Table 11.

12A. *Summer Surface Temperature*				
1	2	3	4	5
A	44.085	0.686	0.828	3.998
B	25.050	0.877	0.936	2.542
C	− 0.294	0.885	0.939	2.506
D	−21.929	0.885	0.938	2.549
AB	−32.653	0.910	0.951	2.297
A^2	−21.847	0.926	0.959	2.119
AC	− 5.426	0.936	0.964	2.012

Table 12 (*cont.*)

12A. Summer Surface Temperature

1	2	3	4	5
BD	4.166	0.938	0.964	2.013
CD	34.139	0.940	0.965	2.014
AD	24.849	0.943	0.966	2.016
B^2	−10.097	0.944	0.966	2.030
D^2	− 2.756	0.944	0.965	2.068
C^2	2.064	0.944	0.965	2.109
BC	− 1.074	0.944	0.964	2.154
K	6.385			

12B. Winter Surface Temperature

1	2	3	4	5
A	49.451	0.790	0.889	3.730
B	20.608	0.896	0.945	2.673
D	− 9.811	0.905	0.949	2.599
C	− 9.645	0.910	0.952	2.568
AB	−37.258	0.949	0.972	1.963
A^2	−22.953	0.967	0.982	1.617
CD	19.132	0.975	0.986	1.443
C^2	10.480	0.977	0.987	1.404
BC	4.695	0.977	0.987	1.414
D^2	4.029	0.978	0.987	1.433
B^2	− 5.130	0.978	0.987	1.458
AD	11.415	0.978	0.986	1.486
BD	7.374	0.978	0.986	1.505
AC	1.673	0.978	0.986	1.537
K	1.342			

12C. Surface Salinity

1	2	3	4	5
D	− 3.170	0.277	0.526	0.857
A	6.861	0.425	0.644	0.777
B	2.965	0.676	0.815	0.594
C	1.442	0.709	0.833	0.573
A^2	− 4.674	0.739	0.849	0.552
AC	− 4.343	0.772	0.867	0.526
AB	− 2.995	0.797	0.880	0.505
AD	3.024	0.800	0.879	0.511
D^2	1.115	0.802	0.878	0.519
BC	− 1.101	0.803	0.876	0.527
B^2	− 0.862	0.803	0.874	0.537
BD	1.140	0.804	0.871	0.548
CD	0.954	0.804	0.869	0.559
C^2	− 0.159	0.804	0.866	0.571
K	33.549			

Table 13. Observed and Estimated Surface Temperatures and Salinity for Sixty-One Core Tops

No: Locality code number. Columns 1: Observed. 2: Estimated. 3: Residual (because of truncation, residuals may be off by ±0.1)

NO.	SUMMER SURFACE TEMPERATURE			WINTER SURFACE TEMPERATURE			AVERAGE SURFACE SALINITY		
	1	2	3	1	2	3	1	2	3
1	2.0	5.9	-3.9	0.0	0.4	-0.4	33.7	33.5	-0.2
2	5.0	6.1	-1.1	-1.0	0.5	-1.5	33.5	33.5	-0.0
3	5.5	5.9	-0.4	1.0	0.1	0.8	34.0	33.5	-0.4
4	7.0	5.8	1.1	0.5	-0.1	0.6	34.0	33.6	-0.6
5	7.0	9.5	-2.5	2.0	2.6	-0.6	34.0	34.1	-0.1
6	10.5	8.5	1.9	0.5	0.7	-0.2	34.7	34.4	0.2
7	11.0	5.8	5.1	1.0	0.1	0.8	33.5	33.5	-0.0
8	10.0	11.1	-1.1	6.0	4.7	1.2	35.0	35.0	-0.0
9	13.0	14.3	-1.3	5.0	8.0	-3.0	34.0	34.7	-0.7
10	12.0	12.4	-0.4	7.5	7.7	-0.2	35.2	35.4	-0.1
11	14.0	14.7	-0.7	7.5	8.5	-1.0	35.0	34.7	0.2
12	14.5	14.8	-0.3	9.0	10.3	-1.3	35.5	35.4	0.0
13	15.0	14.9	0.0	10.0	9.8	-0.1	35.5	35.4	0.0
14	14.5	15.7	-1.2	11.0	11.0	-0.0	34.7	34.7	-0.0
15	16.0	14.2	1.7	11.0	8.6	2.3	35.5	35.2	0.2
16	18.0	19.4	-1.4	13.0	15.3	-2.3	35.0	35.4	-0.4
17	20.0	19.7	0.2	13.5	13.1	0.3	36.0	35.8	0.1
18	18.0	18.0	-0.0	15.5	14.2	1.2	35.5	35.1	0.3
19	19.0	15.8	3.1	15.0	11.5	3.4	35.0	34.7	0.2
20	16.5	23.2	-4.7	17.0	18.5	-1.5	35.7	36.2	-0.4
21	21.5	23.6	-2.1	16.0	16.7	-0.7	36.5	36.4	-0.0
22	21.0	22.2	-1.2	16.5	17.2	-0.7	35.7	35.9	-0.2
23	21.0	23.2	-2.2	16.7	18.3	-1.6	35.7	36.1	-0.4
24	24.0	23.1	0.8	16.0	15.9	0.0	36.5	36.3	0.1
25	24.0	21.9	2.0	16.0	16.3	-0.3	36.2	35.9	0.2
26	23.0	22.3	0.6	18.0	17.8	0.1	36.0	36.0	-0.0
27	24.0	24.9	-0.9	18.0	17.7	0.2	36.0	36.6	-0.6
28	23.0	23.3	-0.3	19.3	17.7	1.2	35.7	36.2	-0.5
29	23.0	24.2	-1.2	19.0	18.6	0.5	37.0	36.4	0.5
30	24.0	26.0	-2.0	19.0	21.5	-2.5	36.5	36.4	0.0
31	25.0	24.7	0.2	18.5	19.8	-1.3	36.2	36.6	-0.3
32	26.0	23.4	2.5	18.0	18.0	-0.0	36.5	36.2	0.2
33	26.0	24.8	1.1	19.0	18.0	0.9	36.7	36.6	0.1
34	26.0	23.2	2.7	19.0	17.8	1.1	36.2	36.2	0.4
35	25.0	23.5	1.4	20.0	18.3	1.6	36.5	36.2	0.2
36	26.0	25.1	0.8	20.5	19.4	1.0	36.6	36.6	0.3
37	26.0	27.3	-1.3	20.5	21.1	-0.6	37.0	37.0	-0.0
38	24.5	25.4	-0.9	22.0	22.4	-0.4	36.7	35.9	0.7
39	27.0	24.2	2.7	20.0	18.8	1.1	36.7	36.3	0.4

(40–61)	Col1	Col2	Col3	Col4	Col5	Col6	Col7	Col8	Col9
40	0.3	36.9	37.0	0.8	20.1	21.0	0.1	26.0	26.2
41	0.7	36.2	37.0	-1.0	23.5	22.5	-1.7	26.7	25.0
42	0.1	35.8	36.0	-1.1	22.6	21.5	-0.9	25.5	26.5
43	-0.9	36.2	35.2	-1.1	23.1	22.0	-0.0	26.3	26.2
44	-0.5	36.6	37.2	-0.5	23.0	23.0	-0.3	26.3	26.0
45	0.1	35.5	35.7	-0.3	23.3	22.5	-0.2	26.2	27.0
46	-0.9	36.1	35.2	0.5	21.9	23.5	1.2	25.7	27.0
47	-0.5	36.5	35.0	0.1	23.3	24.0	1.3	25.6	27.5
48	0.1	35.5	35.7	-0.8	24.8	25.0	0.2	27.2	27.0
49	0.2	36.2	30.5	-0.1	24.6	25.0	0.3	26.6	27.0
50	0.5	35.6	35.7	-0.7	25.7	25.5	-0.5	27.5	27.0
51	0.5	36.4	37.0	0.3	24.6	26.0	-0.4	27.4	27.0
52	-0.3	35.8	35.7	-0.6	26.1	24.5	0.0	26.9	29.0
53	0.0	36.1	36.0	-0.1	26.1	25.0	0.9	27.9	28.5
54	-0.5	37.0	36.2	0.0	24.4	26.0	-2.1	26.8	27.5
55	-0.8	35.8	36.5	1.0	23.9	26.0	1.0	27.4	27.5
56	-0.7	35.7	35.0	2.1	23.8	26.5	0.7	26.7	27.0
57	-1.1	36.1	36.5	0.6	25.3	26.5	0.4	27.4	27.0
58	-0.5	35.7	35.2	0.1	25.8	26.5	-0.2	27.9	28.0
59	-0.0	35.6	35.7	1.8	24.6	26.5	0.2	26.7	27.0
60	0.3		36.0	0.2	26.2		-1.0	28.0	28.0
61				0.6	25.6		0.4	27.5	

Table 14. Estimated Surface Temperatures (°C) and Salinity (per mil)
for Core V12–122

Column 1: Depth in core (cm). 2: Estimated summer surface temperature. 3: Estimated
winter surface temperature. 4: Estimated average surface salinity.

1	2	3	4
0	27.5	25.7	35.6
10	27.6	25.7	35.8
20	26.1	23.9	35.9
30	25.1	22.5	35.8
40	25.7	22.7	36.2
50	25.6	22.7	36.1
60	24.8	22.1	36.4
70	24.6	21.1	36.3
80	25.7	22.9	36.2
90	25.1	21.2	36.2
100	25.6	21.7	36.4
110	25.3	21.8	35.8
120	25.8	22.4	36.0
130	25.5	22.2	36.0
140	25.0	22.2	35.8
150	25.1	21.9	36.1
160	26.0	21.9	36.3
170	24.8	21.8	36.0
180	25.2	21.9	36.3
190	26.4	23.0	36.1
200	25.9	22.6	36.1
210	26.7	23.5	36.2
220	26.9	24.2	35.9
230	27.3	25.1	35.8
240	26.8	23.8	36.1
250	25.8	22.9	35.9
260	26.3	23.1	35.9
270	26.5	23.5	36.0
280	26.3	23.1	35.9
290	26.6	24.0	35.9
300	26.2	22.9	36.6
310	26.2	22.4	36.4
320	25.9	22.5	36.0
330	24.9	21.1	36.5
340	26.3	22.5	36.5
350	26.4	22.7	36.2
360	25.1	22.6	36.3
370	27.1	24.0	36.2
380	26.8	23.5	36.2
390	26.7	23.0	36.3
400	25.0	21.6	36.2
410	27.2	23.9	36.3
420	26.8	23.8	36.1
430	26.6	23.3	36.2
440	26.3	23.1	36.0
450	27.2	24.7	36.0
460	26.9	23.9	36.1
470	25.7	22.1	36.1
480	25.0	22.0	35.9
490	27.2	24.6	36.0
500	27.7	25.8	35.9
510	26.4	22.9	36.1

520	26.5	23.5	35.8
530	25.7	23.0	36.0
540	25.3	22.1	36.2
550	25.4	21.7	35.9
560	25.3	22.8	35.6
570	26.5	23.6	35.6
580	24.7	22.1	35.6
590	26.2	23.0	36.3
600	26.0	22.6	36.3
610	26.0	22.5	36.2
620	26.7	22.3	36.4
630	24.9	22.5	35.9
640	26.5	23.6	35.9
650	25.9	23.1	35.9
660	26.4	22.5	36.2
670	25.3	22.4	35.8
680	26.6	22.6	36.3
690	27.7	23.7	36.7
700	26.5	22.8	36.2
710	27.3	23.1	36.5
720	26.7	21.4	36.6
730	26.5	22.6	36.3
740	26.7	23.8	35.8
750	25.4	22.0	36.2
760	26.6	23.2	36.2
770	27.1	23.9	36.2
780	25.7	24.0	36.0
790	27.2	25.2	36.0
800	27.1	24.2	36.1
810	26.2	23.4	36.1
820	26.1	22.7	36.1
830	25.6	21.7	36.1
840	26.8	22.7	36.4
850	26.7	23.2	36.3
860	25.6	21.9	36.4
870	26.8	22.7	36.4
880	25.9	22.2	36.1
890	24.0	19.9	36.0
900	25.4	21.6	35.9
910	23.8	20.1	36.0
920	25.8	21.9	36.1
930	26.3	22.2	36.2
940	26.5	23.2	36.1
950	25.3	22.8	36.2
960	25.3	22.8	35.9
970	25.4	22.5	36.0
980	25.1	22.1	36.2
990	25.6	22.5	36.2
1000	25.3	22.4	36.0
1010	23.1	20.0	36.1
1020	24.3	20.2	36.0
1030	23.1	20.9	35.6
1040	24.4	20.5	36.0
1050	25.9	22.6	36.0
1060	25.9	22.5	35.8
1070	23.7	20.2	36.1
1080	26.1	23.1	36.0
1090	25.6	21.8	36.3

Table 15. Digital Model of Hypothetical Core Data

No.	Component 1	Component 2	Signal	Differential sedimentation	Randomness	Distorted signal
1	100	100	200	150	40	190
2	98	90	188	200	−20	180
3	96	8	104	160	−16	144
4	91	8	99	98	68	166
5	84	90	174	174	−60	114
6	71	100	171	150	40	190
7	58	90	148	61	60	121
8	44	8	52	40	−4	36
9	30	8	38	46	−66	−20
10	21	90	111	115	96	211
11	11	100	111	114	12	126
12	5	90	95	103	−76	27
13	2	8	10	74	−7	67
14	2	8	10	10	68	78
15	5	90	95	4	−12	−8
16	11	100	111	55	−40	15
17	21	90	111	104	−8	96
18	30	8	38	113	0	113
19	44	8	52	111	−4	107
20	58	90	148	50	−48	2
21	71	100	171	40	0	40
22	84	90	174	60	52	112
23	91	8	99	156	24	180
24	96	8	104	174	−52	122
25	98	90	188	98	56	154
26	100	100	200	150	68	218
27	98	90	188	200	84	284
28	96	8	104	160	68	228
29	91	8	99	98	28	126
30	84	90	174	174	−16	158

31	77	100	171	150	32	182
32	58	90	148	61	−4	57
33	44	8	52	40	0	40
34	30	8	38	46	−20	26
35	21	90	111	115	−40	75
36	11	100	111	114	88	202
37	5	90	95	103	64	167
38	2	8	10	74	−7	67
39	2	8	10	10	−12	−2
40	5	90	95	4	−64	−60
41	11	100	111	55	−48	7
42	21	90	111	104	−48	56
43	30	8	38	113	12	125
44	44	8	52	111	56	167
45	58	90	148	50	108	158
46	71	100	171	40	−20	20
47	84	90	174	60	−52	8
48	91	8	99	156	12	168
49	96	8	104	174	−44	130
50	98	90	188	98	4	102
51	100	100	200	150	−12	138
52	98	90	188	200	24	224
53	96	8	104	160	92	252
54	91	8	99	98	28	126
55	84	90	174	174	−8	166
56	71	100	171	150	−28	122
57	58	90	148	61	12	73
58	44	8	52	40	80	120
59	30	8	38	46	8	54
60	21	90	111	115	−92	23
61	11	100	111	114	8	122
62	5	90	95	103	24	127
63	2	8	10	74	−59	15
64	2	8	10	10	−8	2
65	5	90	95	4	28	32

(*cont.*)

Table 15 (cont.)

No.	Component 1	Component 2	Signal	Differential sedimentation	Randomness	Distorted signal
66	11	100	111	55	68	123
67	21	90	111	104	32	136
68	30	8	38	113	−12	101
69	44	8	52	111	48	159
70	58	90	148	50	28	78
71	71	100	171	40	−12	28
72	84	90	174	60	−60	0
73	91	8	99	156	−60	96
74	96	8	104	174	60	234
75	98	90	188	98	−28	70
76	100	100	200	150	32	182
77	98	90	188	200	0	200
78	96	8	104	160	−32	128
79	91	8	99	98	52	150
80	84	90	174	174	40	214
81	71	100	171	150	8	158
82	58	90	148	61	64	125
83	44	8	52	40	−12	28
84	30	8	38	46	64	110
85	21	90	111	115	36	151
86	11	100	111	114	36	150
87	5	90	95	103	0	103
88	2	8	10	74	21	95
89	2	8	10	10	20	30
90	5	90	95	4	48	52
91	11	100	111	55	16	71
92	21	90	111	104	−92	12
93	30	8	38	113	−44	69
94	44	8	52	111	−56	55
95	58	90	148	50	116	166
96	71	100	171	40	24	64
97	84	90	174	60	−100	−40
98	91	8	99	156	32	188
99	96	8	104	174	36	210
100	98	90	188	98	−40	58

Table 16. Percentages of Benthic Foraminifera in Core Top Samples

Sample no.	Percent total foraminifera	Sample no.	Percent total foraminifera
1	0.26	36	0.39
2	2.51	37	0.33
3	0	38	0.25
4	1.40	39	2.24
5	5.57	40	0.92
6	4.98	41	9.07
7	1.21	42	2.37
8	0.44	43	0.52
9	17.63	44	0.33
10	0.78	45	1.07
11	1.40	46	0.13
12	1.45	47	1.70
13	0.75	48	0.44
14	1.73	49	0.65
15	0	50	0
16	0.24	51	0.56
17	0.65	52	3.76
18	5.76	53	0
19	0.15	54	0.47
20	1.23	55	5.99
21	1.82	56	0.78
22	1.24	57	0.16
23	0.82	58	15.15
24	1.49	59	0.18
25	0.61	60	1.23
26	1.55	61	0.61
27	0.67		
28	0.51		
29	0.23		
30	0.82		
31	0.50		
32	0.80		
33	0.53		
34	2.17		
35	0.26		

Table 17. Summary of Communalities for Core Tops and Core V12–122

N is the number of samples averaged; h^2 the average communality for an interval, i.e. the percent of the total original sum of squares explained by the five-assemblage model; h is the fraction of original data explained.

	N	h^2	h	Range of h^2
Core tops	61	83	91	51–99
Total core V12–122	110	69	83	47–90
Z zone	4	84	92	78–90
Y zone	17	73	85	51–86
X zone	10	69	83	57–80
W zone	5	68	82	51–78
V zone	52	67	82	47–80
U zone	22	66	81	47–84
Subpolar maxima*	10	67	82	54–81

* Ten samples with high values of the subpolar assemblage.

Table 18. Coarse-Fraction and CaCO₃ Data for Caribbean Core V12–122

Depth in core (cm)	Total* CaCO₃ (%)	Coarse† fraction (%)	CaCO₃‡ (% fine fraction)	Zone
10–15	75	—	—	Z
20–25	75	—	—	Z
35–40	67	—	—	Y–Z
73–77	56	—	—	Y
128–131	67	21	58	Y
180–185	59	15	52	Y
215–220	66	26	54	X
275–280	67	32	52	X
310–315	67	17	60	X
315–320	64	16	47	W–X
345–350	53	12	47	W
365–370	58	13	52	V
419–421	56	29	38	V
477–480	61	21	51	V
535–540	71	10	68	V
612–614	61	18	53	V
697–701	60	8	55	V
752–758	63	17	56	V
780–785	60	24	47	V
910–915	56	15	48	U
1061–1065	64	18	56	U

* From Ku and Broecker (1966) and Broecker and Van Donk (1970).
† Weight percent > 74 μ; measured by D. B. Ericson.
‡ Calculated on the assumption that all the coarse fraction is CaCO₃.

Table 19. Coincidence of Major Peaks in *Globorotalia menardii*, Estimated Winter Temperature, and O¹⁸ Curves in Caribbean Core V12–122

Depth (cm)	G. menardii*	Winter† temperature	O¹⁸‡	Zone
0–10		X	X	Z
230–300	X	X	X	X
430			X	V
500	X	X	X	V
650	X		X	V
790	X	X	X	V
970			X	U
1060	X			U

* Greater than 9 percent in Table 2.
† Greater than 24°C according to estimates plotted in Figure 30.
‡ Less than 1.5 per mil on Figure 33.

181

Cesare Emiliani

6. THE AMPLITUDE OF PLEISTOCENE CLI-
MATIC CYCLES AT LOW LATITUDES AND
THE ISOTOPIC COMPOSITION OF
GLACIAL ICE

Research during the past 120 years in the middle and high northern latitudes (especially in Europe and North America) has proved the repeated occurrence of major glaciations. In their monumental work (1199 pages), Penck and Brückner (1909) concluded that four great glaciations (Günz, Mindel, Riss, and Würm) had occurred since the "beginning" of the Pleistocene. These conclusions received such universal acceptance that they became almost a gospel and are still widely accepted and widely quoted in the literature. Evidence of numerous additional glaciations revealed by a host of subsequent workers (see Zeuner 1959 for a summary of the classical literature) was generally forced into Penck and Brückner's scheme, usually by the workers themselves, who assigned them the secondary role of substages within the four major glaciations. Occasional findings suggesting the occurrence of an apparently exorbitant number of glaciations (e.g. the eleven weathering horizons found at Paks by Scherf, 1936) were generally ignored and rarely quoted.

The geochemical work of Arrhenius (1952) on a suite of deep-sea cores from the equatorial Pacific and the isotopic work by the present writer on cores from the Atlantic and adjacent seas (Emiliani, 1955, 1958, 1964, 1966a) revealed the occurrence of numerous, quasiperiodic variations in isotopic parameters with almost constant amplitudes. These variations were shown, by absolute dating, to be strictly related to the glacial/interglacial cycles of the northern latitudes, and led to the conclusion that approximately eight major glaciations had occurred during the past 400,000 years and that, more generally, some twenty glaciations per million years may have occurred during the past few million years and may continue to occur during the next few million years (Emiliani 1966b). It would appear that the numerous glacial episodes summarized by Zeuner (1959, p. 54) were in fact major glaciations of approximately similar magnitude.

Although the clearest picture of climatic changes during the more recent portion of the Pleistocene is that provided by oxygen isotopic analysis of deep-sea cores, a clear picture has been obtained also by appropriate micropaleontological analysis of these cores (Lidz 1966; Emiliani 1969; Ruddiman, in press; Imbrie, this volume).

The amplitude of O^{18}/O^{16} variations in the calcium carbonate of the tests of planktonic foraminiferal species of shallow habitat (*Globigerinoides sacculifera* and *G. rubra*) in deep-sea cores from low latitudes in the Atlantic and the Caribbean is about 1.8 per mil. While this amplitude, if entirely due to temperature, would indicate a glacial/interglacial variation of about 8°C at these low latitudes, the present writer estimated (Emiliani, 1955, 1966a) that about 30 percent of the amplitude mentioned is due to concomitant glacial/interglacial variations in the oxygen isotopic composition of sea water. By contrast some authors (Olausson 1965; Shackleton 1967; Dansgaard and Tauber 1969) have suggested that the water effect is much greater and may account either largely or entirely for the amplitude observed. Their arguments are based on models attempting to estimate the oxygen isotopic composition of the ice of the Pleistocene ice caps. The purpose of the present paper is to estimate the glacial/interglacial amplitude of temperature variations at low latitudes by means other than the O^{18}/O^{16} ratios; and from this to estimate the oxygen isotopic composition of glacial ice and provide an explanation for this composition.

CONTINENTAL EVIDENCE AT MIDDLE AND LOW LATITUDES

The classical methods for estimating past temperatures are both geological and paleontological, the more reliable ones being those based on accurate analysis of fossil faunas and floras. Many such estimates exist in the literature, mainly for Europe and North America.

In the middle latitudes, lowering of nivation sculptures (about 1200 meters in both the Alps and the Rocky Mountains), the southern extension of frozen soils, and the pollen evidence, all indicate a glacial/interglacial temperature range of 6 to 10°C (yearly mean) (Emiliani and Flint 1963). At lower latitudes, the 10° of latitude displacement toward the equator of the boundary between the westerlies and the trade wind belt in Africa north of the equator (Fairbridge 1964), the lowering of nivation sculptures on Mount Kenya (Flint 1959, 1963) and in New Guinea (Gentilli 1961), forest biogeography in East Africa (Moreau 1952), and pollen work in both South America and Southeast Asia (Van der Hammen 1961; Tsukada 1967) indicate a glacial/interglacial range of 5–10°C. While most of this evidence pertains to environments at some altitude above sea level, there is some evidence from land deposits at low altitudes also indicating marked temperature variations. Thus Watts (1969) found *Fagus, Acer, Tilia, Carpinus,* and even some *Picea,* together with more abundant *Quercus, Carya, Pinus,* and *Liquidambar,* in a section 10 to 13 meters below the sediment surface at Mud Lake, Florida. Although Watts does not present temperature estimates, this assemblage, which represents a glacial age, could not have lived in Florida without a temperature decrease of several degrees centigrade in both yearly and summer means.

A complicating factor in studying continental Pleistocene climates at middle and low latitudes is that not only temperature but also precipitation changed during the Pleistocene. These two factors, equally importantly, although in different ways, affecting the continental ecology, cannot be easily separated by· the usual methods of paleontological analysis. Thus, in their palynological study of two long cores from the lake bed upon which Mexico City is built, Sears and Clisby (1955) stress the close relationship between moisture and temperature without separating the two factors.

While it is often stated in the classical literature that the climate of southern Europe and elsewhere in the direction of the equator from the ice caps became much more humid, and only secondarily colder, during glacial ages, Gentilli (1961), Bonatti (1963, 1966), Fairbridge (1964), Galloway (1965), Van der Hammen et al. (1965), and Damuth and Fairbridge (1970) concluded that climate in these areas was cold and dry.

The conclusions obtained from the continental evidence are summarized in Table 1.

THE OCEANIC EVIDENCE

The precipitation or humidity factor, important in evaluating the continental evidence, is insignificant in the open ocean, because the ocean is wet anyway. This factor is thus eliminated, and an attempt can be made to estimate temperature alone from the changes in the microfaunas.

The composition of planktonic foraminiferal faunas in deep-sea cores from the Atlantic and adjacent seas varied markedly in response to the glacial/interglacial cycles of the continents. Phleger et al. (1953) published detailed analyses of the microfaunas of the North Atlantic deep-sea cores collected by the Swedish Deep-Sea Expedition of 1947–48. The average percentages of the different species during glacial stage 2 and interglacial stage 5 in deep-sea cores 234, 246, 280, and 288 are shown in Table 2. The locations, depth of the water, and core lengths are shown in Table 3, together with climatic data.

If the pertinent data of Table 2 are compared with Figures 3–19 in Phleger et al. (1953), striking latitudinal displacements are noticed. The absence of *Globorotalia menardii* during stage 2 indicates a latitude not lower than about 20°, while the continued abundance of *Globigerinoides sacculifera* and *Globoquadrina eggeri* indicates a latitude not higher than about 20°. This evidence alone indicates that an oceanic climate similar to that now occurring at 20°N latitude prevailed in the equatorial Atlantic during the glacial ages. The equatorial February–August surface temperature range, now approximately 27–25°C, would have been reversed during glacial ages to approximately 20–24°C, showing the climatic effect of the ice caps transmitted through the hydroatmosphere and overcoming the effect of

Table 1. Amplitude of Glacial-Interglacial Temperature Variations at Low Latitudes
Estimated by Methods Other than Oxygen Isotope Analysis

(Data from the authors mentioned)

Method	Amplitude (°C)
Continental Evidence	
Lowering of nivation sculptures on Mt. Kenya (Flint)	
Lowering of nivation sculptures in New Guinea (Gentilli)	
Biogeography of East Africa (Moreau)	5–10
Pollen analysis, Colombia (Van der Hammen)	
Pollen analysis, Florida (Watts)	
Pollen analysis, Taiwan (Tsukada)	
Oceanic Evidence	
Frequency of *Globorotalia menardii, G. inflata,* and *Globigerinoides sacculifera,* equatorial and North Atlantic cores (Phleger et al.)	5 (yearly max) 3 (yearly min) 4 (yearly av)
Frequency of *Pulleniatina obliquiloculata,* equatorial and North Atlantic cores (Phleger et al.)	5
Frequency of *Pulleniatina obliquiloculata,* Caribbean cores (Emiliani)	8
Frequency of planktonic foraminifera, eastern equatorial Atlantic (Schott)	5–9
Latitudinal displacement of Coccolithophoridae, Atlantic Ocean (McIntyre)	10 (yearly max) 5–6 (yearly min) 8 (yearly av)
Frequency of *Pulleniatina obliquiloculata* and *Sphaeroidinella dehiscens,* eastern equatorial Pacific, 90–94°W (Arrhenius)	> 2 < 5
Frequency of *Pulleniatina obliquiloculata* and *Sphaeroidinella dehiscens,* eastern equatorial Pacific, 130–135°W (Arrhenius)	> 3 < 9
Frequency of *Pulleniatina obliquiloculata,* central Pacific Ocean, 150°W (Brotzen)	< 8
Frequency of different pelagic foraminiferal faunas, South Pacific (Blackman)	3–4
Pore concentration in *Globoquadrina eggeri* (Wiles)	
Caribbean	6
Equatorial Pacific	3.5
Conclusions	
Caribbean	7–8
Equatorial Atlantic	5–6
Equatorial Pacific	3–4

Table 2. Average Percentages of Different Species of Pelagic Foraminifera
During Stages 2 (a) and 5 (b) in Four Atlantic Cores*

(Data from Phleger et al. 1953)

	234		246		280		288	
	a	b	a	b	a	b	a	b
Globigerina bulloides	2.5	3.7	5.3	1.5	16.7	20.8	46.0	37.3
G. pachyderma	0.1	0	0.6	0	1.9	1.9	6.7	3.7
Globigerinoides conglobata	1.5	1.8	0.2	1.0	0.2	0.9	0	0
G. rubra	24.5	17.2	27.0	36.0	11.3	12.5	1.9	3.3
G. sacculifera	42.0	25.0	18.3	23.5	1.3	0.9	0.3	0
Globoquadrina eggeri	19.5	21.7	22.6	6.0	9.7	11.8	13.0	14.3
Globorotalia inflata	4.0	1.1	2.5	0.4	54.3	34.2	13.3	13.7
G. menardii	0	16.5	0.7	6.5	0	0.1	0	0.1
G. truncatulinoides	0.8	0.6	9.0	3.0	3.0	9.5	0.8	8.7
Hastigerinella aequilateralis	1.5	1.0	1.0	3.0	0.7	0.5	0.2	0.1
Orbulina universa	0.8	1.4	0.2	0.5	1.1	0.6	0	0.4
Pulleniatina obliquiloculata	0	3.2	1.3	5.0	0	0.2	0	0

* Stage 2: core 234, 20–60 cm; core 246, 70–100 cm; core 280, 50–150 cm; core 288, 130–230 cm. Stage 5: core 234, 210–320 cm; core 246, 450–470 cm; core 280, 400–530 cm; core 288, 560–690 cm. Core locations and other pertinent data are given in Table 3.

Table 3. Data on Deep-Sea Cores Discussed in Text[*]

Core no.	Lat.	Long.	Length (cm)	Depth (m)	Temperature Max. (°C)	Min. (°C)
234	5°45'N	21°43'W	3577	673.5	29.4	24.4
246	0°48'N	31°28'W	3210	870.0	29.4	23.9
280	34°57'N	44°16'W	4256	1002.5	27.8	15.6
288	41°08'N	31°50'W	2684	1334.0	25.6	12.8
38	2°52'S	89°50'W	3225	405.0	29.0	18.0
39	2°44'S	92°45'W	3600	528.0	29.0	18.0
40	1°04'N	93°20'W	2820	1017.0	29.0	20.0
58	6°44'N	129°28'W	4440	991.0	28.5	25.0
59	3°05'N	133°06'W	4370	972.0	28.5	23.0
62	3°00'S	136°26'W	4510	1479.0	27.0	23.5
74	0°29'S	151°33'W	4545	1346.0	28.5	24.5
75	0°20'N	150°36'W	4405	872.0	28.5	24.0
P6702-1	5°00'N	103°00'W	3159	740.0	29.0	24.0
A179-4	16°36'N	74°48'W	2965	690.0	30.0	24.0
V19-47	17°00'S	111°12'W	3422	450.0	26.0	23.0

[*] Lengths of cores 234 and 280 refer to the undisturbed Pleistocene sections. Temperatures are yearly maxima and minima (from Oceanographic Atlas of the North Atlantic Ocean. Section II. Physical Properties, U.S. Naval Oceanographic Office, Publ. no. 700, 1967; and Monthly Charts of Mean, Minimum, and Maximum Sea Surface Temperature of the North Pacific Ocean, U.S. Naval Oceanographic Office, Special Publ. SP-123, 1969).

the sun's double passage. The glacial/interglacial temperature range would thus be about 5°C (yearly minimum) or 3°C (yearly maximum), for an average of 4°C.

The pelagic foraminiferal species most limited to warm waters is *Sphaeroidinella dehiscens* (cf. Bradshaw 1959). This species, however, is generally scarce in the Atlantic cores (Phleger et al. 1953) and, as a consequence, frequency noise is large. Better represented is *Pulleniatina obliquiloculata*. This species is best adapted to a depth habitat of 25–50 meters in the equatorial Atlantic (Jones 1969) and 100–200 meters in the Sargasso Sea (Bé 1960). At these depths, yearly temperature ranges from 18 to 24°C (Bé 1960, Fig. 11; Muromtsev, 1963, pp. 586–92). In the Sargasso Sea, where extensive monthly observations have been made (Bé 1960), *Pulleniatina obliquiloculata* has been found to be a winter species, best adapted to a temperature of about 18–20°C. Should temperature at the optimum depth habitat of *P. obliquiloculata* decrease below 18°C, the species could conceivably follow the upward movement of the 18°C isotherm.[1] However, should even surface temperature decrease below 18°C, *P. obliquiloculata* could not survive. This is borne out by the lower limit of 18°C in surface temperature published by Bradshaw (1959, Fig. 40) for this species.

Pulleniatina obliquiloculata is absent during much of the glacial ages in the Caribbean and equatorial Atlantic cores (see Phleger et al. 1953; Emiliani 1964, Fig. 6, curves F; 1969, Fig. 1, middle curve), showing that glacial temperatures in these cores decreased to at least 18°C. By comparison with modern temperatures, glacial/interglacial temperature ranges of 8°C and 5°C are indicated for the Caribbean and the equatorial Atlantic respectively.

McIntyre and Bé (1967) have demonstrated the latitudinal distribution of modern coccolithophorid species in the Atlantic, and deduced temperature ranges and temperature optima for each species. Applying this information to postglacial and glacial sediments of the North Atlantic, McIntyre (1967) was able to reconstruct a paleoisotherm map showing that glacial isotherms were displaced 10–20° of latitude toward the equator, corresponding to a temperature decrease of 5–6°C below the February mean at low latitudes, or about 10°C below the August mean, or about 7.5°C below the yearly mean. These figures, especially those for the yearly and August means, are higher than those obtained above from the pelagic foraminiferal faunas.

Summarizing the micropaleontological evidence discussed above, one may conclude that glacial surface temperatures were at least 8°C lower than today in the Caribbean and at least 5°C lower than today in the equatorial Atlantic.

In the suite of cores from the eastern equatorial Pacific described by Arrhenius (1952), considerable variations in the carbonate percentage and other concomitant parameters were noticed, but it is difficult or impossible to estimate temperature variations from these oscillations. The mechanism proposed by Arrhenius

---

1. Such vertical migrations, however, do not seem to have occurred (see Emiliani 1954; 1955, p. 546).

(1952) to explain the observed carbonate oscillations, namely increased carbonate productivity related to increased atmospheric and oceanic circulations during glacial ages, appears correct. Because of the pattern of vertical and horizontal circulation in the area where the cores described by Arrhenius were collected, interglacial temperatures could conceivably be lower than glacial temperatures at some core locations. In particular, a glacial displacement of the surface isotherms toward the equator could conceivably raise equatorial temperatures by bringing closer to the equator the high-temperature belt now at 5 to 10°N. While the large amount of postdepositional solution in these cores makes it impossible to estimate past temperatures on the basis of faunal composition, the occurrence and relative abundances of *Sphaeroidinella dehiscens* and *Pulleniatina obliquiloculata*, the two species most restricted to warm waters, provide some pertinent information.

According to the comprehensive work of Bradshaw (1959), *S. dehiscens* is absent at temperatures below 23°C while *P. obliquiloculata* is absent at temperatures below 18°C. Both species are markedly resistant to postdepositional solution (Berger 1968, Table 1), so that their absence, when other more soluble species remain present, cannot be ascribed to postdepositional solution. In the long cores 58, 59, 62, *S. dehiscens* is frequent below a stratigraphic level estimated to be about 400,000 years old (at 400, 1000, and 570 cm below top respectively; see Emiliani 1966b, Fig. 7), while it becomes rare or absent above this level. Although the general scarcity of *S. dehiscens* above the 400,000-year level makes random effects important, the conclusion may be drawn that temperature at the location of cores 58, 59, and 62 was often below 23°C, a decrease of 2 to 4°C with respect to the temperatures prevailing today at the location of the three cores (Table 3). Contrary to *S. dehiscens*, *P. obliquiloculata* remains generally well represented during the past 400,000 years in all three cores, showing that temperature did not decrease at any time below 18°C. One may conclude, on the evidence of the Pacific cores under discussion, that the glacial/interglacial temperature range was greater than 2 to 4°C but smaller than 9°C.

Additional evidence is provided by cores 38, 39, 40 (cf. Arrhenius 1952, Plates 2.38, 2.39, 2.40). The yearly surface temperature minimum at the location of the three cores is 18°C (cores 38 and 39) and 20°C (core 40). While *S. dehiscens* is entirely absent, as expected, *P. obliquiloculata* in different core layers is either very rare or absent. The conclusion is that the temperature minimum must have been often below 18°C, indicating a glacial/interglacial amplitude greater than 1 to 3°C.

In core P6702-1, *P. obliquiloculata* was found to be frequent to abundant throughout the core, while *S. dehiscens* was often absent (Emiliani, unpublished data). Again, absence of *S. dehiscens* indicates a temperature decrease of at least 2°C below the present yearly minimum of 24°C at the location of the core. A glacial/interglacial temperature range greater than 2°C but smaller than 5°C is deduced.

In cores 74 and 75 from the central equatorial Pacific described by Brotzen and Dinesen (1959), *P. obliquiloculata* is frequent throughout, indicating a glacial/interglacial temperature range smaller than 8°C (present yearly range at the location of the two cores being 24 to 28°C; see Table 3). A "small" cooling of the surface water in the Timor Sea during the last glacial age was inferred by Van Andel et al. (1967) on the basis of pelagic foraminiferal changes.

Blackman (1966) and Blackman and Somayajulu (1966) presented evidence showing remarkable glacial/interglacial faunal changes in deep-sea cores from the southeastern Pacific. Various micropaleontological criteria, including relative abundance of different species and coiling directions, were studied in considerable detail. The results indicate a 5–6° displacement at 45°S of the high-latitude fauna toward the equator during glacial ages, and a very pronounced displacement of *Globorotalia inflata* into the equatorial region (Blackman 1966, p. 169). At the longitude of 110°W, the yearly temperature ranges from 8°C to 13°C at 45°S, and from 11°C to 17°C at 40°S. Thus, a decrease of 5° latitude implies a temperature decrease of 3–4°C in both yearly minima and yearly maxima, and in averages. As noted before, it is difficult to estimate the glacial/interglacial temperature range closer to the equator because of the complex system of horizontal and vertical circulation. However, according to Blackman (1966), an intensification of the Peru current could account for the displacement of *G. inflata* toward the equator. This would entail, necessarily, some temperature decrease.

The evidence from all Pacific cores discussed above, taken together, indicates a glacial/interglacial temperature range greater than 2°C but smaller than 6°C.

Pore counts by Wiles (1967) on *Globoquadrina eggeri* specimens from the Caribbean core A179-4 and the eastern equatorial Pacific cores V19-47 show a marked glacial/interglacial variation in pore concentration in both cores. There is a remarkable correlation between pore concentration variations in core A179-4 and O^{18}/O^{16} ratios as measured in shells of *Globigerinoides sacculifera* (Wiles 1967, Fig. 1; see also Fig. 1 of this paper). An excellent correlation also exists between pore concentration variations in core A179-4 and those in core V19-47 (Fig. 1). The difference in pore concentrations between stage 2 (the temperature minimum of the last glacial age taken as base of reference) and the other stages is shown in Table 4. As may be seen from this table, pore concentrations are generally smaller in the Pacific than in the Atlantic. The yearly range of the surface temperature at the location of the two cores is given in Table 3. The yearly average is 26.5°C at the site of the Caribbean core and 24.5°C at the site of the Pacific core. To this difference of 2.0°C corresponds a difference in pore concentration of 1.5. From the difference in pore concentration of 3.5 (core A179-4) and 1.6 (core V19-47) between the present maximum and the glacial extreme of stage 2, glacial/interglacial surface temperature ranges of 6°C and 2.5°C may be calculated for the Caribbean and the eastern equatorial Pacific respectively. However, the Pacific value should probably be larger because the top of core V19-47 appears slightly truncated (Wiles 1967, Fig. 2; this paper, Fig. 1). If the pore concentration value for stage 1 in this core were the same as that of stage 5 (as it is for the

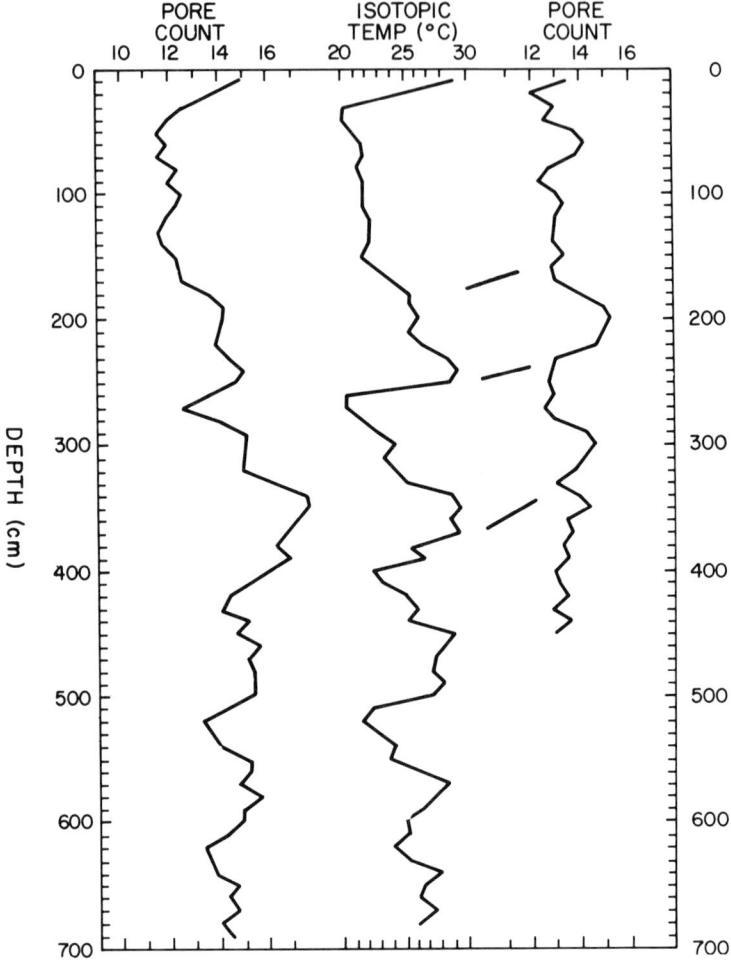

Fig. 1. Left curve: core A179-4, pore concentration in *Globoquadrina eggeri*. Middle curve: core A179-4, isotopic temperatures. Right curve: core V19-47, pore concentrations in *Globoquadrina eggeri* (from Wiles 1967)

Caribbean core) the difference in pore concentration values between stages 2 and 1 would be 2.1 instead of 1.6, bringing the glacial/interglacial surface temperature range to 3.5°C. This is about two-thirds as large as that of the Caribbean, confirming conclusions previously reached on the basis of isotopic data (Emiliani 1966a, p. 119). It should be noticed that the above argument assumes that the Caribbean and Pacific populations of *Globoquadrina eggeri* reacted in similar ways to temperature as far as pore concentration is concerned.

The conclusions obtained from the above review of the micropaleontological evidence are shown in Table 1.

Table 4. Pore Concentrations in *Globoquadrina eggeri*
from Caribbean Core A179-4 and Pacific Core V19-47 *

Stage no.	Core A179-4	Core V19-47
1	3.5	1.6
3	1.0	0.9
5	3.5	2.1
6	1.0	0.4
7	6.0	3.2

* The difference between various stages and stage 2 is taken as base of reference.

THE OXYGEN ISOTOPIC COMPOSITION OF GLACIAL ICE

The nonisotopic foraminiferal evidence presented above can be taken to indicate that glacial surface temperatures were about 7 to 8°C lower than today in the Caribbean, about 5 to 6°C lower than today in the equatorial Atlantic, and about 3 to 4°C lower than today in the equatorial Pacific. The isotopic ranges measured on the foraminiferal shells, in combination with the temperature values mentioned, lead to an estimate of the glacial/interglacial isotopic range of the sea water in different areas (Table 5). As may be seen, the isotopic range may have been as small as zero in the Caribbean, where the increased O^{18}/O^{16} ratio in sea water during glacial ages may have been balanced by an increased influx of fresh water from the glaciated Andes and other ranges of northern South America.

Greater glacial/interglacial temperature ranges are expected at higher latitudes because of the greater proximity to the ice caps; and smaller ranges are expected for deeper water. In the bottom water, the maximum glacial/interglacial temperature range is that measured isotopically on the benthonic foraminiferal shells, amounting to 4°C for the Atlantic (Emiliani 1958), and 2°C for the Pacific (Emiliani 1955). It is certain, however, that a portion of this range was due to glacial/interglacial variations in the oxygen isotopic composition of the sea water. As previously discussed (Emiliani 1966a) the boundary case is provided by assigning these ranges in their entirety to oxygen isotopic variations of the sea water.

The overall glacial/interglacial range for the entire ocean is likely to lie between the value obtained for the equatorial Atlantic (0.6 per mil) and that obtained for the equatorial Pacific (0.3 per mil). Thus, a value of 0.5 per mil does not seem unreasonable. This value, which coincides with that previously obtained on a different basis (Emiliani 1966a), leads to an estimate of −15 per mil for the average isotopic composition of glacial ice (cf. Emiliani 1966a), a value which is considerably higher than those measured in Antarctica (−20 to −50 per mil) and Greenland (−20 to −40 per mil) (see Emiliani 1966a, pp. 119–20 for the references).

Dansgaard et al. (1969) have found that glacial ice, still preserved in the

Table 5. Glacial-Interglacial Ranges in Various Regions[*]

	δO^{18} ‰ (Foraminifera)	Temperature (°C)	δO^{18} ‰ (Surface sea water)
Caribbean	1.8	7–8	0
Equatorial Atlantic	1.8	5–6	0.6
Equatorial Pacific	1.2[†]	3–4	0.3

[*] Isotopic composition of surface sea water derived from the measured isotopic composition of planktonic foraminifera and the surface temperature estimated from nonisotopic evidence.

[†] Estimated; see Emiliani 1966, Table 4.

bottom layers of the Greenland ice cap, is 10 per mil lighter than postglacial ice. On this basis, Dansgaard and Tauber (1969) estimate that Antarctic and Greenland ice during glacial ages, corresponding to 29×10^6 km³ of water, had an average isotopic composition 5 per mil lighter than today. Using this estimate and the previous estimate of +0.5 per mil for the average oxygen isotopic composition of sea water during glacial ages, and assigning an isotopic value of −40 per mil to the Siberian ice (equivalent to 4×10^6 km³ of water; see Dansgaard and Tauber 1969, Table 1), an average isotopic composition of −9 per mil is obtained for the North American and European ice sheets. This value, much heavier than that assumed by Dansgaard and Tauber (1969, Table 1), is not at all unexpected. During glacial ages, in fact, the North American and European ice sheets extended well into the middle latitudes, and were in close proximity to the oceanic areas supplying moisture (northeastern Pacific, Gulf of Mexico, North Atlantic, Mediterranean). Under these conditions, most of the moisture is expected to have fallen directly onto the ice sheets as snow without undergoing the extensive process of fractional precipitation so apparent today for the high-latitude ice masses of Greenland and Antarctica. Supporting this contention is the observed aridity of the intervening areas (Bonatti 1963, 1966; Fairbridge 1964; Galloway 1965; Van der Hammen et al. 1965; Damuth and Fairbridge 1970). An isotopic composition of −30 per mil for the North American and European ice sheets, as advocated by Dansgaard and Tauber (1969), would require a vast amount of precipitation in the intervening areas, an amount for which no evidence exists.

Summary. A detailed analysis of the paleontological evidence both on land and at sea shows that the amplitude of the glacial/interglacial temperature range was 7–8°C in the Caribbean, 5–6°C in the equatorial Atlantic, and 3–4°C in the equatorial Pacific. These values and the glacial/interglacial isotopic range of 1.8 per mil measured in the shells of pelagic foraminifera indicate that the average oxygen isotopic composition of sea water during glacial ages was +0.5 per mil and that of glacial ice was −15 per mil. The average oxygen isotopic composition of the North American and European ice caps was about −9 per mil, a value considerably heavier than that assumed by some but in agreement with the expected pattern and isotopic composition of precipitation on these areas during glacial ages.

Acknowledgments. Financial support was provided by the National Science Foundation (grants GA–10082, GA–15226, GA–4569). Contribution no. 1218 from the School of Marine and Atmospheric Sciences, University of Miami, Miami, Florida.

REFERENCES

Arrhenius, G., Sediment cores from the East Pacific, *Swedish Deep-Sea Exped., 1947–1948, Repts.,* 5, fasc. 1, 227 pp., 1952.

Bé, A. W. H., Ecology of Recent planktonic foraminifera. Part 2, Bathymetric and seasonal distributions in the Sargasso Sea off Bermuda, *Micropaleontology,* 6, 373, 1960.

Berger, W. H., Planktonic foraminifera: Selective solution and paleoclimatic interpretations, *Deep-Sea Res.,* 15, 31, 1968.

Blackman, A., Pleistocene stratigraphy of cores from the southeast Pacific Ocean, 200 pp., Ph.D. thesis, University of California, San Diego, 1966.

Blackman, A., and B. L. K. Somayajulu, Pacific Pleistocene cores: Faunal analyses and geochronology, *Science,* 154, 886, 1966.

Bonatti, E., Stratigrafia pollinica dei sedimenti postglaciali di Baccano, lago craterico del Lazio, *Atti Soc. Toscana Sci. Nat., Ser. A,* 70, 13 pp., 1963.

Bonatti, E., North Mediterranean climate during the last Würm glaciation, *Nature,* 209, 984, 1966.

Bradshaw, J. S., Ecology of living planktonic foraminifera in the north and equatorial Pacific Ocean, *Contrib. Cushman Found. Foram. Res.,* 10, pt. 2, 25, 1959.

Brotzen, F., and A. Dinesen, On the stratigraphy of some bottom sections from the central Pacific, *Swedish Deep-Sea Exped., 1947–1948, Repts.,* 10, fasc. 4, 43, 1959.

Damuth, J. E., and R. W. Fairbridge, Equatorial Atlantic deep-sea arkosic sands and ice-age aridity in tropical South America, *Bull. Geol. Soc. Am.,* 81, 189, 1970.

Dansgaard, W., and H. Tauber, Glacier O^{18} content and Pleistocene ocean temperatures, *Science,* 166, 499, 1969.

Dansgaard, W., S. J. Johnson, J. Møller, and C. C. Langway, Jr., One thousand centuries of climatic record from Camp Century on the Greenland ice sheet, *Science,* 166, 377, 1969.

Emiliani, C., Depth habitats of some species of pelagic Foraminifera as indicated by oxygen isotope ratios, *Am. J. Sci.,* 252, 149, 1954.

Emiliani, C., Pleistocene temperatures, *J. Geol.,* 63, 538, 1955.

Emiliani, C., Paleotemperature analysis of core 280 and Pleistocene correlations, *J. Geol.,* 66, 264, 1958.

Emiliani, C., Paleotemperature analysis of the Caribbean cores A254–BR–C and CP–28, *Bull. Geol. Soc. Am.,* 75, 129, 1964.

Emiliani, C., Paleotemperature analysis of the Caribbean cores P6304–8 and P6304–9, and a generalized temperature curve for the past 425,000 years, *J. Geol.,* 74, 109, 1966a.

Emiliani, C., Isotopic paleotemperatures, *Science,* 154, 851, 1966b.

Emiliani, C., A new paleontology, *Micropaleontology,* 15, 265, 1969.

Emiliani, C., and R. F. Flint, The Pleistocene record, in *The Sea,* edited by M. N. Hill, vol. 3, p. 888, Interscience, New York, 1963.

Fairbridge, R. W., African ice-age aridity, in *Problems in Palaeoclimatology,* edited by A. E. M. Nairn, p. 356, Interscience, London–New York, 1964.

Flint, R. F., Pleistocene climates in eastern and southern Africa, *Bull. Geol. Soc. Am.*, *70*, 343, 1959.

Flint, R. F., Pleistocene climates in low latitudes, *Geograph. Rev.*, *53*, 123, 1963.

Galloway, R. W., Late Quaternary climates in Australia, *J. Geol.*, *73*, 603, 1965.

Gentilli, J., Quaternary climates of the Australian region, *Ann. N. Y. Acad. Sci.*, *95*, art. 1, 465, 1961.

Jones, J. I., Planktonic foraminifera as indicator organisms in the eastern equatorial Atlantic current system, *Symp. on Oceanog. and Fisheries Resources of the Tropical Atlantic, Abijan, Ivory Coast*, 1966, p. 213, UNESCO, Paris, 1969.

Lidz, L., Deep-sea Pleistocene biostratigraphy, *Science*, *154*, 1448, 1966.

McIntyre, A., Coccoliths as paleoclimatic indicators of Pleistocene glaciations, *Science*, *158*, 1314, 1967.

McIntyre, A., and A. W. H. Bé, Modern Coccolithophoridae of the Atlantic Ocean. I. Placoliths and Cyrtoliths, *Deep-Sea Res.*, *14*, 561, 1967.

Moreau, R. E., Africa since the Mesozoic: With particular reference to certain biological problems, *London Zool. Soc., Proc.*, *121*, 869, 1952.

Muromtsev, A. M., The principal hydrological features of the Atlantic Ocean, 838 pp. (in Russian), State Oceanographic Institute, Moscow, 1963.

Olausson, E., 1965. Evidence of climatic changes in North Atlantic deep-sea cores, with remarks on isotopic paleotemperature analysis, in *Progress in Oceanography*, edited by M. Sears, vol. 3, p. 221, 1965.

Penck, A., and E. Brückner, *Die Alpen im Eiszeitalter*, 1199 pp., Tauchnitz, Leipzig, 1909.

Phleger, F. B., F. L. Parker, and J. F. Peirson, North Atlantic Foraminifera, *Swedish Deep-Sea Exped., 1947–1948, Repts., 7*, fasc. 1, 122 pp., 1953.

Ruddiman, W. F., Pleistocene sedimentation in the equatorial Atlantic. I. Stratigraphy and paleoclimate, *Bull. Geol. Soc. Am.*, in press.

Scherf, E., 1936. Versuch einer Einteilung des ungarischen Pleistozäns auf moderner polyglazialistischer Grundlage, *Intern. Quatern. Congr., III, Acta*, Vienna, 1936, p. 237, 1936.

Schott, W., Foraminiferenfauna und Stratigraphie der Tiefsee-Sedimente im Nordatlantischen Ozean, *Swedish Deep-Sea Exped., 1947–1948, Repts., 7*, 355, 1966.

Sears, P. B., and K. H. Clisby, Palynology in southern North America, Part IV. Pleistocene climate in Mexico, *Bull. Geol. Soc. Am.*, *66*, 521, 1955.

Shackleton, N., Oxygen isotope analyses and Pleistocene temperatures re-assessed, *Nature*, *215*, 15, 1967.

Tsukada, M., Vegetation in subtropical Formosa during the Pleistocene glaciations and the Holocene, *Palaeogeograph., Palaeoclimatol., Palaeoecol.*, *3*, 49, 1967.

Van Andel, Tj. H., G. R. Heath, T. C. Moore, and D. F. R. McGeary, Late Quaternary history, climate, and oceanography of the Timor Sea, northwestern Australia, *Am. J. Sci.*, *265*, 737, 1967.

Van der Hammen, Th., The Quaternary climatic changes of northern South America, *Ann. N. Y. Acad. Sci.*, art. 1, 676, 1961.

Van der Hammen, Th., T. A. Wijmstra, and W. H. van der Molen, Palynological study of a very thick peat section in Greece, and the Würm-glacial vegetation in the Mediterranean region, *Geol. Mijnb.*, *44*, 37, 1965.

Watts, W. A., A pollen diagram from Mud Lake, Marion County, north-central Florida. *Bull. Geol. Soc. Am.*, *80*, 631, 1969.

Wiles, W. W., 1967. Pleistocene changes in pore concentration of a planktonic forami-
niferal species from the Pacific Ocean, in *Progress in Oceanography,* edited by M.
Sears, vol. 4, p. 153, 1967.
Zeuner, F. E., *The Pleistocene Period,* 447 pp., Hutchinson, London, 1959.

G. Wollin, D. B. Ericson, and M. Ewing

7. LATE PLEISTOCENE CLIMATES RE-CORDED IN ATLANTIC AND PACIFIC DEEP-SEA SEDIMENTS

It is generally agreed that the last ice age ended about 11,000 years ago when the continental ice sheets disappeared from the more temperate parts of northwestern Europe and northeastern America. It is convenient to speak of the last 11,000 years as postglacial time even though this term may be a trifle euphemistic. It can hardly be said, for example, that postglacial time has come to Antarctica and Greenland or even to much of Alaska. For this reason we prefer to regard this important but as yet unsatisfactorily named climatic event not as a separate epoch but as a part of the Pleistocene.

Deevey and Flint (1957), Wiseman (1967), Dansgaard et al. (1969), and others have found evidence of major climatic variations in postglacial time. One of the aims of this article is to present a detailed study of the climate during this time as recorded in deep-sea sediment cores.

Under normal conditions the rate of sediment accumulation in the deep oceans varies from 1.5 to 3 cm in 1000 years. In consequence, the last 11,000 years are normally represented by a layer of sediment no thicker than 15 to 30 cm. Because burrowing animals mix together the upper 3 or 4 cm of sediment as it accumulates, records of climatic events of duration less than about 2000 years are normally unresolvable. However, exceptional cores contain records of postglacial climatic events which are resolvable and suitable for study. Because of peculiarities of local bottom topography and action of gentle bottom currents, the rate of accumulation of the fine fraction of normal deep-sea sediment, consisting mostly of clay particles, may be exceptionally fast. Presumably such exceptional rates of accumulation depend upon concentration of fine sediment transported or wafted by gentle bottom currents from an adjacent rise to a depression in the sea floor. This kind of rapid deposition is entirely different from the virtually instantaneous deposition by a turbidity current. Rapid accumulation in depressions should, we believe, be orderly, continuous, and at a constant rate. Apparently, wafting by deep currents is controlled mainly by currents of velocity too low to carry the rela-

Table 1. Geographical Location and Water Depth of the Cores

Core	Latitude	Longitude	Water depth (m)
SP9-3	53°52′N	21°06′W	2740
A167-8	33°13′N	73°39′W	4655
A179-4	16°36′N	74°48′W	2965
A179-15	24°48′N	75°55′W	3110
A180-47	15°19′N	17°55′W	2195
V9-30	06°01′N	36°39′W	4890
V12-122	17°00′N	74°24′W	2730
V15-32	03°15′S	82°30′W	2855
V21-146	37°04′N	163°02′E	3965
RC8-92	31°33′S	108°30′W	2710
RC8-93	29°22′S	105°14′W	3160
RC8-94	27°17′S	102°05′W	3075

tively large, heavy shells of foraminifera; these reach the site of deposition by settling vertically from the upper layers of water.

Lamont-Doherty Geological Observatory of Columbia University has a unique collection of deep-sea sediment cores from all the oceans. It contains more than 5000 cores, obtained in more than fifty expeditions. About fifteen cores in the collection contain postglacial sections that range from about 200 to 60 cm long and are suitable for detailed study of climate. The results from our study of seven of these cores are included in this article.

All the cores discussed here consist of foraminiferal lutite, a mixture of fine mineral particles from the continents and particles of calcium carbonate secreted by planktonic organisms. Table 1 gives the geographical locations and depths of water for the cores.

Two main conclusions were drawn from the study of the cores: (1) there were about twenty climatic oscillations in postglacial time; (2) comparison of Pleistocene climatic records (defined by variations in abundance of planktonic foraminifera in cores from the Pacific and from the Atlantic) suggests that times of warm surface water in some regions of the Pacific were partly synchronous with times of cool water in the Atlantic.

METHODS OF FAUNAL ANALYSIS

Ericson (1959), Bandy (1960), and Jenkins (1967) have presented evidence from the areal distribution of the dextral and sinistral forms of the planktonic foraminifer, *Globigerina pachyderma,* that the dextral form is tolerant of higher temperature than the sinistral. We have found that the coiling[1] of *Globorotalia truncatuli-*

---

1. We follow the gastropod convention in defining the direction of coiling. That is, a shell is right coiling if, when viewed from above with the dorsal side up, the chambers have been added in clockwise direction.

noides in the North Atlantic has responded to the climatic changes of the last 75,000 years, but unlike *Globigerina pachyderma*, left coiling dominated during times of warm climate and vice versa. Apparently the rule that sinistral races tolerate lower temperatures does not always hold true. Thus, for the first time we are using coiling ratios of both *G. pachyderma* and *Globorotalia truncatulinoides* as climatic indicators.

A third climatic indicator we are using is based on variations in abundance of the *Globorotalia menardii* group. The method generally employed in foraminiferal studies is to determine percentages of species in the total number of tests in the sample. This is time-consuming and, in the case of climatic studies, to some extent wasteful in that it necessarily involves counting all species, even though some of the most abundant are of little or no significance from the standpoint of climate.

In deep-sea sediments of particle-by-particle deposition, the material coarser than 74 micrometers in particle diameter consists almost entirely of the tests of planktonic foraminifera. Therefore, the ratio of the number of tests of a particular species to the weight of the material coarser than 74 micrometers may be substituted for percentages. Experience shows that plots of variation in this ratio from level to level in sediment cores closely parallel curves for variation in percentages of species calculated in the conventional way. The method has several advantages: (1) it results in a great saving of time; (2) it makes it possible to plot variations in the relative abundance of particular climatically sensitive species without having to count other species; (3) this in turn makes it possible to count larger samples, thereby increasing the statistical validity of the data.

The first step is to weigh to the nearest milligram the washed sample to be analyzed. Then the tests of the particular species to be studied are counted, and the ratio of the count to the weight (in milligrams) of the washed sample is calculated. This ratio, which may be called the frequency of the species, is, then, an index of productivity with respect to the total productivity of all planktonic species at the particular core level being studied.

Our counts of the *Globorotalia menardii* group and of *G. truncatulinoides*, *G. scitula*, *G. inflata*, *Pulleniatina obliquiloculata*, *Sphaeroidinella dehiscens*, and *Globigerinella aequilateralis* indicate that the *Globorotalia menardii* group is the most sensitive climatic indicator in the cores. High abundance of the *G. menardii* group indicates warm climate. This group includes the four subspecies *Globorotalia menardii menardii*, *G. m. tumida*, *G. m. flexuosa*, and *G. m. fimbriata*.

Lidz (1966), after having studied the foraminifera in a core from the Caribbean, concluded that "The *Globorotalia menardii* group, used by Ericson and associates to delineate Pleistocene temperatures, has been found to be an unreliable indicator of temperature. The major drawback in using the *G. menardii* group is that its specific and subspecific components appear to have considerably different temperature habitats." We believe that Lidz's conclusion is misleading.

The major component of the group, *G. m. menardii*, according to Lidz displays the best positive correlation with the curve of oxygen isotope analyses (O^{18}/O^{16} ratios) among the twenty-two species and subspecies counted in his study. Lidz's own curve of abundance of G. m. flexuosa is also in good agreement with the climatic curve based on isotopic ratios.

Another member of the group, G. m. fimbriata, is a strictly tropical form. It is always a very minor component of the G. menardii group—so much so that its inclusion in the group cannot have an appreciable effect on the reliability of the group as a whole.

Schott (1953) and Phleger et al. (1953) have shown that G. m. tumida is a tropical to midlatitude form in the Atlantic and adjacent seas. Bradshaw (1959), whom Lidz cites in support of his conclusion that the G. menardii group "have considerably different temperature habitats," in discussing the geographical range of G. m. tumida in the Pacific, writes: "This species is found only in the tropical regions. It appears to be less tolerant of cold temperatures than G. menardii and is not found so far north as that species." In our opinion Lidz's phrase "considerably different" implies rather more than Bradshaw actually says. In spite of Bradshaw's coupling of cold temperatures with G. menardii (G. m. menardii in our classification) in the quotation above, his chart shows clearly that G. menardii is hardly a cold-water form. There is nothing about its distribution in the Pacific that is inconsistent with what is known about its distribution in the Atlantic.

Recently J. Imbrie and N. G. Kipp of Brown University have counted the tests of thirty species and subspecies of planktonic foraminifera in a Caribbean core. They have then plotted by computer the counts as percentages of the total planktonic population. John van Donk of the Lamont-Doherty Geological Observatory has determined oxygen isotope ratios in the tests of Globigerinoides ruber and G. sacculifer in the same core. In Imbrie's opinion (Imbrie and Kipp, this volume) the curve of variation in abundance of the G. menardii group is in closer agreement with the curve of variation of the isotopic ratios than is that of any other single species or subspecies.

CLIMATIC INTERPRETATION OF THE ATLANTIC SEDIMENTS

Studies by Cushman and Henbest (1940), Phleger (1942), Ovey (1950), Schott (1953), Wiseman (1954), Parker (1958), Ericson et al. (1964), Ericson and Wollin (1968), and others, of the planktonic foraminifera in deep-sea sediment cores, have shown that the relative abundances of the species which are sensitive to temperatures vary from level to level. These investigators agree that the variations record shifts in the geographical ranges of the species and that these shifts were a consequence of the climatic changes of the late Pleistocene. Like Schott, we believe that the Globorotalia menardii group is the most sensitive and reliable of the climatic indicators.

Climate curves based on the variations in abundance of the G. menardii group, coiling ratios of Globigerina pachyderma, and coiling ratios of Globoro-

talia truncatulinoides in cores from the Atlantic and the Caribbean are shown in Figures 1–5. We have plotted variations in the coiling of *G. truncatulinoides* on a horizontal scale running from zero percentage of left coiling on the left to higher percentages of left coiling on the right in order to conform to the convention of plotting rising temperatures from left to right. Variations in coiling of *Globigerina pachyderma* are plotted in the usual way, that is, with left dominance on the left side of the diagrams and vice versa. The age determinations shown in the figures were made by the radiocarbon, protactinium-ionium, protactinium, and thorium-230 methods.

There is good agreement among the three methods used for climatic interpretation. All methods indicate that postglacial time started about 11,000 years ago.

We find evidence, particularly in core A180-47 (Fig. 1), of about twenty short climatic fluctuations superimposed upon the long-term climatic trend of postglacial time. The fact that in many cases several counts of the temperature-sensitive species at successive levels define single changes in the trends of the curves of Figure 1, in our opinion proves their reality, as opposed to merely random variations in abundance. We suggest that these minor fluctuations of about 550 years' duration correspond to the astronomic-eustatic oscillations cited by Fairbridge (1961).

The curves in Figures 1, 2, 3, and 4 indicate that there was a major shift from warm to colder climate about 3000 years ago in the Atlantic. Figures 4 and 5 show evidence of three intervals of mild climate during the Wisconsin glacial stage. According to variations in the coiling of *Globorotalia truncatulinoides* (Fig. 4), which give the finest resolution, the most recent of these had two temperature peaks. Core A179-4 (Fig. 5) fails to show this interval probably because of poor resolution. In view of expectable inaccuracies of dating and probable variations in rate of accumulation from level to level in the cores we can only estimate that these mild interludes occurred about 25,000, 40,000, and 65,000 years ago.

Our Atlantic and Caribbean climate curves are in good agreement with the climate curve of Dansgaard et al. (1969) obtained from oxygen isotope analyses of an ice core from the Greenland ice sheet.

COMPARISON OF PACIFIC AND ATLANTIC CLIMATES

The Pleistocene climatic record as defined by variations in abundance of planktonic foraminifera in sediment cores from the Atlantic and adjacent seas is rather clear. Micropaleontological studies by Cushman and Henbest (1940), Ovey (1950), Phleger et al. (1953), Schott (1953), Wiseman (1954), Parker (1958), Ericson et al. (1964), and others on many cores from the Atlantic, the Caribbean, and the Mediterranean revealed the occurrence of alternating warmer and colder stages. Correlations can be made among the various cores and between climatic zones in the cores and Pleistocene stages on the continents. In addition, oxygen

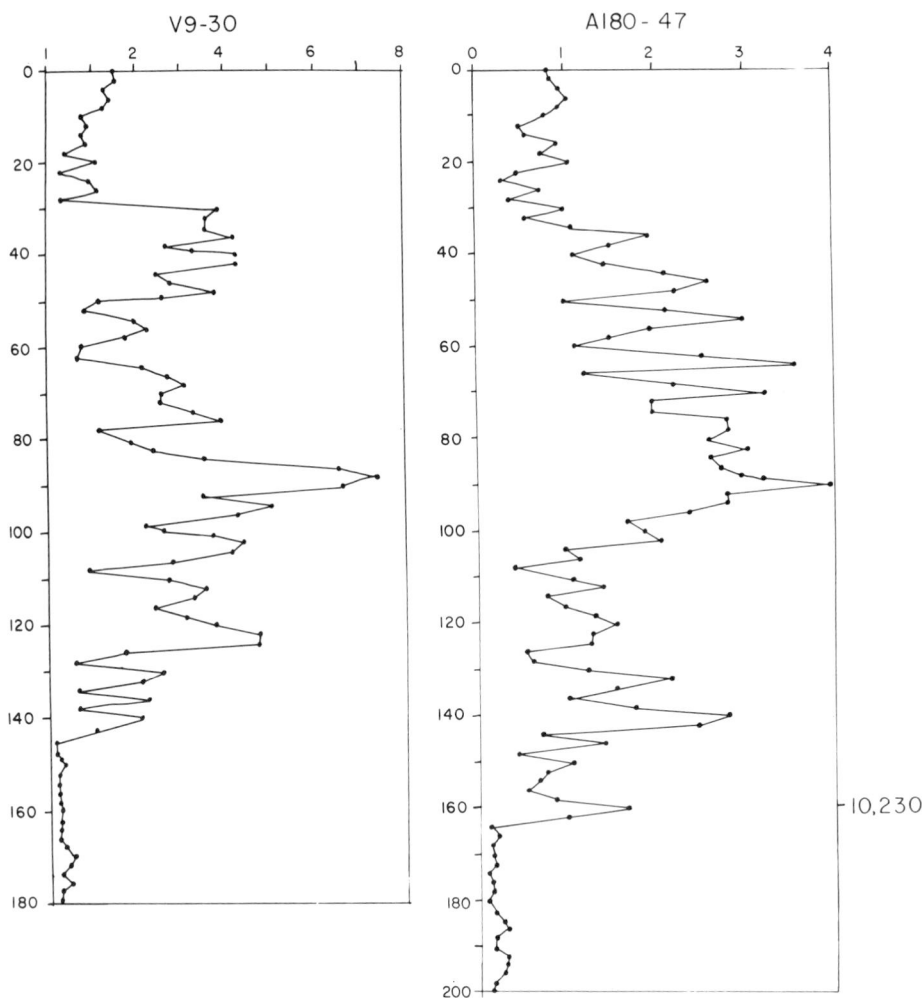

Fig. 1. Climate curves based on variations in the frequency of the *Globorotalia me-nardii* group in two cores from the Atlantic. The units at the top of the columns are ratios of the number of tests of the *G. menardii* group to the total weight in milligrams of the tests of all species of foraminifera in the samples. Since the magnitudes of these ratios differ from core to core, we have plotted the ratios for each core on a different scale in order to show the climatic changes more clearly. The number to the right of core A180–47 is a radiocarbon age. The horizontal line opposite the age indicates the mid-depth of the sample used for dating. Numbers to the left of the columns are depths in cores (in centimeters).

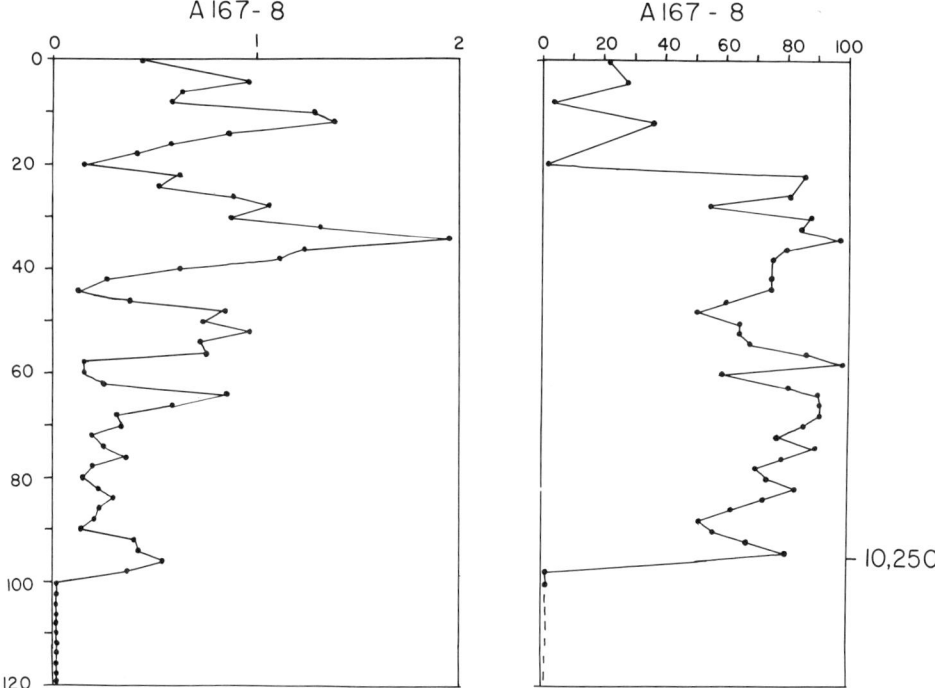

Fig. 2. Climate curves in a core from the Atlantic. The curve to the left is based on variations in the frequency of the *Globorotalia menardii* group. The units at the top of the left
column are ratios of the number of tests of the *G. menardii* group to the weight in milligrams of the tests of all species in the washed samples. The climate curve to the right is
based on percentages of left-coiling *G. truncatulinoides*. The number to the right is a
radiocarbon age. The horizontal line opposite the age indicates the mid-depth of the
sample used for dating. Numbers to the left are depths in the core (in centimeters).

isotope analyses by Emiliani (1966) of deep-sea cores from the Atlantic and the
Caribbean have yielded temperature records that correlate closely with each
other.

Although the Pacific Pleistocene climatic record is less clear than the Atlantic,
studies of climates in deep-sea cores have been interpreted to mean that the temperature variations of the surface waters of both oceans were synchronous. Our
study seems to indicate that during the last million years temperature fluctuations
in at least part of the Pacific have not been in step with temperature fluctuations
of the Atlantic.

So far, only relatively few cores from the Pacific have been analyzed micro-
paleontologically. Brotzen and Dinesen (1959) published climate curves based on
variations in abundance of planktonic foraminifera in two cores from the central
Pacific. In their opinion the climatic zones in these cores correspond directly to
similarly defined zones in a core from the Atlantic. However, according to their

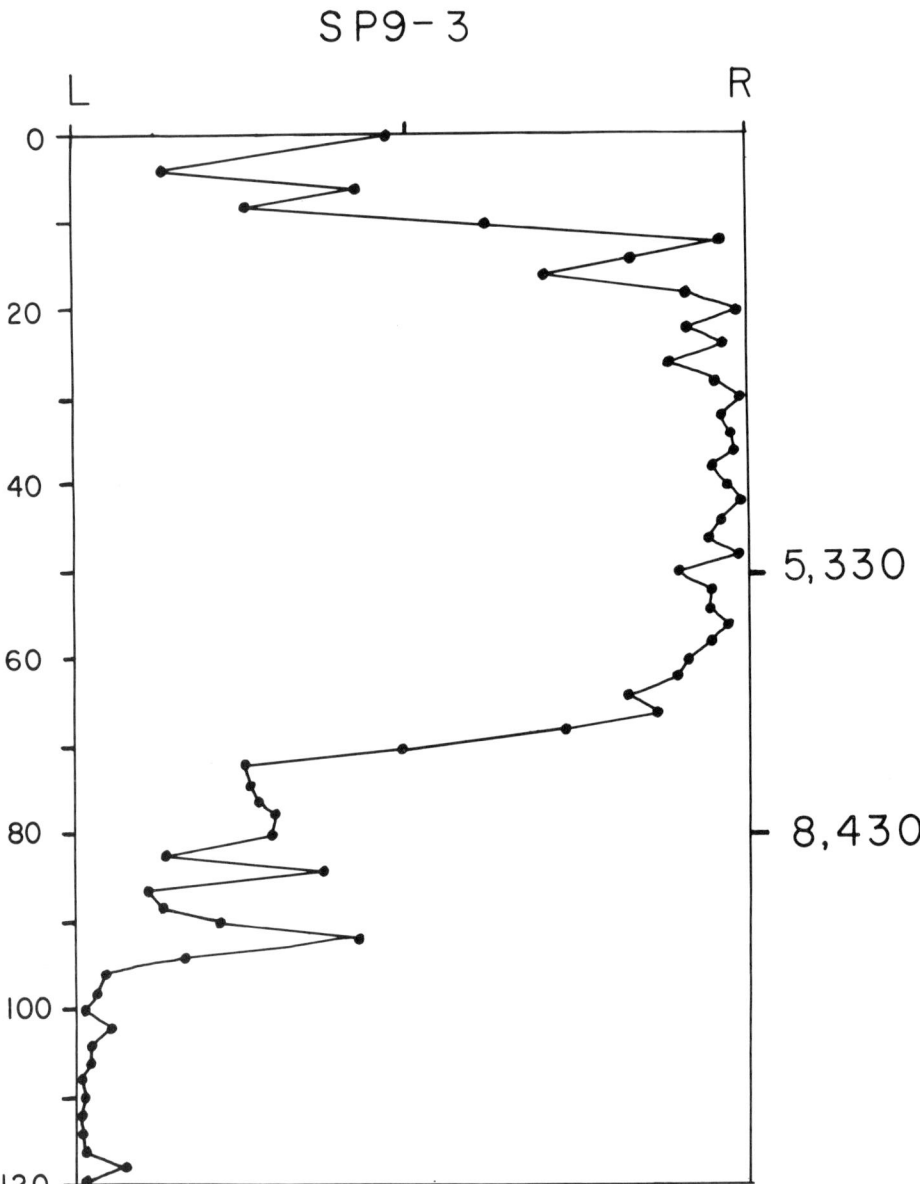

Fig. 3. Climate curve based on changes in the direction of coiling of *Globigerina pachyderma* in a core from the Atlantic. The scale runs from 100 percent left coiling at the lefthand margin of the column to 100 percent right coiling at the righthand margin. Numbers to the right of the column are radiocarbon ages. The horizontal lines opposite the ages indicate the mid-depths of the samples used for dating. Numbers to the left of the column are depths in the core (in centimeters).

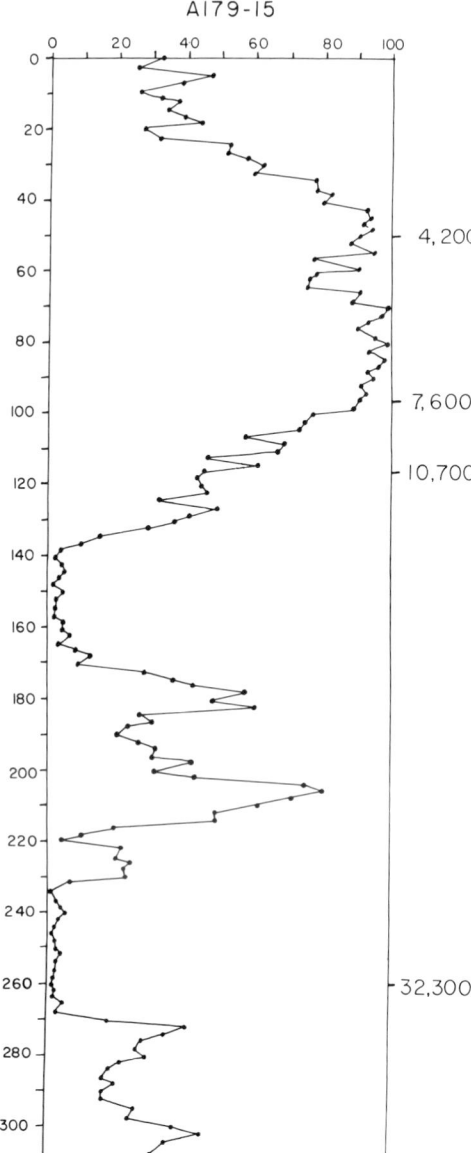

Fig. 4. Climate curve based on percentages of left-coiling *Globorotalia truncatulinoides* in a core from the Atlantic. Numbers to the right of the column are radiocarbon ages. The horizontal lines opposite the ages indicate the mid-depths of the samples used for dating. Numbers to the left of the column are depths in the core (in centimeters).

chronology of events in the Pacific, a maximum of warm conditions occurred at a time they estimate to have been between 18,000 and 20,000 years ago, after which temperature declined to a minimum about 7000 years ago. Since then temperature has risen steadily. But students of the climatic record in Atlantic cores are in accord that a temperature minimum is indicated at the level corresponding to about 18,000 to 20,000 years B.P. Thus, this study by Brotzen and Dinesen seems

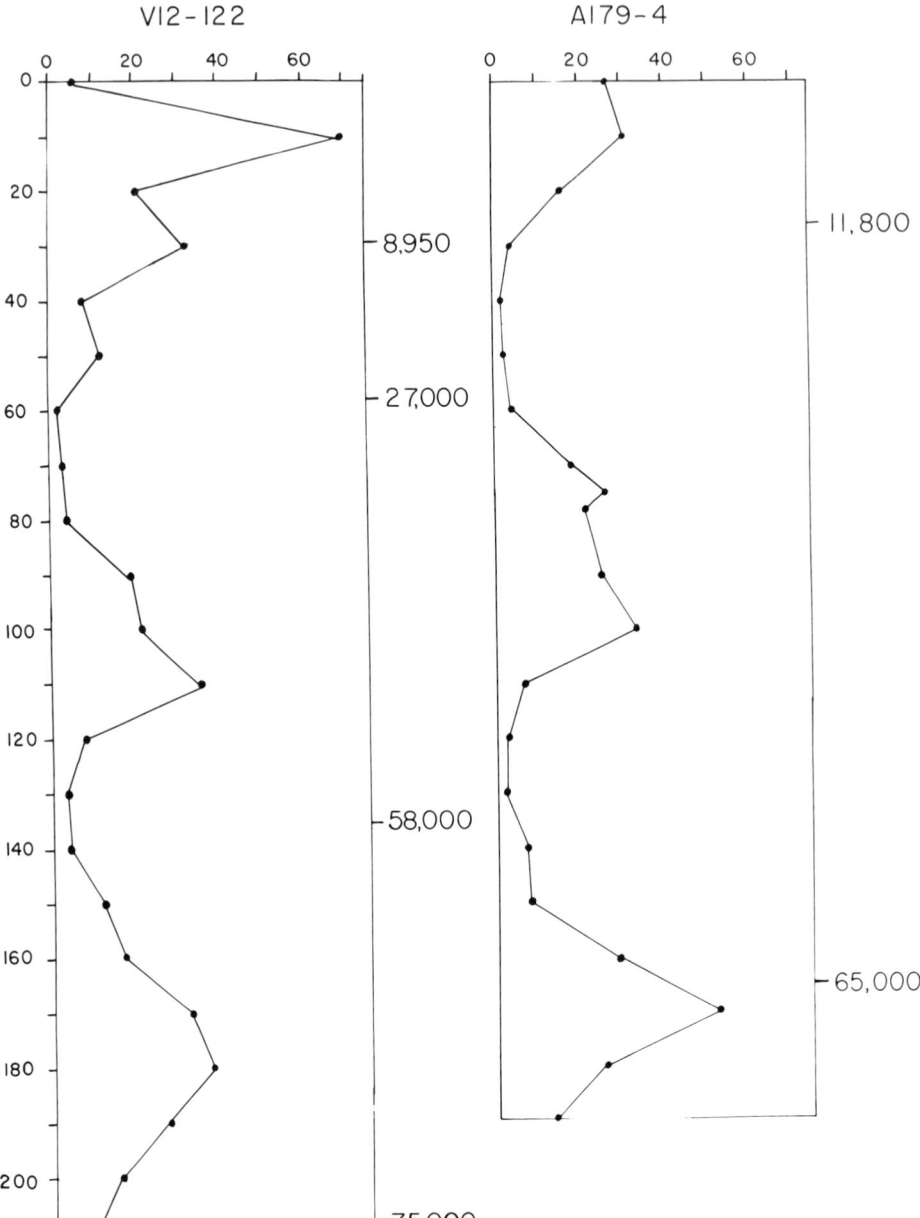

Fig. 5. Climate curves based on percentages of left-coiling *Globorotalia truncatuli-noides* and radiochemical age determinations in two cores from the Caribbean. Numbers to the left are depths in cores (in centimeters). The age of 58,000 was determined by Sackett (1965) by the protactinium method. Ku and Broecker (1966) determined the age of 75,000 by the thorium-230 method. The age of 65,000 was determined by Rosholt et al. (1961) by the protactinium-ionium method. The rest of the ages are radiocarbon dates. The horizontal lines opposite the ages indicate the mid-depths of the samples used for dating.

to show that, at least during the last 20,000 years, changes in temperature of the surface water of the two oceans have been more nearly opposed than synchronous.

Blackman and Somayajulu (1966) studied the foraminifera in two cores from stations near 20°S latitude and 560 and 890 km from the coast of Peru. They concluded that faunal zones representing alternations of warm and cool climates extending back some 200,000 years corresponded to similar zones in cores from the Atlantic. Blackman and Somayajulu's chart of the distribution of temperature-sensitive species of foraminifera in the southeast Pacific shows that the region in which their two cores were taken is exceptional in that a high-latitude assemblage is directly adjacent to the equatorial assemblage. Presumably the Peru current, a northward flowing offshoot of the Antarctic circumcurrent, is responsible for this exceptional condition. We surmise that conditions in this area during the Pleistocene have been dominantly influenced by growth and decay of the Antarctic continental ice sheet. If so, the sequence of climatic events in this particular area may have conformed with the sequence of the Atlantic instead of with that of the southeast Pacific in general, where direct insolation may have been the primary determinant of the temperature of surface water.

Hays et al. (1969) have found a correlation between variations in carbonate content in equatorial Pacific cores and the evidence of temperature fluctuations of surface waters of the Atlantic established by Emiliani (1955, 1961, 1966) and Ericson and Wollin (1968). Their correlation is based on oxygen isotope analyses and on variations in foraminifera respectively. From the assumption that sediment layers rich in carbonate were laid down during times when surface waters in the Pacific were cool, Hays and colleagues concluded that the temperature fluctuations of the waters of the Pacific and Atlantic were synchronous during the last 400,000 years.

Having studied cores from the North Atlantic and equatorial Pacific, Olausson (1967) concluded that the rate of accumulation of calcium carbonate in the North Atlantic was lower during the glacial stages and higher during the interglacial stages and that the opposite was true in the equatorial Pacific. According to Olausson a carbonate minimum in the North Atlantic is synchronous with a maximum in the equatorial Pacific.

Geochemical work by Arrhenius (1952) on cores from the equatorial Pacific revealed the occurrence of numerous alternating low-carbonate and high-carbonate layers. Arrhenius suggested that the low-carbonate sections were deposited during interglacial stages and the high-carbonate ones during glacial stages.

We agree with Arrhenius that during glacial stages atmospheric circulation would be particularly vigorous, and that this would lead to much upwelling of deep, nutrient-rich water with consequent increased productivity of planktonic organisms among which would be the lime-secreting planktonic foraminifera and coccolithophoridae. The deduction that, if the calcareous layers had accumulated during glacial stages, the surface water of the equatorial Pacific must have been cooler at times of copious carbonate precipitation, seemed to be confirmed when

Emiliani (1955) published oxygen isotope analyses of samples from the layers of maximum and minimum carbonate in three of the cores studied by Arrhenius; the layers of maximum carbonate contained slightly larger oxygen-18 ratios, implying slightly lower temperature.

However, the small differences in the isotope ratios between high- and low-carbonate stages led Emiliani to conclude that no more than approximately 40×10^6 km^3 of ice were formed during the glacial stages. He based this conclusion on the fact that a larger amount of ice would have changed the isotopic composition of sea water to such an extent as to make it necessary to apply a correction to the paleotemperatures, which would make temperatures during the ice ages significantly higher than temperatures during the interglacials, a condition which he considered unacceptable. Emiliani's estimate of ice formed seems to be excessively conservative; according to Donn et al. (1962) the volume of ice formed during the last glacial stage was between 70 and 84×10^6 km^3.

Emiliani also assumed that the delta value of ocean water during the glacial stages could not have been greater than +0.4 per mil because a larger delta value would mean that surface waters of the equatorial Pacific were warmer during the glacial stages. However, Dansgaard and Tauber (1969), basing their reasoning on oxygen isotope analyses of an ice core from the Greenland ice sheet, have concluded that the mean delta value of ocean water during the last glacial stage cannot have been less than +1.2 per mil.

Emiliani (1955) concluded that the temperature fluctuations of the waters of the Pacific and Atlantic were synchronous, but this new delta value together with a more realistic estimate of ice formed suggests that his oxygen isotope analyses of the Pacific cores can be interpreted to mean that the temperature variations of the surface water of the two oceans were not synchronous.

Climate curves based on the variations in abundance of the *Globorotalia menardii* group and coiling ratios of *Globigerina pachyderma* in two cores from the Pacific are shown in Figure 6. A comparison with the Atlantic curves shows that they are different.

In Figure 7 the frequency curves for the *G. menardii* group in three Pacific cores are compared with Ericson and Wollin's (1968) generalized climate curve based on variations in the frequency of the *G. menardii* group in cores from the Atlantic. The time scale used in Figure 7 is based on magnetic reversals in Atlantic and Pacific cores (Ericson and Wollin 1970). The climate curves indicate that temperature variations of the surface waters of a part of the Pacific during the last million years were not synchronous with temperature variations in the Atlantic which at times seem to have been opposed to temperature variations in the part of the Pacific from which the cores were raised.

These results are not easily reconciled with the Milankovitch theory of causation of the ice ages (see Emiliani 1955, 1961, 1966), but they harmonize with a modified version of Simpson's (1938) hypothesis. Simpson pointed out that accumulation of ice on the continents implied copious precipitation, which in turn

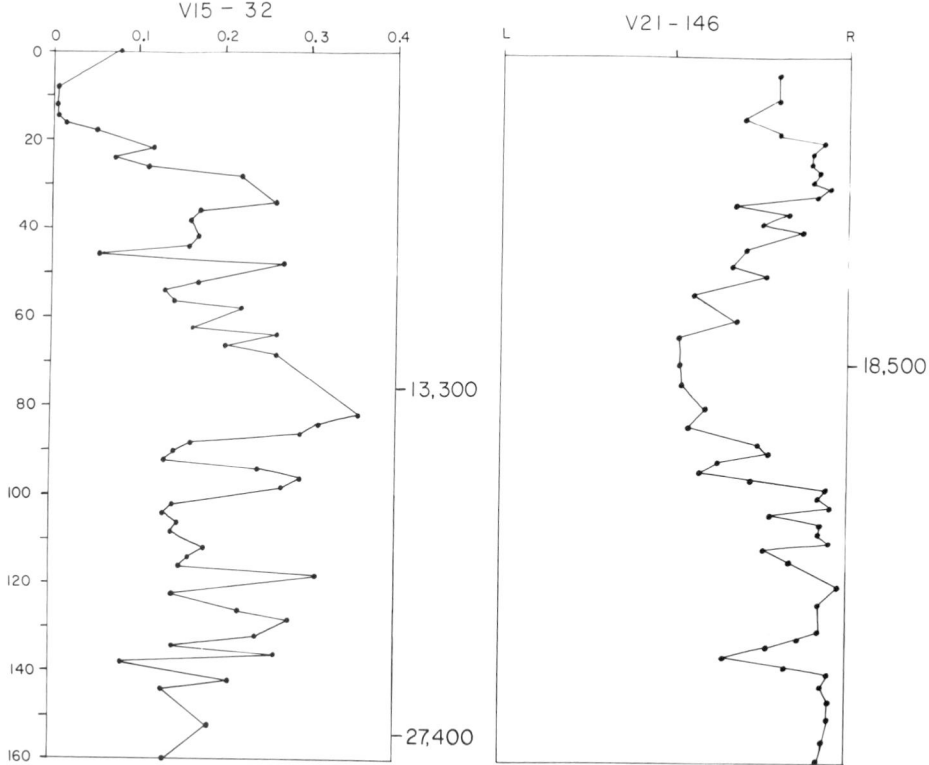

Fig. 6. Climate curves in two cores from the Pacific. The curve for core V15–32 is based on variations in the frequency of the *Globorotalia menardii* group. The units at the top of the left column are ratios of the number of tests of the *G. menardii* group to the weight in milligrams of the tests of all species in the washed samples. The climate curve for core V21–146 is based on changes in the direction of coiling of *Globigerina pachyderma*. The scale runs from 100 percent left-coiling at the lefthand margin of the column to 100 percent right-coiling at the righthand margin. Numbers to the right of the columns are radiocarbon ages. The horizontal lines opposite the ages indicate the mid-depths of the samples used for dating. Numbers to the left are depths in cores (in centimeters).

required copious evaporation from the oceans, a process dependent upon increased energy from the sun.

Apparently the surface waters of the North Atlantic were colder during the ice ages; they were probably cooled by the quantities of adjacent ice. Colder waters do not preclude, however, the possibility of increased insolation at times. Such insolation could have raised the temperature of the surface waters of the equatorial Pacific and thus have caused increase in evaporation in low latitudes to make possible the greater precipitation necessary for the accumulation of ice in high latitudes.

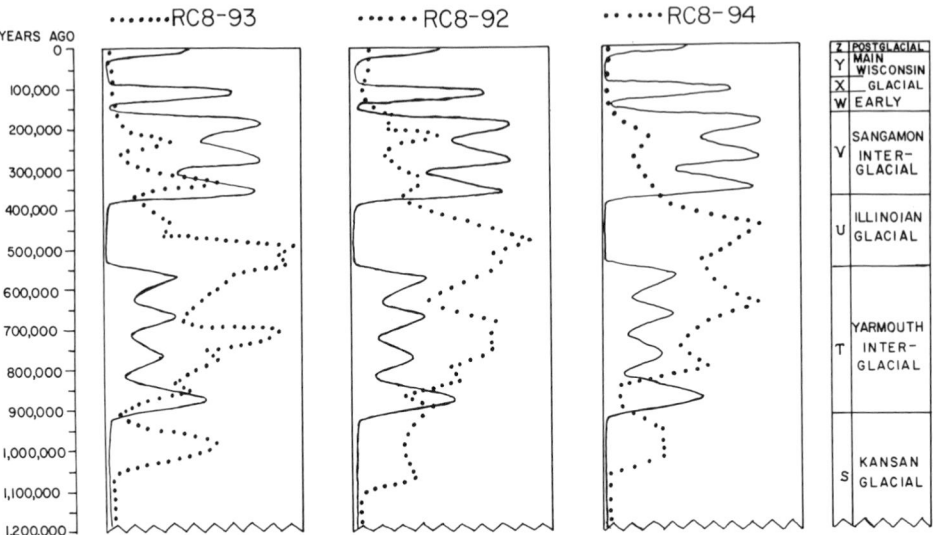

Fig. 7. Comparison of Pacific and Atlantic climate curves. Generalized climate curves (solid lines) are based on variations in the frequency of the *Globorotalia menardii* group in cores from the Atlantic (Ericson et al. 1964; Ericson and Wollin 1968) and the frequency curves of three Pacific cores (dotted lines). The correlation of the climatic zones with glacial and interglacial stages is for the generalized climate curve of the Atlantic only. The climatic zones are designated by letters according to the system devised by Ericson and Wollin (1968). The time scale at left is based on magnetic reversals in the Atlantic and Pacific cores (Ericson and Wollin 1970).

Ericson and Wollin first proposed this idea in 1964. Since then, Olausson (1965) has attempted in a similar way to reconcile the Simpson hypothesis with the seemingly contradictory evidence of lower temperatures in the Atlantic during the ice ages.

Our tentative assumption is that, in contrast to the Atlantic, the Pleistocene climatic record in the Pacific varies from area to area. In order to get a clear picture of the Pleistocene climatic record in the Pacific, many more cores need to be studied.

Summary. The climatic records of the past in this study are based on variations in abundance of the *Globorotalia menardii* group, coiling ratios of *Globigerina pachyderma*, and coiling ratios of *Globorotalia truncatulinoides* in deep-sea sediment cores. Good agreement was found among these three planktonic foraminiferal indicators for interpreting climates. In cores from the Atlantic about twenty climatic oscillations with periods of about 550 years are superimposed on the long-period trend of postglacial time. The climatic records in Atlantic and Caribbean cores indicate that there were maximums of warm conditions about 25,000, 40,000, and 65,000 years ago during the Wisconsin glacial stage.

Comparison of the Pleistocene climatic records of the last million years in cores from the Pacific with similar records in cores from the Atlantic suggests that times of warm surface water in some regions of the Pacific were partly synchronous with times of cool water in the Atlantic.

Acknowledgments. We thank Janet Wollin, Sandella Walker, and Julie McGowan for able assistance in the laboratory investigations. We are grateful to W. S. Broecker, D. L. Thurber, and Hans Suess for radiocarbon determinations. The collection, preservation, and analysis of the cores were supported in large part by grants and contracts with the National Science Foundation (GA–982, GA–3959, and GA–15639) and the Office of Naval Research (contract N00014–67–A–0108–0004). Lamont-Doherty Geological Observatory Contribution no. 1536.

REFERENCES

Arrhenius, G., Sediment cores from the east Pacific, *Swedish Deep-Sea Exped., 1947–1948, Repts.,* 5, fasc. 1, 1, 1952.

Bandy, O. L., The geological significance of coiling ratios in the foraminifer *Globigerina pachyderma* (Ehrenberg), *J. Paleontol.,* 34(4), 671, 1960.

Blackman, A., and B. L. K. Somayajulu, Pacific Pleistocene cores: Faunal analyses and geochronology, *Science,* 154, 886, 1966.

Bradshaw, J. S., Ecology of living planktonic foraminifera in the north and equatorial Pacific Ocean, *Contrib. Cushman Found. Foram. Res.,* 10, pt. 2, 25, 1959.

Brotzen, F., and A. Dinesen, On the stratigraphy of some bottom sections from the central Pacific, *Swedish Deep-Sea Exped., 1947–1948, Repts.,* 10, fasc. 3, 43, 1959.

Cushman, J. A., and L. G. Henbest, Geology and biology of North Atlantic deep-sea cores between Newfoundland and Ireland, *U.S., Geol. Surv., Profess. Papers,* 196–A, pt. 2, Foraminifera, 35, 1940.

Dansgaard, W., and H. Tauber, Glacier oxygen-18 content and Pleistocene ocean temperatures, *Science,* 166, 499, 1969.

Dansgaard, W., S. J. Johnson, J. Møller, and C. C. Langway, Jr., One thousand centuries of climatic record from Camp Century on the Greenland ice sheet, *Science,* 166, 377, 1969.

Deevey, E. S., and R. F. Flint, Postglacial hypsithermal interval, *Science,* 125, 182, 1957.

Donn, W. L., W. R. Farrand, and M. Ewing, Pleistocene ice volumes and sea-level lowering, *J. Geol.,* 70(2), 206, 1962.

Emiliani, C., Pleistocene temperatures, *J. Geol.,* 63(6), 1955.

Emiliani, C., Cenozoic climatic changes as indicated by the stratigraphy and chronology of deep-sea cores of *Globigerina*-ooze facies, *Ann. N. Y. Acad. Sci.,* 95(1), 521, 1961.

Emiliani, C., Paleotemperature analysis of Caribbean cores P6304–8 and P6304–9 and a generalized temperature curve for the past 425,000 years, *J. Geol.,* 74(2), 109, 1966.

Ericson, D. B., Coiling direction of *Globigerina pachyderma* as a climatic index, *Science,* 130, 219, 1959.

Ericson, D. B., and G. Wollin, *The Deep and the Past,* Alfred A. Knopf, New York, 1964.

Ericson, D. B., and G. Wollin, Pleistocene climates and chronology in deep-sea sediments, *Science,* 162, 1227, 1968.

Ericson, D. B., and G. Wollin, Pleistocene climates in the Atlantic and Pacific oceans: A comparison based on deep-sea sediments, *Science, 164,* 1483, 1970.

Ericson, D. B., M. Ewing, and G. Wollin, The Pleistocene epoch in deep-sea sediments, *Science, 146,* 723, 1964.

Fairbridge, R. W., Convergence of evidence on climatic change and ice ages, *Ann. N. Y. Acad. Sci., 95*(1), 542, 1961.

Hays, J. D., T. Saito, N. D. Opdyke, and L. H. Burckle, Pliocene-Pleistocene sediments of the equatorial Pacific: Their paleomagnetic, biostratigraphic, and climatic record, *Bull. Geol. Soc. Am., 80,* 1481, 1969.

Jenkins, D. G., Recent distribution, origin, and coiling ratio changes in *Globorotalia pachyderma* (Ehrenberg), *Micropaleontology, 13*(2), 195, 1967.

Ku, T. L., and W. S. Broecker, Atlantic deep-sea stratigraphy: Extension of absolute chronology to 320,000 years, *Science, 151,* 448, 1966.

Lidz, L., Deep-sea Pleistocene biostratigraphy, *Science, 154,* 1448, 1966.

Olausson, E., On the Pleistocene stratigraphy and climate, *Intern. Quatern. Congr., VII, Abstr.,* p. 367, 1965.

Olausson, E., Climatological, geoeconomical and paleooceanographical aspects on carbonate deposition, *Progress in Oceanography, 4,* 245, 1967.

Ovey, C. D., On the interpretation of climatic variations as revealed by a study of samples from an equatorial Atlantic deep-sea core, *Centenary Proc. Roy. Meteorol. Soc.,* 211, 1950.

Parker, F. L., Eastern Mediterranean foraminifera, *Swedish Deep-Sea Exped., 1947–1948, Repts., 8,* fasc. 4, 217, 1958.

Phleger, F. B., Foraminifera of submarine cores from the continental slope, *Bull. Geol. Soc. Amer., 53,* 1395, 1942.

Phleger, F. B., F. L. Parker, and J. F. Peirson, North Atlantic foraminifera, *Swedish Deep-Sea Exped., 1947–1948, Repts., 7,* fasc. 1, 1, 1953.

Rosholt, J. N., C. Emiliani, J. Geiss, F. F. Koczy, and P. J. Wangersky, Absolute dating of deep-sea cores by the Pa^{231}/Th^{230} method, *J. Geol., 69*(2), 162, 1961.

Sackett, W. M., Deposition rates by the protactinium method, *Symposium on Marine Geochemistry, Univ. of Rhode Island, Occasional papers,* no. 3, 29, 1965.

Schott, W., Die Foraminiferen in dem äquatorialen Teil des atlantischen Oceans, *Wiss. Ergeb. Deut. Atlant. Exped. 'Meteor' 1925–27,* Bd. III, T. 3, Lf. 1, B., 43, 1953.

Simpson, G. C., Ice Ages, *Nature, 141,* 591, 1938.

Wiseman, J. D. H., The determination and significance of past temperature changes in the upper layer of the equatorial Atlantic Ocean, *Proc. Roy. Soc. (London), Ser. A, 222,* 296, 1954.

Wiseman, J. D. H., Evidence for recent climatic change in cores from the ocean bed, *Proc. Roy. Meteorol. Soc.,* International Symposium on World Climate from 8000 to 0 B.C., London, p. 83, 1967.

Kenneth Hunkins, Allan W. H. Bé,
Neil D. Opdyke, and Guy Mathieu

8. THE LATE CENOZOIC HISTORY OF THE
ARCTIC OCEAN

The Arctic Ocean is the only ocean which at present supports a permanent ice cover. The ice pack of other oceans is only a seasonal phenomenon; for example, the extensive winter ice cover of the seas surrounding Antarctica shrinks to a narrow fringe around the continent in summer. The Arctic Ocean ice pack covers all the central part of the ocean in the winter, with some open water developing around the shores in summer (Fig. 1). The borders of the pack advance and retreat hundreds of kilometers through each change of seasons. Arctic pack ice averages 3 meters in thickness with a seasonal variation of about 10 percent. This thin veneer of ice has undoubtedly had an important function in controlling climate in the past. Since it fluctuates considerably even today, it is possible that it disappeared completely in the past and may do so again in the future. Removal of the Arctic ice sheet would change the albedo of this ocean, 13 million square kilometers in area, from about 0.6 to 0.1, with profound effects on global climate. Evaporation from the waters of an open Arctic Ocean would be much increased, perhaps leading to major glaciation on the surrounding continents.

The history of the Arctic Ocean climate, and particularly a record of the presence or absence of ice cover, should be preserved in the sediments of the ocean floor. Disappearance of the ice cover would have strongly influenced sedimentation. In an open Arctic Ocean productivity of plant and animal life would be increased, leading to greater skeletal carbonate deposition. Ice-rafted gravels would cease to be deposited in a completely open ocean with no floes, ice islands, or icebergs. A partial ice cover would enhance transport and deposition of rafted material. However, completely rigid ice cover would again inhibit ice rafting as well as productivity.

Sampling of ocean sediments has been an important part of scientific programs at the drifting ice stations which have been maintained in the Arctic Ocean by both the United States and the Soviet Union. The results of early Soviet work were summarized by Belov and Lapina (1960). They discussed the physical, chemical, and biological characteristics of 231 cores representing a wide coverage

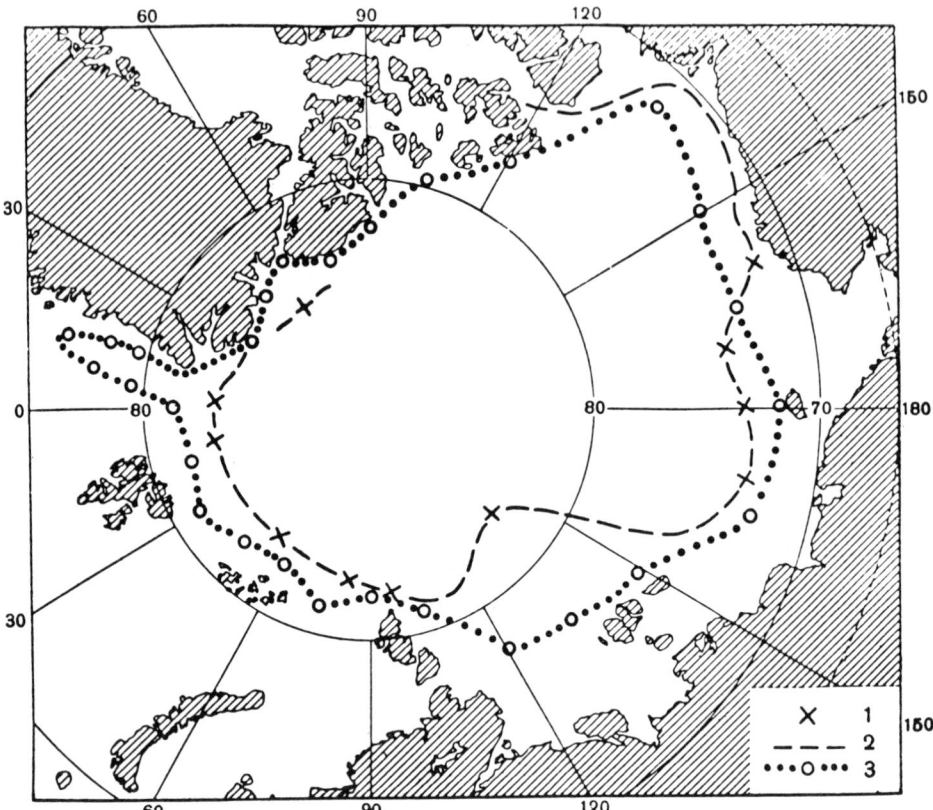

Fig. 1. Extent of the Arctic Ocean ice cover. 1. Points reached by ships in free sailing. 2. Conventional limit of free sailing. 3. Limit of pack ice in August (Gordienko and Laktionov 1969).

of the Arctic basin. Cores collected from U.S. stations, Alpha and T-3, were analyzed by Ericson et al. (1964) with special attention given to the microfauna. Further paleontological descriptions of some of the Alpha cores were reported by Herman (1964, 1969).

The establishment of a time scale within the cores is essential for their climatic interpretation. Radioisotope dating with C^{14} and uranium-series isotopes was described by Hunkins and Kutschale (1967) and by Ku and Broecker (1967). They found a very low rate, about 2 mm per 1000 years, for pelagic sedimentation over the past 70,000 years. The paleomagnetic method of dating based on past reversals of the earth's field has given similar average rates over much longer time spans. Lin'kova (1965) obtained an average rate on the Lomonosov ridge of 1.5 mm per 1000 years for the past 700,000 years. Steuerwald et al. (1968) measured a rate of 2 mm per 1000 years on the Mendeleyev ridge over the same time span for a suite of cores collected by the U.S. Geological Survey.

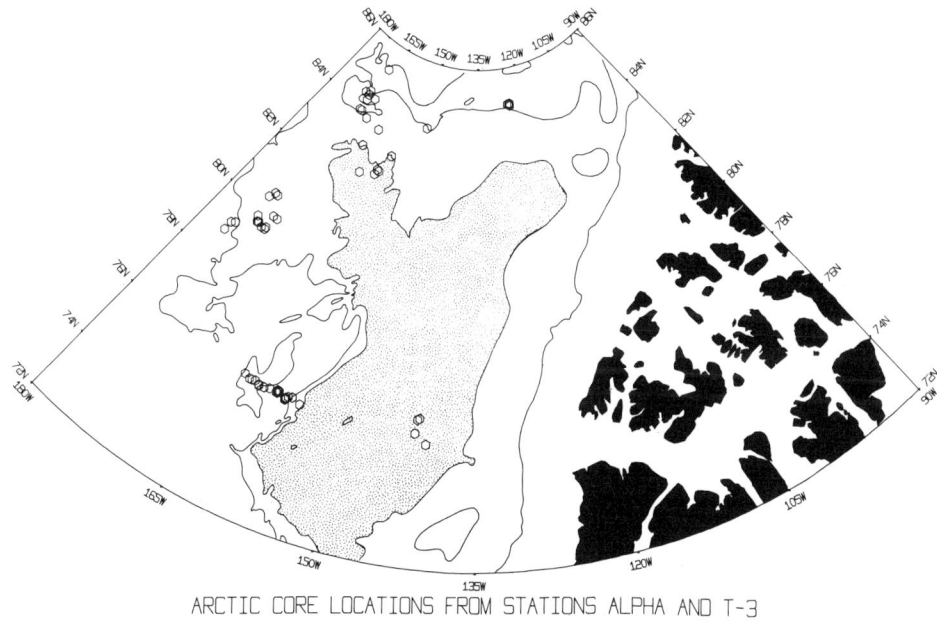

ARCTIC CORE LOCATIONS FROM STATIONS ALPHA AND T-3

SHOWN WITH 2000 AND 3500 METER CONTOURS

Fig. 2. Locations of cores raised in the Arctic Ocean by Lamont-Doherty Geological Observatory from ice stations T-3 and Arlis II. Canada Abyssal Plain is represented by the stippled area enclosed within the 3500-meter contour.

The oxygen isotope composition of both planktonic and benthonic foraminifera from Arctic cores has been determined in order to interpret past conditions (Van Donk and Mathieu 1969). They conclude that the temperature of the water in which the planktonic foraminifera live has not changed significantly over the past 25,000 years.

This paper summarizes past studies and presents new data on cores collected from ice island T-3. Sites in the Arctic Ocean from which cores have been raised by the Lamont group are shown in Figure 2. A piston corer has been used for most of the work and core lengths are usually several meters long. A list of the cores discussed here is given in Table 1.

CORE LITHOLOGY AND SEDIMENTARY PROCESSES

Sediment types resulting from four different processes are recognized in Arctic Ocean cores. These are: (1) normal pelagic sediments accumulated from the slow and continuous settling of mineral particles derived from land and of the hard parts of planktonic organisms; (2) ice-rafted sediments, an unsorted mixture of rock types, which have been carried from shore by ice and later dumped into deep water; (3) turbidity current deposits which have been carried rapidly from

Table 1. Sediment Cores from the Arctic Ocean

Core no.	Lat. N	Long. W	Depth (m)	Core length (cm)
T3/63-1	83°02′	163°34′	3437	275
T3/67-6	79°44′	173°04′	2815	432
T3/67-9	79°38′	172°07′	2237	356
T3/67-10	79°36′	172°21′	2688	310
T3/67-11	79°35′	172°30′	2810	250
T3/67-12	80°21′	173°31′	2867	374
T3/69-03	84°58′	124°50′	1726	205

shallow water and laid down in graded beds in the deepest parts of the basin; (4) volcanic glass shards. Most of the cores raised from Arctic ridges and rises contain the first two kinds of sediment. These cores are most useful for climatic studies since they are undisturbed by turbidity currents. Only on abyssal plains are turbidity current deposits generally encountered. The cores from ridges and rises show color banding with alternating dark brown and light tan layers. Clay-sized material makes up most of the sediment in all Arctic cores. Variable amounts of microfossils are present in the clay matrix, principally foraminifera with the planktonic species *Globigerina pachyderma* predominating. The dark layers generally contain a higher percentage of foraminiferal tests than the light tan layers.

Also present are variable amounts of unsorted material, sands, and gravels which are classified as ice-rafted. A high percentage of gravels and sands has also been noted in dredges and in bottom photographs from the Arctic Ocean (Hunkins et al. 1960; Schwarzacher and Hunkins 1961). The roundness, shape, and striation of the dredged gravels are indicative of glacial origin, derived primarily from sedimentary rocks. The ice-rafted sediments are believed to have been transported by ice islands, which are tabular ice sheets, thicker than sea ice, that have been broken from the Ellesmere ice shelf and are now floating within the pack. Glacial till material has been observed on two: T-3 and Arlis II. Arlis II split while a party was aboard and large quantities of gravel were dumped into the ocean. This provided a graphic illustration of how such material is transported and deposited.

The Upper Foraminifera-rich Layer

Good correlation was found between the various peaks of foraminiferal abundance among cores on the Alpha Cordillera (Ericson et al. 1964). There is a dark brown layer, rich in foraminifera, at the top of the cores and extending to a depth of 10 or 15 cm in all Alpha Cordillera cores examined by Ericson and colleagues. This upper dark layer has been well studied since it correlates between cores and should relate to the most recent climatic events. Lapina and Belov (1960) reported the presence of this layer in all 231 cores from various parts of the Arctic basin. They found that the average thickness ranged from 9.7 cm in the Canada basin to a maximum of 16.3 cm on the crest of the Alpha Cordillera.

The first attempts to determine range in time of this layer were made by the radiocarbon method on *G. pachyderma* tests from Alpha Cordillera dredge samples and showed that the topmost samples were quite young. Results of radiocarbon age-dating on carbonate from arctic cores are summarized in Table 2. A date of 9300 ± 180 years B.P. was obtained for dredge sample L-501, which presumably represents the top few centimeters of sediment. A similar sample from the Chukchi Plateau, L-565 A, gave a date of 4800 ± 700 years B.P. Both dates are within the 11,000-year span of postglacial time. Further confirmation that this upper layer extends to the present was found in the date of 700 ± 100 years B.P. for a dredge sample from the Canada Abyssal Plain (Hunkins and Kutschale 1967). Gravity core samples taken near the same location as this dredge showed that approximately 3 mm of dark brown foraminiferal lutite, like that from the top layer on ridges, overlies a long section of gray unfossiliferous lutite. The gray layer is considered to be a turbidity current deposit and extends to a depth of at least 5 meters. The last turbidity current must have occurred 1400 years ago, double the mean age of 700 years given by the radiocarbon date. Since that time normal pelagic sedimentation has produced the upper 3-mm layer.

A date for the base of this upper layer has not been firmly fixed but it appears to be near the beginning of the classical Wisconsin glaciation. A composite sample from the basal 3 cm of this layer made up from four cores (DSA 3,4,5, and 6) was dated as 25,000 ± 3000 years B.P. Another composite sample, taken with a multibarrel cover, of the basal 1 cm of the layer was too old to date—greater than 30,000 years B.P. Almost the same level was dated from the same multibarrel core by Ingrid Olsson (personal communication) as 12,400 ± 1400 years B.P. (Van Donk and Mathieu 1969). Some of the discrepancies between these dates may be attributed to the difficulty in controlling precisely the sampling level since the rate of deposition is so low that short core intervals span large periods of time.

Dating with uranium series isotopes gives a considerably greater age for the base of the upper layer than do the previous radiocarbon dates. On the basis of eight samples from core T3/63-1, Ku and Broecker (1967) estimated the transition from foraminifera-poor to foraminifera-rich sediments (at 15 cm in this core) to have taken place 70,000 years ago, with a probable error no greater than 30 percent. A similar measurement on T3/66 placed the transition at about 100,000 years ago. These dates appear more acceptable than the radiocarbon dates since they are in better accord with the paleomagnetic dates, as will be shown later. It is likely that the radiocarbon dates are systematically too young because of the reworking and mixing of the sediment by burrowing animals. Any reworking would be expected to bring material from both above and below into a given layer. However, foraminifera are much more abundant above the transition and

Table 2. Summary of C[14] Data on Arctic Deep-Sea Sediments

Lamont no.	Sample type	Location	Depth (cm)	C[14] age (yrs B.P.)	Estimated sed. rate (mm/1000 yrs)
L-960	Combined 2 dredges*	80°32'N, 140°07'W 80°23'N, 140°03'W	top 0.3	700 ± 100	2
L-565A	Trawl‡	77°52'N, 163°W	ca. top 5	4,800 ± 700	—
L-501	Dredge†	84°22'N, 148°51'W	ca. top 5	9,300 ± 180	—
L-508	Combined 4 cores† −3 −4 −5 −6	84°12'N, 168°33'W 84°21'N, 168°49'W 84°28'N, 169°04'W 85°15'N, 167°54'W	7–10	25,000 ± 3,000	3
L-1001	Combined multibarrel core samples T3-63-1S	82°56'N, 155°54'W	7–8	> 30,000	< 3
	T3-63-1S‡	" "	1 + 2 outer 1 + 2 inner 3 + 4 outer 3 + 4 inner 6 whole	6,500 ± 500 8,360 ± 190 10,600 ± 300 13,800 ± 400 12,400 ± 1,200	2.3 1.8 3.3 2.5 4.8
L-1213	T3-66W§	75°41'N, 157°29'W	2.5 6.5 10.5	6,700 ± 250 10,500 ± 850 14,250 ± 1,400	3.7 6.2 7.4

* Hunkins & Kutschale 1967.
† Olson & Broecker 1961.
‡ Ingrid Olsson, unpublished.
§ Van Donk & Mathieu 1969.

hence the mixing would bias the dates toward the more youthful direction. Uranium series and paleomagnetic dates would not be affected by the burrowing since they do not depend upon carbonate material.

The deposition of this upper layer thus must have begun 70,000 to 100,000 years ago and have continued to the present. Its beginning coincides approximately with the commencement of the classical Wisconsin glaciation (80,000 years B.P.). However, there is no marked change in sediment type at the close of the Wisconsin (11,000 years B.P.), as there is in cores from the Atlantic Ocean. In the Arctic Ocean sedimentation continues unchanged from the Wisconsin into the Recent.

PALEOMAGNETIC STRATIGRAPHY

The earth's magnetic field has undergone a number of reversals in polarity in the past, and the dates of these reversals have now been well established by potassium-argon dating of volcanic sequences (Cox 1969). Ocean sediments also retain the polarity of the earth's field at the time they were deposited, thus providing an important method of dating and correlating bottom cores. Lin'kova (1965) found magnetic reversals in five of six cores from the Lomonosov ridge which she examined. The change from normal to reversed polarity, representing the base of the Brunhes normal epoch, occurred at a depth of 110 cm.

Four Mendeleyev ridge cores showed magnetic reversals, corresponding to the base of the Brunhes epoch, at depths ranging from 110 to 165 cm (Steuerwald et al. 1968).

Magnetic inclination and declination were measured on five Lamont cores from the Mendeleyev ridge and on two cores from the Alpha Cordillera. The 1-cm^3 samples were first subjected to an alternating field of 50 oersteds to remove unstable magnetization before measurement with the Lamont spinner magnetometer (Foster 1966). Results for core T3/67-6 are shown in Figure 3. This core was sampled at 5-cm intervals, and the reversal at the base of the Brunhes epoch is clearly evident at a depth between 220 and 225 cm in both the declination and inclination plots. This is one of the best magnetic records from cores run at Lamont. Even so the data exhibit considerable scatter and other cores show even more.

The Brunhes-Matuyama reversal is at 148 cm in T3/63-1 from the Alpha Cordillera, but the data for 69-3 are not considered reliable enough to determine the reversal (Fig. 4). The same reversal is found at depths of 195, 112, and 90 cm in cores 67-9, -11, and -12 (Figs. 5, 6). Data from 67-10 are not considered to be reliable. The presence of ice-rafted pebbles in the cores probably accounts for the somewhat erratic magnetic results. A single pebble, disoriented because of its large size, would be sufficient to invalidate a sample.

The paleomagnetic data are in fairly good accord, as shown in Table 3 where the average sedimentation rates over the Brunhes epoch are presented. The average rates for three investigations by different authors on different cores are in

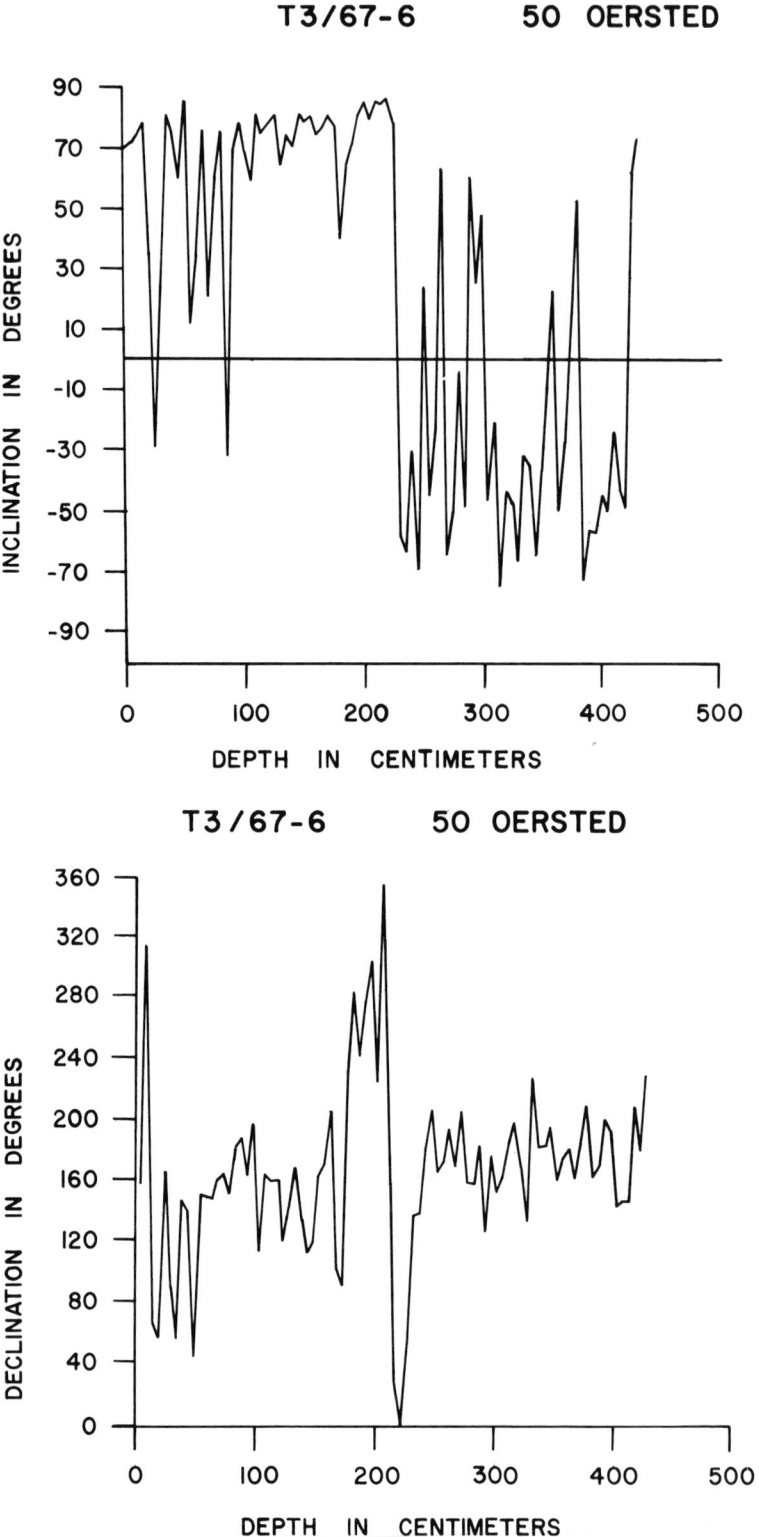

Fig. 3. Upper: Magnetic inclination for core T3/67-6 determined by spinner mag
tometer. Note reversal at about 220 cm. Lower: Magnetic declination for the same c
Note rapid change at about 220 cm.

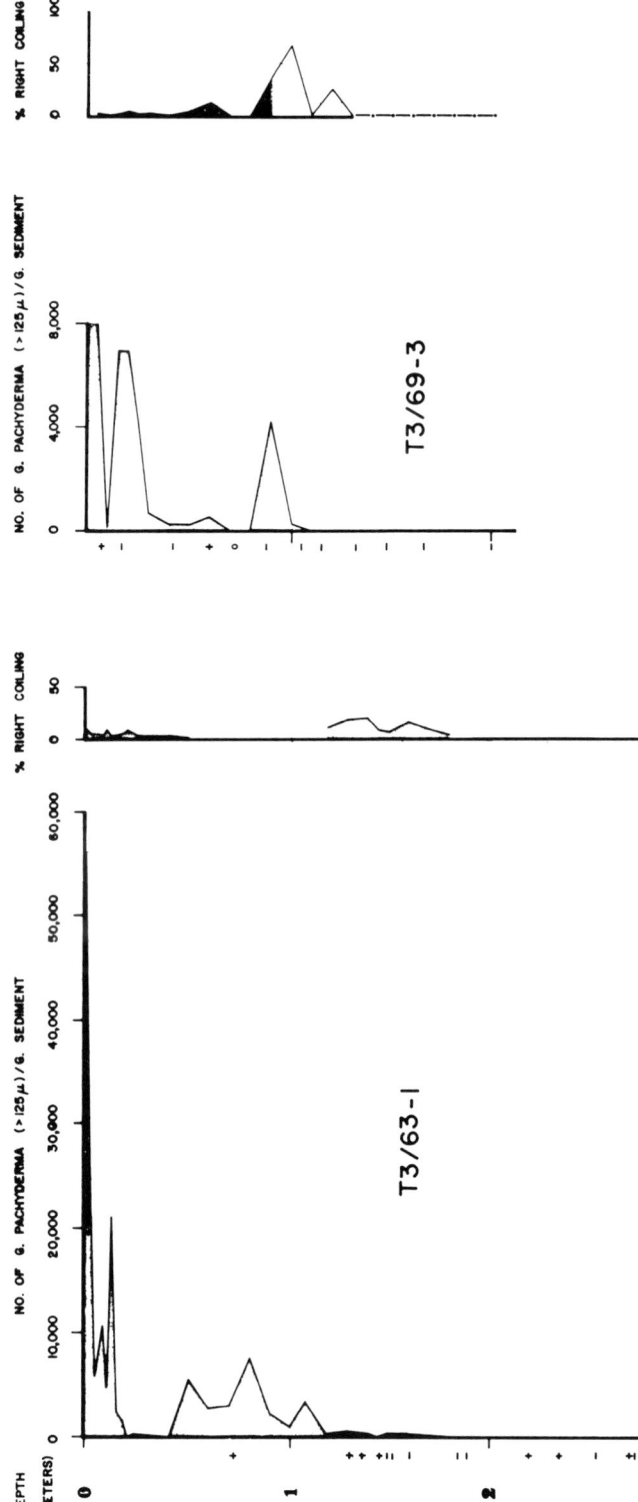

Fig. 4. Fluctuations in abundance of *Globigerina pachyderma* and their coiling direction in cores T3/63-1 and T3/69-3 from the Alpha Cordillera. Normal and reversed magnetic polarity is indicated by + and −. Sampling interval for microfossils was 10 cm.

Fig. 5. Fluctuations in abundance of *Globigerina pachyderma* and their coiling direction in cores T3/67-9 and T3/67-10 from the Mendeleyev ridge. Normal and reversed magnetic polarity are indicated by + and −. Sampling interval for microfossils was 10 cm.

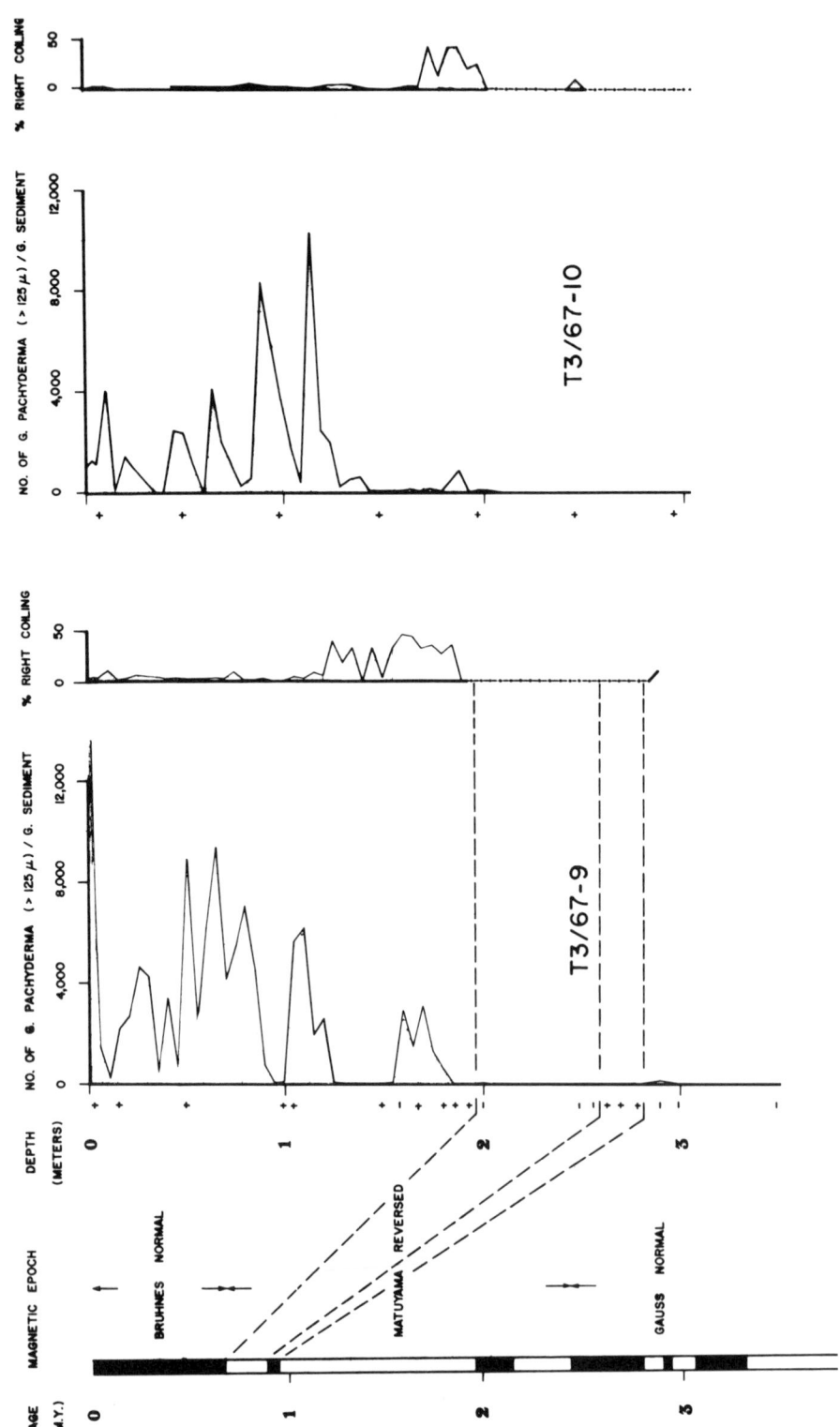

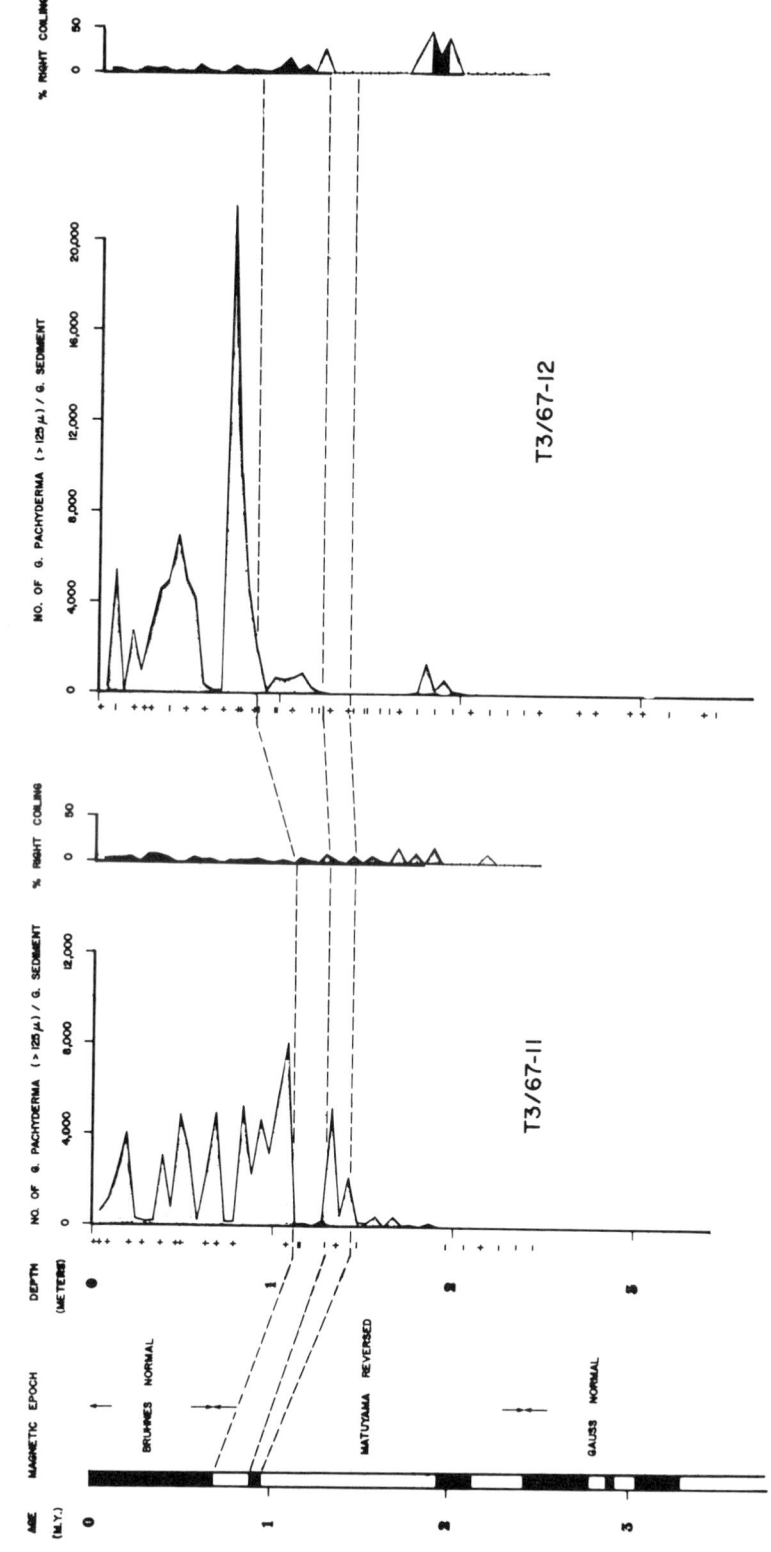

Fig. 6. Fluctuations in abundance of *Globigerina pachyderma* and its coiling direction in cores T3/67-11 and TE/67-13 from the Mendeleyev ridge. Normal and reversed magnetic polarity are indicated by + and −. Sampling interval for microfossils was 5 cm for T3/67-11 and 10 cm for T3/67-12.

Table 3. Average Sedimentation Rates over the Past 700,000 Years in the
Arctic Ocean, Based on Paleomagnetic Reversals

Province	Source	No. of cores	Rate (mm/1000 yrs)
Lomonosov ridge	Lin'kova (1965)	5	1.5
Mendeleyev ridge	Steuerwald et al. (1968)	5	2.0
Mendeleyev ridge	Lamont	4	2.2
Alpha Cordillera	Lamont	1	2.1

quite close agreement. The sedimentation rates over the past 700,000 years lie
within the range of 1.5 to 2.2 mm per 1000 years and they are compatible with
radioisotope dating. This accord in rates over different time intervals suggests that
pelagic sedimentation rates have been relatively constant in the Arctic Ocean for
at least the past 700,000 years.

MICROPALEONTOLOGY

Globigerina pachyderma (Ehrenberg) is the predominant planktonic foraminifer
in the cores listed in Table 1. This species is still living in the Arctic Ocean, where
it has been collected in plankton nets and is known to reach maximum concentrations during the summer months. The specimens in the upper 200 meters are generally immature; their shells are thin-walled and five-chambered and have a large
aperture. Apparently they develop further at greater depth by adding a small final
chamber and secreting a heavy calcite crust to become the mature, thick-walled
individuals found in bottom sediments (Bé 1960). Oxygen isotope observations
support this, since they indicate that about one-half of the carbonate in the *G.
pachyderma* tests was incorporated at depths below 300 meters (Van Donk and
Mathieu 1969).

A wide variation in size and shape of *G. pachyderma* may be found in the
cores, particularly in the upper foraminifera-rich zones (Plate 1). The shells may
have four or five chambers per whorl; the chambers may be hemispherical or coalescent; the shell walls may be thin (ca. 10 μ) or thick (ca. 30 μ), although the
latter are by far more common. Occasionally, thin-walled *G. pachyderma* predominate (e.g. at 195 cm in core 67-10), and this is probably indicative of ideal
sea floor conditions for shell preservation. At times the five-chambered specimens
with large apertures may be more common and larger in size than the compact,
encrusted variety; this is tentatively interpreted to be due to seasonal growth variations.

Globigerina quinqueloba Natland (Plate 2) occurs in small numbers in the
74-125 micron fraction whenever *G. pachyderma* is common. The former resembles *G. pachyderma*, but the presence of spine holes in its shell microstructure indicates that the living specimens possessed elongate spines. *G. quinqueloba* differs
from nonspinose *G. pachyderma* in having a more compressed shell, a prominently

projecting lip over its aperture, and a finer wall texture composed of smaller euhedral calcite crystals.

We have not observed any pteropods or temperate and subtropical planktonic foraminifera, such as *Globorotalia inflata*, *G. crassaformis*, *Globoquadrina dutertrei*, *Globigerinoides* sp., cf. *G. sacculifer*, and *Globigerinoides* sp., cf. *G. ruber*, which were encountered by Herman (1969). We did observe various species of benthic foraminifera and to a lesser extent radiolaria, ostracodes, sponge spicules and, very rarely, diatoms.

The abundance of *Globigerina pachyderma* fluctuates widely within the upper part of the cores corresponding to the Brunhes normal epoch. Foraminiferal tests are rare below the Brunhes-Matuyama boundary (Figs. 3, 4, 5).

Left-coiling *G. pachyderma*, generally associated with cold conditions (Ericson et al. 1964), predominate in these cores. Greater than 90 percent of the *G. pachyderma* at the tops of the cores are left coiling, and this ratio generally persists throughout the Brunhes epoch. However, late in the Matuyama reversed epoch, in the vicinity of the Jaramillo event, there is a zone where planktonic foraminifera are sparse but where the percentage of right-coiling *G. pachyderma* reach as high as 20 to 45 percent.

Core T3/67-11 was sieved into two different size fractions (Fig. 7). Total *G. pachyderma* was more abundant in the coarser fraction (125 μ) but the right-coiling types more prevalent in the finer fraction (74–125 μ). Benthonic foraminifera were most abundant in the Brunhes epoch as were the planktonic foraminifera. There is a weak correlation between the abundance of planktonic and benthonic species (Fig. 8).

Climatic Interpretation

Three climatic indicators were examined in the Arctic cores studied here: abundance of *G. pachyderma*, coiling direction of the same species, and the amount of insoluble ice-rafted material. Because these cores contain only two species of planktonic foraminifera, there is no possibility of using variations in faunal assemblages as an indicator of climatic change.

Since the sedimentation rate has been shown to be relatively constant, the abundance of planktonic foraminifera should be a measure of productivity which, in turn, must be related to the amount of open water. The oscillations in abundance are believed to represent changes in the Arctic pack ice cover from its present condition with some open leads in summer to even more severe conditions with an almost rigid pack throughout the year.

The uppermost level from cores taken on the Alpha Cordillera and described by Ericson et al. (1964) are rich in foraminifera. This deposition was shown by radiocarbon dating to extend to the present and hence to represent present ice conditions (Hunkins and Kutschale 1967). Cores from the Mendeleyev ridge (Figs. 4, 5) sometimes have a barren zone at the top. This may result from current winnowing which has removed a small amount of material from the surface. It

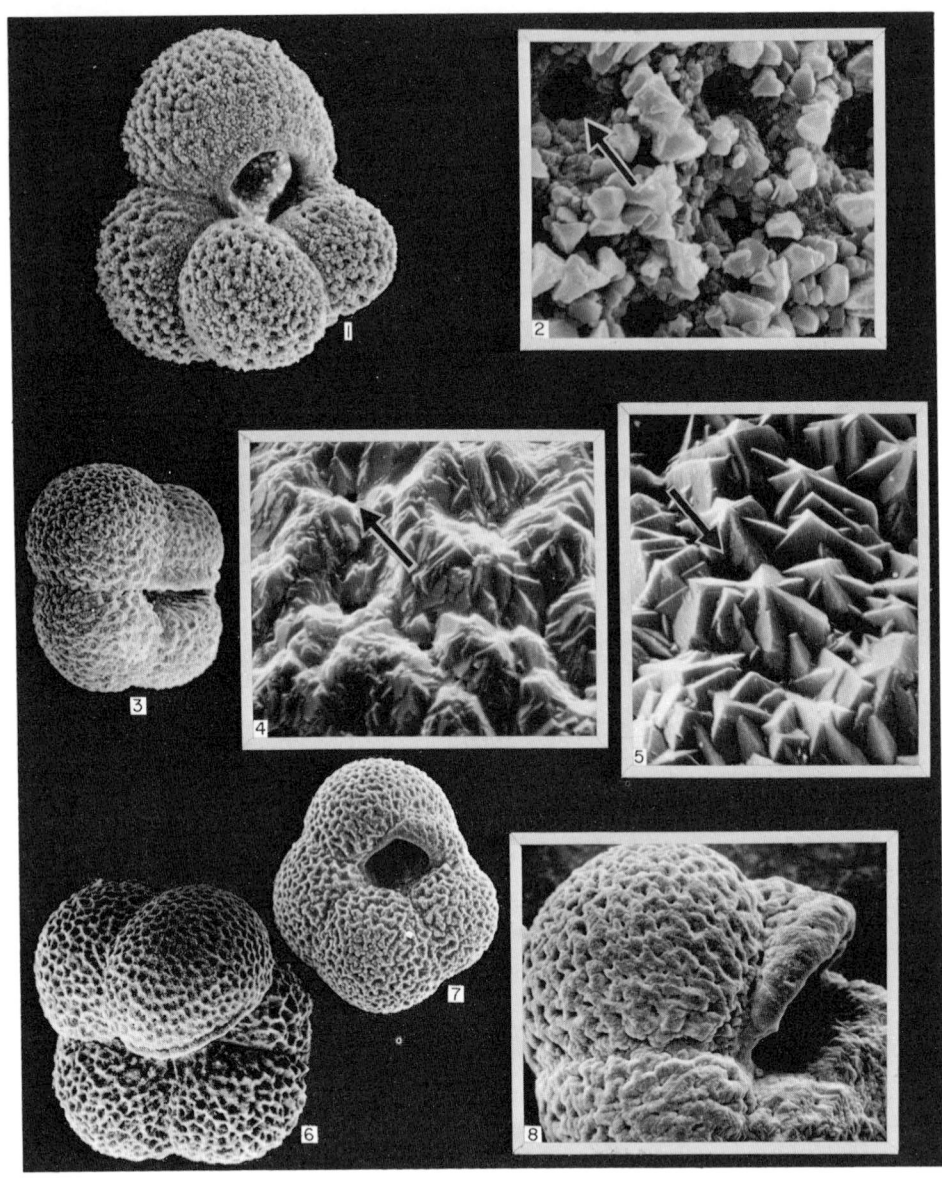

Plate 1. Scanning electron micrographs of typical specimens of foraminifera from Arctic Ocean core samples.

1. *Globigerina pachyderma* (Ehrenberg). Juvenile or early stage with large aperture, five hemispherical chambers, and a relatively thin wall. From 155 cm in core T3/67-11; × 260.

2. Detail of earliest chamber of last whorl of specimen on left. Euhedral calcite crystals of medium size surround pores. Arrow points to pore. *G. pachyderma* does not possess spines and therefore lacks spine holes; × 2600.

3. *Globigerina pachyderma* (Ehrenberg). Typical adult stage with heavily encrusted calcite crust and four coalescent chambers per whorl. From 5 cm in core T3/67-11; × 142.

4. Detail of final chamber of specimen on left, showing incompletely formed calcite crystals surrounding the pores (arrow); × 1420.

5. Detail of antepenultimate chamber of specimen at far left. Euhedral crystals have obliterated the pores (arrow); × 1200.

6. *Globigerina pachyderma* (Ehrenberg). Large, five-chambered variety from 250 cm in core T3/67-11. Polygonal pattern of ridges surrounding pores precedes formation of calcite crust; × 160.

7. *Globigerina pachyderma* (Ehrenberg). Thick-walled adult stage with diminutive final chamber. From 5 cm in core T3/67-11; × 135.

8. Detail of specimen in 7; × 315.

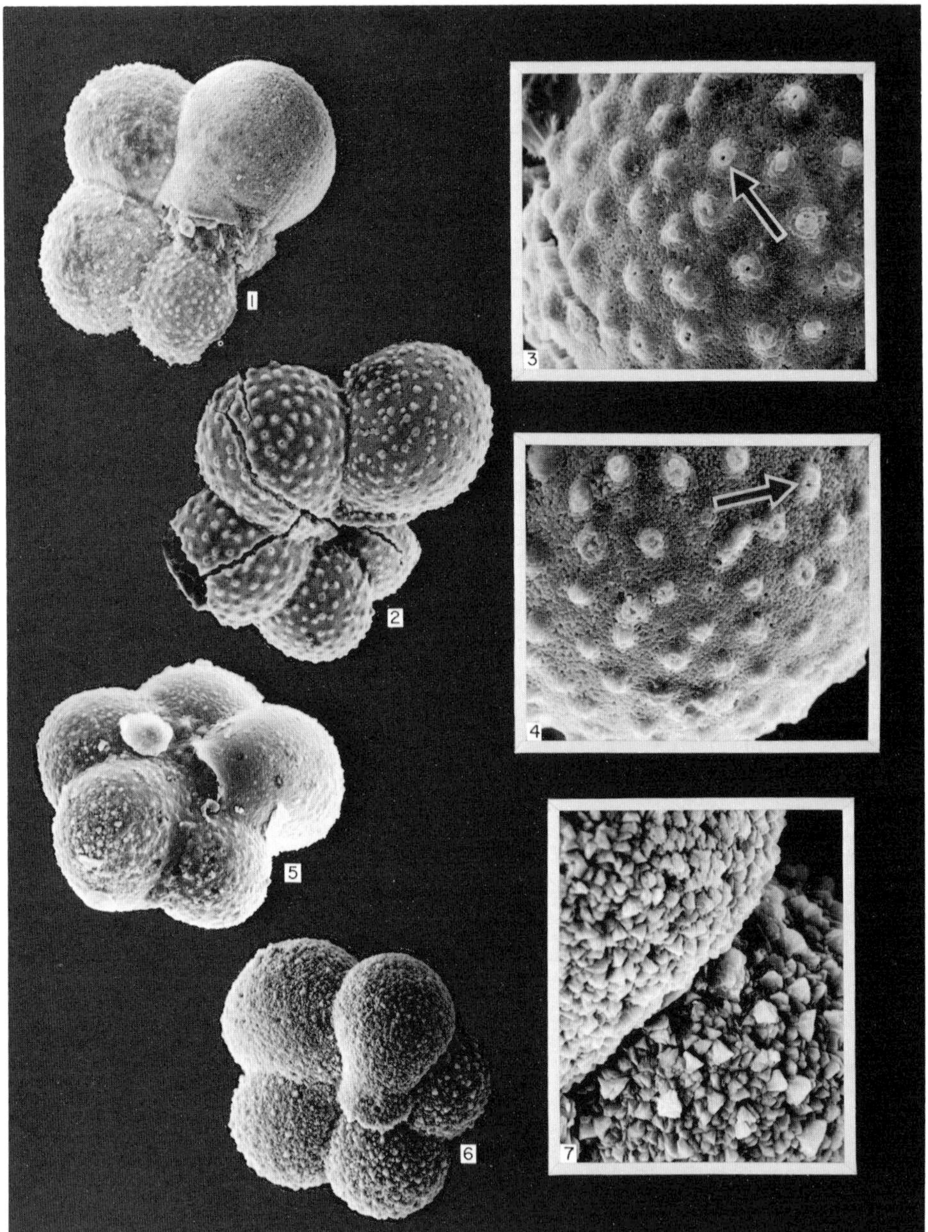

Plate 2. Scanning electron micrographs of typical specimens of foraminifera from Arctic Ocean core samples.

1. *Globigerina quinqueloba* Natland. From 185 cm in core T3/67-10; × 260.

2. *Globigerina quinqueloba* Natland. This laterally compressed five-chambered specimen has spine bases. From 185 cm in core T3/67-10; × 280.

3. Detail of penultimate chamber of specimen in 2, below. The spine holes (arrow) in the centers of the topographic highs are caused by the disappearance of former spine shafts; × 1000.

4. Detail of earliest chamber of specimen in 2, showing more spine holes and spine bases (arrow); × 1000.

5. *Globigerina quinqueloba* Natland. From 185 cm in core T3/67-10; × 260.

6. *Globigerina quinqueloba* Natland. The final chamber has characteristic tear-drop shape and projecting lip over aperture. From 185 cm in core T3/67-10; × 277.

7. Detail of earliest chamber (below) and final chamber (above) of specimen in 6, left. Note the somewhat larger crystals in the earliest chamber; × 1460.

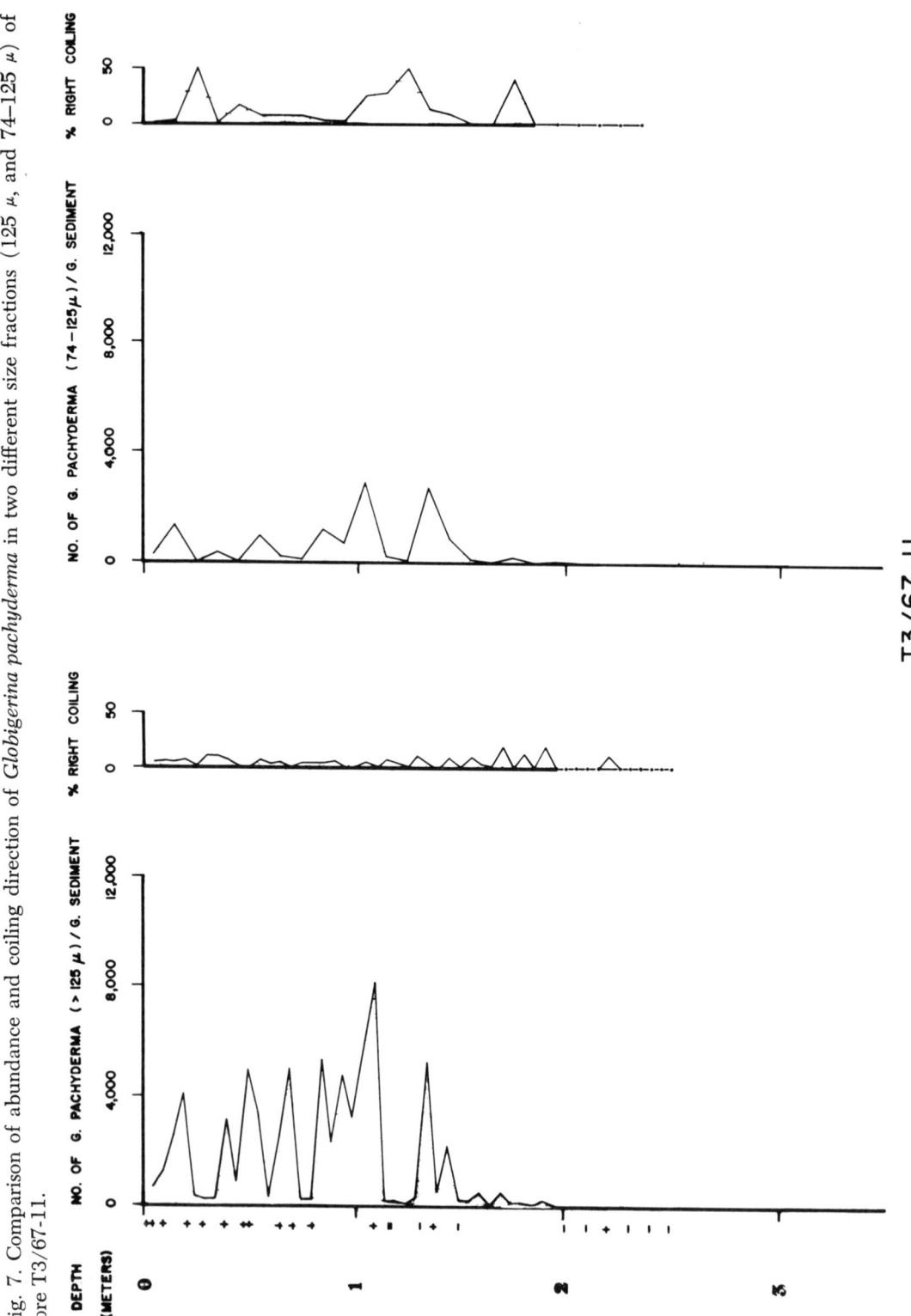

Fig. 7. Comparison of abundance and coiling direction of *Globigerina pachyderma* in two different size fractions (125 μ, and 74–125 μ) of core T3/67-11.

T3/67-11

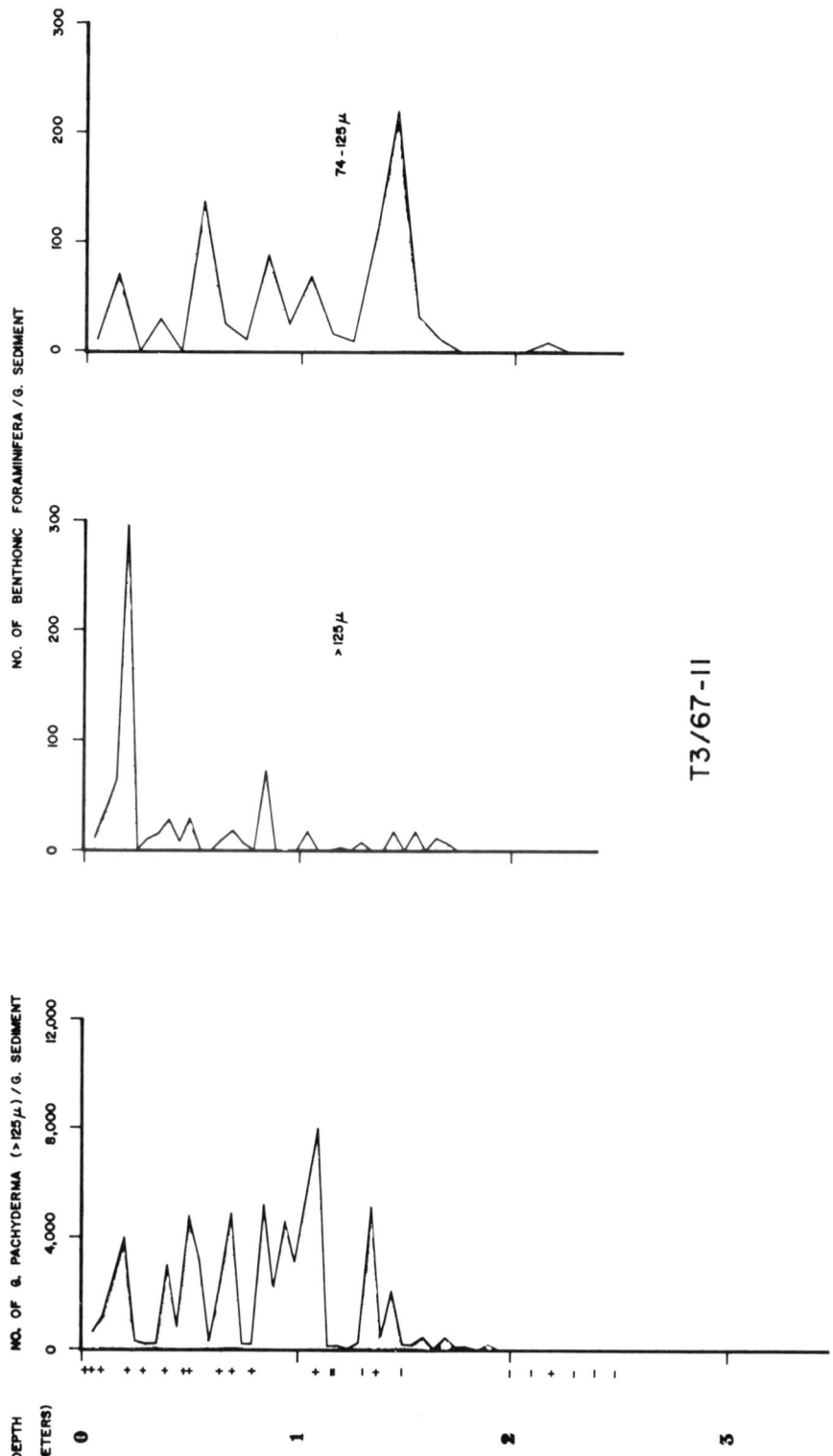

Fig. 8. Abundance of benthonic foraminifera in two different size fractions (125 μ, and 74–125 μ) of core T3/67-11. Abundance of planktonic foraminifera (*Globigerina pachyderma*) are also shown for comparison.

appears that bottom currents are swifter over the Mendeleyev ridge (Hunkins et al. 1969). The Alpha Cordillera cores are taken to be the norm because of the careful radiocarbon dating of their tops. The peaks of abundance within the Brunhes epoch do not correlate clearly from core to core. This is probably due to the difficulties in sampling cores with such slow deposition rates and to slight variations in that rate. However, the number of peaks varies only from four to seven in the cores and this resembles results from other oceans. Hays et al. (1969) noted eight distinct carbonate cycles within the Brunhes epoch for the equatorial Pacific. Kennett (1969) identified seven oscillations of warm and cold climate in Antarctica during the Brunhes epoch on the basis of changes in faunal assemblages. He found only subdued fluctuations in the Matuyama epoch with generally cold conditions and two weak warm periods.

Validity of microfossil abundance as a climatic indicator depends upon the tests having been little affected by solution. The preponderance of thick-shelled G. pachyderma and the lack of aragonitic pteropods and siliceous diatoms attest that carbonate and silica dissolution do take place on the Arctic sea floor. For example, the pteropod Limacina helicina is commonly encountered in the Arctic Ocean waters, but Ericson et al. (1964) have found it only sporadically in certain core horizons. However, the degree to which dissolution alters the Arctic fossil assemblages is not known. In our cores there is a comparatively small proportion of broken shells, which usually are indicators of early and medium stages of dissolution. It is possible that dissolution is so effective as to leave behind only the most resistant, thick-walled foraminifera. Thus, an alternative interpretation of the foraminifera-poor zones is that physicochemical conditions during these intervals were favorable for more complete dissolution of calcareous material. The abundance of foraminifera in the most recent sediments suggests that carbonate solution is not rapid under present bottom conditions.

Coiling direction in Globigerina pachyderma is also an effective climatic indicator in deep-sea cores. Ericson et al. (1964) associate left-coiling G. pachyderma with cold Arctic water and right-coiling specimens with warmer subpolar waters. The change in coiling direction in surface sediments from the Atlantic occurs along a line that coincides with the present 7.2°C isotherm in April. The transition line lies south of Iceland. North of it, left-coiling types predominate; south of it, right-coiling types prevail. Surface samples from Arctic Ocean cores which they examined all showed 97 percent or more of left-coiling specimens. The cores described here generally confirm this, although the percentage may be as low as 90 in some cases. Steuerwald et al. (1968) found less than 80 percent left-coiling types at the top, which is much less than those found by us and by Ericson.

The zone with a higher percentage of right-coiling types at a depth just below the Brunhes-Matuyama boundary in the Lamont cores indicates warmer conditions at that time. This interpretation conflicts with the evidence above based on abundance. Since foraminiferal concentration is low during the Matuyama epoch, it is possible that solution effects were more pronounced prior to about one million years ago. Manganese nodules are much more prevalent during the Matuyama in these cores, indicating different bottom water conditions in that earlier epoch. The presence of manganese nodules and lack of foraminifera suggest conditions like those in other oceans below the carbonate compensation depth.

Large amounts of ice-rafted material would indicate that ice islands moved freely about with enough open water into which debris could be washed or dumped. The coarse fraction of core T3/67-11 was treated with hydrochloric acid in order to remove foraminiferal tests. This treatment also dissolved the manganese nodules. Percentages of soluble and insoluble material are plotted against depth in Figure 9. Correlation is generally poor between soluble material (mostly tests) and insoluble material (mostly quartz grains). A regression dia-

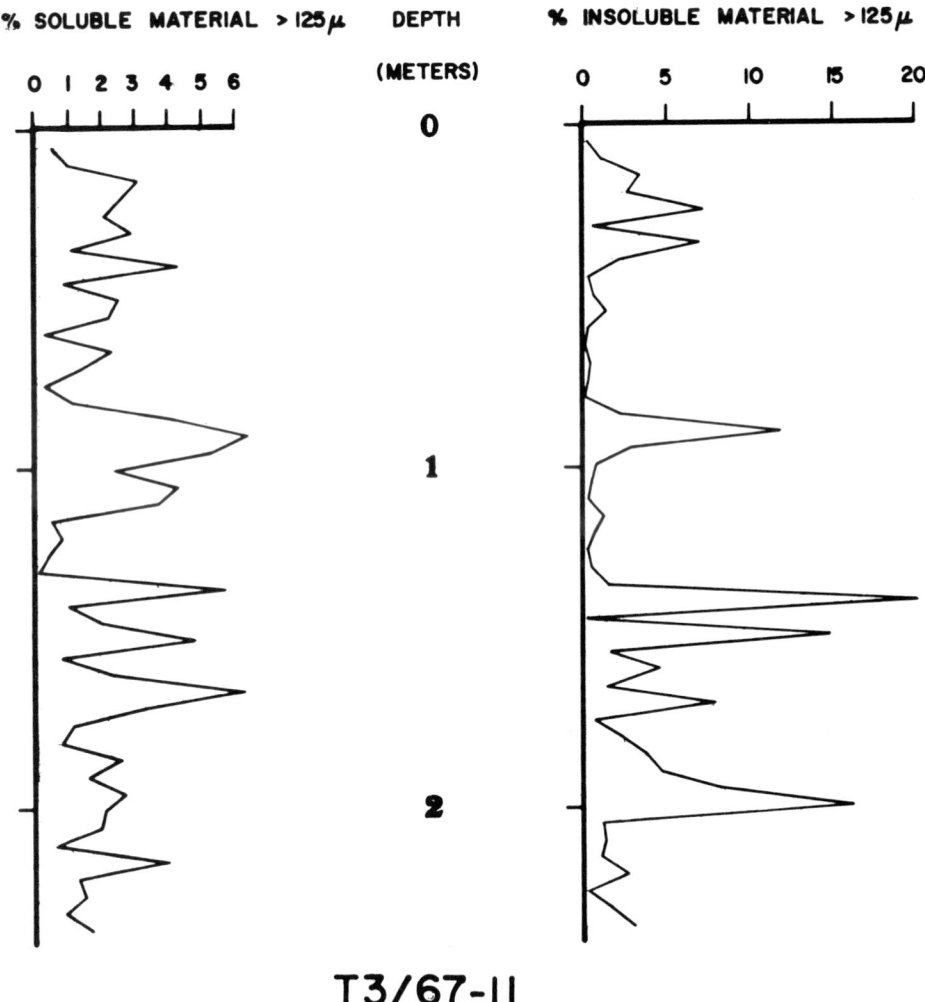

Fig. 9. Percentages of acid-soluble and acid-insoluble material in the coarse fraction of core T3/67-11 expressed in terms of total dry weight of sediment.

gram showed no clear correlation either positive or negative. There is some tendency for the insoluble residue to be higher in the Matuyama epoch, indicating more open conditions, in agreement with evidence for warmer climate from coiling direction.

Ice conditions in the Matuyama epoch, at least back to about 1.5 million years ago, were more uniform, without the wide swings of the Brunhes epoch. The climatic indicators conflict in their evidence as to whether heavy or light ice conditions prevailed then. Ice-rafted debris and coiling direction indicate warmer waters open enough for ice islands to move freely about and deposit material. But low faunal abundance would indicate a more rigid pack. It is possible, however, that the foraminiferal tests have been mostly dissolved by a different bottom water chemistry during the Matuyama epoch, thus rendering this indicator unreliable.

The evidence suggests that the present Arctic ice pack has persisted more or less unchanged from about 80,000 years ago. Prior to that there had been a number of cycles with conditions similar to the present alternating with even colder conditions when the ice pack was extremely tight with no open water even in summer. Most of the cores show about seven cycles through the Brunhes epoch, making each cycle 100,000 years in duration.

Acknowledgments. We are grateful to Dr. Tsunemasa Saito, who furnished the data for core T3/63–1. Werner Tiemann and Grace Irving ably assisted with the laboratory investigations and Joseph Forns helped in the preparation of the scanning electron micrographs.

This work was supported under contract N00014–67–A–0108–0016 with the Office of Naval Research, and a National Science Foundation grant (GA 14,177). Lamont-Doherty Geological Observatory Contribution no. 1556.

References

Bé, A. W. H., Some observations on Arctic planktonic foraminifera, *Contrib. Cushman Found. Foram. Res., 11* (2), 64, 1960.

Belov, N. A., and N. N. Lapina, New data on the stratification of the bottom sediments in the Arctic Basin, *Dokl. Akad. Nauk SSSR, 122,* 1958.

Còx, A., Geomagnetic reversals, *Science, 163,* 237, 1969.

Ericson, D. B., M. Ewing, and G. Wollin, Sediment cores from the arctic and subarctic seas, *Science, 144,* 1183, 1964.

Foster, J., A paleomagnetic spinner magnetometer using a fluxgate gradiometer, *Earth Planetary Sci. Letters, 1,* 463, 1966.

Gordienko, P. A., and A. F. Laktionov, Circulation and physics of Arctic Basin waters, *Ann. Intern. Geophys. Yr., 46,* 94, 1969.

Hays, J., T. Saito, N. Opdyke, and L. Burckle, Pliocene-Pleistocene sediments of the equatorial Pacific: Their paleomagnetic, biostratigraphic and climatic record, *Bull. Geol. Soc. Am., 80,* 1481, 1969.

Herman, Y., Temperate water foraminifera in Quaternary sediments of the Arctic Ocean, *Nature, 201,* 386, 1964.

Herman, Y., Arctic Ocean Quaternary microfauna and its relation to paleoclimatology, *Paleogeograph., Paleoclimatol., Paleoecol., 6,* 251, 1969.

Hunkins, K., and H. Kutschale, Quaternary sedimentation in the Arctic Ocean, in *Progress in Oceanography,* edited by M. Sears, vol. 4, p. 89, Pergamon, Oxford and New York, 1967.

Hunkins, K., M. Ewing, B. Heezen, and R. Menzies, Biological and geological observations on the first photographs of the Arctic Ocean deep-sea floor, *Limnol. Oceanog., 5,* 154, 1960.

Hunkins, K., E. M. Thorndike, and G. Mathieu, Nepheloid layers and bottom currents in the Arctic Ocean, *J. Geophys. Res., 74,* 6995, 1969.

Kennett, J., Foraminiferal studies of Southern Ocean deep-sea cores, *Antarctic J., 4,* 178, 1969.

Ku, T.-L., and W. Broecker, Rates of sedimentation in the Arctic Ocean, in *Progress in Oceanography,* edited by Sears, vol. 4, p. 95 (see Hunkins and Kutschale, above), 1967.

Lapina, N. N., and N. A. Belov, Peculiarities of sediment formation in the Arctic Ocean, in *Recent Sediments of the Seas and Oceans,* Reports of the Congress, 24–27 May 1960, edited by Strahov, Bezrukov, and Yablokov, p. 86, English transl. by U.S. Navy Electronics Lab., 1 Aug. 1965.

Lin'kova, T. I., Some results of paleomagnetic study of Arctic Ocean floor sediments, in *The Present and the Past of the Geomagnetic Field,* p. 279, Nauka Press, Moscow, 1965.

Olson, E., and W. Broecker, Lamont natural radiocarbon measurements VII, *Radiocarbon, 3,* 141, 1961.

Schwarzacher, W., and K. Hunkins, Dredged gravels from the central Arctic Ocean, in *Geology of the Arctic,* edited by G. O. Raasch, vol. 1, p. 666, Univ. of Toronto Press, 1961.

Steuerwald, B. A., D. L. Clark, and J. A. Andrew, Magnetic stratigraphy and faunal patterns in Arctic Ocean sediments, *Earth Planetary Sci. Letters, 5,* 79, 1968.

Van Donk, J., and G. Mathieu, Oxygen isotope compositions of foraminifera and water samples from the Arctic Ocean, *J. Geophys. Res., 74,* 3396, 1969.

Wallace S. Broecker

9. CALCITE ACCUMULATION RATES AND GLACIAL TO INTERGLACIAL CHANGES IN OCEANIC MIXING

Among the many problems to be solved if we are to substantially improve our understanding of the nature of glacial to interglacial climatic changes is the effect of these fluctuations on the patterns and rates of large-scale oceanic mixing. To judge by the difficulties encountered in trying to ascertain the present mode of operation of this vast system, it is extremely unlikely that meaningful paleo-reconstructions will be accomplished by theoretical arguments alone. We will need to have some observational evidence to guide our endeavors. The most important source of such information lies in the carbonate component of deep-sea sediments. To see why this is the case let us consider the factors controlling the rate at which the mineral calcite accumulates on any given region of the sea floor.

1. *The rate at which the element phosphorus is supplied to the sea surface in a given region.* The calcite accumulating on the sea floor is produced by planktonic plants (coccolithophoridae) and animals (foraminifera). The overall productivity of these plants and animals is limited by the availability of phosphorus in the sunlit surface waters. This is attested by the fact that over 95 percent of the phosphorus that is currently added to these waters by upward mixing returns to the deep sea in particulate form. The only way marine organisms could appreciably increase their overall productivity would be to more efficiently recycle the phosphorus within the surface waters before it falls back to the deep sea. The rate of supply of phosphorus to the surface depends only on the manner in which the sea mixes, because of the 200,000-year mean oceanic residence time of phosphorus.

2. *The amount of calcite falling per unit of phosphorus brought to the surface ocean.* Not all marine planktons produce calcite. Those that do, produce it in variable proportions to organic tissue. Thus the amount of this mineral produced per unit of phosphorus fixed by marine organisms will depend on the makeup of the marine life in the given oceanic region. Further, since planktonic animals derive energy from the oxidation of organic tissue, phosphorus must be recycled within the surface water. On the other hand, as surface waters are everywhere

supersaturated with $CaCO_3$, recycling of this substance in all likelihood does not take place. Thus the calcite content of falling particulate debris must be enriched over that found in the living organisms. The magnitude of this enrichment will depend on the number of photosynthesis–oxidation cycles a phosphorus atom goes through before it is trapped in a particle which falls beyond the surface water regime.

3. *The fraction of falling calcite that survives solution and becomes a permanent component of the sediment.* Some of the falling calcite may be destroyed by the benthic animals which feed on the falling planktonic debris, some dissolves in the corrosive waters that are found in the deepest parts of all ocean basins (Li et al. 1969), and some falls victim to the chemical reactions taking place within the pores of the sediment. Red clay owes its existence to the high efficiency at which these processes operate on large regions of the sea floor. In the absence of solution virtually all deep-sea sediments would be carbonate-rich oozes. The important question that arises in connection with this study is to what degree even carbonate-rich sediments have suffered solution loss.

With these three factors in mind let us consider how the pattern of carbonate accumulation rates might change between glacial and interglacial times. First of all, since a complete climatic cycle averages about 100,000 years in length, there should be no large change in the total dissolved phosphorus content of the sea. Even if the rate of phosphorus input were to change appreciably, the long residence time of this element in the sea would largely damp the variations (Fig. 1). If this is the case, then in a given area of the ocean the rate at which calcite accumulates will reflect the mode of oceanic mixing, the ecology of planktonic organisms, and the efficiency of the calcite-destroying processes. If some means could be devised to evaluate the effects of ecology and solution, a handle would be obtained on climatic-induced changes in oceanic mixing.

That the accumulation rate of $CaCO_3$ is related to the rates of oceanic mixing has been widely recognized (Arrhenius 1952; Wiseman 1956). However, these early studies have suffered from a faulty assumption. In the absence of adequate radioactive dating these workers were forced to assume that one of the components accumulating in the sediments of interest arrived at a constant rate regardless of changes in climate. The component selected was the fine-grained silicate detritus. Unfortunately, as shown by Broecker et al. (1958), at least for the equatorial Atlantic this proved to be a bad guess. The rate of accumulation of fine detritus is the least constant of all the components of interest. Turekian (1965) overcame this difficulty by using cores dated by radioisotopic methods. His study was, however, restricted mainly to the equatorial and northern Atlantic and thereby permitted no conclusions about global circulation. Although this study will also prove inadequate, it does take a step forward in that absolute age control is used and the problem is treated on a global scale.

The simplest model for the $CaCO_3$ cycle in the ocean would be as follows. The remains of calcitic and aragonitic particles produced by organisms fall toward the deep-sea floor. Due to its higher solubility, aragonite dissolves more

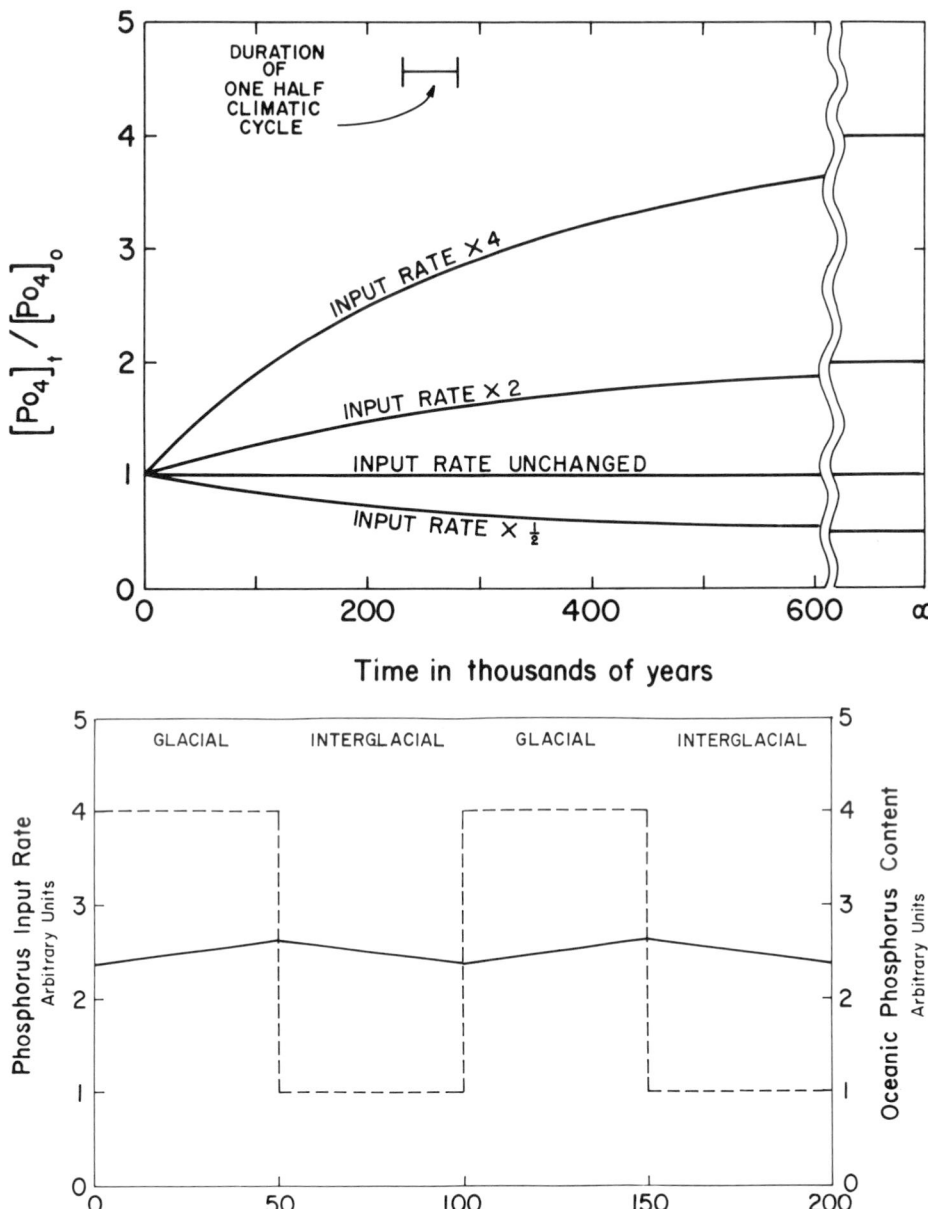

Fig. 1. (A) Response of oceanic phosphorus content to various changes in the rate of phosphorus supply by rivers (other factors held constant). (B) Hypothetical glacial to interglacial phosphorus cycle. Note that despite the large change in input rate (× 4) there is only a small range in phosphorus content (∼10 percent). Also note that the mean phosphorus content for interglacials is identical with that for glacials.

readily than calcite. In the present-day ocean only a very small fraction of the aragonite and about 15 percent of the calcite falling into the deep sea survive. The CaCO$_3$ that survives does so because it falls upon topographic highs which project above the layer of corrosive bottom water (this layer, of course, extends to much shallower oceanic depths for aragonite than for calcite). Assumed that the anomalies in bottom topography and in surface productivity are randomly arranged, one with respect to the other, measurements of calcite accumulation rates on topographic highs should yield a measure of the mean rate of calcite rain from the overlying surface water. Further, by measuring the amount of excess CaCO$_3$ dissolved in deep-sea water (relative to surface water) and the residence time of water in the deep sea, it should be possible to estimate the rate at which CaCO$_3$ is currently dissolving in the deep sea. This flux would include that portion of the calcite and aragonite which fall onto regions of the sea floor below the respective saturation horizons for these phases. The CaCO$_3$ cycle can thus be expressed in the following mathematical form:

$$A_0(R_C + R_A) = A_0 R_S + (A_A R_A + A_C R_C)$$

where A_0, A_A, and A_C are the area of the total sea floor, that covered by aragonite-rich sediments, and that covered by calcite-rich sediments; and R_S, R_A, and R_C are the average rates at which CaCO$_3$ dissolves in deep-sea water, accumulates in aragonite-rich sediments, and accumulates in calcite-rich sediments. $A_0(R_C + R_A)$ is then the total amount of calcite and aragonite falling into the deep sea. $A_0 R_S$ is the amount dissolving. $A_A R_A$ and $A_C R_C$ are the amounts of aragonite and calcite accumulating per unit time. The assumption is made that the average accumulation rate for aragonite and calcite in sediments rich in these minerals typifies the rate at which these minerals fall over the world ocean toward the sea floor.

To test this model we will attempt to estimate the appropriate rates and areas to see whether a meaningful balance can be obtained.

ACCUMULATION RATES

Table 1 summarizes the available data on the average rate at which CaCO$_3$ has accumulated on various parts of the deep-sea floor during the last few hundred thousand years. As in all cases the cores encompass at least one complete climatic cycle, these results should average out any climatic-induced rate changes. Since these sediments are in general devoid of aragonite, the results apply only to the mineral calcite. Although the results are inadequate to permit the definition of precise global accumulation rate distributions, they do suggest that the average rate of accumulation in the Atlantic (0.10 ± 0.03 mole CaCO$_3$/m^2 yr) is greater than that for the Pacific-Indian Ocean (0.05 ± 0.02 mole/m^2 yr).

Because of its scarcity in deep-sea sediments the rate of accumulation of aragonite is very difficult to assess. Where pteropods have been reported, their abundance (if given at all) is usually in the number of specimens per gram of

Table 1. Summary of CaCO₃ Accumulation Rates

Core	Latitude	Longitude	Depth (km)	CaCO$_3$ (%)	Accumulation rate CaCO$_3$ (moles/m^2 yr) [*]	Reference
		CORES DATED STRATIGRAPHICALLY (using Broecker-Van Donk 1970 chronology) [†]				
		North Atlantic				
A164-6	38°N	70°W	3.7	20	0.07	Ericson et al. 1961 / Turekian 1965
A164-15	36°N	69°W	4.5	15	0.05	Ericson et al. 1961 / Turekian 1965
A164-44	34°N	63°W	3.4	75	0.08	Ericson et al. 1961 / Turekian 1965
A180-9	39°N	46°W	4.1	40	0.05	Ericson et al. 1961 / Turekian 1965
280	35°N	44°W	4.3	50	0.16	Emiliani 1958
A180-15	39°N	37°W	4.6	65	0.17	Ericson et al. 1961 / Turekian 1965
A180-16	38°N	32°W	2.3	85	0.20	Ericson et al. 1961 / Turekian 1965
A180-39	26°N	19°W	3.5	70	0.06	Ericson et al. 1961 / Turekian 1965
R5-60	20°N	23°W	3.9	75	0.08	Ericson et al. 1961 / Turekian 1965
		Caribbean				
A172-2	16°N	72°W	3.1	60	0.14	Ericson et al. 1961 / Turekian 1965
A179-4	17°N	75°W	3.0	65	0.10	Ericson et al. 1961 / Turekian 1965
P6304-8	15°N	69°W	3.9	65	0.14	Emiliani 1966
P6304-9	15°N	69°W	4.1	65	0.14	Emiliani 1966

(cont.)

Table 1 (cont.)

Core	Latitude	Longitude	Depth (km)	CaCO$_3$ (%)	Accumulation rate CaCO$_3$ (moles/m² yr)*	Reference
V12-122	17°N	74°W	2.8	60	0.12	Broecker & van Donk 1970
A240-M1	15°N	68°W	4.2	60	0.11	Rosholt et al. 1961
Equatorial Atlantic						
A180-76	0°	26°W	3.5	80	0.23	Ericson et al. 1961 / Turekian 1965
A180-73	0°	23°W	3.7	65	0.16	Ericson et al. 1961 / Emiliani 1955
A180-74	0°	24°W	3.3	80	0.20	Ericson et al. 1961 / Turekian 1965
Equatorial Pacific						
RC11-209	04°N	140°W	4.4	80	0.08	Hays et al 1969
60	02°N	135°W	4.5	70	0.15	Arrhenius 1952
Equatorial Indian						
IC-5	02°N	78°E	4.3	35	0.06	Oba 1969
CORES DATED BY IONIUM						
(based on interpretation by Ku et al. 1968)						
North Atlantic						
ZEP24	27°N	20°W	4.3	62	0.12	Goldberg et al. 1963
ZEP18	23°N	44°W	5.0	62	0.03	Goldberg et al. 1963
ZEP15	21°N	44°W	3.3	78	0.03	Goldberg et al. 1963
ZEP13	20°N	49°W	4.7	53	0.03	Goldberg et al. 1963
V16-21	17°N	48°W	4.0	64	0.03	Ku 1966
Caribbean						
A254-Br-C	16°N	73°W	—	50	0.13	Goldberg et al. 1964

Table 1 (cont.)

Core	Latitude	Longitude	Depth (km)	CaCO$_3$ (%)	Accumulation rate CaCO$_3$ (moles/m² yr) *	Reference
Equatorial Atlantic						
LUS223	12°N	45°W	4.0	61	0.03	Goldberg & Griffin 1964
LUS217	04°N	34°W	3.3	62	0.07	Goldberg & Griffin 1964
V9-29	04°N	35°W	4.7	75	0.04	Ku 1966
V9-27	0°	30°W	4.5	65	0.20	Ku 1966
V9-11	03°S	32°W	4.2	50	0.08	Goldberg & Koide 1962
South Atlantic						
LUS183	20°S	13°W	3.5	89	0.06	Goldberg & Griffin 1964
LUS178	24°S	16°W	4.0	81	0.05	Goldberg & Griffin 1964
LUS168	29°S	09°W	3.9	87	0.09	Goldberg & Griffin 1964
LUS163	31°S	02°E	4.2	77	0.10	Goldberg & Griffin 1964
LUS159	34°S	15°E	4.2	55	0.10	Goldberg & Griffin 1964
Indian Ocean						
MON57G	26°S	74°E	–	74	0.03	Goldberg & Koide 1963
South Pacific						
RC8-93	29°S	105°W	3.2	79	0.03	Ku et al. 1968
V19-54	17°S	114°W	2.9	64	0.06	Unpublished L-DGO data
V21-48	10°S	126°W	3.9	85	0.04	Ku et al. 1968

* Assumes the average density of water-free core material to be 0.8 gm/cm³.

† Rates measured between termination II (127,000 years B.P.) and termination I (11,000 years B.P.). The terminations are identified either from faunal or O^{18} data.

coarse fraction rather than as weight fraction. The unavailability of any reliable estimate for the ratio of aragonite to calcite in the falling carbonate debris will prove to be one of the major limitations in this study.

The fraction of the Atlantic Ocean floor covered by calcite-rich sediments is about 65 percent, and of the Pacific-Indian Ocean about 40 percent (Sverdrup et al. 1942). Aragonite-rich sediments cover a negligible portion of the sea floor.

The solution rate for the last few thousand years can be estimated from the present distributions of total dissolved inorganic carbon, of alkalinity, and of radiocarbon in the world ocean. Although our ability to construct whole-ocean mixing models is still extremely limited, it is possible to obtain separate estimates for the average carbonate solution rates in the deep Atlantic and the deep Pacific-Indian Ocean basins.

Our knowledge of the mixing *patterns* in the ocean stems largely from studies of the distributions of temperature, salinity, dissolved oxygen, and the trace isotope O^{18}. Surface water from the Atlantic is cooled in the Norwegian Sea. It sinks and spills over the topographic barriers running from the British Isles to Greenland into the deep Atlantic basin. Mixing processes at depth carry this water down the Atlantic to the Antarctic. Here it is joined by dense water created by ice formation in the Weddell Sea. The mixture is carried through the deep Indian basin into the Pacific. The steady input of this new abyssal water is balanced by a more or less uniform upwelling of deep water through the main oceanic thermocline into the warm surface ocean. Superimposed on this general upwelling is a diffusive mixing of the warm water overlying the main oceanic thermocline with the cold water underlying it. This combination of advective and diffusive mixing brings phosphorus from the deep to the surface sea.

The absolute *rates* for these processes have been estimated from the oceanic distribution of the cosmic-ray-produced isotope C^{14}. The difference in the abundance of this radionuclide in carbon extracted from the deep waters in the North Pacific (the region most isolated from the atmospheric birthplace of C^{14}) and other more accessible water masses provides a means for quantifying the rates of mixing. As C^{14} is carried by the falling organic debris as well as by the water it is necessary to consider the distribution of dissolved carbon as well as dissolved C^{14}. As pointed out by Lal (1969), taken together these distributions allow the effects of particulate and water transport to be separated.

A model proposed by Broecker and Li (1970) treats the Atlantic (NADW) and Pacific-Indian (PIDW) deep waters as separate well-mixed reservoirs and yields an estimate of the average rate at which particulate $CaCO_3$ falling from the warm surface water of the ocean is being dissolved in each of these two major deep reservoirs. As shown in Figure 2, in order to establish the two particulate fluxes and the six water fluxes necessary to define the transfer of carbon among these reservoirs, eight restrictions on the system are needed. Six of these are achieved from the requirement that the amount of water, carbon, and radiocarbon in each of the reservoirs remain constant with time. The remaining two restrictions are imposed by the salinity–O^{18} relationships. As the latter allow a range

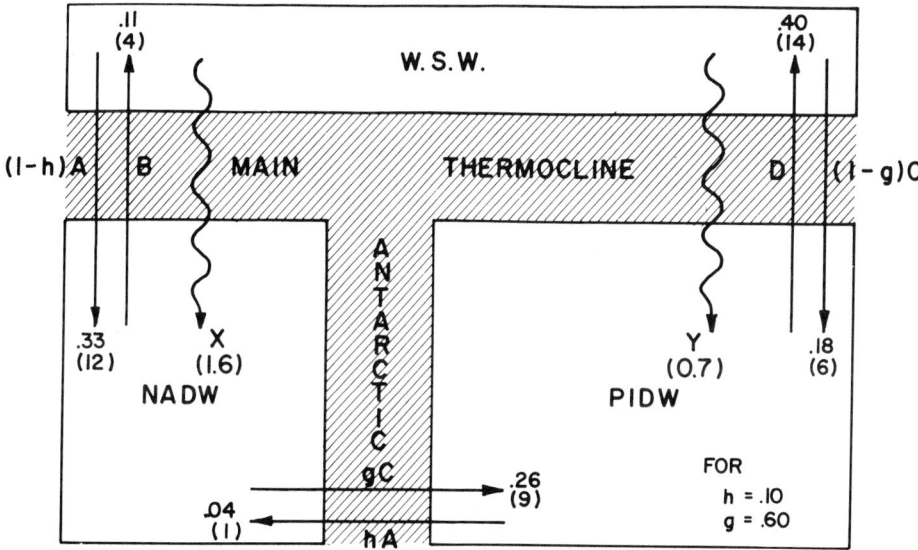

Fig. 2. Three-box oceanic mixing model proposed by Broecker and Li (1970). The straight arrows designate water fluxes and the wavy arrows particle fluxes. The parameter h represents the fraction of NADW supplied by water of the PIDW type. The O^{18}-S diagrams given by Craig and Gordon (1965) demand that h be less than 0.20. The parameter g represents the fraction of PIDW supplied by water of the NADW type. The O^{18}-S diagrams of Craig and Gordon (1965) demand that g be greater than 0.40. The numbers at the ends of the arrows indicate the water fluxes derived for the best estimates of g (i.e. 0.60) and of h (i.e. 0.10). The upper number for each water flux is in units of ocean volumes per 1000 years while that in parentheses is in sverdrups (10^6 m³/sec). The particle fluxes are given in units of moles of carbon per square meter of ocean surface per year.

of possible values, the final results (Table 2) are subject to some uncertainty. It turns out, however, that the particle flux into the Pacific-Indian deep reservoir is nearly independent of this range of choice. The carbon flux lies between 0.6 and 0.7 mole per m² year of carbon falling in particulate form. For the Atlantic the range of possibility is 1.4 to 2.2 moles per m² year.

Measurements of both the change in total dissolved inorganic carbon content (due to the formation and destruction of both organic tissue and $CaCO_3$) and change in alkalinity (due to the formation and destruction of *only* $CaCO_3$) between various oceanic water masses allow the $CaCO_3$ component of the particle flux to be isolated. A plot of A versus ΣCO_2 for samples from throughout the ocean is shown in Figure 3. The slope of lines connecting various water types yields the ratio of the alkalinity change (ΔA) to total dissolved inorganic carbon change ($\Delta \Sigma CO_2$) in going from one water type to another. Since NADW is formed from at least 80 percent surface water the $\Delta A / \Delta \Sigma CO_2$ resulting from the *in situ* destruction of particles is about 0.20 (1 mole $CaCO_3$ per 10 moles particu-

Table 2. Values of the Fluxes for Various Combinations of Permissible
Values for the Parameters g and h*

h	g	A	C	hA	(1-h)A	gC	(1-g)C	B	D	X	Y
					(ocean volumes/1000 yrs)					moles/m² year	
.20	.40	.62	.40	.12	.50	.16	.24	.46	.28	2.2	0.7
.10	.40	.37	.40	.04	.33	.16	.24	.31	.36	1.6	0.7
.00	.40	.26	.40	.00	.26	.16	.24	.10	.40	1.4	0.7
.20	.60	.62	.44	.12	.50	.26	.18	.36	.32	2.2	0.7
.10†	.60†	.37†	.44†	.04†	.33†	.26†	.18†	.11†	.40†	1.6†	0.7†
.00	.60	.26	.44	.00	.26	.26	.18	.00	.44	1.4	0.7
.20	.80	.62	.48	.12	.50	.38	.10	.24	.33	2.2	0.6
.10	.80	.37	.48	.04	.33	.38	.10	.00	.44	1.6	0.6
.00	.80	.26	.48	.00	.26	.38	.10	-.12‡	.48	1.4	0.6

* See Figure 2 for definition of parameters and Broecker and Li (1970) for details.
† Best estimate.
‡ Physically impossible.

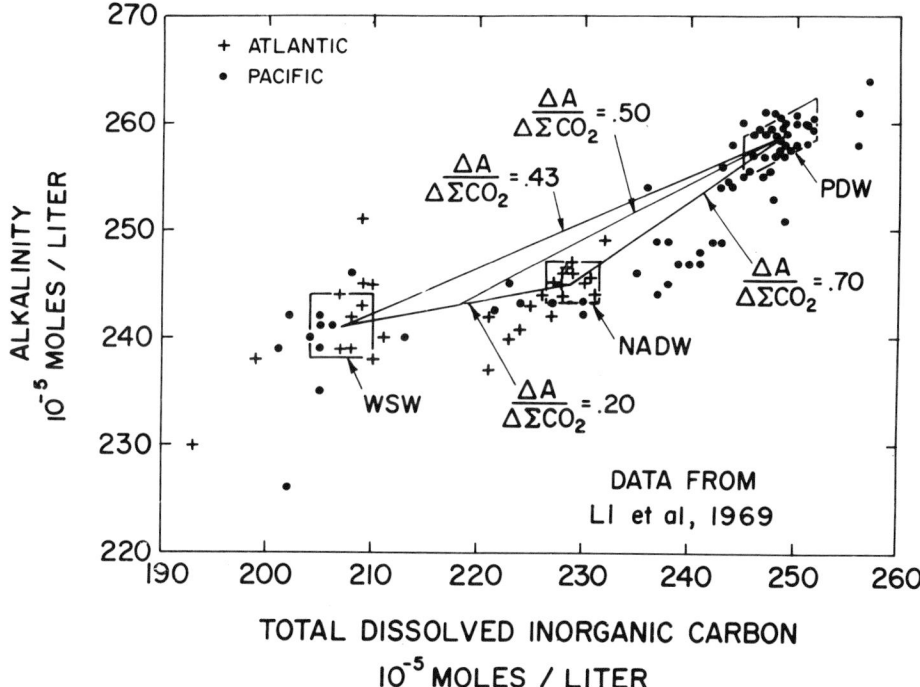

Fig. 3. Plot of alkalinity versus total dissolved inorganic carbon for a wide variety of lo-cations and depths in the world ocean (normalized to a salinity of 35.00 per mil). Sam-ples from the three major water types considered by Broecker and Li (1970) are indi-cated by the ruled boxes. The source of NADW is taken to be WSW and that of PIDW a mixture of equal volumes of NADW and WSW. The slopes of the lines connecting the water mass with its source yield the ratio of organic tissue to CaCO$_3$-derived car-bon released to the water while at depth.

late carbon). The deep Pacific water, on the other hand, comes roughly half from the surface and half from the deep Atlantic. Its *in situ* $\Delta A/\Delta\Sigma CO_2$ must be about 0.50 (1 mole CaCO$_3$ to 4 moles particulate carbon). The flux of CaCO$_3$ destined to dissolve in the deep Pacific is thus about 0.15 mole per m^2 yr and in the deep Atlantic between 0.14 and 0.22 mole per m^2 yr.

If we assume that aragonite is quantitatively unimportant relative to calcite in falling carbonate debris, then the rate of carbonate solution divided by the fraction of the oceanic area covered by calcite-free sediments should equal the rate at which calcite accumulates on topographic highs, i.e.

$$\frac{R_s}{(A_o - A_c)/A_o} = R_c$$

These comparisons are made in Table 3. In the oceans as a whole the actual rate of CaCO$_3$ accumulation is only one-fifth of that predicted if no CaCO$_3$ solution occurs. The lack of agreement could be explained in a number of ways.

Table 3. Summary of CaCO$_3$ Flux Data[*]

(See text for sources)

		Atlantic	Pacific
R_S	Model solution rate	0.18	0.15
$R_C = R_S/(1 - A_C/A_0)$	Model production rate	0.50	0.25
$R_C f$ [†]	Observed accumulation rate	0.07	0.02
f (apparent value)	Model + observed accumulation rates	0.28	0.12

[*] Aragonite component assumed negligible. Fluxes are in the units of moles CaCO$_3$/m^2 yr.
[†] Where f is the fraction of the falling calcite which dissolves.

1. Aragonite is an important component of falling CaCO$_3$ debris.

2. Since the deep-sea solution rate estimates are indicative of only the last 1000 or so years, while the accumulation rate estimates cover a period 100 times as long, possible temporal fluctuations in the rates pose a serious problem. This is especially critical since the last thousand years represent the height of interglaciation while the accumulation rates are averages for an entire climatic cycle. Table 4 shows comparisons of rates for postglacial time with those for the last 100,000 years in those few cores where such information is available.

3. The estimates of accumulation rate averages may not be correct since they are based on relatively few observations and since these observations show considerable geographic variation (see Table 1).

4. The solution rate estimates may be too large since Broecker and Li (1970) in their mixing model are forced by lack of information to neglect the direct input of C^{14} to the deep sea through its cold water "outcrops." If direct addition is taking place at a significant rate, all their water fluxes and particle fluxes would have to be reduced.

5. The scatter in the alkalinity and total dissolved inorganic carbon data (see Fig. 2) permits a fair latitude in assigning the fraction of the excess carbon in deep water derived from solution of CaCO$_3$ (as opposed to the oxidation of organic matter). This uncertainty is agravated by the fact that the compositions for water entering these deep basins are not well defined.

6. CaCO$_3$ solution even in areas of carbonate-rich sediment is appreciable.

As none of these individual points is likely to explain a difference of more than a factor of 2 between the observed and predicted rates, more than one will eventually have to be invoked.

Before going on to the comparison of the actual accumulation rate let us consider the kinds of changes that should be found for various kinds of mixing changes. If, for example, the pattern of oceanic mixing were maintained and the rates changed, then in the absence of ecologic or solution changes, the rates of accumulation should change in proportion to the mixing rates. On the other hand, were both the pattern and rate to change, the resulting accumulation rate change would be complex. Perhaps the most. important element in the mixing cycle is the export of NADW to the Indian and Pacific oceans. Were the intensity of this

Table 4. Comparison of Postglacial Sedimentation Rates
with Those for the Late Glacial, the Entire Last Glacial,
and the Last Interglacial Time

SEDIMENTATION RATE (cm/10³ years)

Core	Last[1] interglacial (127,000– 75,000)	Last[1] glacial (75,000– 11,000)	Late[2] glacial (22,000– 11000)	Post[2] glacial (11,000– 3000)
Caribbean				
A240-M1	2.3	2.1	~4.0°	2.2°
V12-122	2.4	2.4	~3.2†	—
A179-4	1.7	2.4	~4.2‡	~2.4‡
Equatorial Atlantic				
A180-73	3.1	2.8	~4.2§	~2.0§
A180-74	—	4.4	5.1‖	2.3‖

[1] See Table 1.
[2] As determined by C¹⁴ dating.

° Rusnak et al. 1963.
† Ku and Broecker 1966.
‡ Rubin and Suess 1955.
§ Rubin and Suess 1956.
‖ Broecker et al. 1958.

deep water source to change relative to that of waters formed in the Antarctic, the ratio of $CaCO_3$ falling in the Atlantic relative to that in the Pacific would be bound to change. As the factors determining the flux of phosphorus to the surface Atlantic are not well understood, it is not possible to predict even the sense of this change at this time.

Thus, whereas the potential of the approach is clear, any firm conclusions await the resolution of the major points of uncertainty.

COMPARISON OF GLACIAL AND INTERGLACIAL CARBONATE ACCUMULATION RATES

With this background in mind the record of interglacial to glacial changes in $CaCO_3$ accumulation rates will be examined. Although the ideal way to do this would be to take radiocarbon-determined rates in the 8000- to 3000-year time interval (peak of interglaciation) and compare them with those in the 19,000- to 14,000-year interval (peak of glaciation), an insufficient number of cores has been studied in this way to allow sufficient coverage of the ocean (see Table 4 for a summary of the cores where this is possible). An alternative approach is to take cores where the transitions between Emiliani's (1955) stages 1 and 2, 4 and 5, and 5 and 6 have been defined by O^{18} measurements. Based on the absolute chronology of Broecker and Van Donk (1970), these boundaries have ages of 11,000, 75,000, and 127,000 years. The 11,000-year age is based on C¹⁴ data, the 127,000-

year age on Th^{230} and Pa^{231} data for core V12-122 (and supported by paleomagnetic data on other cores), and the 75,000-year age on the hypothesis that the
sharp cooling separating stages 5 and 4 was caused by the summer insolation drop
between the warm peak 82,000 years ago and the cold peak 70,000 years ago in
the astronomical curves (Vernekar 1968). The 52,000-year interval between
127,000 and 75,000 years ago is a period of reduced ice cover (interglacial) while
the 64,000-year interval from 75,000 to 11,000 years is one of substantial ice cover
(glacial) (Broecker and Van Donk 1970). In Figures 4–8 are shown the O^{18}/O^{16}
records for one North Atlantic core, one equatorial Atlantic core, two Caribbean
cores, and one equatorial Indian core (see Table 1 for locations and depths).
The three boundaries of interest are readily identified in the four Atlantic cores.
In the Indian Ocean core the position chosen for the 75,000-year boundary might

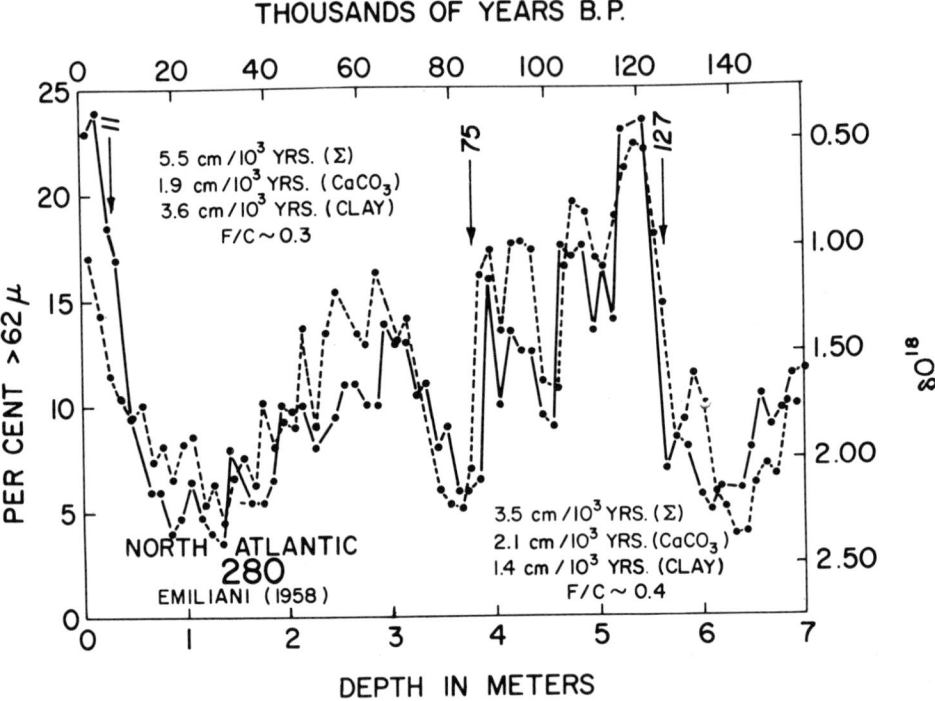

Fig. 4. Plots of the O^{18} (dashed line) and coarse-fraction data (solid line) for core 280
from 34°N in the Atlantic Ocean (Emiliani 1958). The time scale given on the upper
scale is the one that would apply were the sedimentation rate constant and the absolute age of termination II 127,000 years. The average CaCO$_3$ content of the glacial section is taken to be 35 percent and that of the interglacial section 60 percent (based on
data published by Emiliani 1958). The ratio of the foram to coccolith carbonate (F/C)
is obtained by dividing the mean percentage coarse fraction for the interval by the difference between the mean CaCO$_3$ content and the mean percentage coarse fraction.
The term "clay" refers to the bulk noncarbonate fraction of the core.

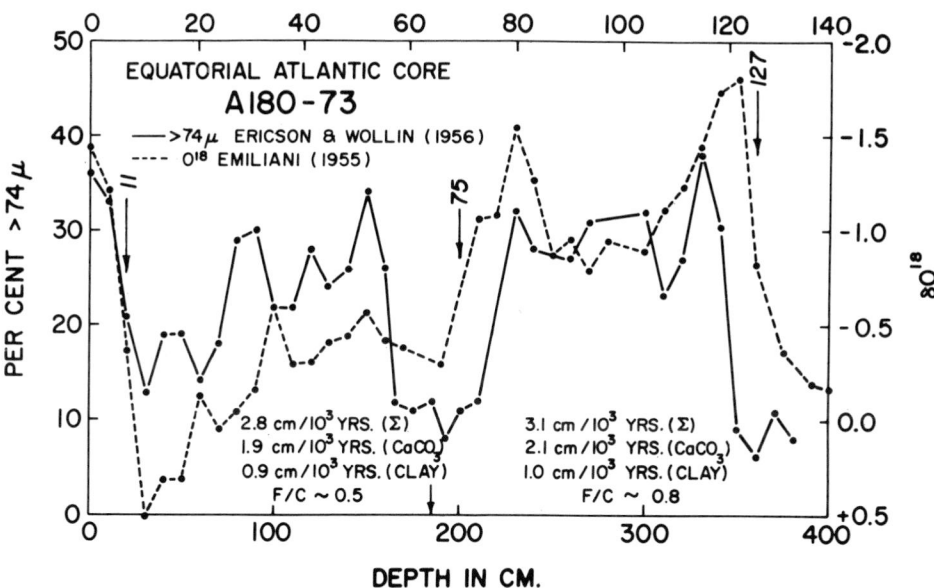

Fig. 5. Plot similar to Figure 4 for core A180–73. The average CaCO₃ content for both the glacial and interglacial intervals is taken to be 68 percent (based on data from Ericson and Wollin 1956). The arrow indicates the point where *Globorotalia menardii* disappears from the faunal assemblage at the end of the interglacial interval.

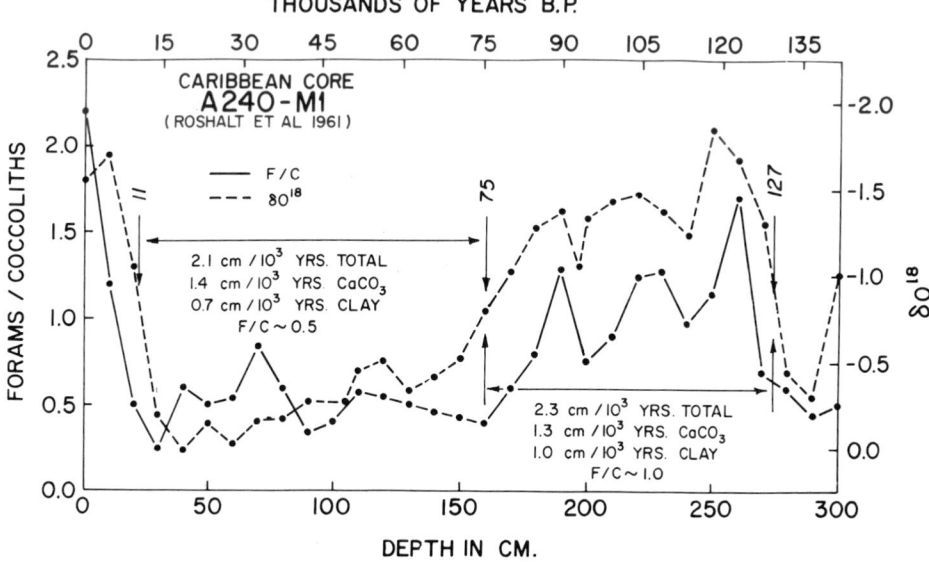

Fig. 6. A plot similar to that in Figure 4 for Caribbean core A240–M1 except that the foram-to-coccolith weight ratio is plotted instead of the coarse fraction percentage.

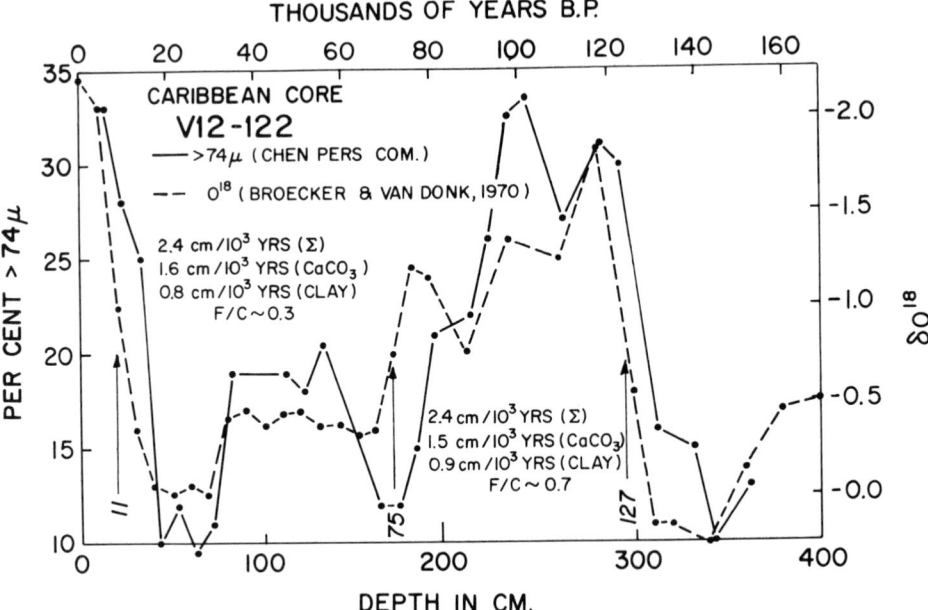

Fig. 7. Plot similar to Figure 4 for Caribbean core V12–122. The average carbonate content of the glacial interval is taken to be 67 percent and that of the interglacial interval 62 percent (Broecker and Van Donk 1970).

be challenged. Although the absolute ages assigned to these boundaries are the subject of some disagreement (Broecker and Ku 1969; and Emiliani and Rona 1969), the core-to-core correlations are firm. If the Broecker-Van Donk (1970) time scale is correct, then in each case the average CaCO₃ accumulation rate is no different during the interglacial interval than during the glacial interval!

Hays et al. (1969) have recently confirmed Arrhenius' (1952) carbonate cycles in cores from the equatorial Pacific. C^{14} and paleomagnetic dating on core RC11-209 suggest that these cycles correspond to those for O^{18} in the Caribbean (Fig. 9). With the same 52,000- and 64,000-year time spans used for the last interglacial and glacial intervals in the Atlantic, the carbonate accumulation rate for the two periods is once again nearly the same.

Table 5 summarizes the accumulation rate data for core 60 also from the equatorial Pacific (Arrhenius 1952). The boundaries were selected in the same way as for RC11-209. In this case the carbonate accumulation rate is more than twice greater during glacial than during interglacial time. The clay rate accumulation is four times lower during glacial than interglacial time and the biogenic silica rates are about the same. Since, as will be discussed below, the interglacial section was subjected to far more intense solution effects than the glacial section, the significance of the higher carbonate accumulation rate in glacial time is open

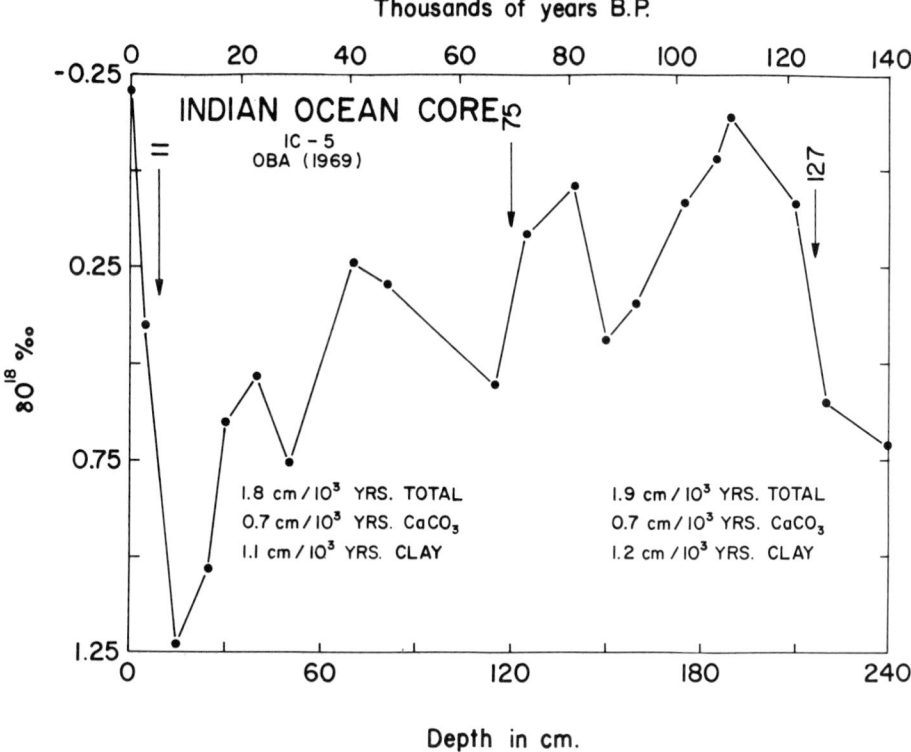

Fig. 8. Plot similar to that in Figure 4 for Indian Ocean core IC–5.

to question. The ratios of the silica accumulation rates may, in this case, prove more indicative of mixing rates than those of carbonate.

The accumulation rate data for the four Atlantic, the one Indian, and the two Pacific cores are summarized in Table 6.

For each of the six cores, coarse-fraction data are also available. When a sieve of 64 or 75 μ is used, the coarse fraction consists almost entirely of the tests of foraminifera shells. Although a substantial portion of the carbonate passing through the sieve represents small juvenile forms and broken tests of foraminifera, the majority is coccolith carbonate. Thus by sieving the core as well as determining its bulk $CaCO_3$ content, a rough estimate of the relative contributions of plant and animal $CaCO_3$ can be obtained. However, since solution of $CaCO_3$ leads to a rapid breakdown of foram shells into small fragments, a variation in this ratio can record solution rather than ecologic changes. For the Atlantic cores in all cases the ratio of coarse to fine $CaCO_3$ is considerably higher during interglacial than during glacial intervals. (This gives rise to the general correlation between the coarse fraction and O^{18} curves noted by Emiliani in 1955.) Thus whereas

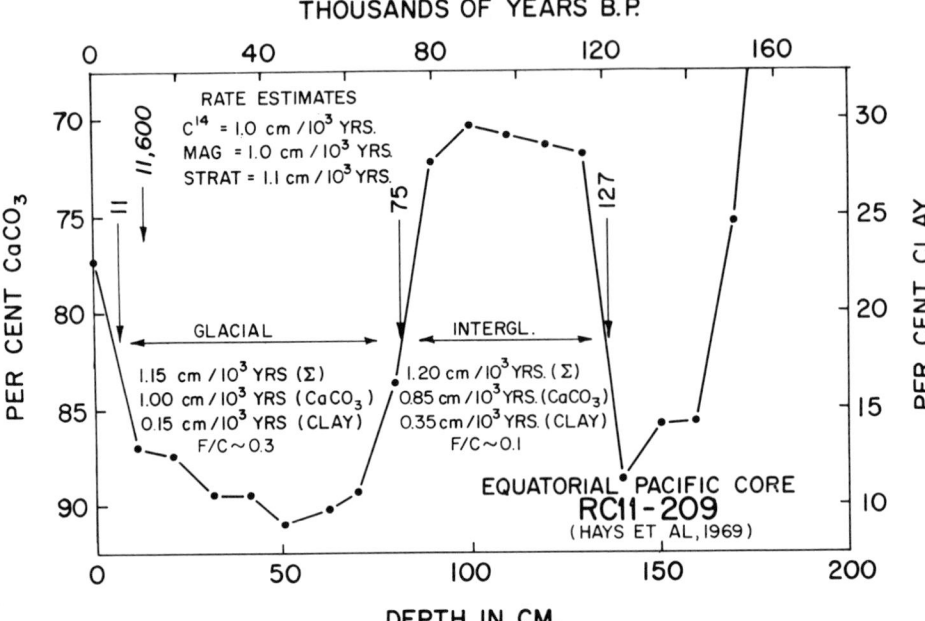

Fig. 9. Plot of percentage CaCO₃ versus depth in equatorial Pacific core RC11–209. The coarse fraction data are summarized in the text. The single radiocarbon date available for this core (11,600 years at a depth of 11 cm) is shown. The Brunhes-Matuyama magnetic boundary appears at a depth of 700 cm in this core (Hays et al. 1969).

the total rate of CaCO₃ accumulation remains nearly constant, the rate at which fine carbonate accumulates is slightly higher during glacial intervals and that for coarse carbonate considerably higher during interglacial intervals in the Atlantic and Caribbean.

Hays et al. (1969) do not give coarse-fraction data for core RC11-209. Four analyses made as part of this study by Guy Mathieu yielded 25 and 32 percent greater than 74 μ for the glacial interval (at 30 and 60 cm depth) and 6 and 7 percent for the interglacial interval (100 and 120 cm). The change in the ratio of coarse to fine carbonate is thus in the opposite sense for the equatorial Pacific than for the Atlantic and Caribbean (see Table 7 for a summary of the glacial–interglacial comparison of fine and coarse accumulation rates for all six cores). This pattern is consistent with that found by Arrhenius (1952) for core 60 and for other cores from the equatorial Pacific.

The problem in interpreting these results is clear. The fine and coarse components give different answers. While the fine component shows almost no glacial to interglacial change in either ocean, the coarse component increases in one ocean and decreases in the other. How much of this change is ecologic, how much due to solution and the physical degradation of the foram shells, and how much to a change in mixing pattern and time scale are the next matters for discussion.

Table 5. Accumulation Rates and Foram Statistics for Core 60
from the Equatorial Pacific

(Based on data given by Arrhenius 1952)

	Last Inter- glacial	Last glacial
Total sediment (cm/10^3 yrs) [*]	1.73	2.73
$CaCO_3$ (cm/10^3 yrs)	1.00	2.32
Clay (cm/10^3 yrs)	0.33	0.08
SiO_2 (cm/10^3 yrs)	0.40	0.33
Foram fragments/whole planktonics	12	4
Planktonics/gm	800	3500
Benthics/gm	200	200
Benthics/planktonics	0.24	0.06

[*] Based on assumptions of age 11,000 years at 15 cm, 75,000 years at 190 cm, 127,000 years at 280 cm. A single Pa^{231}/Th^{230} age of 95,000 years at a depth of 200 cm has been obtained for this core (see Arrhenius 1963). However, the inherent uncertainty in such an age is sufficiently large that it can only be said that it is not inconsistent with the time scale adopted here.

ATTRITION BY SOLUTION

Several indicators of partial carbonate solution have been proposed. Ruddiman and Heezen (1967) and Berger (1968) have summarized efforts to use the differential preservation of individual planktonic species; Arrhenius (1952) and Oba (1969) have summarized the use of the ratio of foram fragments to whole foram tests and of the ratio of benthic to planktonic forams; Chen (1968) has shown the usefulness of alternating occurrence and absence of aragonitic pteropods. The logic behind these studies is that certain types of carbonate (for example, benthic foram shells) are more resistant to solution than others and that the degradation of shells to fragments is the first step in the solution sequence for forams.

For several of the areas considered in this paper good evidence exists that solution effects were more important during *interglacial* than during glacial time. Oba's (1969) results show a strong increase in both the ratio of benthic forams and of foram fragments relative to whole-foram tests in interglacial relative to glacial sections in core IC-5 from the equatorial Indian Ocean. Arrhenius (1952) shows the same pattern in core 60 from the equatorial Pacific. Chen (1968) shows that pteropods were present in the glacial stages of Caribbean cores but absent in interglacial sections. Figure 10 summarizes these data.

In the north and equatorial Atlantic, cores taken from depths less than 4500 meters do not show the high percentages of benthic fauna found in the interglacial Pacific and Indian sediments (Phleger et al. 1953). The presence of

Table 6. Summary of Accumulation Rate Estimates for Carbonate and
Noncarbonate Material in Five Oceanic Areas

Core no.	NONCARBONATE			CARBONATE		
	Glacial	Interglacial	(G/I)	Glacial	Interglacial	(G/I)
	(cm/10³ yr)			(cm/10³ yr)		
North Atlantic						
280	3.6	1.4	2.6	1.9	2.1	0.9
Equatorial Atlantic						
A180-73	0.9	1.0	0.9	1.9	2.1	0.9
Caribbean						
A240-M1	0.7	1.0	0.7	1.4	1.3	1.1
V12-122	0.8	0.9	0.9	1.6	1.5	1.1
Equatorial Pacific						
60	0.4	0.7	0.6	2.3	1.0	2.3
RC11-209	0.15	0.35	0.4	1.0	0.9	1.1
Equatorial Indian						
IC-5	1.1	1.2	0.9	0.7	0.7	1.0

Table 7. Summary of Accumulation Rate Estimates for the Coarse and Fine Contributions to the Carbonate

Core no.	Coarse CaCO$_3$ (cm/10³ yr)			Fine CaCO$_3$ (cm/10³ yr)		
	Glacial	Interglacial	(G/I)	Glacial	Interglacial	(G/I)
North Atlantic						
280	0.4	0.6	0.7	1.5	1.5	1.0
Equatorial Atlantic						
A180-73	0.6	0.9	0.7	1.3	1.2	1.1
Caribbean						
A240-M1	0.5	0.7	0.7	0.9*	0.6	1.5
V12-122	0.4	0.6	0.7	1.2*	0.9	1.3
Equatorial Pacific						
RC11-209	0.25	0.08	3.0	0.75	0.75	1.0
Equatorial Indian						
IC-5	0.4	0.3	1.2	0.3	0.4	0.8

* Includes pteropods.

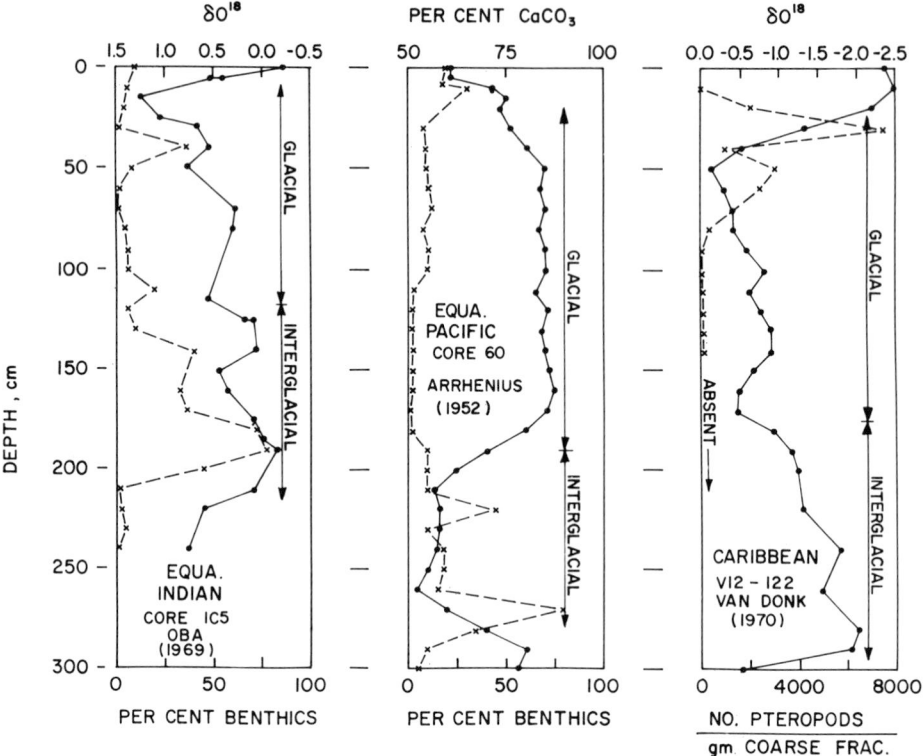

Fig. 10. Evidence for enhanced solution effects during interglacial compared to glacial times. For the Pacific and Indian cores the evidence comes from the great enrichment of benthic forams in the coarse fraction through the selective destruction of planktonic tests and for the Caribbean from the presence of pteropods in glacial age sediments (Chen, personal communication). The solid curves give the O^{18} data and the dashed lines the faunal data.

Table 8. Variation of the Frequency of Benthic Tests
with That of Planktonic Tests in Equatorial Pacific Cores

(Arrhenius 1952; based on many thousand analyses)

Planktonic	Benthics	Ratio
(entire tests/gm)		*(B/P)*
10,000	300	0.03
1,000	150	0.15
100	80	0.80
10	25	2.5
1	10	10

pteropods in the glacial age Caribbean sediments attests to the lack of solution of calcite. Imbrie (personal communication) has found the benthic fauna to be uniformly low in Caribbean core V12-122. Thus no evidence for large solution losses of calcite appears in the Atlantic record.

As for the measurements of accumulation rates made here, the implication is that at least during interglacial times most of the cores used in this study lost more $CaCO_3$ to solution than during glacial times. The actual ratio of interglacial to glacial rain of $CaCO_3$ must be somewhat greater than that measured in the sediments (hence the interglacial rain rate may have been somewhat greater than the glacial rain rate rather than equal to it, as suggested by sediment dating). Finally the enhancement of the coarse relative to the fine $CaCO_3$ fraction in the Indian and Pacific oceans during glacial time reflects, at least in part, the greater deterioration of foram tests deposited during interglacial time.

Measurements by Li et al. (1969) demonstrate that waters in the Atlantic Ocean at depths in excess of 5000 meters are undersaturated with respect to the mineral calcite. Waters below 1000 meters in the Pacific are either close to saturation or undersaturated with respect to calcite. As these authors point out, the difference in level reflects the fact that in the deep sea there is an excess of CO_2 derived from oxidation of organic tissue with respect to $CO_3^=$ derived from calcium carbonate solution. This excess is considerably greater in the Pacific than in the Atlantic (i.e. $[\Delta\Sigma CO_2 - 1/2\Delta A]_P > [\Delta\Sigma CO_2 - 1/2\Delta A]_A$). This leads to a higher carbonate ion content at all levels in the Atlantic than in the Pacific deep waters and the consequent Atlantic to Pacific rise in the $CaCO_3$ compensation level. It thus depicts the present mode of deep-water circulation.

Although the correspondence is not exact, the level where calcite disappears from deep-sea sediments roughly corresponds to the saturation horizon in the deep water. A notable exception is the equatorial Pacific where abundant carbonate material is found well below the presumed saturation horizon. In this region either local saturation of bottom water or the failure of dissolution to match the carbonate rain permits carbonate-rich sediments to accumulate.

Li et al. (1969) also point out that the fraction of the deep-sea floor on which $CaCO_3$ accumulates must be self-adjusting such that this material leaves the ocean at a rate matching its input through continental erosion. Since marine organisms precipitate $CaCO_3$ at a rate several times faster than it is being supplied to the ocean, a balance between input and loss is maintained by solution of carbonate in the deep sea.

The response time of the ocean to changes in the input–output balance is given roughly by the ratio of the carbonate ion content of the ocean ($\sim$400 moles/ m^2) to the average rate of calcium carbonate accumulation (0.02 mole/m^2 yr); hence about 20,000 years. If this time constant is correct, there is time within a given climatic interval for a substantial response in the level of compensation. Thus an increase in the oceanwide rate of $CaCO_3$ accumulation (or a decrease in $CaCO_3$ supply rate by rivers) should be matched by a rise in the saturation horizon. This rise should be recorded especially in those carbonate-bearing sediments lying close to the compensation level.

If the above-mentioned indicators of solution are valid, and if these few cores from key areas typify the world ocean, then the conclusion can be drawn that the carbonate compensation level stood at greater depths during glacial than during interglacial times. This in turn demands that the ratio of $CaCO_3$ input by rivers to $CaCO_3$ precipitation by organisms was higher during glacial than during interglacial times. It is not possible from this evidence alone to say whether this change in glacial time reflects an increase in $CaCO_3$ input to the ocean or a decrease in productivity (and hence a decrease in the average rate of oceanic mixing).

ECOLOGIC EFFECTS

The destruction of foram tests by partial solution in the interglacial intervals of the equatorial Indian and Pacific cores confounds any attempt to use the faunal and floral changes associated with interglacial to glacial climatic changes as indices of ecologic variation. On the other hand, in the Atlantic and Caribbean the large glacial to interglacial change in coarse to fine fraction ratio can only be explained by a change in the ratio of coccoliths to forams living in the surface waters during glacial and interglacial time. If the forams are taken as the best index of overall productivity, then we conclude that productivity was greater during interglacial than glacial times. If, instead, the coccoliths are chosen, no significant difference is found.

This paper is presented more as a guide to future research than as a final report. The potential of the approach described is clear. The conclusions to be drawn from the data in hand must be vague and tentative because of the many remaining loose ends. There is an indication, however, that productivity of carbonate was greater during interglacial than during glacial time. This higher productivity manifests itself as a drop in the calcite compensation level in the equatorial Indian and equatorial Pacific oceans and of the aragonite compensation level in the Caribbean Sea during interglacial times. It also may be reflected by a higher rate of foram test production throughout the Atlantic in interglacial compared to glacial time. These observations could, of course, be attributed to a higher oceanic concentration of phosphorus during interglacial than glacial times, or to an increased $CaCO_3$ supply to the ocean during interglacial time but a better guess is that they reflect more vigorous oceanic mixing during interglacial than during glacial times. The change in mixing rate need not have been very large; the observations require no more than a few tens of percent reduction during the last glacial period. The study clearly rules out large changes in mixing rate as, for example, were called upon by Weyl (1968) in his theory of glaciation.

NOTES OF ADDED CONFUSION

Since the completion of this paper I have become aware of two additional complications. First Peter Weyl at Stony Brook, upon hearing this story, pointed out that the lower $CaCO_3$ productivity called upon during glacial times may reflect merely the decrease of carbonate deposition rate in shallow-water environments resulting from lowered sea level. If so, then there would be no need for a mixing rate change. Second, carbon isotope results obtained by Van Donk (1970) on benthic and planktonic forams raise the suspicion that the phosphorus to carbon ratio in sea water has changed over the course of a single climatic cycle. Perhaps, as suggested by Arrhenius (1967), the residence time of phosphorus in the ocean is only about 10,000 years. Broecker and Van Donk are preparing a separate paper on this subject.

Summary. Among the factors which control the rate at which $CaCO_3$ accumulates on any region of the sea floor are the patterns and rates of large-scale oceanic mixing. If the influence of the other factors of importance can be isolated, then by contrasting $CaCO_3$ accumulation rates between glacial and interglacial intervals at various places on the sea floor it should be possible to place some restrictions on the degree to which oceanic mixing changes with climate. A preliminary evaluation of this scheme is presented in this paper. The first step is to relate the observed rates of $CaCO_3$ accumulation to the present-day mode of oceanic mixing and to ascertain the sensitivity of these rates to changes in this scheme. Next, the rates of accumulation during the last interglacial and the last glacial periods are compared in the central North Atlantic, equatorial Atlantic, Caribbean, equatorial Indian, and equatorial Pacific oceans. The total rate of $CaCO_3$ accumulation is found to remain nearly constant in all these areas, but that of foraminiferal shells is distinctly lower during glacial than interglacial time in the Atlantic and higher during glacial than interglacial time in the Pacific. Finally, means of separating the influences of ecologic and solution effects from those of mixing are discussed.

Acknowledgments. Discussions with K. K. Turekian, J. Imbrie, H. Craig, T. Takahashi, A. McIntyre, C. Chen, P. Weyl, and Y. H. Li have proven very helpful to the author.

Financial support was provided by a grant from the U.S. Atomic Energy Commission, AT (30–1) 2663; and from National Science Foundation grant GA–1346.

Lamont-Doherty Geological Observatory Contribution No. 1565.

REFERENCES

Arrhenius, G., Sediment cores from the East Pacific, *Swedish Deep-Sea Exped., 1947–1948, Repts.,* 5, 1, 1952.

Arrhenius, G., Pelagic sediments, in *The Sea,* vol. 3, chap. 25, 655, Wiley, New York and London, 1963.

Arrhenius, G., Deep sea sedimentation: A critical review of U.S. work 1963–1967, Report of the 14th General Assembly of the International Union of Geodesy and Geophysics.

Berger, W. H., Planktonic foraminifera: Shell production and preservation, 241 pp., Ph.D. thesis, University of California, San Diego, 1968.

Broecker, W. S., and T. L. Ku, Caribbean cores P6304–8 and P6304–9: New analysis of absolute chronology, *Science, 166,* 404, 1969.

Broecker, W. S., and Y. H. Li, Interchange of water between the major oceans, *J. Geophys. Res., 75,* 3545, 1970.

Broecker, Wallace S., and Jan van Donk, Insolation changes, ice volumes, and the O^{18} record in deep-sea cores, *Rev. Geophys. Space Phys., 8,* 169, 1970.

Broecker, W. S., K. K. Turekian, and B. C. Heezen, The relation of deep sea sedimentation rates to variations in climate, *Am. J. Sci., 256,* 503, 1958.

Chen, C., Pleistocene pteropods in pelagic sediments, *Nature, 219,* 1145, 1968.

Craig, H., and L. I. Gordon, Isotopic oceanography: Deuterium and oxygen–18 variations in the ocean and the marine atmosphere, *Symposium on Marine Geochemistry,* University of Rhode Island, Occ. Publ., no. 3, 277, 1965.

Emiliani, C., Pleistocene temperature, *J. Geol., 63,* 538, 1955.

Emiliani, C., Paleotemperature analysis of core 280 and Pleistocene correlations, *J. Geol., 66,* 264, 1958.

Emiliani, C., Paleotemperature analysis of Caribbean cores P6304–8 and P6304–9 and a generalized temperature curve for the past 425,000 years, *J. Geol., 74,* 109, 1966.

Emiliani, C., and E. Rona, Caribbean cores P6304–8 and P6304–9: New analysis of absolute chronology. A reply, *Science, 166,* 1551, 1969.

Ericson, D. B., and G. Wollin, Correlation of six cores from the equatorial Atlantic and the Caribbean, *Deep-Sea Res., 3,* 104, 1956.

Ericson, D. B., M. Ewing, G. Wollin, and B. C. Heezen, Atlantic deep-sea sediment cores, *Bull. Geol. Soc. Am., 72,* 193, 1961.

Goldberg, E. D., and M. Koide, Geochronological studies of deep sea sediments by the ionium/thorium method, *Geochim. Cosmochim. Acta, 26,* 417, 1962.

Goldberg, E. D., and M. Koide, Rates of sediment accumulation in the Indian Ocean, in *Earth Science and Meteoritics,* p. 90, North-Holland, Amsterdam, 1963.

Goldberg, E. D., and J. J. Griffin, Sedimentation rates and mineralogy in the South Atlantic, *J. Geophys. Res., 69,* 4293, 1964.

Goldberg, E. D., M. Koide, J. J. Griffin, and M. N. A. Peterson, A geochronological and sedimentary profile across the North Atlantic Ocean, in *Isotopic and Cosmic Chemistry,* p. 211, North-Holland, Amsterdam, 1963.

Goldberg, E. D., M. Koide, and J. J. Griffin, Ionium/thorium geochronology on Miami core A254–Br–C, in *Recent Researches in the Fields of Hydrosphere, Atmosphere and Nuclear Geochemistry,* Sugswara Festival Volume, Maruzen, Tokyo, 1964.

Hays, J. D., T. Saito, N. D. Opdyke, and L. H. Burckle, Pliocene-Pleistocene sediments of the equatorial Pacific: Their paleomagnetic, biostratigraphic, and climatic record, *Bull. Geol. Soc. Am., 80,* 1481, 1969.

Ku, T. L., Uranium series disequilibrium in deep sea sediments, 157 pp., Ph. D. thesis, Columbia University, New York, 1966.

Ku, T. L., and W. S. Broecker, Atlantic deep-sea stratigraphy: Extension of absolute chronology to 320,000 years, *Science, 151,* 448, 1966.

Ku, T. L., W. S. Broecker, and N. Opdyke, Comparison of sedimentation rates measured by paleomagnetic and the ionium methods of age determination, *Earth Planetary Sci. Letters, 4,* 1, 1968.

Lal, D., Characteristics of large scale oceanic circulation as derived from the distribution of radioactive elements, *Proc. 2nd Intern. Oceanog. Congr.*, North-Holland, Amsterdam, 1969.

Li, Y. H., T. Takahashi, and W. S. Broecker, Degree of saturation of $CaCO_3$ in the oceans, *J. Geophys. Res.*, 74, 5507, 1969.

Oba, T., Biostratigraphy and isotopic paleotemperature of some deep-sea cores from the Indian Ocean, Tohoku University, *Sci. Rept., 2nd Ser. (Geol.)*, 41, 2, 129, 1969.

Phleger, F. B., F. L. Parker, and J. F. Peirson, North Atlantic foraminifera, *Swedish Deep-Sea Exped., 1947–1948, Repts.*, 7, 1, 1953.

Rosholt, J. N., C. Emiliani, J. Geiss, F. F. Koczy, and P. J. Wangersky, Absolute dating of deep-sea cores by the Pa^{231}/Th^{230} method, *J. Geol.*, 69, 162, 1961.

Rubin, M., and H. E. Suess, U.S. Geological Survey radiocarbon dates II, *Science*, 121, 481, 1955.

Rubin, M., and H. E. Suess, U.S. Geological Survey radiocarbon dates III, *Science*, 123, 442, 1956.

Ruddiman, W. F., and B. C. Heezen, Differential solution of planktonic foraminifera, *Deep-Sea Res.*, 14, 801, 1967.

Rusnak, G. A., A. L. Bowman, and H. G. Ostlund, Miami natural radiocarbon measurements II, *Radiocarbon*, 5, 23, 1963.

Sverdrup, H. U., M. W. Johnson, R. H. Fleming, *The Oceans*, 1060 pp., Prentice-Hall, Englewood Cliffs, N.J., 1942.

Turekian, K. K., Some aspects of the geochemistry of marine sediments, in *Chemical Oceanography*, edited by Riley and Skirrow, Vol. 2, p. 81, Academic Press, N.Y., 1965.

Van Donk, J., The oxygen isotope records in deep-sea sediments, Ph. D. thesis, Columbia University, New York, 1970.

Vernekar, A. D., Long-period global variations of incoming solar radiation, in *Research on the Theory of Climate*, vol. 2, 289 pp., report of the Travelers Research Center, Hartford, Conn., 1968.

Weyl, P. K., The role of the oceans in climatic change: A theory of the ice ages, *Meteorol. Monographs*, 8, 37, 1968.

Wiseman, J. D. H., The rates of accumulation of nitrogen and calcium carbonate on the equatorial Atlantic floor, *Advan. Sci.* (London), 12, 579, 1956.

George H. Denton, Richard L. Armstrong,
and Minze Stuiver

10. THE LATE CENOZOIC GLACIAL HISTORY

OF ANTARCTICA

At the present time the planet earth supports approximately 26×10^6 km³ of glacier ice (Flint 1969) which covers about 10 percent of the available land area and is concentrated in the Greenland ice sheet, the Antarctic ice sheet, and smaller glaciers in Arctic and alpine regions. About 2.6×10^6 km³, or 10 percent of the total, resides in the Greenland ice sheet (Bauer 1955), and less than 1.0 percent in small alpine and Arctic glaciers. Thus slightly more than 89 percent of the total, or about 24×10^6 km³ (Table 1) (Bardin and Suyetova 1967), is represented by the huge Antarctic ice sheet, centered roughly over the south pole.

The Antarctic ice sheet greatly influences our environment. It forms a major heat sink and therefore is a significant factor in the heat budget of the earth. Cold saline water formed in the Southern Ocean near Antarctica composes most of the deep water of the world ocean and plays a dominant role in circulation of the oceans. If melted, the Antarctic ice sheet would release enough water to raise sea level by about 55 meters. Substantial melting of the sheet may have caused the high sea levels of Quaternary interglacial ages, and future melting could flood settlements on coastal areas. Although it is unlikely that a large volume of Antarctic ice will melt in the foreseeable future, the distinct possibility remains that the Antarctic ice sheet may undergo regular and catastrophic surges (Wilson 1964, 1969; Weertman 1966), much like many present-day glaciers in Alaska, Canada, Iceland, and Spitsbergen (Meier and Post 1969; Thorarinson 1969; Liestøl 1969). The discharge of large quantities of surging glacier ice from Antarctica into the adjacent Southern Ocean would cause rapid eustatic rises of sea level. For example, a large surge of ice from Wilkes Land could raise sea level by 17 meters in less than 100 years (Hollin 1969, pp. 903–04). Wilson (1964) even postulated that large-scale surges of the Antarctic ice sheet triggered past Quaternary ice ages and may well cause them in future. Moreover, major surges of the Antarctic ice sheet in the Tertiary may have caused changes in both sea level and climate.

For all these reasons it seems pertinent to ask several questions about the

Table 1. Area and Volume of Glacier Ice in Antarctica

(Data from Bardin and Suyetova 1967; Flint 1969; and Mercer 1968a)

Glacier	Area (10^6 km^2)		Volume (10^6 km^3)		Potential effect on sea level† (m)
	With ice shelves	Excluding ice shelves	With ice shelves	Excluding ice shelves*	
Ice Sheet in West Antarctica	3.532	2.297	3.830	3.300 (3.026)	8 − 4 = 4
Ice Sheet in East Antarctica	10.443	10.183	20.201	20.149 (18.477)	51
Entire Antarctic Ice Sheet	13.975	12.480	24.031	23.449 (21.503)	59 − 4 = 55

* Water equivalent given in parentheses and calculated as 91.7 percent of ice volume.

† The sea level effect is based on the volume of the land-based ice sheet, for the ice shelves are floating and thus have no effect on sea level. Moreover, 4 meters must be deducted from the sea level effect of the ice sheet in West Antarctica, because only about 1.5 to 1.8 x 10⁶ km³ of this ice sheet is located above present sea level (Mercer 1968a, p. 81). The actual effect of the Antarctic ice sheet on sea level might be diminished by one-third if the sea floor adjusts isostatically to loads of this magnitude, as suggested by Bloom (1967). The area of the world ocean is taken as 362 x 10⁶ km² (Menard and Smith 1966).

Antarctic ice sheet. What are its present characteristics? Is it growing thicker? When, why, and how did it form? What was its subsequent behavior? How did it affect worldwide Quaternary glaciations? What will occur in the future? In this paper possible answers to these questions are discussed in the light of presently available data, although many gaps exist and much work remains to be done in Antarctica. This review is by no means complete and focuses only on the questions posed above. For a more complete historical survey of the glacial history of Antarctica, see Nichols (1964, 1966).

PRESENT ANTARCTIC ICE SHEET

The present Antarctic ice sheet and floating ice shelves contain about 24×10^6 km^3 of glacier ice spread over an area of nearly 14×10^6 km^2 (Bardin and Suyetova 1967, Tables 1, 4); on the basis of differences in basal and surface topography, it can be divided into ice sheets in East and West Antarctica (Fig. 1).[1] The Transantarctic Mountains separate the two ice sheets except in the area of the Filchner ice shelf.

The ice sheet in East Antarctica overlies a low-relief bedrock surface situated almost entirely above sea level (Bentley et al. 1964). One margin is dammed behind the Transantarctic Mountains, but elsewhere the sheet in East Antarctica either terminates directly in the ocean or is fringed by floating ice shelves, including the Amery ice shelf. All these shelves are protected by embayments or islands; none occurs on the open ocean. Centered roughly over the center of East Antarctica (Fig. 1), the ice sheet forms a huge elliptical dome which reaches maximum surface altitudes of about 400 meters on the interior polar plateau and is some 3600 meters thick (Bentley et al. 1964). The entire sheet in East Antarctica, including shelves, contains about 20.2×10^6 km^3 of glacier ice, or about 83 percent of the total volume of ice in Antarctica; the grounded ice covers a land area of about 10.2×10^6 km^2 (Bardin and Suyetova 1967, Tables 2–4).

The ice sheet in West Antarctica overlies a rugged bedrock floor, much of which is now well below sea level and, with several exceptions, would remain so if the ice sheet were removed and the continent were allowed to adjust isostatically (Bentley and Ostenso 1961, pp. 892–95). These exceptions include the four high-mountain areas that probably formed the original island groups in West Antarctica (Giovinetto 1964, p. 150). The surface topography of the sheet in West Antarctica is very irregular, due in part to the rugged bedrock floor. Floating ice shelves, all of which occur in protected embayments or are buttressed on islands, fringe the sheet. The Ross and Filchner ice shelves, the two largest,

1. Antarctica is divided geographically into East and West Antarctica. The basic dividing line is the 0°–180° meridian. These divisions have nothing to do with compass directions in any given part of Antarctica. For example, in the Ross ice shelf area, one goes west into East Antarctica and east into West Antarctica. East Antarctica includes all of Antarctica between longitudes 0° and 180°; West Antarctica lies between longitudes 180° and 360°.

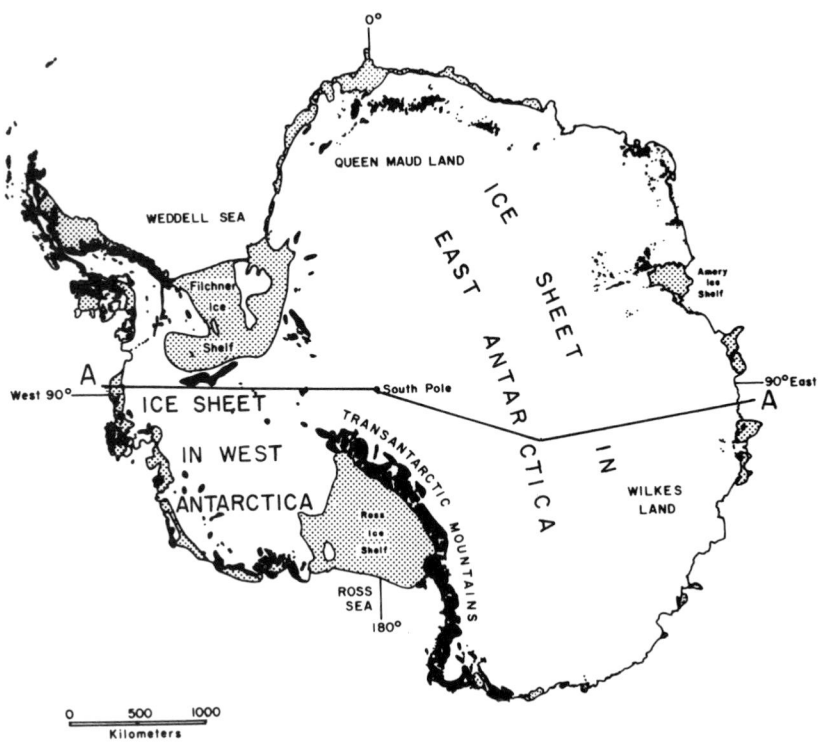

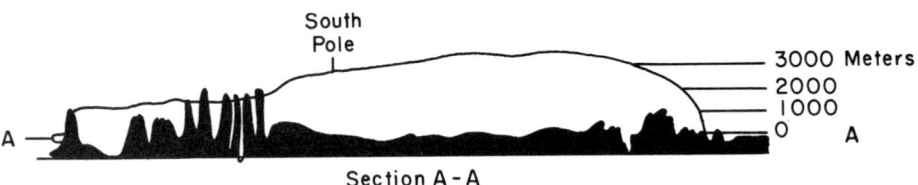

Fig. 1. Antarctic ice sheet. Ice shelves indicated in stipple; exposed bedrock shown in black (after Bentley 1965).

occupy major embayments that reach far inland; the Ross ice shelf is about 0.53×10^6 km² in area and the Filchner ice shelf about 0.40×10^6 km² (Swithinbank and Zumberge 1965, p. 199).

From a mass-balance viewpoint the Antarctic ice sheet is unique. Nearly the entire surface forms a huge area of accumulation (Giovinetto 1964), with rates

Table 2. Mass Balance of Antarctic Ice Sheet and Ice Shelves

(Data from Bull, in press, and Loewe 1967)

	Mass input $(10^6$ g. $yr^{-1})$	Mass output $(10^6$ g. $yr^{-1})$
Positive balance on accumulation zone	208	
Negative balance on ablation zone		1
Calving		145
Bottom melting of ice shelves		20
Excess of estimated input over output:	42 x 10^6 g. yr^{-1}	

of accumulation varying from more than 40 grams per cm² per year near the coast to less than 10 grams per cm² per year over vast areas of the interior (Bull, in press). Nearly all loss of ice from Antarctica occurs by calving of icebergs into the surrounding ocean. Thus the lateral extent of the ice sheet is determined by sea level which controls the position of the grounding line where glacier ice begins to float, thus becoming subject to calving (Hollin 1962, p. 179). Without the confining influence of the peripheral Southern Ocean, the ice sheet probably would cover a much larger area. The most accurate mass-balance figures for the present Antarctic sheet have been calculated by Bull (in press) and are given in Table 2. Although the calculation indicates a positive mass balance, the results must be treated with great caution because of possible large errors in ablation terms (Bull, in press). If real, the positive mass balance suggests that the entire ice sheet is growing thicker, on the average, by 3 cm per year.

Because of difficulties inherent in determining the overall mass balance of the ice sheet, attention has been focused on the budgets of individual drainage systems within the sheet (Fig. 2). The four systems that have been examined in some detail include the western part of the Ross ice shelf drainage system (Giovinetto et al. 1966), the Lambert Glacier system (Budd et al. 1967), part of a system along the Adelie coast (Lorius 1962), and part of the system in Wilkes Land (Budd 1966). All show a budget surplus, although the surplus for the last two is very close to the possible error in measurements. From the results of Giovinetto and others (1966), Hollin (in press) calculated that the surface level of the western part of the Ross ice shelf drainage system is rising at a rate of 3 cm per year. In sum, despite acknowledged uncertainties, a growing consensus suggests that much of the ice sheet is gaining volume. Other glaciological methods of determining whether the sheet is becoming thicker, such as strain measurements and core analyses, have given ambiguous results (Hollin, in press).

ORIGIN OF ANTARCTIC ICE SHEET

Evidence

The inception of glaciation on Antarctica, and the consequent spread of ice sheets over 98 percent of the continent with concomitant effects on worldwide climate,

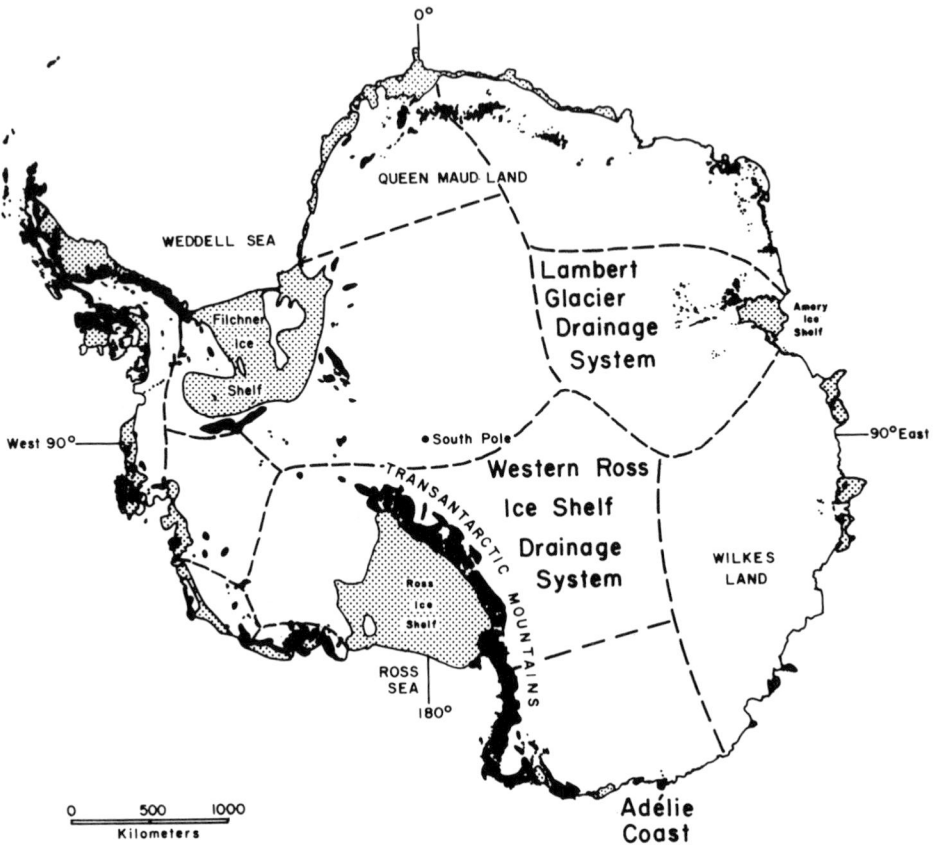

Fig. 2. Individual drainage basins of the Antarctic ice sheet (after Giovinetto 1964).

sea level, Quaternary glaciation, ocean circulation, and ecology in Antarctica, is a subject of interest to several disciplines. The present knowledge of the origin of the ice sheet is far from complete and must be pieced together from deep-sea cores from the Southern Ocean and from scattered stratigraphic sections in West Antarctica, East Antarctica, and Australia.

Tertiary stratigraphic sections in Antarctica are rare and have been studied only in reconnaissance. The best-known occur on islands near the Antarctic Peninsula in West Antarctica. Here rocks of Eocene, Oligocene, and Miocene age crop out on King George Island at lat. 62°S (Fig. 3). Numerous fossil leaves and wood fragments contained in these rocks indicate that a rich and varied flora, including many species of deciduous and coniferous trees, occupied the Antarctic Peninsula through much of the first half of the Tertiary (Adie 1964, pp. 129–35). Similarly, leaf and wood remains occur in the Seymour Island Series exposed on Seymour Island at lat. 64°S near the Antarctic peninsula (Adie 1964, p. 141). Dusén (1908) and Gothan (1908) described a rich, warm-climate flora of seventy

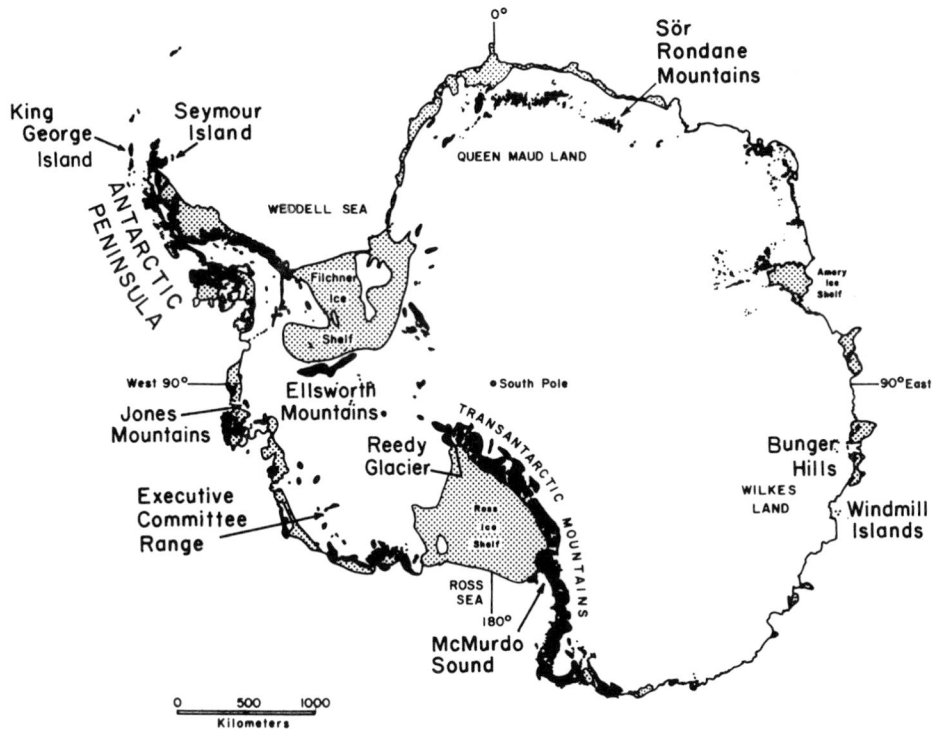

Fig. 3. Index map of Antarctica.

species from the lower part of this series; although Dusén (1908) considered these rocks to be Eocene and Oligocene, recent work has shown them to be early Miocene (Adie 1964, pp. 141, 148; Cranwell 1959, p. 1783). The upper part of the Seymour Island Series consists of littoral marine deposits with a rich invertebrate fauna also of early Miocene age (Sharman and Newton 1894, 1898; Adie 1964, p. 148). These marine deposits contain, in addition, the remains of fossil penguins that are of greater size than those now inhabiting the continent (Wiman 1905a, 1905b; Marples 1953). Cranwell (1959) examined pollen grains from a sample from this series and concluded that the pollen assemblage is dominated by conifers and southern beeches. From the evidence assembled above it appears that the climate of the Antarctic Peninsula may have been incompatible with extensive Cenozoic glaciation through at least the early Miocene; however, more study is necessary to assemble detailed Tertiary paleoclimatic data before this conclusion can be considered firm.

Tertiary rocks also occur as erratics in glacial drift on the lower slopes of volcanic islands in McMurdo Sound in East Antarctica at lat. 78°S (Fig. 3). These rocks, derived from beneath the Ross Sea or the ice sheet in West Antarctica (Cranwell et al. 1960, p. 700), have yielded pollen and microplankton assemblages of probable Eocene age (Cranwell et al. 1960; McIntyre and Wilson 1966, pp. 318–21; Wilson 1967, p. 82; Cranwell, written communication, 1969). The pollen assemblages suggest a dominance of several species of southern beeches, with lower abundances of conifers and ferns, and suggest that a temperate climate dominated this area of Antarctica during part of the Eocene.

The Tertiary stratigraphic sections of southern Australia are pertinent to the question of early Antarctic glaciation, for East Antarctica and Australia were contiguous through the early Tertiary (Sproll and Dietz 1969). The orientation and ages of magnetic anomalies on the ocean floor between the two continents indicate that East Antarctica and Australia separated about 40 m.y. ago and have since moved to their present positions (Heirtzler et al. 1968; LePichon 1968). The impression from sea floor structures around Antarctica and from paleomagnetic measurements from both continents is that Australia moved north while Antarctica remained near the south pole (Irving 1964; Creer 1967). Several areas in southern Australia expose Tertiary stratigraphic sections (Brown et al. 1968). No glacial deposits occur in these sections, and the fossil flora and fauna within the rocks suggest temperate climates throughout Eocene, Oligocene, and perhaps Miocene time, although there are indications of at least local cooling in the Miocene. Thus it appears that vast portions of the East Antarctic-Australia continent did not support large ice sheets prior to 40 m.y. ago, and in fact were marked by temperate climates. Furthermore, biological evidence and δO^{18} measurements of Tertiary fossil oysters and clams indicate that Australia suffered a sharp drop in temperature in the late Miocene (Gill 1961, pp. 333–42; Dorman 1966). This is compatible with formation of continent-wide ice sheets on Antarctica in the late Miocene or early Pliocene.

Indirect evidence of the volume of polar ice bodies can be obtained from Tertiary sea level curves, if sea level changes caused by polar ice sheets can be separated from changes due to tectonic events. In a review of Tertiary sea levels Tanner (1968) suggested that a steady decline of 70 to 100 meters began in middle Miocene and ended by the late Pliocene; he attributed it primarily to buildup of polar ice sheets. Moreover, Tertiary sea level positions do not provide any compelling evidence for the existence of large polar ice sheets prior to the middle Miocene, although reliable data are scarce for this time interval. Unfortunately, insufficient detail is available to refine these very general conclusions, although suggestions have been made about fluctuations of sea level during its general decline of Miocene and Pliocene time. On the basis of foraminifera assemblages, Kennett (1967, 1968) inferred low sea level and cool climate in the late Miocene in New Zealand, and Bandy (1968) recognized low sea levels in the late Miocene and middle Pliocene in California. Webb and Tessman (1967) suggested, from stratigraphic and vertebrate paleontologic evidence in Florida, that sea level was close to its present value by the middle Pliocene. In sum, the history of Tertiary sea level as presently interpreted suggests that large ice sheets did not build up in polar regions until after the middle Miocene. These sea level data, although suggestive, are very preliminary and must be treated with caution.

Another type of evidence for inception of extensive glaciers on Antarctica concerns late Miocene changes in pelagic sedimentation in the equatorial Pacific. These changes involved increased grain size, decreased sediment sorting, and increased carbonate fraction, as well as influx of chlorite, kaolinite, and pyroxene (Heath 1969a, 1969b). This change in sedimentation may well be related to increased Antarctic bottom water accompanying formation of the Antarctic ice sheet.

The earliest evidence of glaciers on Antarctica occurs as ice-rafted quartz grains with surface textures diagnostic of glaciation in lower Eocene, middle Eocene, and Oligocene sediments in deep-sea cores from the Southern Ocean (Geitzenauer et al. 1968; Rex and Margolis 1969; Margolis and Kennett, in press). These cores are dated by enclosed foraminifera. Although they have been attributed to continent-wide ice sheets, the quartz grains indicate only that calving glaciers existed on an adjacent continent; however, the size, distribution, and character of the inferred glaciers cannot be interpreted from deep-sea cores, as detailed in a later section. As mentioned previously, the paleoclimatic data derived from Tertiary rocks on the Antarctic Peninsula, in the McMurdo Sound region, and in southern Australia point to a temperate climate, inconsistent with large ice sheets, over vast portions of the Antarctic-Australian continent through much of the early and middle Tertiary. Moreover tillites of pre-Miocene age have not yet been found in Antarctica or southern Australia, and the history of early and middle Tertiary sea level positions affords no evidence of large polar ice sheets. Therefore, until more data are available, it seems safest to interpret the glacial quartz grains as the products not of widespread ice sheets but of relatively restricted calving glaciers. Coastal glacier complexes, including small ice caps, may have occupied highlands or mountain ranges on the Antarctic Peninsula, on islands in what is now West Antarctica (Rutford, ms., 1969, pp. 254–55), on the Transantarctic Mountains, on sub-Antarctic islands, or in southern Patagonia. Lack of Tertiary paleobotanical data from Queen Maud Land, however, leaves open the possibility of early Tertiary ice sheets in that region of Antarctica.

The oldest direct evidence of widespread glaciation on Antarctica occurs in the Jones Mountains near the coast of West Antarctica at 74°S latitude (Rutford et al. 1968) (Fig. 3). Here a major erosional unconformity truncates a granitic basement of Mesozoic age. This widespread, low-relief erosion surface is marked by glacial striations, grooves, and chattermarks; it is overlain by lenses of tillite, which in turn are associated with volcanics. Although preliminary K/Ar dates of the volcanic rocks were ambiguous, redating has shown that the volcanics and associated glacial deposits are at least 7 m.y. old (Rutford et al. 1970). They are thus late Miocene or early Pliocene in age, depending on whether the Miocene-Pliocene boundary is placed at 13 m.y. (Kulp 1961) or at 7 m.y. (Funnell 1964). Clasts incorporated within the tillite lenses include rock types that could have been derived only from the Ellsworth Mountains (Rutford, written communication, 1970); topographic and geographic factors in the Jones Mountains also argue for a substantial ice sheet in West Antarctica by the late Miocene or early Pliocene (Rutford, ms. 1969, pp. 250–51). Additional suggestions of Miocene-Pliocene glaciation in West Antarctica occur in the Executive Committee Range, where possible glacial grooves occur on volcanic rocks about 6 m.y. old (Doumani 1964, pp. 668, 670).

Evidence of the origin and age of the vast ice sheet in East Antarctica is rare, probably because most of the critical data are buried beneath glacier ice. Much of the available evidence, therefore, comes from the peripheral Transantarctic Mountains. At the present time the huge ice sheet in East Antarctica is dammed by the Transantarctic Mountains (Bentley et al. 1964). A number of glacial valleys, which cut across the Transantarctic Mountains, are filled with long outlet glaciers that head in the ice sheet in East Antarctica and terminate either in the Ross Sea or in the Ross ice shelf. Probably most of the glacial valleys were carved by outlet glaciers from the ice sheet in East Antarctica, although some may have been cut by glaciers fed from local ice sheets located on presently ice-buried mountains on the inland flank of the Transantarctic Mountains.

Taylor, Wright, and Victoria valleys, which cross the Transantarctic Mountains in the McMurdo Sound region (Figs. 3, 4), are of special interest because they are now ice-free except for small alpine glaciers on their walls. They cross the mountains from the ice sheet on the west to McMurdo Sound on the east. Taylor and Wright Upper glaciers, which are small tongues of the ice sheet, spill over bedrock thresholds and occupy the western ends of Taylor and Wright valleys. The altitude of the ice sheet surface is not sufficient to allow large quantities of ice to flow over the thresholds that occupy the valley heads (Bull et al. 1962, p. 71; Bull 1966, p. 177). However, these glaciers formerly flowed through the valleys when the ice sheet west of the valleys attained higher surface altitude.

Taylor and Wright valleys exhibit classical features of glacial erosion, both head in an area where the ice sheet is dammed to considerable thickness immediately behind the steep western flank of the mountains (Crary and Van der Hoeven 1961), and both were carved by outlet glaciers from the ice sheet and not by local alpine glaciers or intermontane ice sheets (Nichols 1961; Bull et al. 1962, p. 72; Denton et al. 1969, p. 269, 1970, p. 16; Calkin and others 1970, p. 22). Numerous basaltic lava flows occur in association with glacial features in Taylor and Wright valleys. K/Ar dates of lava flows in critical stratigraphic positions show that these valleys had been glaciated extensively and had attained essentially their present profiles prior to about 4 m.y. ago (Denton et al. 1969, pp. 269, 277, 1970, p. 20). In order to erode these valleys through the Transantarctic Mountains, the ice sheet in East Antarctica must have attained a full-bodied stage, with surface levels at least locally higher than now, during or before the Pliocene epoch.

Farther north, on islands and peninsulas in the northwestern portion of the Ross Sea, Hamilton and Armstrong (1969) have reported ice-contact breccias up to 7.4 m.y. old that may indicate more extensive ice in late Miocene and early Pliocene time than exists today in that region.

The occurrence of ice-rafted sediments in deep-sea cores from the floor of

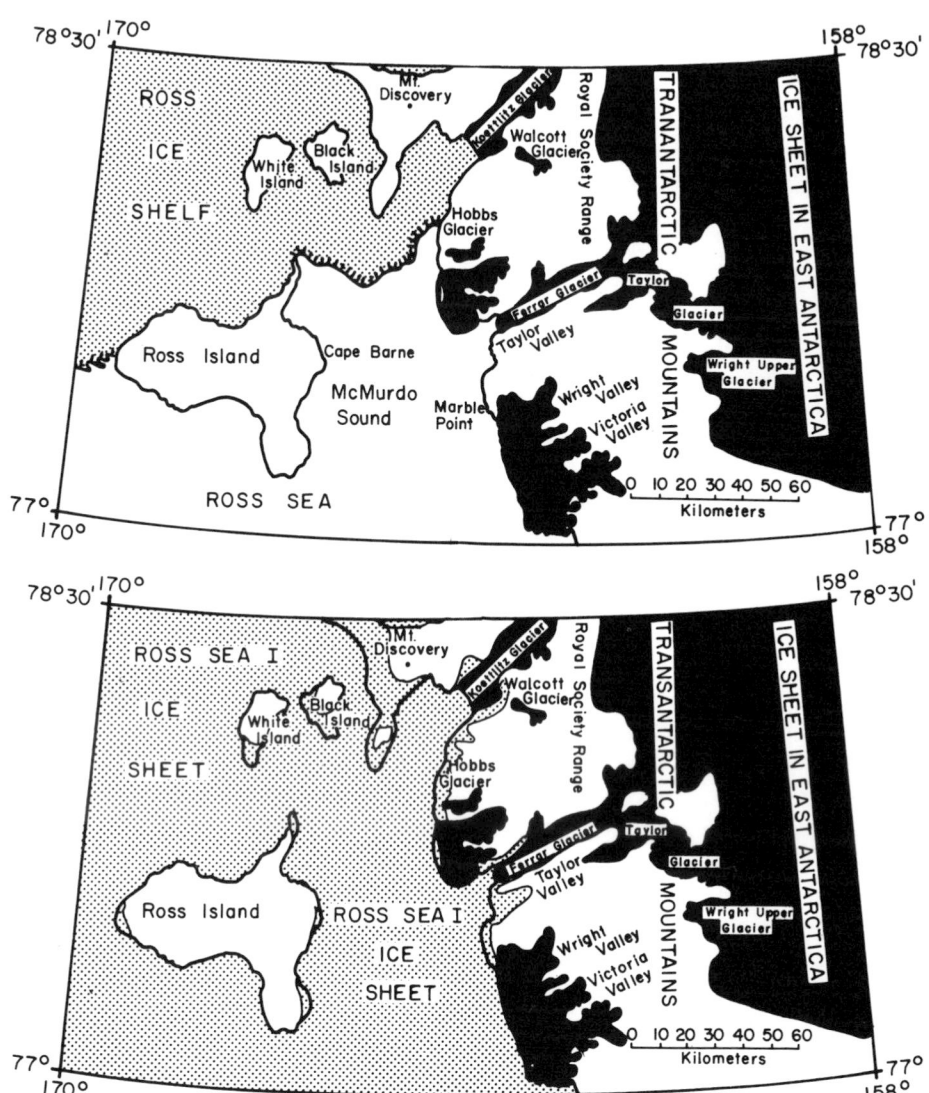

Fig. 4. McMurdo Sound, southern Victoria Land, Antarctica (south is at the top). Upper: Index map; only a few pertinent alpine glaciers are shown in the Transantarctic Mountains. Lower: The extent of the Ross Sea I ice sheet in McMurdo Sound. In places this ice sheet exceeded 1000 meters in thickness. The outer limits of Ross Sea II, III, and IV ice sheets projected short distances beyond the borders of the Ross Sea I ice sheet. The present-day configuration is shown for the ice sheet in East Antarctica.

the Southern Ocean supplements evidence of ancient glaciation preserved on the Antarctic continent. The early and middle Tertiary cores, with enclosed ice-rafted sediments have been mentioned previously. The younger cores are dated by magnetic-reversal stratigraphy, a method which is at present limited to about the last 4 or 5 m.y. Ice-rafted sediments occur continuously from the top of most cores to a depth beyond the range of dating. The major conclusion derived from these cores is that glaciers with calving termini have occupied coastal Antarctica for more than 5 m.y. (Goodell et al. 1968, p. 41). The extent of the glaciers, however, cannot be determined directly from evidence in deep-sea cores.

The accumulated evidence suggests that temperate, perhaps cool, climates characterized much of Antarctica throughout the early and middle Tertiary. Probably during that time glaciers occurred in coastal mountains, and local ice caps may have mantled highlands, but there is no compelling evidence that extensive ice sheets covered the continent. However, the available paleoclimatic evidence is restricted areally, and ice sheets might have existed in Queen Maud Land (Fig. 3). In any case, by the late Miocene a large ice sheet existed in West Antarctica, and by at least the middle Pliocene the ice sheet in East Antarctica had attained a full-bodied condition, at times surpassing its present size. Events accompanying growth of these ice sheets during the Miocene and Pliocene included sea level lowering of about 55 meters (Table 1), climatic cooling, and changes in oceanic circulation.

Theory

Any attempt to reconstruct the sequence of events in the formation of the Antarctica ice sheet must consider both the glaciology of polar ice sheets and the preglacial topography of Antarctica. The subglacial bedrock topography suggests that preglacial Antarctica consisted of the two distinct provinces of West Antarctica and East Antarctica (Fig. 5) (Bentley et al. 1964; Bentley 1965, pp. 262–69). Preglacial West Antarctica consisted of four groups of mountainous islands located off the coast of East Antarctica and surrounded by deep water (Giovinetto 1964); these preglacial island groups now compose the four highland areas of West Antarctica, namely, the Ellsworth Mountains, the Jones Mountains, the Executive Committee Range, and the base of the Antarctic Peninsula. The formation of the ice sheet in this environment in West Antarctica involved a series of interconnected events (Bentley and Ostenso 1961). Extensive glaciers first formed on highlands on the four island groups (Giovinetto 1964, p. 150). These glaciers eventually reached sea level and fed floating ice shelves, which subsequently spread between the islands and East Antarctica. The ice shelves could exist only because of the protection afforded by the island groups. Eventually these ice shelves thickened and grounded, forming the ice sheet in West Antarctica. Thus a critical phase in the formation of the ice sheet in West Antarctica necessarily involved vast ice shelves (Bentley and Ostenso 1961, p. 895; Mercer 1968a, p. 81).

At present, ice shelves fringe about 45 percent of the continental periphery

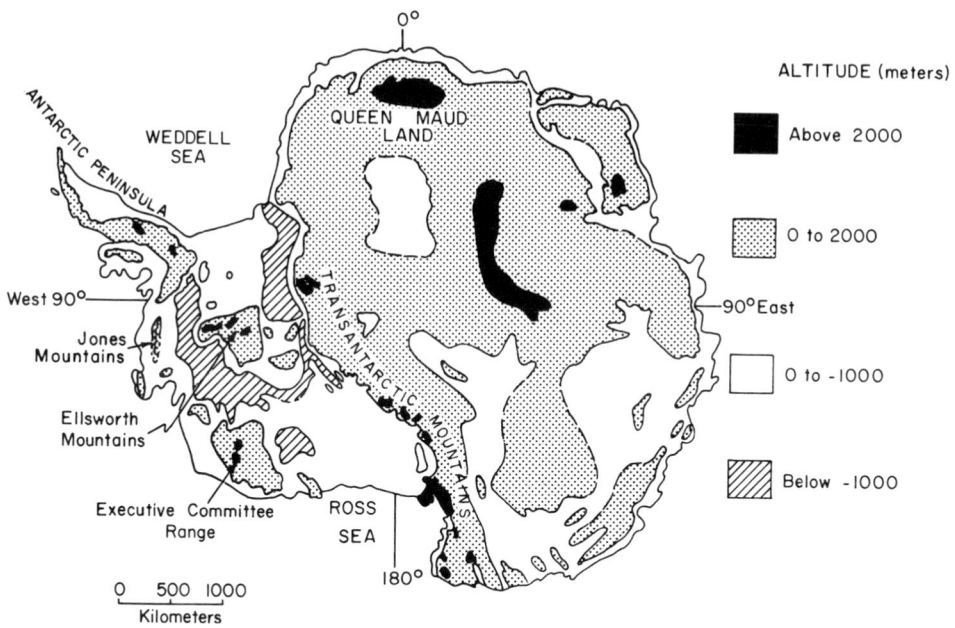

Fig. 5. Subglacial bedrock topography of Antarctica (adapted from Soviet Subglacial Map of Antarctica 1966, from Bentley 1966, and from Schopf 1969).

(Bardin and Suyetova 1967, Table 5). They consist of floating ice tongues that range in thickness from about 200 meters near their seaward edges to 1300 meters at the grounding line where the shelves merge into land-based ice sheets (Crary et al. 1962; Swithinbank and Zumberge 1965, pp. 199–200. The upper surfaces of ice shelves are nearly flat, with altitudes ranging from about 30 meters near the seaward edge to about 100 meters at the grounding line far inland. Ice shelves are fed by glaciers and by direct surface snowfall. For mechanical reasons floating ice shelves apparently cannot form on the open ocean; rather, they require protected areas with lateral constraints (Swithinbank and Zumberge 1965, p. 216; Weertman 1966). All known ice shelves are composed predominantly of polar ice, with ice temperatures well below the pressure-melting point except near the ice–water interface (Robin and Adie 1964, p. 105). Moreover, ice shelves apparently require extreme polar climates, as is well illustrated on the northern Antarctic Peninsula where recent increase of the mean temperature of the warmest month to values above 0°C has caused drastic retreat of the shelves. Farther south, where the mean temperature of the warmest month is still negative, shelves have not retreated (Mercer 1968b). In addition, ice shelves are relatively thin and cannot survive if large surface areas are in the ablation zone. They are therefore found only where snowlines are very low. Indeed the surfaces of modern Antarctic ice shelves are generally areas of net accumulation (Swithinbank and Zumberge 1965, pp. 206–

13). Therefore, in view of the very special conditions required for their existence, the early ice shelves in West Antarctica must have required polar climates as well as the existence of offshore island groups.

The events outlined above had led to an ice sheet in West Antarctica by at least 7 m.y. ago (Rutford et al. 1970). It is doubtful that large Tertiary ice sheets existed in West Antarctica prior to the middle Miocene, for fossil floras found in early and middle Tertiary sedimentary rocks in West Antarctica are incompatible with the polar conditions and low snowlines required for ice shelves. Therefore early Tertiary glaciers in West Antarctica were probably confined to highlands on the initial island groups.

The subglacial bedrock topography suggests that preglacial East Antarctica consisted of a high continent situated almost wholly above sea level (Bentley 1965, pp. 263–66). Probably a high bedrock plateau occupied the interior, and the Transantarctic Mountains may have flanked the eastern margin of the continent. Mercer (1968a, p. 80–81) postulated that the ice sheet in East Antarctica should have formed earlier than the ice sheet in West Antarctica, for an ice sheet based nearly entirely above sea level does not require an early ice shelf phase with severe climatic conditions. Furthermore, the land-based ice sheet in East Antarctica could have consisted entirely of temperate ice, whereas the ice shelf phase in West Antarctica required polar ice. Thus, considering the polar conditions that must have existed in West Antarctica, it is reasonable to argue that an ice sheet also formed in East Antarctica by the late Miocene or early Pliocene, especially in view of the long history of glaciation that probably occurred in Taylor and Wright valleys prior to 4 m.y. ago (Denton et al. 1970, pp. 16, 20).

The mode of origin of the ice sheet in East Antarctica differed significantly from that of the ice sheet in West Antarctica because an ice shelf phase was not required to construct an ice sheet on a base located almost entirely above sea level. Mercer (1968a, p. 80) postulated that the earliest glaciers formed when the Transantarctic Mountains were uplifted above existing snowline; they then flowed onto adjacent lowlands to build up piedmont glaciers which subsequently expanded and coalesced to form the ice sheet. Flint (1943, 1957, pp. 314–18) previously outlined a similar sequence for the growth of the Greenland, Laurentide, and Scandinavian ice sheets.

Alternative possibilities exist for the sequence of glacial events leading to the presence of a huge ice sheet in East Antarctica. As suggested by Calkin (1964, p. 24) and Mercer (1968a, pp. 79–80), some of the first Antarctic glaciers probably occupied the high Transantarctic Mountains, if we assume that those mountains existed at the time. However, it does not necessarily follow that the initial glaciers flowed down the mountain flanks and coalesced to form large piedmont glaciers that eventually expanded into ice sheets. To form an ice sheet in East Antarctica by this mechanism, large piedmont glaciers would have had to be situated on the rather low landward flank of the mountains, not on the flank that at present is adjacent to the Ross Sea. The long valley glaciers necessary to feed such piedmont ice bodies should have carved an extensive system of glacial valleys on the interior flank of the mountains, as today in southern Alaska, where long valley glaciers with numerous tributaries flow down the southern (seaward) flank of the St. Elias Mountains to form the Bering and Malaspina piedmont glaciers. No such extensive complex of major valleys appears to exist on the exposed interior flank of the Transantarctic Mountains, although perhaps a valley system lies buried by ice or has been removed by erosion. Most major glacial valleys drain toward the Ross Sea, where early piedmont glaciers may well have formed near a precipitation source. Topographic evidence in the Transantarctic Mountains therefore does not strongly support the theory that large valley glaciers fed piedmont glaciers on the low interior flank of the mountains.

The subglacial topography of Antarctica shows that the ice sheet in central East Antarctica is underlain by a high bedrock plateau (Fig. 5) (Bentley 1965, pp. 263–66; Schopf 1969, Fig. 1). Large areas of the plateau presently attain altitudes exceeding 2000 meters; if the ice were removed, isostatic adjustment would increase these values to 2500 or 3000 meters. In the absence of the ice sheet, the plateau area would be surrounded by lower terrain, except in the area of the Transantarctic Mountains. In such a case, moisture-laden air masses, associated with cyclonic storms like those that now circle Antarctica, might have penetrated far inland to provide precipitation for building glaciers on the interior plateau. The topography of the plateau would have been suitable for the generation of highland ice caps such as those now on Baffin Island, Devon Island, and other islands in the Canadian Arctic archipelago. Subsequent nonequilibrium expansion of similar early Antarctic ice caps over the entire land mass of East Antarctica could have been triggered either by a feedback effect, caused by the influence of the ice caps on local climate, or by a slight decline in mean annual summer temperature. The plateau area in central East Antarctica seems a likely source for the original glaciers that eventually expanded and merged to form the ice sheet in East Antarctica. If so, stratigraphic and chronologic evidence for the origin of the ice sheet in East Antarctica cannot be examined, for it would be buried beneath several kilometers of glacier ice.

Duration and Cause of the Late Cenozoic Glaciation

The earth has experienced at least three long intervals during which major ice sheets covered continental areas. They occurred during the late Precambrian about 600 m.y. ago, during the late Paleozoic between 300 and 250 m.y. ago, and during the Late Cenozoic; in addition, Ordovician and Devonian glacial deposits occur in Africa. Late Cenozoic glaciations have long been thought to encompass only the last 2 to 3 m.y. However, it has now been documented that major glaciers existed in West Antarctica by the late Miocene or early Pliocene (Rutford et al. 1968, 1970); in East Antarctica by the Pliocene (Denton et al. 1969, 1970) or perhaps the late Miocene (Hamilton and Armstrong 1969); and in the coastal mountains of southern Alaska by the late Miocene (Miller 1953, 1957; Plafker and Miller 1957; Hopkins 1967; Bandy et al. 1969; Denton and Armstrong 1969). Major glaciers have therefore occupied the earth since the late Miocene, for the Antarctic ice sheet probably has remained intact since its inception and major Alaskan glaciations have occurred at numerous intervals since the late Miocene. In addition the Greenland ice sheet may well have formed long before the Quaternary. Therefore, the Late Cenozoic ice age has already lasted approximately 10 m.y., in the sense that major ice sheets and glaciers have occurred on the earth's surface through that length of time.

Theories of ice ages must explain two basic events: (1) the occurrence of long-term glaciation like that in the late Paleozoic or Late Cenozoic, and (2) fluctuations within the long-term intervals of glaciation. The formation of the huge Antarctic ice sheet, and possibly the Greenland ice sheet, with associated worldwide temperature depression, provided a background for the rather minor temperature fluctuations that repeatedly created and destroyed the Laurentide and Scandinavian ice sheets. Hence the initiation of the Cenozoic glacial age perhaps can be explained if we can answer the question: What caused the formation of the Antarctic ice sheet? Several theories of ice ages, such as those proposed by Tanner (1965) and Donn and Ewing (1966), suggest that the origin of the Antarctic ice sheet can be explained simply as a consequence of Antarctica being placed over the south pole through the mechanisms of polar wandering or continental drift. The available paleomagnetic measurements from Antarctica, however, suggest that the continent has been near or over the south pole throughout the Cenozoic (Blundell 1962), although the data are not sufficiently precise to estimate closely the position of the pole. Evidence cited previously suggests that, despite a near-polar position, large ice sheets probably did not form on Antarctica until the late Miocene. If so, a near-polar position is a necessary though not a sufficient condition, for the growth of ice sheets in Antarctica. This situation is not unique, for geologic history records several periods when location of continents in polar regions was not sufficient to cause the formation of large ice sheets (Cox 1968). What, then, caused the Antarctic ice sheet to form long after the continent achieved a near-polar position? A unique answer is not clear. Perhaps it involved worldwide climatic change caused by extraterrestrial factors. Perhaps it involved uplift of the Transantarctic Mountains (Mercer 1968a) or of central East Antarctica. Perhaps expansion of the ice sheet was related to migration of the south pole from the periphery of the continent to its present central position. Alternatively, it may have been the result of a series of unrelated events elsewhere in the world that depressed temperature sufficiently to trigger the growth of ice sheets in Antarctica and other regions (Crowell and Frakes 1970). An event that may well have caused sufficient gradual temperature depression to trigger glacier growth on Antarctica was the Tertiary migration of the north pole to its present isolated position, with consequent formation of a more extensive cover of sea ice on the Arctic Ocean.

FLUCTUATIONS OF THE ANTARCTIC ICE SHEET

Evidence

The Antarctic ice sheet has not remained unchanged through the long interval since its inception and subsequent expansion over the continent. Some fluctuations, especially those in the Ross and Weddell seas, have been major, whereas some, such as recognized altitude changes of the surface of the ice sheet in East Antarctica, may have been minor. In addition the sheet may have undergone massive and periodic surges into the adjacent Southern Ocean, which may even have triggered worldwide Quaternary ice ages. Therefore the Antarctic ice sheet cannot be considered a constant factor in equations of Quaternary ice ages and sea level fluctuations. The following discussion treats (1) glacier fluctuations in East Antarctica, (2) glacier fluctuations in West Antarctica, and (3) the record of glacial events in the Transantarctic Mountains which separate the ice sheets in the east and west. The glacial history of the Antarctic Peninsula is omitted because present-day glaciers on the peninsula are not contiguous with the major ice sheets; a detailed review of glacier fluctuations in this area is given by Mercer (1962).

The ice sheet in East Antarctica covers nearly the entire available land base, and the rather small ice shelves that fringe much of its periphery occupy almost all available protected areas afforded by embayments and islands. Under present conditions, therefore, the lateral extent of the ice sheet in East Antarctica is determined mainly by topography and the confining influence of the peripheral Southern Ocean. However, evidence from the few small ice-free areas and nunataks near the edge of the continent points to past increases in area and volume of the ice sheet. Much of this evidence has been catalogued in detail by Mercer (1962).

The scattered ice-free areas on the coast of East Antarctica, with the exception of the Transantarctic Mountains, were covered formerly by an expanded ice sheet. In addition, glacial features on isolated nunataks and mountain ranges, which occur 100 to 300 km inland from the coast around the periphery of East Antarctica, suggest that the surface of the ice sheet stood 300 to 800 meters higher in the past (Mercer 1962; Grindley 1967). These figures may be somewhat misleading; much of the suggested increase in thickness may have been apparent rather than real, because nunataks could have been isolated as the ice sheet lowered its entire base by erosion over a long time interval (Trail 1964, p. 150; Rutford, ms. 1969, p. 266). Except for the McMurdo Sound region, mentioned below, the ages of former expansions of the ice sheet in East Antarctica are not documented closely. It is not known, for example, if a general expansion of the ice sheet in East Antarctica occurred during the late Wisconsin (Würm) glaciation of the northern hemisphere. No definitive evidence of such expansion exists, except on the Windmill Islands located off the coast of Wilkes Land (Fig. 3). Those islands have been overrun by the ice sheet on at least one occasion

(Hollin and Cameron 1961, pp. 840–41; Cameron 1964, pp. 3–4, 32–35). The removal of ice during the last major recession resulted in a series of emerged shoreline features reaching 30 meters above present sea level. A sample of coralline algae from a beach 23 meters above present sea level yielded a C^{14} date of about 6040 years (M-1052) (Cameron 1964, p. 4), suggesting that the shoreline features may have formed as a result of ice recession during late Wisconsin (Würm) time. The ice margin here has recently readvanced very close to the shoreline features and apparently is now stable (Cameron 1964, p. 4). However, in the Bunger Hills in Wilkes Land (Fig. 3) the presence of highly weathered bedrock and glacial drift over the entire ice-free area, including terrain adjacent to the present ice margin (Shumskiy 1957, p. 60), suggests that the last expansion of the ice sheet here may be very ancient. Present stability of ice fronts in the Sør Rondane Mountains (Van Autenboer 1964, pp. 95–96) and in Queen Maud Land (Fig. 3) (Schytt 1961) is suggested by the presence of large lichens on ice-free terrain adjacent to ice margins.

In West Antarctica evidence of fluctuations of the ice sheet is restricted mainly to nunataks. The altitude of apparently higher surface ice levels measured on nunataks ranges from about 30 meters far inland to about 600 near the coast (Mercer 1962). The ages of high-level deposits are not known. Furthermore, the values of high-surface levels may not be real, for the nunataks may have been stranded as the ice sheet eroded its base.

The Transantarctic Mountains trend almost entirely across the continent and separate the ice sheets in East and West Antarctica. Ice-free areas in the mountains have the potential of recording fluctuations of both ice sheets. The most productive area in the Transantarctic Mountains is the McMurdo Sound region (Fig. 4). This region incorporates some of the most extensive ice-free terrain on the continent, including Taylor and Wright valleys which transect the Transantarctic Mountains from the ice sheet in East Antarctica to McMurdo Sound. Pioneer work on the glacial history of the McMurdo Sound region was carried out by Péwé (1960), who first described multiple glaciation in Antarctica, by Nichols (1961), and by Bull and others (1962). Subsequent contributions to mapping and radiometric dating were made by Calkin (1964), Calkin and others (1970), Wilson (1967), Armstrong and others (1967), Denton and Armstrong (1968), and Denton and others (1969, 1970).

The ice-free terrain in the McMurdo Sound region contains a unique and datable record of former expansions of three major glacier systems: the Ross Ice Shelf, the ice sheet in East Antarctica, and independent alpine glaciers in the Transantarctic Mountains (Fig. 4). Because fluctuations of these three glacier systems were not synchronous, they must be treated separately, as shown in Table 3. The glaciations within each system are numbered from youngest to oldest in McMurdo Sound, in Taylor Valley, and in upper Wright Valley (Denton et al. 1970; Calkin et al. 1970); glaciations in lower Wright Valley have been labeled with geographic names (Nichols 1961, in press; Calkin et al. 1970).

On at least four occasions the Ross ice shelf expanded to form an ice sheet

Table 3. Schematic Correlation Chart and Chronology of Glacial Events in Taylor Valley and in McMurdo Sound

The tentative correlation of glacial events in Taylor and Wright valleys is given in Figure 6. The K/Ar dates given here are rough averages of numerous age determinations made over a period of several years. The dating is still in progress; thus the averages given here and in previous papers have changed and will change slightly as new dates become available (after Denton et al. 1970).

Taylor Glaciations (ice sheet in East Antarctica west of Taylor and Wright valleys)	Ross Sea Glaciations (Ross ice shelf)	Alpine Glaciations
Taylor I	4450 yrs B.P. (I-627; Marble Point) * 5900 yrs B.P. (L-462; Hobbs Glacier) † 6100 yrs B.P. (Y-2401; Hobbs Glacier) 9490 yrs B.P. (Y-2399; Hobbs Glacier)	Alpine I 12,200 yrs B.P. (I-3019; Hobbs Glacier) ‡
	Ross Sea I 34,800 yrs B.P. (no laboratory number given; Cape Barne) § > 47,000 yrs B.P. (Y-2641; Cape Barne same locality as sample dated 34,800 yrs B.P.) > 49,000 yrs B.P. (Y-2642; Cape Barne)	Alpine II K/Ar dates; 2.1 to 0.4 m.y. (Walcott Glacier area)
Taylor II	Ross Sea II Ross Sea III Ross Sea IV	

(cont.)

Table 3 (cont.)

Taylor Glaciations (ice sheet in East Antarctica west of Taylor and Wright valleys)	Ross Sea Glaciations (Ross ice shelf)	Alpine Glaciations
Taylor III	K/Ar dates; 3.1 to 1.2 m.y. (Walcott Glacier area)	
K/Ar dates; 1.6 to 2.1 m.y. (Taylor Valley)		K/Ar dates; 2.1. m.y. (Taylor Valley) Alpine III
Taylor IV		K/Ar dates; 3.5 m.y. (Taylor Valley)
K/Ar dates; 2.7 to 3.5 m.y. (Taylor Valley) and 3.7 m.y. (Wright Valley) Taylor(s) V		

* Nichols (1968, p. 471); Olson and Broecker (1961, p. 150). The C^{14} date given in the chart and text is corrected. The uncorrected date is 5650 ± 150 yrs. B.P. (L-627).
† Péwé (1960).
‡ Black and Bowser (1969).
§ Wilson (in press).

grounded on the floor of the Ross Sea (Denton et al. 1970, pp. 16–20). During such Ross Sea glaciations ice sheets in McMurdo Sound attained thicknesses of more than 1000 meters (Fig. 4), and their outlet glaciers pushed westward into the mouths of ice-free valleys on the west coast of McMurdo Sound and dammed large lakes in Taylor Valley. Strandlines of the former lakes extend along the valley walls; those of Ross Sea I age occur up to about 310 meters in altitude, whereas those of Ross Sea II age reach 400 meters. The surfaces of all existing lakes in Taylor Valley are less than 70 meters above sea level and are permanently frozen except for narrow marginal moats of open water that form for a few weeks during the warmest part of each summer. However, despite their relatively high altitudes the large lakes in Taylor Valley during Ross Sea glaciations I and II had marginal open water during part of the summer, for their shorelines are commonly composed of stratified sediments that required open water for formation. Therefore the mean annual temperatures of summer months during Ross Sea glaciations I and II must have been close to present-day values and may well have been slightly higher.

Several K/Ar and C^{14} dates place limits on the ages of Ross Sea glaciations and suggest that all are younger than 1.2 m.y. Pertinent C^{14} dates suggest that the Ross Sea glaciation I may have been contemporaneous with the late Wisconsin (Würm) glaciation of the northern hemisphere (Denton et al. 1970, pp. 16–20). The recessional phase of the Ross Sea I glaciation correlates closely with the rapid rise of sea level at the end of the late Wisconsin (Würm) glaciation. During that retreat, open water reached McMurdo Sound some 3000 to 4500 years ago when the Ross ice shelf attained approximately its present configuration; since then the Ross shelf in the McMurdo area has not receded south of its present position (Denton et al. 1970, p. 19).

During each Ross Sea glaciation, the present Ross ice shelf was transformed into a grounded ice sheet, with concurrent increases in thickness and surface slope (Fig. 4). Consequently, profiles of outlet glaciers flowing across the Transantarctic Mountains from the ice sheet in East Antarctica to the Ross ice shelf adjusted to achieve mechanical equilibrium with the Ross ice sheet. One of the outlets, the Reedy Glacier, has been studied in detail (Mercer 1968c). This glacier has undergone several variations in thickness in response to changes in the shelf; the variations were much greater in lower portions than in the upper reaches of the glacier. At least three former high ice levels, each corresponding to a Ross Sea glaciation, have been recognized (Mercer 1968c). Similar thickening, presumably in response to a Ross Sea glaciation, has been recognized along the lower Beardmore Glacier, another outlet glacier that crosses the Transantarctic Mountains (Oliver 1964).

The ice sheet in East Antarctica is dammed to considerable thickness on the steep west flank of the Transantarctic Mountains in the McMurdo Sound region. Small tongues from the ice sheet spill over bedrock thresholds and occupy the heads of Taylor and Wright valleys. On at least five former occasions increases of altitude of the surface of the ice sheet in East Antarctica have caused major advances in Taylor and Wright valleys (Fig. 6) (Péwé 1960; Denton et al. 1970; Calkin et al. 1970). All five advances were related to the sheet in East Antarctica and were not associated with local alpine glaciers. In Taylor Valley the advances are termed Taylor glaciations (Denton et al. 1970); in Wright Valley they are called Wright Upper glaciations (Calkin et al. 1970). Most of the K/Ar dates associated with these advances are restricted to Taylor Valley, so that correlations between glacial events in the two valleys are based on subjective criteria such as weathering and relative position of drift bodies (Calkin et al. 1970, p. 22). As previously mentioned, both Taylor and Wright valleys were carved prior to 4 m.y. ago during Taylor glaciation(s) V by glacier tongues from the ice sheet in East Antarctica. While the valleys were being carved the ice sheet must have been full-bodied, with surface altitudes slightly higher than today's. Because the four subsequent advances of Taylor and Wright glaciers have not substantially modified the valleys, probably the glaciers that carved the valleys were wet-based, whereas subsequent glaciers were dry-based. The advances of Taylor glaciations V, IV, and III reached the mouth of Taylor Valley, whereas the advance of Taylor glaciation II was considerably less extensive. Wright Upper glaciations IV, III, and II were confined to the western half of Wright Valley. The ages of pertinent drift sheets in Taylor Valley are bracketed by K/Ar dates of associated lava flows and by C[14] dates of related Ross Sea drift bodies (Table 3).

Of great interest is the behavior of the ice sheet in East Antarctica during

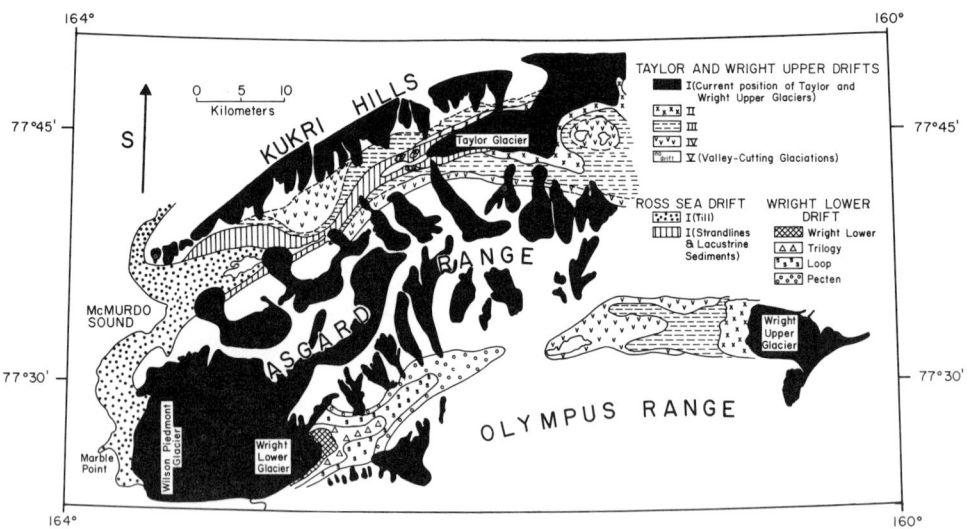

Fig. 6. Schematic map showing some of the glacial units in Taylor and Wright valleys. Many units, including drift deposited by alpine glaciers and by older Ross Sea ice sheets, cannot be included at this scale. The correlation of drift sheets in Taylor and Wright valleys is tentative. (Map of Taylor Valley from Denton et al. 1970; map of Wright Valley from Calkin et al. 1970, and Nichols, in press.)

Quaternary glaciations, particularly through Wisconsin (Würm) and post-Wisconsin time. Radiometric dates and geologic relations in both Taylor and Wright valleys show that the latest increase in surface altitude of the ice sheet in this area, called Taylor glaciation I and Wright Upper glaciation I, is probably current, for the surface now occupies its maximum level in a very long time. (Denton and Armstrong 1968; Denton et al. 1969, pp. 272, 277, 1970, p. 19; Calkin et al. 1970, p. 24). This is in accord with the present positive mass budgets for this drainage system of the ice sheet in East Antarctica (Giovinetto et al. 1966). Correlation of ice sheet deposits with dated drift near McMurdo Sound indicates, in fact, that the ice sheet here is now at its maximum height since before the entire Wisconsin (Würm) glaciation as defined elsewhere in the world (Denton et al. 1969, p. 277, 1970, p. 19). These glacial-geologic data are in accord with the conclusion reached by Wilson (1967, p. 157) from geochemical data that Taylor and Wright Upper glaciers did not advance eastward through the valleys during Wisconsin (Würm) time.

In summary, the chronology of glaciation in the McMurdo Sound region suggests (1) that Ross Sea glaciations probably were contemporaneous with major northern hemisphere glaciations, (2) that the ice sheet in East Antarctica attained a full-bodied stage more than 4 m.y. ago, (3) that the recognized changes in surface altitude of the ice sheet in East Antarctica in the McMurdo Sound region were not synchronous with major northern hemisphere glaciations despite, (4) that the ice sheet in this area was smaller than now during Wisconsin (Würm) time, and (5) that the ice sheet in the McMurdo Sound area now occupies its maximum surface altitude since some time before the Wisconsin (Würm) glaciation of the northern hemisphere.

Most of the evidence from ice-free areas on Antarctica relates to former expansions of the ice sheet. Any indication of major interglacials is buried beneath ice and must be sought elsewhere. The persistent occurrence of ice-rafted sediments in numerous deep-sea cores from the Southern Ocean suggests that Antarctica has not experienced a major interglacial in at least the last 5 m.y. (Goodell et al. 1968, p. 60). The high sea levels of Quaternary interglacial ages, however, suggest recession of the Antarctic and Greenland ice sheets. For example Mercer (1968b) postulated that the possible sea level rise of 6 meters about 120,000 years ago was due to disintegration of the ice sheet in West Antarctica. However, Quaternary interglacial sea levels cannot be related quantitatively to recession of polar ice sheets until tectonic (Flint 1966) and isostatic (Bloom 1967) factors can be distinguished from eustatic sea level changes.

Theory

The Antarctic ice sheet is somewhat unique because it is centered near a pole, has an extreme climatological environment, lacks a surface ablation zone, and is fringed entirely by the Southern Ocean. The reaction of such an ice sheet to temperature changes is difficult to predict. Quite possibly glacial changes in Antarctica may not parallel those elsewhere, despite apparent synchroneity in late Pleistocene temperature changes in Antarctica and the northern hemisphere (Epstein et al. 1970). Three main mechanisms have been proposed to explain fluctuations of the Antarctic ice sheet. All three could have interacted continuously, although any one might have been dominant at a given time or locality. Alternatively, the ice sheet may have responded to mechanisms or events hitherto unrecognized. A definitive answer will not emerge until theoretical and field studies are far more advanced than at present.

At the turn of the nineteenth century Scott (1905) proposed that the Antarctic ice sheet fluctuated out of phase with northern hemisphere glaciations. This hypothesis, expanded by Gow (1965) and by Markov (1969), postulates that worldwide interglacial conditions would allow relatively warm and moist air to penetrate Antarctica, greatly increasing accumulation and causing expansion of the ice sheet. Conversely, intervals of worldwide glaciation would drastically reduce accumulation and lead to ice recession. Such a hypothesis requires that Quaternary temperature trends in Antarctica mirror those elsewhere, an assumption that may be justified (Epstein et al. 1970). However, Nye (1959) showed that ice sheet profiles depend primarily on the plastic properties of ice and that they are less dependent on other factors such as accumulation. For example, if accumulation on the interior of Antarctica were reduced by a factor of 2, the thickness of the ice sheet would decrease by less than 10 percent. However, even such relatively small changes would be sufficient to explain some of the recorded ice sheet fluctuations.

The possibility that changes in surface altitude of the ice sheet are due to accumulation changes can be tested in the McMurdo Sound area. The current increase in surface level of the ice sheet in that region should be limited to the last 17,000 years, if it is due to increased accumulation during a relatively warm interval, including the Hypsithermal, subsequent to the maximum of the late Wisconsin (Würm) glaciation of the northern hemisphere. In fact, this increase in surface altitude, although it may be caused partly by increased accumulation during the last 17,000 years, appears to be a long-term event, for the ice sheet now occupies its maximum surface altitude since some time prior to the Wisconsin (Würm) glacial age. Such extended buildup may have resulted from a combination of increased accumulation since the late Wisconsin (Würm), superimposed on increased ice thickness caused by reduced plasticity of the ice sheet during a Wisconsin (Würm) cold interval. Again, this mechanism assumes that Quaternary temperature trends in Antarctica paralleled those in the northern hemisphere.

A second hypothesis proposes that sea level is the dominant factor in determining the area and volume of the Antarctic ice sheet which is basically a huge accumulation area, with most ablation occurring as icebergs calving from the periphery of the continent. Hollin (1962) suggested, therefore, that the ice sheet has the capacity to expand over a much larger land base but is restricted from doing so by the present position of relative sea level on the Antarctic continent. However, during Quaternary glaciations eustatic sea level oscillations would have alternately expanded and contracted the area above sea level in Antarctica, leading to concurrent ice sheet fluctuations. The resultant ice volume variations would be enormous in the case of ice shelves but would be less drastic for a grounded ice sheet such as that in East Antarctica (Fig. 7). This sea level hypothesis requires that the Antarctic ice sheet fluctuate in phase with major Quaternary glaciations in the northern hemisphere, not because of climatic events but because of mechanical adjustments of the ice sheet to fluctuation of sea level. Quaternary sea level changes were controlled to a large extent by the history of the Laurentide ice sheet in North America. For example, the Laurentide ice sheet caused about 60 percent of the sea level variation within each Quaternary glacial age (Flint 1969). Therefore, climatically induced volume changes of the Laurentide ice sheet may have had a major influence on the size of the Antarctic ice sheet through the link of sea level changes.

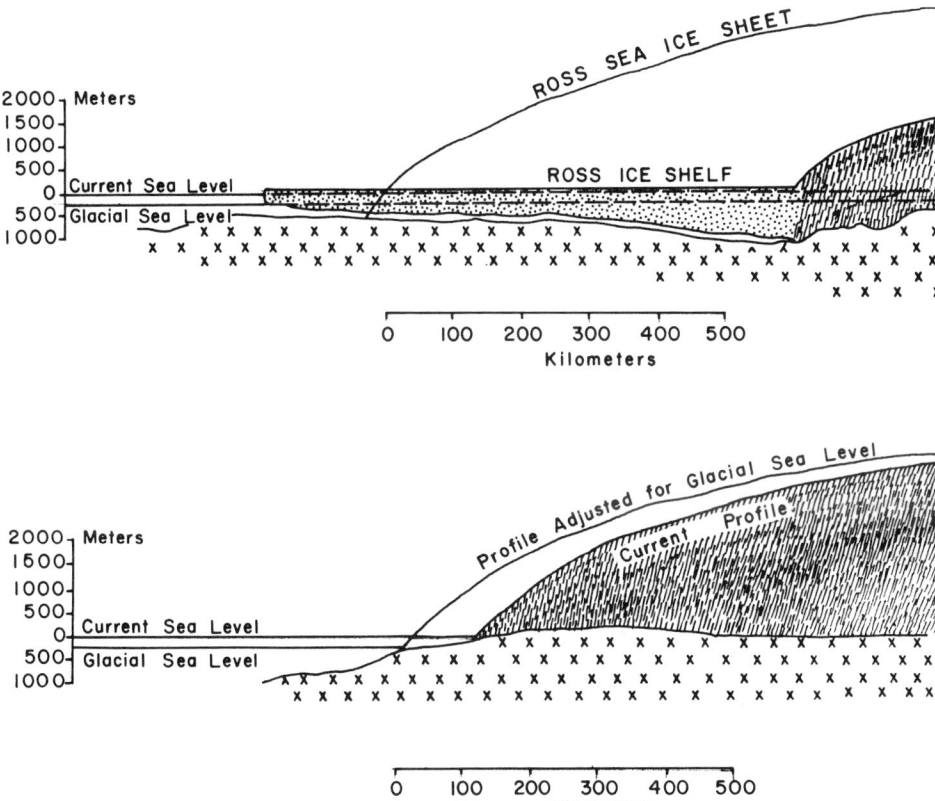

Fig. 7. Idealized profile changes in an Antarctic ice shelf and grounded ice sheet caused by a sea level drop of 130 meters during a Quaternary ice age (after Hollin 1962). Upper: Profile changes of an Antarctic ice shelf. The dimensions given are those measured on the Ross ice shelf along meridian 168°W by Crary and Van der Hoeven (1961). The terminal position of the Ross Sea ice sheet was chosen arbitrarily. Lower: Profile changes of grounded ice sheet.

The sea level mechanism adequately explains the huge changes in ice volume that occurred in the Ross Sea during each Ross Sea glaciation, for only a minor depression of sea level would be required to transform the Ross ice shelf into a grounded ice sheet (Hollin 1962, pp. 186–87; Denton et al. 1970, pp. 16, 20). It further explains why C[14]-dated Ross Sea events appear to correspond with Quaternary glaciations, and why the recession of the Ross Sea I ice sheet correlates closely with the rise of sea level during the closing phase of the late Wisconsin (Würm) glaciation. The alternative explanation that Ross Sea glaciations were controlled by climate appears unlikely, however, because alpine glaciers in the McMurdo Sound region diminished in size during Ross Sea glaciations (Denton et al. 1969, pp. 276–77).

Grounded glacier ice adjacent to the Ross ice shelf would have adjusted mechanically to the existence of Ross Sea ice sheets. Probably the same situation obtained near the Filchner ice shelf. Thus large areas of the ice sheet in West Antarctica, as well as the lower lengths of outlet glaciers that drain through the Transantarctic Mountains into the Ross ice shelf, probably became thicker during Ross Sea glaciations. In East Antarctica, however, field evidence bearing on whether the ice sheet fluctuated in phase with Quaternary glaciations, as required by the sea level mechanism, is contradictory. The time when ice-free areas adjoining the ice sheet in coastal areas of East Antarctica were last overrun by glacier ice is unknown except for the Windmill Islands and the Bunger Hills. As previously mentioned, evidence from the Windmill Islands favors a late Wisconsin (Würm) expansion of the ice sheet, whereas evidence from the Bunger Hills argues against such an expansion. Furthermore, the thick ice sheet dammed behind the Transantarctic Mountains in McMurdo Sound, as previously noted, is now at its highest surface altitude since the Wisconsin (Würm) glaciation of the northern hemisphere; this history of interior surface level variations does not agree with the chronology required by the sea level mechanism according to Hollin (1962). A possible answer is that the sea level mechanism exerts the dominant control on the lateral extent of the ice sheet, whereas an alternative mechanism, such as changes in rate of accumulation, dominates thickness variations in the interior.

A third hypothesis postulates that the Antarctic ice sheet surges periodically, that the ice discharged into the Southern Ocean during each surge forms a vast ice shelf extending north to the Antarctic Convergence, and that the albedo effect of this ice shelf cools the earth and triggers a worldwide Quaternary ice age, which terminates when the ice shelf on the Southern Ocean disintegrates (Wilson 1964, 1969). Such a sequence may have caused each Quaternary ice age. If so, a glacier surge from Antarctica should have occurred immediately before each Quaternary ice age. The Antarctic ice sheet then would have built up to attain a maximum volume at the end of an interglacial age, when a surge would have triggered the next Quaternary ice age. This sequence of events should have left a geologic record (Fig. 8). The changes of surface level of the interior of the ice sheet would not have been in phase with Quaternary ice ages but would have followed the cycle depicted in Figure 8. Data from the McMurdo Sound region suggest that a limited

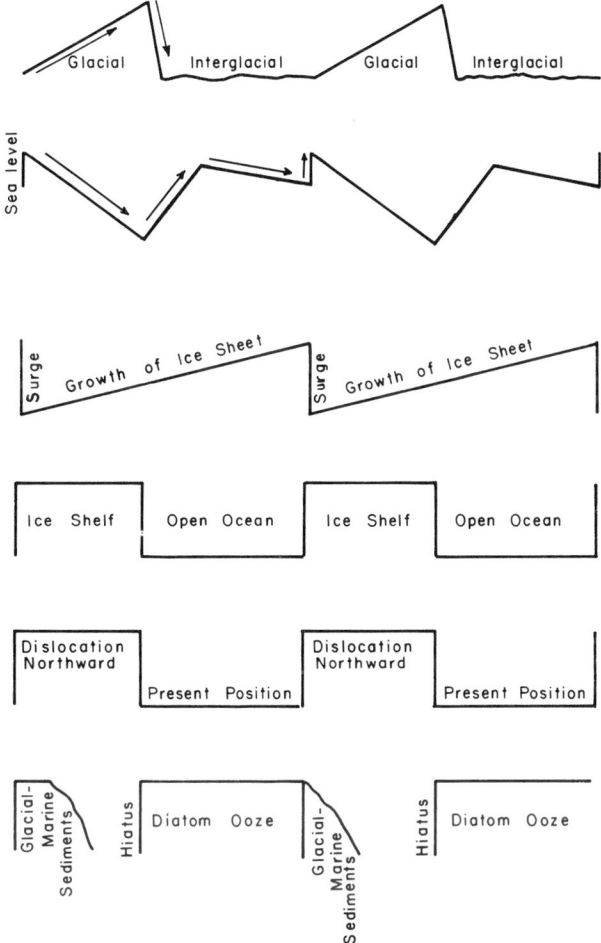

Fig. 8. Schematic diagram showing hypothetical events that would occur if massive surges of the Antarctic ice sheet caused Quaternary ice ages, as suggested by Wilson (1964). These events should have left marks in the geologic record that could be used to test Wilson's hypothesis.

Quaternary glacial-interglacial cycle (Broecker and Van Donk 1970)

Sea level changes (Hollin 1965)

Surface level changes of interior of ice sheet in East Antarctica

Southern Ocean

    a. Presence of ice shelf on Southern Ocean

    b. Antarctic Convergence

    c. Sedimentation in present zone of siliceous diatom ooze

part of the ice sheet may partially fill these requirements; however, fieldwork is required in many areas in Antarctica before definite conclusions are possible. Furthermore, a massive surge of the Antarctic ice sheet would have caused a rapid rise of eustatic sea level, which would have been followed by a gradual decline in sea level as northern hemisphere ice sheets developed during the ensuing Quaternary ice age (Hollin 1969). The sea level curve that would result from this sequence is shown in Figure 8. To date, however, the search for such a sea level record has produced ambiguous results (Hollin 1969, pp. 907–08).

Finally, if large-scale glacier surges formed vast ice shelves around Antarctica, the record should be preserved in deep-sea cores from the floor of the Southern Ocean. The present distribution of sediments there is shown in Figure 9. Large-scale surges should have three possible effects on this distribution of ocean floor sediments (Fig. 8). First, the basal sliding involved in a surge should vastly in-

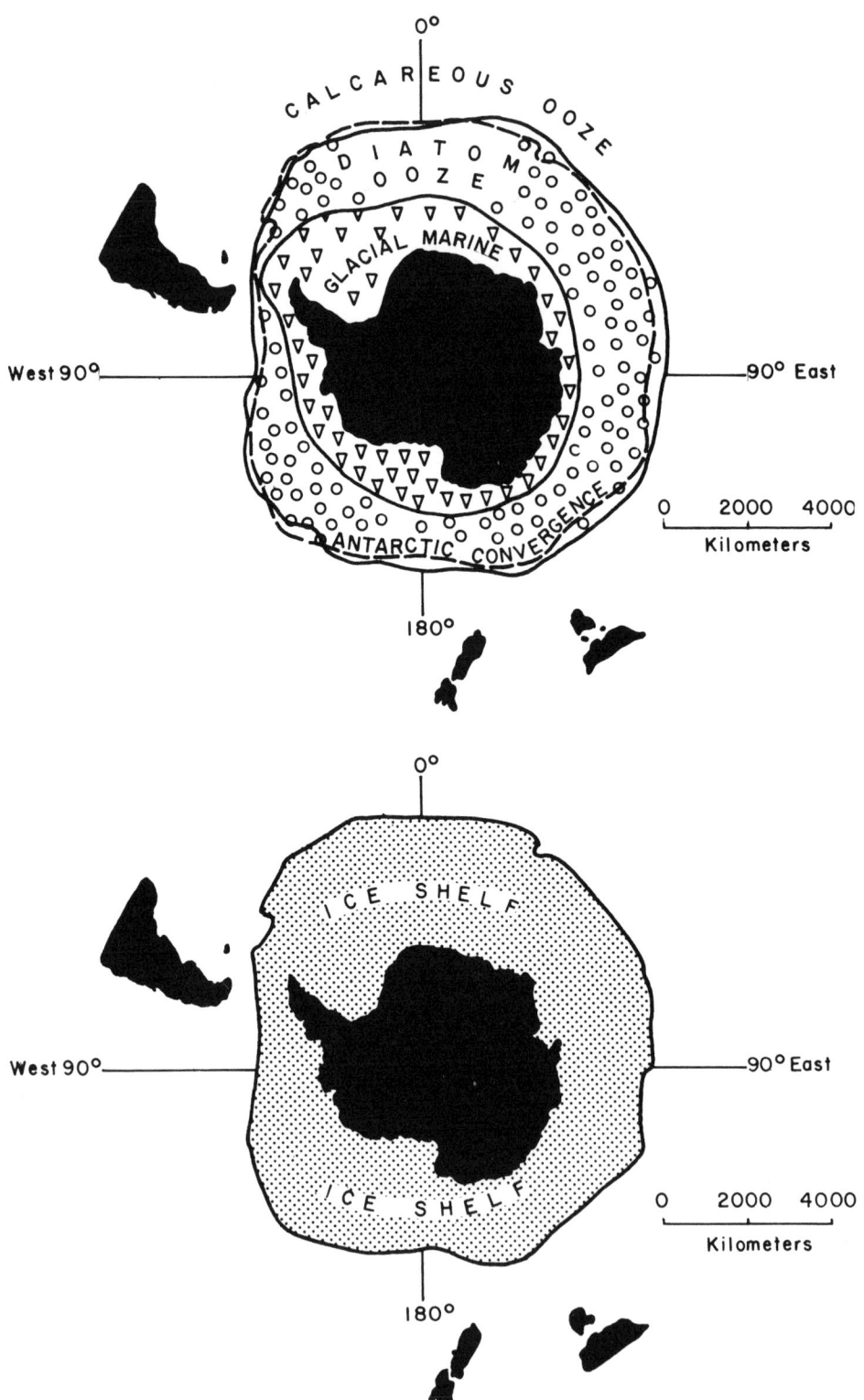

Fig. 9. The Southern Ocean; Antarctica is shown in black. Upper: Present sediment distribution on floor of Southern Ocean (after Hays 1967). Lower: Extent of possible ice shelf generated by a massive surge of the Antarctic ice sheet (after Wilson 1964).

crease the amount of ice-rafted, glacial-marine sediment delivered to the Southern Ocean. The initial ice shelf would be composed of glacier ice that surged from the continent with substantial amounts of basal morainal material. Some of these morainal sediments would have been present in the ice before the surge, as is now the case in West Antarctica (Gow et al. 1968), and additional material would have been incorporated during the surge. Thus the zone of glacial-marine sedimentation surrounding the continent would have extended north to the Antarctic Convergence during the early stages of the ice shelf. Morainal material would not have been deposited during the later stages of the shelf, because by then most of the ice would have been derived from direct snow accumulation on the shelf. Available records of deep-sea cores from the Southern Ocean reveal that widespread layers of glacial-marine sediments were not deposited at the intervals required by the hypothesis.

Furthermore the existence of the vast ice shelf postulated by the surge hypothesis probably would have caused major changes in the position of the Antarctic Convergence during each Quaternary glaciation (Fig. 8). Such major dislocations apparently did not occur during the Quaternary as Hays (1967) reported evidence only for minor migrations of the Antarctic Convergence. The existence of a continuous ice shelf around Antarctica during a Quaternary ice age also would have prohibited the presence of large foraminifera populations in the Southern Ocean and thus would have curtailed deposition of forminiferal ooze on the ocean floor. However, sedimentation rates in the Southern Ocean are so low that the resulting hiatus might be indiscernible; furthermore, hiatuses in siliceous-ooze deposits cannot be documented by C^{14} dates. In summary, data from deep-sea cores seem to argue against massive surges from Antarctica with large associated ice shelves in the Southern Ocean. However, they do not eliminate the possibility of smaller surges of individual drainage systems within the ice sheet.

Deep-sea Cores and Antarctic Glaciation

Several types of evidence from deep-sea cores from the Southern Ocean contribute to knowledge of the extent of glaciation on Antarctica. These include faunal changes and the presence and amount of ice-rafted sediments. Changes in these parameters through time, which is determined by magnetic-reversal stratigraphy, often have been related directly to the extent of glaciation in Antarctica. Seldom are these assumed direct relationships entirely justified.

As mentioned previously, the occurrence of ice-rafted sediments in deep-sea cores from the Southern Ocean reflects nothing more than presence of calving glaciers on adjacent continents. These sediments do not necessarily imply large ice sheets but may well be related to active coastal mountain glaciers and highland ice caps. Likewise the assumption that variations in amount of ice-rafted sediment through time reflect changes in the extent of glaciation in Antarctica (Goodell and Watkins 1968) may not always apply. Variations in the amount of ice-rafted sediment deposited in the Southern Ocean also may result either from fluctuation of ice shelves fringing the continent or from changes in the activity index (Meier 1961) of the ice sheet or of coastal glaciers. Antarctic ice shelves are fed both by glacier ice from land and by surface accumulation of snow; ablation of ice shelves occurs by bottom melting and by calving of icebergs (Zumberge and Swithinbank 1962, pp. 201–04; Swithinbank and Zumberge 1965, pp. 206–14). The glacier ice that enters ice shelves from land may carry considerable basal morainal material. As it moves toward the seaward edge of the shelf, however, the ice is melted from below and is replaced by snowfall from above. In most cases, therefore, icebergs from the seaward edges of shelves are clean, for the original land ice has been ablated from below and the enclosed basal sediment dropped beneath the shelves. At present, ice shelves produce 62 percent of the volume of icebergs annually discharged into the Southern Ocean (Bardin and Suyetova 1967, Table 5). A significant amelioration of climate could cause drastic reduction of Antarctic ice shelves, with consequent increase in the amount of ice-rafted sediment delivered directly to the open ocean. Conceivably then, an increase of ice-rafted sediment in deep-sea cores could reflect climatic amelioration and ice shelf retreat, rather than climatic deterioration and more extensive glaciation.

Increased amount of ice-rafted sediment could result also from change in the activity index of the ice sheet, an event that also might be caused by climatic amelioration. Warnke (1968) suggested that the gradual decline of ice-rafted sediment delivered to the Southern Ocean through the last 3 m.y. may reflect a general cooling trend that caused the ice sheet to become frozen to its base. Some doubt is cast on this suggestion, however, by the recent core hole drilled to the base of the ice sheet in West Antarctica; the core revealed that basal ice there was at the pressure-melting point and that it contained significant amounts of morainal material (Gow et al. 1968). Moreover, calculated temperature gradients suggest that basal ice may be at the pressure-melting point in much of the interior of the Antarctic ice sheet (Paterson 1969, pp. 178–93). Basal temperatures in ice sheets, of course, depend not only on climate but also on ice thickness, rate of accumulation, and rate of movement (Robin 1955).

Stratigraphic and faunal zones in cores from the floor of the Southern Ocean are described by Opdyke and others (1966), Hays (1967), Hays and Opdyke (1967), Goodell and others (1968), and Goodell and Watkins (1968), among many. The present distribution of sediment beneath the Southern Ocean appears to have been established only about 2 m.y. ago (Opdyke et al. 1966, p. 356; Hays 1967, p. 130), but the relation of this event to glaciation on Antarctica is not clear. Minor displacements of sediment zones on the floor of the Southern Ocean during the Quaternary have been related to north–south migrations of the 0°C surface water isotherm. Migrations of this isotherm, in turn, have been assumed to mirror the extent of glaciers on Antarctica (Goodell et al. 1968, p. 107). Although they may have been induced by climatic change, migrations cannot be linked automatically to the extent of Antarctic glaciers because the potential effects of climatic fluctuations on the Antarctic ice sheet are not understood fully.

In sum, the application of data from deep-sea cores to the history of glaciers on Antarctica is limited. Full utilization of data must await detailed knowledge of the possible chain of processes between Antarctic glaciation and changes in sedimentation and faunal distribution in the Southern Ocean. At present the most useful tools are the direct evidence of coastal glaciers afforded by ice-rafted sediment, and the indirect evidence of climatic events provided by faunal changes.

What Does the Future Hold?

The glacier ice contained in the Antarctic ice sheet poses a potential threat to our environment because of its possible influence on sea level, oceanic circulation, and climate. Changes in area and volume of the ice sheet might influence world climate considerably. An extensive surge of the ice sheet into the adjacent Southern Ocean would inundate vast coastal areas and could trigger a new ice age. Likewise, melting of the land-based part of the sheet would have drastic results in coastal areas.

The ice sheet in West Antarctica may be vulnerable to climatic amelioration because it is based largely below sea level. Mercer (1968b) suggested that it is in delicate mechanical balance with the fringing ice shelves, which in turn require low temperatures for survival. Recent events on the Antarctic Peninsula suggest that ice shelves can exist only when the average January temperature is less than 0°C (Mercer 1968b, p. 219). Since this temperature is about −4°C at the edges of the Ross and Filchner ice shelves, a modest temperature increase might cause disintegration of these shelves, which in turn could lead to flotation of the adjacent grounded ice sheet. Conceivably the process could continue until the entire portion of the ice sheet grounded below sea level was consumed. The critical factor in this mechanism is the temperature near the edge of the ice shelves, not the temperature on the interior of the ice sheet. A minor increase in temperature from either natural or human causes might lead to disintegration of the ice sheet in West Antarctica, with an attendant rise of sea level of about 4 meters (Mercer 1968a, 1968b).

The land-based ice sheet in East Antarctica contains 83 percent of the ice on Antarctica (Bardin and Suyetova 1967) and is not vulnerable to the mechanism proposed by Mercer (1968b). Furthermore, the map of surface temperatures over the ice sheet (Bentley et al. 1964) suggests that a considerable increase in mean annual temperature, perhaps more than 15°C, would be required to produce a zone of ablation wide enough to diminish the area of the ice sheet, unless the increased precipitation associated with warmer air masses over it should more than compensate for increased ablation. Apparently, therefore, the huge volume of ice locked up in East Antarctica is not threatened by temperature variations such as those that elsewhere characterized the glacial-interglacial cycles of the Quaternary epoch. The problem of getting rid of the ice sheet appears more difficult than the original problem of forming it, because an ice sheet over the south pole tends to perpetuate itself by influencing Antarctic climates to produce such exceedingly low mean annual temperatures. Probably the ice will remain as backdrop for future Quaternary ice ages as long as the Antarctic continent is situated over the south pole.

The possibility remains that part of the Antarctic ice sheet may surge into the surrounding Southern Ocean. Some of the necessary conditions for a surging are met: several drainage basins of the ice sheet are growing thicker, large parts of the ice sheet may be floored with a thin layer of water, and theoretical considerations suggest that a major surge is possible. However, other evidence argues against a possible surge, particularly some of the geologic data mentioned previously. Further advances in theoretical glaciology and additional fieldwork in Antarctica are needed before this question of future surges of the Antarctic ice sheet can be answered.

Summary. The Antarctic ice sheet, which at present contains about 24×10^6 km^3 of glacier ice, covers 98 percent of Antarctica and is almost entirely fringed by open ocean or floating shelf ice. On the basis of surface and subsurface topography the sheet can be divided into two parts, separated by the Transantarctic Mountains. The ice sheet in East Antarctica forms a huge dome that overlies a low-relief bedrock floor situated almost wholly above sea level, whereas the smaller sheet in West Antarctica has irregular surface topography and rests on a rugged bedrock floor far below sea level. Removal of the Antarctic ice sheet thus would leave a continent in East Antarctica and several archipelagoes in West Antarctica.

Although it occupied a polar or near-polar position throughout the Cenozoic, Antarctica apparently did not support continent-wide ice sheets until late Miocene time, for early and middle Tertiary rocks from West Antarctica, East Antarctica, and southern Australia, which was contiguous with East Antarctica until about 40 m.y. ago, contain fossil faunas and floras incompatible with widespread polar ice sheets. However, the presence of ice-rafted quartz grains in deep-sea cores from the Southern Ocean points to the existence of calving glaciers on Antarctica during lower Eocene, middle Eocene, and Oligocene time. In view of the paleobotanical evidence from the continent, these quartz grains probably represent relatively small complexes of mountain glaciers and highland ice caps with calving termini. The possibility of large early Tertiary ice sheets in Antarctica remains open, however, until more data are available. In any case, by 7 m.y. ago in the late Miocene or early Pliocene, a large ice sheet existed in West Antarctica, and by at least 4 m.y. ago in the middle Pliocene the huge ice sheet in East Antarctica had attained a full-bodied condition. Formation of the entire Antarctic ice sheet was accompanied by a decline in sea level of about 55 meters.

Subsequent fluctuations of the Antarctic ice sheet have involved major changes around the periphery, especially in ice shelf areas, and minor changes of surface altitude on the central portion of the ice sheet. The most complete glacial record comes from the McMurdo Sound region, where the huge ice sheet in East Antarctica is dammed to considerable thickness west of the Transantarctic Mountains and the Ross ice shelf occurs east of the mountains. Past changes in surface altitude of the ice sheet in East Antarctica have caused repeated invasions of Taylor and Wright valleys. At least five invasions, each of successively lesser extent, have been identified and numbered separately from youngest to oldest for each valley, although the numbering system implies possible correlation. Advances of Taylor glaciation V and Wright Upper glaciation V shaped the valleys to their present profiles. Advances of Wright Upper glaciations IV, III, and II were confined to western Wright Valley. Advances of Taylor glaciations V, IV, and III reached McMurdo Sound, whereas the advance of glaciation II was much less extensive. Taylor glaciation I and Wright Upper glaciation I are current, and the ice sheet in East Antarctica west of Taylor and Wright valleys is presently at its maximum surface height since before Wisconsin (Würm) time as defined in the northern hemisphere. Substantial ice recession separated each advance. Lava flows deposited in Taylor Valley between glaciations V and IV range from about 2.7 to 3.5 m.y. old, and volcanic cones in a similar stratigraphic position in Wright Valley are about 3.7 m.y. old. Lava flows separating drifts of Taylor glaciations IV and III are between 1.6 and 2.1 m.y. old. These data indicate that the ice sheet of East Antarctica had attained a full-bodied stage prior to nearly 4 m.y. ago and has since undergone several changes of surface altitude. The recorded changes in surface level of the ice sheet in East Antarctica were not synchronous with northern hemisphere Quaternary glaciations, despite indications of synchronous temperature changes.

Elsewhere, along the coast of East Antarctica, the ice sheet has undergone peripheral thickening of 300 to 800 meters, perhaps accompanied by a lateral expansion(s) of about 100 km. The ages of former advances over coastal areas are not known, except on the Windmill Islands where the last recession may have occurred shortly before 6000 years ago. In West Antarctica the ice sheet has varied in apparent thickness from 30 meters far inland to 600 near the coast.

The Ross ice shelf has expanded into a grounded ice sheet on at least four occasions. All four Ross Sea glaciations are younger than 1.2 m.y. Shells deposited in McMurdo Sound during the recession between Ross Sea glaciations II and I give C^{14} ages of > 47,000 years B.P. (Y-2641) and >49,000 years B.P. (Y-2642). Algae resting on Ross Sea I drift provide a minimum date of 9490 years B.P. (Y-2399) for the initial stages of withdrawal of the Ross Sea I ice sheet. During the final stages of recession, seasonally open water existed in McMurdo Sound by 3000 to 4500 years B.P. The Ross ice shelf in the McMurdo region has not retreated south of its present position during the last several thousand years. These data indicate that the withdrawal phase of Ross Sea glaciation I coincided closely with the rapid rise of sea level during late Wisconsin (Würm) time, and that the Ross Sea glaciations most probably corresponded with worldwide Quaternary glaciations.

The fluctuations of the unique Antarctic ice sheet, which is situated in an extremely cold polar environment and is fringed nearly entirely by the Southern Ocean, are not related in a simple manner to temperature changes, even though such changes were probably synchronous between Antarctica and the northern hemisphere. Three major hypotheses attempt to explain fluctuations of the Antarctic ice sheet. The first suggests that possible increases in accumulation during Quaternary interglacial ages caused expansions of the ice sheet, whereas recessions resulted from decreased accumulation in Antarctica during Quaternary glaciations. The second hypothesis suggests that the relative position of sea level on the Antarctic continent controlled the lateral extent of the ice sheet. Lower sea levels of Quaternary glaciations thus would have been accompanied by expansions of the Antarctic ice sheet. The third hypothesis postulates that periodic surges of the Antarctic ice sheet triggered individual worldwide Quaternary glaciations through the albedo effects of vast ice shelves generated on the Southern Ocean. The field evidence that bears on these hypotheses suggests that sea level controlled fluctuations along the periphery of the ice sheet, especially in ice shelf areas, but that an alternative mechanism, such as periodic surging or changes in accumulation rate, controlled surface level oscillations in interior East Antarctica. Evidence from deep-sea cores from the Southern Ocean suggests, however, that possible surges did not form large ice shelves and thus did not trigger Quaternary ice ages.

Acknowledgments. This paper represents ideas accumulated for nearly a century by numerous workers, most of whom are cited here or by Nichols (1964, 1966). We particularly wish to acknowledge many fruitful discussions about Antarctic glacial history with P. E. Calkin, L. M. Cranwell, H. W. Borns, Jr., Colin Bull, J. D. Hays, J. T. Hollin, Wibjörn Karlén, J. H. Mercer, R. L. Nichols, J. F. Nye, T. L. Péwé, R. H. Rutford, I. G. Speden, and A. T. Wilson. An early draft of this paper was substantially improved by R. F. Flint and A. L. Washburn; this effort is greatly appreciated.

Fieldwork and radiometric dating by the present writers was financed by NSF grants GA-1156 and GA-4034 to the American Geographical Society, and NSF grants GA-1157 and GP-4879 to Yale University. The University of Maine afforded support while the paper was being written. The U.S. Navy and the Office of Antarctic Programs, National Science Foundation, provided logistic support in the field.

The first writer owes a great debt to Professor R. L. Nichols of Tufts University for introducing him to Antarctic fieldwork, for suggesting the mapping and radiometric dating program in the McMurdo Sound region, and for constantly encouraging the ensuing fieldwork. The basis for much of what is known about Antarctic glacial geology was laid by important pioneer work in the McMurdo Sound region during the I.G.Y. by R. L. Nichols and T. L. Péwé. Many field workers, including the present writers, are deeply indebted to these men for their excellent work, which provided the background and stimulation for all subsequent research into the glacial history of the McMurdo Sound region.

References

Adie, R. J., Geologic history, in R. Priestley, R. J. Adie, and G. de Q. Robin, eds., *Antarctic Research*, p. 118, Butterworths, London, 1964.

American Commission on Stratigraphic Nomenclature, Code of Stratigraphic Nomenclature, *Bull. Am. Assoc. Petrol. Geologists, 45*, 645, 1961.

Armstrong, R. L., Warren Hamilton, and G. H. Denton, Glaciation in Taylor Valley, Antarctica, older than 2.7 million years, *Science, 159*, 187, 1967.

Bandy, O. L., Cycles in Neogene paleoceanography and eustatic changes, *Paleogeograph., Paleoclimatol., Paleoecol., 5*, 63, 1968.

Bandy, O. L., E. Ann Butler, and R. C. Wright, Alaskan Upper Miocene marine glacial deposits and the *Turborotalia pachyderma* datum plane, *Science, 166*, 607, 1969.

Bardin, V. I., and I. A. Suyetova, Basic morphometric characteristics for Antarctica and budget of the Antarctic ice cover, Tokyo Nat. Science Mus. Jare Scientif. Repts., Spec. Issue 1, 92, 1967.

Bauer, Albert, Über die in her heutigen Vergletscherung der Erde als Eis gebundene Wassermasse, *Eiszeitalter Gegenwart, 6*, 60, 1955.

Bentley, C. R., The land beneath the ice, in T. Hatherton, ed., *Antarctica*, p. 259, Reed and Reed, Wellington, 511 pp., 1965.

Bentley, C. R., and N. A. Ostenso, Glacial and subglacial topography of West Antarctica, *J. Glaciology, 3*, 882, 1961.

Bentley, C. R., R. L. Cameron, C. Bull, K. Kojima, and A. J. Gow, *Physical Character-*

istics of the Antarctic Ice Sheet, American Geographical Society, Antarctic Map Folio Series, folio 2, 1964.

Black, R. F., and C. J. Bowser, Salts and associated phenomena of the termini of the Hobbs and Taylor glaciers, Victoria Land, Antarctica, Comm. Snow and Ice, General Assembly of Bern, Sept.–Oct. 1967, p. 227, 1969.

Bloom, A. L., Pleistocene shorelines: A new test of isostasy, *Bull. Geol. Soc. Am., 78,* 1477, 1967.

Blundell, O. J., Paleomagnetic investigations in the Falkland Islands Dependencies, British Antarctic Survey Scientific Reports, 39, 24 pp., 1962.

Broecker, W. S., and Jan van Donk, Insolation changes, ice volumes and the O^{18} record in deep sea cores, *Reviews of Geophysics and Space Physics, 8,* 169, 1970.

Brown, D. A., K. S. W. Campbell, and K. A. W. Crook, *The Geological Evolution of Australia and New Zealand,* 409 pp., Pergamon, Oxford, 1968.

Budd, W. F., Glaciological studies in the region of Wilkes, Eastern Antarctica, 1961, Australian National Antarctic Research Expeditions, Scientific Reports, Series A (IV) Glaciology, Pubn. no. 88, 152 pp., 1966.

Budd, W. F., I. H. Landon Smith, and E. R. Wishart, The Amery Ice Shelf, in H. Oura, ed., *Physics of Snow and Ice,* vol. 1, pt. 1, p. 447, Sopporo, Institute of Low Temperature Science, Hokkaido University, 1967.

Bull, Colin, Climatological observations in ice-free areas of southern Victoria Land, Antarctica, p. 177, in M. J. Rubin, ed., *Studies in Antarctic Meteorology: Am. Geophys. Union, Antarctic Res. Ser., 9,* 231, 1966.

Bull, Colin, Snow accumulation in Antarctica, in *Proceedings of the Symposium on Antarctica,* American Association for the Advancement of Science, in press.

Bull, Colin, B. C. McKelvey, and P. N. Webb, Quaternary glaciations in southern Victoria Land, Antarctica, *J. Glaciology, 4,* 63, 1962.

Calkin, P. E., Geomorphology and glacial geology of the Victoria Valley system, southern Victoria Land, Antarctica, Ohio State University, Institute of Polar Studies, Report no. 10, 66 pp., 1964.

Calkin, P. E., R. E. Behling, and Colin Bull, Glacial history of Wright Valley, southern Victoria Land, Antarctica, *Antarctic J. U.S., 5,* 22, 1970.

Cameron, R. L., Glaciological studies at Wilkes Station, Budd Coast, Antarctica, in Malcolm Mellor, ed., *Antarctic Snow and Ice Studies,* Am. Geophys. Union, Antarctic Res. Ser., 2, 1, 1964.

Cox, Allan, Polar wandering, continental drift, and the onset of Quaternary glaciation, *Meteorol. Monographs, 8,* 112, 1968.

Cranwell, L. M., Fossil pollen from Seymour Island, Antarctica, *Nature, 184,* 1782, 1959.

Cranwell, L. M., H. J. Harrington, and I. G. Speden, Lower Tertiary microfossils from McMurdo Sound, Antarctica, *Nature, 186,* 700, 1960.

Crary, A. P., and F. G. Van der Hoeven, Sub-ice topography of Antarctica, Long. 160° W to 130° E, IUGG Symposium on Antarctic Glaciology, Helsinki, 1960, Int. Assn. Sci. Hydrol. Publ. 55, 1961.

Crary, A. P., E. C. Robinson, H. F. Bennett, and W. W. Boyd, Glaciological regime of the Ross Ice Shelf, *J. Geophys. Res., 67,* 2791, 1962.

Creer, K. M., A synthesis of world-wide palaeomagnetic data, in S. K. Runcorn, ed., *Mantles of the Earth and Terrestrial Planets,* p. 351, Interscience, New York, 1967.

Crowell, J. C., L. A. Frakes, Phanerozoic glaciation and the causes of ice ages, *Am. J. Sci.*, *268*, 193, 1970.

Denton, G. H., and R. L. Armstrong, Glacial geology and chronology of the McMurdo Sound region, *Antarctic J. U.S.*, *3*, 99, 1968.

Denton, G. H., and R. L. Armstrong, Miocene-Pliocene glaciations in southern Alaska, *Am. J. Sci.*, *267*, 1121, 1969.

Denton, G. H., R. L. Armstrong, and M. Stuiver, Histoire glaciaire et chronologie de la région due détroit de McMurdo, sud de la Terre Victoria, Antarctide; note préliminaire, *Rev. Geograph. Phys. Geol. Dyn.*, *11*, 265, 1969.

Denton, G. H., R. L. Armstrong, and Minze Stuiver, Late Cenozoic glaciation in Antarctica: The record in the McMurdo Sound region, *Antarctic J. U.S.*, *5*, 15, 1970.

Donn, W. L., and Maurice Ewing, A theory of ice ages III, *Science*, *152*, 1706, 1966.

Dorman, F. H., Australian Tertiary paleotemperatures, *J. Geol.*, *74*, 49, 1966.

Doumani, G. A., Volcanoes of the Executive Committee Range, Byrd Land, in R. J. Adie, ed., *Antarctic Geology*, p. 666, North-Holland, Amsterdam, 1964.

Dusén, P., Über die Tertiäre Flora der Seymour Insel, *Wiss. Ergeb. Schwedische Südpolarexpedition 1901–1903*, *3* (3), 1, 1908.

Epstein, Samuel, R. P. Sharp, and A. J. Gow, Antarctic Ice Sheet: Stable isotope analyses of Byrd Station cores and interhemispheric climatic implications, *Science*, *168*, 1570, 1970.

Flint, R. F., Growth of the North American ice sheet during the Wisconsin age, *Bull. Geol. Soc. Am.*, *54*, 325, 1943.

Flint, R. F., *Glacial and Pleistocene Geology*, 553 pp., Wiley, New York, 1957.

Flint, R. F., Comparison of interglacial marine stratigraphy in Virginia, Alaska, and Mediterranean areas, *Am. J. Sci.*, *264*, 673, 1966.

Flint, R. F., The position of sea level in a glacial age, paper given at INQUA Congr., VII, Paris, 30 August–5 September 1969, 1969.

Funnell, B. M., The Tertiary period, in W. F. Harland, A. G. Smith, and B. Wilcock, eds., The Phanerozic Time Scale, *Geol. Soc. London Quart. J.*, v. 120s, 179, 1964.

Geitzenauer, K. R., S. V. Margolis, and D. S. Edwards, Evidence consistent with Eocene glaciation in a South Pacific deep-sea sedimentary core, *Earth Planetary Sci. Letters*, *4*, 173, 1968.

Gill, E. D., The Climates of Gondwanaland in Kainozoic time, in A. E. M. Nairn, ed., *Descriptive Palaeoclimatology*, p. 332, Interscience, New York, 1961.

Giovinetto, M. B., The drainage systems of Antarctica: Accumulation, in Malcolm Mellor, ed., *Antarctic Snow and Ice Studies*, Am. Geophys. Union, Antarctic Res. Ser., *2*, 127, 1964.

Giovinetto, M. B., E. S. Robinson, and C. W. M. Swithinbank, The regime of the western part of the Ross Ice Shelf drainage system, *J. Glaciology*, *6*, 55, 1966.

Goodell, H. G., and N. D. Watkins, The paleomagnetic stratigraphy of the Southern Ocean: 20° West to 160° East longitude, *Deep-Sea Res.*, *15*, 89, 1968.

Goodell, H. G., N. D. Watkins, T. T. Mather, and S. Koster, The Antarctic glacial history recorded in sediments of the Southern Ocean, *Paleogeograph., Paleoclimatol., Paleoecol.*, *5*, 41, 1968.

Gothan, W., Die fossilen Hölzer von der Seymour- und Snow Hill-Insel, *Wiss. Ergeb. Schwedische Südpolarexpedition, 1901–1903*, *3* (8), 1, 1908.

Gow, A. J., The ice sheet, in T. Hatherton, ed., *Antarctica*, p. 221, Reed and Reed, Wellington, 511 pp., 1965.

Gow, A. J., H. T. Veda, and D. E. Garfield, Antarctic Ice Sheet: Preliminary results of first core hole to bedrock, *Science, 161*, 1011, 1968.

Grindley, G. W., The geomorphology of the Miller Range, Transantarctic Mountains; with notes on the glacial history and neotectonics of East Antarctica, *New Zealand J. Geol. Geophys., 10*, 557, 1967.

Hamilton, W. B., and R. L. Armstrong, Late Tertiary glaciation of Antarctica: *U.S., Geol. Surv., Profess. Papers*, 650A, p. 214, 1969.

Hays, J. D., Quaternary sediments of the Antarctic Ocean, in Mary Sears, ed., *Progress in Oceanography*, vol. 4, p. 117, *The Quaternary History of the Ocean Basins*, Pergamon, Oxford, 1967.

Hays, J. D., and N. D. Opdyke, Antarctic radiolaria, magnetic reversals, and climatic change, *Science, 158*, 1001, 1967.

Heath, G. R., Carbonate sedimentation in the Abyssal equatorial Pacific during the past 50 million years, *Bull. Geol. Soc. Am., 80*, 689, 1969a.

Heath, G. R., Mineralogy of Cenozoic deep-sea sediments from the equatorial Pacific Ocean, *Bull. Geol. Soc. Am., 80*, 1997, 1969b.

Heirtzler, J. R., G. D. Dickson, E. M. Herron, W. C. Pitman III, and Xavier LePichon, Marine magnetic anomalies, geomagnetic field reversals and motions of the ocean floor and continents, *J. Geophys. Res., 73*, 2119, 1968.

Hollin, J. T., On the glacial history of Antarctica, *J. Glaciology, 4*, 173, 1962.

Hollin, J. T., Wilson's theory of ice ages, *Nature, 208*, 12, 1965.

Hollin, J. T., Ice-sheet surges and the geological record, *Can. J. Earth Sci., 6*, 903, 1969.

Hollin, J. T., Is the Antarctic Ice Sheet growing thicker?, in press.

Hollin, J. T., and R. L. Cameron, IGY glaciological work at Wilkes Station, Antarctica, *J. Glaciology, 3*, 833, 1961.

Hopkins, D. M., Quaternary marine transgressions in Alaska, in D. M. Hopkins, ed., *The Bering Land Bridge*, p. 47, Stanford Univ. Press, Stanford, 1967.

Irving, E., *Paleomagnetism and Its Application to Geological and Geophysical Problems*, 399 pp., Wiley, New York, 1964.

Kennett, J. P., Recognition and correlation of the Kapitean Stage (Upper Miocene, New Zealand), *New Zealand J. Geol. Geophys., 10*, 1051, 1967.

Kennett, J. P., Paleo-oceanographic aspects of the foraminiferal zonation in the upper Miocene–lower Pliocene of New Zealand, in Committee Mediterranean Neogene Stratigraphy Proc. IV Session, Bologna, 1967, *Giorn. Geol.* (2), *35*, no. 3, 143, 1968.

Kulp, J. L., Geologic time scale, *Science, 133*, 1105, 1961.

LePichon, Xavier, Sea-floor spreading and continental drift, *J. Geophys. Res., 73*, 3661, 1968.

Liestøl, Olav, Glacier surges in West Spitsbergen, *Can. J. Earth Sci., 6*, 895, 1969.

Loewe, F., The water budget in Antarctica, in T. Nagata, ed., *Proceedings of the Symposium on Pacific-Antarctic Sciences*, p. 101, Eleventh Pacific Science Congress, Tokyo, 1966: Tokyo, National Science Museum, 1967.

Lorius, C., Contribution to the knowledge of the Antarctic Ice Sheet: A synthesis of glaciological measurements in Terre Adélie, *J. Glaciology, 4*, 79, 1962.

Margolis, S. V., and J. P. Kennett, Antarctic glaciation during the Tertiary recorded in Subantarctic deep-sea cores, *Science*, in press.

Marini, M. A., M. F. Orr, and E. L. Coe, Surviving macromolecules in Antarctic seal mummies, *Antarctic J. U.S.*, *2*, 190, 1967.

Markov, K. K., The Pleistocene history of Antarctica, in T. L. Péwé, ed., *The Periglacial Environment*, p. 263, 1969.

Marples, B. J., Fossil penguins from the mid-Tertiary of Seymour Island, Falkland Islands Dependencies Survey Scientific Reports, 5, 15 pp., 1953.

McIntyre, O. J., and G. J. Wilson, Preliminary palynology of some Antarctic Tertiary erratics, *New Zealand J. Botany*, *4*, 315, 1966.

Meier, M. F., Mass budget of South Cascade Glacier, 1957–60, *U.S., Geol. Surv., Profess. Papers, 424-B*, p. 206, 1961.

Meier, M. F., and Austin Post, What are glacier surges?, *Can. J. Earth Sci.*, *6*, 807, 1969.

Mercer, J. H., Glacier variations in the Antarctic, Am. Geograph. Soc., Glaciological Notes, no. 11, 5, 1962.

Mercer, J. H., The discontinuous glacio-eustatic fall in Tertiary sea level, *Paleogeograph., Paleoclimatol., Paleoecol.*, *5*, 77, 1968a.

Mercer, J. H., Antarctic ice and Sangamon sea level, Internatl. Assoc. Sci. Hydrology, Publication no. 79, p. 217, 1968b.

Mercer, J. H., Glacial geology of the Reedy Glacier area, Antarctica, *Bull. Geol. Soc. Am.*, *79*, 471, 1968c.

Miller, D. J., Late Cenozoic marine glacial sediments and marine terraces of Middleton Island, Alaska, *J. Geol.*, *61*, 17, 1953.

Miller, D. J., Geology of the southeastern part of the Robinson Mountains, Yakataga district, Alaska, *U.S., Geol. Surv., Oil and Gas Inv. Map OM 187*, 1957.

Nichols, R. L., Multiple glaciation in the Wright Valley, McMurdo Sound, Antarctica, Pacific Sci. Congr., 10th, Honolulu, Abstracts of Papers., p. 317, 1961.

Nichols, R. L., The present status of Antarctic glacial geology, in R. J. Adie, ed., *Antarctic Geology*, p. 123, Interscience, New York, 1964.

Nichols, R. L., Geomorphology of Antarctica, in J. C. F. Tedrow, ed., Antarctic Soils and Soil Forming Processes, *Am. Geophys. Union, Antarctic Res. Ser.*, *8*, 1, 1966.

Nichols, R. L., Coastal geomorphology, McMurdo Sound, Antarctica, *J. Glaciology*, *7*, 449, 1968.

Nichols, R. L., Glacial geology of the Wright Valley, Antarctica, in *Proceedings of the Symposium on Antarctica*, American Association for the Advancement of Science, in press.

Nye, J. F., The motion of ice sheets and glaciers, *J. Glaciology*, *3*, 493, 1959.

Oliver, R. L., The level of former glaciation near the mouth of the Beardmore Glacier, in R. J. Adie, ed., *Antarctic Geology*, p. 138, Interscience, New York, 1964.

Olson, E. A., and W. S. Broecker, Lamont natural radiocarbon measurements VII, *Radiocarbon*, *3*, 141, 1961.

Opdyke, N. D., B. Glass, J. D. Hays, and J. Foster, Paleomagnetic study of Antarctic deep-sea cores, *Science*, *154*, 349, 1966.

Paterson, W. S. B., *The Physics of Glaciers*, 250 pp., Pergamon, Oxford, 1969.

Péwé, T. L., Multiple glaciation in the McMurdo Sound region, Antarctica—a progress report, *J. Geol.*, *68*, 498, 1960.

Plafker, George, and D. J. Miller, Reconnaissance geology of the Malaspina district, Alaska, *U.S., Geol. Surv., Oil and Gas Inv. Map OM 189*, 1957.

Rex, R. W., and S. V. Margolis, Surface features on sand grains from Antarctic continental shelf and deep-sea cores, *Antarctic J. U.S.*, *4*, 168, 1969.

Robin, G. de Q., Ice movement and temperature distribution in glaciers and ice sheets, *J. Glaciology*, 2, 523, 1955.

Robin, G. de Q., and R. J. Adie, The ice cover, in Raymond Priestley, R. J. Adie, and G. de Q. Robin, eds., *Antarctic Research*, p. 100, Butterworths, London, 1964.

Rutford, R. H., The glacial geology and geomorphology of the Ellsworth Mountains, West Antarctica, 311 pp., Ph.D. dissertation, University of Minnesota, Minneapolis, 1969.

Rutford, R. H., C. Craddock, and T. W. Bastien, Late Tertiary glaciation and sea-level changes in Antarctica, *Paleogeograph., Paleoclimatol., Paleoecol.*, 5, 15, 1968.

Rutford, R. H., Campbell Craddock, R. L. Armstrong, and C. M. White, Tertiary glaciation in the Jones Mountains, Summaries of Papers, SCAR/IUGS Symposium on Antarctic Geology and Solid Earth Geophysics, Oslo, Norway, 6–15 August 1970, 1970.

Schopf, J. M., Ellsworth Mountains: Position in West Antarctica due to sea-floor spreading, *Science, 164*, 63, 1969.

Schytt, Valter, Glaciology II. Blue ice-fields, moraine features and glacier fluctuations: Norwegian-British-Swedish Antarctic Expedition, 1949–52, Scientific Results (Oslo, Norsk Polarinstitutt), v. 4, E, p. 181, 1961.

Scott, R. F., Results of the National Antarctic Expedition I, *Geograph. J., 25*, 353, 1905.

Sharman, Georges, and E. T. Newton, Notes on some fossils from Seymour Island, *Trans. Roy. Soc. Edinburgh, 37*, 707, 1894.

Sharman, Georges, and E. T. Newton, Notes on some additional fossils collected at Seymour Island, Graham's Land, by Dr. Donald and Captain Larsen, *Trans. Roy. Soc. Edinburgh, 22*, 58, 1898.

Shumskiy, P. A., Glaciological and geomorphological Reconnaissance in the Antarctic in 1956, *J. Glaciology, 3*, 56, 1957.

Soviet Subglacial Map of Antarctica, Map 66, in V. G. Bakaev, and others, eds., Atlas Antarktiki, vol. 1 of Sovekskaya Antarticheskaya Ekspeditsiya, 1966.

Sproll, W. P., and R. S. Dietz, Morphological continental drift fit of Australia and Antarctica, *Nature, 222*, 345, 1969.

Swithinbank, C. W. M., and J. H. Zumberge, 1965, The ice shelves, in T. Hatherton, ed., *Antarctica;* p. 199, Reed and Reed, Wellington, 511 p.

Tanner, W. F., Cause and development of an ice age, *J. Geol., 73*, 413, 1965.

Tanner, W. F., Tertiary sea level symposium—introduction, *Paleogeograph., Paleoclimatol., Paleoecol.*, 5, 7, 1968.

Thorarinsson, Sigurdur, Glacier surges in Iceland, with special reference to the surges of Brúarjökull, *Can. J. Earth Sci., 6*, 875, 1969.

Trail, D. S., The glacial geology of the Prince Charles Mountains, in R. J. Adie, ed., *Antarctic Geology*, p. 143, Interscience, New York, 1964.

Van Autenboer, T., The geomorphology and glacial geology of the Sør-Rondane, Dronning Maud Land, in R. J. Adie, ed., *Antarctic Geology*, p. 81, Interscience, New York, 1964.

Warnke, D. A., Comments on a paper by H. G. Goodell and N. D. Watkins, The paleomagnetic stratigraphy of the Southern Ocean: 20° West to 160° East longitude, *Deep-Sea Res., 15*, 723, 1968.

Webb, S. D., and N. Tessman, Vertebrate evidence of a low sea level in the Middle Pliocene, *Science, 156*, 379, 1967.

Weertman, J., Effect of a basal water layer on the dimensions of ice sheets, *J. Glaciology*, 6, 191, 1966.

Wilson, A. T., Origin of ice ages: An ice shelf theory for Pleistocene glaciation, *Nature*, 201, 147, 1964.

Wilson, A. T., The lakes of the McMurdo dry valleys, *Tuatara*, 15, 152, 1967.

Wilson, A. T., The climatic effects of large-scale surges of ice sheets, *Can. J. Earth Sci.*, 6, 911, 1969.

Wilson, A. T., Radiocarbon age of a raised marine deposit on Cape Barne, Ross Island, Antarctica, in press.

Wilson, G. J., Some new species of lower Tertiary dinoflagellates from McMurdo Sound, Antarctica, *New Zealand J. Botany*, 5, 57, 1967.

Wiman, C., Voraüfige Mitteilung über die alttertiaren Vertebraten der Seymour Insel, *Bull. Geol. Inst. Univ. Upsala*, 6, pt. 2, 249, 1905a.

Wiman, C., Über die alttertiaren Vertebraten der Seymour Insel, *Wiss. Ergeb. Schwedische Sudpolarexpedition, 1т01–1903*, 3 (1), 1, 1905b.

Zumberge, J. H., and C. W. M. Swithinbank, The dynamics of ice shelves, *Am. Geophys. Union, Antarctic Reser., Monograph 7*, 197, 1962.

Stephen C. Porter

11. FLUCTUATIONS OF LATE PLEISTOCENE ALPINE GLACIERS IN WESTERN NORTH AMERICA

Fluctuations of alpine glaciers in western North America during the late Pleistocene are widely recorded by end moraines and associated meltwater sediments within and beyond major mountain systems. Morphologic, stratigraphic, and chronologic studies of these deposits have been carried out for nearly a century, but much of the resulting body of data is descriptive in character and of limited value in attempting to compare details of regional glacier behavior. During the past two decades, however, many glacial-geologic studies of the North American Cordillera have emphasized careful, detailed mapping, semiquantitative methods of relative dating, and establishment of absolute chronologies based on radiometric dates, thereby permitting limited interregional comparison of glacial successions and tentative evaluation of certain widely held assumptions regarding synchrony of Pleistocene glacier fluctuations.

My objectives in this paper are to review critical data bearing on the chronology of glacier fluctuations in western North America and evaluate the degree to which synchronous behavior of late Pleistocene glaciers can currently be demonstrated. Any such attempt at synthesis necessarily relies heavily on data generated by other workers whose interpretations generally reflect, to varying degrees, their own methodology and philosophy. I have recast some of these data for the purpose of comparison, and have accepted only those interpretations and correlations for which the evidence appears reasonably compelling. All radiocarbon dates cited in the text are given in radiocarbon years before the present (B.P.).

REVIEW OF REGIONAL CHRONOLOGIES

The aim of this review is to compare regional patterns of glacier behavior without attempting to infer the nature of climatic variations which may have generated them, for such inferences are a hazardous undertaking at best. Consequently, only evidence bearing directly on glacier fluctuations will be considered. Although pollen analyses, soils, and pluvial lake stratigraphies have been used rather extensively in the establishment and correlation of glacial chronologies, especially in regions where radiometric dates are scarce or absent, in general such data cannot be confidently used to infer details of glacier history. Plants, soil-forming processes, and pluvial lakes may have responded to changes of climate that also caused fluctuations of alpine glaciers, but there is no compelling evidence indicating that such responses were synchronous, except in a very gross sense. The response of a glacier to a change in climatic regime, which involves a complex of as yet poorly understood processes, may ultimately lead to a fluctuation of the glacier terminus and generation of some sort of lasting geologic evidence of that fluctuation. However, the delay between the initial environmental change and the ultimate dynamic response of the glacier terminus may be measured in tens, hundreds, or even thousands of years, depending on the flow characteristics and size of the glacier (Kamb 1964; Meier 1965). It is therefore dangerous to assume that changes in flora, lakes, and soil formation were necessarily in phase with variations of glacier termini. For reasonably detailed evaluation of past glacier behavior, only radiometric dates and semiquantitative relative age criteria that are directly related to glacial deposits can provide adequate temporal control.

Interregional comparisons of glacial successions traditionally have been made through the use of correlation charts which depict major stratigraphic units in columnar form, calibrated either to relative or absolute time. Where time–stratigraphic units are employed, charts require placement of arbitrary boundaries between units, generally a highly subjective undertaking in view of the time-transgressive nature of most glacial deposits. Where radiometric dates are few, as is often the case, arbitrarily placed boundaries may be quite misleading if used for detailed comparison and correlation.

A far more suitable and graphic means of data presentation is the time–distance curve, which depicts terminal fluctuations of glaciers through time. Such curves effectively illustrate the time-transgressive nature of glacier variations and, when arrayed in parallel succession, permit more objective comparison of the gross pattern of first-order glacier advance and retreat, as well as comparison of detailed fluctuations reflecting second-order readvances and recessional intervals. Furthermore, where temporal control is adequate, they also permit evaluation of mean rates of advance and retreat, and of synchroneity or nonsynchroneity of glacial maxima in different regions. The construction of detailed time–distance curves, however, requires a good knowledge of the position of former glacier margins and adequate dating control, preferably through absolute ages. Regrettably, in relatively few alpine regions is such information yet available. The best data have come from coastal mountain ranges of northwestern North America where organic matter is widely and often abundantly preserved in the stratigraphic record, thereby providing a reasonably large number of radiocarbon dates. Elsewhere, organic matter at critical horizons has rarely been found, and a general paucity of radiocarbon dates has led to the development of relative chronologies commonly based on time-dependent weathering, mass-wasting, and erosional parameters. In no cases have relative and absolute time scales been calibrated over a sufficiently long period to allow independent use of relative dating methods for detailed correlation. Moreover, it appears unlikely that relative time based on such criteria will have a simple linear relationship to radiometrically controlled time scales. Many factors contribute to this, not the least of which is the probable variation in weathering rate and intensity through time brought about by changes of climate. Time–distance curves employing relative age criteria are therefore not as useful as those calibrated with radiometric ages, but they can emphasize gross regional patterns of first-order glacier fluctuations.

Relatively few studies in western North America have generated sufficient data for construction of detailed time–distance curves, and the quality and completeness of the data vary considerably from region to region. Glacial successions selected for review include those from the Brooks Range of arctic Alaska, the northeast St. Elias Mountains of Yukon Territory, the Puget-Fraser Lowland of western Washington and British Columbia, the central Cascade Range of Washington, the east slope of the Sierra Nevada in California, and the Sawatch Range in the Rocky Mountains of Colorado (Fig. 1). Although an elaborate glacial chronology based on postulated pulsatory climatic oscillations has been proposed for Cook Inlet in south-central Alaska, insufficient radiocarbon control and a lack of relative age data preclude construction of detailed time–distance curves for that region. In western North America only that part of the late Pleistocene encompassing the Wisconsin glacial age is sufficiently well represented by radiometric dates and relative age data to permit realistic interregional comparisons, and this inquiry is therefore restricted to that interval. Post-Wisconsin glacier fluctuations in the North American Cordillera have been reviewed recently in another paper (Porter and Denton 1967).

Brooks Range

Four successively younger and less extensive ice advances during the Itkillik glaciation, designated Banded Mountain, Anayaknaurak, Antler Valley, and Anivik Lake, represent the last major expansion of glaciers on the north side of the Brooks Range (Porter 1964, 1967). A three- or fourfold subdivision of Itkillik drift, based on moraines, associated outwash trains, stratigraphy, and morphologic contrasts, has also tentatively been recognized along several major north-draining valleys through field reconnaissance and photo interpretation. A radiocarbon age of 13,270 ± 160 years for a sample of organic matter beneath Anayaknaurak till provides a close limiting date for that advance. Banded Mountain till underlies the dated horizon and probably represents the maximum expansion of glaciers during late Wisconsin time (Fig. 2b). The Anivik Lake advance at Anaktuvuk Pass occurred before 7241 ± 95 years ago; a somewhat tenuous correlation of radiocarbon-dated pollen zones from Umiat, north of the range, with nondated pollen-bearing lake sediments in the Chandler Valley inferred to be contemporaneous with Anivik Lake drift, suggests that the advance may have culminated close to 8300 years ago (Porter 1964).

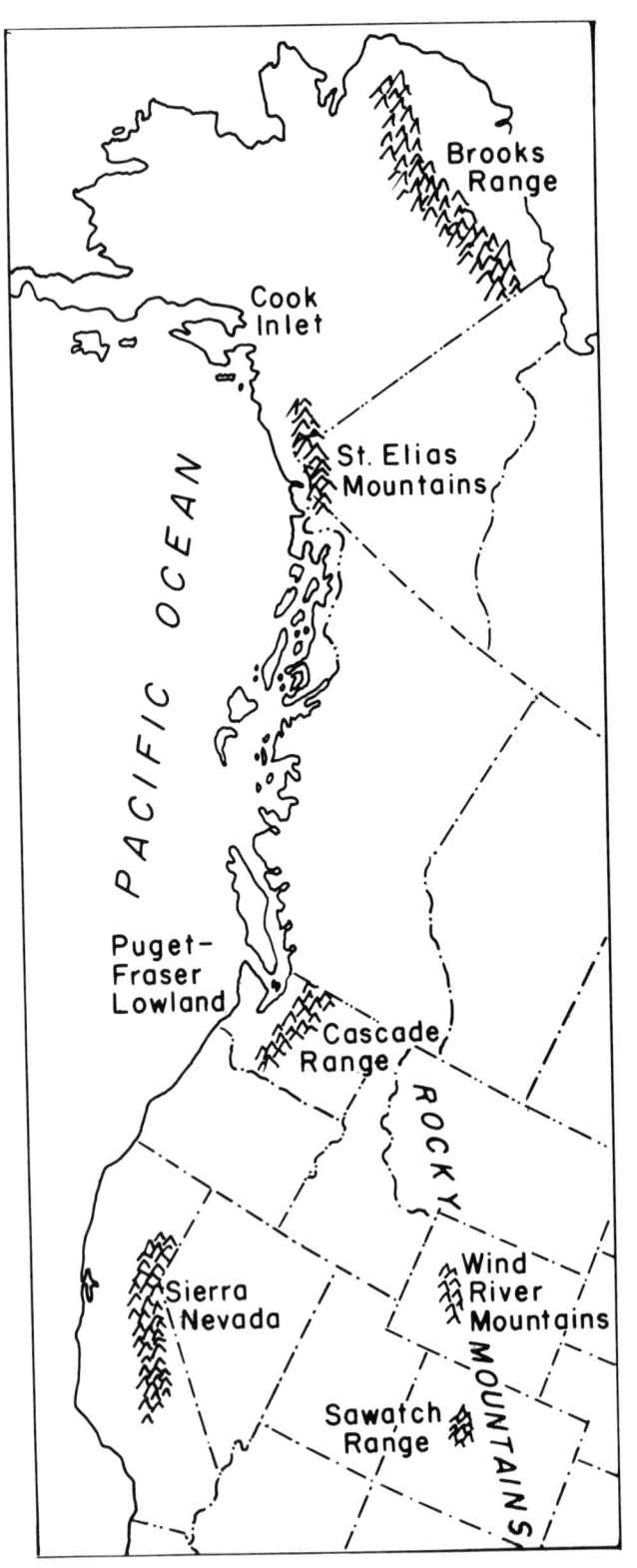

Fig. 1. Western North America, showing areas from which chronologies have been selected for comparison.

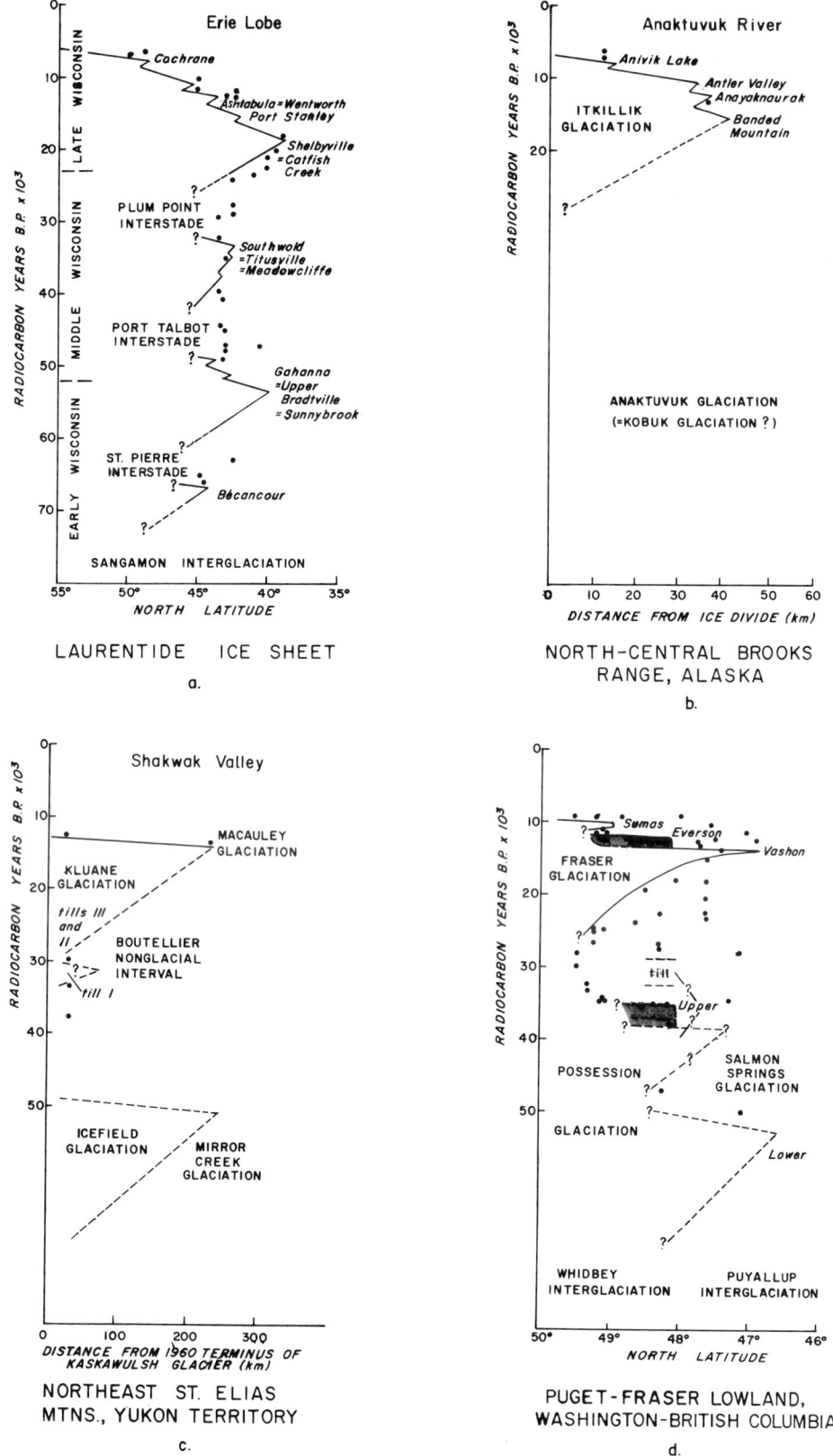

Fig. 2. Absolute time–distance diagrams for the Laurentide ice sheet and for alpine glaciers in northwestern North America. Solid circles indicate finite radiocarbon dates. Shaded areas indicate times of glacial-marine sedimentation.

Drift of the Anaktuvuk glaciation lies beyond Itkillik drift along the lower Anaktuvuk River, and although it was considered pre-Wisconsin in age by Detterman and colleagues (1958) and possibly Nebraskan by Karlstrom (1957), it may prove to be early Wisconsin in its type area (Porter 1964). However, in the absence of radiometric dates, such a correlation cannot yet be convincingly demonstrated. Preliminary studies of the lower Anaktuvuk River area suggest that Anaktuvuk drift may be divisible into two separate bodies having somewhat similar surficial characteristics but different distributional patterns.

Hamilton (1969) mapped a succession of drifts on the south side of the Brooks Range which bears a striking resemblance to the Wisconsin sequence on the Arctic slope. Moraines built during two successive advances of the Kobuk glaciation lie well beyond the mountain front and have been tentatively correlated with the early Wisconsin. In two exposures organic matter associated with till of the earlier advance was dated as > 34,300 years and > 38,000 years old. The Itkillik glaciation in the Alatna Valley included four main advances, designated Siruk Creek, Chebanika, Helpmejack, and Range Front, and is broadly regarded as late Wisconsin in age. In terms of gross pattern and morphology, the moraines closely resemble those of Itkillik age along the Anaktuvuk River. Dates of 8990 ± 80 years and 8205 ± 195 years on organic matter overlying Chebanika drift provide minimum limiting ages for the second advance, but the samples may not closely date the time of glacier recession from the Chebanika moraine. A date of 6760 ± 90 years on wood fragments above drift of the Range Front advance is similar to limiting dates for the Anivik Lake advance at Anaktuvuk Pass and suggests that major trunk glaciers of late Itkillik age in the two drainages may have largely wasted away by 6000–7000 years ago.

St. Elias Mountains

The northeast St. Elias Mountains generated a vast glacier system during the late Pleistocene, with piedmont lobes terminating some 100 to 200 km beyond the mountain front. Stratigraphy and chronology of Wisconsin age drift sheets in the Shakwak Valley near Kluane Lake have been investigated by Denton and Stuiver (1967), while terminal drift farther north has been studied and mapped by Krinsley (1965) and Rampton (1969).

Icefield drift, the older of the two uppermost Pleistocene drifts in the Shakwak Valley, includes two bodies of outwash separated by till, together with associated lake sediments and ice-contact stratified drift (Denton and Stuiver 1967). Samples of organic matter taken from the till sheet and from near the base of ice-contact stratified drift deposited during stagnation of the Icefield glacier are more than 49,000 years old. Organic silt close to the base of the upper outwash unit is 37,700 + 1500/− 1300 years old, and probably was deposited shortly after the glacier started to retreat. Organic matter near the top of the same outwash unit has an age of 30,100 ± 600 years. The interval encompassing these dates, named the Boutellier nonglacial interval, was a time when the Shakwak Valley may have been largely or entirely ice-free.

Kluane drift overlies Icefield drift and includes three till sheets (designated I, II, and III), two outwash bodies, associated lake sediments, and ice-contact stratified drift. Kluane till I was deposited after 33,400 ± 800 years ago, whereas tills II and III are less than 30,100 years old. Although till I may be related to the same advance that deposited the two younger tills, possibly it represents a separate advance which culminated between 33,400 and 30,100 years ago (Fig. 2c). Basal organic matter from bog sediments resting on Kluane ice-contact stratified drift in the Shakwak Valley is 12,500 ± 200 years old and provides a minimum date for deglaciation of the valley.

End moraine systems north and east of the northern St. Elias Mountains and the Wrangell Mountains, initially mapped by Bostock (1952) and subsequently remapped by Krinsley (1965), recently have been reinterpreted by Rampton (1969). The terminal moraine of the youngest drift sheet, designated Macauley by Rampton and approximately equivalent to the Donjek moraine of Krinsley, probably represents the maximum extent of ice during the Kluane glaciation. Radiocarbon samples from deposits beneath and beyond Macauley drift have ages of more than 35,000–42,000 years. Basal organic matter from a bog formed on terminal Macauley drift is 13,660 ± 180 years old, which is a minimum age for culmination of the maximum Kluane ice advance (Fig. 2c). Ice in the White River drainage basin retreated to a position close to the present margin of the Russell glacier by 11,270 ± 200 years ago, the recession essentially paralleling that along the Shakwak Valley system farther south.

Puget-Fraser Lowland

The best-dated late Pleistocene glacial sequence in western North America is found in the Puget-Fraser Lowland of western Washington and southwestern British Columbia. The stratigraphic succession, which has evolved largely over the last two decades, has been discussed by Crandell (1965), Armstrong and others (1965), and Easterbrook and others (1967). The abundance of radiocarbon dates from this region makes it possible to construct a reasonably detailed time–distance curve (Fig. 2d), but because few dates are available for sediments more than 35,000 years old, the chronology of glacier fluctuations shown on the lower part of the curve cannot be considered well established.

Two major advances of the Puget lobe recognized in the southern Puget Lowland probably encompass all or most of Wisconsin time. The most extensive advance, which took place during the Salmon Springs glaciation, has not been dated. Crandell (1963) recognized stratigraphic evidence of two Salmon Springs advances in the southeastern Puget Lowland which were separated by a nonglacial interval during which peat was deposited. The peat has a radiocarbon age of about 50,000 years, but possibly this represents only a minimum age (Easterbrook et al. 1967). Robert J. Carson (personal communication, 1970) has mapped two outwash terraces of Salmon Springs age along the Chehalis River in central-west Washington, the principal meltwater channel beyond the Puget lobe, and the highest of these can be traced to terminal Salmon Springs drift. Outwash comprising a lower terrace is inferred to pass beneath terminal sediments of the next most extensive drift sheet which was deposited during the Fraser glaciation.

In the northern Puget Lowland, drift occupying the same stratigraphic position as Salmon Springs drift has been designated Possession (Easterbrook et al. 1967). Peat beds interstratified with Possession drift on Whidbey Island have finite radiocarbon ages of 46,700 (+3000/−1800), 34,900 (3000/−2000), and 27,200 (+1000/−900) years. The local and regional stratigraphic relationships suggest that a glacial advance, represented by a thin till, occurred some time between about 27,000 and 35,000 years ago, and that possibly an earlier advance took place between about 38,000 and 47,000 years ago (Fig. 2d). Either of these postulated advances may have been the one that deposited late Salmon Springs drift in the southeastern Puget Lowland. A still earlier advance occurred prior to 47,000 years ago and may be equivalent to the early Salmon Springs advance recognized farther south which culminated prior to 50,000 years ago. It is uncertain whether the greatest extent of the Salmon Springs glacier occurred during the early or late advance. Glacial-marine drift apparently related to a late phase of the Salmon Springs–Possession glaciation has a radiocarbon age of about 38,100 ± 1500 years (Donald D. Biederman, personal communication, 1970). These inferred correlations should be viewed as very tentative, for they are based on only a few dates which have large standard deviations. Nevertheless, they appear to be consistent with evidence currently at hand.

Numerous dates from sediments above Salmon Springs and Possession drifts suggest that nonglacial conditions characterized most of the interval from about 34,000 to 24,000 years ago in the northern part of the lowland, and from about 25,000 to 15,000 years ago in the southern lowland. Nonglacial conditions were terminated transgressively as the Puget lobe advanced during the Fraser glaciation, reaching its maximum extent about 14,000 years ago. Glacial recession was rapid and by about 13,000 years ago the southern and central parts of the lowland were largely deglaciated. However, floating ice persisted in the northern lowland for an additional 1500 years, as indicated by a large number of radiocarbon dates from uplifted glacial-marine sediments. A late readvance or, alternatively, grounding of floating ice, occurred during the Sumas stade sometime between 11,000 ± 900 and 9920 ± 760 years ago, after which the glacier retreated rapidly into the mountains of southern British Columbia.

Cascade Range

Only a few valleys in the Cascade Range of Washington have been studied in any detail, and of these the upper Yakima River valley contains one of the best-preserved and most complete records of late Pleistocene glaciation (Porter 1965a, 1965b, 1969a, 1969b). In this drainage basin three major drift sheets have been recognized, the youngest two of which (Kittitas and Cle Elum) are broadly correlated with the Salmon Springs and Fraser drifts, respectively, of the Puget Lowland. Unlike the glacial sediments in the Lowland, however, those in most east-draining valleys of the central Cascades contain very little associated organic matter; consequently, subdivision and correlation of drift sheets has been based largely on relative age criteria. Although a variety of parameters has proved suitable for relative age determinations, the most widely applicable and easily quantified parameter is thickness of weathering rinds on clasts of fine-grained basalt which are common to all tills and outwash bodies except those of the latest advance (Hyak) of the last glaciation. Statistical evaluation of weathering-rind data suggests that rind thickness is time-dependent and increases progressively, but probably nonlinearly, with increasing age of the drift (Fig. 3a). Morphologic, stratigraphic, pedologic, and weathering data have led to subdivision of the Kittitas drift into two units of lesser stratigraphic rank, Virden and Indian John, which reflect two distinct ice advances. The Cle Elum drift has been subdivided into five units designated, from oldest to youngest, Tillman, Ronald, Nelson, Lakedale, and Hyak, each recording a glacier advance or stillstand of sufficiently large magnitude or long duration to have resulted in the construction of a distinct outwash train and related moraines. Tracing of alpine drift across the Cascade crest and into the lower end of a west-draining valley that contains deposits of the Puget lobe tentatively suggests that drift of Lakedale age may have been deposited about the time the Puget lobe last reached its maximum position (ca. 14,000 years ago). Stratigraphic evidence from at least five valleys on the west side of the Cascades indicates that during the Fraser glaciation alpine glaciers reached their greatest extent and had begun to retreat before the Puget lobe achieved its maximum stand. Near Mount Rainier, this early episode of alpine glacier advance has been designated Evans Creek (Crandell 1963; Armstrong et al. 1965). In the upper Yakima drainage basin the Tillman, Ronald, and Nelson advances may have occurred during Evans Creek time; however, the maximum advances of different valley glaciers in the Cascades may not have been synchronous. A radiocarbon age of 7140 ± 95 years for basal bog sediments overlying drift of Hyak age and underlying volcanic ash of the Mazama eruption (ca. 6700 years old) provides a minimum limiting date for disappearance of many Cle Elum glaciers along the crest of the range.

Sierra Nevada

The glacial succession first delineated by Blackwelder (1931) along the eastern flank of the Sierra Nevada has, with minor modifications, been used as the standard time–stratigraphic sequence for the mountain range by subsequent workers. The current status of stratigraphic units has been reviewed by Wahrhaftig and Birman (1965) and Bateman and Wahrhaftig (1966). Three successively younger glaciations, the Tahoe, Tenaya, and Tioga, are commonly correlated with the Wisconsin glaciation, but the relationship of the Tenaya to the other two has not been satisfactorily resolved. Wahrhaftig and Sharp (1965) suggested, on the basis of surficial weathering characteristics, that the time interval between the Tahoe and Tenaya advances may have been longer than that separating the maximum stands of the Tenaya and Tioga glaciers, and both Morrison (1965) and Janda (1967) considered the Tenaya to be the earliest stade of the Tioga glaciation. Although most workers in the Sierra currently appear to regard the Tenaya as closer in age to the Tioga than to the Tahoe, weathering data from different localities are

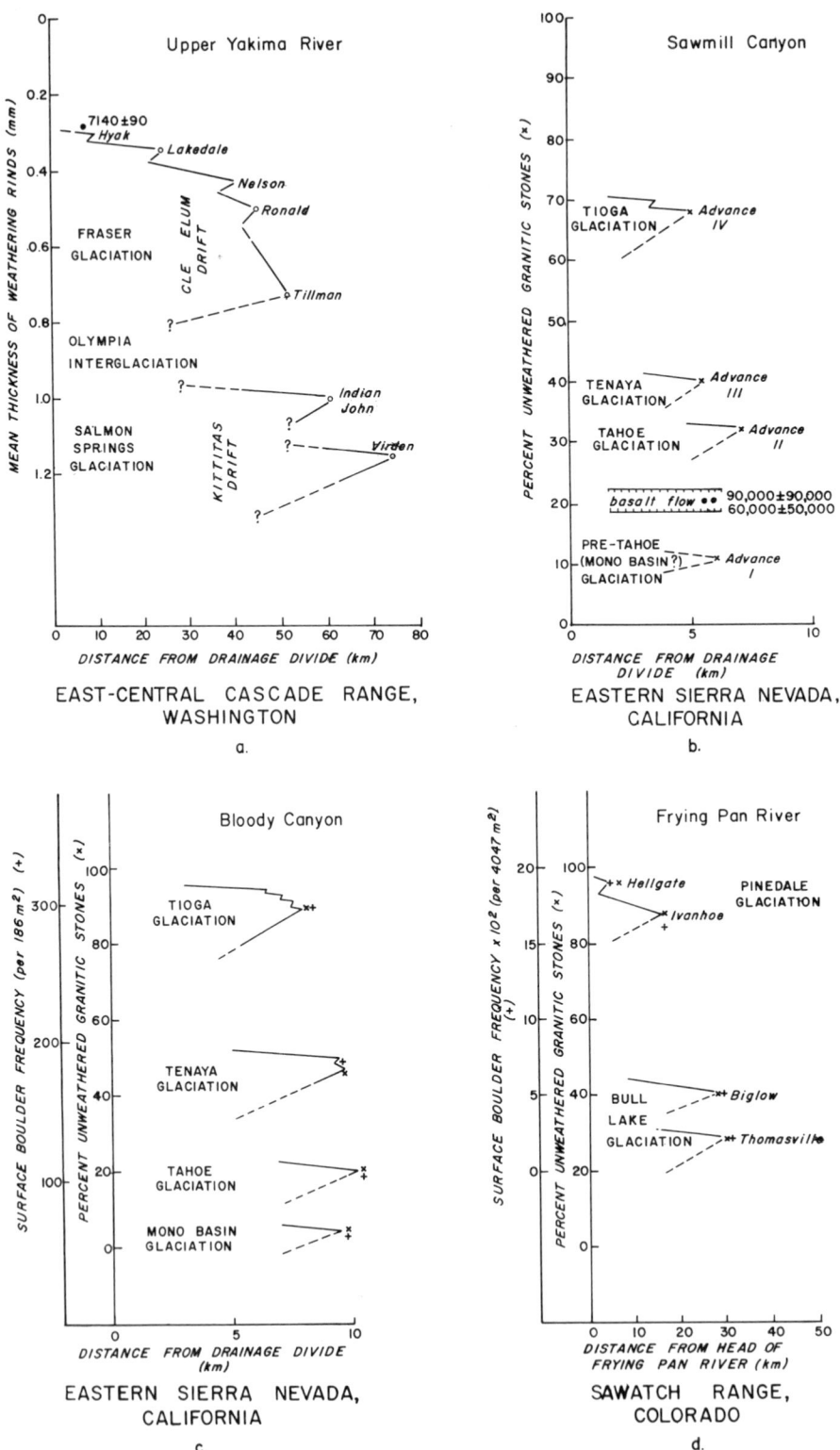

Fig. 3.   Relative time–distance diagrams for alpine glaciers in western United States. Solid circles indicate radiometric dates.

often contradictory, and some data can be interpreted as indicating that the Tahoe/Tenaya interval was the shorter of the two nonglacial intervals (Figs. 3b, 3c). Because differences in degree of weathering reflect the combined effects of many variables, and because the rate of weathering probably varied with time as environmental conditions changed, estimates of difference in absolute age among the three drifts based on weathering characteristics may be unreliable.

Evidence for multiple advances of Tahoe and Tioga glaciers has been reported (Kesseli 1941; Birkeland 1964; Birman 1964; Rahm 1964; Sharp 1969), but there is no general agreement on how the major drift sheets should be subdivided. Tills possibly representing a late Tioga advance have been described by Birkeland (1964) and Birman (1964), but the interpretations are controversial and as yet no absolute dates for the tills are available.

A pre-Tahoe ice advance, designated the Mono Basin glaciation by Sharp and Birman (1963), was generally less extensive than the Tahoe, and evidence of it is rarely encountered. Although Sharp and Birman suggested that Mono Basin drift possibly correlates with Illinoian deposits of the midcontinent, P. W. Birkeland and R. J. Janda compared it with pre-Tahoe drifts elsewhere in the Sierra Nevada and, on the basis. of moraine morphology and soil development, concluded that the Mono Basin may be closer in age to Tahoe drift than to pre-Tahoe tills with well-developed soils (Wahrhaftig and Sharp 1965). They further suggested the possibility that Mono Basin drift was deposited by the earliest Wisconsin ice advance. Smith (1968) also favors an early Wisconsin age for the Mono Basin glaciation; however, he estimates that the Wisconsin began about 130,000 years ago, some 60,000 years earlier than the date favored by a number of other workers (Broecker et al. 1958; Flint 1963; Emiliani 1966; Frye et al. 1968; Dansgaard et al. 1969). In view of the obvious differences of opinion and lack of definitive age criteria, correlation of the Mono Basin glaciation with early Wisconsin or older deposits in the midcontinent remains uncertain.

Only four absolute dates have been obtained that bear directly on probable Wisconsin-age drifts of the Sierra Nevada. A radiocarbon date of 9990 ± 800 years B.P. obtained from sediments near the base of Osgood Swamp, a small lake dammed by a terminal moraine of Tioga age in the Lake Tahoe Basin, provides a minimum date for ice recession near the crest of the range (Adam 1967). The date is close to a stratigraphic boundary marking a change from glacial to post-glacial flora. as determined by study of fossil pollen. Potassium-argon dates of 90,000 ± 90,000 and 60,000 ± 50,000 years obtained by Dalrymple (1964) on basalt underlying drift of probable Tahoe age in Sawmill Canyon and a date of 150,000 ± 30,000 years on pre-Tahoe basalt near Sonora Pass indicate that the Tahoe glaciation probably is less than about 100,000 years old and support assignment of a Wisconsin age to the Tahoe/Tenaya/Tioga complex. Because of large standard deviations, the dates still leave considerable doubt as to the true age of the Tahoe drift sheet and do not permit close correlation with other established chronologies.

Semiquantitative relative age criteria have been used rather extensively in the Sierra for subdivision and correlation of Mono Basin and younger drifts. Granite-weathering ratios and surface boulder frequency are among the most widely employed parameters, and can be used to construct time–distance curves for several glaciated drainage basins. Data from Sawmill Canyon, reported by Dalrymple (1964), give evidence of five glacial advances (Fig. 3b). The oldest till (advance I), which lies beneath the basalt dated 60,000–90,000 years old, was regarded as probably pre-Tahoe in age. Granite-weathering ratios derived for this drift may not be significant because the till was once buried by the basalt and later exhumed. Drift of the youngest (advance V) is confined to a small cirque and was considered probably Hilgard (late Tioga?) in age. Weathering data from moraines at the mouth of Bloody Canyon in the east-central Sierra Nevada (Fig. 3c) were reported by Sharp and Birman (1963). No special significance can be attached to the multiple Tenaya and Tioga moraines they mapped, for in an adjacent valley multiple Tioga moraines were not mapped and Tenaya drift apparently was not recognized.

Indirect evidence of the age of Sierran glacial advances comes from studies of fluctuations in the level of Searles Lake, which occupied a desert basin east of the Sierra, as revealed by surface and subsurface stratigraphy and radiocarbon dates (Flint and Gale 1958; Stuiver 1964; Smith 1968). During high stands of the lake, water was derived largely by overflow from Owens Lake which in turn received runoff directly from glaciers along the east side of the Sierra and, during the Tahoe glaciation, from Mono Lake to the north. The last 40,000 years of the record have been dated by radiocarbon and include two intervals of high water inferred to have been contemporaneous with glacial advances in the Sierra. The dates, when adjusted for apparent radiocarbon deficiency of the lake waters (Broecker and Kaufman 1965), indicate that the uppermost mud unit (Parting Mud) recovered in cores from the basin was deposited between ca. 8500 and ca. 22,000 years B.P. This unit is regarded by Smith (1968) as broadly correlative with Tioga drift in the nearby mountain range. An older mud unit (Bottom Mud) is more than about 30,000 years old and is considered probably equivalent in age to the Tahoe and Mono Basin drifts.

Radiocarbon dates related to lake clays in the southern San Joaquin Valley demonstrate a similar chronology of fluctuations of late Pleistocene lakes which received meltwater from Sierran glaciers (Croft 1968). Lacustrine clay deposition began in the Tulare and Buena Vista–Kern lake basins shortly after 26,780 ± 600 years ago and terminated before 14,000 years ago. A second shorter lacustrine interval, marked by a date of 9040 ± 300 years, may have lasted from about 10,000 to 8000 years ago. If waxing and waning of lakes coincided approximately with the growth and retreat of valley glaciers in the nearby Sierra Nevada, then the Tioga glaciation may have lasted approximately from 25,000 to 8000 years ago.

Rocky Mountains

Richmond (1965) has reviewed the glacial succession in the Rocky Mountains of western United States, adopting the standard nomenclature of the Wind River Mountains of Wyoming for the region as a whole. The Bull Lake and Pinedale glaciations generally are regarded as broadly encompassing Wisconsin time, but too few radiometric dates are yet available to allow close regional and interregional correlations. In the Rocky Mountains of southern Canada radiocarbon ages for organic matter lying beneath till of the last major glaciation (presumably Pinedale) range from about 25,000 to more than 37,000 years (Richmond 1965). In Yellowstone National Park, wood from lake sediments deposited during and following recession of a Bull Lake ice cap and prior to generation of a Pinedale ice cap is more than 42,000 years old. Dates on marl and mollusk shells collected from the bottom of kettles formed in early and middle Pinedale outwash indicate that the maximum Pinedale advance occurred before 11,330 ± 330 years ago in the Wind River Mountains, while peat in recessional ice-contact stratified drift indicates that the Pinedale maximum in the Yellowstone Park area occurred before 13,140 ± 700 years ago. Pinedale drift in the Wasatch Range of Utah is inferred to be approximately contemporaneous with radiocarbon-dated sediments and strandlines of Lake Bonneville which record pluvial conditions between about 25,400 and 11,300 years ago (Richmond 1964, 1965). However, the correlation is indirect because the lake sediments are not in contact with the drift. Dates for postglacial peat deposits in the Rockies show that Pinedale glaciers in several ranges had largely disappeared by 6000–7000 years ago. Despite the aforementioned dates, in no single glaciated drainage basin within the Rocky Mountain system is there sufficient radiometric age control to construct a detailed absolute time–distance curve.

In the absence of critical radiometric dates, various relative age parameters have been used to differentiate late Pleistocene drifts. Soil-stratigraphic units, rather widely employed for subdivision and comparison of glacial stratigraphic successions throughout the Rockies, are useful for broad correlations but do not provide sufficiently precise information to make detailed evaluations of glacier fluctuations. Because in most studies relative age criteria have been treated qualitatively, few published data are available for construction of relative time–distance curves. An exception is Nelson's (1954) work in the Frying Pan River drainage basin in the Sawatch Range of central Colorado. Nelson recognized six drifts, four of which he considered Wisconsin in age. Two of these, the Thomasville and Biglow, he equated with Bull Lake drifts, while two younger ones, Ivanhoe and Hell Gate, he considered Pinedale equivalents (Fig. 3d). Semiquantitative measurements of nine morphologic and weathering parameters showed consistent variations reflecting differences in relative age of the drift sheets. Similar studies, currently in progress in other Rocky Mountain valleys, should generate data suitable for construction of additional time–distance curves (Peter Birkeland, personal communication, 1970).

COMPARISON OF REGIONAL CHRONOLOGIES

When the time–distance curves depicting late Pleistocene glacier fluctuations in western North America are arrayed in parallel succession and compared, certain gross similarities and differences are immediately apparent.

The pattern of each succession is broadly comparable in having several distinct first-order maxima and minima. The curves based on absolute age indicate that these major fluctuations each covered a span of 10^4 years or longer. In most parts of the Cordillera, two glaciations are broadly equated with the Wisconsin (Kobuk-Itkillik, Icefield-Kluane, Salmon Springs-Fraser, Bull Lake-Pinedale). The Sierra Nevada is the only notable exception, having as many as four drifts of possible Wisconsin age represented; however, in many drainage basins only the Tahoe and Tioga drifts are recognized. In each of these regions the most recent first-order glacial advance generally was less extensive than one or more earlier advances, leading to formation of two or more belts of moraines which increase in age away from the source of ice.

The extent of ice recession during first-order minima is poorly known in most alpine regions. Pleistocene glaciers in the northeast St. Elias Mountains were not much more extensive than at present during the Boutellier nonglacial interval, and in western Washington the Puget Lowland was largely free of ice during the long Olympia interglaciation. In both cases, however, glacier lobes were supplied from sizable mountain ice caps, and deglaciation of lowland areas may not have been accompanied by extensive deglaciation of adjacent mountains. Elsewhere, data are largely lacking. Minimum values for ice recession in the Cascade Range can be determined from stratigraphic evidence, but the actual extent of deglaciation between the Salmon Springs and Fraser advances remains unknown.

The pattern of second-order fluctuations is highly variable, not only from region to region but within any given region. Furthermore, without stratigraphic information, it generally is not possible to say whether a given second-order event involved a halt or a readvance of a glacier terminus. Subdivision of a glacial stratigraphic record into units of stadial or substage rank can be a highly subjective undertaking, and because subdivisions of alpine successions often are based on different criteria, the resulting chronologies may not be closely similar. In the Brooks Range, a fourfold subdivision of the last glaciation has been made both on the north and south sides of the range, but this subdivision cannot be recognized in all valleys. Deposits of Fraser age in the Cascade Range have been variously subdivided into three, four, or five units of lesser stratigraphic rank in different drainage basins, whereas in the Puget Lowland only two stadial episodes have been determined for the Puget lobe. Early and late advances of Salmon Springs glaciers are recognized both in the Puget Lowland and in the adjacent Cascade Range and Olympic Mountains, but not every alpine valley provides evidence of the twofold subdivision. Similarly, no unanimity of opinion exists as to how the Tioga and Tahoe glaciations of the Sierra Nevada should be subdivided. In the Rocky Mountains, second-order fluctuations of Pinedale and Bull Lake glaciers have been recognized by some workers, who generally have based their subdivision of the record on morphologic evidence. Richmond (1965) has recognized a regional threefold subdivision of the Pinedale glaciation and a twofold subdivision of the Bull Lake, the youngest stade of which he considers multiple. However, evidence for each of the different units is not present in all areas and precise correlations are hampered by a general lack of radiocarbon dates.

The absence of a uniform and widely recognized second-order subdivision of alpine drifts is not surprising. Because late Pleistocene glaciers in western North

America ranged considerably in size and in climatic environment, their activity and response rates must have varied at least as much as those of modern glaciers throughout this broad region. For a given second-order climatic fluctuation, termini of small glaciers are likely to have responded within a relatively short interval of time (perhaps $\leqslant 10^2$ years) whereas the response time for long valley glaciers and piedmont lobes may have been an order of magnitude greater. Consequently, there is little reason to expect synchronous advance and retreat of all glacier termini in response to second-order climatic fluctuations throughout a region of subcontinental size. Furthermore, local factors may have influenced glaciers in such a way that the relative magnitude of a climatically induced advance may have varied in different drainage basins, even for glaciers of approximately the same size, so that a given terminal fluctuation might be recorded in one area but not another.

At least some of the second-order alpine advances or halts which punctuated the recessional phase of the last glaciation may represent isostatic effects, rather than a response of glaciers to worldwide changes of climate. In heavily glaciated alpine regions, a rapid decrease in glacier volume during recession could lead to isostatic uplift of a mountain range, thereby raising glaciers to higher altitudes and, at the same time, resulting in a relative lowering of equilibrium lines. The net effect of a relatively sudden isostatic movement, therefore, might be to halt or reverse the trend toward deglaciation and possibly lead to the deposition of one or more moraines and associated valley trains. Such physical evidence could easily be misinterpreted as having climatic significance, which would be true only in the sense of a local climate change that affected glaciers in the region subject to uplift. Such effects might generate a succession of moraines that could be broadly correlative within a given mountain range but have no widespread regional significance. Glacier advances generated by such a mechanism might not be synchronous in detail unless the response rates and sizes of the glaciers were comparable.

An additional complication arises from the fact that not all short-lived advances are controlled by climate. Surges, now a widely recognized glacier phenomenon in certain alpine regions, appear to be unrelated to climatic fluctuations, yet they can generate deposits which are essentially indistinguishable from those made by a glacier advancing in response to a change of climate. How many second-order advances, which have been formalized through stratigraphic names and have found their way into numerous correlation charts, were merely local surges having no special climatic significance? Clearly, until the nature of a glacier advance can be effectively demonstrated as having been climatically or not climatically induced, considerable caution should be taken in attaching regional significance to second-order terminal fluctuations.

COMPARISON OF CORDILLERAN AND LAURENTIDE CHRONOLOGIES

The standard time–stratigraphic nomenclature for North America has evolved along the southern margin of the vast region formerly covered by the Laurentide ice sheet. Because many workers in the Cordillera have attempted to correlate alpine drift sheets with those in north-central United States and often refer to certain alpine glaciations as "late Wisconsin" or "early Wisconsin," it may be instructive to compare chronologies developed in the two regions by examining time–distance curves.

Radiocarbon dates now span most of the Wisconsin glacial interval for the region of the Erie lobe in Ohio and adjacent parts of Ontario, and a rather detailed time–distance curve can be constructed (Fig. 2a). The curve is based largely on work summarized by Goldthwait and others (1965) and Dreimanis and others (1966), with modifications resulting from unpublished information supplied by Alexis Dreimanis (personal communication, 1969). Confidence in the accuracy of the curve is greatest for the younger portion, which is controlled by numerous radiocarbon dates. The older part is less well controlled and many of the dates have large standard deviations.

The stratigraphic succession has been subdivided somewhat arbitrarily into three major time–stratigraphic units, designated early, middle, and late Wisconsin, following Goldthwait and others (1965). Major first-order advances of the ice sheet margin culminated about 18,000–20,000 years ago during the late Wisconsin (ca. 14,000 years ago elsewhere along the drift border), and between about 50,000 and 55,000 years ago during the early Wisconsin. An advance of possible first-order magnitude also occurred between about 30,000 and 35,000 years ago in the Lake Ontario and Lake Erie basins which may have been approximately contemporaneous with advances in Illinois between 26,900 and 32,600 years ago (Kempton and Hackett 1968), and in Wisconsin about 29,000 to 32,000 years ago (Frye et al. 1965). A still earlier advance, also possibly of first-order magnitude and as yet not closely dated, is represented by the Bécancour till in the St. Lawrence Lowland which is more than 67,000 years old.

Significant intervals of glacier recession occurred during the Plum Point and Port Talbot interstades, and possibly during the St. Pierre interstade, but as yet the extent of deglaciation at these times is now known. Independent evidence suggests that middle Wisconsin deglaciation may have been more extensive than generally assumed. Data bearing on the position of sea level during the Wisconsin glaciation suggests relatively high stands of the sea during the middle Wisconsin. Milliman and Emory (1968), citing evidence collected from the Atlantic continental shelf of the United States, suggest that sea level was near the present level about 30,000 to 35,000 years ago, while Hoyt and others (1968) have described evidence for submergence of coastal Georgia about 48,000 to 40,000 and about 30,000 to 25,000 years ago. Veeh and Chappel (1970) provide evidence from New Guinea which also points to a marine transgression between about 50,000 and 35,000 years ago during which sea level stood within about 25 meters of its present level. These dated intervals correspond rather closely with the Port Talbot and Plum Point interstades. Because large-magnitude fluctuations of sea level were chiefly a function of changes in volume of upper middle-latitude continental ice sheets, these inferred high sea level stands imply substantial volume loss and terminal recession of the Laurentide ice sheet. Consequently, the Plum

Point and Port Talbot intervals may have been times of extensive deglaciation in north-central North America and, because of the widespread effects of the continental ice sheets on world climate, times of extensive deglaciation in the Cordillera as well.

Also of considerable interest in this regard is a curve of oxygen isotope variations in a Greenland ice core which appears to record major global oscillations of climate during the Wisconsin glacial age (Dansgaard et al. 1969). The curve displays two long intervals of first-order magnitude characterized by relatively high δO^{18} values, possibly correlative with the Plum Point and Port Talbot interstades of central North America. The curve is not controlled by radiocarbon dates, but ice flow considerations suggest that these intervals lasted from about 29,000–35,000 and 41,900–49,000 years ago. The intervening period, during which δO^{18} values were low, may correspond approximately to the interval during which the Laurentide ice sheet advanced to its middle Wisconsin maximum.

Second-order fluctuations of the Laurentide ice sheet may have little time–stratigraphic significance beyond any given terminal lobe. Not only was the time of the maximum late Wisconsin advance different along different segments of the glacier margin, but also several significant second-order advances, such as the Port Huron and Valders, appear to have been confined to certain sectors or lobes, instead of being regional. For reasons given above, considerable uncertainty also exists as to the climatic significance of the numerous moraines that record second-order terminal fluctuations in the region of the Great Lakes (Flint et al. 1959). Many, if not all, may owe their existence to isostatic effects resulting from deglaciation or, as has been suggested by Prest (1969), to major surges. Furthermore, the regional or worldwide significance of certain long-accepted interstadial climatic oscillations has been seriously questioned in recent years (Cushing 1967; Davis 1967; Mercer 1969). Consequently, little confidence can be placed in attempts to equate second-order glacier fluctuations between the midcontinent and the Cordillera, if for no other reason than that the response time of a small alpine glacier will be so vastly different from that of a continental ice sheet. Although precise dating may indicate that certain secondary maxima were reached simultaneously in the two regions and therefore are correlative, *sensu stricto*, the cause of the advances may be unrelated, both physically and temporally.

If we compare the time–distance curve constructed for the Laurentide ice sheet with those from western North America, certain correlations and generalizations can be made. For the alpine successions closely controlled by radiocarbon dates, it is apparent that the most recent first-order advance is broadly equivalent to the late Wisconsin advance of the Laurentide glacier. One also can reasonably infer that the last major glacial cycle in such areas as the Cascade Range, the Sierra Nevada, and the Rocky Mountains was contemporaneous, in the broad sense, with the late Wisconsin. As yet the dating is too imprecise, however, to demonstrate the degree of synchrony of glacial maxima (Fig. 4). In the Brooks Range the last maximum advance culminated before 13,270 years ago, in the St. Elias Mountains probably shortly before 13,660 years ago, in the Puget Lowland

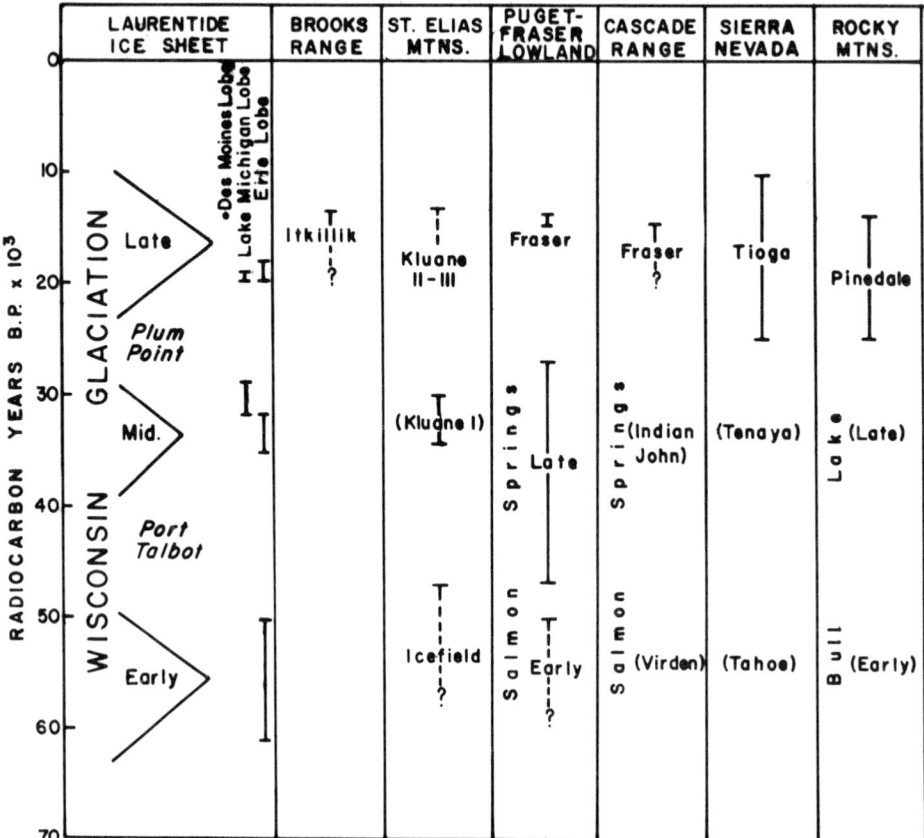

Fig. 4. Times of culmination of first-order late Pleistocene glacier advances in North America. Dashed lines indicate uncertain lower limit for maximum. Placement of bracketed units is speculative.

close to 14,000 years ago, in the Cascade Range some time before 14,000 years ago, in the Sierra Nevada between 9990 and about 25,000 years ago, and in the Rocky Mountains some time between about 13,000 and 25,000 years ago. Earlier first-order maxima are too poorly dated to make reliable correlations. However, several radiocarbon dates from the Yukon and from Washington suggest that the Icefield glaciation and the early episode of the Salmon Springs glaciation occurred during early Wisconsin time. In these same two regions tills are present in the stratigraphic succession which may have been deposited within the interval marked by an advance of the Laurentide ice sheet during middle Wisconsin time.

If the first-order advances of the Laurentide ice sheet depicted in Figure 4 reflect a response of the glacier to major oscillations of climate, then such oscillations may also have generated a significant response in alpine glaciers. Accordingly, one might speculate on the possible regional significance of the pattern of pre-late Wisconsin first-order advances in the Cordillera which as yet lack radio-carbon control. In the Brooks Range, the Cascade Range, and the Rocky Mountains an inferred pre-late Wisconsin and post-Sangamon glaciation is divided into two separate but essentially similar phases. If these phases were broadly correlative from region to region, could they reflect the same climatic oscillations that generated first-order advances of the Laurentide ice sheet during early and middle Wisconsin time (Fig. 4)? In the Northern and Middle Rocky Mountains, both the lower and upper drifts of Bull Lake age are overlain by a mature soil indicating nonglacial conditions (Schultz and Smith 1965). Were these two weathering intervals possibly contemporaneous respectively with Port Talbot and Plum Point interstades? In the Sierra Nevada, might the enigmatic Tenaya advance, which preceded the late Wisconsin Tioga glaciation, record a middle Wisconsin climatic oscillation? Such an interpretation would not be inconsistent with much of the weathering data (Figs. 3b, 3c). The fact that the Tenaya advance has not been recognized in some Sierran drainage basins, and that the late Bull Lake and late Salmon Springs advances are not recorded in all alpine valleys of the Rockies and the Cascades, suggests that if a middle Wisconsin advance is represented in the Cordillera, it frequently was less extensive than the maximum late Wisconsin advance. The crux of the problem lies in determining the relative importance or magnitude of different alpine advances, and distinguishing first-order advances, which might have broad regional climatic significance, from those of secondary rank possibly having no regional or climatic significance. In areas where radiometric dates are lacking, judgments are necessarily based on quantitative or semiquantitative measurements of soil parameters and other weathering phenomena.

Conclusions

It should be apparent from the preceding survey that remarkably few details are known about the chronology of late Pleistocene glacier fluctuations in the North American Cordillera. In part, this reflects a paucity of studies in critical areas where either radiocarbon dates or semiquantitative relative age data can be obtained. The two areas which at present provide the most thoroughly dated alpine glacial successions are not truly representative of conditions throughout much of the Cordillera, for they involve massive piedmont ice lobes which were fed by mountain ice caps. Because small alpine glaciers are likely to respond more rapidly to changes of climate than are large glacier systems, potentially they can provide a better index of former climate. Additional detailed studies of Pleistocene alpine glaciers of modest size are therefore badly needed. Past experience suggests that the opportunity for obtaining good radiocarbon control is better in northwestern North America than in most other parts of the Cordillera, largely because of the abundance of organic sediments associated with glacial drift sheets and the greater likelihood of their preservation. Further research in this region, with emphasis on dating the record of early and middle Wisconsin advances, should prove especially fruitful. At the same time, quantitative and semiquantitative studies of glacial deposits in other alpine areas can provide important information bearing on problems of local and regional correlation.

The existing data show quite clearly that widespread synchrony of alpine glacier variations cannot be demonstrated at present with any degree of confidence. Although the last major advance of alpine glaciers appears to have been broadly synchronous throughout most of western North America and to correlate in a general way with the late Wisconsin advance of the Laurentide ice sheet, possible equivalence of earlier first-order advances remains uncertain. Second-order advances not only lack regional uniformity of pattern but are too poorly dated to demonstrate any degree of synchrony. Furthermore, theoretical considerations dictate against the likelihood of synchronous terminal response of glaciers to a given climatic oscillation, and many second-order fluctuations may be totally unrelated to climatic events. Considerable caution must therefore be excercised in inferring former fluctuations of climate from glacial geologic data.

Summary. Time–distance curves depict the position of a glacier terminus through time and may provide information on relative magnitude of first- and second-order terminal fluctuations, times of glacial maxima and minima, and mean rates of advance and retreat. Curves constructed for late Pleistocene alpine glaciers show first-order fluctuations, having a time span of 10^4 years or longer, which appear broadly synchronous throughout western North America, and correlate in a gross way with first-order fluctuations of the southern margin of the Laurentide ice sheet during the Wisconsin glaciation. Second-order fluctuations, having a time span of 10^3 years or less, show no uniformity of pattern from region to region. Furthermore, climatic control of such fluctuations cannot be convincingly demonstrated, and many advances inferred from glacial-geologic evidence may, in fact, represent either response of glaciers to isostatic effects or to random surges not climatically induced. Even where climate was the controlling factor, second-order terminal fluctuations are not likely to have been precisely synchronous regionally because of the considerable range in environment, size, and response rates of Cordilleran glaciers.

REFERENCES

Adam, D. P., Late-Pleistocene and Recent palynology in the central Sierra Nevada, California, in E. J. Cushing and H. E. Wright, Jr., eds., *Quaternary Paleoecology,* Yale Univ. Press, p. 275, New Haven, 1967.

Armstrong, J. E., D. R. Crandell, D. J. Easterbrook, and J. B. Noble, Late Pleistocene stratigraphy and chronology in southwestern British Columbia and northwestern Washington, *Bull. Geol. Soc. Am.,* 76, p. 321, 1965.

Bateman, P. C., and Clyde Wahrhaftig, Geology of the Sierra Nevada, in Geology of northern California, *Calif. Div. Mines Geol. Bull., 190,* 107, 1966.

Birkeland, P. W., Pleistocene glaciation of the northern Sierra Nevada, north of Lake Tahoe, California, *J. Geol., 72,* 810, 1964.

Birman, J. H., Glacial geology across the crest of the Sierra Nevada, *Geol. Soc. Am., Spec. Papers, 75,* 80 pp., 1964.

Blackwelder, Eliot, Pleistocene glaciation in the Sierra Nevada and Basin ranges, *Bull. Geol. Soc. Am., 42,* 865, 1931.

Bostock, H. S., Geology of northwest Shakwak Valley, Yukon Territory, *Geol. Surv. Canada, Mem., 267*, 54 pp., 1952.

Broecker, W. S., and Aaron Kaufman, Radiocarbon chronology of Lake Lahontan and Lake Bonneville II: *Bull. Geol. Soc. Am., 76*, 537, 1965.

Broecker, W. S., K. K. Turekian, and B. C. Heezen, The relation of deep sea sedimentation rates to variations in climate, *Am. J. Sci., 256*, 503, 1958.

Crandell, D. R., Surficial geology and geomorphology of the Lake Tapps quadrangle, Washington, *U.S., Geol. Surv., Profess. Papers, 388-A*, 84 pp., 1963.

Crandell, D. R., The glacial history of western Washington and Oregon, in H. E. Wright, Jr., and D. G. Frey, eds., *The Quaternary of the United States*, Princeton, Princeton Univ. Press, p. 341–353, 1965.

Croft, M. G., Geology and radiocarbon ages of late Pleistocene lacustrine clay deposits, southern part of San Joaquin Valley, California, in Geological Survey Research, 1968, *U.S., Geol. Surv., Profess. Papers, 600-B*, B151–B156, 1968.

Cushing, E. J., Late-Wisconsin pollen stratigraphy and the glacial sequence in Minnesota, in E. J. Cushing and H. E. Wright, Jr., eds., *Quaternary* Paleoecology: New Haven, Yale Univ. Press, p. 59–88, 1967.

Dalrymple, G. B., Potassium-argon dates of three Pleistocene interglacial basalt flows from the Sierra Nevada, California, *Bull. Geol. Soc. Am., 75*, 753, 1964.

Dansgaard, W., S. J. Johnsen, J. Møller, and C. C. Langway, Jr., One Thousand centuries of climatic record from Camp Century on the Greenland Ice Sheet, *Science, 166*, 377, 1969.

Davis, M. B., Late-glacial climate in northern United States; a comparison of New England and the Great Lakes region, in E. J. Cushing and H. E. Wright, Jr., eds., *Quaternary Paleoecology*: New Haven, Yale Univ. Press, p. 11–43, 1967.

Denton, G. H., and M. Stuiver, Late Pleistocene glacial stratigraphy and chronology, northeastern St. Elias Mountains, Yukon Territory, Canada, *Bull. Geol. Soc. Am., 78*, 485, 1967.

Detterman, R. L., A. L. Bowsher, and J. T. Dutro, Jr., Glaciation on the Arctic Slope of the Brooks Range, northern Alaska, *Arctic, 11*, 43, 1958.

Dreimanis, A., J. Terasmae, and G. D. McKenzie, The Port Talbot interstade of the Wisconsin glaciation, *Can. J. Earth Sci., 3*, 305, 1966.

Easterbrook, D. J., D. R. Crandell, and E. B. Leopold, Pre-Olympia Pleistocene stratigraphy and chronology in the central Puget Lowland, Washington, *Bull. Geol. Soc. Am., 78*, 13, 1967.

Emiliani, C., Isotopic paleotemperatures, *Science, 154*, 851, 1966.

Flint, R. F., Status of the Pleistocene Wisconsin Stage in central North America, *Science, 139*, 402, 1963.

Flint, R. F., and W. A. Gale, Stratigraphy and radiocarbon dates at Searles Lake, California, *Am. J. Sci., 256*, 689, 1958.

Flint, R. F., R. B. Colton, R. P. Goldthwait, and H. B. Willman, Glacial map of the United States east of the Rocky Mountains, Geol. Soc. Am., 1959.

Frye, J. C., H. B. Willman, and R. F. Black, Outline of glacial geology of Illinois and Wisconsin, in H. E. Wright, Jr., and D. G. Frey, eds., *The Quaternary of the United States*, Princeton, Princeton Univ. Press, p. 43–61, 1965.

Frye, J. C., H. B. Willman, M. Robin, and R. F. Black, Definition of Wisconsinan Stage, *U.S. Geol. Surv., Bull., 1274-E*, 22 pp., 1968.

Goldthwait, R. P., A. Dreimanis, J. L. Forsyth, P. F. Karrow, and G. W. White, Pleistocene deposits of the Eire Lobe, in H. E. Wright, Jr., and D. G. Frey, eds., *The Quaternary of the United States*, Princeton, Princeton Univ. Press, p. 85–97, 1965.

Hamilton, T. D., Glacial geology of the lower Alatna Valley, Brooks Range, Alaska, in S. A. Schumm and W. C. Bradley, eds., United States contributions to Quaternary Research, *Geol. Soc. Am., Spec. Papers, 123,* 181–223, 1969.

Hoyt, J. H., V. J. Henry, Jr., and R. J. Weimer, Age of late-Pleistocene shoreline deposits, coastal Georgia, in R. B. Morrison and H. E. Wright, eds., *Means of Correlation of Quaternary Successions*, p. 381, Univ. Utah Press, Salt Lake City, 1968.

Janda, R. J., Pleistocene history and hydrology of the upper San Joaquin River, California, *Dissertation Abstr., 27,* 1967.

Kamb, B., Glacier geophysics, *Science, 146,* 353, 1964.

Karlstrom, T. N. V., Tentative correlation of Alaskan glacial sequences, 1956, *Science, 125,* 73, 1957.

Kempton, J. P., and J. E. Hackett, The late-Altonian (Wisconsinan) glacial sequence in northern Illinois, in R. B. Morrison and H. E. Wright, Jr., eds., *Means of Correlation of Quaternary Successions*, p. 535, Univ. Utah Press, Salt Lake City, 1968.

Kesseli, J. E., Studies in the Pleistocene glaciation of the Sierra Nevada, California, *Univ. Calif. Pub. Geography, 6,* 315, 1941.

Krinsley, D. B., Pleistocene geology of the south-west Yukon Territory, Canada, *J. Glaciology, 5,* 385, 1965.

Meier, M. F., Glaciers and climate, in H. E. Wright, Jr., and D. G. Frey, eds., The Quaternary of the United States, p. 795, Princeton Univ. Press, Princeton, 1965.

Mercer, J. H., The Allerød oscillation; a European climatic anomaly?, *Arctic Alpine Res., 1,* 227, 1969.

Milliman, J. D., and K. O. Emory, Sea levels during the past 35,000 years, *Science, 162,* 1121, 1968.

Morrison, R. B., Quaternary Geology of the Great Basin, in H. E. Wright, Jr., and D. G. Frey, eds., *The Quaternary of the United States*, p. 265, Princeton Univ. Press, Princeton, 1965.

Nelson, R. L., Glacial geology of the Frying Pan River drainage, Colorado: *J. Geol., 62,* 325, 1954.

Porter, S. C., Late Pleistocene glacial chronology of north-central Brooks Range, Alaska, *Am. J. Sci., 262,* 446, 1964.

Porter, S. C., Late Wisconsin alpine glaciation of east-central Cascade Range, Washington, in Abstracts for 1965, *Geol. Soc. Am., Spec. Papers, 87,* 131, 1965a.

Porter, S. C., Day 5—Yakima to Seattle, in C. B. Schultz and H. T. U. Smith, eds., Guidebook for Field Conference J, Pacific Northwest, INQUA Congr., VII, Nebraska Acad. Sci., p. 34, 1965b.

Porter, S. C., Pleistocene geology of Anaktuvuk Pass, central Brooks Range, Alaska, Arctic Inst. North America Tech. Paper 18, 100 pp., 1967.

Porter, S. C., Relative dating of alpine drift sheets using weathering rinds, in Abstracts for 1968, *Geol. Soc. Am., Spec. Papers, 121,* 1969a.

Porter, S. C., Pleistocene geology of the east-central Cascade Range, Washington, *Guidebook for 3d Pacific Coast Friends of the Pleistocene Field Conf.,* 54 pp., 1969b.

Porter, S. C., and G. H. Denton, Chronology of neoglaciation in the North American Cordillera, *Am. J. Sci., 265,* 177, 1967.

Prest, V. K., Retreat of Wisconsin and Recent ice in North America, *Geol. Surv. Canada, Map 1257A*, 1969.

Rahm, D. A., Glacial geology of the Bishop area, Sierra Nevada, California, in Abstracts for 1963, *Geol. Soc. Am., Spec. Papers*, 76, 221, 1964.

Rampton, V. N., Pleistocene geology of the Snag-Klutlan area, southwestern Yukon Territory, Canada, 237 pp., Ph.D. dissertation, University of Minnesota, 1969.

Richmond, G. M., Glaciation of Little Cottonwood and Bells Canyons, Wasatch Mountains, Utah, *U.S., Geol. Surv., Profess. Papers*, 454-D, 41 pp., 1964.

Richmond, G. M., Glaciation of the Rocky Mountains, in H. E. Wright, Jr., and D. G. Frey, eds., *The Quaternary of the United States*, p. 217, Princeton Univ. Press, Princeton, 1965.

Schultz, C. B., and H. T. U. Smith, eds., Guidebook for Field Conference E, northern and middle Rocky Mountains, INQUA Congr., VII, Nebraska Acad. Sci., 129 pp., 1965.

Sharp, R. P., Semiquantitative differentiation of glacial moraines near Convict Lake, Sierra Nevada, California, *J. Geol.*, 77, 68, 1969.

Sharp, R. P., and J. H. Birman, Additions to the classical sequence of Pleistocene glaciations, Sierra Nevada, California, *Bull. Geol. Soc. Am.*, 74, 1079, 1963.

Smith, G. I., Late-Quaternary geologic and climatic history of Searles Lake, southeastern California, in R. B. Morrison and H. E. Wright, Jr., eds., *Means of Correlation of Quaternary Successions*, p. 293, Univ. Utah Press, Salt Lake City, 1968.

Stuiver, Minze, Carbon isotopic distribution and correlated chronology of Searles Lake sediments, *Am. J. Sci.*, 262, 377, 1964.

Veeh, H. H., and John Chappell, Astronomical theory of climatic change: Support from New Guinea, *Science*, 167, 862, 1970.

Wahrhaftig, Clyde, and J. H. Birman, The Quaternary of the Pacific Mountain system in California, in H. E. Wright, Jr., and D. G. Frey, eds., *The Quaternary of the United States*, p. 299, Princeton Univ. Press, Princeton, 1965.

Wahrhaftig, Clyde, and R. P. Sharp, Sonora Pass Junction to Bloody Canyon, in C. B. Schultz and H. T. U. Smith, eds., Guidebook for Field Conference I, northern Great Basin and California, INQUA Congr., VII, Nebraska Acad. Sci., p. 71, 1965.

Barrie C. McDonald

12. LATE QUATERNARY STRATIGRAPHY AND DEGLACIATION IN EASTERN CANADA

Two quite separate aspects of late Quaternary history with special reference to eastern Canada are here discussed: widespread stratigraphic evidence for two major Wisconsin interstades, together with the paleogeographic inferences that they permit; and selected aspects of the late Wisconsin deglaciation. No attempt has been made to document a complete regional history with its complexity of attendant problems, as this has been the subject of recent attention by Prest (1969, and in press).

The discussion concentrates on that part of Canada south of about 66°N and east of about 104°W (Fig. 1). Although the discussion has been confined largely to Canada, abundant related data exist in the adjacent regions of the United States (cf. Wright and Frey 1965).

WISCONSIN STADES AND INTERSTADES

Evidence for significant Wisconsin interstades in Ontario and in the midwestern United States was presented a decade ago in two fundamental papers (Dreimanis 1960; Frye and Willman 1960). Each paper indicated that at least one major glacial phase and a subsequent interstade were represented in the post-Sangamon but "pre-classical" Wisconsin sediments. Since 1960, stratigraphic sections from North America and elsewhere have been correlated to these increasingly well-documented type areas. Dreimanis (1960) referred the interstade between about 50,000 and 23,000 B.P. to the "mid-Wisconsin"; earlier Wisconsin events were referred to "early Wisconsin." This usage is followed here, and events postdating about 23,000 B.P. are referred to "late Wisconsin."

Recent stratigraphic studies in widely separated areas of eastern Canada have added a considerable body of new data that permit regional reconstruction of the areal extent and succession of Wisconsin glacial phases and intervening interstades. With this reconstruction as an objective, the sequence of events in widely separated "control areas" is discussed, along with evidence from some neighboring areas, followed by an attempt at regional correlation. The strati-

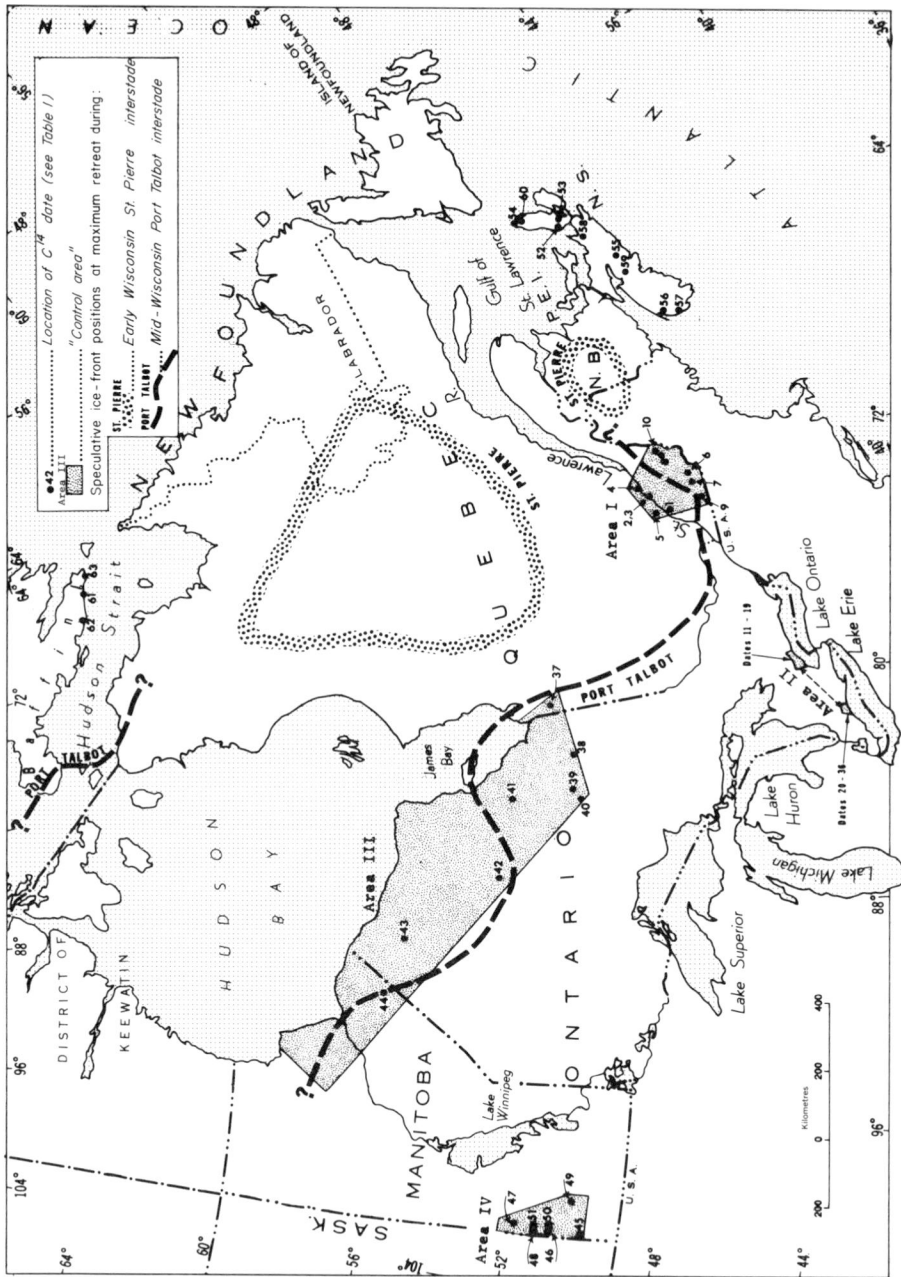

Fig. 1. Locations of radiocarbon dates, control areas, and speculative ice front positions.

Table 1. Radiocarbon Dates

(See Fig. 1 for locations)

No.	C^{14} years B.P.	Lab. no.[a]	Material	Stratigraphic unit	Reference
			I Southeastern Quebec		
	St. Lawrence Lowlands				
1	67,000 ± 1000[b]	Gro-1711	wood	St. Pierre beds	Dreimanis 1960
2	65,300 ± 1400[c]	GrN-1799	wood	St. Pierre beds	Muller 1964
3	64,000 ± 2000	Gro-1766	wood	St. Pierre beds	Dreimanis 1960
4	>44,470	Y-463	wood	St. Pierre beds	E. S. Deevey, written comm., 1957
5	>30,840[d]	Y-255	wood	St. Pierre beds	Preston et al. 1955
	Appalachian region				
6	>54,000	Y-1683	fine plant	Massawippi formation	McDonald 1967
7	>41,500	GSC-507	fine plant	Massawippi formation	McDonald 1967
8	>40,000	GSC-1084	fine plant	Massawippi formation	Shilts 1969
9	>28,000	GSC-655	wood	late-glacial gravel, sample reworked	Lowdon and Blake 1968
10	>20,000	GSC-1137	fine plant	Gayhurst formation?	Shilts 1969
			II Southern Ontario		
	Toronto vicinity				
11	>46,000[e]	L-409	wood	Don formation	Karrow 1967
12	>54,340 ± 500[f]	GrN-4817	wood	Scarborough formation	Karrow 1969
13	>50,800	GrN-4237	wood	Scarborough formation	Karrow 1969
14	>49,700[g]	GSC-203	wood	Scarborough formation	Dyck et al. 1965
15	>53,000[h]	GSC-1228	peat	Thorncliffe formation	A. A. Berti, pers. comm., 1970
16	44,600 ± 190[i]	GrN-4454	peat ball in gravel	Thorncliffe formation	Karrow 1969

(cont.)

Table 1 (*cont.*)

No.	C^{14} years B.P.	Lab. no.[a]	Material	Stratigraphic unit	Reference
17	> 41,000	GSC-629	peat	Thorncliffe formation	Karrow 1969
18	32,000 ± 690	GSC-1221	fine plant	Thorncliffe formation	A. A. Berti, pers. comm., 1970
19	28,300 ± 600	GSC-1082	fine plant	Thorncliffe formation	A. A. Berti, pers. comm., 1970
	North shore of Lake Erie				
20	47,700 ± 1200	GSC-217	wood from peat balls	modern shore at Port Talbot site	Dyck et al. 1965
21	47,600 ± 400[j]	GrN-2597 / GrN-2601	gyttja	Port Talbot beds	De Vries and Dreimanis 1960
22	46,700 ± 1400	GrN-2570	gyttja	Port Talbot beds	De Vries and Dreimanis 1960
23	46,400 ± 940	GSC-993-2	peat ball	Plum Point beds, beach facies	Lowdon and Blake 1970
24	45,800 ± 1200	GrN-4427	peat	Port Talbot beds?	Dreimanis et al. 1966
25	45,100 ± 1000	GrN-2619	peat ball	modern shore at Port Talbot site	De Vries and Dreimanis 1960
26	44,400 ± 1200	GrN-2580	wood	Southwold till, sample reworked	De Vries and Dreimanis 1960
27	43,400 ± 1300[k]	GrN-4800	wood	Port Talbot beds	Dreimanis, in press
28	42,700 ± 1200	GrN-4799	wood	Port Talbot beds	Dreimanis, in press
29	> 40,000	GSC-770	wood	Plum Point beds, beach facies	Lowdon and Blake 1970
30	38,000 ± 1500	GrN-4272	peat ball in gravel	late-glacial gravel, sample reworked	Dreimanis et al. 1966
31	34,000 ± 500	GrN-4238	peat ball in gravel	late-glacial gravel, sample reworked	Dreimanis et al. 1966
32	33,000 ± 1500	L-370A	gyttja	Port Talbot beds	Dreimanis et al. 1966

Table 1 (cont.)

No.	C^{14} years B.P.	Lab. no.[a]	Material	Stratigraphic unit	Reference
33	28,200 ± 1500[l]	L-185B	wood	Catfish Creek till	Broecker and Kulp 1957
34	27,500 ± 1200	W-177	wood	Catfish Creek till	Rubin and Suess 1955
35	27,470 ± 130	GrN-2625	wood	Catfish Creek till	De Vries and Dreimanis 1960
36	24,600 ± 1600	L-217B	wood	Catfish Creek till	Broecker and Kulp 1957
			III *Hudson Bay Lowland*		
37	> 42,000	Y-1165	peat	Missinaibi beds	Stuiver et al. 1963
38	> 43,600	GSC-435	wood	Missinaibi beds	Lowdon et al. 1967
39	> 53,000[m]	Gro-1435	wood	Missinaibi beds	Terasmae and Hughes 1960
40	> 42,000	Gro-1921	peat	Missinaibi beds	Terasmae, pers. comm., 1968
41	> 54,000	GSC-1185	peat	Missinaibi beds	Coll. 1968 by R. Wilkins; unpub.
42	> 35,800	GSC-83	wood	Missinaibi beds	Dyck and Fyles 1963
43	> 41,000	GSC-1011	peat	Missinaibi beds	McDonald 1969
44	> 37,000	GSC-892	wood	Missinaibi beds	McDonald 1969
			IV *Southwestern Manitoba*		
45	> 42,000	GSC-750	wood	unnamed sediments below Shell till	Lowdon and Blake 1968
46	> 41,000	GSC-678	wood	unnamed sediments below Shell till	Klassen 1969
47	> 37,760	GSC-284[n]	peat	Roaring River clay	Klassen et al 1967
48	> 34,000	GSC-676	peat	Roaring River clay	Klassen 1969
49	> 31,300	GSC-297	grass	Roaring River clay	Dyck et al. 1966
50	> 30,000	GSC-218	wood	Roaring River clay?	Dyck et al. 1965
51	37,700 ± 1500[o]	GSC-653	charcoal	unnamed sediments between Lennard and Minnedosa tills	Klassen 1969

(cont.)

Table 1 (*ccnt.*)

No.	C^{14} years B.P.	Lab. no.[a]	Material	Stratigraphic unit	Reference
			V Neighboring Areas		
	Maritime Provinces of Eastern Canada				
52	>51,000[p]	GSC-370	wood	intertill organic beds	Mott and Prest 1967
53	>44,000	GSC-290	wood	intertill organic beds	Mott and Prest 1967
54	>38,300	GSC-283	wood	intertill organic beds	Mott and Prest 1967
55	>33,800	GSC-33	wood	from lower of two tills	Dyck and Fyles 1963
56	>39,000	GSC-887	marine shells	marine clay	Lowdon and Blake 1970
57	>38,000	GSC-695	marine shells	till	Grant 1968
58	$33,700 \begin{smallmatrix} +2300 \\ -1800 \end{smallmatrix}$	I-3236	wood	intertill silt	MacNeill 1969
59	$33,200 \begin{smallmatrix} +2000 \\ -1700 \end{smallmatrix}$	I-3237	wood	subtill sediments	MacNeill 1969
60	20,300 ± 400	I-2438	peat	subtill sediments	Buckley and Willis 1970
	Hudson Strait area				
61	34,800 ± 1100	GSC-426	marine shells	till	Blake 1966
62	30,200 ± 1500	GSC-414	marine shells	till	Blake 1966
63	>25,900	GSC-468	marine shells	postglacial marine sand	Blake 1966

[a] Laboratories: Gro, GrN = Groningen; GSC = Geological Survey of Canada; I = Isotopes, Inc.; L = Lamont; S = Saskatchewan; W = United States Geological Survey; Y = Yale University.

[b] Another sample from same location (Pierreville): >29,630 (Y-256).

[c] Other samples from same location (St. Pierre-les-Becquets): >44,000 (L-396A); >40,000 (L-190A); >40,000 (W-189); >30,840 (Y-242).

[d] Another sample from same location (Les Vieilles Forges): >29,630 (Y-254).

[e] Another sample from Don formation: >30,000 (W-121).

[f] Although reported by the laboratory as finite, Karrow (1969) regards this date as minimal because of contamination by modern rootlets. Other samples from same location (Seminary section, Scarborough): >52,000 (Gro-2555); >40,000 (L-522B).

[g] Another sample from same location (Woodbridge cut): >49,000 (GSC-729).

[h] Two other samples, from the same peaty lens as GSC-1228, have been dated at 48,000 ± 1400 (GSC-534) and 38,900 ± 1300 (GSC-271). The presence in the peat of delicate *Dryas* leaves suggests that the material had not been reworked. It is possible that the finite dates resulted from contamination of those samples by modern rootlets.

Notes to Table 1 (*cont.*)

[i] Other samples from same location (Markham): $>34,000$ (W-194); $>23,000$ (S-26).

[j] Other samples from same location (Port Talbot): $>40,000$ (L-370A); $>39,000$ (L-217A); $>38,000$ (L-185A); $>34,000$ (S-46); $>32,000$ (W-100); $>29,500$ (L-440); $>25,000$ (S-7).

[k] Dates 27 and 28 (from the same location) indicate that a date reported by Dreimanis et al. in 1966 (33,400 ± 500, GrN-4397, but reported there erroneously as GrN-4238) was in error due to contamination (Dreimanis, in press).

[l] Dates 33 to 36 are from pieces of wood that were removed from Catfish Creek till stratigraphically overlying the Plum Point member of the Port Talbot interstadial sediments.

[m] Other samples from same location (Missinaibi River): $>42,600$ (L-396B); $>38,000$ (W-241); $>38,000$ (W-242); $>30,840$ (Y-270); $>29,630$ (Y-269).

[n] This date was erroneously labeled as GSC-286 in Klassen et al. (1967).

[o] Because subjacent marl was dated at 28,220 ± 380 (GSC-711; Lowdon and Blake 1968), charcoal was recollected from the same bed as GSC-653 and has been dated at 23,700 ± 290 (GSC-1279; Klassen, unpublished).

[p] Other samples from same location (Hillsborough): $>38,000$ (W-157); $>21,000$ (Y-232).

graphic control areas (Fig. 1) are: I, southeastern Quebec; II, southern Ontario; III, Hudson Bay Lowland; IV, southwestern Manitoba.

Pertinent radiocarbon dates are listed in Table 1 and located on Figure 1; composite stratigraphic columns for the control areas are shown on Figure 2. All the control areas are underlain by Paleozoic and/or Mesozoic strata and are separated by areas of Precambrian shield where Quaternary stratigraphy is generally poorly preserved.

Stratigraphic Record

Southeastern Quebec. Stratigraphic studies in the St. Lawrence Lowlands (Terasmae 1958; Gadd 1960, and in press) and in the Quebec Appalachians to the southeast (McDonald 1967; Shilts 1969; McDonald and Shilts, in press) have reported on the late Quaternary stratigraphy in southeastern Quebec. A till stratigraphy that indicates a sequence of Wisconsin ice flow directions broadly similar to that recorded in the Quebec Appalachians has been reported from northern Vermont (Stewart and MacClintock 1964).

The earliest glacial phase for which there is direct evidence is represented in the St. Lawrence Lowlands by the Bécancour till and in the Appalachians by its correlative, the Johnville till. Till fabrics and till lithology indicate that glaciers during this episode flowed from the north and northwest. An early Wisconsin age for this glacial event is suggested by the presence beneath Johnville till of gravel deposited by through-flowing streams which require that the Lowlands were glacier-free; this gravel is judged by its unusually deep weathering to represent the Sangamon interglacial interval. The presence in this gravel of erratics from the Precambrian shield provides indirect evidence for yet an earlier glaciation.

The St. Pierre peat beds, overlying the Bécancour till in the Lowlands, represent the best-known exposures in eastern North America of sediments of the earliest Wisconsin interstade. They are represented in the Appalachians by peat beds and plant-bearing lake sediments of the Massawippi formation. Radiocarbon dates, reported as "finite," would place the duration and age of this interstade at a few thousand years and about 65,000 B.P. respectively (Table 1). Spore and pollen assemblages in both these units indicate a climate cooler than the present and vegetation similar to the northern boreal forest. Peat beds near sea level in the Lowlands and stream gravel in the Appalachians indicate that both the St. Lawrence Lowlands and the Appalachians were ice-free and that through-flowing drainage was established to the Gulf of St. Lawrence.

In the St. Lawrence Lowlands a single till, the Gentilly till, overlies the St. Pierre beds and is overlain in turn by marine sediments of the Champlain Sea, suggesting that from about 60,000 to about 12,000 B.P. glaciers occupied this region. In the Appalachians, however, this time span is represented by the Chaudière and Lennoxville tills and intervening glaciolacustrine sediments of the Gayhurst formation. The basal part of the Chaudière till, representing renewed glaciation after the St. Pierre interstade, was deposited by ice flowing from east-northeast across the Appalachians. This flow direction is documented by till fabrics, till lithology, and boulder trains. Such an ice flow direction may have been responsible for the early blocking of the St. Lawrence River, necessary to permit the deposition of a thick sequence (25 meters) of proglacial-lake sediments that are exposed beneath Gentilly till. During the Chaudière glacial phase the glacier dispersal center influencing southeastern Quebec shifted westward. Till fabrics, till lithology, and boulder trains indicate that by the end of the Chaudière phase, and for the rest of the Wisconsin, ice flow in this area was from the northwest.

Graded couplets of the silt–clay laminas, interpreted as varves, indicate that the Gayhurst sediments separating the Chaudière and Lennoxville tills record at least 4000 years of deposition. Finely divided plant material, judged from field evidence as being from probable Gayhurst sediments, has been dated at > 20,000 B.P. (Table 1). It is proposed (McDonald and Shilts, in press) that this unit represents a significant mid-Wisconsin interstade, and that the associated ice front, which blocked the northwest-flowing drainage of the Appalachians, lay at the edge of the St. Lawrence Lowlands along the Appalachian front.

The entire late Wisconsin is represented in the Appalachians by Lennoxville till and in the Lowlands by the upper part of the Gentilly till.

Southwestward from the southeastern Quebec control area, a considerable body of stratigraphic data was gathered in excavations for the St. Lawrence Seaway by MacClintock and Stewart (1965). The stratigraphy recorded there extends to the vicinity of Montreal (Prest and Hode Keyser 1962). The sequence can be summarized briefly: (1) ice flowing up the St. Lawrence Valley from the northeast deposited the Malone till; (2) a partial deglaciation followed, perhaps of relatively short duration, during which the ice front retreated in the St. Lawrence Valley to the vicinity of Montreal; glacial-lake sediments were deposited in the valley, but through-flowing drainage was not established to the Gulf of St. Lawrence; (3) ice flowing from the northwest deposited the Fort Covington till. No radiocarbon dates were obtained from these sediments. The entire sequence may correlate with the Gentilly till. Glacial-lake sediments separating the Malone and Fort Covington tills would then be correlative with the Gayhurst sediments of the Quebec Appalachians and represent a significant mid-Wisconsin interstade.

Southern Ontario. This control area includes the well-documented exposures at Toronto (Coleman 1933; Terasmae 1960; Karrow 1967, 1969) and those near Port Talbot on Lake Erie (Dreimanis 1960, and in press; Dreimanis et al. 1966). These exposures are classic; careful field examination over many years, and the large number of radiocarbon dates (Table 1) from these sections, have given this area fundamental importance as a type area for Quaternary stratigraphy in eastern North America. Correlation between Toronto and Port Talbot (Fig. 2) has been drawn from a chart prepared by Dreimanis (in press).

Sediments of the Sangamon interglacial interval overlie an older till (York) at Toronto and are represented by the richly fossiliferous Don formation. On the basis of palynology and paleontology Terasmae (1960) estimated that the annual mean temperature during deposition of the Don beds was 5°F warmer than at present. Overlying the Don beds are fossiliferous deltaic sediments of the Scar-

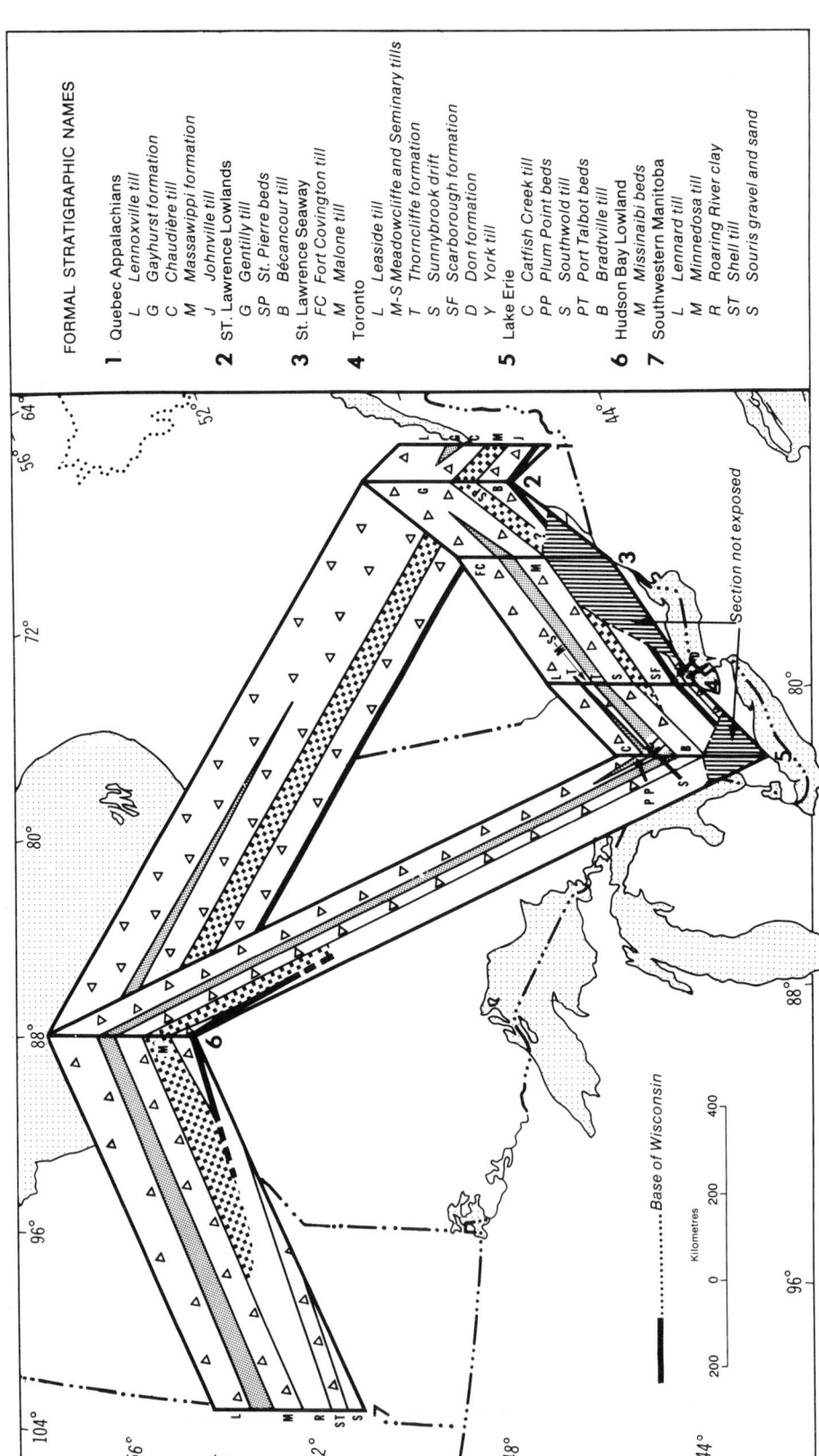

FORMAL STRATIGRAPHIC NAMES

1 Quebec Appalachians
   L   Lennoxville till
   G   Gayhurst formation
   C   Chaudière till
   M   Massawippi formation
   J   Johnville till

2 ST. Lawrence Lowlands
   G   Gentilly till
   SP St. Pierre beds
   B   Bécancour till

3 St. Lawrence Seaway
   FC Fort Covington till
   M   Malone till

4 Toronto
   L   Leaside till
   M-S Meadowcliffe and Seminary tills
   T   Thorncliffe formation
   S   Sunnybrook drift
   SF   Scarborough formation
   D   Don formation
   Y   York till

5 Lake Erie
   C   Catfish Creek till
   PP Plum Point beds
   S   Southwold till
   PT Port Talbot beds
   B   Bradtville till

6 Hudson Bay Lowland
   M   Missinaibi beds

7 Southwestern Manitoba
   L   Lennard till
   M   Minnedosa till
   R   Roaring River clay
   ST Shell till
   S   Souris gravel and sand

Fig. 2. Proposed Quaternary correlations, eastern Canada. (Bases of vertical columns mark locations of stratigraphy shown.)

borough formation. They are interpreted by Karrow (1969) to have formed in a lake that was dammed by ice in the St. Lawrence River valley to a level about 50 meters above present Lake Ontario. The suggested proximity of glaciers to Toronto is compatible with pollen and spore data that indicate an annual mean temperature during Scarborough time that was 10°F lower than present (Terasmae 1960). Subsequent establishment of an open drainage system, presumably through the St. Lawrence Valley, permitted deep dissection of the Scarborough beds and local accumulation of channel gravel and sand.

The first Wisconsin glaciers to reach Toronto and Lake Erie flowed into southern Ontario from the east-northeast. This glacial phase is represented at Toronto by Sunnybrook till and in the Lake Erie basin by Bradtville till.

Subsequent deglaciation marked the beginning of the Port Talbot interstade, a major nonglacial interval representing the mid-Wisconsin (Dreimanis et al. 1966). Sediments of this interstade include the Port Talbot and Plum Point beds of the Lake Erie basin and the Thorncliffe formation of Toronto. The sediments contain abundant plant material; radiocarbon dates (Table 1) indicate that the interstade began at least 50,000 B.P. and lasted, with some interruption, until the beginning of the late Wisconsin glaciation, about 23,000 B.P. The youngest date reported from sediments of this interval is 22,800 ± 450 (GSC–816), on wood from the buried St. Davids gorge near Niagara Falls, 15 km south of Lake Ontario (Hobson and Terasmae 1969). Pollen and spores indicate that the climate during the Port Talbot interstade was cooler than the present, and that the vegetation of southern Ontario resembled that of the present northern boreal forest. The presence within the Port Talbot sequences, both in the Lake Erie basin and at Toronto, of till and ice-rafted sediment indicates that during at least part of this interstade glaciers were in the southern Ontario region. On the basis of far-traveled lithologic indicators, Dreimanis (1960) reported that these tills were derived from source areas west of those of the till stratigraphically below the Port Talbot.

The Port Talbot interstade ended with the advance to southern Ontario of late Wisconsin ice. Despite a complex pattern of lobation resulting in part from a local center of ice dispersal in the Lake Ontario basin, lithology of the late Wisconsin tills continued to reflect a more westerly provenance than the early Wisconsin till.

Hudson Bay Lowland. Quaternary geologic and palynologic data from the southeastern part of the Hudson Bay Lowland were discussed by Terasmae (1958) and by Terasmae and Hughes (1960), and new geologic data from the entire Lowland were reported by McDonald (1969). A detailed study of Quaternary Stratigraphy in the Lowland south of James Bay is currently underway (Skinner 1970). The Hudson Bay Lowland has a particular importance to Quaternary history because of the following: (1) the unusually excellent and abundant exposures of Quaternary sediments are surrounded on all sides by several kilometers of Precambrian terrain, where good exposures that would facilitate correlation with the other control areas are not abundant; (2) due to its location near the center of former ice sheets, ice flow directions associated with the various till units can provide useful data relevant to the growth and spread of these ice sheets; and (3) owing to its location near the center of former ice sheets, nonglacial events here must correlate with significant nonglacial events in areas farther south.

The Missinaibi beds, intertill nonglacial sediments including stream deposits, and peat over richly fossiliferous marine strata, are widespread in the Hudson Bay Lowland and provide a useful stratigraphic marker. All radiocarbon dates on the peat beds (Table 1) are "greater than" dates. Pollen and spores indicate that, during deposition of the Missinaibi beds, vegetation was northern boreal forest, similar to that in the area today. However, because the boreal forest zone is so broad, the sensitivity of the pollen record here to climatic change is low.

The rank of Missinaibi beds is still questionable. Because floral elements indicating a climate warmer than that of today were absent, and because the Missinaibi time interval appears to be relatively short, Terasmae (1958) proposed an interstadial rank for these beds and suggested their correlation with the St. Pierre beds of the St. Lawrence Lowlands. McDonald (1969) considered the Missinaibi beds to be of interglacial rank because the marine strata and deposits of throughflowing streams at low altitude required that Hudson Bay and Hudson Strait be ice-free, indicating a severe diminution of any contemporary ice sheet. If the Missinaibi beds are wholly correlative with the St. Pierre beds, it is strange that marine strata, present in the Hudson Bay Lowland to altitudes of about 75 meters above present sea level, have not been observed in the excellent exposures in the St. Lawrence Lowland. Barring the possibility that the Hudson Bay Lowland was uplifted tectonically in early Wisconsin time, and perhaps since, it is possible that: (1) marine strata associated with the buried peat exist also in the St. Lawrence Lowlands but have yet to be observed; or (2) isostatic depression from the preceding glaciation was more pronounced in the Hudson Bay Lowland than in the St. Lawrence Lowlands; or (3) marine strata underlying buried peat in the Hudson Bay Lowland are truly interglacial and record a higher sea level. If the last is true, the overlying peat may still be a St. Pierre equivalent, although in that case the absence of a till between the marine strata and peat would indicate that the earliest Wisconsin glacial phase did not extend westward to the Hudson Bay Lowland.

The Missinaibi beds directly overlie till at a few localities. On the Missinaibi River (Terasmae and Hughes 1960; 50°17′N, 82°40′W) a complex sequence of several till units and interstratified sediments underlie the Missinaibi beds. Insufficient data are in hand with which to indicate reliably the provenance of these tills. In the central Hudson Bay Lowland, the lower till appears from till fabrics to have been deposited from ice flowing due south, perhaps from a dispersal center located over Hudson Bay.

Two distinct tills overlie the Missinaibi beds throughout the Lowland. In the northwestern and southeastern parts of the Lowland the tills are separated by glacial-lake sediments, apparently devoid of organic material. There is no evidence of subaerial exposure, nor are there marine sediments between the tills. In the central part of the Lowland the stratified-sediment unit is absent and the two tills are in contact with each other. This may indicate that the central part of the Lowland was occupied by ice continuously from Missinaibi time until final deglaciation about 7500 B.P. It is probable that the break between the two uppermost tills here correlates with the Port Talbot interstade. Apparently, although the northwestern and southeastern parts of the Hudson Bay Lowland were temporarily ice-free, glaciers continued to occupy Hudson Bay. The effect of ice in Hudson Bay, instead of the moderating effect of the sea, may be compatible with a cooler climate in southern Ontario.

Stone orientations and striated boulder pavements from tills in the central and northwestern parts of the Lowland indicate that the till immediately overlying the Missinaibi beds was deposited by ice flowing from about N40°E. However, the uppermost till was deposited by ice flowing from between N25°E and N45°W. Sparse data from the southeastern part of the Lowland indicate that this shift in ice flow direction may also be recorded there. Thus it appears that at least temporary glacier dispersal centers were located over Hudson Bay.

Southwestern Manitoba. A Quaternary stratigraphy documented with radiocarbon dates, and consistent with radiocarbon-controlled stratigraphy farther west in Saskatchewan, has recently been presented for southwestern Manitoba (Klassen et al. 1967; Klassen 1969; see also Christiansen 1968a, b).

The Sangamon interglacial interval may be represented in Manitoba by the Roaring River clay. Radiocarbon dates from this unit are "greater than" dates, and paleontologic study has indicated deposition of the unit during a cool–warm–cool climatic sequence. The clay overlies Shell till, the deep weathering of which has been attributed to interglacial conditions (Klassen 1969).

Two Wisconsin glacial phases are recorded in Manitoba by the Lennard and Minnedosa tills. These tills are separated by stratified sediments containing a thin till unit that may be related to the Minnedosa glacial phase. A radiocarbon date of 37,000 ± 1500 (Table 1; see also footnote o, Table 1) from the stratified sediments indicates their correlation with the mid-Wisconsin Port Talbot interstade.

The Minnedosa till, underlying the mid-Wisconsin stratified unit, was deposited by ice flowing from the northeast. The Lennard till, overlying the stratified unit and representing the entire late Wisconsin, was deposited by ice flowing from the northwest.

Neighboring areas. Although a detailed Quaternary stratigraphy is not available for the Maritime Provinces of eastern Canada, evidence for Wisconsin interstades in Nova Scotia is presented by Mott and Prest (1967) and Livingstone (1968); radiocarbon dates are listed in Table 1. Pollen study of a series of intertill stratified sediments indicates the existence of a climate cooler than the present and vegetation similar to the northern boreal forest. Because of this and the date of > 51,000 B.P. from the Hillsborough site, these nonglacial sediments were correlated with the St. Pierre beds of the St. Lawrence Lowlands. Wood in the lower of two tills near Milford, Nova Scotia (date no. 55, Table 1), may have been derived from a correlative unit. If so, there could be two post-St. Pierre tills present in this area. Three finite radiocarbon dates in the range 33,700 to 20,300 B.P. (Table 1) have been obtained from plant material in and beneath till in Nova Scotia. These indicate that at least parts of the Maritimes were ice-free also during the Port Talbot interstade, and were subsequently overrun by late Wisconsin glaciers.

Two "finite" dates of 34,800 and 30,200 B.P. (Table 1) have been obtained from marine shells in till on the south coast of Baffin Island (Blake 1966). The till was deposited from an ice tongue flowing generally eastward in Hudson Strait. Numerous other marine-shell dates, "finite" in the range of 30,300 to 42,400 B.P., have been reported from other widely separated localities in the Canadian Arctic Archipelago. Because of the small amount of contamination needed to make very old shells produce a "finite" date, these dates should be regarded with caution until their reliability can be corroborated by dates from associated plant material.

Regional History

An attempt to correlate stratigraphy from the control areas has been made by means of a fence diagram (Fig. 2). This is a simplified framework built on an imperfect terrestrial record. Many of the radiocarbon dates are "greater than," thereby providing unsatisfactory control of early Wisconsin chronology. It is even questionable whether the three "finite" dates reported from the St. Pierre beds are truly finite. It would be encouraging if laboratories other than Groningen would produce similar dates, and it would be encouraging also to see some of these attempts reported as "greater than" at the 70,000 B.P. level of activity. As noted by Broecker (1965), without checks on different chemical fractions of the same sample these ages should be regarded as minimal.

Stratigraphic data indicate that two widespread Wisconsin interstades, the early Wisconsin St. Pierre and the mid-Wisconsin Port Talbot interstades, separate three major Wisconsin glacial phases. In southwestern Manitoba, unaffected by the earliest Wisconsin glacial phase, sediments of the St. Pierre interstade directly overlie sediments of interglacial age. Although on Figure 2 the Missinaibi beds of the Hudson Bay Lowland have been shown wholly correlative with the St. Pierre beds of the St. Lawrence Lowland, it must be kept in mind that the lower marine member of the Missinaibi beds may be interglacial. If the latter is true, the correlation shown on Figure 2 would have to be modified to show a greater similarity between the sequence in the Hudson Bay Lowland and that in southwestern Manitoba. No deep weathering was observed on the till beneath the Missinaibi beds. Until more diagnostic data become available, perhaps from a detailed paleoecologic study of the Missinaibi marine fauna, the rank of the Missinaibi beds will remain in doubt. The most useful criteria permitting identification of the interglacial episode preceding the Wisconsin are intense weathering zones, and flora that indicate a climate significantly warmer than that of today. Lack of evidence of glaciers, for example in the case of strata assigned to the St. Pierre interstade, does not necessarily indicate interglacial conditions.

However, the physical character of nonglacial sediments can lead to some inferences regarding the extent of ice retreat. A speculative ice front position at maximum retreat during the early Wisconsin St. Pierre interstade is shown on Figure 1. Peat beds at low altitude in the St. Lawrence Lowland, and peat, stream gravel, and marine strata in the Hudson Bay Lowland, indicate that during the St. Pierre interstade the St. Lawrence River, Hudson Bay, and Hudson Strait were all ice-free. The main area of ice during the St. Pierre interstade is shown in Quebec because it is apparent that the earliest Wisconsin glacial phase had its greatest effect in eastern Canada, and also because in southeastern Quebec the earliest Wisconsin glaciers flowed from the north and northwest. A small area of ice is also shown in the Maritime region, because early post-St. Pierre glaciers flowed from the east-northeast across southeastern Quebec, suggesting that this was an important accumulation area at that time.

A speculative ice front position at maximum retreat during the mid-Wisconsin Port Talbot interstade is also shown on Figure 1. During the Port Talbot interstade, glaciers continued to occupy the St. Lawrence Lowlands and Hudson Bay. Stratigraphic data indicate, however, that the Quebec Appalachians, part of the St. Lawrence River valley, southern Ontario, part of the Hudson Bay Lowland, southwestern Manitoba, and at least part of the Maritimes were ice-free. If marine-shell dates from the Arctic Archipelago are proven to be reliable, part of Hudson Strait was also ice-free.

Both post-St. Pierre glacial phases were more extensive than the earliest Wisconsin glacial episode, and they affected areas farther west. The early Wisconsin ice sheets grew in eastern Canada. Evidence of ice flow directions from the Quebec Appalachians suggests that, in early post-St. Pierre time, a major dispersal center existed in the Maritime region, perhaps over northern Maine and central New Brunswick. Reliable stratigraphic data, ideally with radiocarbon control, are needed from this region in order to document these events better.

Evidence of ice flow directions indicates that, for each of the control areas discussed, the mid-Wisconsin Port Talbot interstade was accompanied by a westward shift of the centers of glacier outflow. The nature of this shift, whether episodic or continuous, is unknown, as are the number, shape, and areal extent of these dispersal centers. This shift was recognized in southern Ontario and Quebec by Dreimanis (1960). In the Appalachians of southeastern Quebec, this shift occurred during the Chaudière glacial phase. The glacial members of the mid-Wisconsin sequence in southern Ontario also show evidence of this shift. It is tempting to speculate that the Port Talbot interstade and the westward migration of glacier dispersal centers are related to the same cause.

Late Wisconsin Deglaciation

Deglaciation following the late Wisconsin glacial phase was characterized, with a few notable local exceptions, by the more or less regular backwasting of an active ice front, interrupted by relatively short-lived and commonly local readvances. Reconstruction of the chronology of the deglaciation has been attempted recently

(Bryson and Wendland 1967; Bryson et al. 1969; Prest 1969). The amount of interpretation involved has resulted in marked differences in the reconstructions. Figures for this discussion are based largely on Prest (1969) because his data are more complete and because they were presented on a scale suitable for measurement.

As a basis for discussion, deglaciation is depicted by a plot of area occupied by late Wisconsin glaciers versus time (Fig. 3); relevant glacier areas are listed in Table 2. The area at 18,000 B.P. is highly interpretive; over large areas, ice lay at the late Wisconsin maximum about 15,000 B.P., and this position has also been accepted as valid for 18,000 B.P. The times when large areas separated from the main ice sheet are identified, and the rates of deglaciation within these areas are shown. Deglaciation was most rapid between 10,000 and 7,000 B.P.; the rate of deglaciation during this period had increased by 50 percent over the rate between 14,000 and 10,000 B.P. The final remnants in Labrador and west of Hudson Bay disappeared about 6,000 B.P. The area presently occupied by glaciers is shown as constant since about 8,000 B.P. In fact, this area has fluctuated, but data on the fluctuations are incomplete, and the scale of fluctuation is probably very minor relative to the scale of Figure 3.

Although Figure 3 could provide a very crude guide to the shape of a eustatic sea level curve, the basic assumptions involved in the extension of these data to ice volumes (cf. Donn et al. 1962) are so large that they make a rigorous treatment unreliable at best. Some dangers inherent in a simple-function extrapolation of glaciated areas to ice volumes are: (1) a simple surface profile for the ice sheet must be assumed, yet, throughout the whole late Wisconsin and especially after about 13,500 B.P. when the ice sheet was disintegrating into separate masses (Fig. 3), the surface of the ice sheet probably had an uneven topography; (2) topography of the base of the ice sheet was highly variable, being essentially that of the partially rebounded present-day topography; and (3) ice front positions are poorly known in places, especially where they stood in the sea and where they were overridden by later ice. In any case, the data and assumptions would have to be extended to include all glaciated regions of the world; Figure 2 includes only Canada and those glaciated parts of the United States south of Canada that were coterminous with areas occupied by the main continental ice sheet. Consideration must be given also to changes in ice volume in Antarctica and Greenland during the late Wisconsin. Calculations from data in Table 2 indicate that Figure 3 accounts for only about 57 percent of the late Wisconsin decrease in glaciated area (82 percent if the figures of Shumskiy are accepted).

Ice front positions at times of maximum retreat during the St. Pierre and Port Talbot interstades were postulated on Figure 1. From analogy with extent of glacier cover at similar stages of late Wisconsin deglaciation, areas of glacier cover during these times would be equivalent to about 0.5×10^6 km^2 for St. Pierre, and to about 5.2×10^6 km^2 for Port Talbot.

Deglaciation in the Maritime Provinces of eastern Canada was accompanied, between about 13,500 and 11,000 B.P., by the isolation of residuals of glacier ice

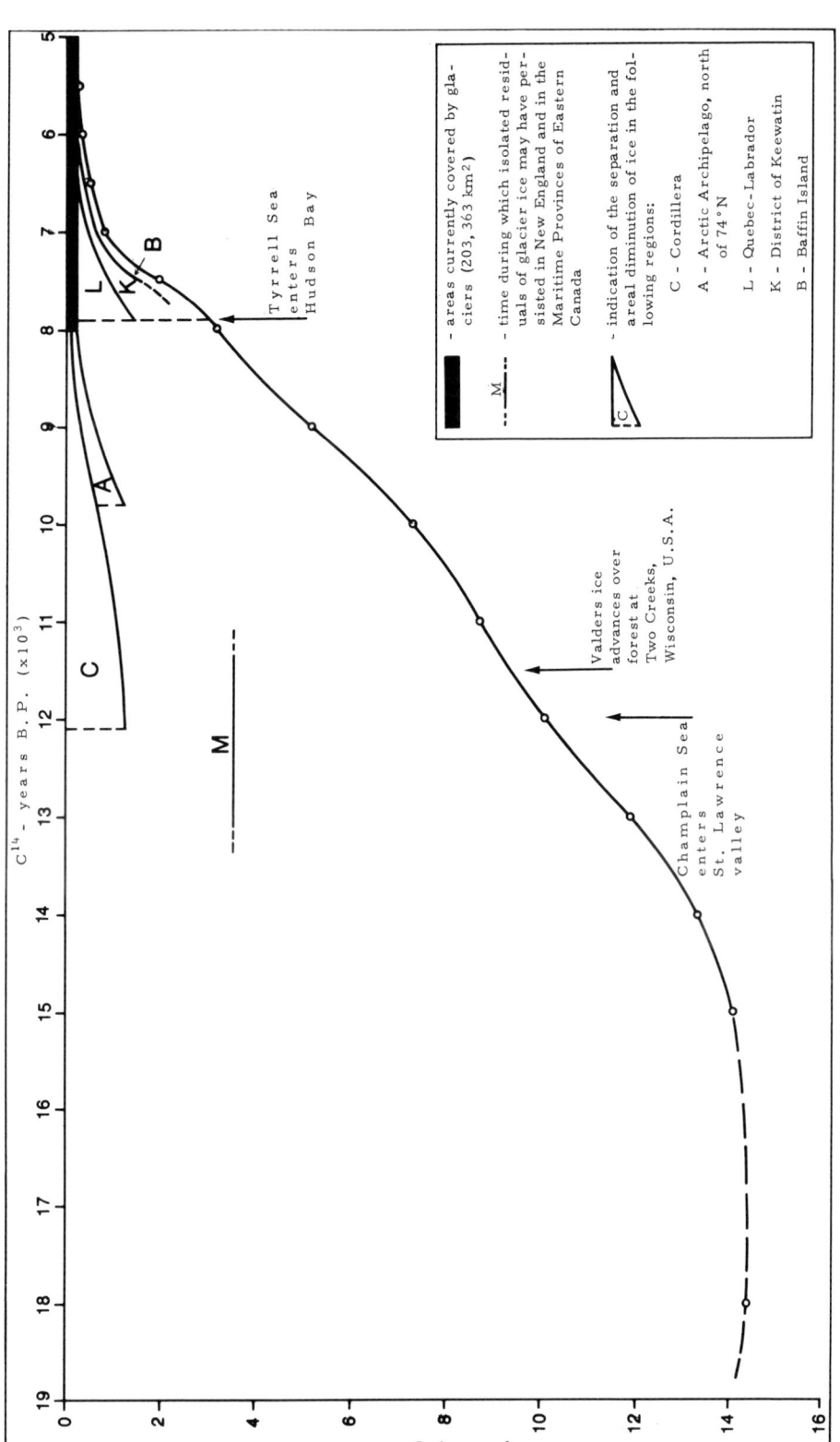

Fig. 3. Variation with time of late Wisconsin glacier areas of North America. (Not included are past and presently glaciated areas of Greenland and Alaska, and those glaciated areas of Mexico and the U.S. Cordillera that were not coterminous with areas occupied by the main continental ice sheet.)

Table 2. Selected Areas of Past and Present Glaciation

Time	Area ice-covered ($km^2 \times 10^6$)	Reference
Comparison of Glaciated Areas of the World to Those of North America		
Quaternary maximum		
World	45.236[a]	Flint 1957
North America	15.735	Flint 1957
Late Wisconsin maximum		
World	40.049[b]	Flint 1957
North America	14.825	Flint 1957
Existing glacier areas		
World	15.872[c]	Flint 1957
Canada		
Cordilleran	0.050	W. E. S. Henoch, unpub. ms., 1967
Arctic Archipelago	0.153	G. Falconer, J. Falconer, and R. Kellerhals, unpub. ms., 1958
Labrador	0.000024	W. E. S. Henoch, unpub. ms., 1967
Retreat of Late Wisconsin Ice Sheet in North America[d]		
18,000 C^{14} years B.P.	14.379	
15,000	14.038	
14,000	13.325	
13,000	11.908	
12,000	10.119	
11,000	8.780	
10,000	7.347	
9,000	5.207	
8,000	3.200	
7,500	1.987	
7,000	0.813	
6,500	0.503	
6,000	0.307	
5,500	0.225	
present day	0.203	Falconer et al., unpub. ms., 1958

[a] Shumskiy et al. 1964, report this area as 55 x 10⁶ km². The figure in Flint (1957) should probably be adjusted upward by about 0.6 x 10⁶ km², in keeping with more recent estimates of Antarctic areas.

[b] Shumskiy et al. 1964, report this area as 33.6 x 10⁶ km². The figure in Flint (1957) should probably be adjusted upward by about 0.6 x 10⁶ km², in keeping with more recent estimates of Antarctic areas.

[c] Shumskiy et al. 1964, report this area as 16.215 x 10⁶ km². The area given by Flint (1957), 14.972 x 10⁶ km², has been adjusted upward by 0.9 x 10⁶ km² to accommodate a more recent estimate of the area of the Antarctic ice sheet (Bentley et al. 1964).

[d] The data presented here were measured from the map of Prest (1969) by the present author using an Aristo polar planimeter. Alterations were made in areas where recent data and the author's interpretations are in conflict with ice front configurations shown by Prest. Not included are past and presently glaciated areas of Greenland and Alaska, and those glaciated areas in Mexico and the U.S. Cordillera that were not coterminous with areas occupied by the main continental ice sheet.

on many of the islands and high areas. Many workers have argued that these residuals remained active for a time, or were subsequently reactivated, and that they supported radial glacier outflow (cf. Flint 1951; Hickox 1962; Borns and Hagar 1965; Brookes 1969; Prest and Grant 1969). Although this may be true, few existing arguments for the postulated activity are compelling. If a large glacier residual remained on the highlands of western Maine, it was not sufficiently active to flow into Quebec (McDonald and Shilts, in press). More field study in the Maritime region, involving stratigraphic control and numerous criteria for establishing flow directions, is necessary.

A controversial aspect of deglaciation in eastern Canada has been the importance and extent of the Cochrane readvance (see Hughes 1965). Prest (1969) attributes it to a large-scale southward surge of part of the ice sheet affecting most of the eastern Hudson Bay area. Because Cochrane till could not be identified in the Hudson Bay Lowland north of 50°20′N (McDonald 1969), the present author agrees with Hughes (1965) that the Cochrane episode is of very limited areal extent and of local importance only.

Before about 12,500 to 11,000 B.P. much of the meltwater discharging southward from the wasting ice sheet entered the Mississippi and Hudson river drainage systems, and large volumes of sediment and fresh water were carried southward to the sea via the Gulf of Mexico and the Hudson River estuary. As the ice retreated north past the Mississippi River drainage divide and north into Quebec, large settling basins, in the form of glacial lakes and estuaries situated along the southern margin of the ice sheet, trapped much of the sediment. After retreat of the Valders ice from the Lake Superior and Lake Michigan basins, meltwater discharge flowed generally eastward through the Great Lakes area and into the Champlain Sea occupying the St. Lawrence Lowlands. After about 7,900 B.P., Hudson Bay also functioned as an efficient sediment trap. There is much discussion in the current literature about local changes in rates of sedimentation in the North Atlantic, and of the possible influence of a dilution of sea water on observed faunal and isotopic variations (Broecker 1965; Olausson 1965; Shackleton 1967; Mercer 1969). The details of the late Wisconsin deglaciation should be taken into consideration. It is probable that the development of settling basins on the continent and the method of disintegration of the ice sheet had an influence on local rates of sedimentation and ocean salinity that would be recognizable in the deep-sea sediments adjacent to the continent. Because of locations of meltwater and sediment discharge, the effects of deglaciation should be more noticeable in the Arctic and Atlantic ocean basins than in the Pacific basin.

Summary. Stratigraphic and radiocarbon data from four "control areas" have been used to reconstruct the areal extent and succession of Wisconsin glacial phases and intervening interstades in eastern Canada between Quebec and Manitoba. Three important Wisconsin glacial phases are represented, separated by the early Wisconsin St. Pierre interstade and by the mid-Wisconsin Port Talbot interstade. Limits of glacier retreat during the St. Pierre and Port Talbot interstades are postulated. Influence of the earliest Wisconsin glacial phase was concentrated in eastern Canada and did not extend west to Manitoba. Both post-St. Pierre glacial phases were more extensive than the earliest Wisconsin glacial episode, and they affected areas farther west. The early Wisconsin ice sheets grew in eastern Canada, but evidence of ice flow directions between Manitoba and Quebec indicate that the mid-Wisconsin Port Talbot interstade was accompanied by a westward shift of the centers of glacier outflow. Some characteristics of late Wisconsin deglaciation are discussed, the influences of which may be recognizable in deep-sea sediments bordering the continent.

Acknowledgments. I would like to thank A. A. Berti, A. Dreimanis, R. W. Klassen, and R. Wilkins for permission to quote unpublished radiocarbon dates, and G. Falconer and W. E. S. Henoch for permission to use unpublished measurements of present glacier areas. Discussions of glacier areas with C. S. L. Ommanney are appreciated. I am particularly grateful to W. Blake, Jr., V. K. Prest, and W. W. Shilts for stimulating discussions and for offering abundant constructive criticism of the manuscript. Permission for publication was given by the Director, Geological Survey of Canada.

REFERENCES

Bentley, C. R., R. L. Cameron, C. Bull, K. Kojima, and A. J. Gow, Physical characteristics of the Antarctic ice sheet, *Antarctic Map Folio Series, Folio 2*, Am. Geograph. Soc., 1964.

Blake, W., Jr., End moraines and deglaciation chronology in northern Canada, with special reference to southern Baffin Island, 31 pp., *Geol. Surv. Canada, Paper 66–26*, 1966.

Borns, H. W., Jr., and D. J. Hagar, Late-glacial stratigraphy of a northern part of the Kennebec River valley, western Maine, *Bull. Geol. Soc. Am., 76*, 1233, 1965.

Broecker, W. S., Isotope geochemistry and the Pleistocene climatic record, in *The Quaternary of the United States*, edited by H. E. Wright, Jr., and D. G. Frey, p. 737, Princeton University Press, Princeton, N.J., 1965.

Broecker, W. S., and J. L. Kulp, Lamont natural radiocarbon measurements IV, *Science, 126*, 1324, 1957.

Brookes, I. A., Late-glacial marine overlap in western Newfoundland, *Can. J. Earth Sci., 6*, 1397, 1969.

Bryson, R. A., and W. M. Wendland, Radiocarbon isochrones of the retreat of the Laurentide ice sheet, 25 pp., *Univ. of Wisconsin, Dept. of Meteorology, Tech. Rept. 35*, 1967.

Bryson, R. A., W. M. Wendland, J. D. Ives, and J. T. Andrews, Radiocarbon isochrones on the disintegration of the Laurentide ice sheet, *Arctic Alpine Res., 1*, 1, 1969.

Buckley, J. D., and E. H. Willis, Isotopes' radiocarbon measurements VIII, *Radiocarbon, 12*, 87, 1970.

Christiansen, E. A., A thin till in west-central Saskatchewan, Canada, *Can. J. Earth Sci., 5*, 329, 1968a.

Christiansen, E. A., Pleistocene stratigraphy of the Saskatoon area, Saskatchewan, Canada, *Can. J. Earth Sci., 5*, 1167, 1968b.

Coleman, A. P., The Pleistocene of the Toronto region, 69 pp., *Ontario Dept. Mines, 41*, pt. 7, 1933.

De Vries, H., and A. Dreimanis, Finite radiocarbon dates of the Port Talbot interstadial deposits in southern Ontario, *Science, 131,* 1738, 1960.

Donn, W. L., W. R. Farrand, and M. Ewing, Pleistocene ice volumes and sea-level lowering, *J. Geol., 70,* 206, 1962.

Dreimanis, A., Pre-classical Wisconsin in the eastern portion of the Great Lakes region, North America, *Intern. Geol. Congr., 21st, Copenhagen, 4,* 108, 1960.

Dreimanis, A., The last ice age in the eastern Great Lakes region, North America, *Communications from VIII INQUA Congress,* Paris, in press.

Dreimanis, A., J. Terasmae, and G. D. McKenzie, The Port Talbot interstade of the Wisconsin glaciation, *Can. J. Earth Sci., 3,* 305, 1966.

Dyck, W., and J. G. Fyles, Geological Survey of Canada radiocarbon dates II, *Radiocarbon, 5,* 39, 1963.

Dyck, W., J. G. Fyles, and W. Blake, Jr., Geological Survey of Canada radiocarbon dates IV, *Radiocarbon, 7,* 24, 1965.

Dyck, W., J. A. Lowdon, J. G. Fyles, and W. Blake, Jr., Geological Survey of Canada radiocarbon dates V, *Radiocarbon, 8,* 96, 1966.

Flint, R. F., Highland centers of former glacial outflow in northeastern North America, *Bull. Geol. Soc. Am., 62,* 21, 1951.

Flint, R. F., *Glacial and Pleistocene Geology,* 553 pp., Wiley, New York, 1957.

Frye, J. C., and H. B. Willman, Classification of the Wisconsinan stage in the Lake Michigan glacial lobe, 16 pp., *Illinois State Geol. Surv., Circ., 285,* 1960.

Gadd, N. R., Surficial geology of the Bécancour map-area, Quebec, 34 pp., *Geol. Surv. Canada, Paper 59-8,* 1960.

Gadd, N. R., Pleistocene geology of the Central St. Lawrence Lowland, *Geol. Surv. Canada, Mem., 359,* in press.

Grant, D. R., Recent submergence in Nova Scotia and Prince Edward Island, *Geol. Surv. Canada, Paper 68-1A,* 162, 1968.

Hickox, C. F., Jr., Late Pleistocene ice cap centered on Nova Scotia, *Bull. Geol. Soc. Am., 73,* 505, 1962.

Hobson, G. D., and J. Terasmae, Pleistocene geology of the buried St. Davids Gorge, Niagara Falls, Ontario: Geophysical and palynological studies, 16 pp., *Geol. Surv. Canada, Paper 68-67,* 1969.

Hughes, O. L., Surficial geology of part of the Cochrane district, Ontario, Canada, in *International Studies on the Quaternary,* edited by H. E. Wright, Jr., and D. G. Frey, *Geol. Soc. Am., Spec. Papers, 84,* 535, 1965.

Karrow, P. F., Pleistocene geology of the Scarborough area, 108 pp., *Ontario Dept. Mines, Geol. Rept., 46,* 1967.

Karrow, P. F., Stratigraphic studies in the Toronto Pleistocene, *Proc. Geol. Assoc. Can., 20,* 4, 1969.

Klassen, R. W., Quaternary stratigraphy and radiocarbon chronology in southwestern Manitoba, 19 pp., *Geol. Surv. Canada, Paper 69-27,* 1969.

Klassen, R. W., L. D. Delorme, and R. J. Mott, Geology and paleontology of Pleistocene deposits in southwestern Manitoba, *Can. J. Earth Sci., 4,* 433, 1967.

Livingstone, D. A., Some interstadial and postglacial pollen diagrams from eastern Canada, *Ecol. Monographs, 38,* 87, 1968.

Lowdon, J. A., and W. Blake, Jr., Geological Survey of Canada radiocarbon dates VII, *Radiocarbon, 10,* 207, 1968.

Lowdon, J. A., and W. Blake, Jr., Geological Survey of Canada radiocarbon dates IX, *Radiocarbon*, *12*, 46, 1970.

Lowdon, J. A., J. G. Fyles, and W. Blake, Jr., Geological Survey of Canada radiocarbon dates VI, *Radiocarbon*, *9*, 156, 1967.

MacClintock, P., and D. P. Stewart, Pleistocene geology of the St. Lawrence Lowland, 152 pp., *N.Y. State Museum Sci. Serv., Bull.*, *394*, 1965.

MacNeill, R. H., Some dates relating to the dating of the last major ice sheet in Nova Scotia, *Maritime Sediments*, *5*, no. 1, 3, 1969.

McDonald, B. C., Pleistocene events and chronology in the Appalachian region of southeastern Quebec, Canada, Ph.D. dissertation, Yale University, New Haven, 1967.

McDonald, B. C., Glacial and interglacial stratigraphy, Hudson Bay Lowland, in *Earth Science Symposium on Hudson Bay*, edited by P. J. Hood, p. 78, *Geol. Surv. Canada, Paper 68-53*, 1969.

McDonald, B. C., and W. W. Shilts, Quaternary stratigraphy and events, southeastern Quebec, *Bull. Geol. Soc. Am.*, in press.

Mercer, J. H., The Allerød oscillation: A European climatic anomaly?, *Arctic Alpine Res.*, *1*, 227, 1969.

Mott, R. J., and V. K. Prest, Stratigraphy and palynology of buried organic deposits from Cape Breton Island, Nova Scotia, *Can. J. Earth Sci.*, *4*, 709, 1967.

Muller, E. H., Quaternary section at Otto, New York, *Am. J. Sci.*, *262*, 461, 1964.

Olausson, E., Evidence of climatic changes in North Atlantic deep-sea cores, with remarks on isotopic paleotemperature analysis, in *Progress in Oceanography*, edited by M. Sears, vol. 3, p. 221, Pergamon, New York, 1965.

Prest, V. K., Retreat of Wisconsin and Recent ice in North America, *Geol. Surv. Canada, Map 1257A*, 1969.

Prest, V. K., Quaternary geology of Canada, in *Geology and Economic Minerals of Canada, Econ. Geol. Series, No. 1*, edited by R. J. W. Douglas, Geol. Surv. Canada, Ottawa, in press.

Prest, V. K., and J. Hode Keyser, *Surficial Geology and Soils, Montreal Area, Quebec*, 36 pp., Dept. Public Works, City of Montreal, 1962.

Prest, V. K., and D. R. Grant, Retreat of the last ice sheet from the Maritime Provinces: Gulf of St. Lawrence region, 15 pp., *Geol. Surv. Canada, Paper 69-33*, 1969.

Preston, R. S., E. Person, and E. S. Deevey, Yale natural radiocarbon measurements II, *Science*, *122*, 954, 1955.

Rubin, M., and H. E. Suess, U.S. Geological Survey radiocarbon dates II, *Science*, *121*, 481, 1955.

Shackleton, N., Oxygen isotope analyses and Pleistocene temperatures re-assessed, *Nature*, *215*, 15, 1967.

Shilts, W. W., Pleistocene geology of the Lac Mégantic region, Southeast Québec, Canada, Ph.D. dissertation, Syracuse University, Syracuse, 1969.

Shumskiy, P. A., A. N. Krenke, and I. A. Zotikov, Ice and its changes, in *Research in Geophysics*, *2*, edited by Hugh Odishaw, p. 425, M.I.T. Press, Cambridge, Mass., 1964.

Skinner, R. G., Quaternary stratigraphy, Moose River basin, *Geol. Surv. Canada, Paper 70-1A*, 186, 1970.

Stewart, D. P., and P. MacClintock, The Wisconsin stratigraphy of northern Vermont, *Am. J. Sci.*, *262*, 1089, 1964.

Stuiver, M., E. S. Deevey, and I. Rouse, Yale natural radiocarbon measurements VIII, *Radiocarbon, 5,* 312, 1963.

Terasmae, J., Contributions to Canadian palynology: Part II, Non-glacial deposits in the St. Lawrence Lowlands, Quebec; Part III, Non-glacial deposits along Missinaibi River, Ontario, *Geol. Surv. Canada, Bull., 46,* 13, 1958.

Terasmae, J., A palynological study of Pleistocene interglacial beds at Toronto, Ontario, *Geol. Surv. Canada, Bull., 56,* 23, 1960.

Terasmae, J., and O. L. Hughes, A palynological and geological study of Pleistocene deposits in the James Bay Lowlands, Ontario, 14 pp., *Geol. Surv. Canada, Bull., 62,* 1960.

Wright, H. E., Jr., and D. G. Frey, eds., *The Quaternary of the United States,* 922 pp., Princeton University Press, Princeton, N.J., 1965.

Arthur L. Bloom

13. GLACIAL-EUSTATIC AND ISOSTATIC CONTROLS OF SEA LEVEL SINCE THE LAST GLACIATION

This appraisal of glacial-eustatic and isostatic controls of sea level is restricted to the events of the latest 18,000 radiocarbon-dated years of Pleistocene and Holocene time. The restriction is required because: (1) adequate data are available only for this short time span, and (2) the probability of large and unevaluated tectonic controls increases with time. Evaluation of sea floor spreading, plate tectonics, and continental drift is left to the appropriate specialists. Only eustatic changes in sea level caused by climate-controlled transfers of water masses within the hydrologic cycle, and isostatic changes in response to those transfers, have been considered.

A general discussion of the climatic and tectonic influences on sea level during the Late Cenozoic glacial ages must be deferred until the current revolution in global tectonics and sea floor spreading has subsided. It would be foolhardy to infer anything about glacial-eustatic control of sea level during one of the early Pleistocene glaciations, for instance, in the face of evidence that the ocean basins are widening at rates of up to 16 cm per year (Bullard 1969). In the 100,000 or so years since the last interglacial high sea level, a vigorous spreading rate of 10 cm per year would have created a prism of new ocean basin 10 km wide and 4 km deep along the crest of the midocean ridges which, according to some authorities, total 65,000 km in length. Following this logic, the meltwater of the latest deglaciation would have returned to an oceanic basin that had grown by 2.6×10^6 km^3 while the water had been temporarily residing on the continents as glacier ice. The water equivalent of the recently melted ice sheets was estimated in the range of 38–44×10^6 km^3 by Donn et al. (1962) and at about 43×10^6 km^3 by Flint (1969). Thus, the spreading of the ocean basins since the last interglacial could accommodate about 6 percent of the returned meltwater, and the postglacial shorelines would be almost 8 meters lower than the interglacial shorelines of 100,000 years ago. Obviously, theories of sea floor spreading must have a major impact on the study of Late Cenozoic sea levels.

As we view the new and remarkably mobile earth model, we need to be reminded that at least the mass of the earth and its fundamental geoid form have been constant in Late Cenozoic time. The mean radius of the earth may have changed because of the volumetric changes associated with the conversion of dense rocks into less dense sediments, or water into glacier ice, but isostasy always works to minimize such changes. All the ocean basins cannot widen or deepen unless some other parts of the world become narrower or higher—unless, of course, the earth is an expanding spheroid. Although the Darwin rise of the western equatorial Pacific may have been subsiding at 3 cm per 1000 years throughout the Cenozoic era, an equally large area of the East Pacific rise may have been rising at a comparable rate (Menard 1964, pp. 118, 138). Menard (this volume) seems to doubt that vertical movements of the ocean floor in the last 10 million years have had any great cumulative effect on sea level.

The venerable term "eustatic" was applied to those changes of sea level, or vertical displacements of the sea surface, that are simultaneous over the entire world (Daly 1934, p. 41). Changes in the volume of the ocean basins, or changes in the amount of water temporarily abstracted from the sea by glaciers (but excluding floating ice shelves), should cause a uniform worldwide, or eustatic, change of sea level. Eustatic changes can thus be subdivided into tectonic and climatic categories, excluding such other minor controls as volumetric expansion of the ocean water with temperature or salinity changes, and the effects of semipermanent ocean currents or high- and low-pressure atmospheric systems on the regional heights of the sea surface. These and many other possible causes of sea level fluctuations were summarized and evaluated by Fairbridge (1966, pp. 479–85).

The most significant cause of eustatic sea level fluctuations in the Late Cenozoic era has been the growth and shrinkage of continental ice sheets. Since Maclaren (1842) proposed his ingenious test of Professor Agassiz's glacial theory, it has been accepted that growth of glaciers on land must cause sea level to fall. The hydrosphere of the earth is a closed system in all significant respects, and the oceans and modern glaciers constitute respectively about 97.2 and 2.2 percent of the total near-surface water inventory. The only other important reservoir in the earth's hydrologic cycle is ground water. Comprising only 0.6 percent of the total water budget, and much of it buried more than 0.8 km under the continental landscapes, ground water is not able to respond significantly to changes in either oceanic or glacier ice volumes.

Isostatic fluctuations of sea level carry a special connotation in the context of Late Cenozoic glacial ages. The emerged, tilted, late-glacial and postglacial marine and lacustrine shorelines of regions recently deglaciated are ample proof that the earth's surface assumes a basin-like form under a load of glacier ice. The total postglacial isostatic uplift should equal approximately one-third of the former ice cap thickness, but the observed uplift rarely exceeds one-tenth of the ice thickness as measured or inferred from other evidence. The form of radiocarbon-dated postglacial uplift curves is exponential, with initial uplift at rates as high as 9 meters per century (Washburn and Stuiver 1962). It is very likely that an un-known but significant fraction of the total isostatic recovery was accomplished be-fore the landscape became ice-free. By the time the sea invaded the valleys in the rising land surface and etched reference marks on the valley sides, the recovery was already underway at a maximum rate. Subsequently, the rate of recovery has decreased exponentially with a half-life on the order of 1000 years.

The tilted, emerged shorelines of glaciated regions are well explained by the theory of postglacial isostatic uplift. It has not been emphasized, however, that the emerged shorelines of the Baltic Sea, Hudson Bay, and similar regions ac-tually record a water volume that has been poured back into the world ocean af-ter the temporary late-glacial or postglacial submergence. This water, decanted from the glaciated continental regions by isostatic uplift, becomes a contribution to the eustatic rise of sea level. To thoroughly intermingle the necessary terms, postglacial emergence of some regions, because of isostatic uplift, has added to the eustatic rise of sea level and has contributed to the submergence of stable coasts elsewhere.

Isostatic response to crustal loads may influence sea level in yet another way. If the layer of sea water returned to the world ocean by glacier melting is heavy enough, the ocean floor might be isostatically downwarped beneath it. Again, as in the predicted response of glacial isostasy, the maximum downwarp should be about one-third of the depth of the added water layer. On a stable continental block, the glacial and interglacial shorelines might therefore record postglacial submergence of only two-thirds the actual depth of the returned water. However, a small, steep-sided island rising from the deep ocean floor like a dipstick might move with the downwarped ocean bottom and record the true eustatic change in sea level (Bloom 1967). There is evidence that some islands in the eastern Caro-line group have acted as Pleistocene dipsticks (Bloom 1970).

THE DEGLACIAL HEMICYCLE OF THE LAST 18,000 YEARS

Farrand (1965) introduced the phrase "deglacial hemicycle" for that time be-tween the initial withdrawal of ice from the outermost moraines to the time of complete disappearance of the continental ice sheets. Farrand's analysis of the latest deglacial hemicycle clearly anticipated many of the conclusions of this con-tribution. Restricted by very limited data, he was nevertheless able to make im-portant inferences about the rate at which the ice sheets disappeared and sea level rose. He noted that the fluctuations shown by some postglacial sea level curves are greatly exaggerated. He also noted that the deglacial hemicycle ended 7000 years ago, and thereby avoided involvement in the controversy about eu-static sea level in the last 7000 years. I do not intend to avoid that interesting mat-ter.

The general technique which Farrand used, and which is also adopted here, is to measure the area of the ice caps at known intervals of time, estimate the average ice thickness at each time, and convert the resulting ice volumes into equivalent water volumes. The change in water volume, divided by the area of the world ocean, should give the glacial-eustatic change of sea level for each time interval. The technique requires precise maps of ice margin positions at successive times during deglaciation ("equicesses" was the term introduced by DeGeer in 1905 for geochronologically determined lines of ice recession, according to J. Lundqvist 1965, p. 163). Progressively more refined equicessal maps have been prepared for portions of the Scandinavian ice sheet (Sauramo 1958, Figs. 142, 143, 149, 153, 154; J. Lundqvist 1965, Fig. 14) but usually at very small scales. Prest (1969) prepared the first map at a scale larger than a text figure of "speculative ice-marginal positions during recession of the last ice-sheet complex" in North America. For the first time we can now visualize the progress, and measure the rate, of deglaciation in North America. We can hope that a committee of European Pleistocene scientists will compile a similar map for the Weichsel ice sheet. Obviously such maps must be speculative, but when they are drawn by the scientists who have been closely involved with the actual fieldwork, they are of great value to Pleistocene studies.

The concept of relating the deglacial hemicycle to eustatic rise of sea level is illustrated first by an abstraction (Fig. 1). Suppose the radius r of a circular ice sheet decreased at a constant rate. The area of the ice sheet would decrease in proportion to r^2 (Fig. 1, solid line). The curve is a parabola, with the most rapid loss of area during the initial reduction of radius. If the thickness of the circular ice sheet remained constant, and it melted as a disk of constantly decreasing radius, the graph of decreasing area with time would also be the graph of decreasing volume with time, or of the equivalent glacial-eustatic rise of sea level. If, however, thickness decreased in linear proportion to radius, volume would decrease in proportion to r^3 (Fig. 1, dashed line), and the glacial-eustatic rise of sea level would be even more rapid initially but correspondingly slower during the final phases of deglaciation.

Ice volumes decreasing in proportion to r^2 or r^3 are but two possibilities for a model of deglaciation and eustatic rise of sea level. Figure 2 illustrates a wider range of possibilities, in which ice volume is proportional to the radius r raised to progressively smaller powers.

If the thickness of a shrinking circular ice sheet decreased rapidly at first, and then less rapidly (Fig. 2a) so that thickness h was always in proportion to the radius r raised to some power greater than 1, then the ice volume at any time during deglaciation would be proportional to hr^2 or to r^3. If the thickness of a circular ice sheet was always in linear proportion to the radius (Fig. 2b), then volume would be proportional to radius cubed, as was illustrated in Figure 1. If thickness remained constant ($h \propto r^0$) during marginal retreat (Fig. 2d), then volume would be proportional to area, or to the radius squared. This possibility was also illustrated in Figure 1. An intermediate case (Fig. 2c) is added to clarify the trend of the series. Note that Figure 2d illustrates a limiting case, for if thickness is made proportional to radius raised to a power less than zero (a negative exponent) then thickness would increase as radius decreased. No field observa-

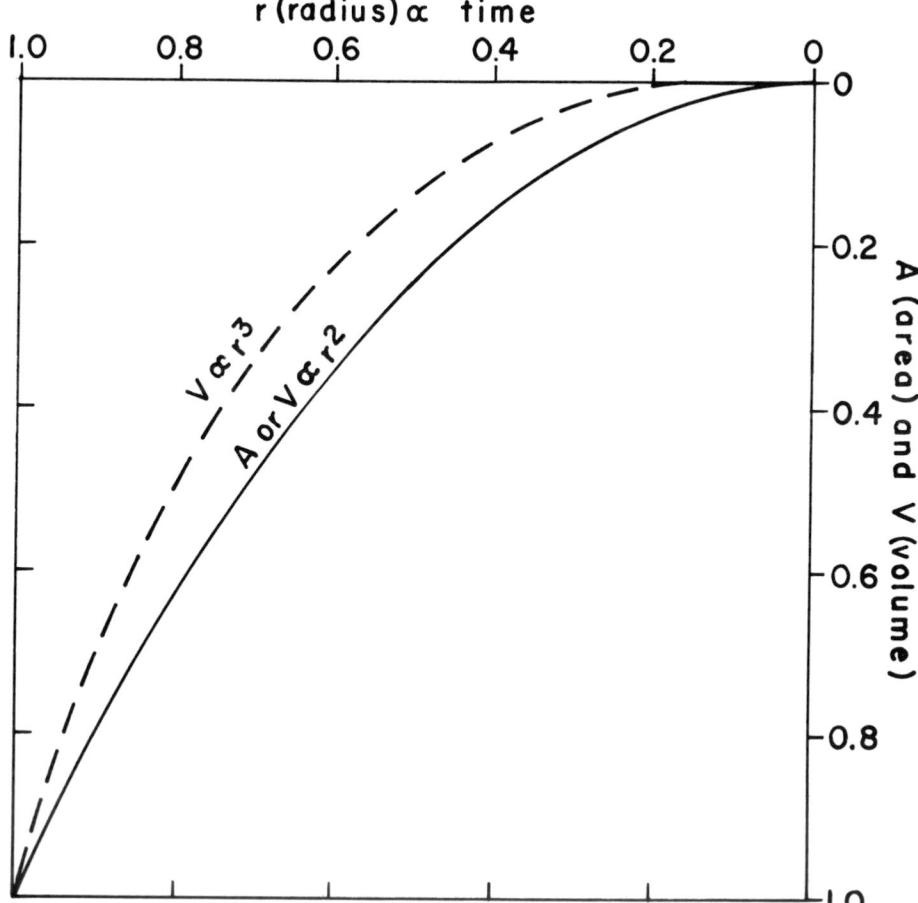

Fig. 1. Ratio of decreasing area and volume of a circular ice cap with constantly decreasing radius. If thickness remains constant, volume is proportional to radius squared, or area (solid line); if the ice sheet thins in linear proportion to the decreasing radius, volume is proportional to radius cubed (dashed line).

tions or theory support such a model of deglaciation. Figure 2b does not necessarily illustrate the other limiting case for the series, but Figures 2a and 2b suggest that volume cannot be proportional to radius raised to a power much larger than 3 lest the initial thinning be so rapid that the ice sheet would be "dead" while it still covered a substantial area.

    The two deglaciation curves of Figure 1 enclose a field within which the glacial-eustatic sea level curve should lie. The lower curve is a limit; the upper curve is not a limit but is probably close to it. Ghosts of old debates about the mode of deglaciation, whether by "backwasting" or "downwasting," flutter over Figures 1 and 2, but the scales of radius and thickness need to be kept in mind.

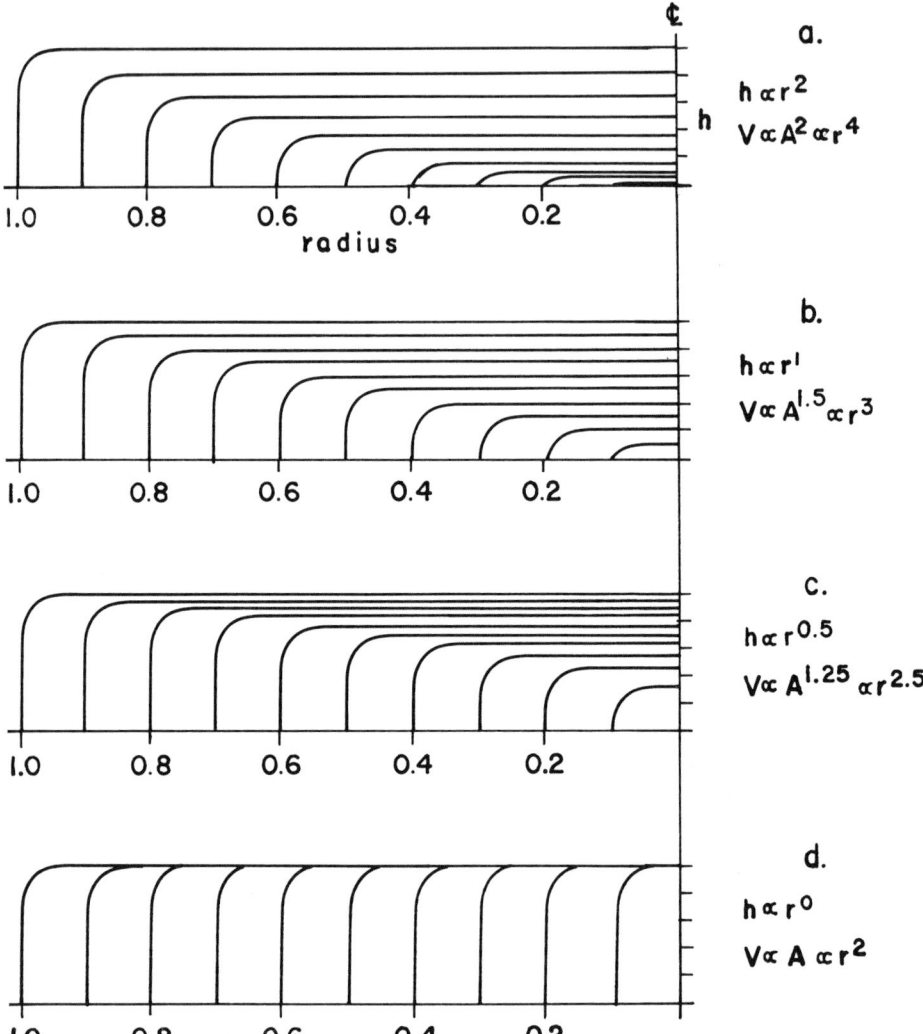

Fig. 2. Profiles of an ice cap in various modes of deglaciation. A circular plan view is assumed. Vertical exaggeration is about × 100. (a) Thickness decreases rapidly at first, then progressively more slowly. (b) Thickness decreases in linear proportion to radius. (c) Thickness decreases slowly at first, then progressively more rapidly. (d) Thickness remains constant.

The diameters of the ice sheets were up to 1000 times greater than their thicknesses. If thinning was always proportional to marginal retreat, a stage would have been reached when the glaciers were too thin to flow. Intuitively one would suppose that, at least during the final stages of deglaciation, thickness must have been nearly constant at some low value while the remnant, stagnant ice caps melted. The implication of Figure 1 for eustatic sea level studies is that at least during the final part of the deglacial hemicycle, the sea level curve ought to look more like the lower curve than the upper one.

The next problem is to replace the assumed linear decrease of radius in Figure 1 with an actual time scale, and to compile the progress of deglaciation of the several ice sheets during the last 18,000 years. The areas of the Laurentide (including the Appalachian and Queen Elizabeth Islands ice) and Cordilleran (east of 141°W long.) ice sheets, shown in Table 1, were measured directly from Prest's (1969) map. Areas were measured in square inches from tracings of the appropriate equicesses. From the map scale of 1:5 million, one square inch was calculated to equal 16,129 km², but that conversion gave areas for the Province of Saskatchewan and Hudson Bay that were respectively 96.3 and 97.2 percent of the areas listed by Goode's *World Atlas* (12th ed.) (Espenshade 1964). A conversion constant of 1 in.² $= 16,815$ km² gave areas correct to within 1 percent for Saskatchewan and Baffin Island and was used to calculate all the areas of Laurentide and Cordilleran ice. Shrinkage of the paper on which the maps are printed requires scale calibration with one or more known areas on each map that is so used.

Norway, Sweden, and Finland each have national atlas map sheets on a scale of 1:2 million that show various equicesses (see Table 1, footnotes). Unfortunately, the projections differ slightly, and various adjustments had to be made for small areas of overlap and gap on the three maps. However, the total area of the three countries, 1,110,000 km², is very close to the measured area of glaciation at 10,150 years B.P., if the ice margin is drawn straight across the Baltic Sea from the central Swedish moraines to the Salpausselkä moraines, and the areas of the Scandinavian ice sheet in Table 1 are probably correct to within a few percent. No clear consensus could be found in the literature concerning either the chronology or extent of Scandinavian ice southwest of Norway on the North Sea floor. Therefore, the Weichsel maximum area of the Scandinavian ice sheet in Table 1, which includes only the Norwegian Atlantic continental shelf ice, is much smaller than previous measurements of 4.25×10^6 km² (Flint 1957, p. 53), or 3.66×10^6 km² (Donn et al. 1962, p. 210). The latter measurement also included Great Britain, which was ignored in compiling Table 1.

Also omitted from Table 1 are Iceland, the European Arctic islands, all the ice sheets of the Soviet Union, and the changes in ice areas of the Greenland and Antarctic ice sheets. The island ice caps were too small to greatly affect the totals, and generally retreated in synchroneity with the larger Scandinavian ice sheet. According to Kind (1965, p. 269; 1969, p. 94) the coldest interval of the Sartansk glaciation of Siberia was between 25,000 and 13,000 B.P. By 11,800 B.P. a warming trend had begun, and the climatic optimum of the Holocene was between 8500 and 4500 B.P. Farrand (1965) reported correlation of a Sartansk retreatal stade with the Salpausselkä moraines of Finland, but shortly thereafter the Siberian ice sheet had broken into small upland remnants. Flint (1957, p. 53) estimated the Siberian ice sheet at 2.17×10^6 km² during the last major glaciation. Donn et al. (1962) estimated its maximum area in the last 30,000 years as 1.32×10^6 km².

Table 1. Areas (in 10^6 km^2) of the Major Ice Sheets That Melted during
the Last Deglacial Hemicycle

Time (yrs B.P.)	Lauren- tide	Cordilleran (E of 141°W long.)	Alaskan (W of 141°W long.)	Scandi- navian
18,000	≤ 11.89[a]	< 1.62	< 0.63	2.63[h]
15,000	11.79	1.62[c]	0.63[f]	2.05[i]
12,500	9.69	1.30	—	—
12,000	8.59[b]	—	—	—
11,800	8.75	1.16[d]	—	—
10,500	7.49	0.90	—	1.10[j]
9,000	4.54	0.28	—	—
8,500	—	0.22[e]	—	0.19[k]
8,000	2.73	—	—	nil[l]
7,000	0.34	nil	0.05[g]	—
6,500	nil	—	—	—
0	0.15	0.03	0.05[g]	0.005[m]

[a] At maximum extent, regardless of age (Prest 1969); boundary variously dated from 18,000 to 13,500 B.P.

[b] By subtracting areas of known Valders advance (0.16 x 10^6 km^2) from area at 11,800 B.P.

[c] Cordilleran maximum, boundary variously dated from 17,000 to 13,500 B.P.

[d] Area within 11,500 B.P. equicess.

[e] Youngest equicess shown by Prest (1969).

[f] Maximum late Pleistocene advance of Coulter et al. (1965); labeled "ca. 13,500" by Prest (1969).

[g] Present glacier area (Flint 1957); Goldthwait (1967) reported that 7000 B.P. glacier areas and reforestation were comparable to today's.

[h] Last Weichsel maximum: "C"line of Denmark; Hamburg and Brandenburg moraines; does not include North Sea ice.

[i] By extrapolating areas between C line and D line of Denmark (Mörner 1969, p. 108) and between Frankfurt and Pomeranian moraines of Germany around remaining 85 percent of ice sheet perimeter and subtracting from area at 18,000 B.P.

[j] Ra-Central Swedish-Salpausselkä moraines and margin inferred by Sauramo (1958) for 10,150 B.P. (Donner 1965, p. 238).

[k] Area at bipartition, 8750 B.P. (Sauramo 1958, p. 454; Donner 1965, p. 238).

[l] Swedish glaciers disappeared entirely in climatic optimum, 8000–5000 B.P. (J. Lundqvist 1965, p. 172).

[m] Present area of 5000 km^2 (Flint 1957); J. Lundqvist (1965, p. 172) reported that all modern Swedish glaciers date from 2500 B.P. or younger.

Base maps: Coulter et al. (1965); Holtedahl and Andersen (1960); G. Lundqvist (1953); Okko (1960); Prest (1969); Stampfuss (1936).

The Siberian ice sheet should be included in Table 1, but its addition would only increase the maximum glaciated area, not the rate of deglaciation expressed as percentage of the maximum.

Flint (1957) tabulated the existing and "last-maximum" areas of the Greenland ice sheet as 1.73×10^6 km^2 and 2.16×10^6 km^2 respectively. It seems to have lost an area of 0.43×10^6 km² during the deglacial hemicycle. By 9000–8500 B.P., deglaciation of coastal northeast Greenland was well under way (Washburn and Stuiver 1962), but the time of last maximum glaciation has not been determined.

Events of the last deglacial hemicycle in Antarctica are largely unknown. According to Denton (Denton et al. 1969; Denton, this volume) the large East Antarctic ice sheet has been little changed throughout Late Cenozoic time. The Taylor I episode is now in progress, with the ice at a maximum surface level where it has been since early Wisconsin time. The West Antarctic ice sheet, in the vicinity of the Ross Sea at least, has fluctuated largely in harmony with the northern hemisphere glaciers. Deglaciation from the Ross I episode was in progress by 9400 B.P. and continued until about 3000 B.P.

The synchroneity of retreat of the West Antarctic ice sheet with the northern hemisphere ice sheets is controlled by eustatic fluctuation of sea level. When the northern hemisphere glaciers were at their maximum extent, the lowered sea level caused the Antarctic floating ice shelves to ground, whereby they could thicken greatly and assume a parabolic or elliptical profile. Hollin (1962, p. 192) estimated the increased volume of Antarctic ice because of the increased area to the grounding line as 2.5–8.5×10^6 km³, or the equivalent of 6–21 meters of additional sea level lowering. He favored the lower value. Mercer (1968, p. 81) estimated that only about 1.5–1.8×10^6 km³ of the West Antarctic ice sheet represents water abstracted from the ocean, since the rest is grounded ice that is displacing sea water today. He calculated that the loss of the entire present West Antarctic ice sheet would cause a eustatic sea level rise of 4.0–4.5 meters, but hypothesized that the West Antarctic ice sheet may have formed as long ago as the Pliocene-Pleistocene boundary. A conservative conclusion is that while most of the Antarctic ice sheet is very old and has been largely passive during the last 18,000 years, a peripheral belt of uncertain volume has calved into the sea and contributed perhaps 6 meters to the latest glacial-eustatic rise of sea level.

The data of Table 1 are shown graphically in Figure 3, along with a summation graph of the total glacier area at successive time intervals. At more than 14,000 B.P., only a possible range can be given for the total area. For instance, at 18,000 B.P. the Cordilleran and Alaskan glaciers were not at their maximum area, and much of the maximum Laurentide perimeter is younger than 18,000 B.P. as well. But at numerous places around the perimeter of the Laurentide ice sheet in the western United States, and in the Puget Sound lobe of the Cordilleran ice sheet, older Wisconsin drift has been reported, which suggests that the 18,000-year equicess must be fairly close to the younger maximum border. The minimum area at 18,000 B.P. for the ice sheets of Table 1 was 14.52×10^6 km², but the actual area was probably closer to the maximum total of 16.77×10^6 km².

The summation of areas at 15,000 B.P. also must be expressed as a range of 16.09–15.46×10^6 km², because the Alaskan glaciers were at less than their maximum advance until 13,500 B.P. However, areas of the larger ice sheets at 15,000 B.P. can be measured reasonably well, and the actual area was probably close to

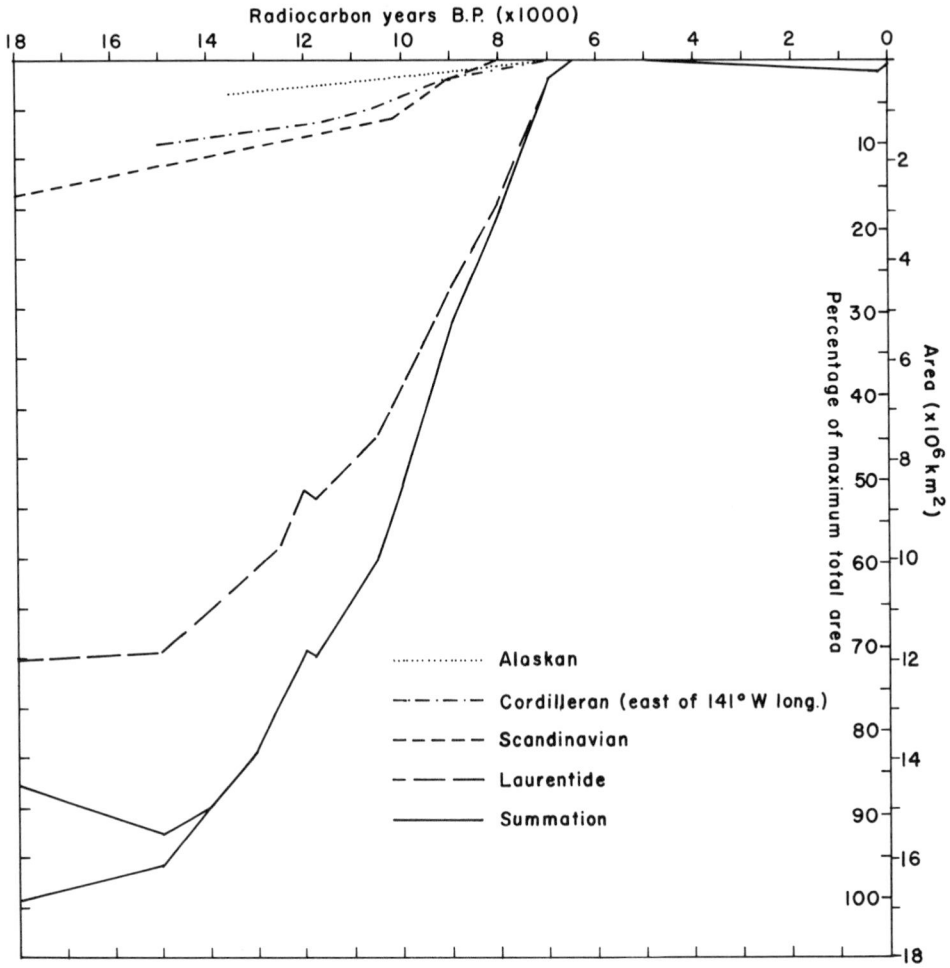

Fig. 3. Areas of major ice sheets that melted in the last 18,000 years. See Table 1 for sources of data.

the maximum of the possible range at that time. It is possible that the last glaciation reached its maximum total area at a time nearer 15,000 than 18,000 B.P.

Of the many minor fluctuations of the ice borders, only the Laurentide Valders readvance at 11,800 B.P. is large enough to show graphically. The area of the Valders readvance was measured by drawing straight lines across the proximal ends of the Valders lobes in the Lake Winnipeg lowland and the Lake Superior-Lake Michigan lowlands. The areas thus measured are probably near the high end of the range of probability. It is true that the technique of measuring the areas inside successive equicesses tends to mask local readvances, but in the course of reviewing the literature of ice margin positions, I was struck with the multitude of minor readvances that have been proposed at almost random times during the deglacial hemicycle. For instance, even if the Salpausselkä moraines of Finland represent pauses in a previously steady retreat of the ice margin, they formed while the much larger Laurentide ice sheet was accelerating its retreat, and they produce no noticeable inflection in the summation curve at 10,150 B.P. (Fig. 3). The rapid calving of Fennoscandian ice into the Gulf of Bothnia between 10,000 and 9000 B.P. does appear as a steepening of the slope on the summation curve, but the even more extensive calving of the heart of the Laurentide ice sheet into the Tyrrell Sea between 9000 and 8200 B.P. is lost in the generally rapid deglaciation of 10,000–7000 B.P. There is no well-documented evidence of synchronous advances along substantial segments of the various ice sheets that would warrant any additional crenulations on the summation curve in Figure 3. The steady decrease in glaciated area suggests that the corresponding eustatic rise of sea level was also a nonpulsating event.

The most striking feature of Figure 3 is the dissimilarity of the various curves with the hypothetical area curve of Figure 1. Obviously, the deglacial hemicycle was not characterized by glacier margins retreating at a constant rate. On the contrary, the loss of area was quite slow until about 14,000 B.P., accelerated to 12,000 B.P. whereupon the Valders readvance caused a small but definite increase in total area, and then went at a nearly constant rate to total deglaciation by about 6500–7000 B.P. The percentage scale of Figure 3 gives some useful visualizations: 88 percent of the total glaciated area was still ice-covered at 14,000 B.P.; at 12,000 B.P., just prior to the Valders readvance, 70 percent of the total ice area remained. Half the ice was gone by 10,000 B.P., and the remaining half disappeared in little more than 3000 years. Deglaciation seems to have been a process that was sluggish at first, then rapid until the final melting of the last upland remnants.

THICKNESS AND WATER VOLUME OF FORMER ICE SHEETS

Glacial-eustatic rise of sea level is obviously related to the decreasing water volume in the ice sheets during the deglacial hemicycle. While the areas of the ice sheets at successive times can now be measured to a precision of perhaps ± 5 percent, the average thicknesses of the ice sheets at successive times can rarely be measured at all. The regional relief of mountain ranges known to have been overtopped gives only minimum values for ice sheet thickness, but some ice sheets had nunataks protruding above them, from which true thicknesses can be inferred. Modern ice sheets, and theoretical models based on flow laws of perfectly plastic ice, can also be used as analogs of former ice sheets, although the ancient glacier regimens were not the same as those of modern ice sheets, nor is the deformation of glacier ice duplicated exactly by laboratory observations. Flint (1969) realistically evaluated the only available approach to ice thickness: "Average thickness was estimated by inspection, including consideration of minimum values imposed by the geologic record. The implied margin of error is thought to be ± 25%."

A few examples of the thickness of modern and former ice sheets, which indicate a reasonable convergence of evidence, follow. Holtzscherer (in Holtzscherer and Robin 1954) reported the thickness of the Greenland ice sheet along several seismic traverses. The maximum thickness he measured in Greenland was 3300 meters, in the area of the central dome. The east–west profile across the central dome (at about 70°–72°N lat.) gave an average thickness of 2300 meters. Along a more northerly seismic traverse, at approximately 79°N latitude, the mean thickness was only about 1500 meters. Airborne radar sounding of the northern Greenland ice sheet along 77°N latitude also gave a maximum ice thickness of 3300 meters (Walker et al. 1968). Holtzscherer estimated that the mean ice thickness for the whole of Greenland must be about 1600 meters. Fristrup (1966) estimated that the average thickness is 1500 meters.

The thickest glacier ice yet found on earth is near Byrd Station, Antarctica, where 4300 meters of ice extend from 1800 meters above sea level to 2500 meters below sea level (American Geographic Society 1962). Nearby, the deep bore hole at Byrd Station (Gow et al. 1968) penetrated 2164 meters of ice to bedrock. Fristrup (1966) estimated an average ice thickness for Antarctica of 2500 meters.

Concerning former ice sheets, Robin (1964) estimated that the Scandinavian ice sheet may have reached a maximum central thickness of 3.5 km, if full isostatic adjustment beneath the ice sheet was achieved. At the time of the Ra moraines, about 10,000 B.P., outlet glaciers at least 1300 meters thick and 30 km long entered the sea in southern Norway (Andersen 1960, p. 406). Accumulation and glacier flow were rapid, although the final retreat of the Scandinavian ice margin began shortly thereafter. At the time of bipartition, 8750 B.P., when the Scandinavian ice sheet broke into two residual ice caps, the southern remnant over the Norwegian-Swedish frontier ceased moving when it was still so thick that the final movements were nearly unaffected by local relief (Gjessing 1966, p. 138). When the ice surface lowered to less than 1800 meters above sea level, it was below the firn limit, and nourishment ended. Depending on the bedrock altitude at the time, the relict ice sheet must have been at least 1500 meters thick (Strøm 1956, p. 747).

Many nunataks protruded through the Cordilleran ice sheet in British Columbia and the state of Washington. Canadian erratics have been found on the flanks of Mt. Baker in Washington to an altitude of 1600 meters. From these and similar related kinds of evidence, the Puget Sound lowland ice lobe of the last major glaciation is inferred to have been about 1700–1800 meters thick on its eastern flank against the Cascade Range (Easterbrook 1969, p. 2282). Near its southern limit, about 200 km south of Mt. Baker, the Puget lobe was only about 400 meters thick (Crandell 1965, p. 345).

In southern British Columbia, 60–70 km north of Vancouver, Mt. Garibaldi was an active volcano in late Pleistocene time (Mathews 1952). It erupted during the deglacial hemicycle and built a cone on the wasting ice sheet. From the distribution of volcanic ejecta, Mathews concluded that the Cordilleran ice sheet reached a maximum altitude of 2200 meters about 17 km north of Mt. Garibaldi,

and 1700 meters about 21 km to the south. The Squamish Valley west of Mt. Garibaldi is only slightly above sea level, so these altitudes are representative of ice
thicknesses on the western slope of the Cordilleran ice sheet.

The Laurentide ice sheet so completely dominates the graph of total glacier
area (Fig. 3) that its thickness is a critical value in estimating total ice volume.
Unfortunately, few facts about its thickness are known. The southern quadrant of
this great ice sheet ranged in thickness from only 460 meters against the Cypress Hills in semiarid southeastern Alberta (Westgate 1968) to more than 1500
meters near Mt. Washington, New Hampshire (Flint 1957, p. 319).

In the absence of direct measurements about the thickness of the Laurentide ice sheet, we must turn to analogy and theory. Hollin (1962) found that the
profile of the Antarctic ice sheet for 375 km inland from Mirnyy Station, approximately along 90°E longitude, almost exactly fits the parabolic equation $h = 4.7 \sqrt{d}$,
where h is the ice thickness in meters, and d is the distance in meters from the
edge of the ice sheet. Hollin's parabola also is the profile of a perfectly plastic ice
sheet with a yield stress of 1.0 bar, so it fits the current theories of glacier flow
quite well. Inland beyond 375 km from the edge of the Antarctic ice sheet, the
actual profile becomes much flatter than the theoretical parabola. The equation
predicts that at the center of the North American Laurentide ice sheet of 18,000
years B.P., which had an approximate radius of 2000 km, the ice was 6.6 km thick,
but such an extrapolation is not valid. Most other estimates of maximum thickness
are only one-third to one-half as great. The same formula more reasonably predicts a maximum ice thickness over central New York State of 1500–2100 meters,
approximately 100–200 km north of the Wisconsin drift border. The local relief of
central New York is only 600 meters, so there is no way of checking the predicted
thickness, but the intense local erosion of the Finger Lakes troughs implies an ice
thickness comparable to that of the Greenland or Scandinavian ice sheet.

From the numerous measurements and estimates, an upper limit of about
3.5 km can be proposed for the thickness of a large ice sheet unless it is grounded
far below sea level and confined by rugged mountains, as seems to be the case for
the thickest part of the West Antarctic ice sheet. Average thicknesses of 1.5 to 2.5
km seem reasonable, but there is no direct evidence to support these values. Assigning an average thickness to each of the ice sheets in Table 1 and Figure 3 at
each stage of deglaciation would obviously be little more than a guessing game.
Instead, the relationships inferred from Figures 1 and 2 are further exploited in
Figure 4. If thickness stayed nearly constant with time, the percentage loss of area
would be equivalent to the percentage loss of volume. If thickness was always
proportional to radius, then $v \propto a^{1.5}$. For simplicity, Figure 4 reproduces as a solid
line the summation curve of ice sheet areas from Figure 3, expressed as percentage of the maximum total area of 16.77×10^6 km². The dashed line is the graph
of $v \propto a^{1.5}$, expressed as percentage of the maximum but undetermined ice volume. The maximum volume is assumed to be at the time of maximum area, but no
attempt has been made to estimate the actual water volume of the maximum glaciation.

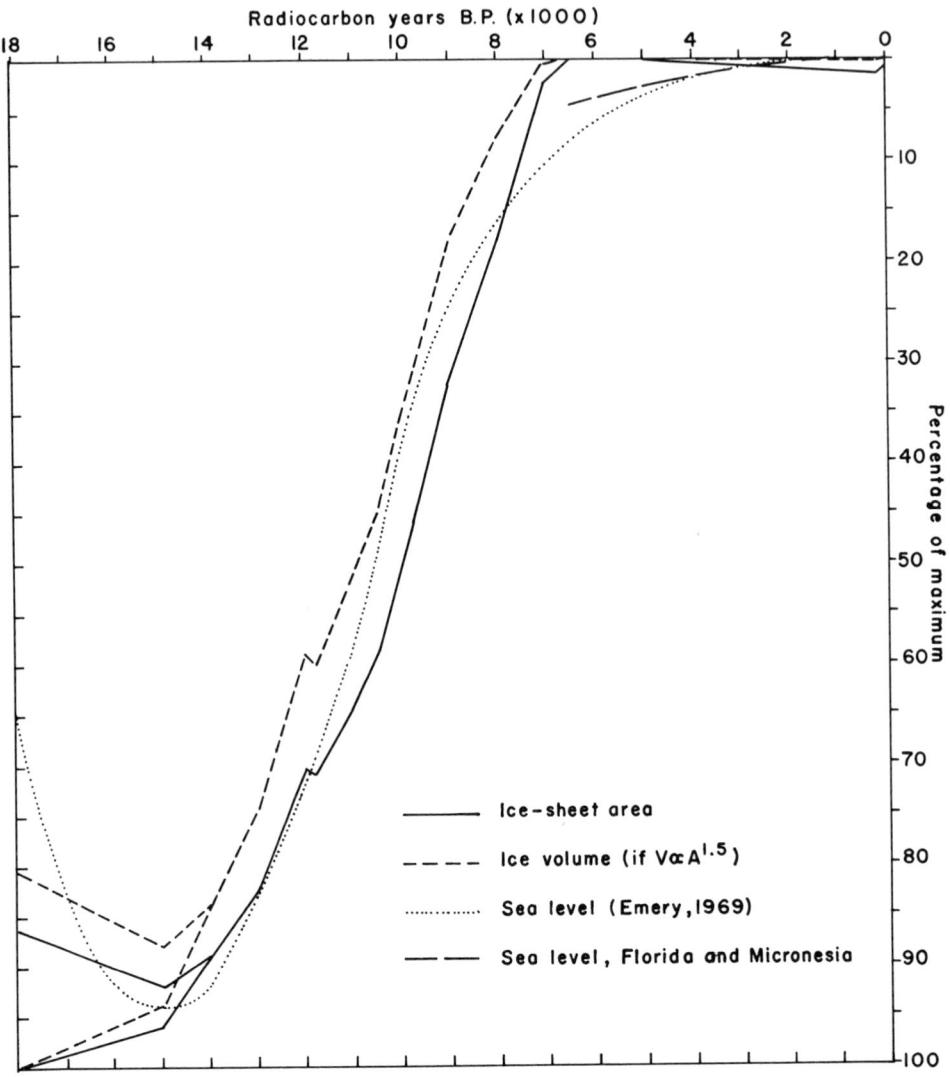

Fig. 4. Ice sheet area, volume, and sea level lowering, all expressed as percentage of maximum during the last 18,000 years.

DEGLACIATION AND RISING SEA LEVEL

As in the hypothetical examples illustrated by Figures 1 and 2, at any time during the deglacial hemicycle the percentage of remaining ice volume should be equal to or less than the percentage of remaining ice area. The ice sheet area and relative-volume graphs of Figure 4 should give the lower, and possibly also the upper, limits of the glacial-eustatic sea level curve for the deglacial hemicycle. Two well-documented sea level curves are added to Figure 4 for comparison. One covers only the last 6500 years, and discussion of it will be deferred to a later part of this paper. The other is the most recent complete summary curve by Emery (1969) and is shown as a dotted line on Figure 4. It is based primarily on about fifty radiocarbon-dated samples of shallow-water salt-marsh peat, oolites, or mollusk shells from the outer continental shelf off the United States. All the samples imply a shallow marine environment at the time of deposition but were collected as deep as 130 meters. The deepest sample is about 15,000 years old, and Emery concluded that the sea was then at its lowest level of the last glaciation. That conclusion is neither definitely supported nor disproved by the area and volume curves of Figure 4, but the predominance of radiocarbon ages of 15,000 years or less for the outermost border of the two North American ice sheets strongly suggests that he is correct.

The sea level curve drawn by Emery is a smooth, sigmoidal curve through data points that show a progressively greater range of depth with increasing age. On Figure 4 his curve has been converted to the percentage of remaining rise of sea level at any time in the deglacial hemicycle, with 130 meters taken as 100 percent. His smoothed curve shows a maximum lowering of only 125 meters, or about 94 percent of the total, but that is obviously the result of a simplification in drafting.

From 18,000 to about 16,000 B.P., Emery's sea level curve does not fit the ice sheet area or relative-volume curves. For the sea to have been as high as he suggested at 18,000 B.P., the ice sheets must have been only as large as they were later, at 11,000–13,000 B.P. That seems highly unlikely, in view of the large number of internally consistent radiocarbon dates that document the Laurentide ice advance from 25,000 to 18,000 B.P., across Ohio, for example (Goldthwait 1958). However, only a few radiocarbon-dated samples support the sea level curve in that time range, and perhaps the only significant feature of the older portion of the curve is that it has a minimum at 15,000 B.P.

The fit of Emery's sea level curve and the summation curve of ice sheet areas is excellent between about 16,000 and 12,000 B.P. It might be argued that any set of decreasing percentages would show similar parallel trends, but the inherent errors of compilation and simplification considered, it is impressive that the two curves from entirely independent sources show such similarity. It can be inferred that until about 12,000 B.P. the ice sheets maintained approximately constant thicknesses, and the volume of water being returned to the sea was proportional to the peripheral loss of ice area.

From 12,000 to 8000 B.P., sea level rose steeply, although ice sheet area actually increased for about 200 years, then for about 1500 years decreased less rapidly than in earlier time. From Figure 3 it is obvious that the changes in total ice sheet area in this time range were dominated by the Valders readvance and subsequent retreat of the Laurentide ice margin. The smaller ice sheets show no important area changes. Perhaps the divergence in Figure 4 of the ice sheet area curve and the sea level curve indicates that a dramatic thinning of the Laurentide ice sheet began with the Valders readvance and continued for about the next 2000 years, even though the ice margin did not retreat as fast as it did earlier and later. Major climatic changes at Valders time have been inferred by many palynologists and geologic oceanographers. Perhaps the peculiarly restricted Valders readvances, reported only in the Lake Winnipeg, Lake Superior, and Lake Michigan basins, record marginal surges into suitably oriented lowlands while most of the vast ice sheet simply thinned but did not expand. Especially in the 11,000 to 9000 B.P. time range in Figure 4, Emery's sea level curve is very close to the relative ice volume curve that was derived from an assumption that thickness is proportional to radius. Glacial-eustatic rise of sea level in post-Valders time is most nearly explained by a constantly thinning Laurentide ice sheet, whereas in pre-Valders time frontal retreat with nearly uniform ice thickness is the best explanation.

For the last 8000 years, Emery's smooth, sigmoid sea level curve departs notably from the graphs of ice sheet area and relative volume that enclose it prior to that time. By 6500 B.P. the Laurentide, Cordilleran, and Scandinavian ice sheets were gone, yet according to Emery's interpretation, 8 percent, or about 10 meters, of the postglacial coastal submergence was yet to be completed (Fig. 4). The discrepancy is not due to the overgeneralization of a draftsman, for even the most diverse published opinions about Holocene sea level history agree that in the last 7000 years most coasts have submerged 8 to 12 meters. Clearly, something other than the melting of the northern hemisphere ice sheets must have caused the continuing submergence.

One additional factor in the submergence is the isostatic downwarping of coasts (Bloom 1967). The rapid rise of sea level prior to 7000 B.P. placed an ever-increasing weight of water on the continental margins. Although the entire deep ocean floor received a uniform addition of water load during the deglacial hemicycle, each coastal segment received a different amount of load, depending on the near-shore bottom profile, the map configuration of the coast, and other variables. By using an averaging technique to determine the regional water load along the Atlantic coast of the United States (Bloom 1967) it was demonstrated that coastal localities adjacent to deep water have submerged as much as 5 meters more in postglacial time than localities adjacent to shallow water. The agreement between the differential submergence and the average water load adjacent to each locality is excellent, justifying the hypothesis that isostatic response of coasts to the postglacial rise of sea level contributed a significant amount to the observed submergence.

If all coasts are warped downward to varying degrees by the superjacent water load of the postglacial rise of sea level, where can we measure the true amount of the glacial-eustatic component without the isostatic increment? Some places are in narrow estuaries that extend far inland, away from the massive water load on the shelves. Other useful coasts are fronted by broad, shallow continental shelves. Unfortunately, these coasts yield records of only the last few meters of total submergence during the last few thousand years. If they have submerged any more than a few meters, they have probably been deformed. The Everglades coast of southwestern Florida, on the feather-edge of the postglacial submergence across a broad limestone shelf, gives the best U.S. record of the last 5000 radiocarbon years, with a total submergence of less than 4 meters (Scholl et al. 1969).

Other good places to measure the glacial-eustatic component of sea level rise are on the outer continental shelf, or on small islands that rise from the deep ocean floor. These places will be warped downward by the water load around them, but because the entire deep ocean floor is depressed, the volume of the ocean basin increases and sea level with reference to an island, or to a hypothetical buoy moored in deep water, should not change because of the isostatic deformation. Whether the isostatic adjustment is rapid or slow will not influence the vertical interval between glacial and interglacial waterlines on the island, or between the present sea surface and a glacial age intertidal peat bed now on the outer continental shelf. For this reason, the portion of Emery's curve of sea level rise prior to 7000 or 8000 B.P., which is based on samples dredged from the continental shelves, is probably a curve of purely glacial-eustatic submergence, whereas the part of his curve younger than 5000 years is based on samples from coastal sites that are very likely downwarped, and it probably shows isostatically produced excess submergence.

Another submergence curve in Figure 4, graphing the final 4.5 percent (about 5.8 meters) of the postglacial rise of sea level in the last 6500 years, diverges tangentially away from Emery's curve. This is the submergence curve for the Florida Everglades for the last 5000 radiocarbon years, but the portion between 5000 and 6500 B.P. is based on six radiocarbon-dated peat samples from Truk and Ponape, in the eastern Caroline Islands of the central Pacific Ocean (Bloom 1970). These two islands, 730 km apart on the crest of the ancient Darwin rise, have identical submergence histories for the last 6500 years. They are excellent candidates for the role of Pleistocene dipsticks. The sea level curve compiled from Florida and Micronesia data should be free from the effect of isostatic warping and should be a reliable measure of eustatic sea level rise in the last 6500 years, just as Emery's curve should be good for ages greater than 7000 or 8000 B.P. That the two segments do not join is due to the progressively greater isostatic displacement of progressively older and deeper coastal samples. A true glacial-eustatic sea level curve should continue to rise parallel to the deglaciation curves, probably between the relative-volume and ice area curves, until 7000–6500 B.P. Then, with completion of the deglacial hemicycle in most regions, the rate of sea level rise should have decreased sharply. The composite curve of sea level changes in Florida and the eastern Caroline Islands demonstrates that the remaining 4.5 percent (5.8 meters) of sea level rise has been very slow. A good synthesis of the glacial-eustatic component of sea level rise can be made by continuing Emery's curve steeply upward from its position at 9000 B.P., when 25 percent of the rise remained, to a point at 7000 B.P., when only 5 percent of the rise was incomplete. From 7000 B.P. to the present, the Florida-Micronesia curve should show only the glacial-eustatic component. The vertical difference between such a synthetic or composite curve and the smooth curve drawn by Emery is primarily a measure of the postglacial isostatic deformation on various parts of the U.S. Atlantic coast.

EUSTATIC SEA LEVEL DURING THE LAST 6500 YEARS

Even if the Florida-Micronesia curve of Figure 4 records the true glacial-eustatic rise of sea level, and the youngest part of Emery's curve is based on samples that have been isostatically lowered, the problem still remains to account for a water volume of 2.10×10^6 km^3 that must have been returned to the sea since 6500 B.P., in order to produce the remaining 5.8 meters of observed submergence. Three sources for the water can be proposed. First, if the ice caps that survived the deglacial hemicycle, including those on Greenland, Antarctica, and the high-Arctic islands of Canada and Eurasia, continued to melt throughout the Hypsithermal interval, they could have supplied some or all of the necessary water volume. Second, postglacial isostatic rebound, especially around Hudson Bay and the Baltic Sea, has elevated areas that were below sea level in late-glacial time. The elevated shorelines of the late-glacial marine limit are evidence of the volume of water that was decanted back into the sea by postglacial isostatic uplift. Third, pluvial lakes may have persisted into the Hypsithermal interval, and their gradual evaporation may have provided a small increment to the volume of water in the ocean. In principle, a fourth possible source of the water returned to the sea in postglacial time might have been a decrease in the ground water of regions that were formerly cooler and more moist. That source is not evaluated here but is not regarded as significant.

Surviving Ice Caps

Baffin Island retained a considerable ice cap later than the general Laurentide deglaciation of 6500 B.P. (Prest 1969). Andrews (1968, p. 128) estimated that the water equivalent of the Baffin Island ice cap was 1.76×10^5 km^3 as late as 6700–7000 B.P. The subsequent shrinkage and fragmentation, of which the Barnes ice cap is a surviving remnant, is not entirely deciphered, but as an upper limit the entire water volume that remained at 6700–7000 B.P. could have raised sea level only 0.48 meter, less than 10 percent of the observed amount.

The Greenland ice sheet is estimated to have lost an area of 4.3×10^5 km^2 during the deglacial hemicycle (Flint 1957). If that area had the average thickness of 1500 meters, 6.5×10^5 km^3 of ice (or 6.0×10^5 km^3 of water) has returned to the sea. The eustatic effect would be a sea level rise of 1.65 meters. However, the fiords of northeast Greenland were ice-free by 8500–9000 B.P. (Washburn and Stuiver 1962), and it is hard to believe that anything more than a small fraction of the total volume of water released by the melted Greenland ice has returned to the sea since 6500 B.P.

A similar argument can be made for Antarctica, but with even fewer supporting facts. Hollin (1962) calculated the water loss of the Antarctic ice sheet to be equivalent to 6 to 21 meters of sea level rise in approximately the last 18,000 years, caused primarily by calving around the periphery of the ice sheet, as sea level rose in response to the melting of the northern hemisphere ice sheets. He favored an estimate near the lower limit of 6 meters. But if the loss of Antarctic ice was in any way proportional to the glacial-eustatic rise of sea level, then all but a few percent of the total had returned to the sea by 6500 B.P. Nevertheless, the history of the Antarctic ice sheet during the Hypsithermal interval remains a problem for further research.

Isostatic Decantation

The amount of water that has been decanted out of Hudson Bay and the Baltic basin can be inferred from the height of the uplifted shorelines. The area of the late-glacial Tyrrell Sea in the Hudson Bay lowland was measured from the Glacial Map of Canada (Prest et al. 1968). Using the measured area of the Tyrrell Sea (1.505×10^6 km^2) and the area of Hudson Bay (0.826×10^6 km^2, exclusive of Foxe Basin and Hudson Strait) as the bases of the frustrum of a cone, and the average elevation of the Tyrrell Sea shorelines as 200 meters, the volume of water that has been decanted is 2.30×10^5 km^3, enough to cause a rise in sea level of 0.63 meter. The uplift has been in progress since at least 8200 B.P., probably most rapid initially, so only a small part of this water can be used to account for the continued rise of sea level after 6500 B.P.

The Baltic Sea region can be evaluated in a similar way. The Litorina Sea had an area only slightly larger than the present Baltic Sea, and reached its maximum extent about 7000 B.P. (Donner 1965, pp. 239–40). The tilted shorelines are near present sea level southeast of Finland but are more than 120 meters above sea level at the northern end of the Gulf of Bothnia. The area of the present Baltic Sea used as equivalent to the Litorina Sea, and an average depth of 50 meters assumed, the water volume of the Litorina Sea would have caused a eustatic rise of sea level of only 5.8 cm. This is a small contribution indeed, but at least it was all returned to the world ocean within the last 7000 years.

Other coastal regions also have emerged in postglacial time but, the small contribution of the Litorina Sea to eustatic sea level considered, no other areas seem worth further evaluation. At the very most, less than a meter of sea level rise in the last 8000 years can be attributed to isostatic decantation of formerly submerged deglaciated regions.

Desiccation of Pluvial Lakes

Many of the arid regions of the world have internal drainage, and in the closed basins of those regions lakes accumulated during Pleistocene periods of increased precipitation, decreased evaporation, or lowered temperature. The pluvial lakes in the Great Basin of the western United States were at their most recent maxima approximately during the time of greatest expansion of the North American ice sheets. The correlation of pluvial and glacial climates on other continents is less certain. If many pluvial lakes were at their maximum extent 18,000–15,000 years

B.P., and survived beyond the end of the deglacial hemicycle, their postglacial desiccation is an additional source of water for the eustatic rise of sea level. However, the following survey of the literature on late Pleistocene pluvial lakes demonstrates that they could not have been significant contributors to the rising sea level of postglacial time. It can be regarded as a scholarly exercise in futility, useful only in the instruction of others.

Table 2 summarizes the data and approximations that were used in compiling the paleovolumes of some major late Pleistocene pluvial lakes. For many of the lakes only length, width, and depth have been reported, so the rectilinear volume may be several times as large as the actual volume of the lake. Conversely, for some of the surviving lakes the present area, as reported in standard atlases, was used even though the area would be larger with any increase in depth. Many other pluvial lakes are known, but those in Table 2 are among the largest and are representative of the others.

The draining of Lake Bonneville by the Snake River flood has been variously inferred at 18,000, 30,000 (Malde 1968), and 12,000 B.P. (Broecker and Kaufman 1965). No estimates of the lake's history permit more than a small fraction of the water in the lake (which at its maximum held the equivalent of about 3 cm of sea level) to have returned to the sea in postglacial time. The remaining lakes of the Great Basin (Lake Bonneville represented 44 percent of the total Pleistocene lake area) might have contributed about 2 cm to the rise of sea level. However, Searles Lake had evaporated to the stage of salt precipitation by 10,000 B.P., having earlier been a freshwater lake (Flint and Gale 1958). Lake Lahontan had a similar history. Throughout the arid southwestern United States, a climate similar to to-

Table 2. Late Pleistocene Pluvial Lakes

Name	Area (km²)	Av. depth (m)	Volume (km³)	Sea level equiv. (m)
Bonneville[a]	5.0×10^4	305	1.016×10^4	0.028
Bonneville[b]	—	—	1.58×10^3	0.004
Gt. Basin, excl.[c] Bonneville	6.39×10^4	100	6.39×10^3	0.018
Caspian Sea[d]	3.94×10^5	26	1.05×10^4	0.028
Aral Sea[e]	6.85×10^4	50	3.43×10^3	0.009
Sinkiang[f]	1×10^5 ?	100+ ?	?	0.02 to 0.03 ?
Dieri, Australia[f]	1.034×10^5	50	5.18×10^3	0.014
Victoria, Africa[g]	6.95×10^4	20	1.39×10^3	0.004
Lisan (Dead Sea)[h]	3.74×10^3	190	3.25×10^2	nil

[a] Total basin volume (Crittenden 1963).
[b] Snake River flood (Malde 1968).
[c] Snyder et al. (1964).
[d] Fedorov and Leontiev [1953].
[e] Kes [1960].
[f] Flint (1957).
[g] Bishop and Trendall (1967).
[h] Neev and Emery (1967).

day's had developed by 7500–7000 B.P. (Mehringer 1967). At most, a few milli-
meters of sea level rise in the last 7000 years could be attributed to the pluvial
lakes of the western United States.

Evaporation of all the other large Pleistocene lakes listed in Table 2 would
collectively add a maximum of about 10 cm to the level of the sea. Obviously, desiccation of pluvial lakes has not contributed significantly to the postglacial rise of
sea level. The insignificance of the process implies a similar minor role for the
ground water lost from regions that were formerly more humid or cooler.

Evaluation

A striking feature of the many published submergence curves is that in spite of
the diverse interpretations of the authors, most of the curves show that sea level
was between 8 and 12 meters below present level at about 7000 B.P. (Fairbridge
1961, 1968; Coleman and Smith 1964; Shepard and Curray 1967; Jelgersma 1967;
Emery 1969). As the rate of sea level rise just prior to 7000 B.P. was on the order
of 10 meters per 1000 years, but decreased to only 1 to 2 meters per 1000 years
thereafter (or oscillated several meters above and below present sea level, or
reached present level and did not change further), the coincidence at 7000 B.P. is
exceptional.

Almost all the data used for compiling the sea level history of the last 7000
years and 10 meters of submergence have been collected from tidal marshes or
lagoons along the continental coasts of North America, Europe, or Australia. Each
coastal site is at the edge of the water load that was imposed on the continental
margins by the glacial-eustatic rise of sea level, and each one has been differen-
tially affected by the isostatic adjustment to that load. The maximum isostatic response of the crust or mantle to the water load would be about one-third of the
glacial-eustatic sea level rise; hence if the deep ocean floor had responded continuously to the glacial-eustatic water load, but the edges of the continents did
not begin to respond until the end of the deglacial hemicycle, then some coasts
may be submerged four-thirds the amount of true glacial-eustatic rise of sea level.
The intersection of most submergence curves 8–12 meters below present sea level
at 7000 B.P. would then imply that the glacial-eustatic component of sea level rise
after that time is about 6–9 meters, which is in good agreement with the Florida-
Micronesia curve that should be unaffected by isostatic depression.

We still must account for the nonisostatic submergence of 5.8 m on "dipstick"
islands in the last 6500 years. Continued melting of Arctic island ice caps and the
Greenland ice sheet could contribute a maximum of 2.1 meters to the sea level
rise, and probably contributed only a tenth of that amount in postglacial time. Decantation of formerly submerged deglaciated regions by postglacial isostatic uplift might have contributed a maximum of 1 meter to the postglacial rise of sea
level, but again, the actual contribution was probably less than the maximum. Desiccation of pluvial lakes offers only a trivial 10 cm. All together, these could have
contributed only 1 to 3 meters to the observed submergence of nearly 6 meters.
Peripheral calving of the Antarctic ice sheet may be the source for the remaining

3–5 meters of sea level rise that followed completion of the deglacial hemicycle in the northern hemisphere. It is necessary, but unsatisfactory, to conclude this review with the admission that the last 4 to 6 percent of the total sea level rise in the last 15,000–18,000 years is as yet unaccounted for, although glacial-eustatic and isostatic controls adequately account for the remaining 94–96 percent.

Summary. The melting of ice sheets between about 18,000 and 6500 B.P. provides a theoretically sound and demonstrably adequate cause for a corresponding glacial-eustatic rise of sea level of about 130 meters. The maximum total area of the North American and Scandinavian ice sheets was 16.77×10^6 km^2 or less, some time between 18,000 and 15,000 B.P. The areas of these ice sheets at various times in the deglacial hemicycle can be measured on recent maps of equicesses, or geochronologically determined lines of ice recession. Other glaciated regions, such as Siberia, Greenland, Antarctica, and various high-latitude islands, are assumed to have lost all or part of their ice sheets in broad synchroneity with the melting of the Laurentide and Cordilleran ice sheets of North America and the Scandinavian ice sheet of Europe. The Laurentide ice sheet complex, with a maximum area of 11.89×10^6 km^2, dominated the history of the deglacial hemicycle.

The thickness, and therefore the volume, of former ice sheets can only be estimated. Theories of the deformation of glacier ice, and geophysical exploration of existing ice sheets, only suggest that the former ice sheets could have been several kilometers thick. The relative change in volume can be inferred from two limiting conditions, however. If the ice sheets remained at a constant average thickness during their marginal recession, the percentage of remaining ice volume at any time in the deglacial hemicycle would be proportional to the percentage of remaining area. If the ice sheets thinned in some constant proportion to their decreasing radii, then the percentage of remaining volume would be proportional to the three-halves power of the percentage of remaining area. The second condition requires that at any stage of deglaciation, the percentage of remaining volume be less than the percentage of remaining area.

A well-documented graph of the postglacial rise of sea level, compiled by Emery primarily from the radiocarbon ages of shallow-water deposits now submerged as much as 130 meters on the outer continental shelf of the U.S. Atlantic coast, shows excellent correspondence with the decreasing area and relative volume of the ice sheets. From a maximum low sea level at 15,000 B.P. until Two Creeks time at 12,000, the curve of rising sea level coincides with the curve of decreasing ice sheet area. After Valders time, the percentage of remaining sea level rise decreased rapidly, while the percentage of remaining ice sheet area increased slightly and then decreased less rapidly for about 2000 years. The rise of sea level approximates the relative-volume change of the ice sheets in this interval, perhaps reflecting a fundamental change in the dynamics of the Laurentide ice sheet in Valders time, from frontal retreat to rapid thinning.

For the last 8000–9000 years, coasts have been submerging at a decelerating rate, even though by 6500 B.P. the deglacial hemicycle was over for the North American and Scandinavian ice sheets. Many coasts have continued to submerge 8 to 12 meters in postglacial time. Isostatic downwarping of coastal regions under the superjacent water load of the glacial-eustatic rise of sea level can account at most for one-fourth of the observed postglacial submergence; therefore a glacial-eustatic rise of 6 to 9 meters remains to be explained. Submergence of 5.8 meters in the last 6500 years is observed on the shallow coast of Florida and on small Micronesian islands, areas that should not be deformed by local isostatic adjustment. Final removal of remnant ice sheets in Hypsithermal time, decantation of water from isostatically uplifted inland and marginal seas, and evaporation of water from pluvial lakes together can account for only one-sixth to one-half of the observed postglacial submergence, so an additional source of meltwater for a glacial-eustatic rise of sea level of about 3 to 5 meters must be sought. The Antarctic ice sheet could have supplied that much water, but its history for the last 7000 years is not yet known.

REFERENCES

American Geographical Society, *Antarctica* (map), 1:5,000,000, New York, 1962.

Andersen, B. G. Sørlandet, in *Geology of Norway*, edited by O. Holtedahl, Norg. Geol. Undersokelse, no. 208, 203, 1960.

Andrews, J. T., Late-Pleistocene history of the Isortoq Valley, north-central Baffin Island, Canada, *Mélanges de Geographie offerts à M. Omer Tulippe, 1,* 118, 1968.

Bishop, W. W., and A. F. Trendall, Erosion-surfaces, tectonics and volcanic activity in Uganda, *Geol. Soc. London Quart. J., 122,* 385, 1967.

Bloom, A. L., Pleistocene shorelines: A new test of isostasy, *Bull. Geol. Soc. Am., 78,* 1477, 1967.

Bloom, A. L., Paludal stratigraphy of Truk, Ponape, and Kusaie, Eastern Caroline Islands, *Bull. Geol. Soc. Am., 81,* 1895, 1970.

Broecker, W. S., and A. Kaufman, Radiocarbon chronology of Lake Lahontan and Lake Bonneville II, Great Basin, *Bull. Geol. Soc. Am., 76,* 537, 1965.

Bullard, E., The origin of the oceans, *Sci. Am., 221,* 66, 1969.

Coleman, J. M., and W. G. Smith, Late recent rise of sea level, *Bull. Geol. Soc. Am., 75,* 833, 1964.

Coulter, H. W., D. M. Hopkins, T. N. V. Karlstrom, T. L. Péwé, C. Wahrhaftig, and J. R. Williams, Map showing extent of glaciations in Alaska, *U.S. Geol. Survey Miscell. Geol. Inv. Map* I–415, 1965.

Crandell, D. R., Glacial history of western Washington and Oregon, in *The Quaternary of the United States,* edited by H. E. Wright Jr. and D. G. Frey, p. 341, Princeton Univ. Press, Princeton, 1965.

Crittenden, M. D., Jr., Effective viscosity of the Earth derived from isostatic loading of Pleistocene Lake Bonneville, *J. Geophys. Res., 68,* 5517, 1963.

Daly, R. A., *The Changing World of the Ice Age,* 271 pp., Yale Univ. Press, New Haven, 1934.

Denton, G. H., R. L. Armstrong, and M. Stuiver, Histoire glaciaire et chronologie de la région du Détroit de McMurdo, Sud de la Terre Victoria, Antarctide, Note préliminaire, *Rev. Geograph. Phys. Geol. Dyn.,* ser. 2, *11,* 265, 1969.

Donn, W. L., W. R. Farrand, and M. Ewing, Pleistocene ice volumes and sea-level lowering, *J. Geol., 70,* 206, 1962.

Donner, J. J., The Quaternary of Finland, in *The Quaternary,* edited by K. Rankama, vol. 1, p. 199, Wiley-Interscience, New York, 1965.

Easterbrook, D. J., Pleistocene chronology of the Puget Lowland and San Juan Islands, Washington, *Bull. Geol. Soc. Am., 80,* 2273, 1969.

Emery, K. O., The continental shelves, *Sci. Am., 221,* 106, 1969.

Espenshade, E. B., ed., *Goode's World Atlas,* 12th ed., 288 pp., Rand McNally, Chicago, 1964.

Fairbridge, R. W., Eustatic changes in sea level, in *Physics and Chemistry of the Earth,* vol. 4, p. 99, Pergamon, New York, 1961.

Fairbridge, R. W., Mean sea level changes, long term—eustatic and other, in *Encyclopedia of Oceanography,* edited by R. W. Fairbridge, p. 479, Reinhold, New York, 1966.

Fairbridge, R. W., Holocene, postglacial or recent epoch, in *Encyclopedia of Geomorphology,* edited by R. W. Fairbridge, p. 525, Reinhold, New York, 1968.

Farrand, W. R., The deglacial hemicycle, *Geol. Rundschau, 54,* 385, 1965.

Fedorov, P. V., and O. K. Leontiev, On the history of the Caspian Sea in the late and post Khvalin time [1953] (abstract), in *Annotated Bibliography of Quaternary Shorelines,* edited by H. G. Richards and R. W. Fairbridge, p. 175, Acad. of Nat. Sci., Spec. Pub. 6, Philadelphia, 1965.

Flint, R. F., *Glacial and Pleistocene Geology,* 553 pp., Wiley, New York, 1957.

Flint, R. F., The position of sea level in a glacial age, paper presented at Eighth Congress, International Association for Quaternary Research, Paris, France, September, 1969.

Flint, R. F., and W. A. Gale, Stratigraphy and radiocarbon dates at Searles Lake, California, *Am. J. Sci., 256,* 689, 1958.

Fristrup, B., *The Greenland Ice Cap,* 312 pp., Univ. of Washington Press, Seattle, 1966.

Gjessing, J., Deglaciation of southeast and east-central Norway, *Norsk Geog. Tidsskr., 20,* 133, 1966.

Goldthwait, R. P., Wisconsin age forests in western Ohio. I, Age and glacial events, *Ohio J. Sci., 58,* 209, 1958.

Goldthwait, R. P., Evidence from Alaskan glaciers of major climatic changes, *Royal Meteorological Society Proceedings of the International Symposium on World Climate from 8000 to 0* B.C., p. 40, 1967.

Gow, A. J., H. T. Ueda, and D. E. Garfield, Antarctic ice sheet: Preliminary results of first core hole to bedrock, *Science, 161,* 1011, 1968.

Hollin, J. T., On the glacial history of Antarctica, *J. Glaciology, 4,* 173, 1962.

Holtedahl, O., and B. G. Andersen, Glacial map of Norway (1:2,000,000), in *Geology of Norway,* edited by O. Holtedahl, Norg. Geol. Undersokelse, no. 208, pt. 2 (maps), Oslo, 1960.

Holtzscherer, J. J., and G. de Q. Robin, Depth of polar ice caps, *Geograph. J., 120,* 193, 1954.

Jelgersma, S., Sea-level changes during the last 10,000 years, *Royal Meteorological Society Proceedings of the International Symposium on World Climate from 8000 to 0* B.C., p. 54, 1967.

Kes, A. S., Fluctuations of the Aral Sea level [1960] (abstract), in *Annotated Bibliography of Quaternary Shorelines,* edited by H. G. Richards and R. W. Fairbridge, p. 178, Acad. of Nat. Sci., Spec. Pub. 6, Philadelphia, 1965.

Kind, N. V., The radiocarbon chrononlogy of the last glaciation and of the postglacial interval in Siberia (abstract), in *Abstracts,* p. 269, Seventh Congress, International Association for Quaternary Research, Boulder, Colorado, 1965.

Kind, N. V., Les variations climatiques et les glaciations en Siberie au Quaternaire Supérieur (abstract), in *Abstracts,* p. 94, Eighth Congress, International Association for Quaternary Research, Paris, France, 1969.

Lundqvist, G., Glacial striae, terminal moraines and lines of ice recession (map), in *Atlas över Sverige*, edited by M. Lundqvist, pls. 21–22, Stockholm, 1953.

Lundqvist, J., The Quaternary of Sweden, in *The Quaternary*, edited by K. Rankama, vol. 1, p. 139, Wiley-Interscience, New York, 1965.

Maclaren, C., The glacial theory of Professor Agassiz, *Am. J. Sci.*, Ser. 1, *42*, 346, 1842.

Malde, H. E., The catastrophic late Pleistocene Bonneville flood in the Snake River Plain, Idaho, *U.S., Geol. Surv., Profess. Paper*, *596*, 1968.

Mathews, W. H., Mount Garibaldi, a supraglacial Pleistocene volcano in southwestern British Columbia, *Am. J. Sci.*, *250*, 81, 1952.

Mehringer, P. J., Jr., The environment of extinction of the Late-Pleistocene megafauna in the arid southwestern United States, in *Pleistocene Extinctions—The Search for a Cause*, edited by P. S. Martin and H. E. Wright, Jr., p. 247, Yale University Press, New Haven, 1967.

Menard, H. W., *Marine Geology of the Pacific*, 271 pp., McGraw-Hill, New York, 1964.

Mercer, J. H., The discontinuous glacio-eustatic fall in Tertiary sea level, *Paleogeograph., Paleoclimatol., Paleoecol.*, *5*, 77, 1968.

Mörner, N., The late Quaternary history of the Kattegatt Sea and the Swedish west coast, 487 pp., *Sveriges Geol. Undersokn.*, Ser. C, no. 640, 1969.

Neev, D., and K. O. Emery, The Dead Sea, 147 pp., *Israel Geol. Surv. Bull.*, no. 41, 1967.

Okko, V., Minerogenic deposits and stages of deglaciation (map), in *Suomen Kartasto* (Atlas of Finland), edited by L. Aario, pl. 4, Helsinki, 1960.

Prest, V. K., Retreat of Wisconsin and Recent ice in North America, *Geol. Surv. Canada, Map 1257A*, 1969.

Prest, V. K., D. R. Grant, and V. N. Rampton, Glacial map of Canada, *Geol. Surv. Canada, Map 1253A*, 1968.

Robin, G. de Q., Glaciology, *Endeavour*, *23*, 102, 1964.

Sauramo, M., Die Geschichte der Ostsee, 522 pp., *Suomalaisen Tiedeakatemian Toimituksia (Annales Academiae Scientiarum Fennicae)*, Ser. A-III, no. 51, 1958.

Scholl, D. W., F. C. Craighead, Sr., and M. Stuiver, Florida submergence curve revised: Its relation to coastal sedimentation rates, *Science*, *163*, 562, 1969.

Shepard, F. P., and J. R. Curray, Carbon-14 determination of sea level changes in stable areas, in *Progress in Oceanography*, vol. 4, p. 283, Pergamon, New York, 1967.

Snyder, C. T., G. Hardman, and F. F. Zdenek, Pleistocene lakes in the Great Basin, *U.S. Geol. Surv. Miscell. Geol. Inv. Map I-416*, 1964.

Stampfuss, R., Karten zur Vorgeschichte, Karte 1, Eiszeitalter (Die altere Steinzeit, etwa 40,000–12,000 v. d. Yr.) 1:4,000,000, Verlag F. E. Wachsmuth, Leipzig, 1936.

Strøm, K., The disappearance of the last ice sheet from central Norway, *J. Glaciology*, *2*, 747, 1956.

Walker, J. W., D. C. Pearce, and A. H. Zanella, Airborne radar sounding of the Greenland Ice Cap: Flight 1, *Bull. Geol. Soc Am.*, *79*, 1639, 1968.

Washburn, A. L., and M. Stuiver, Radiocarbon-dated postglacial delevelling in northeast Greenland and its implications, *Arctic*, *15*, 66, 1962.

Westgate, J. A., Surficial geology of the Foremost-Cypress Hills Area, Alberta, 122 pp., *Res. Council Alberta (Can.), Geol. Div., Bull.*, *22*, 1968.

K. O. Emery, Hiroshi Niino, and Beverly Sullivan

14. POST-PLEISTOCENE LEVELS OF THE EAST CHINA SEA

The particular field of interest of this paper is the continental shelf, which was indirectly influenced by Pleistocene glaciation in areas far beyond those that were covered by glaciers. The lowering of sea level caused by the locking of water in glaciers exposed nearly all the shelves of the world, converting them into extensions of the land. For many thousands of years the shelves were exposed to subaerial processes that developed soils, streams that eroded channels and deposited now-submerged deltas, and ocean waves and currents that eroded sea cliffs and deposited beaches, spits, and offshore islands which protected lagoons and estuaries. The topographic features were the same as those of the present shores and coastal plains. In fact, many of the present features themselves are atop a continental shelf of the future after the remnants of glacier ice melt and cause a further rise of sea level to perhaps 60 meters above the present level.

During the period of glacially lowered sea level the continental shelf became covered with forests and grasses, evidence of which has been preserved as submerged freshwater peat (Emery et al. 1967). Herbivorous animals foraged on the shelf and moved across it to what are now offshore islands, and even traveled between Asia and North America. Remains of these animals, particularly of the larger ones, have been dredged from many shelves of the world; prominent among them are the teeth and larger bones of mammoth, mastodon, bison, elk, musk-ox, moose, horse, tapir, and giant ground sloth (Whitmore et al. 1967). Carnivores and scavengers must also have been present, although their remains have not yet been reported. At the same time, the shore zone was populated by many kinds of mollusks and soft-bodied animals, and the overlying water contained fish and plankton. Only the hard shells of mollusks are readily preserved, and many shells of shallow-water oysters and clams have been dredged from far out on the continental shelf.

Birds must have fed upon shore plants and animals, and they nested within and flew over the grasslands and forests. Similarly, man lived on the former belt of land at the edge of the continent where he found supplies of mammals, birds, fish, and mollusks. To date, however, man's bones have not been reported from the continental shelf, because they are fragile and easily destroyed and because at least the Americas were sparsely populated by man during most of the latest time of low sea level.

The plant remains preserved in peats and the bones and shells of animals that are resistant to weathering are so similar to those of living forms that they do not serve as indicators of dates, other than of broadly Pleistocene or Late Cenozoic time. However, the plant and animal remains found on the shelf at mid-latitudes generally are typical of colder climates than those that prevail today near the dredging localities, and they serve as excellent materials for determining dates by their carbon-14 content. These dates, when combined with the depths from which the remains of shallow-water or shore plants and animals were dredged, provide much information on the former position of shore zones and thus of sea levels during late or post-Pleistocene time.

Sea level curves that span as long a time as 30,000 years have three main features, each of which has been discussed by many investigators. One is the nature of the curve during the past 13,000 years. All investigators appear to agree that rise of sea level before about 3000 years ago was faster than at present; however, they have differed in their interpretation of earlier data points. A few investigators have drawn their sea level curves from point to point, but the scatter of the large number of points shown in Figure 1 suggests that the points are not reliable enough by age or depth to justify such a method. Also, there appear to be differences between the average or best-fit lines for different regions. The most complete data through 1960 are for the shelf off Texas in the western Gulf of Mexico (Curray 1960; Shepard 1960). Subsequent studies of similar materials on the Atlantic shelf of the United States (Emery and Garrison 1967) yielded a curve that diverged from the one for the western Gulf. At 13,000 years ago, sea level was about 105 meters below the present level for the Atlantic shelf, whereas it was only 55 meters below the present level for the western Gulf shelf. In order to try to determine which of the two sets of data points might depict eustatic conditions, Milliman and Emery (1968) compiled published dates and depths for shallow-water materials on shelves of the Caribbean Sea, Mexico, Panama, Australia, southern California, Nigeria, Argentina, Bahama Islands, and western Florida. Although only a few dates were available from any one shelf, the general assemblage of data points was more like that of the sea level curve for the Atlantic than for the western Gulf of Mexico (Fig. 1), suggesting that the Gulf shelf had been raised in post-Pleistocene time.

The second feature of the sea level curves is that if they go back far enough in time they should show a sea level minimum. Some curves with only a few data points suggest that a minimum occurred about 19,000 years ago, a time selected to correspond with the date of maximum glacier advance in the United States. However, the cluster of data points in Figure 1 indicates that the sea level minimum was more nearly 15,000 than 19,000 years ago.

Third, the data points of Figure 1 indicate that sea level was higher 25,000 to 30,000 years ago than 15,000 years ago. Some workers believe that sea level 30,000

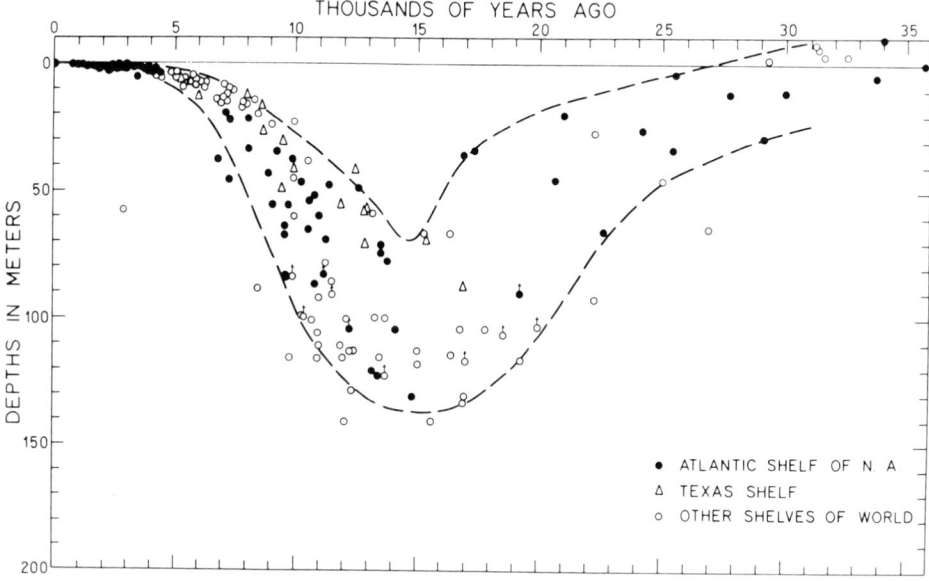

Fig. 1. Compilation of published dates and depths of the present continental shelves of the world on the basis of shallow-water shells, oolites, salt-marsh peat, wood, coralline algae, and coral—mostly from Milliman and Emery (1968). A few of the very many more published dates and depths for peats in modern salt marshes have been added to show the sea levels of the period from 8000 years ago to the present. The dashed lines serve as an envelope to enclose most of the data points.

years ago may even have been higher than at present. However, the scatter of data points within the envelope of dashed lines in Figure 1 supports only a high sea level, not necessarily one higher than at present. The broadening of the envelope with past time may be due to many local changes of level produced by diastrophism and isostatic rebound as well as to the longer time during which shells may have been shifted about the sea floor by currents.

A better approach to estimating eustatic changes of sea level during the past 30,000 years may be that of collecting new samples and determining many new dates for a single shelf that is believed to be at least as stable as the shelves of the Atlantic and western Gulf of Mexico. An area that seems particularly suitable is the shelf in the East China Sea that is at about the same latitude as much of the Atlantic coast of the United States and was not subject to heavy glaciation during the Pleistocene epoch. Studies of rocks and sediments from the floor of the East China Sea (Emery and Niino 1967) revealed several samples that contained shallow-water shells from relatively great depths on this shelf. Additional similar samples were obtained in collections of the Tokaiku Fisheries Experimental Station, the Yamaguchi Fisheries Experimental Station (mainly collected prior to 1930), and from dredgings during 1967 and 1969 aboard the research ships of the Tokyo University of Fisheries. Thirty-two samples from that region (Table 1) indicate

Table 1. Submerged Relict Species in the East China Sea*

Sta.	N Lat.	E Long.	Depth (m)	Land Species	Elephas namadicus Makiyama	Bison occidentalis Lucas	Brackish Water Species	Corbicula japonica Prime	Ostrea gigas Thunberg	Marine Species (0–10 m)	Haliotis sp.	Anadara subcrenata (Lischke)	Arca inflata Reeve	Solens sp.	Mactra chinensis Philippi	Paphia euglypta (Philippi)	Saxidomus purpuratus (Sowerby)	Trapezium sp.	Meretrix lamarkii Deshayes	Mercenaria stimpsoni (Gould)	Rapana thomasiana Crosse	Purpura luteostoma (Holten)	Marine Species (10–30 m)	Mytilus coruscum Gould	Mytilus sp.	Natica janthostoma Deshayes	Pecten albicans (Schröter)	Pecten yessoensis (Jay)	Tonna luteostoma (Küster)	Astarte sp.	Pecten laqueatus Sowerby	Macoma calcarea (Gmelin) (thick walled)	Cardium sp.	Barbatia sp.
1	35°30'	130°30'	210		x																													
2	to	to	220		x																													
3	36°30'	132°00'	250		x																													
4			280		x																													
5	35°29'	131°18'	125																				A			x							
6	35°27'	131°18'	118																				A			x							
7	35°27'	130°36'	194						x							x											x	x	A	A		
8	35°20'	130°25'	150																											A		
9	35°10'	130°29'	162																											x		
10	34°48'	129°13'	192		x																											
11	34°42'	129°16'	207															A											x		
12	34°37'	129°13'	195																											x		
13	34°40'	129°16'	219																											A		
14	34°37'	129°13'	192																											x		
15	34°15'	129°07'	120																													
16	33°49'	128°40'	115																													
17	33°32'	128°56'	146						x																		x			x		
18	33°03'	128°49'	117						x		x		x	x		x				x		x			x						
19	32°30'	124°04'	42			x																							x		
20	32°18'	127°56'	166																													
21	32°03'	128°29'	272																													
22	32°02'	128°02'	150						x																							
23	31°00'	125°31'	64					A	A		A			A																	
24	29°22'	126°18'	112						A						x		x			x					x	x	x			
25	29°20'	126°18'	112						x						x		x			x					x	x	x			
26	29°10'	126°10'	102					A	A						x		x			x					x	x	x			
27	28°45'	126°14'	110						A		A			A																
28	28°29'	125°13'	62						x																						
29	28°21'	126°06'	110						x					x	x		x			x					x	x	x			x	x
30	26°30'	123°00'	140						A																						
31	27°06'	123°30'	142										x	x																	
32	25°44'	123°18'	190						A			A								x		x		x	x		x	x		x	x

that past sea levels were lower than at present. As shown by Figure 2, the samples are in a belt that occupies the seaward 200 to 500 km of the shelf. A dotted pattern depicts the areas where future finds of relict shells are most likely to be made: seaward of the large area of muddy sediment that was and is being deposited by the Yangtze River off Shanghai and the Hwang-ho River that is north of the area shown by the map; seaward of modern deposits along the coasts of

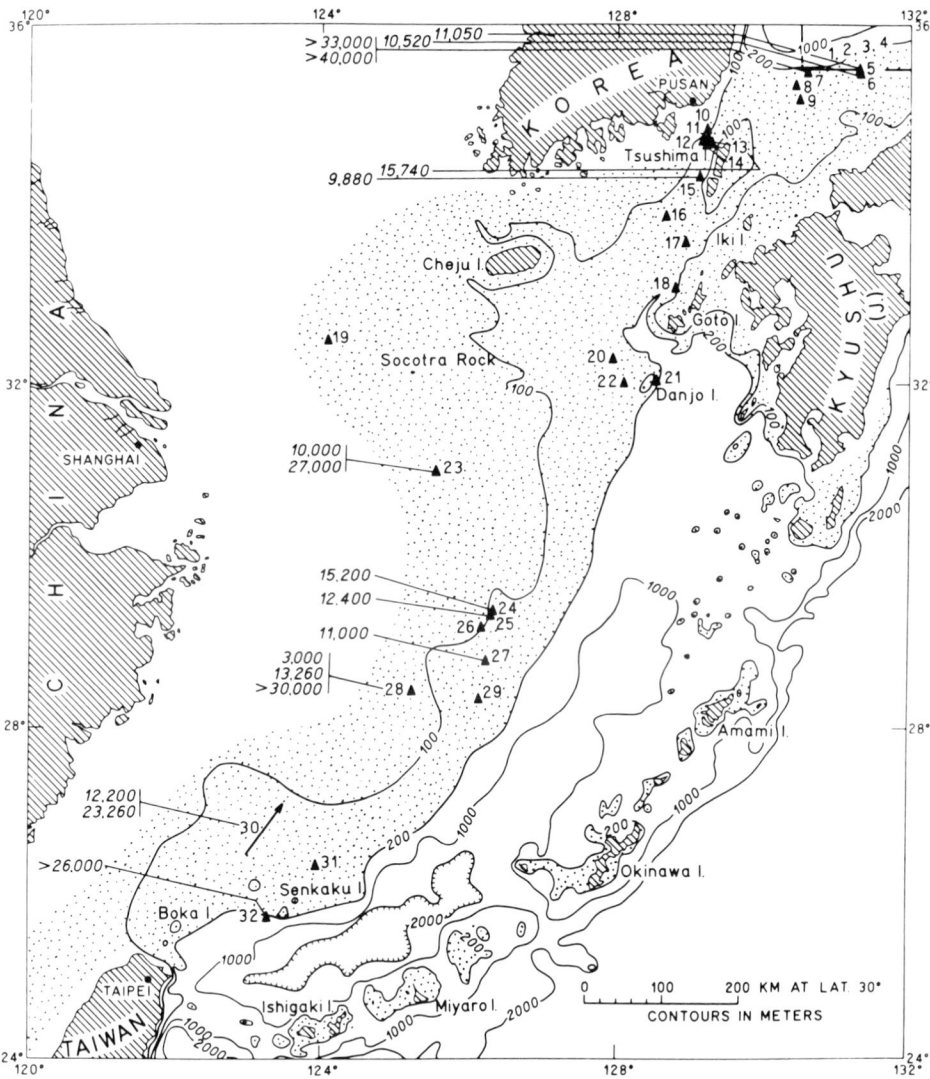

Fig. 2. Positions of samples for which data are presented in Table 1. Dotted pattern shows the areas in which future finds of relict faunas are most likely to be made. Numbers with leaders show radiocarbon ages in years.

South Korea and the large Japanese islands; around the small islands between Kyushu and Taiwan; and the area shallower than the 200-meter contour.

Five of the samples contain teeth of a Pleistocene elephant, and one had the lower jaw of a Pleistocene species of bison. The presence of these bones on the sea floor merely indicates that the shelf at some unknown time during the Pleistocene or early Holocene had been exposed so that land mammals could traverse it, a fact that previously had been known from the presence of Asian Pleistocene elephants, tapirs, bison, and deer on the Japanese islands. The bones provided no precise dates for the lowered sea level.

Twenty-six samples contain mollusk shells that were grouped into the following environments: brackish, marine 0 to 10 meters depth, and marine 10 to 30 meters depth (Table 1). Most samples contained too few shells that were unweathered or present in large enough quantity to justify radiocarbon dating. Twelve samples were selected as the most appropriate, and seventeen dates on various species were obtained (Table 2). Two were dated at National Taiwan University, Taipei, and the others at the U.S. Geological Survey in Washington. The resulting dates and depths are shown as large solid dots on Figure 3, with the same enveloping dashed lines as on Figure 1. For completion, four radiocarbon dates of peat from modern salt marshes of the west coast of Korea are included; these dates were published by Park (1969).

The dates and depths from the shelf and the Korean salt marshes follow the general sea level envelope that was developed for the Atlantic shelf of the United States by Milliman and Emery (1968), at least for the period between 15,000 years ago and the present, except in two respects. The simpler of these exceptions is for the several samples from deeper than about 140 meters. Their depths are greater than the generally accepted maximum sea level lowering of the late Pleistocene epoch, and the samples were dated in the light of that knowledge. All were on steep slopes bordering the continental shelf, whose depth is about 140 meters here. We believed that the dates would be greater than about 15,000 years, supporting the concept that large-scale reworking occurred at the time of lowest sea level. The measured dates vindicated this belief (for samples 7, 13, 32). Except for these dates, the others of Table 2 were added to Figure 1 to complete its worldwide coverage.

The second exception is a much more complicated one. For samples from stations 23, 28, and 30, shells of several different species of pelecypods were separately dated. Differences between the resulting dates differ by a factor of 2 or more within individual dredge samples. In samples 23 and 28 the older dates are for ten to fifteen specimens of *Corbicula japonica*, a brackish-water form. Even though the walls of these shells are very thin, transfer of old carbon into the shell is very unlikely. One might rationalize that, although most of the shells that were dated are from the latest transgressive movement of the shore, *Corbicula* is from the previous regressive stage of sea level. Still unexplained is the problem of the very young age of a composite of *Mactra chinensis* in sample 28. As the original samples were small, checking of the dates by duplicate runs is impossible. The var-

Table 2. Radiocarbon-Dated Samples

Station no.	Depth (m)	Species	Depth range (m)	Dated sea level below present (m)	Radio-carbon lab. no.	Radio-carbon age
5	125	Mytilus corscum	10–30	105	W2214	11,050 ± 600
6	118	Mytilus corscum	10–30	98	W2215	10,520 ± 600
7	194	Astarte sp.	10–30	140?	W2340	> 33,000
13	219	Macoma calcarea	10–30	140?	W2342	> 40,000
15	120	Macoma calcarea	10–30	140?	W2343	15,740 ± 400
23	64	Mercenaria stimpsoni	0–10	115	W2338	9,880 ± 350
		Mactra chinensis	0–10	59	W2217	10,000 ± 600
		Corbicula japonica	brackish	64	W2216	27,000 ± 1000
24	112	Ostrea gigas	brackish	112	W2036	15,200 ± 850
25	112	Ostrea gigas	brackish	112	NTU-22	12,400 ± 500
27	110	Ostrea gigas	brackish	110	NTU-38	11,000 ± 700
28	62	Mactra chinensis	0–10	57	W2220	3,000 ± 300
		Anadara subcrenata	0–10	57	W2254	13,260 ± 600
		Corbicula japonica	brackish	64	W2219	> 30,000
30	140	Ostrea gigas	brackish	140	W2237	12,200 ± 400
		Mactra chinensis	0–10 }	130	W2341	23,260 ± 600
		Pecten albicans	10–30 }			
32	190	Arca inflata	0–10	140?	W2360	> 26,000

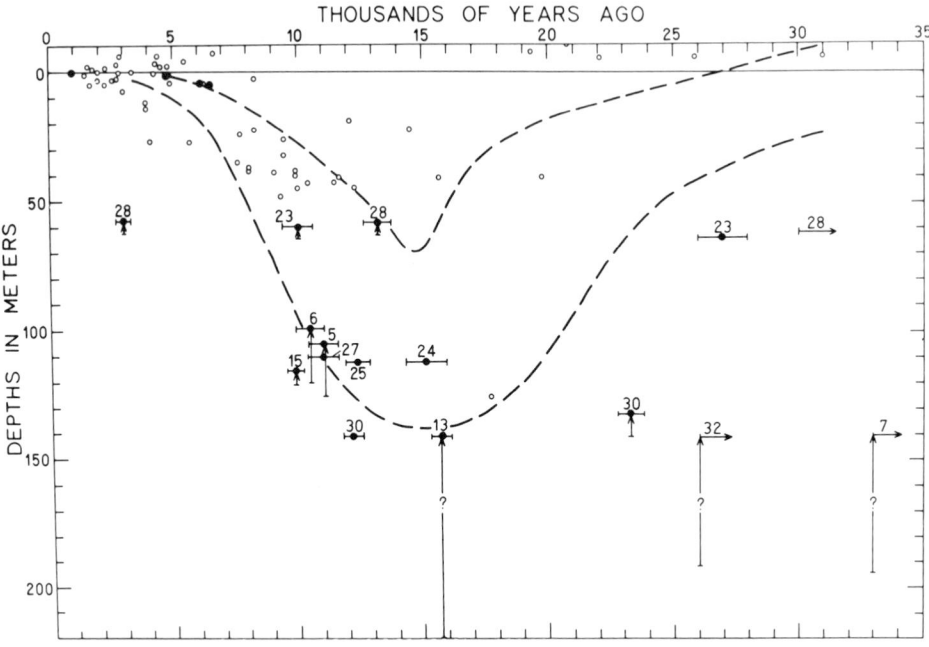

Fig. 3. Depths and ages of radiocarbon-dated shallow-water relict shells from the floor of the East China Sea (Table 2) and adjacent areas. The solid circles at the upper left indicate ages and depths of peat in salt marshes on the west coast of Korea (Park 1969). For comparison, the open circles are for samples from off Japan (Kagami 1962; Hoshino et al. 1967; Hattori 1967; Fujii 1967, 1969); because of known tectonic activity of Japan these points have a wide scatter. The dashed lines denote the envelope for most data points on the world's continental shelves, taken from Figure 1.

iations in radiocarbon age of different species in the same samples points up the need to make similar checks of ages in different species of other sea floor samples that are used for dating purposes.

Other radiocarbon dates are available from around the islands of Japan. Kagami (1962) dated *Corbicula japonica* from 40 to 50 meters depth off Sakata in northern Japan at 10,000 ± 250 years. Hoshino et al. (1967) dated a collection of *Glycymeris (Veletuceta) yessoensis* and *Limopsis tajimae* from 260 meters in Enoura Bay (just west of Tokyo) at 15,900 ± 300 years, and Hattori (1967) provided four dates indicative of past sea levels at Sendai Bay, northeast of Tokyo. Most comprehensive of all are studies by Fujii (1967, 1969), who listed a total of 52 dates for samples of unidentified shell, coral, peat, wood, and charcoal. These have ages of 1400 to more than 31,000 years and are shown with the other dates from Japan as open circles on Figure 3. Elevation corrections by Fujii for tectonism are omitted in Figure 3; the intense tectonic activity and submarine landsliding (Shepard 1933) in Japan require the indicated depths of ancient sea levels there to be considered less reliable than those from the East China Sea. Nevertheless, most of the Japanese data points between 13,000 years ago and the present lie within the enveloping dashed lines that were transferred from Figure 1.

The conclusion appears to be clear that the continental shelf between Japan, Korea, and China was exposed subaerially when sea level was low during early Holocene time. Probably it was alternately exposed and submerged several times during the Pleistocene epoch of glacial and interglacial stages. Accordingly, quite different faunas left their fossil remains on the shelf, and the shelf alternately served as a barrier and as an avenue of migration between the Asian mainland and both Japan and Taiwan. Radiocarbon dates are still inadequate to prove or disprove high sea levels about 30,000 years ago in the East China Sea, and dates of that general age from Japan are suspect owing to known diastrophic activity in Japan. Existing dated samples appear to favor a minimum sea level about 15,000 years ago, in accordance with dated samples from most of the rest of the world's continental shelves.

Summary. Radiocarbon ages of shore or shallow-water animals and plants from many continental shelves of the world show that sea level was about 130 meters below the present level about 15,000 years ago and probably at an intermediate level about 30,000 years ago. Only a few data points are available for most shelves other than those off the Atlantic and the western Gulf coasts of the United States, and the indicated sea level curve is different for these two shelves. A third shelf, in the East China Sea, was selected for intensive sampling and radiocarbon dating. The seaward half of this shelf has a large expanse of somewhat calcareous sandy sediment that contains the remains of animals that lived there during glacial times of low sea level. Included are bones of large terrestrial mammals, shells of brackish-water mollusks, and shells of shallow-water marine mollusks. Radiocarbon dating was attempted for seventeen sets of shells from twelve different dredge stations. Four sets were too old to be dated by radiocarbon (more than 26,000 years). Data points for ten of the remaining thirteen sets lie within or near the age and depth range of similar data points for the Atlantic shelf of the United States. Other data points from Japan fit less well, probably because of changes in land elevation due to active diastrophism.

Woods Hole Oceanographic Institution, Woods Hole, Mass., Contribution no. 2441 (K.O.E.). Published by permission of the Director, U.S. Geological Survey.

REFERENCES

Curray, J. R., Sediments and history of Holocene transgression, continental shelf, northwest Gulf of Mexico, in F. P. Shepard, F. B. Phleger, and Tj. H. van Andel, eds., *Recent Sediments, Northwest Gulf of Mexico*, p. 221, Amer. Assoc. Petroleum Geologists, Tulsa, Okla., 1960.

Emery, K. O., and L. E. Garrison, Sea levels 7,000 to 20,000 years ago, *Science, 157*, 684, 1967.

Emery, K. O., and Hiroshi Niino (1967) Stratigraphy and petroleum prospects of Korea Strait and the East China Sea: Report of Geophysical Exploration, Geol. Survey of Korea, vol. 1, p. 249; Also in Econ. Commission for Asia and the Far East, Committee for Co-ordination of Joint Prospecting for Mineral Resources in Asian Off-shore Areas, Geol. Survey of Japan, Tech. Bull. 1 (1968) p. 13.

Emery, K. O., R. L. Wigley, Alexandra S. Bartlett, Meyer Rubin, and E. S. Barghoorn, Freshwater peat on the continental shelf, *Science, 158,* 1301, 1967.

Fujii, Shoji, Postglacial deposits and their carbon-14 datings in the Japanese Islands, *The Quaternary Research*, vol. 6, p. 192, 1967.

Fujii, Shoji, Sea level changes in Japan during the past 11,000 years, *INQUA Congr., VII, Abstr.,* Paris, p. 198, 1969.

Hattori, Mutsuo, Recent sediments of Sendai Bai, Miyagi Prefecture, Japan, *Sci. Rept. Tohoku Univ., Sendai, Second Ser.* (Geol.), *39,* 1, 1967.

Hoshino, Michihei, Masatake Nishihara, and Takeshi Aoki, Absolute dating of mollus-can shells collected from the sea bottom of the mouth of Enoura Bay, *Chikyu Kagaku* [Earth Science], *21,* no. 6 (in Japanese), 1967.

Kagami, H., Modal size distribution of the shelf sediments around Honshu, Japan, Univ. Tokyo, Geol. Inst., Doctoral dissertation, 1962.

Milliman, J. D., and K. O. Emery, Sea levels during the past 35,000 years, *Science, 162,* 1121, 1968.

Park, Yong Ahn, Submergence of the Yellow Sea coast of Korea and stratigraphy of the Sinpyeongcheon Marsh, Kimje, Korea, *J. Korea Geol. Surv., 5,* 57, 1969.

Shepard, F. P., Depth changes in Sagami Bay during the great Japanese earthquake, *J. Geol., 41,* 527, 1933.

Shepard, F. P., Rise of sea level along northwest Gulf of Mexico, in F. P. Shepard, F. B. Phleger, and Tj. H. van Andel, eds., *Recent Sediments, Northwest Gulf of Mexico,* p. 338 (see Curray, above), 1960.

Whitmore, F. C., Jr., K. O. Emery, H. B. S. Cooke, and D. J. P. Swift, Elephant teeth from the Atlantic continental shelf, *Science, 156,* 1477, 1967.

T. van der Hammen, T. A. Wijmstra,
and W. H. Zagwijn

15. THE FLORAL RECORD OF THE LATE CENOZOIC OF EUROPE

A glance at a vegetation map will convince us that before the arrival of Neolithic man, Europe was almost entirely covered by forests. Only on the northernmost tip, near the Arctic Ocean, is there a narrow strip of treeless vegetation or tundra (Fig. 1). Some 15,000 to 20,000 years ago, during the coldest part of the last glacial, this picture was entirely different. An inland ice sheet covered the northern part of the continent and glaciers covered the Alps. The whole area between the southern edge of the inland ice and the Alps was either completely devoid of vegetation or covered with tundra or cold-steppe vegetation. The entire North Sea basin was dry, connecting Great Britain with the continent.

This last period was called the Würm glacial in Penck and Brückner's classical studies of glaciation in the Alps, and Weichsel glacial was proposed for the probably contemporaneous last glacial of northern Europe. There is little or no doubt about this correlation. However, the chronology of earlier glaciations is more complex. The correlation of the Alpine Riss and Mindel with the Saale and Elster glaciations seems probable, but the correlation of the Alpine Günz and earlier glaciations with phenomena in northern Europe and the Mediterranean is problematic. Moreover, the time interval covered by the three last and best-known glacials and interglacials may be only a fifth to a tenth of the total duration of the Quaternary. Therefore, if we do not want to lose ourselves in a terminology based on vague correlations, it seems necessary to establish a stratigraphic nomenclature not primarily based on moraines in mountainous areas but on sedimentary sequences in basins outside these glaciated regions.

In the Netherlands there are extensive deposits from glacial and interglacial times. They can be studied in surface exposures, but especially in deep bore holes. Study of the stratigraphic and climatic sequences of the last million years in detail requires a continuous record of the changing vegetation. Pollen grains are the most suitable fossils to reflect these changes. Long, continuous pollen diagrams are like a four-dimensional (although somewhat distorted) picture of the vegetation throughout time.

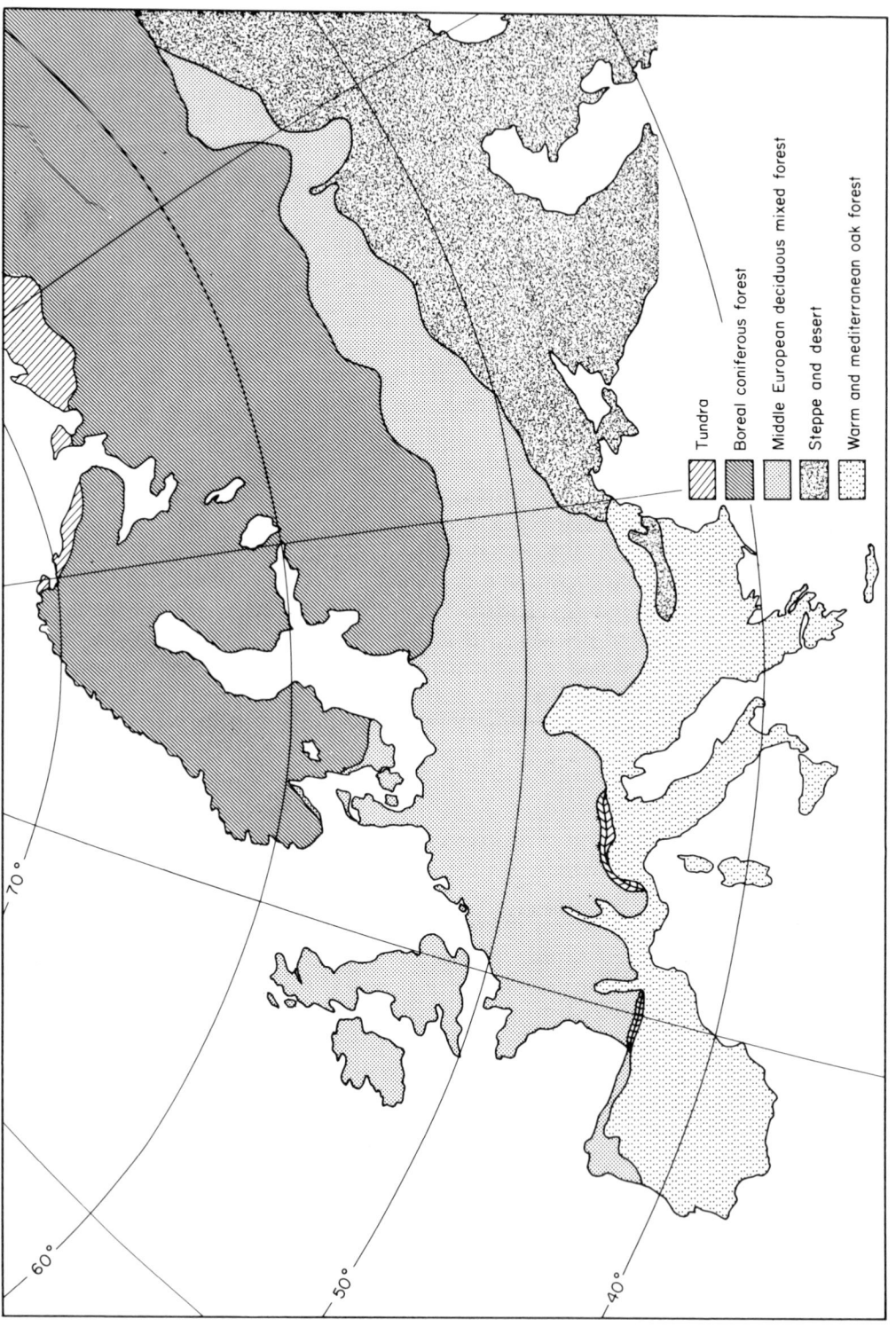

Fig. 1. Principal vegetation zones of Europe.

Unfortunately, the pollen record is not continuous in northwestern Europe because, especially in the middle and upper Pleistocene, sediments suitable for pollen preservation, like peat and clay, were principally formed during the interglacials and interstadials. Sediments from glacial time are often sands or are otherwise unsuitable for pollen studies. Hence the total picture of the floral sequence must be gradually assembled and for the older and deeper-lying deposits we often cannot be sure the picture is complete. For example, between the cold Menapian and the Elsterian (Mindel) glacial we formerly placed only one interglacial, the Cromerian. Recently, however, two more interglacials in this interval were established by means of pollen analysis of several new sections (Andersen 1965; West and Wilson 1966; Zagwijn unpublished). Therefore our knowledge of the Pleistocene sequence for northwestern Europe may still be incomplete; on the other hand, correlation with glacial phenomena is facilitated because of the proximity of the land ice.

To be sure, if all existing interglacials, glacials, and interstadials were to be represented in our stratigraphical table, we would need a very long, continuous series of lake sediments. These types of deposits are apparently not present north of the Alps but have been found farther south in the northern Mediterranean area (Sercelj 1967; Wijmstra 1969; Florschütz et al. 1970; etc.). Because of these very fortunate findings, we now possess continuous pollen diagrams, representing up to some 500,000 years of vegetation history. The future may even provide us with a total coverage of a major part of the entire Quaternary. The floral record of the Late Cenozoic of Europe is probably the most complete we have. As presented here, it is based on the latest results of the jigsaw puzzle in northwestern Europe and partly on still unpublished long, continuous pollen diagrams from southern Europe.

First we will discuss the history of climate and vegetation of about the last 120,000 years, as an example of the interglacial-glacial cycle. Then we will try to give a summary of what we know of the history of the last 10 million years. Finally we will discuss briefly the possibility of long-distance correlation of the European climatic–vegetational sequence with that of other continents.

The Last Interglacial-Glacial Cycle

Northwestern Europe (Netherlands). The upper Quaternary sequence in the Netherlands is now rather well known. It comprises the type localities for the Eemian and for the Amersfoort, the Moershoofd, Hengelo, and Denekamp interstadials (Zagwijn 1961; Van der Hammen et al. 1967; Vogel and Van der Hammen 1967). The type localities for the Brørup, Bølling, and Allerød interstadials are in Denmark (Iversen 1947; Andersen 1961; etc.) and that for the Odderade interstadial is in Germany (Averdieck 1967).

The pollen diagram presented for the Netherlands (Fig. 2, middle) is a composite diagram with ample stratigraphical and C^{14} control. It is based on diagrams from Amsterdam, Amersfoort, Moershoofd, Hengelo, Elsloo, and Halder

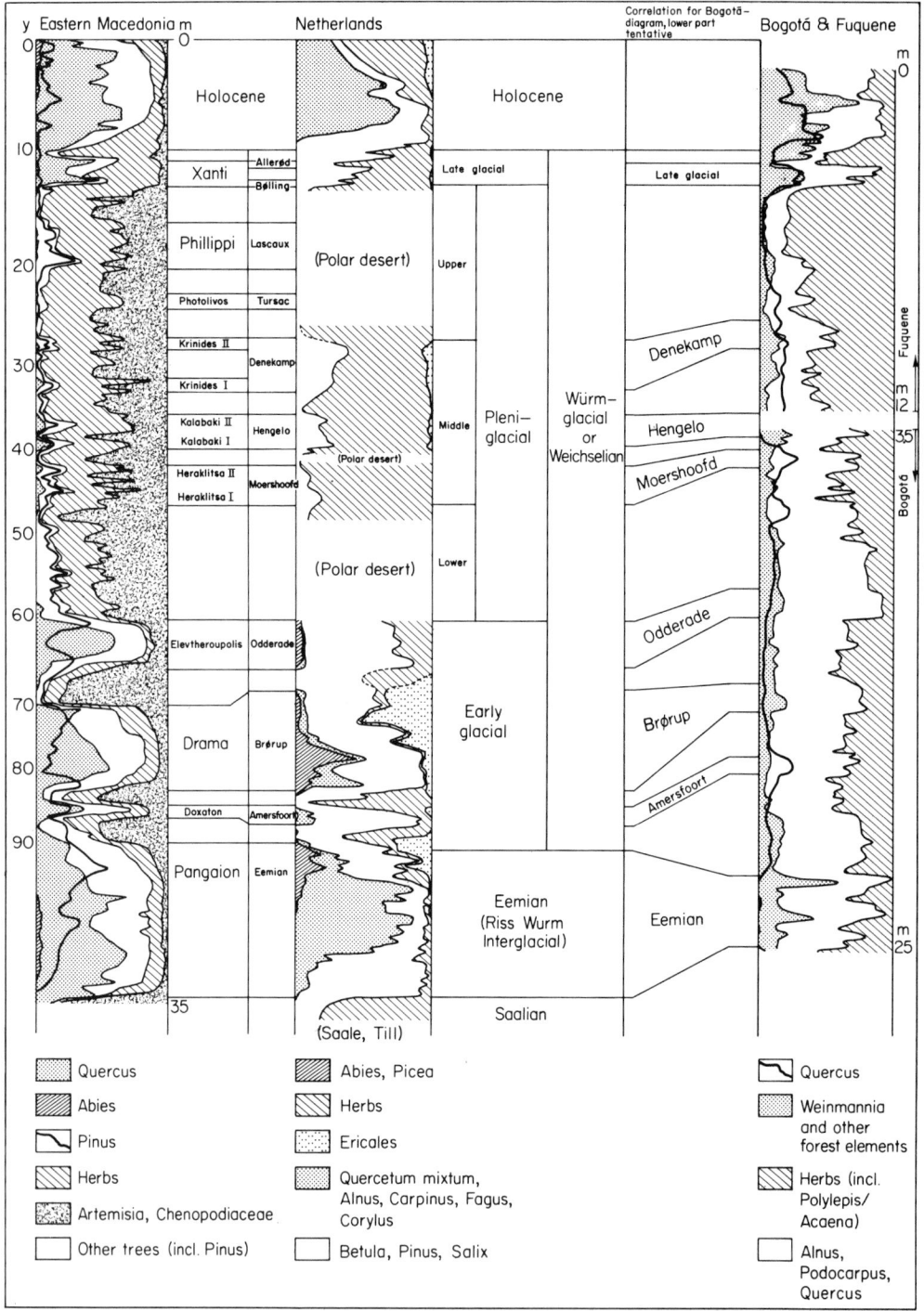

Fig. 2. Pollen diagrams for the last interglacial (Eemian), last glacial (Weichselian), Holocene from Philippi (northeastern Macedonia), the Netherlands, and Bogotá, Colombia (redrawn after Wijmstra 1969; Zagwijn, parts unpublished; and Van der Hammen et al., parts unpublished).

(Zagwijn 1961; Van der Hammen et al. 1967; and partly unpublished material). The vertical scale is adjusted to the continuous diagram from Philippi (Fig. 2, left; Wijmstra 1969). There is no pollen record from the coldest intervals; vegetation was apparently either completely absent or extremely scarce then, and there is ample evidence of heavy frost action (e.g. frost wedge polygons). These intervals have been indicated on the diagram as "polar desert." Some relevant C^{14} dates are indicated. The diagram shows cumulatively the percentage of the pollen of the principal ecological groups of plants. At the right side are those indicating open vegetation (tundra, heather), at the left the temperate (to boreal) forest (deciduous oak forest, fir, and spruce), and in the middle the subarctic (to boreal) pine-birch-willow forest or shrub.

The diagram starts above the Saale (Riss) till with the very end of this glacial period, reflected by high herb percentages (tundra). The Eemian starts with the immigration of subarctic-boreal birch and pine forest, soon replaced by a deciduous oak forest, indicating temperate conditions. In the second half of the interglacial there is a marked extension of hornbeam (*Carpinus*). In the later part fir (*Abies*) appears and spruce (*Picea*) begins a relatively important role. At the same time there is an increase of the pine-birch group and of heath. These facts indicate that soils are becoming increasingly acid toward the end of the interglacial; moreover, temperatures decline. With a marked rise of herbs the Eemian ends, indicating that a vegetation poor in trees became dominant. With this rise begins the first "stadial" of the early Weichselian glacial. Then follow the Amersfoort interstadial, the second "stadial," the Brørup interstadial, the third "stadial," and the Odderade interstadial. The stadials are represented as maxima of the herb group, indicating that the vegetation became a tundra or a "park tundra." The interstadials show a minor increase of the temperate-boreal elements in the case of the Amersfoort and Odderade interstadials and a major increase in the Brørup. The last one, therefore, was apparently the warmest interstadial of the early Weichselian. Both the Brørup and Odderade interstadials differ from all other Weichselian interstadials and the Eemian by the presence of pollen grains of the *Picea omorikoides* type. An older C^{14} date for the Amersfoort interstadial was 63,500 ± 900, but it seems probable that this date must now be considered minimal.

After the Odderade interstadial the Pleniglacial starts with extremely cold conditions; vegetation was very scarce or absent, resulting in a polar desert. This lower Pleniglacial lasted until the beginning of the Moershoofd interstadial, which seems to have started shortly after 50,000 B.P. After a short polar desert phase, two more interstadials follow, the Hengelo and Denekamp, dated respectively around 38,000 and 30,000 B.P. During the following upper Pleniglacial that lasts until the beginning of the late glacial, extremely cold polar desert conditions again prevail.

As a whole, the climate during the middle Pleniglacial seems to have been generally less cold and more humid than the lower and upper Pleniglacial. Although there was apparently one short polar desert phase in the middle Pleniglacial, most of the time the vegetation was more tundra-like. Macro-remains (leaves) of tundra plants like *Salix polaris, S. herbacea, Saxifraga oppositifolia,* etc. are mostly found in this interval, while signs of sedimentation from running water are frequent. Tundra-peat layers are frequent, representing the interstadials. Although the increase of birch pollen in the Moershoofd interstadial is very slight, it is somewhat more important in the other two interstadials. Probably this pollen is of *Betula nana* (dwarf birch), and the vegetation during the interstadials could be defined as a shrub tundra.

Artemisia pollen also seems to rise slightly during the middle Pleniglacial interstadials, but a marked rise in *Artemisia* percentage is not found until the beginning of the late glacial. It seems to be the first indication of an amelioration of the climate after the upper Pleniglacial polar desert and probably dates back to 13,000–14,000 B.P. (Van der Hammen and Vogel 1966). Around 12,400 B.P. there is immigration of the first large birches, resulting in a park tundra during the Bølling interstadial. From 12,000 to 11,800 B.P. there is a short, rather cold interruption and in many places an open tundra-like vegetation returns ("Older Dryas time"). Characteristic of the late glacial as a whole, but especially of the interval between 13,000 and 11,800, is a relatively high percentage of pollen from plants that may be called steppe elements: *Artemisia,* Chenopodiaceae etc.

The next interstadial is the well-known Allerød, lasting from 11,800 to 10,900 B.P. First a closed birch forest establishes itself and then pine immigrates for the first time since the end of the Odderade interstadial. The end of the Allerød interstadial is marked by a rise of the herb group and of heath pollen (*Empetrum*) and a decline of pine. Forest fires apparently occurred in the dead pine forest over great areas in Europe, attested by the frequent occurrence of charcoal in beds of this age. The vegetation becomes partially open again (park landscape); the increase of Ericales may indicate leaching of the soil during Allerød time and eventually a greater humidity. This colder period (Younger Dryas time) lasted until ca. 10,000 B.P. Closed pine-birch forest then establishes itself at the beginning of the Holocene. In some places a short, minor cooling seems to be reflected, around 10,000 (Friesland fluctuation; Behre 1967), but temperate elements soon immigrate and by 8,000 B.P. the mixed deciduous oak forest is dominant. In the later Holocene there is immigration of beech (*Fagus*) and hornbeam (*Carpinus*), but by that time human influence on the forests becomes increasingly important, resulting in an increase of the herb and heath groups.

In Figure 3 a schematic representation is given of the glacial-interglacial cycle in northwestern Europe (after Iversen 1958, and others; see also Andersen 1967). An interglacial begins with the replacement of the open (tundra) vegetation by pine-birch forest. Soils are still alkaline, shade is increasing. Further amelioration of the climate results in the replacement of the pine forest by temperate deciduous trees. First there is a maximum of hazel (*Corylus*), but gradually an oak forest, with local elms (*Ulmus*) and limes (*Tilia*), becomes dominant. There is a further increase of shade, and a slightly acid forest soil is formed. Hornbeam then appears,

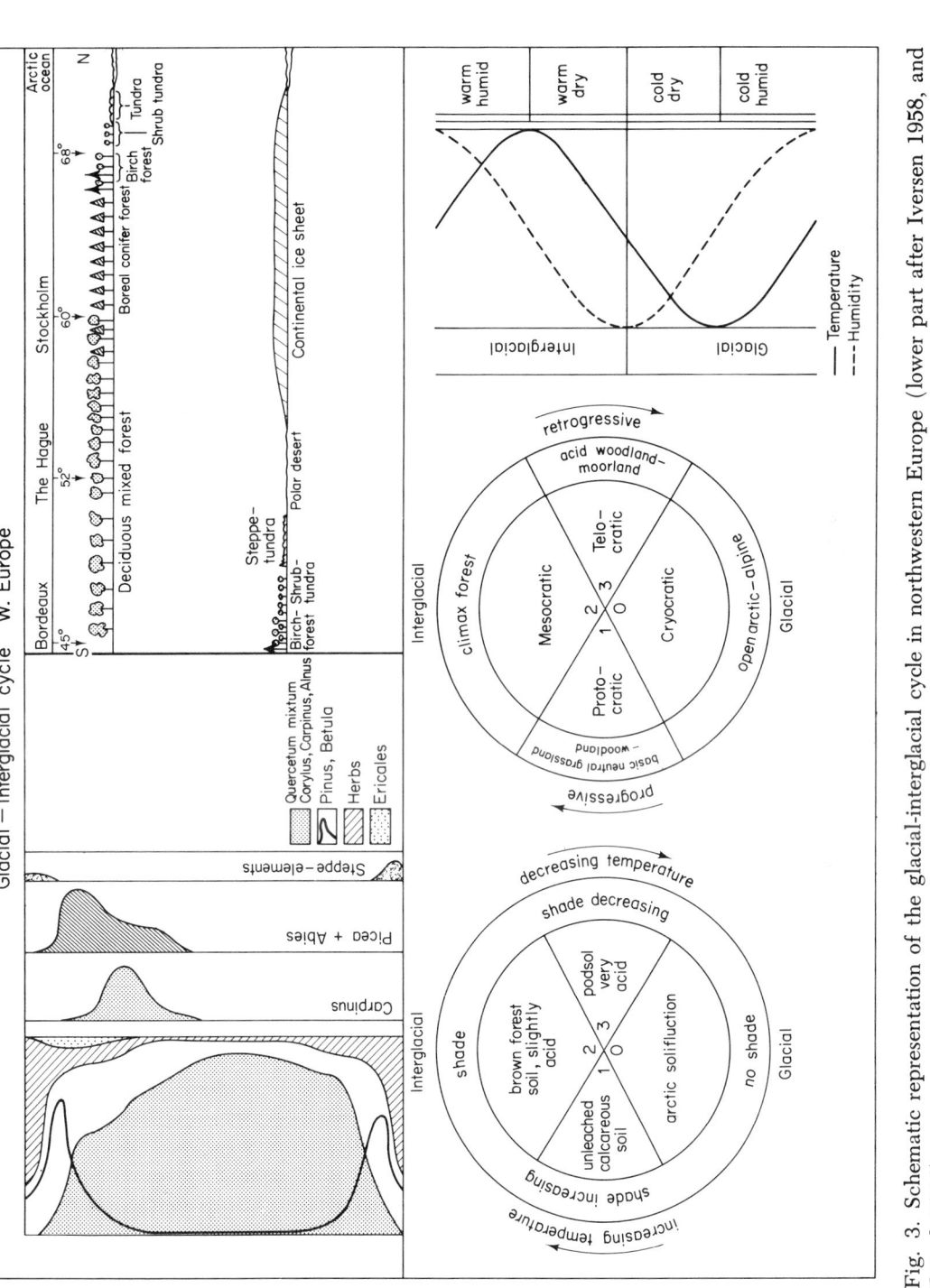

Fig. 3. Schematic representation of the glacial-interglacial cycle in northwestern Europe (lower part after Iversen 1958, and Gricuk 1964).

reflected as a clear zone in the pollen diagram. At the same time fir (*Abies*) and spruce (*Picea*) begin to show, often resulting in well-marked zones in the pollen diagrams—spruce, especially, appears above the hornbeam zone. During that time soils are becoming more acid, shade decreases again, and there is an increase of heath. The temperature also begins to decrease, finally resulting in pine-birch forest. Because of further decrease of temperature this forest is then replaced by an open herb (tundra) vegetation.

Extreme full-glacial conditions cause the complete or almost complete disappearance of vegetation and the establishment of a polar desert in a zone immediately south and southwest of the continental ice sheet. So-called steppe elements are more frequent in the late-glacial open vegetation than in the early glacial. This may be due in part to soil conditions, but there seems to be no doubt that the climate was relatively dry during the later part of the glacials and the earlier part of the interglacials. The climate was relatively humid during the later interglacials and the earlier glacials.

Mediterranean area (eastern Macedonia). The upper Quaternary sequence in the Mediterranean area is well known from a few places, where continuous series of lake sediments covering that period were found. The best and most complete diagram known at present is from the Tenaghi Philippon (ca. 50 meters above sea level) in eastern Macedonia (Wijmstra 1969). As it is well dated by a series of C^{14} analyses, it may easily be compared with the sequence north of the Alps. Figure 2 (left) is based on a continuous 35 meters of sediments. The figures at the left indicate thousands of years. They are based on eleven C^{14} dates for the upper 50,000 years and on extrapolation down to 90,000 years. The diagram shows, cumulatively, the percentage of the total of pollen grains (excluding marsh and aquatic plants) for different ecological groups. At the extreme right are the open vegetation members, represented by a group of so-called steppe elements (*Artemisia* and Chenopodiaceae) and a group of herbs (principally grass). The whole area left of these two groups represents the total percentage of forest elements. From the right, oak and the fir are represented. The white area between the curve for herb and steppe elements and the oak curve represents pine and other trees like hornbeam and elm. To clarify the total picture, the pine pollen percentage is also indicated as a separate curve (not cumulative), to be read from the left axis. The lowest part of the diagram shows a very high percentage of steppe elements—trees must have been almost totally lacking. It should represent the very end of the penultimate glaciation (Saalian, Riss).

The next interval, called Pangaion interglacial, is characterized by the dominance of forest and should correspond to the Eemian. The early part of this interglacial is characterized by the pollen of *Pistacia* and *Quercus ilex* (a Mediterranean evergreen oak). The main part of the interglacial, however, is dominated by pollen of deciduous oaks. The lower part shows a maximum of the curve for elm and the upper part a maximum of hornbeam. Fir is present in the middle part. The last part of the interglacial shows a decrease of oak and an increase of pine. Here again pollen grains of *Pistacia* and *Quercus ilex* are found. The early Weichselian starts then with a rapid increase of open vegetation, especially steppe elements.

The general course of the diagram during the early Weichselian is very similar to that of northwestern Europe, and there seems to be little doubt that the Doxaton, Drama, and Elevtheroupolis forest periods correspond to the Amersfoort, Brørup, and Odderade interstadials respectively. The "stadials" are then represented by open steppe vegetation. The Drama interstadial is clearly the warmest of the three, like the Brørup, showing an elm maximum in the lower and a hornbeam maximum in the upper part. All three interstadials show a pine maximum at the base and at the top, and an oak maximum in the middle.

Some 60,000 years ago a long steppe period begins, which lasts until around 14,000 years ago (the beginning of the late glacial). It corresponds to the Weichselian Pleniglacial. This long, almost treeless steppe period is interrupted only by some minor increases of oak and pine. They are the Heraklitsa, Kalabaki, and Krinides interstadials, corresponding in time to the Moershoofd, Hengelo, and Denekamp. A very small fluctuation (Photolivos) and another, principally characterized by an increase of pine (Philippi interstadial), have not yet been established from the Netherlands; but they correspond in time respectively to the Tursac and Lascaux interstadials described (from caves) in France (Leroi Gourhan 1965, 1968).

Electron microscopic studies of the Chenopodiaceae pollen of this "Pleniglacial" steppe period were carried out by Smit and Wijmstra (unpublished). Among the genera were *Kochia* and *Eurotia*, which occur today on some relict sites of steppe vegetation in middle Europe as well as in Asian cool and cold steppe vegetations like that in the Pamir. We might therefore consider the Mediterranean Pleniglacial steppes as an extension to the west of the cool-to-cold steppe zone of Central Asia (see also Fig. 1).

The beginning of the late glacial is marked by an increase of oak and a marked fall of the curve for steppe elements. The late glacial ends with a marked rise of herbs; this period should correspond to the Younger Dryas time. The Xanthi interstadial should represent both Allerød and Bølling interstadials. *Pistacia* and *Quercus ilex* are characteristic of the relatively open late-glacial vegetation. The beginning of the Holocene, some 10,000 years ago, is marked by an increase of oak and a decline of the herbs. The lower part is characterized by a maximum of elm and the upper part by hornbeam and fir. The final rise of the herb group and of *Pistacia* clearly indicates the influence of man on the vegetation (forest clearing).

In Figure 4 a comparison is given of the early glacial to Recent part of the Philippi diagram (elevation 48 meters) with one from Ioannina in western Macedonia (Bottema 1967) at an elevation of ca. 500 meters. There is a fair agreement between the general course of the curves. However, while the Pleniglacial in Philippi is represented by almost treeless steppe, at Ioannina there is still a fair representation of oak during that time. Beech is still equally represented in that interval in Ioannina, while at Philippi there is only beech in the early glacial inter-

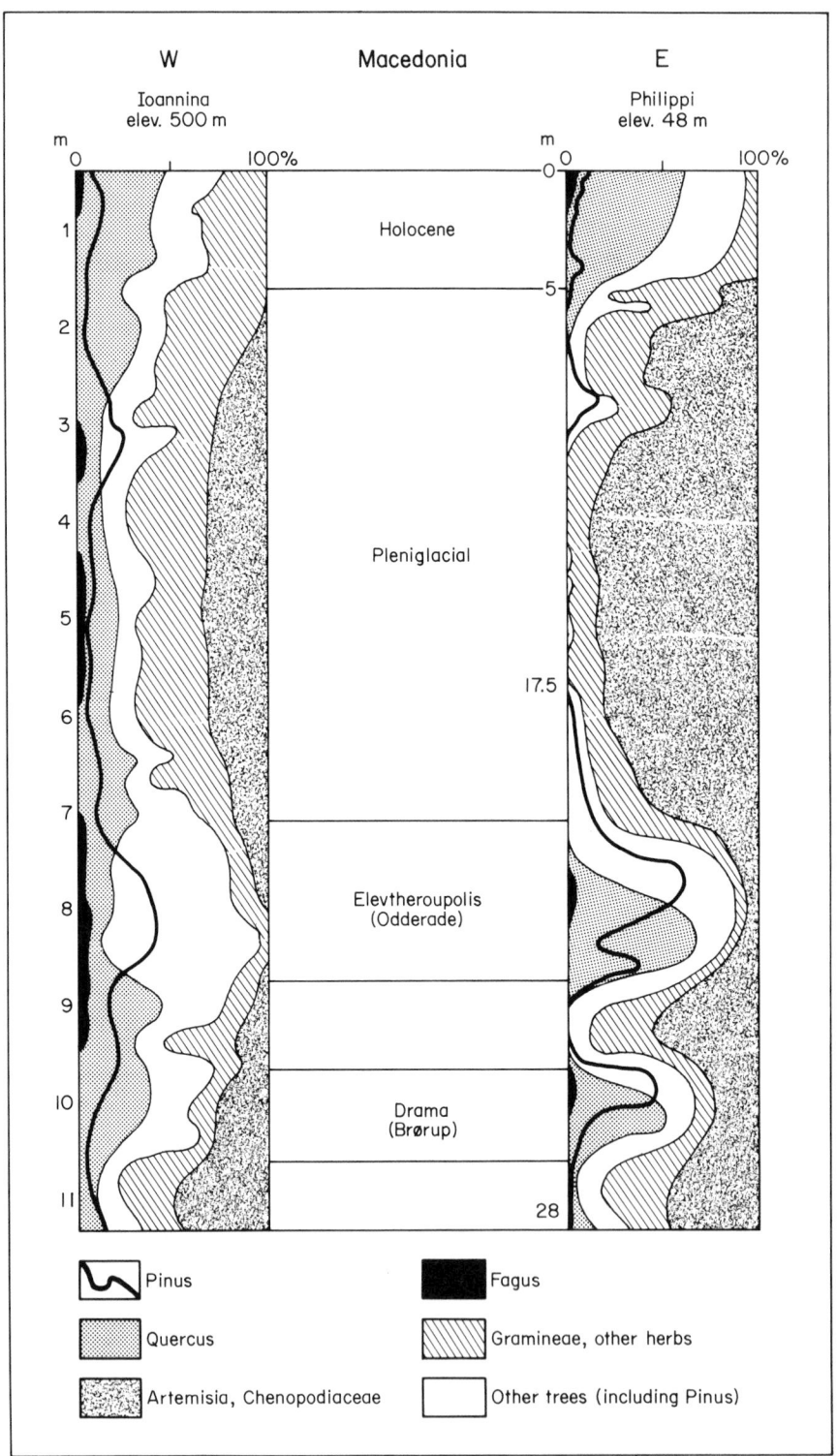

Fig. 4. Pollen diagrams representing early glacial to Recent, from western Macedonia (Ioannina, elevation ca. 500 meters) and eastern Macedonia (Philippi, elevation 48 meters; redrawn after Bottema 1967, and Wijmstra 1969).

stadials. These facts seem to show that during the Pleniglacial there was locally a (probably narrow) montane forest belt at around 500 meters in the Balkan mountains, intercalated between a lower cool steppe belt and a higher alpine (steppe?) belt. This situation is very similar to that in some parts of the central Asian highlands (Pamir etc.), where a narrow montane forest belt may occur between cool and cold steppe-vegetation with *Eurotia*. It may also be compared to the Ceja de Montana montane forest belt between a lower steppe zone and the higher puna zone in Peru and with similar successions in the southwestern United States (Walter 1968). This local montane forest belt may have been a refugium during glacial time for a number of European forest species.

Figure 5 is a schematic representation of what we now think was the glacial-interglacial vegetation cycle in the Mediterranean area. Soil development has not yet been clearly detected. Although temperature and shade play an important role, the factors dominating the general picture are the changes in humidity. The most astonishing fact is that these changes, reflected at the transition from forest to steppe and vice versa, are perfectly synchronous with the changes based principally on temperature in northwestern Europe (forest-tundra and vice versa). The only explanation seems to be that changes in humidity depend directly on changes in temperature as a primary cause for changes of the world's climatic pattern as a whole.

Figure 6 is a schematic representation of the changes of vegetation during an interglacial and a glacial period in a south–north section through Europe. The interglacial situation can be directly compared with the vegetation map (Fig. 1). The continent was almost entirely covered with forest. During full-glacial time very little forest was left, because of lowered temperature (tundra and polar desert in northwestern Europe) or because of extreme dryness (steppe in Mediterranean area). Forest refuges existed only locally, as in a local montane zone (ca. 500-meter elevation) in the Balkan mountains.

Extinction and Evolution

It becomes more and more evident that the Quaternary changes of climate in many places of the world caused a considerable speciation in both fauna and flora (see e.g. the interesting article of Haffer 1969 for the avifauna). The same factor may be at least partially responsible for pre-Quaternary Cenozoic phases of accelerated evolution (Van der Hammen 1961, 1964). However, most obvious and striking in the European Late Cenozoic is the successive extinction of a great number of floral taxa. Most of these are directly related to the present-day flora of East Asia and eastern North America. They belong to a holarctic flora that during the upper Tertiary existed in the northern hemisphere and in large areas of North America, Europe, and Asia. The successive disappearance of these elements from Europe during the Late Cenozoic and their existence today in two almost opposite areas of the globe (East Asia and eastern North America) has been known for many years. Excellent papers on the seed floras of the European Tertiary and lower Pleistocene were published by Reid and others (e.g. Reid and

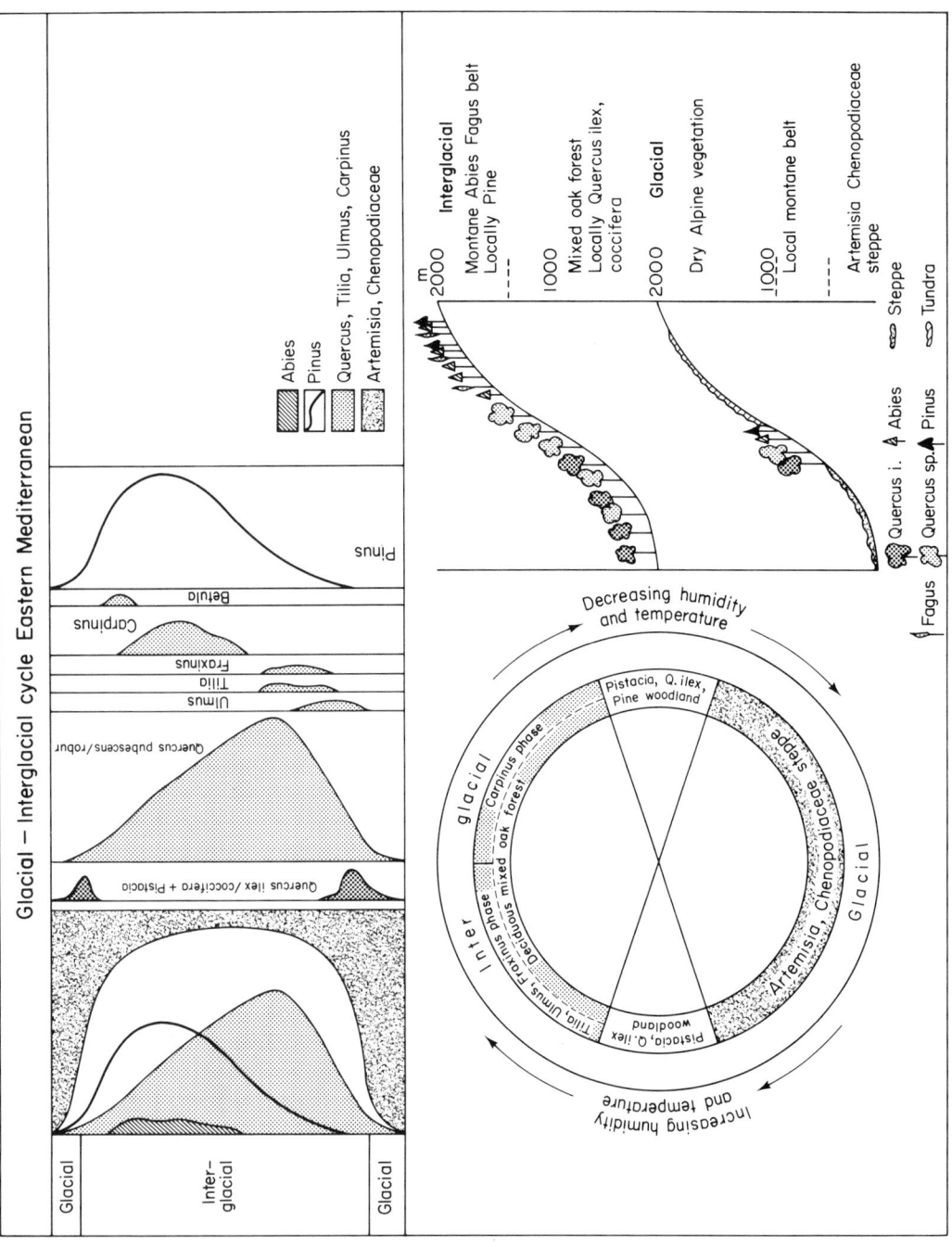

Fig. 5. Schematic representation of the glacial-interglacial cycle in Macedonia (Mediterranean area)

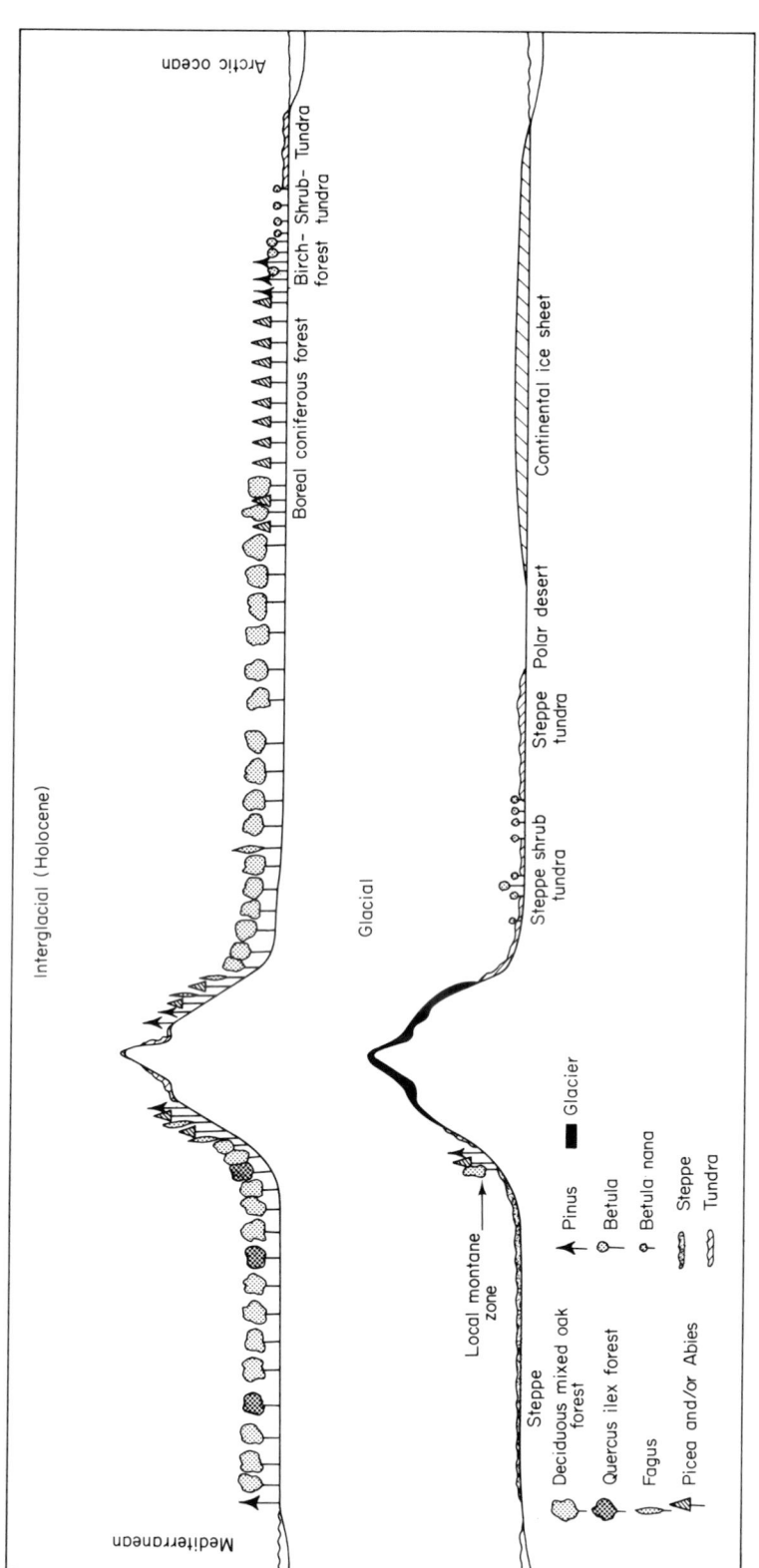

Fig. 6. Schematic representation of the vegetation during an interglacial and during a glacial, in a south–north section through Europe.

Reid 1907, 1910, 1915). With some additions (e.g. Florschütz 1925; Van der Hammen 1951b; Zagwijn 1959), these publications are still a major source of information on the subject.

The present disjunct areas of many of the eastern American–eastern Asian taxa were explained years ago by the theory of Asa Gray: the Pleistocene continental glaciations, combined with east–west chains of glaciated mountains (Pyrenees, Alps, and Carpathians, Caucasian Mountains, etc.) were responsible for the extinction of these elements in Europe and western Asia; but the mountain chains and valleys of East Asia and eastern North America, running northwest, permitted these temperate to warm temperate elements to migrate southward to warmer areas and to come back during the interglacial intervals. This theory still holds, although it may need to be amplified in view of new information (i.e. the phenomenon of the dry steppes in the Mediterranean and associated local montane zone during glacial times). It is clear that speciation must have taken place at the same time, so that typical European floral elements became more and more abundant in our area. However, because of the extremely rich woody floras of eastern Asian affinity as compared to the European, the impression of successive extinction is strongly dominant, especially in studies of the pollen flora.

Table 1 gives a clear illustration of this successive extinction and change since the middle Miocene in northwestern Europe. Comparison with the curve at the left side shows a clear relation to cold or cool intervals. After almost every cooler period, some taxa do not return. A few examples of disappearance related to a cooler phase are the following.

At the end of	*Extinct genera*
Middle Miocene	*Libocedrus, Metasequoia, Pandanus, Castanopsis, Mastixia*
Upper Miocene	*Cinnamonum, Clethra, Engelhardtia, Coriaria,* and the Fagaceae *Tricolpites henrici, T. microhenrici,* etc.
Brunssumian	*Corylopsis, Cunninghamia, Elaeagnus, Glyptostrobus, Palmae, Rhus, Symplocos*
Reuverian	*Aesculus, Diospyros, Liquidambar, Nyssa, Pseudolarix, Styrax, Zelkova* (and probably *Sequoia* and *Taxodium*)
Lower Tiglian	*Fagus, Liriodendron*
Upper Tiglian	*Actinidia, Euryale, Magnolia, Phellodendron*
Waalian	*Carya, Castanea, Juglans, Ostrya, Pterocarya, Tsuga*
Cromer I	*Eucommia, Celtis, Parthenocissus*
Holsteinian	*Azolla*
Eemian	*Brasenia, Dulichium*

Besides the influence of the relatively short cold intervals, there is a general tendency toward temperature decrease which is in part responsible for the disappearance from northwestern Europe of many elements found in or related to the warm-temperate-subtropical "evergreen forest" of southeastern China and its replacement by elements of the temperate "mixed mesophytic forest" of southeastern and eastern China.

Chronostratigraphy and Dating

Table 1 gives an excellent basis for stratigraphic correlation, especially of somewhat larger intervals, as has been done by using the appearance and disappearance of mammals like mastodonts, elephants, rhinoceros, mice, etc. A more refined "climatic" chronostratigraphy is based, at least for the Quaternary, on the succession of cool or cold and warmer periods, as reflected by the changing vegetation (and faunal associations). This has led to the climatic chronostratigraphical subdivisions for the Quaternary, as indicated in the left upper part of Table 1 (Van der Vlerk and Florschütz 1953; Zagwijn 1957, 1960, 1963b).

Recent studies have shown that longer subdivisions, like the Menapian, Waalian, Eburonian, and Tiglian, might be subdivided into a number of colder–warmer cycles of the same order of time as the well-known glacial-interglacial cycles of the upper part of the Quaternary. Their amplitude, however, may be less, so that the glacial-interglacial cycle appears to be longer in the lower Quaternary. This might be caused by the fact that these cycles are superposed on another cycle of longer duration, causing e.g. the generally cold Menapian, warm Waalian, and cold Eburonian. However, further research is needed to prove or disprove this supposition.

Most important, especially for correlation of widely separated sections in Europe and with other continents, is absolute dating. C^{14} dating has solved the main problems of correlation within the Weichselian and Holocene (see above); K/Ar dating has contributed to the dating of parts of the lower Quaternary in central France, but correlation of the latter with the sequence in the Netherlands, for example, is not yet precise (e.g. Lumley 1969; Glangeaud 1969). However, the data available for the Netherlands at least indicate that our correlation is a reasonable approximation. In northwestern Europe the possibilities for K/Ar dating are not very great, as volcanic materials are scarce. An exception is the dating of the Eifel volcanics and its relation to the Rhine terraces by Frechen and Lippolt (1965). By means of stratigraphic correlations based on petrology (the first occurrence of an augite association in the base of the formation at Urk) it seems possible to apply their date of ca. 400,000 years to approximately the base of the Cromerian III, perhaps correlative to the Cromerian interglacial at its type site. Dating by paleomagnetic reversals now seem to be one of the most promising techniques. Work has been done in central France (see e.g. Lumley 1969; Glangeaud 1969) and in eastern Europe (Demek and Kukla 1969). For the Netherlands the first data, based on measurements of clays, were published by Van Montfrans and Hospers (1969). It seems probable that they found the first reversal (Brunhes/Matuyama) in the lower part of the "Cromerian complex" (somewhere between I and II). The lower part would then date 700,000 years B.P. Normal magnetization was found by Van Montfrans and Hospers in the upper Waal-

Table 1. Chronostratigraphical Chart of the Interval Miocene to Recent for Northwestern Europe

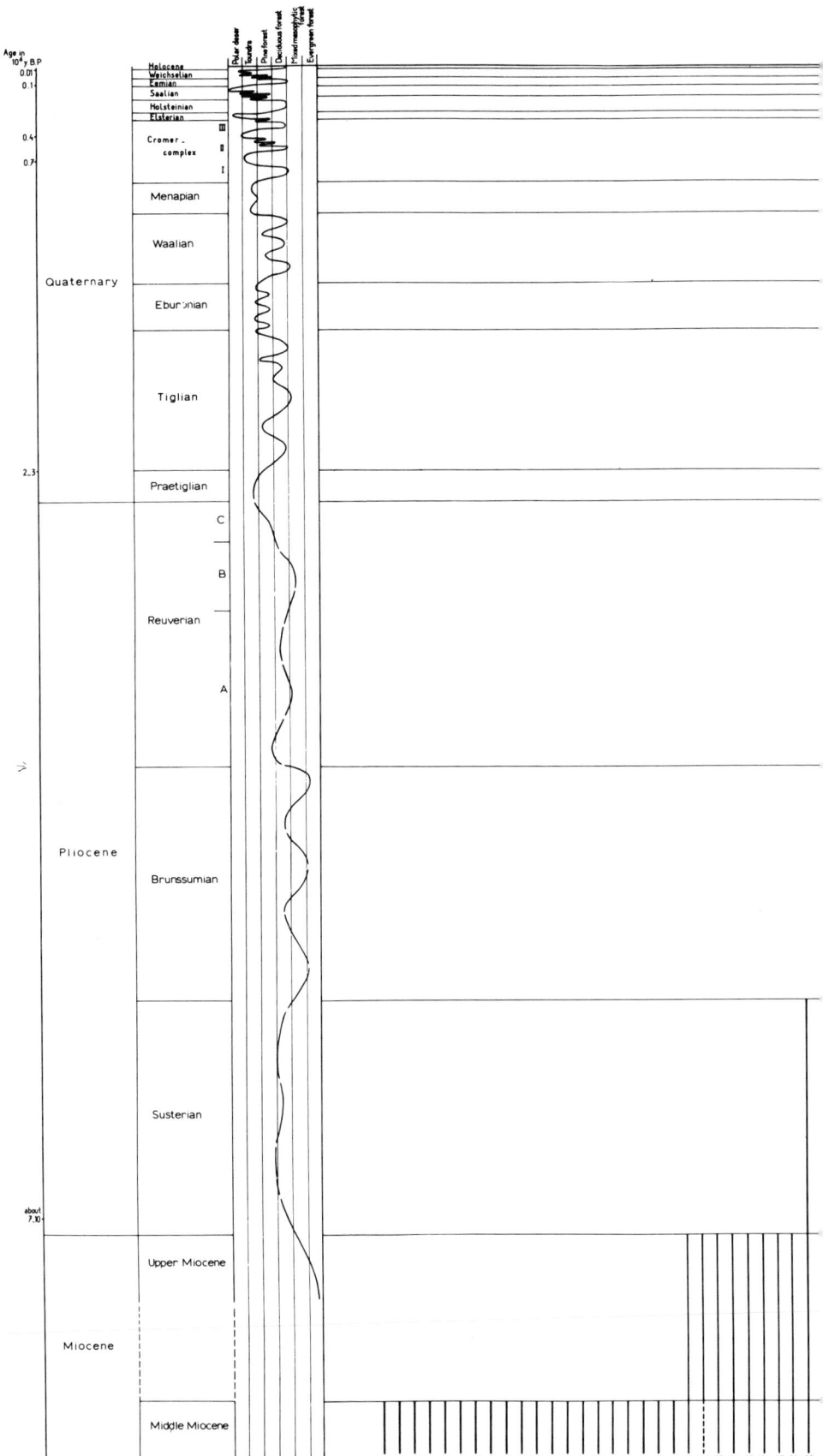

The curve indicates the general changes of vegetation; the successive extinction (or appearance) of a number of important taxa is shown (after Zagwijn, unpublished).

1	m*	*Libocedrus*	47	m + p	*Aesculus*
2	m	*Elatides†*	48	m	*Diospyros*
3	m	*Tetraclinis*	49	m	*Epipremnum*
4	m	*Metasequoia*	50	m	*Halesia*
5	m	Bromeliaceae	51	m	*Karwinskia*
6	m	*Pandanus*	52	p	*T. liblarensis fallax*
7	m	*Laurocarpum†*	53	m + p	*Liquidambar*
8	m	*Asimina*	54	m	*Meliosma*
9	m	*Castanopsis*	55	m + p	*Nyssa*
10	m	*Magnoliaespermum†*	56	m	*Pistacia*
11	m	*Ailanthus*	57	m	*Pseudolarix*
12	m	*Durania†*	58	p	*Sciadopitys*
13	m	*Ganitrocera†*	59	m + p	*Sequoia*
14	m	*Koelreuteria†*	60	m	*Stewartia*
15	m	*Mastixia*	61	m	*Styrax*
16	m	*Nephelium*	62	m	*Taxodium*
17	m	*Pallioporia†*	63	m + p	*Zelkowa*
18	m	*Sphenotheca†*	64	m + p	*Fagus*
19	m	*Tectocarya†*	65	m	*Liriodendron*
20	p	*Platycarya*	66	m	*Proserpinaca*
21	m	*Cinnamomum*	67	m	*Actinidia*
22	m	*Coriaria*	68	m	*Euryale*
23	p	*Clethra (T. exactus)*	69	m	*Magnolia*
24	p	*T. henrici*	70	m + p	*Phellodendron*
25	p	*T. microhenrici*	71	m + p	*Carya*
26	p	*T. villensis*	72	p	*Castanea*
27	p	*T. pseudocingulum*	73	m	*Decodon*
28	p	*Engelhardtia*	74	m + p	*Juglans*
29	p	*Cyrilla (T. brühlensis)*	75	m + p	*Ostrya*
30	m	*Alangium*	76	m + p	*Pterocarya*
31	m	*Berchemia*	77	m	*Tsuga*
32	m	*Corylopsis*	78	m + p	*Eucommia*
33	m	*Cunninghamia*	79	p	*Celtis*
34	m	*Cyclocarya*	80	p	*Parthenocissus*
35	p	*Elaeagnus*	81	m	*Azolla*
36	p	*T. edmundi (Cissus?)*	82	m	*Brasenia*
37	m	*Fothergilla*	83	m	*Dulichium*
38	m	*Glyptostrobus*	84‡	m	*Abies*
39	m	*Martyia†*	85	m	*Aldrovanda*
40	m	*Orixa*	86	p	*Picea*
41	m + p	Palmae	87	m	*Staphylea*
42	m	*Rhus*	88	m	*Salvinia*
43	m	*Schizandra*	89	m	*Trapa*
44	m	*Spirematospermum†*	90	m	*Vitis*
45	m + p	*Symplocos*	91	p	*Buxus*
46	m	*Torreya*			

* m = macrofossils; p = pollen.
† = extinct.
‡ Nos. 84–91 are still extant in Europe, but extinct in the Netherlands.

ian to lower Menapian. This is suggestive of the presence near the Waalian-Menapian boundary of the Jaramillo event and would date the Waalian-Menapian around 0.9 to 1 million years. However, Van Montfrans and Hospers consider that insufficient data are available to enable them to state this with any certainty.

Normal magnetization was also found in the upper Eburonian and near the boundary of the Tiglian and Eburonian. Nothing more is known at the moment and we cannot place the base of the Pleistocene more definitely than probably somewhere between 2 and 3 million years B.P.

There is no doubt that the Reuverian and Brunssumian belong to the Pliocene. The Susterian is sometimes placed in the lowermost Pliocene and sometimes in the upper Miocene. As we think this last interpretation more probable, we suggest an age of approximately 7 to 10 million years for its base; this again is no more than a guess, based on the known dates for the age of the Miocene-Pliocene boundary.

THE UPPER AND MIDDLE PLEISTOCENE SEQUENCE (CROMERIAN COMPLEX TO RECENT)

The Eemian to Recent sequence, discussed above, provides the model for the next older glacial and interglacial sequences, which are similar in general features. In the upper to middle Pleistocene sequence it is not easy to distinguish one inter-glacial from another on a palynologic basis. Only the Cromerian complex I inter-glacial seems to be distinguishable by the presence of *Eucommia*, one of the last "Tertiary relicts" in the Pleistocene. On the other hand, there are certain differences in the pollen zonation of the interglacials, based on different successions and quantitative changes, which may be used only if a complete pollen diagram of the entire interglacial deposit is available. The sequence in northwestern Europe is reconstructed with the help of guide horizons (like the Saalian drift deposits), superposition, sedimentary petrological data, etc.

As far as we know, the Saalian glacial has several rather warm interstadials (Andersen 1965) and presents the most advanced land–ice border. The Holsteinian is a widely known interglacial, often represented by relatively thick clay layers containing remnants of the aquatic fern *Azolla filiculoides* (Neede clays; Florschütz 1935). Below the deposits of the Elsterian glacial we find the thick sequence that we call the Cromerian complex. As has been mentioned above, in the Netherlands there seem to be at least three interglacials in this interval (Zagwijn et al., in press), of which the upper one (indicated in Table 1 as III) perhaps corresponds to the English-type Cromerian, as defined by West and Wilson (1966). The interglacial II of the complex, originally found at Westerhoven (Zagwijn and Zonneveld 1956), corresponds to the Danish Harreskovian (Andersen 1965). Interglacial I, with *Eucommia*, is equivalent to the interglacial of Osterholz, Germany (Grüger 1968). However, all these correlations still lack confirmation, and for the moment we prefer to use the neutral indications I, II, and III and the term Cromerian complex for the whole interval between Menapian and Elsterian (see Table 1).

A unique, continuous pollen diagram from the eastern Mediterranean (Philippi, eastern Macedonia) is presented in Figure 7 (from Wijmstra, in preparation). The diagram is based on the analysis of a series of lake sediments and peat 120 meters thick. The upper 35 meters correspond, on a smaller scale, to the first diagram of Figure 2. The C^{14} dates versus depth for the last 50,000 years lie approximately on a straight line. Extrapolation of these data gives the plausible date of 90,000 years for the top of the Eemian and 400,000 years for the lower part of the section. The gradual downward increase of compaction taken into account, a date around 500,000 years seems to be more probable. At any rate, the lower part would reach into the Cromerian complex. Since we know the glacial-interglacial succession for the area from the upper part (steppe forest), we can interpret the diagram in terms of glacials, interglacials, and interstadials with local names. These have been correlated with the northwestern European sequence (Fig. 7). Thus the lower part of the diagram should correspond to the upper part of the Cromerian complex, already postulated from extrapolation of the rate of sedimentation of the upper part. This diagram represents the most complete and continuous registration of upper and middle Pleistocene climatic sequences on the continent that we now have. Moreover, it is known that the total thickness of the Quaternary sediments in the basin is almost 300 meters. This material will be cored in the near future and it seems probable that within a few years we will have a continuous section representing a major part of the Quaternary.

With the data now available it seems possible to make a close calculation of the average duration of a glacial-interglacial cycle in the upper and middle Pleistocene. There are five cycles between the base of the Cromerian complex and the base of the Holocene, representing approximately 750,000 years. This would mean 150,000 years for one cycle. This figure seems to be too high in comparison with current estimates of the base of the Eemian (ca. 130,000 years) but agrees rather well with estimations of 300,000 years for the base of the Holsteinian. It seems, therefore, that the length of a glacial-interglacial cycle would be in the order of 120,000 to 150,000 years. As for correlation with the Alpine sequence, it seems probable that all four glaciations (Würm-Riss-Mindel-Günz) fall within the middle and upper Pleistocene.

THE LOWER PLEISTOCENE SEQUENCE (PRAETIGLIAN TO MENAPIAN)

In the upper half of this sequence (see Table 1) the cold Eburonian, the warm Waalian, and the cold Menapian have been described from the Netherlands (Zagwijn 1960). As far as can be established by pollen analysis, during the coldest parts of the Eburonian and Menapian, a tundra-like landscape existed in that country. As the polar-desert stage had apparently not been reached, it seems that the extremely cold conditions of the maxima of the later glaciations were not reached. On the other hand, the time represented by both the Eburonian and

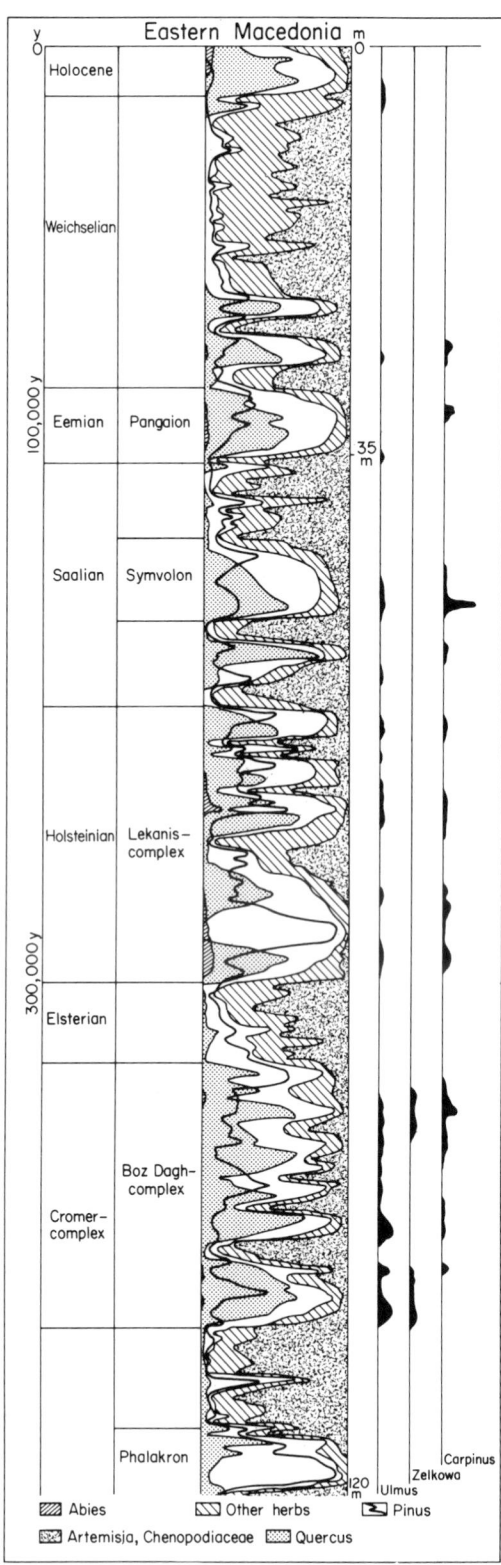

Fig. 7. Pollen diagram of a 120-meter section from Philippi (eastern Macedonia), representing the interval Cromerian complex to Recent (after Wijmstra, unpublished).

Menapian should be a multiple of the length of the later glacials. Equally, the warm Waalian should have lasted a multiple of the later interglacials. However, some minor climatic changes have been deduced from the available pollen data, which could be of the same duration as the later glacials and interglacials. More detailed pollen data are certainly needed, but for the time being it seems that in this part of the Pleistocene the influence of a longer cold–warm cycle seems to dominate over the still recognizable glacial-interglacial cycle of the order of the upper and middle Pleistocene (see curve in Table 1).

The lower half of the sequence comprises the Praetiglian and Tiglian (Van der Vlerk and Florschütz 1953). The Praetiglian, as defined by Zagwijn (1960), has always been considered to be the base of the Pleistocene. Although, as we will see, cooler phases occur farther down in the Late Cenozoic, the Praetiglian is the first time that a tundra-like open vegetation prevailed in the Netherlands. The long Tiglian represents as a whole a warm interval; however several cool intervals are recognized, when the temperate deciduous forest was replaced by boreal pine forest (Zagwijn 1960, 1963a).

The cold of the Praetiglian had a serious influence on the flora. Many elements of the rich "mixed mesophytic forest" of Chinese type disappeared and never came back in northwestern Europe (see above and Table 1). Although a number of floral elements disappeared at the beginning of the cold Eburonian, the influence of the long, cold Menapian was much more serious. Most of the Tertiary relicts still present in the Waalian, like *Carya*, *Castanea*, *Juglans*, and *Tsuga*, disappear. All data available at this moment (mammal faunas, paleomagnetic data, etc.) seem to show that most of the lower Pleistocene sequence from northwestern Europe corresponds to the upper Villafranchian of Italy (or middle to upper Villafranchian of France).

The Praetiglian and the base of the Tiglian are represented in the uppermost part of Figure 9. The deterioration of the climate at the transition Reuverian-Praetiglian (Pliocene-Pleistocene) is very clearly registered. The "tertiary elements" show a sudden fall, immediately followed by a replacement of the temperate deciduous forest elements by pine, representing the boreal forest. Then there is a marked rise of open herb vegetation, representing park-tundra conditions.

A diagram from Senèze in central France (Fig. 8 redrawn after Elhai 1969) is shown as an example of a pollen diagram from a long lower Pleistocene sequence. It is based on the analysis of 120 meters of sediments. The floral composition during the forest phases is of the Tiglian to Waalian type in northern Europe. The cool phases, however, are represented here by a very high percentage of steppe elements (*Artemisia*, Chenopodiaceae), giving the diagram a typical Mediterranean aspect. It seems probable that the section between 45 and 120 meters represents part of the Tiglian, including a marked cool steppe interval. It is possible that the Eburonian and Waalian are represented in the upper 45 meters, but without further stratigraphical data this interpretation cannot be positive.

The similarity of this diagram (excluding the presence of the Tertiary relict elements) to the upper-middle Pleistocene diagram of Philippi (Fig. 7) is strik-

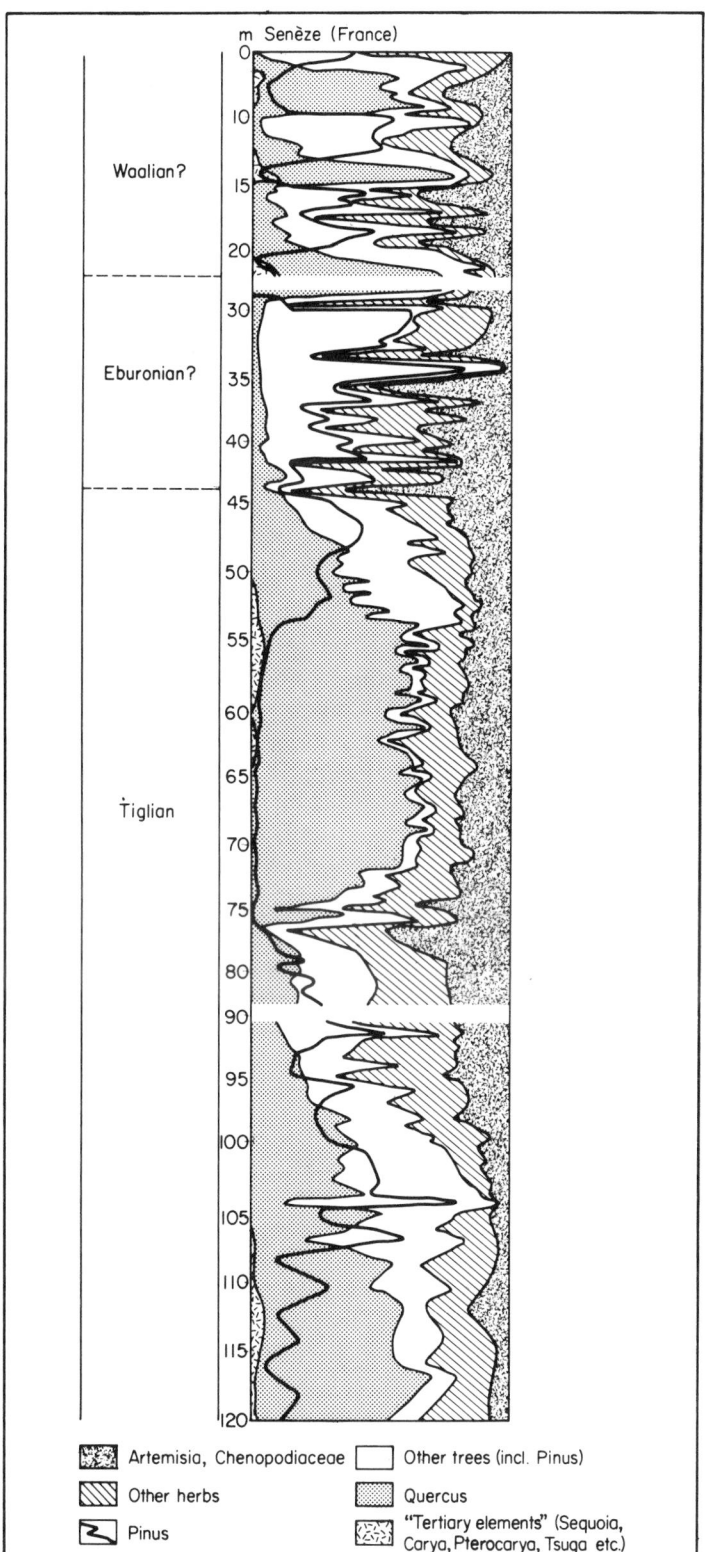

Fig. 8. Pollen diagram of a deep section from Senèze (Massif Central), France, representing a part of the lower Pleistocene (Tiglian to Waalian?; redrawn after Elhai
1969).

ing. As both cores are 120 meters long, it is easy to compare the thicknesses of the interglacial forest phases and the glacial steppe phases. The thicknesses are of the same order of magnitude. This at least indicates that the climatic fluctuations of the lower Pleistocene may be of the same approximate duration as the later glacial-interglacial cycle. However, this indication must be confirmed by more data.

In Italy, several thick lower Pleistocene sections have been described. The best known is that of Leffe (Lona 1950; Lona and Follieri 1957). At least part of the Tiglian and possibly the Eburonian, Waalian, and Menapian (see Zagwijn 1957) seem to be represented. Detailed palynological study of these long sections, combined with paleontological and paleomagnetic surveys, might considerably increase our knowledge of the climatic-vegetational succession of the lower Pleistocene and refine our curve.

THE MIOCENE AND PLIOCENE SEQUENCE

During the Miocene, warm temperate to subtropical conditions prevailed in western and northwestern Europe. Especially abundant are elements related to the flora of the evergreen forest in southeastern China today (Wang 1961). From Table 1, however, a phase of remarkably strong extinction at the middle-upper Miocene boundary is apparent. In addition there are some palynological indications of a cooling climate at that time. Comparison with the Late Cenozoic coincidence of cool or cold phases with extinction of species indicates that there was indeed an interval of cooler climate at that boundary.

Pollen diagrams from parts of the interval from the uppermost Miocene to the lower Tiglian are illustrated in Figure 9 (redrawn after Zagwijn 1960). Between the definite upper Miocene deposits and the Pliocene Brunssumian lies the Susterian. The age of the Susterian (a thickish series of predominantly rather coarse sediments) is not certain (lowermost Pliocene or uppermost Miocene). However, an upper Miocene age seems acceptable. Many elements disappear from northwestern Europe at the base of that interval and a good number of them never returned (Table 1). A pollen diagram from part of the Susterian (Fig. 9) shows that pine dominates. The flora is very poor in species and everything seems to indicate that a cool-temperate to boreal climate prevailed. An estimated age for the Susterian could be somewhere between 6 and 10 million years. Pollen diagrams, representing parts of the Pliocene Brunssumian and Reuverian are shown in Figure 9 and may be compared with the corresponding intervals in Table 1.

During this time the dominant vegetation is that of the Chinese mixed mesophytic forest type (Wang 1961), corresponding to a temperate climate with rainfall throughout the year. However, there is no doubt that there are several cooler phases during both the Brunssumian and Reuverian. The coolest phase is again related to local extinction of a number of elements, and pine forest dominates. This coolest phase corresponds to the base of the Reuverian and could be interpreted as boreal. The other cooler phases correspond to a cooler temperate climate—there is an increase of pine but it does not dominate.

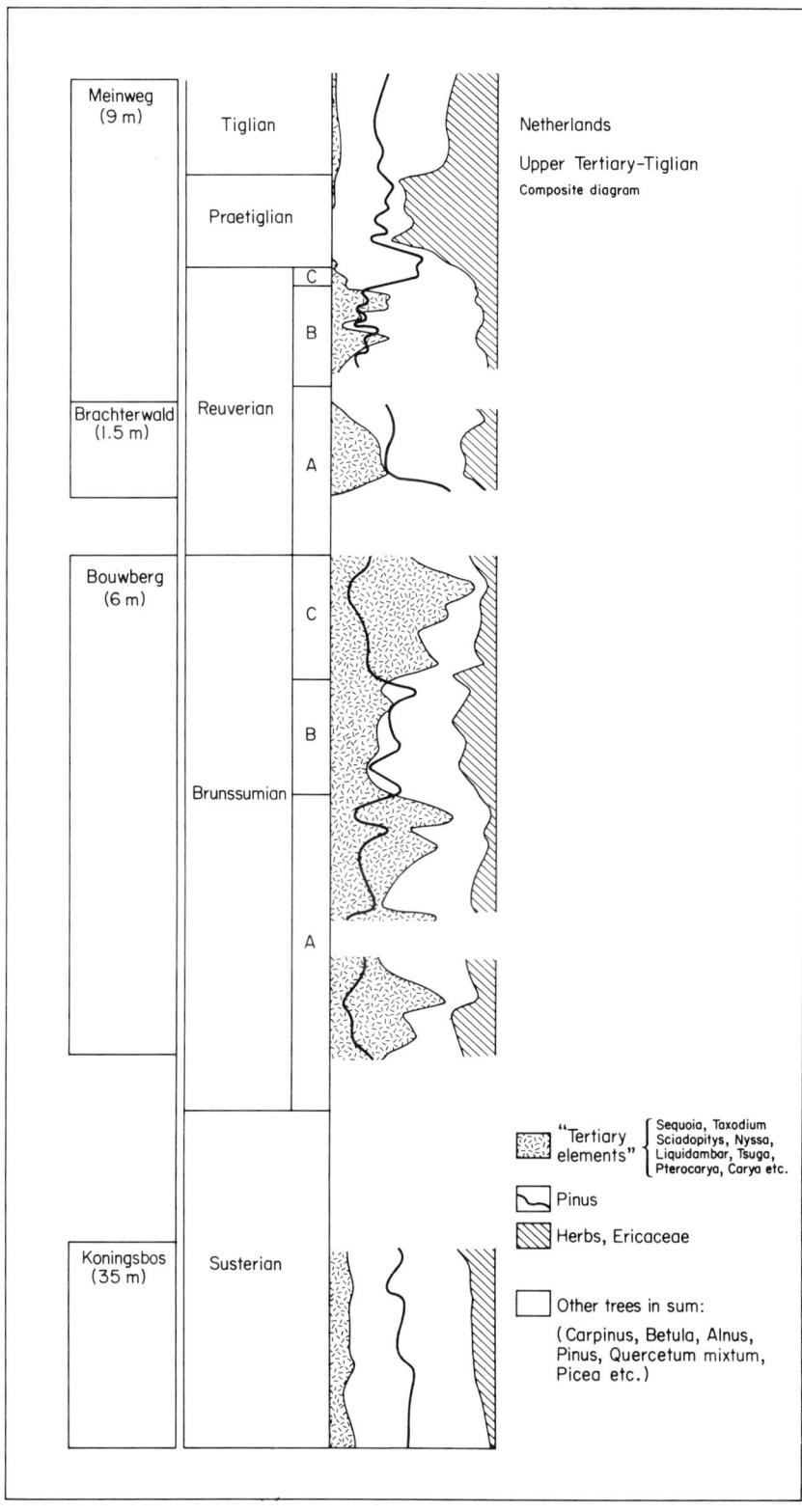

Fig. 9. Pollen diagrams from the Netherlands, representing parts of the interval Susterian to Tiglian (redrawn after Zagwijn 1960).

To verify the climatic character of the changes visible in the diagrams, one of these intervals (from the Brunssumian) was studied in detail and another type of pollen diagram was constructed (Fig. 10; redrawn after Zagwijn 1967). Only those elements were used in the pollen sum that have a clear climatic meaning; all elements of the marsh- and swamp-forest were omitted from the sum. This diagram shows a clear succession from a warm temperate forest with deciduous and evergreen broadleaves (zone II) to a montane-boreal type of forest with *Pinus*, *Picea*, and *Tsuga*. This can be interpreted only in terms of a cooling climate. Zone I also shows a dominance of pine with some hemlock that also should be of the montane-boreal type.

Another example from the Pliocene, but from the western Mediterranean (Spain), is the pollen diagram from the 25-meter core of old lake sediments from Villaroya (Fig. 11; redrawn after Remy 1958). The diagram shows the presence of Tertiary–lower Pleistocene elements (*Tsuga, Carya, Pterocarya*) besides *Zelkova* and *Liquidambar*. These two elements disappeared from northwestern Europe at the Plio-Pleistocene boundary. *Zelkova* is still found in the middle Pleistocene of the eastern Mediterranean (Philippi). The presence in the upper part of the diagram of a continuous curve of *Liquidambar*, however, is strongly indicative of a pre-Pleistocene age. Moreover, in the sediments above the lake deposits,

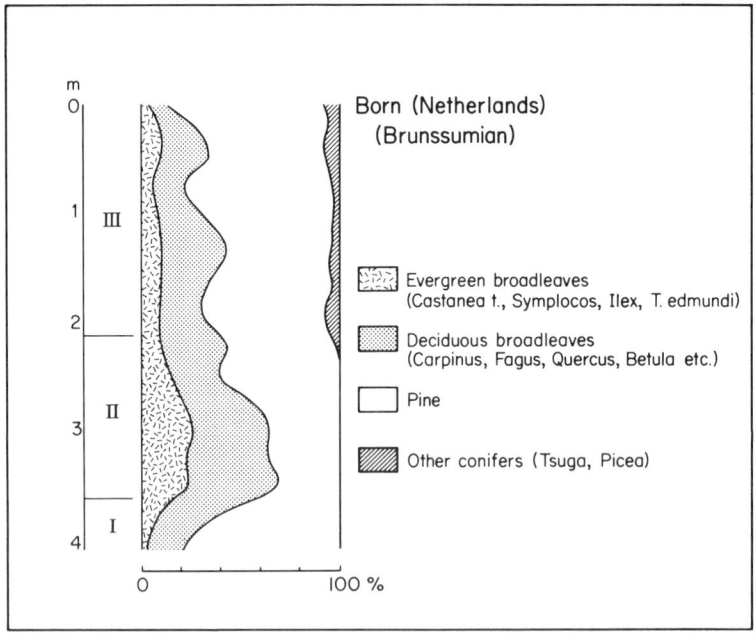

Fig. 10. Pollen diagram from Born (Netherlands), representing part of the Brunssumian. Species from the marsh or swamp forest have been excluded from the sum (redrawn after Zagwijn 1967).

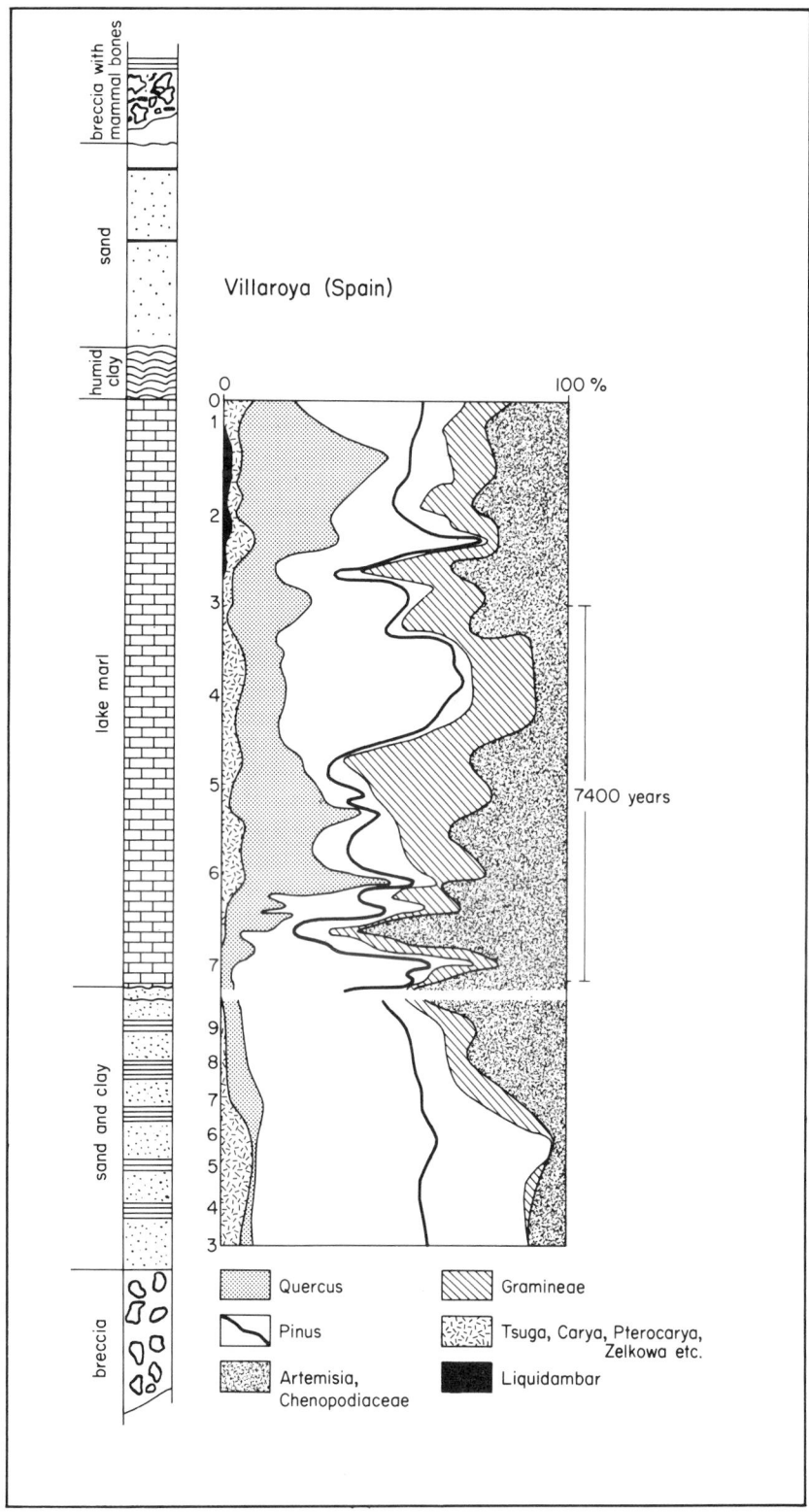

Fig. 11. Pollen diagram of a probably lower Villafranchian section from Villaroya, Spain (redrawn after Remy 1958).

a vertebrate fauna was found that according to Remy (1958) shows certain affinities with the fauna of the lower Villafranchian. An upper Pliocene age therefore seems probable.

This diagram is again of the typical Mediterranean type, with a phase showing high percentages of steppe elements (*Artemisia* and Chenopodiaceae). At the highest maximum of the steppe elements, the Tertiary elements have practically disappeared for a short interval. The steppe phase indicates drier conditions. An interesting additional fact is that a thin annual layering was found in part of the lake sediments. They represent 7400 years for this interval, as indicated in Figure 11. This shows that time length of the climatic deterioration was at least of the same order of magnitude as some of the cold steppe phases in the middle and upper Pleistocene of the Philippi diagram (Fig. 7).

Many of the facts mentioned above show clearly that climatic changes occurred in the Pliocene and even in the Miocene of Europe. Other facts, pointing in the same direction and not all mentioned here, can be found in the scientific literature (e.g. Altehenger 1958; etc.). These changes may be compared with the Pleistocene climatic changes, although they never reach the extreme cold conditions of the later Pleistocene glaciations. The climate in general during the Miocene and Pliocene changed from a subtropical to relatively warm temperate character, and the cool phases do not reach conditions colder than those corresponding to boreal-montane vegetation types. Very little is known of the duration of the cooler periods. The Susterian seems to represent a considerable lapse of time, but the cooler phases of the Brunssumian must have been much shorter. The shortest registered periods may be the warm phase (zone II) of Born (Fig. 10) and the dry (cool?) phase of Villaroya (Fig. 11) and might be of the order of magnitude of the later Pleistocene fluctuations. Changes of climate, related to periods of extinction and evolution, will probably play an increasingly important role in pre-Quaternary stratigraphy (Van der Hammen 1961, 1964a).

POSSIBILITIES OF LONG-DISTANCE CORRELATION

The possibility of correlation of the known European sequence with other continents, by means of long-sequence, continuous pollen diagrams, depends on the contemporaneity of climatic change over all or large parts of the globe. The only way to verify this is by means of direct or indirect absolute dating. Carbon-14 dates of the climatic change at the transition of the last glacial to the Holocene from many places in the world have shown that this change invariably takes place between about 13,000 and 10,000 years ago. Even minor climatic fluctuations in this interval, like the Allerød interstadial, are detectable and contemporaneous in places as far apart as Europe, Asia, Africa, and South America. Since it is impossible within the scope of this article to review even part of the most relevant data, we will give only one example of a possible correlation between Europe and South America.

In Figure 2, at the right, is a diagram from the Eastern Cordillera of Colombia (after Van der Hammen and Gonzalez 1960 and Van Geel, unpublished). It is composed of two parts; the upper part is from the Laguna de Fuquene, the lower from the Sabana de Bogotá; both are from an elevation of ca. 2580 meters. The diagram is based on a total of about 35 meters of lake sediments. Both the thickness and the supposed time interval represented are the same as those of the lake series of Philippi (left). At the right part of the diagram is the total of open vegetation elements (principally grasses). At the left is the total of trees. Two important elements of the Andean forest are indicated by individual curves. The dividing line between the percentage of open vegetation elements and forest elements is, generally speaking, a reflection of the vertical movements of the altitudinal forest line, determined principally by changes of temperature and partly by changes of humidity. The upper part of the diagram is C^{14}-dated (the last 40,000 years). These dates show that the registered changes of climate are, within the limits of the method, contemporaneous with similar changes of climate in Europe. These correspond e.g. to the Allerød–Younger Dryas time boundary, to the Denekamp interstadial, and to the Hengelo interstadial. The lower part of the section, below 40,000 B.P., shows two major interstadials above the interval of interglacial type at the bottom of the diagram; it seems very probable now that they correspond to the Odderade and Brørup interstadials. In summary we may say there is ample evidence that the climatic fluctuations of the last 40,000 years in Europe and South America are contemporaneous. It seems most probable from the comparison of the diagrams of Figure 2 that the same holds for the entire Eemian-Weichselian-Holocene sequence.

Figure 12 is a hitherto unpublished diagram of a 200-meter section from the same place (for a less detailed diagram of this section see Van der Hammen 1964b). The uppermost part of this section corresponds to the diagram given in Figure 2, but on a smaller scale. Since we are able to interpret the upper part of Figure 2, we can recognize this sequence of glacials, interglacials, and major interstadials. Knowing this, we have tried to make a correlation with Europe (indicated at the right side of the diagram). Presumptions for this correlation are that the section is continous and that we may deduce from the proved correlation (C^{14}-dated) of the upper part of the section that the other major climatic changes are also contemporaneous. The following facts support the given interpretation. The interval that is to be correlated with the Menapian is indeed a very long, cold period (interrupted only by short minor fluctuations) which had a profound influence on the specific composition of the flora. Immediately after this cold period a good number of new elements appear, some of which immigrated from the north. The long Menapian cold in Europe also had a profound influence on the vegetation, but there it caused the definite disappearance from the area of a number of "Tertiary relicts." Finally, dates calculated by extrapolation on the base of the sedimentation rate of the upper part of the section give figures that correspond reasonably well for correlated intervals.

Correlations like the one suggested in Figure 12 need confirmation. If no material for direct dating (e.g. volcanic ash layers for K/Ar) was available, direct

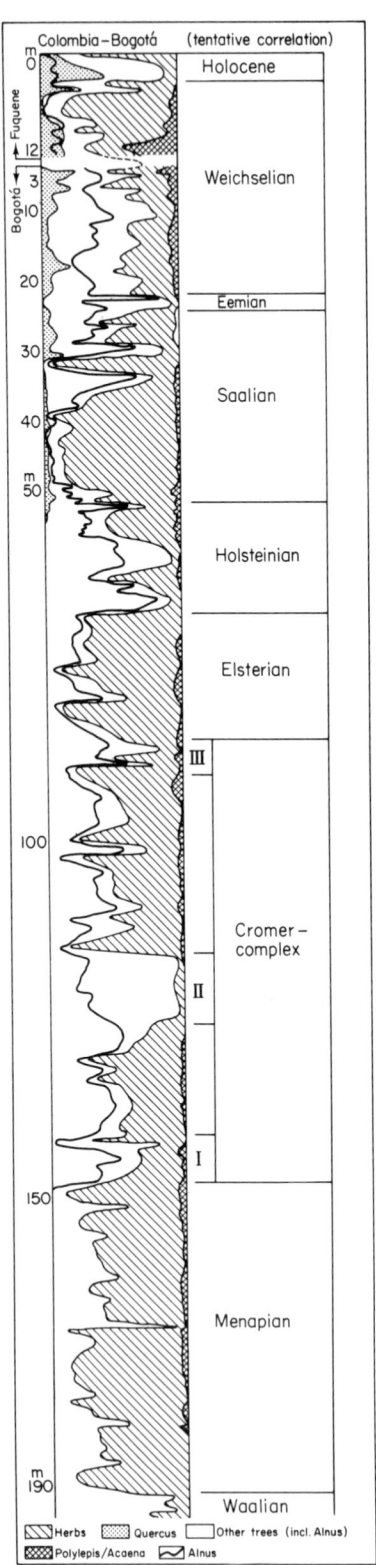

Fig. 12. Pollen diagram representing the interval Waalian (?) to Recent, from Colombia (Bogotá and Fuquene). The correlation with the European chronostratigraphical sequence below the Eemian level is tentative.

dating was impossible until recently. Measurements of paleomagnetic reversals, however, can be applied to sedimentary clays, and this seems to provide us with the necessary tool to confirm or revise palynological correlations of long, continuous sections. If, for instance, the first magnetic reversal were found somewhere between 130 and 140 meters, our correlation would be confirmed as basically right. We hope to make paleomagnetic studies of parallel cores of this section in the near future. By that time we also hope to have more paleomagnetic data from the northwestern European sequence and also from a deeper new section from Philippi and other places in the Mediterranean area.

Final Remarks

We have seen that the total picture of the history of vegetation in Europe north of the Alps has been built up from many isolated pieces representing minor parts of the time-stratigraphical sequence. Continuous sequences of considerable time length from the Mediterranean area have completed the European picture for the last half-million years and we expect soon to be able to complete it for the last few millions of years. It seems probable that more climatic changes comparable to the later Quaternary glacial-interglacial cycle (but possibly with a minor amplitude) than we know of at this moment may be found in the earlier Quaternary. Climatic changes also play an important role in the pre-Quaternary Late Cenozoic. At least some of the cool periods seem to be of the same time length as the later glacials. The cold periods are also apparently responsible for the successive elimination of now exotic or extinct species both in the Pliocene and the Quaternary.

Recently discovered very long lake sequences and modern dating techniques leave little doubt that within the next few years considerable progress will be made toward completing and refining our knowledge of the Late Cenozoic floral and climatic sequence.

Climatic changes, reflected by the changing flora and vegetation, have proved to be a reliable basis for chronostratigraphy. Comparisons of C^{14}-dated pollen diagrams from the last glacial and Holocene sequences from widely separated areas have been sufficiently encouraging to support the probability of long-distance correlations on the basis of climatic change, as reflected by changing vegetation. However, such correlations should be undertaken only for continuous sequences with at least *some* fix-points determined (e.g. by paleomagnetic reversals or absolute dating).

Summary. This article outlines our present knowledge of the history of the Late Cenozoic flora, vegetation, and climate of Europe. Frenzel (1967, 1968) provides a good recent survey of existing literature. It is based principally on the latest results of palynological and stratigraphical studies in northwestern Europe and the Mediterranean area.

Besides the known upper to middle Pleistocene glacials and interglacials, many climatic changes have been found in the middle and lower Pleistocene and also in the upper Tertiary. Successive extinctions of now exotic species, principally of East Asian–eastern North American affinity, are related to cool or cold phases, both in the upper Tertiary and in the Quaternary.

Cold phases in the Pleistocene are represented in northwestern Europe by a polar desert or a tundra. The same phases in the Mediterranean area are represented by the extension of cool steppes. In some places (e.g. the Balkan mountains) a montane forest zone with oak, beech, etc. was present above these steppes at an elevation of about 500 meters. During warm phases both northwestern Europe and the northern Mediterranean area were covered with forest.

The possibilities of long-distance correlation of known vegetational climatic successions in other continents is discussed and illustrated with an example from northern South America.

References

Altehenger, P. A., Klimaschwankungen im Pliozän von Wallensen (Hils), *Eiszeitalter Gegenwart, 9,* 104, 1958.

Andersen, S. T., Vegetation and its environment in Denmark in the Early Weichselian Glacial (Last Glacial), *Danmarks Geol. Undersoegelse II, 75,* 1, 1961.

Andersen, S. T., Interglacialer og interstadialer i Danmarks Kvartaer, *Medd. Dansk. Geol. Foren., 15* (4), 486, 1965.

Andersen, S. T., Interglaziale Pflanzensukzessionen aus Dänemark und ihr Verhältnis zu Umweltfaktoren, in R. Tüxen, *Pflanzensoziologie und Palynologie,* p. 106, The Hague, 1967.

Averdieck, F. R., Die Vegetationsentwicklung des Eem-Interglazials und der Frühwürm-Interstadiale von Odderade, Schleswig-Holstein, in C. O. Gripp, *Frühe Menschheit und Umwelt,* Vol. II, pp. 101–125, Köln-Graz, 1967.

Behre, K. E., The Late-glacial and Early Postglacial history of vegetation and climate in northwestern Germany. *Rev. Palaeobotany Palynology, 4,* 149, 1967.

Bottema, S., A Late Quaternary pollen diagram from Ioannina, northwestern Greece, in E. S. Higgs et al., The climate, environment and industries of Stone Age Greece: Part III. *Proc. Preh. Soc.,* Vol. 33, 1967.

Demek, J., and J. Kukla, *Periglazialzone, Löss und Paläolithikum der Tschechoslowakei.* Tschechosl. Acad. Wissensch., Geogr. Inst., Brno, 1969.

Elhai, H., La flore sporo-pollinique du gisement Villafranchien du Senèze (Massif Central, France), *Pollen Spores, 11,* 1, 127, 1969.

Florschütz, F., On *Pseudolarix kaempferi Gord.* from the clay of Reuver, *Rec. Trav. Bot. Neerl., 22,* 269, 1925.

Florschütz, F., Over Azolla en de ouderdomsbepaling van interglaciale zoetwaterafzettingen in Nederland, *Geol. Mijnb., 14,* 12, 1935.

Florschütz, F., J. Menendez Amor, and T. A. Wijmstra, Palynology of a deep section in southern Spain, in press, 1970.

Frechen, J., and H. J. Lippolt, Kalium-Argon-Daten zum Alter des Laacher Vulkanismus, der Rheinterrassen und der Eiszeiten, *Eiszeitalter Gegenwart, 16,* 5, 1965.

Frenzel, B., *Die Klimaschwankungen des Eiszeitalters,* Brunswick, 1967.

Frenzel, B., *Grundzüge der Pleistozänen Vegetationsgeschichte Nord-Eurasiens,* Wiesbaden, 1968.

Glangeaud, L., Massif Central et bordure Mediterranéenne, Livret-Guide de l'excur-
sion A9, Report, VIII Internat. Congr. Quat., Paris, 1969.

Gricuk, V. P., Comparative study of the interstadial flora of the Russian Plain, Report,
VI Internat. Congr. Quat., Warsaw 1961, 2, 395, 1964.

Grüger, E., Vegetationsgeschichtliche Untersuchungen an Cromerzeitlichen Ablagerungen im nördlichen Randgebiet der deutschen Mittelgebirge, Eiszeitalter Gegenwart, 18, 204, 1969.

Haffer, J., Speciation in Amazonian forest birds, Science, 165, 131, 1969.

Hammen, T. van der, Late-glacial flora and periglacial phenomena in the Netherlands,
Leidse Geol. Mededel., 17, 71, 1951a.

Hammen, T. van der, A contribution to the palaeobotany of the Tiglian, Geol. Mijnb.,
7, 242, 1951b.

Hammen, T. van der, Upper Cretaceous and Tertiary climatic periodicities and their
causes, Ann. N.Y. Acad. Sci., 95 (1), 440, 1961.

Hammen, T. van der, Paläoklima, Stratigraphie und Evolution, Geol. Rundschau, 54,
428, 1964a.

Hammen, T. van der, A pollen diagram from the Quaternary of the Sabana de Bogotá
(Colombia) and its significance for the geology of the northern Andes, Geol.
Mijnb., 43, 113, 1964b.

Hammen, T. van der, and E. Gonzalez, Upper Pleistocene and Holocene climate and
vegetation of the Sabana de Bogotá (Colombia, South America), Leidse Geol.
Mededel., 25, 261, 1960.

Hammen, T. van der, and J. C. Vogel, The Susacá interstadial and the subdivision of
the late-glacial, Geol. Mijnb., 45, 33, 1966.

Hammen, T. van der, G. C. Maarleveld, J. C. Vogel, and W. H. Zagwijn, Stratigraphy,
climatic succession and radiocarbon dating of the last glacial in the Netherlands,
Geol. Mijnb. 45, 79, 1967.

Iversen, Johs., Plantevaekst, Dyreliv og klima i det senglaciale Danmark, Geol. Foren.
Stockholm Forh., Stockholm, 69, 1947.

Iversen, Johs., The bearing of glacial and interglacial epochs on the formation and extinction of plant taxa. Uppsala Univ. Arsskrift, p. 210, 1958.

Leroi Gourhan, A., Les analyses polliniques sur les sédiments des grottes, Bull. Assoc.
Franc. Étude Quaternaire, 3 (2), 145, 1965.

Leroi Gourhan, A., L'Abri du facteur à Tursac, Gallia Prehistoir, 2, 123, 1968.

Lona, F., Contributi alla storia della vegetatione e del clima nella val Padana: Analisi
pollinica del giacimento Villafranchiano di Leffe (Bergamo), Atti Soc. Sci. Nat.,
89, 123, 1950.

Lona, F., and M. Follieri, Successione pollinica della serie superiore (Günz-Mindel) di
Leffe (Bergamo), Veröff. Geobot. Inst. Rübel Zürich, 34, 86, 1957.

Lumley, H. de, Les civilisations préhistoriques en France; correlations avec la chronologie quaternaire, Etudes francaises sur le Quaternaire, 1969.

Montfrans, H. M. van, and J. Hospers, A preliminary report on the stratigraphical position of the Matuyama-Brunhes geomagnetic field reversal in the Quaternary sediments of the Netherlands, Geol. Mijnb., 48, 565, 1969.

Reid, C., and E. M. Reid, The fossil flora of Tegelen-sur-Meuse, near Venlo, in the Province of Limburg, Verhandel. Koninkl. Ned. Akad. Wetenschap., Afdel. (2e sectie),
Natuurk., Sec. II, 13, 6, 1907.

Reid, C., and E. M. Reid, A further investigation of the Pliocene flora of Tegelen, *Verslag Koninkl. Akad. Wetenschap.*, *19*, 192, 1910.

Reid, C., and E. M. Reid, The Pliocene floras of the Dutch-Prussian border. *Meded. Rijksopsp. v. Delfstoffen*, Vol. 6, 1915.

Remy, H., Zur Flora und Fauna der Villafranca-Schichten von Villaroya, Prov. Logrono, Spanien, *Eiszeitalter Gegenwart*, *9*, 83, 1958.

Sercelj, A., Quartäre Vegetationsgeschichte Jugoslaviens auf Palynologischer Grundlage, in R. Tüxen, *Pflanzensoziologie und Palynologie*, p. 87, The Hague, 1967.

Vlerk, I. M. van der, and F. Florschütz, The Palaeontological base of the subdivision of the Pleistocene in the Netherlands. *Verhandel. Koninkl. Ned. Akad. Wetenschap., Afdel. Natuurk, Sec. II, 20* (2), 1, 1953.

Vogel, J. C., and T. van der Hammen, The Denekamp and Paudorf interstadials, *Geol. Mijnb.*, *46*, 187, 1967.

Walter, H., Die Vegetation der Erde in öko-physiologischer Betrachtung. Vol. II: *Die gemässigten und arktischen Zonen*, Jena, 1968.

Wang, Chi-Wu, The forests of China, Maria Moors Cabot Foundation, vol. 5, Harvard University, 1961.

West, R. G., and D. G. Wilson, Cromer Forest Bed series, *Nature*, *209*, 497, 1966.

Wijmstra, T. A., Palynology of the first 30 metres of a 120 m deep section in northern Greece, *Acta Botan. Neerl.*, *18* (4), 511, 1969.

Zagwijn, W. H., Vegetation, climate and time-correlations in the Early Pleistocene of Europe. *Geol. Mijnb.*, n.s., *19*, 233, 1957.

Zagwijn, W. H., Zur stratigraphischen und pollenanalytischen Gliederung der pliozänen Ablagerungen im Roertal-Graben und Venloer-Graben der Niederlande, *Fortschr. Geol. Reinland. Westfalen*, *4*, 5, 1959.

Zagwijn, W. H., Aspects of the Pliocene and Early Pleistocene vegetation in the Netherlands, *Mededel. Geol. Sticht., Ser. C, 3*, 1, 5, 1, 1960.

Zagwijn, W. H., Vegetation, climate and radiocarbon datings in the Late Pleistocene of the Netherlands. Part I: Eemian and Early Weichselian, *Mededel. Geol. Sticht.*, n.s., *14*, 15, 1961.

Zagwijn, W. H. Pollenanalytic investigations in the Tiglian of the Netherlands, *Mededel. Geol. Sticht.*, n.s., *1*, *16*, 49, 1963a.

Zagwijn, W. H., Pleistocene stratigraphy in the Netherlands, based on changes in vegetation and climate, *Verh. Kon. Ned. Geol. Mijnb. Genootsch., Geol. serie, 21* (2), 173, 1963b.

Zagwijn, W. H., Ecologic interpretation of a pollen diagram from Neogene beds in the Netherlands, *Rev. Palaeobotany Palynology*, *2*, 173, 1967.

Zagwijn, W. H., and J. I. S. Zonneveld, The Interglacial of Westerhoven, *Geol. Mijnb.*, n.s., *18*, 37, 1956.

Zagwijn, W. H., H. M. van Montfrans, and J. G. Zandstra, Subdivision of the "Cromerian" in the Netherlands pollen analysis, paleomagnetism, sedimentary petrology, *Geol. Mijnb.*, in press, 1971.

H. E. Wright, Jr.

16. LATE QUATERNARY VEGETATIONAL
HISTORY OF NORTH AMERICA

The Wisconsin ice sheet eliminated the forest cover of Canada and the northern part of the United States. To what extent was the vegetation of unglaciated areas affected by the associated climatic changes? What was the progress of the revegetation of the deglaciated terrain during the retreat, especially at times when the ice margin temporarily readvanced? What have been the subsequent shifts in vegetation throughout postglacial time? What vegetational changes result from climatic change?

These have been the central questions for recent pollen investigations in North America. Contrasts as well as similarities with the well-known northwest European vegetational history are immediately apparent, and they provide valuable insight into regional climatic history and biogeography, as well as into the effectiveness of pollen analysis as a paleoecological technique. Because some of the problems in North America are different from those in Europe, several of the usual procedures of historical pollen analysis have been modified or amplified; examples follow:

1. European vegetational history was first worked out principally from examination of exposed section in peat bogs, but the absence of extensive peat workings in the United States led to the investigation of lake sediments instead. The homogeneity of many organic lake sediments, in comparison with peat, makes it possible to estimate the concentration of pollen grains per unit volume of sediment, rather than simply the percentage of different types (Davis 1967a, 1969a). When combined with close-interval radiocarbon dating, the concentration can be converted to absolute pollen influx (API), expressed as number of grains per unit area per year. Recovery of undisturbed cores of annually laminated organic sediments from deep-water lakes provides a time scale of even greater accuracy (Craig 1970). The API technique has made it possible to determine, for example, that the high percentages of spruce pollen in the late-glacial herb pollen zone result from long-distance pollen transport in a tundra environment of low local pollen production, rather than the local occurrence of spruce trees, and one of the perplexing problems in the reconstruction of late-glacial vegetation is thus essen-

tially solved. The technique might be profitably applied to any problems where differential pollen production may result in serious discrepancies between percentages of pollen in the sediment and percentages of plants in the vegetation of the area—for example, the problem of the areal extent of Neolithic forest clearance in western Europe.

The utility of lake sediments for pollen analysis has led to the development and perfection of coring devices that permit the recovery of nearly continuous cores of many different types of sediment, even in deep lakes (Wright et al. 1965). Wide-diameter corers have been designed to provide material for radiocarbon dating. Large cores are also useful for stratigraphic plant-macrofossil analysis—a complementary technique that actually preceded pollen analysis in northern Europe because of the availability of exposed peat sections that could be easily sampled in bulk. Macrofossil analysis of lake sediments in some of the American investigations has been quantified and treated in the same manner as pollen analysis, so that the fossil pollen and seed flora combined from a site may number well over a hundred taxa and thus may constitute an appreciable percentage of the total flora of the area (Watts and Winter 1966). The rich information about the aquatic and wet-ground plants supplied by such analyses provides an additional dimension to paleoecological reconstruction, for the succession of these plants may reflect the changing limnology and hydrology of the lake and its basin.

2. Because the vegetation of the forested portions of North America has been relatively little disturbed by man, compared to that of most of Europe, it is possible to determine what type of pollen rain is produced by known vegetation of reasonably natural areas, and then to use this information to control the vegetational reconstructions made from pollen diagrams of premodern stratigraphic sections. For this purpose the pollen content of samples of surface sediment, moss polsters, or even surface soil has been determined on many transects across both forested and treeless regions in North America (Davis 1967b; Wright 1968a; Lichti-Federovich and Ritchie 1968; McAndrews and Wright 1970).

Surface-sample studies have also been applied to local areas in order to determine the nature of pollen dissemination from areas of mapped vegetation of different types, and to search for the pollen record of relatively minor taxa that are diagnostic of particular vegetation types. The local, extralocal, and regional components of the pollen rain can thus be related to the known local, extralocal, and regional vegetation, and this information can be subsequently applied to the interpretation of the vegetational succession from the pollen stratigraphy at a particular site (Janssen 1966, 1967).

The search for modern analogs for past vegetation in North America has been only partly successful, and it is now realized that some past vegetation types may have no modern analog, even though the range of climatic and other environmental conditions may be superficially similar. The best example of this situation concerns the late-glacial spruce pollen zone for interior North America from Saskatchewan to Ohio (Wright 1968a). The spruce zone contains no appreciable amount of pine pollen, despite the fact that today the boreal spruce forest from

Labrador to Alberta contains much jack pine, whose pollen shows up in surface samples in appreciable quantity. Furthermore, the spruce zone contains modest pollen percentages of oak and other temperate deciduous trees, unlike the surface samples from the modern spruce forest. So the late-glacial spruce pollen zone must represent a vegetation that no longer exists to any extent. The basic cause may in fact be a late-glacial climate different from that in the boreal forest region today, or the situation may represent variable rates of migration of major tree types following an epoch of pronounced climatic change (Wright 1968a).

The problem of modern analogs has also risen in the interpretation of post-glacial diagrams for southeastern United States (Watts 1970), and it must now be realized that forest vegetation is subject to modification and reconstitution to assemblages that may lack certain expectable major constituents, or may contain others that are not expectable on the basis of their distribution and association today. The situation makes it particularly important that the ecological tolerances and ranges of major forest trees be worked out, so that their occurrence in the past may reveal more explicitly the nature of past conditions.

The search for modern analogs for past vegetation has never been an integral part of European studies, largely because it has been assumed that the pollen yield from unnatural vegetation would only confuse the interpretation of pollen diagrams. Studies now are being made, however, of the details of pollen dispersal from known vegetation (Tauber 1965), and at least one recent project has involved extensive analyses of surface samples in a local area (Birks 1969).

3. Even though the forest vegetation of North America has not been vastly disturbed, compared to that of northwestern Europe, one must admit that extensive lumbering and forest clearance has taken place in certain areas. But because this disturbance dates from only one or two centuries ago, presumably under climatic conditions similar to those of the present, the effect on the pollen rain can be accurately determined. In contrast, the effective disturbance of the European forest vegetation started with the Neolithic forest clearance more than 5000 years ago, and the vegetation that prevailed before that time may have reflected a climate different from today's climate in that area.

The technique for studying the extent of vegetational disturbance brought about by the white man in North America was first systematically used by McAndrews (1966), who prepared pollen diagrams for short cores of lake sediment in a transect across several vegetational belts. The time of first disturbance was easily detected in all the cores by the abrupt rise in pollen percentages of *Ambrosia* (ragweed) and Chenopodiineae, both of which represent common agricultural weeds. Concomitantly the percentages of pine decreased slightly and the values of birch increased, as a result of lumbering and the spread of secondary growth of birch over cleared areas. The changes in the tree pollen curves are relatively small, however, compared to the earlier changes that reflected natural modifications in the vegetation probably in response to climatic change.

This paper attempts to summarize the present status of knowledge concerning the late Quaternary vegetational history of North America. Most of the review is concerned with late Wisconsin and post-Wisconsin vegetation history in north-central and eastern United States, as relatively little is known at present about vegetational changes in western North America, or about the earlier time intervals anywhere in North America. The review is not intended to be exhaustive or bibliographic, for part of the subject has been summarized elsewhere (Ogden 1965; Davis 1965; Whitehead 1965; Cushing 1965; Martin and Mehringer 1965; Heusser 1965). The main emphasis here will be on the results of several more recent investigations representing some of the new approaches outlined above. Several of these studies have not been fully published, and I am obliged to various authors for providing information as well as for discussion of several problems. John Birks, W. A. Watts, and E. J. Cushing kindly read a draft of the manuscript.

Modern Vegetation

The major vegetational regions of North America (Fig. 1) in general appear to reflect the modern climatic patterns, but the local flora is the product of a long history of plant migrations in response to climatic and edaphic change. A very gross subdivision of the vegetation in Canada recognizes the tundra, forest/tundra transition, and boreal conifer forest (Rowe 1959). The boreal forest is characterized by *Picea glauca* (white spruce), *P. mariana* (black spruce, confined to bogs in the south), *Pinus banksiana* (jack pine), and *Abies balsamea* (balsam fir), with *Betula papyrifera* (white birch) and *Populus tremuloides* (aspen) especially on areas subject to cutting or burning.

Through the Great Lakes area there is a transitional conifer/hardwood forest containing *Pinus strobus* (white pine), *P. resinosa* (red pine), and (except in Minnesota) *Tsuga canadensis* (hemlock), along with the species of the boreal forest to the north and species of the hardwood forest to the south. Many of these trees (or closely related species) extend down the Appalachian Mountains, which reach elevations of 2000 meters in North Carolina.

The hardwood forest changes its composition from west to east. *Quercus* spp. (oak) are present throughout, but *Carya* spp. (hickory) are more common in the west, whereas *Fagus grandifolia* (beech) is more common to the east. On richer soils, *Ulmus americana* (American elm), *Tilia americana* (basswood), *Acer saccharum* (sugar maple), and *Betula lutea* (yellow birch) usually are dominant. *Castanea dentata* (chestnut) was important in the east before it was almost completely destroyed by disease a few decades ago.

In the lowlands of southeastern United States, the tree flora is tremendously rich. *Quercus* (>40 spp.) and *Pinus* (>10 spp.) are common almost everywhere, but *Liriodendron tulipifera* (tulip tree), *Magnolia* spp., *Nyssa sylvatica* (black gum), *Acer rubrum* (red maple), *Carya* (>10 spp.), *Carpinus caroliniana* (hornbeam), *Liquidambar styraciflua* (sweet gum), *Taxodium distichum* (bald cypress), and a host of other trees are common in many areas.

The deciduous forest is bordered on the west by prairie. In the transition between the two, a mosaic of prairie openings in the forest occurs, especially in the

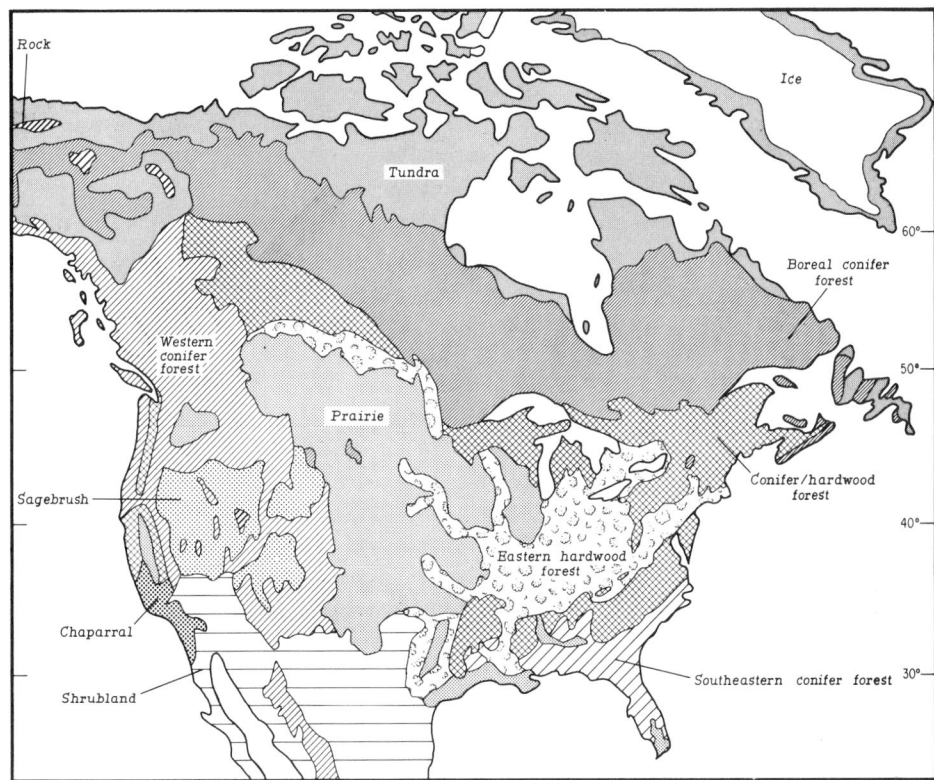

Fig. 1. Vegetation map of United States.

so-called prairie peninsula, which extends eastward from Iowa to Indiana. A *Quercus* savanna is also present in the transitional region—at least it was before the intensive agriculture and fire protection of the last hundred years largely destroyed it.

The prairie, dominated by Gramineae, extends westward to the Rocky Mountains. Westward the prairie plants become shorter and less densely grown, as the precipitation decreases, and *Artemisia* (sage) becomes more common (McAndrews and Wright 1970). Many of the tree species that do occur in the prairie are confined to stream valleys or the north sides of hills. The Black Hills, which are now dominated by *Pinus ponderosa*, a Rocky Mountains species, also contain outlying stands of *Picea glauca* and *Betula glandulosa* (dwarf birch), which are disjunct from the boreal forest far to the north and are probably relics of a more extensive Pleistocene distribution.

In the Rocky Mountains the vegetation is dominated by conifers, which are arranged in a clear altitudinal zonation. But the species involved in the zonation change from one major mountain range to the next, partly as a result of slightly different climatic and edaphic relations, and partly as a result of the complexities of migration during the Pleistocene. In the southern Rocky Mountains of Colorado and New Mexico, as well as in the Colorado plateaus to the west, the lowest tree zone consists of a savanna or woodland of *Pinus edulis* (pinyon) and *Juniperus* spp., with sagebrush interspersed. This grades upward generally to a *Pinus ponderosa* zone, with *Quercus gambelii* common in the lower part and *Pseudotsuga menziesii* (Douglas-fir), *Pinus flexilis* (limber pine), and *Populus tremuloides* (aspen) in the upper part. At still higher elevations comes the spruce/fir forest, dominated by *Picea pungens* (blue spruce), *P. engelmanni* (Engelmann spruce), *Abies lasiocarpa* (alpine fir), and *A. concolor* (white fir), along with *Pinus albicaulis* (whitebark pine).

In the Middle Rocky Mountains of Wyoming and even in much of Colorado this conifer zonation is not followed in detail, because some tree species are inexplicably absent and others are present. For example, on Grand Mesa in west-central Colorado, ponderosa pine does not occur, and the mountain slopes above the pinyon/juniper woodland at the base are covered by oak and other large shrubs, with Douglas-fir in some north-facing slopes below the spruce/fir zone, especially where fire has been common.

In the Northern Rocky Mountains other conifer species occur, especially in the western part of the area. *Larix occidentalis* (western larch) occurs in the ponderosa zone, and *Thuja plicata* (western red cedar), *Tsuga heterophylla* (western hemlock), and *Pinus monticola* (western white pine) are common associates of Douglas-fir.

The mountains and coastal region of California have an entirely different and highly varied group of conifers, including the redwoods and many species of pine, but the coastal ranges of the Pacific Northwest are more closely related in their tree flora to the Northern Rocky Mountains. *Picea sitchensis* (sitka spruce) is confined to coastal slopes in British Columbia and Alaska, where it occurs with *Tsuga mertensiana* (mountain hemlock), *Chamaecyparis nootkatensis* (Alaska yellow cedar), and *Abies amabilis* (amabilis fir).

On the drier sides of the Canadian cordillera, in Alberta, the western tree flora is reduced, but it is supplemented by elements of the boreal forest of the Canadian interior. In Alberta *Picea engelmanni* of the mountains hybridizes with *P. glauca* of the plains, and *Pinus contorta* with *P. banksiana*. Northward along the piedmont the pines drop out in southwestern Yukon, and the forested parts of interior Alaska have only *Picea glauca* and *P. mariana*, along with *Betula papyrifera* (white birch), *Populus* spp., and shrubs like alder and willow.

CENTRAL AND EASTERN NORTH AMERICA

Pre-Wisconsin History

Detailed pollen diagrams for early and middle Quaternary sites are so few that little can be said about the vegetational history before the time of Wisconsin glaciation. The Don beds near Toronto, presumably of Sangamon age, indicate temperate forest with *Liquidambar* and other trees that do not now reach so far north (Terasmae 1960). Pollen diagrams for two nearby sections or organic sediments beneath Wisconsin till in eastern Indiana (Fig. 2) illustrate an interglacial cycle starting with a conifer zone, proceeding to a hardwood zone dominated by *Quercus* (oak) and *Carya* (hickory), and terminating with a conifer zone (Kapp and Gooding 1964).

Of special interest is a diagram from near Vandalia in southern Illinois for the sediments of a still existent lake on Illinoian drift (E. Grüger 1971). The sequence starts with a conifer zone, as the manifestation of a late-glacial boreal forest terminating the Illinoian glaciation. Then follows a hardwood zone that indicates a rich south-temperate deciduous forest for Sangamon time. Parts of this zone are characterized by maxima of Taxodiaceae/Cupressaceae that probably indicate the development of *Taxodium* (bald cypress) around the lake.

In central Florida and southern Georgia (Watts 1969, 1971) pollen profiles have been studied that are believed to date to the Sangamon interglacial (Fig. 3). They show a vegetational development similar to that of the Holocene sediments in the upper part of the profiles.

Early Wisconsin

The vegetation history from the end of the Sangamon interglacial until the time of the maximum Wisconsin glaciation is recorded in only two areas in central and eastern North America. In southwestern Ontario are intertill peats and other or-

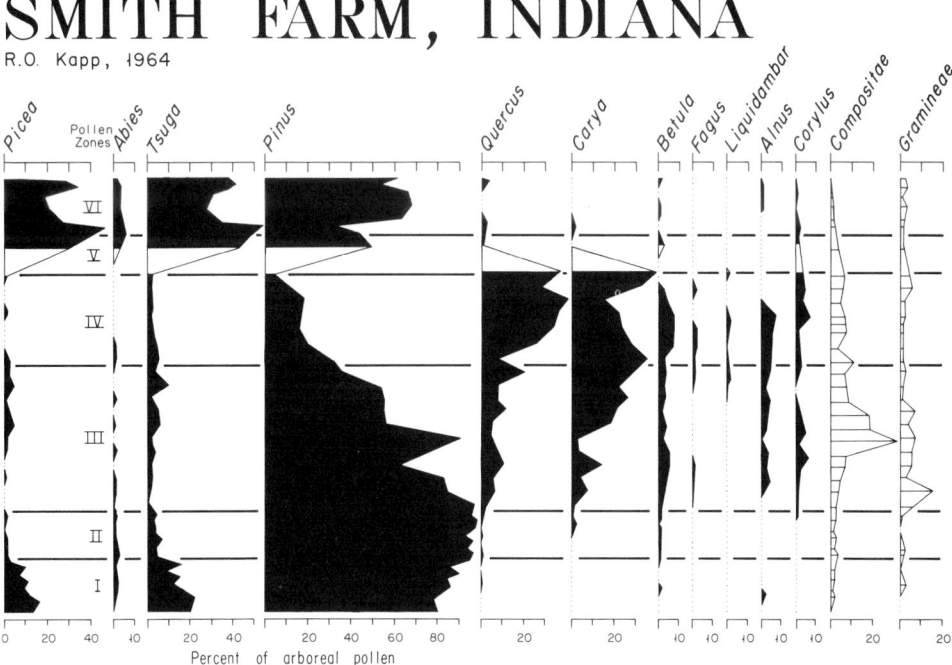

Fig. 2. Pollen diagram from Sangamon deposits at Smith Farm site, western Indiana (redrawn from Kapp and Gooding 1964).

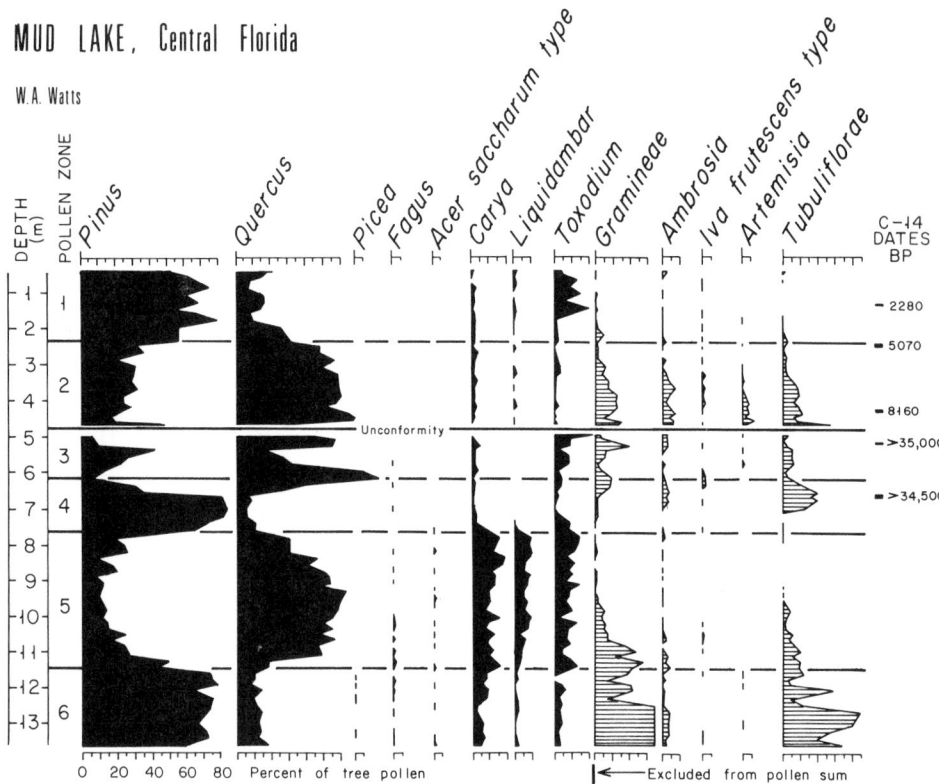

Fig. 3. Pollen diagram from Mud Lake, central Florida (redrawn from Watts 1969).

ganic sediments assigned to the Port Talbot and Plum Point interstadials of early Wisconsin time, with radiocarbon dates of 48,000 to 24,000 years ago. Pollen counts indicate that the vegetation was dominated by *Picea* and *Pinus*, whereas today the vegetation is deciduous forest (Dreimanis et al. 1966).

In southern Illinois at the Vandalia site on Illinoian drift, the pollen sequence continues upward from the Sangamon, above a discontinuity of unknown duration (E. Grüger 1971). The rich south-temperate deciduous forest of Sangamon time changed to a predominantly treeless vegetation, but with a few oak trees. The pollen assemblage for this interval resembles more that of prairie than of tundra, so the climate was probably dry rather than simply cold. It should probably be correlated with the early Wisconsin, at which time the ice sheet existed in northeastern Illinois, some 40,000 years ago (Kempton and Hackett 1968).

The pollen diagram for the Vandalia site proceeds upward from this herb assemblage to a pine zone, with increasing percentages of spruce. Radiocarbon dates from other sites with similar pollen assemblage suggest assignment of this zone to the Farmdalian interstadial (28,000–22,000 years ago). Pollen analysis of buried Farmdalian organic silts farther north in Illinois shows higher values of spruce (E. J. Cushing, unpublished). The species of pine involved has not been determined, but it is probably *Pinus banksiana,* which is now associated with spruce throughout the modern boreal forest except in the northernmost part next to the tundra.

The general sequence of events for the early Wisconsin in the Great Lakes region may be summarized as follows. During the long period of cooling that marked the advance of the Laurentide ice sheet (70,000–20,000 years ago), the deciduous trees of the Sangamon interglacial in the Great Lakes area were eventually replaced by conifer forest. As the climate became even colder and drier during the maximum of Wisconsin glaciation, pine was eliminated from the Midwest forest, although it survived in the Appalachians and was even dominant in the southern Appalachians. Although ice margin fluctuations are recognized for the early Wisconsin in both southwestern Ontario and northern Illinois, the climatic changes do not appear to have been sufficiently great to alter the composition of the conifer forest significantly. The trend in southern Illinois implies gradual replacement of pine by spruce as the Lake Michigan lobe slowly advanced and finally reached its maximum about 20,000 years ago.

Main Wisconsin

Much of the terrain of Europe north of the Alps during the last glacial maximum supported a tundra or, close to the ice, even a cold rock desert. Treeless conditions even dominated in northern Italy and Greece (Frank 1969; Wijmstra 1969; Van der Hammen et al., this volume). In America, however, the few full-glacial records available indicate that much of the area south of the ice sheet was covered with boreal forest rather than tundra. Two factors may have been important in causing this difference. First, the area is much farther south—the ice limit in Illinois is 39°, whereas in Germany it is 52°. Second, the Alps with their large ice cap were in just the right position to reinforce the semipermanent area of atmospheric high pressure that spread out from the Scandinavian ice sheet, so that most of the westerly flow of warm air that might have entered Europe was diverted south of the Alps. No such southern mountain mass existed in North America; the north–south orientation of the Appalachians provides no real barrier to air movement and has relatively little effect on the climate of adjacent areas.

Full-glacial vegetational records are sparse in central and eastern United States, partly because lakes and bogs are absent or inconspicuous south of the glacial border. Martin's (1958) claim for full-glacial tundra (C^{14} date of 13,500 years ago) in northeastern Pennsylvania is weakened by his postulation that much of the arboreal pollen was deposited secondarily and did not represent the contemporaneous vegetation. More substantial evidence for tundra comes from a marsh site in the Allegheny Plateau of western Maryland at an elevation of about 810 meters: the high percentages of herbs and the low values for pollen influx (as determined from radiocarbon dates of 18,500 to 12,500 years ago) imply a treeless vegetation, at least at higher elevations (J. A. Maxwell, personal communication). Farther south in the Appalachian Mountains and the Atlantic

Coastal Plain there are several sites covering the full-glacial time range of the Wisconsin (ca. 23,000–14,000 years ago). For example, in the Atlantic Coastal Plain of North Carolina, the pollen assemblage for sediments dated as about 35,000 to 12,000 years ago is dominated by *Pinus* and *Picea* (Whitehead 1967). Pollen diagrams for the Shenandoah Valley in Virginia (Fig. 4) indicate a similar assemblage, for the time range 20,000 to 9,500 years ago (Craig 1969). In the piedmont of northern Georgia both pollen and plant macrofossil analyses of sediments 23,000 to 13,500 years old point to a forest of *Pinus banksiana* (jack pine), with enough *Picea mariana* (black spruce) to yield 5 percent spruce pollen (Fig. 5), and with an aquatic and semiaquatic macroflora of boreal character, like that found today in northern New England and the Great Lakes region (Watts 1970). Inasmuch as the distance today from the boreal forest (except for the montane forests) is 1200 km, a considerable climatic change is implied. Just where the temperate deciduous forests were located at this time is difficult to say, be-

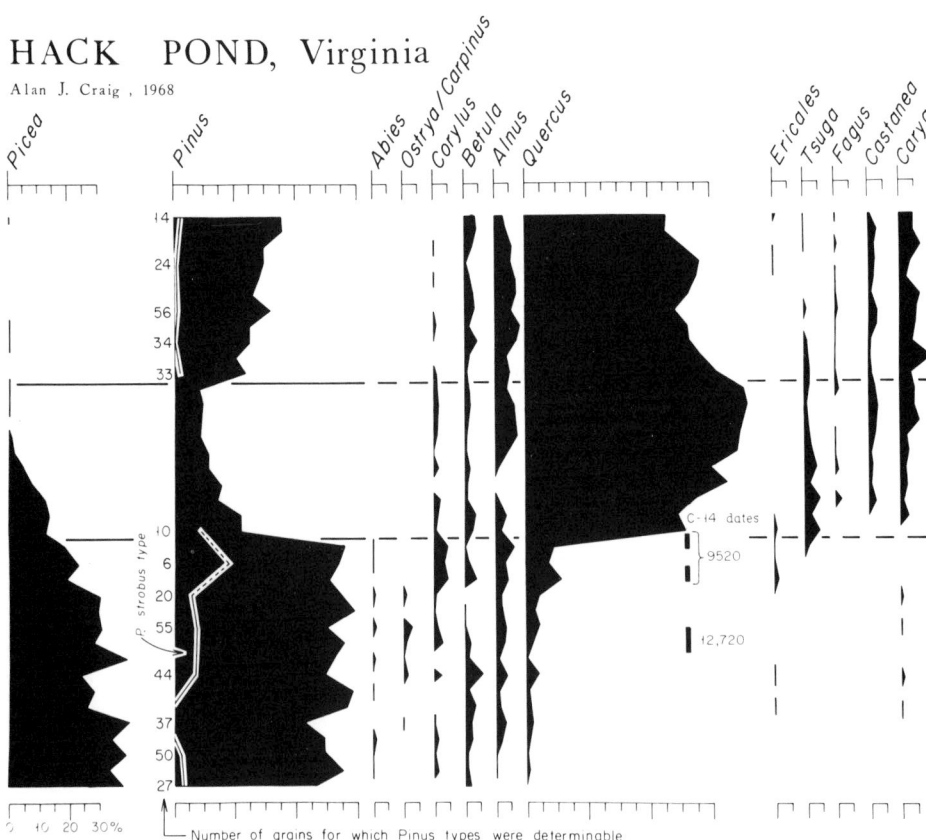

Fig. 4. Pollen diagram for Hack Pond, Shenandoah Valley, Virginia (redrawn from Craig 1969).

QUICKSAND POND
BARTOW COUNTY, GEORGIA
W. A. Watts, 1967

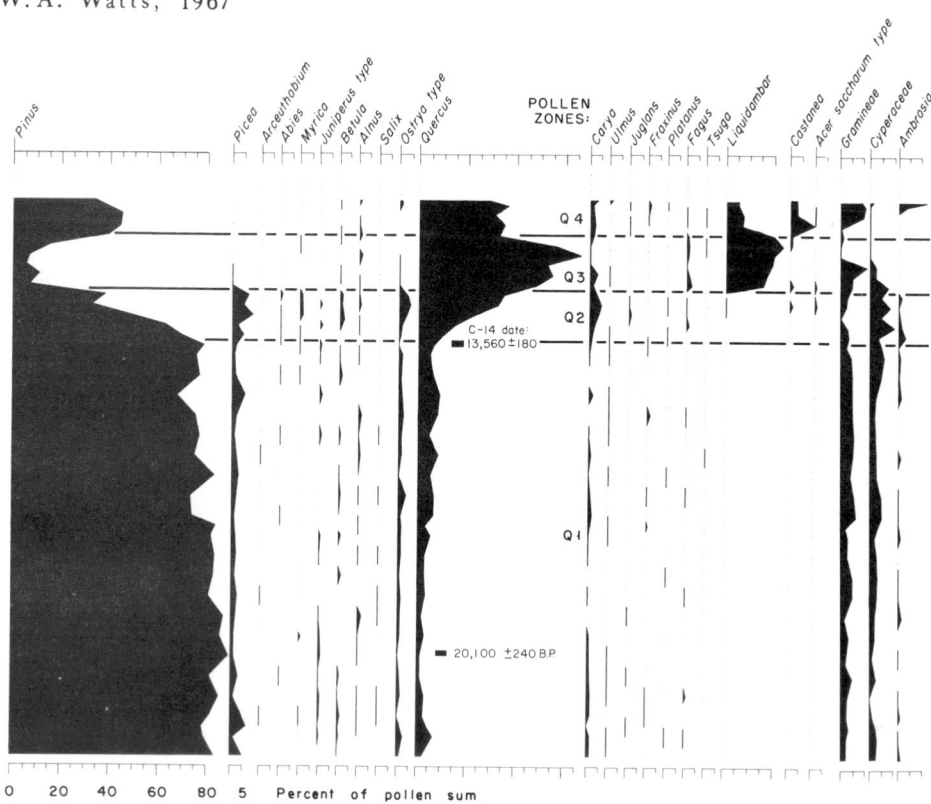

Fig 5. Pollen diagram for Quicksand Pond, northwestern Georgia (redrawn from Watts 1970a).

cause no sites farther south cover the relevant time range. Sites so far studied in Florida (Fig. 3) and southern Georgia contain an unconformity that probably represents the entire Wisconsin interval (Watts 1969, 1971). Apparently the lakes were dry because the water table in the highly permeable Florida aquifer was greatly lowered as a reflection of sea level depression.

Diagrams for the full glacial in northern Georgia are marked by 10 percent pollen of *Quercus*, *Carpinus* type, and other temperate deciduous trees (Fig. 5), and it seems likely that the boreal forest did not completely overwhelm the deciduous forest during the full glacial, but rather that patches of temperate trees remained on favorable sites. In any case it seems certain for the Appalachian

Highlands and the Atlantic Coastal Plain that the full-glacial vegetation was vastly different from today's during the time of maximum Wisconsin glaciation, and that the view of Braun (1950) to the contrary is now untenable. Taxonomic evidence recently cited by Iltis (1965) in defense of the Braun hypothesis must be explained in other ways—at least as far as it applies to the vegetation (as distinct from the flora).

In the Middle West the case is much the same. The pollen spectra for full-glacial time in southern Illinois indicate a spruce/pine assemblage, as in the Appalachian Mountains, but with oak values reaching 30 percent (E. Grüger 1971). In northeastern Kansas, about 500 km to the west, new pollen diagrams imply a spruce forest for the interval 24,000 to at least 16,000 years ago (probably 12,000), with no pine and little oak (J. Grüger 1971). Thus unlike the boreal spruce forest of today, which has a fairly uniform composition across Canada from the oceanic east to the continental west, the boreal forest of full-glacial time showed differences. In the Appalachian Mountains it was dominated by *Picea* and *Pinus banksiana*, with less spruce to the south. Westward in Illinois, pine and spruce both occurred in abundance, although the pine was not determined positively as *P. banksiana*. Still farther west, in Kansas, pine was completely absent, as it was in late-glacial time throughout the Middle West (including Illinois).

The localities in Illinois and Kansas are 300 km south of the southern limit of the Wisconsin ice sheet, and there is no proof yet that tundra did not exist in the intervening area. Reports of permafrost features in central Indiana (Wayne 1967), central Illinois (Frye and Willman 1958), and western Wisconsin (Black 1965) suggest that tundra may have occurred there. The paucity of such features in the United States compared to Europe, however, indicates that permafrost and tundra were probably absent or localized over most of the area at the time of glacial maximum, and a periglacial environment such as is visualized for most of Europe was indeed restricted. It will be seen that tundra did develop farther north on deglaciated landscape in New England and northern Minnesota at the time of ice retreat, but positive evidence for its presence at the time of the ice maximum is limited.

The great expanse of sand dunes in northern Nebraska, dated as full glacial because they grade southeastward into loess of Wisconsin age, implies a treeless landscape—even completely barren of vegetation—but this situation presumably reflects the strength of periglacial winds and abundance of sand rather than frigidity of temperature (Wright 1970). Whether or not the boreal spruce forest extended to the Rocky Mountains at this time can be determined only after the discovery of suitable pollen sites in this very large area.

The southern limit of boreal conifer forest for the full glacial is also unknown. Fossils of subarctic mammals in Wisconsin deposits of southwestern Kansas (Hibbard and Taylor 1960) imply an extension of boreal conditions to the southern Great Plains. The report of *Picea* pollen in undated sediments in east Texas (Potzger and Tharp 1943), however, failed to find strong confirmation in reexamination of the deposit (Graham and Heimsch 1960). Also, a deposit with spruce cones in Louisiana (Brown 1938), presumed to date from Wisconsin time, yielded, on recollection, a radiocarbon date of 7240 years ago (Bender et al. 1967). The occurrence of pine in west Texas and eastern New Mexico during full-glacial time (Hafsten 1964) probably related more to eastward expansion of mountain forests than to southern extension of the boreal forest from the northern plains. The southern limit of boreal conifer forest during a full-glacial time was probably somewhere in south-central United States, perhaps extending westward from Georgia. It thus may have formed a latitudinal belt as broad as it is today—1000 km from Hudson Bay to the Great Lakes. The nature of the full-glacial vegetation belts to the south is almost completely unknown.

Late Wisconsin

For the present purpose, late Wisconsin time is taken to include the fluctuating retreat of the Wisconsin ice sheet until the time marked by the abrupt decimation of the boreal forest, i.e. from about 14,000 to about 11,000 years ago. The record of late-glacial and postglacial vegetation was first studied in the glaciated areas, where lakes and bogs are abundant. A record comparable to that of Europe was anticipated, i.e. a basal herb assemblage representing the late-glacial tundra, followed by various tree pollen zones for the postglacial. The early pollen diagrams of Potzger (e.g. 1946, 1953) throughout the Great Lakes region and the Northeast served to outline grossly the postglacial vegetational sequence. Most of his counts did not include herb pollen, however, so it was not possible to identify the basal tundra zone that was expected on the basis of European experience. Deevey (1951) was so convinced that tundra must have been present that he challenged not only Potzger's nonidentification of herb pollen but his coring techniques, pointing out that the tundra zone should be expected in the mineral sediments beneath richly organic sediments, and that Potzger did not adequately core these mineral sediments (Potzger 1953).

It turns out that many sites lack a mineral-rich herb zone at the base merely because the lake or bog did not come into existence until long after active glacial ice left the area, as a result of the persistence of buried stagnant ice (Florin and Wright 1969). Even where radiocarbon dates show that stagnant ice did not persist, however, a herb zone with tundra-type macrofossils has been identified only in New England and in northeastern Minnesota.

In New England the tundra lasted until about 12,000 years ago (Davis 1969b). Early evidence for tundra was based on the high percentages of nonarboreal pollen (NAP) (Deevey 1951), but the plant macrofossils of tundra plants found at comparable horizons at other sites made the case more convincing (Argus and Davis 1962), as do the counts of absolute pollen influx (API) at Rogers Lake in southern Connecticut (Davis 1969b). The API work indicates that pollen deposition occurred at an extremely slow rate in the herb zone, allowing the influx of tree pollen from distant forests to reach values of 30 percent on the percentage plot (Fig. 6).

Late-glacial tundra landscape also occurred in northeastern Minnesota from the time of local ice retreat to as late as 10,000 years ago. Here the record lies primarily in the high values of NAP (notably Cyperaceae, *Artemisia, Ambrosia,*

ROGERS   LAKE , CONNECTICUT

Margaret B. Davis, 1967

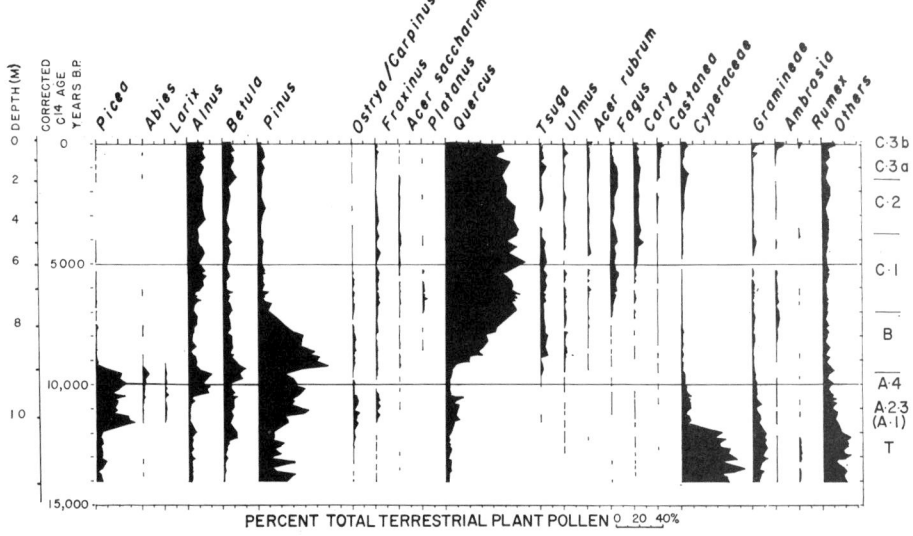

PERCENT TOTAL TERRESTRIAL PLANT POLLEN 0 20 40%

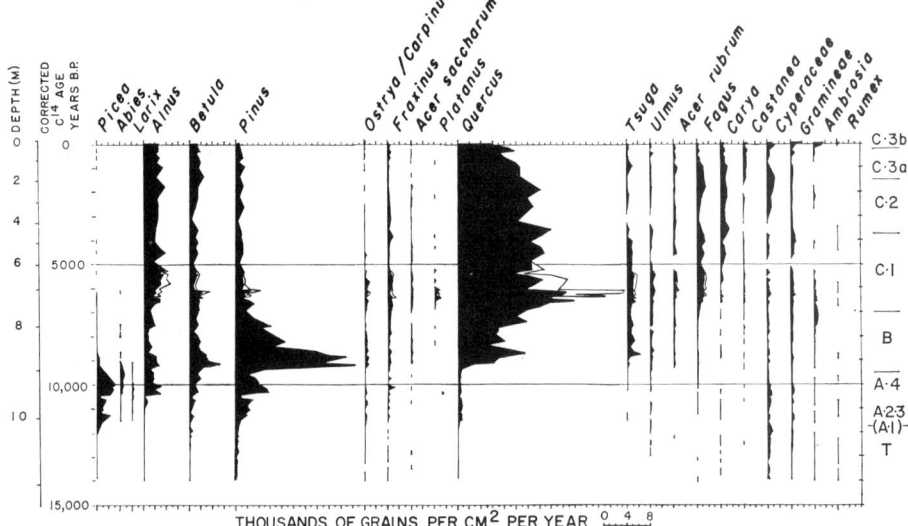

THOUSANDS OF GRAINS PER CM² PER YEAR 0 4 8

Fig. 6. Pollen diagrams showing percentages (above) and absolute pollen influx (below) for Rogers Lake, Connecticut (from Davis 1967a).

Gramineae, and *Salix*) and in macrofossils of *Dryas integrifolia, Vaccinium uligi-nosum* var. *alpinum, Salix herbacea*, and other arctic types (Fries 1962; Baker 1965; Watts 1967). An API diagram for one of the sites indicates low pollen influx at this time (Craig 1970). The pollen sequence at some of these sites suggests that the tundra had a closing phase of *Betula glandulosa* (dwarf birch), prior to the brief development of a boreal *Picea* forest about 10,000 years ago.

Farther south in Minnesota, tundra was apparently absent from the late-gla-cial landscape, and the spruce forest extended to the edge of the ice—in fact it probably covered a broad zone of stagnant ice in the terminal area. Many sections of late-glacial sediments have at their base a layer of terrestrial plant detritus (in-cluding its associated terrestrial diatoms) that represents either the development of superglacial forest duff *in situ* or its inwashing into shallow pools formed as buried ice began to melt out (Florin and Wright 1969).

Two special problems exist in the interpretation of late-glacial pollen assem-blages—the occurrence of the relatively high pollen percentages for thermophilous plants, and the relations between pollen curves and ice margin fluctuations.

Quercus, Ostrya/Carpinus, and other thermophilous deciduous types exceed a total of 20 percent in the late-glacial spruce/oak pollen zone A-2-3 in southern New England (Davis 1969b). In Minnesota, *Fraxinus* pollen reaches 17 percent at one site, and the pollen of other thermophilous deciduous trees totals 7 percent at the same levels. Grains of *Ambrosia* type, *Typha latifolia, Humulus lupulus*, and other thermophilous plants are common. Three explanations are generally offered for the apparently anomalous association of boreal and temperate plants.

1. Redeposition of pollen from reworking of earlier interglacial or interstadial sediments by the last ice sheet; Anderson (1954) favored this explanation for a site in southern Michigan, where the counts of thermophilous tree pollen ex-ceeded 35 percent. Neither Anderson nor anyone else, however, has discovered such pollen grains in the tills that provided the inorganic sediments to the lake. In fact, in the only careful investigation to date in America on this subject, Cush-ing (1964) identified conifer grains and other microfossils derived from Creta-ceous rock fragments in the local till but no thermophilous Pleistocene types.

2. Long-distance transport from stands of thermophilous plants. This expla-nation probably applies to all occurrences of thermophilous pollen types in the herb pollen zone, for the local pollen production is low for these horizons. It might also apply for modest occurrences of a few percent in the spruce zone. For example, grains of *Ephedra* and *Sarcobatus*, plants that now grow only in the southwestern and western arid regions, must certainly have blown in to midwest-ern sites from afar (Maher 1964). As far as *Quercus, Ulmus, Ostrya* type, and *Corylus* are concerned, their full-glacial or late-glacial refugia are not known, so one cannot estimate the distance of long-distance transport. Pollen analyses from northeastern Kansas (J. Grüger 1971) show a spruce pollen zone with lower rather than higher percentages of these thermophilous types than in the Minne-sota sites, implying that no major refuge for them existed to the south and that the explanation of long transport from that direction may not be valid (Fig. 7).

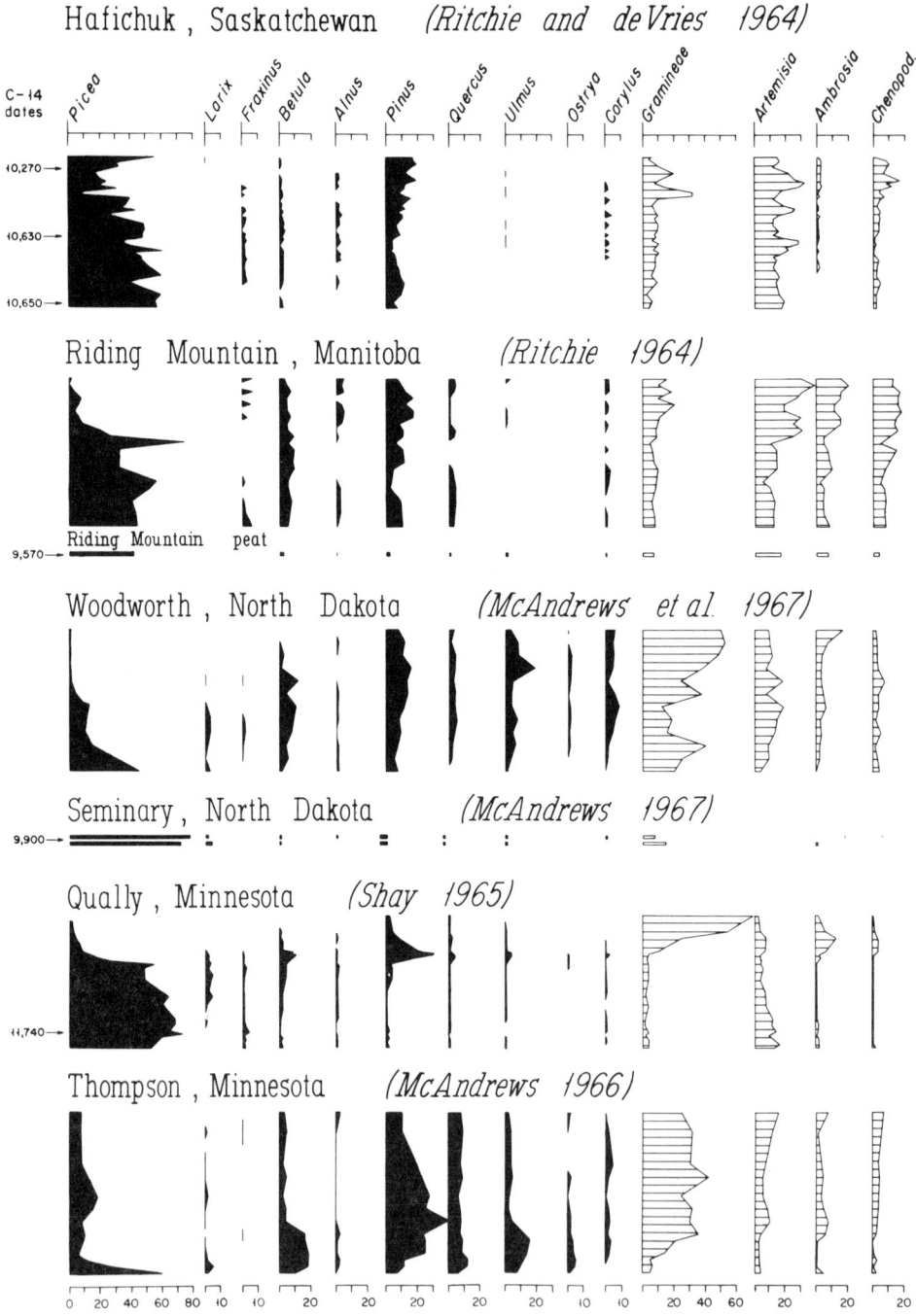

Hafichuk, Saskatchewan *(Ritchie and de Vries 1964)*

Riding Mountain, Manitoba *(Ritchie 1964)*

Woodworth, North Dakota *(McAndrews et al. 1967)*

Seminary, North Dakota *(McAndrews 1967)*

Qually, Minnesota *(Shay 1965)*

Thompson, Minnesota *(McAndrews 1966)*

0 20 40 60 80 10 10 20 20 20 20 20 20 10 10 20 40 60 20 20 20

Percent of total pollen

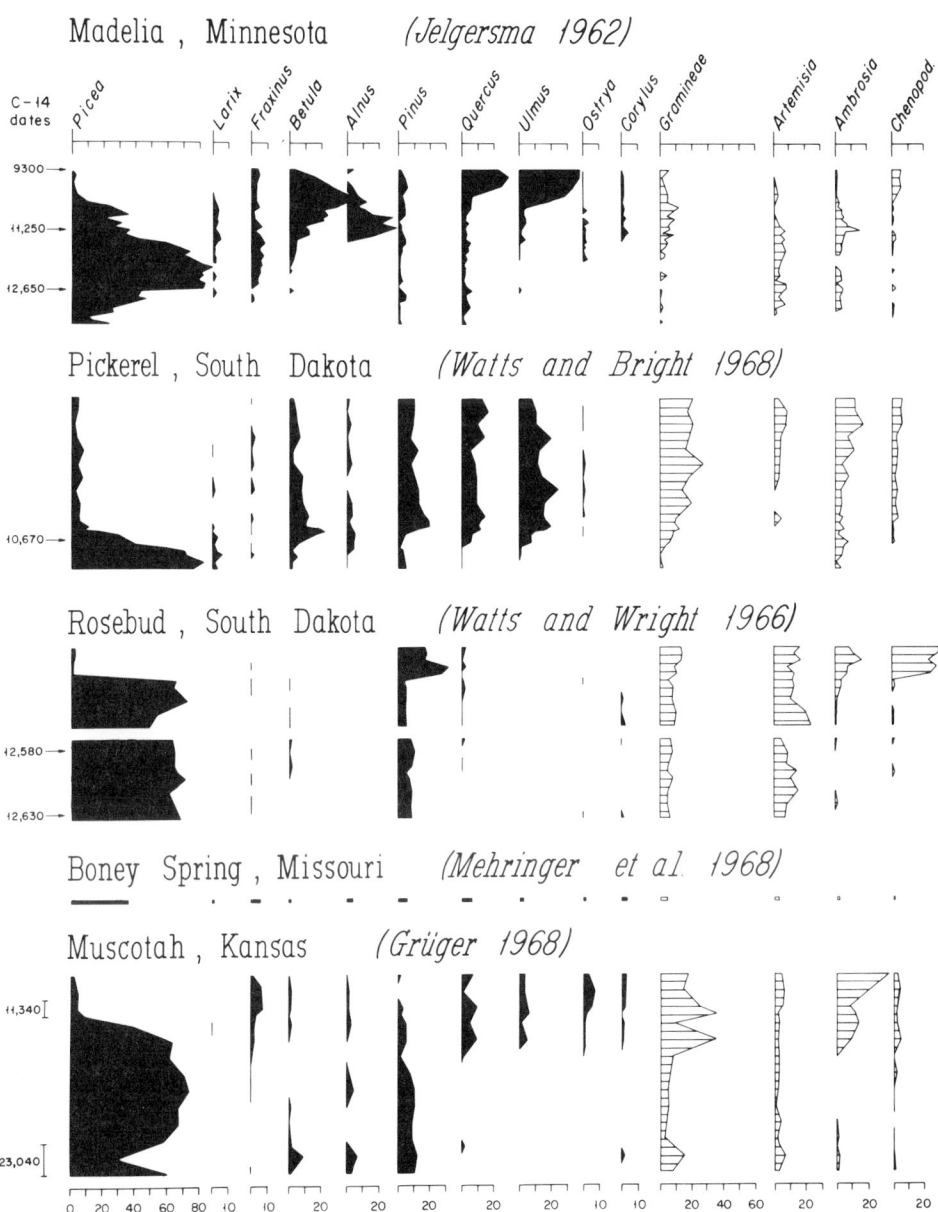

Fig. 7. Abbreviated pollen diagrams from sites on the northern and central Great Plains (from Wright 1969).

On the other hand the oak pollen percentages for contemporaneous sediments in southern Illinois are sufficiently high to imply a potential source of oak pollen for long transport to the north. Perhaps the most convincing argument against distant transport is that surface samples from the boreal forest today contain very low pollen percentages of thermophilous types, even relatively close to the deciduous forest (Wright 1968a).

3. Occurrence of thermophilous trees in the same forest as *Picea*. At present, the boreal spruce forest of Canada contains in its southern part (for example in southern Manitoba) *Fraxinus pennsylvanica*, *F. nigra*, *Quercus macrocarpa*, and other generally temperate forms, but not in sufficient quantity to register appreciably in the modern pollen rain in this region (Lichti-Federovich and Ritchie 1968). On the other hand, the late-glacial spruce forest of the Great Lakes region differed from the modern boreal forest of Canada in another major feature—the apparent absence of *Pinus*. Whereas the spruce pollen zone of New England has 20 to 40 percent pine pollen and relatively high rates of pine pollen deposition (Davis 1969b)—and thus presumably *Pinus banksiana* in the vegetation—the pine pollen values west of the Appalachian Mountains (Fig. 7) are generally less than 2 percent, all of which can be attributed to long-distance transport (Wright 1968a). The modern boreal forest, on the other hand, contains *P. banksiana* throughout its extent, except close to the northern treeline, and it yields pine pollen values generally of 10 to 50 percent.

The weight of evidence seems to point toward a late-glacial spruce forest for the Great Lakes region and New England that contained thermophilous elements in sufficient quantity to total 10 to 25 percent in the pollen rain. Certainly ash must be included in this category, and probably oak and ironwood, if not the minor components. Pine definitely was absent, however, all the way from Manitoba to Ohio—or at least it occurred in extremely small numbers. Macrofossil analysis may ultimately confirm the presence of thermophilous trees, although analysis of most late-glacial lake sediments so far has yielded no remains of thermophilous types (or of pine). The only pertinent report is that by Rosendahl (1948) for a find of *Fraxinus pennsylvanica* along with *Picea mariana* in sediments 11,300 years old in western Minnesota (Wright et al. 1963).

Subdivision of the late-glacial pollen zones to match fluctuations in the ice margin has been attempted ever since radiocarbon analysis provided the means for independent dating of pollen and glacial sequences. The interpretation of the late Wisconsin glacial sequence in the Great Lakes region became somewhat stabilized when the stratigraphic position, importance, and dating of the Two Creeks forest bed in northeastern Wisconsin was worked out, and this horizon, along with the Valders glacial stadial that followed, has served as the basis for a standard climatic curve for late Wisconsin time. Because a comparable ice margin fluctuation in Scandinavia as far as the central Swedish (Salpausselkä) moraine has long been shown to match the Older Dryas/Allerød/Younger Dryas fluctuation in the pollen sequence (Fries 1965; Tauber 1970), first dated by the varve chronology and then confirmed by radiocarbon analysis, the transatlantic correlation of these two sequences has generally been accepted, and the search for a fluctuation in the pollen curves to match the Two Creeks/Valders fluctuation has been vigorous (but loose) throughout the country. It immediately became apparent that the Allerød organic layer intercalated between clays at many European sites had no litho-stratigraphic equivalent in America, probably because the region was already well forested and not subject to solifluction and the introduction of clay into the lakes. In New England, attempts to correlate changes in the spruce curve with glacial fluctuations led to confusion (Davis 1965), especially when few C^{14} dates were available. The matter was made more difficult by lack of knowledge of where the Valders ice border was in New England—or in adjacent Canada. The most recent interpretation, based in part on API counts, is that the late-glacial pollen sequence in New England can be interpreted as a straight record of warming climate, without evidence for climatic reversal (Davis 1967b). Thus the rise in *Quercus* pollen at the base of the spruce zone (Fig. 6) is said to represent an advance of oak trees toward (but not into, according to Davis) the *Picea/Pinus* forest, thus indicating a warming trend. The subsequent fall in oak pollen percentage does not mean a retreat of oak trees but a great influx of pine pollen.

In the western Great Lakes region, identification of a climatic reversal in the pollen sequence was made by West (1961) in eastern Wisconsin, where sites a few kilometers beyond the Valders drift border show a small pollen maximum of *Artemisia* (sage) near the base (West 1961; Schweger 1969). Such a maximum does not occur in the diagram for a site on the Valders drift itself. The *Artemisia* rise here is assumed to record a local expansion of tundra-like openings in the spruce forest. A similar interpretation was made of a rise in nonarboreal pollen at Madelia (Fig. 7) in south-central Minnesota (Jelgersma 1962). The same reasoning was used in interpreting a fluctuation of *Artemisia* (implying cold conditions) and *Fraxinus* (warm) at Kirchner Marsh in southeastern Minnesota, and radiocarbon dates provide some general support for a Two Creeks/Valders correlation there (Wright et al. 1963). The very high pollen values of *Artemisia* in the late-glacial tundra sites of northeastern Minnesota (20–50 percent) support the correlation of an *Artemisia* maximum with a cold interval (Fig. 8, Weber Lake). On the other hand, as Cushing (1967) points out, many *Artemisia* species are dominantly prairie plants, and the high *Artemisia* pollen values may indicate a warm and dry climate rather than a cold one. It may not be possible to resolve this difference in interpretation, partly because no large modern analogs to freshly deglaciated landscape in temperate regions can be studied today. *Artemisia* species occur in the present tundra, but they are not common, perhaps because so much of the tundra has a peat mat unfavorable for growth of a genus that generally favors bare, well-drained (and thus locally xeric) soils. In fact, an edaphic condition favoring the full-glacial and late-glacial expansion of *Artemisia* may have been provided by the extensive outwash plains that were actively forming at this time.

In Minnesota the question has been complicated by the fact that the history

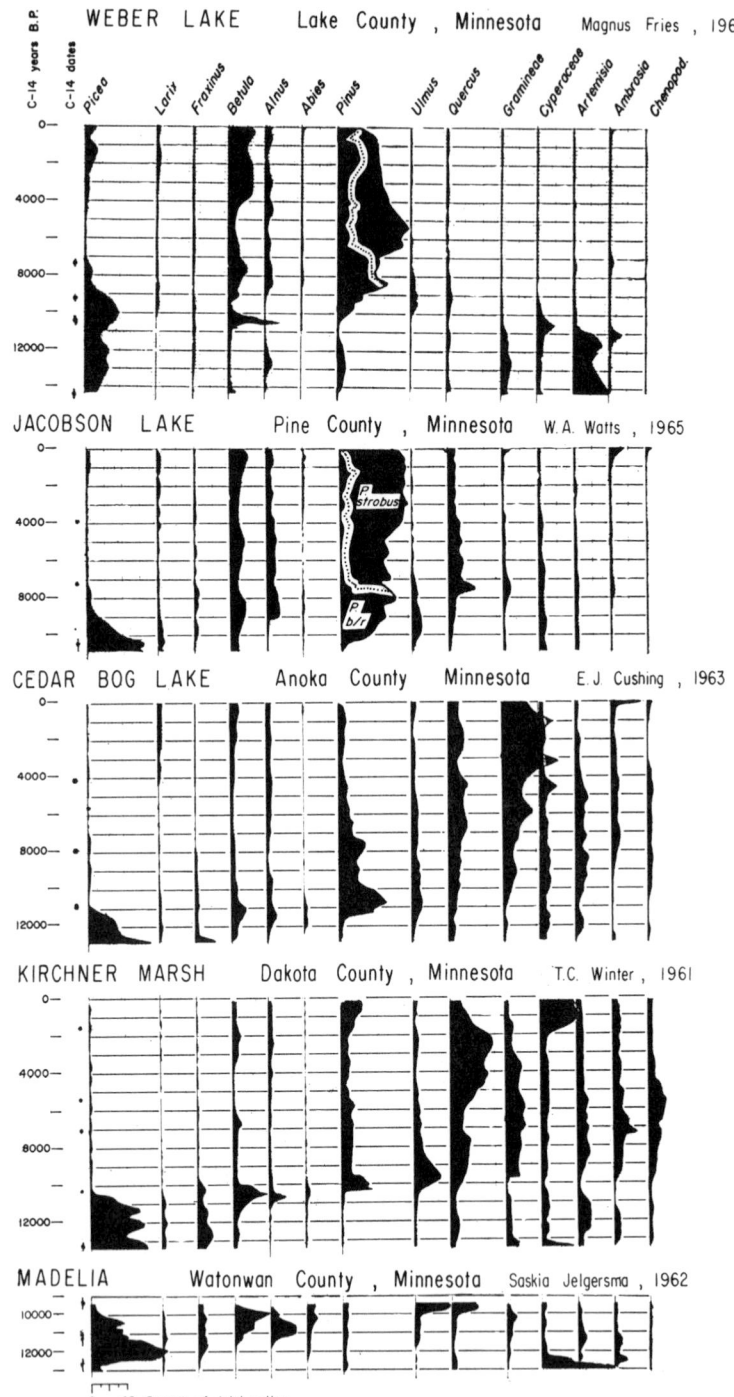

Fig. 8. Abbreviated pollen diagrams from five sites from northeastern (top) to south-central Minnesota (from Wright 1968a).

of late Wisconsin ice margin fluctuations (Fig. 9) is more complex than previously assumed, and that adjacent ice lobes did not necessarily fluctuate synchronously (Wright and Ruhe 1965; Wright and Watts 1969). It is possible that some of the ice lobe fluctuations in the Great Lakes region—perhaps even the famous Valders fluctuation of the Lake Michigan lobe—resulted from sporadic glacial surges rather than from regional climatic fluctuations (Wright 1969). This raises the entire question about a simplified climatic control on ice margin fluctuation, as well as the relation of vegetational changes to both glacial and climatic changes.

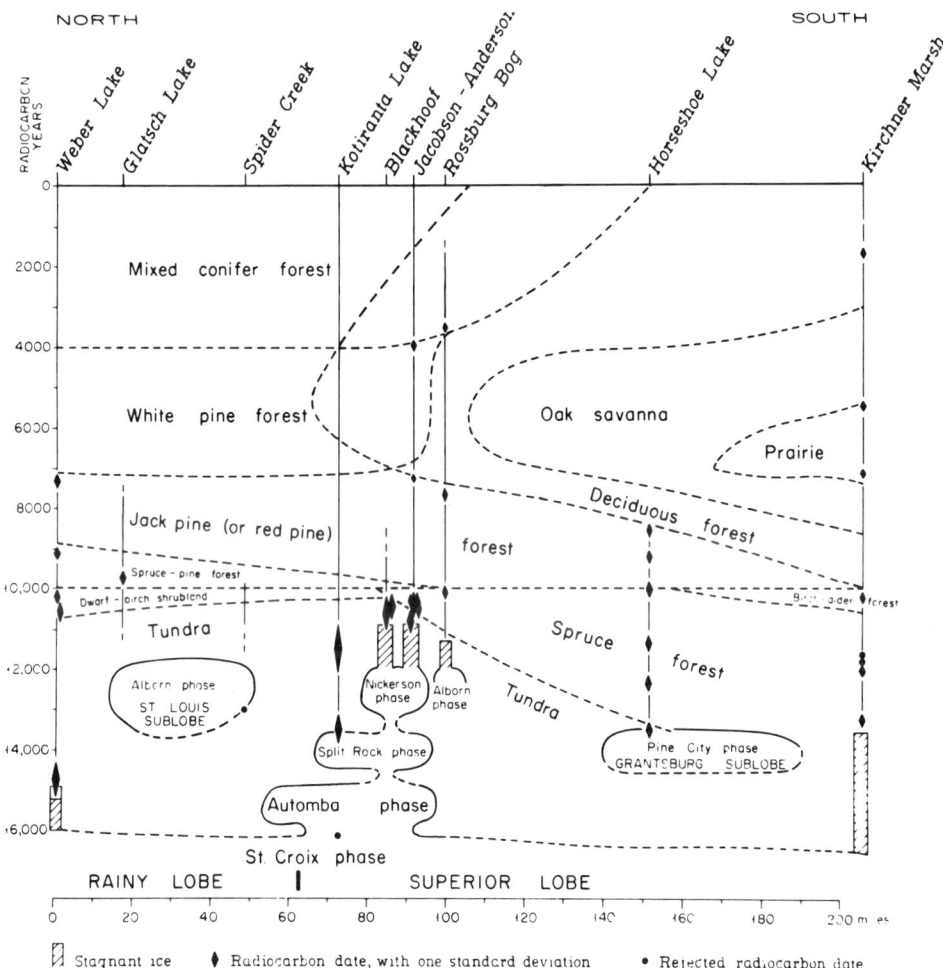

Fig. 9. Diagrammatic cross section from north (left) to south in eastern Minnesota showing development of vegetation types. Based on carbon-dated pollen profiles at the sites indicated. Patterns at the base show the times and areas of development of various Wisconsin glacier lobes (from Wright and Watts 1969).

Although the question is not unsolvable, many more detailed pollen diagrams, with numerous radiocarbon dates, must be completed before the relationships can be considered.

Post-Wisconsin

The late-glacial/postglacial boundary in the pollen sequence is conveniently taken as the top of the spruce zone. This is an extremely sharp boundary that is found from Manitoba to New England (Fig. 7). The *Picea* values commonly drop from 50 to 5 percent in a few decimeters of sediment, equivalent to only a few hundred years. The *Picea* communities were replaced principally by *Betula*, *Pinus*, or *Quercus*, depending on the area and the availability of seed sources.

In New England (Fig. 6), pine and birch (and oak?) were already present in the late-glacial spruce forest, so that when the climate became too warm for the regeneration of spruce the gaps created by wind throw or fire were filled primarily with birch and pine, giving rise to the pine pollen zone (zone B). Whereas the pine pollen involved in the spruce zone that preceded was mostly *Pinus banksiana/resinosa* (Davis 1969b), the pine that so rapidly expanded 9000 years ago at the spruce fall was *P. strobus*, whose pollen may be distinguished by the verrucae on the furrow.

In Ohio, the end of the *Picea* zone is equally sharp, but the spruce was replaced by oak rather than pine or birch (Ogden 1966). All three of the pine species, which presumably had their full-glacial refuges in the Appalachian Highlands, must have moved north of Lake Erie in their westward migration, for by this time it may have been too warm for their development in Ohio, in competition with temperate deciduous trees.

In Minnesota, a network of pollen diagrams enables the detailed tracing of events in the transformation of the late-glacial spruce forest to its successor (Wright 1968a). The sequence may be followed from southwest to northeast (Fig. 8). Part of the succession is represented diagrammatically in Figure 9, which represents a north–south transect in eastern Minnesota. At Madelia, on the edge of the modern prairie in south-central Minnesota, the spruce forest started its deterioration as early as 12,000 years ago, for the trend of increasing warmth first affected its outer portions. Gaps in the forest were taken by birch and alder, for pine was probably still restricted to the Appalachian Highlands and New England and was thus not locally present to provide a seed source. Spruce persisted in reduced amount for almost 2500 years, until elm and oak finally entered the scene in quantity from the south, about 9500 years ago.

In southeastern Minnesota (as exemplified by the Kirchner Marsh pollen diagram) the *Picea* forest did not begin to decline until about 11,000 years ago, and the deterioration lasted about 700 years (Fig. 9). *Betula*, already in the forest, took over a major share of the coverage. Pine (as *Pinus banksiana/resinosa*) did not arrive until almost 1000 years later, and the abrupt rise of pine pollen to 20 percent may represent only the close approach (rather than the actual existence) of pine trees. Elm and then oak arrived about 9500 years ago.

Farther north, e.g. at Horseshoe Lake, near Cedar Bog Lake, the spruce fall and the pine rise come at the same horizon, probably representing not much more than a century of complete transformation of the forest composition, about 10,000 years ago. Birch was hardly involved, being overshadowed by pine, at least in the pollen rain.

Still farther north the decline of the spruce forest started later and lasted longer. At Weber Lake, northwest of Lake Superior, shrub-birch tundra still existed 10,000 years ago. Spruce then invaded from the south but immediately began a slow decline, not reaching low values until about 9000 years ago.

The very abrupt rise in pine pollen in eastern Minnesota about 10,000 years ago requires comment, because the virtual absence of pine pollen at lower levels implies very few pine trees in the vicinity to provide the seed source for rapid spread when the spruce forest deteriorated. Although detailed diagrams are not available across the Great Lakes region to trace the pine migration from its Appalachian refuge, the movement must have been extremely rapid. In late-glacial time pine spread into New England along with spruce, but it did not make any progress westward into the flourishing spruce forest of Ohio. Westward migration was blocked also by ice itself (and by its proglacial lakes) until final withdrawal of the ice in the Lake Michigan basin at the end of the Valders stadial about 11,000 years ago. This event also marked the accelerated climatic changes that decimated the spruce forest throughout the area, and pine could then spread across the northern Great Lakes area, where the climate (probably not much different from today's) was favorable for its vigorous growth. It rapidly covered an area about the same as its modern distribution, reaching central but not western Minnesota. *Pinus banksiana* was most likely to have been the species involved in the first expansion, in view of its relatively rapid regeneration time and its occurrence in more northerly regions today. But *P. resinosa* also occurred in Minnesota at least as early as 8200 years ago, as indicated by finds of needles and seeds; *P. strobus* does not appear to have been involved.

Subsequent changes in the forest composition of early postglacial time show the dynamic character of the vegetation under an initial impulse of climatic change. The trend toward a warmer, drier climate in Minnesota, so clearly the cause for the demise of the spruce forest caused the pine and birch components to be replaced in turn by elm, oak, and other thermophilous deciduous trees. Whereas the pines had clearly reached Minnesota from the east, the deciduous types moved in from the south. In Ohio, oak replaced spruce as the dominant pollen type 10,500 years ago (Ogden 1966). *Quercus* and its associates reached southern Minnesota about 9500 years ago and spread rapidly northward.

The warming trend is better seen by the subsequent advance of the prairie. When the spruce forest gave way to deciduous forest near Pickerel Lake in northeastern South Dakota 11,000 years ago, there were already signs of some prairie openings, as indicated by the high pollen percentages of Gramineae (Watts and Bright 1968). Prairie openings moved slowly eastward; at Kirchner Marsh in southeastern Minnesota, the early dominance of *Ulmus* was gradually replaced (about 9000 years ago) by *Quercus*, along with Gramineae and other prairie types, indicating already the development of an oak savanna (Fig. 9). This vegetation type reached east-central Minnesota (Jacobson Lake) about 7200 years ago, when *Quercus* and Gramineae pollen rose abruptly at the expense of *Pinus banksiana/resinosa* type (Wright and Watts 1969).

Meanwhile, true prairie was not far behind. Its widespread development is not well dated at Pickerel Lake, but it can probably be placed at about 9500 years ago. It reached Kirchner Marsh, 500 km to the east, about 7200 years ago, at the same time that oak savanna reached Jacobson Lake (Fig. 9). The arrival of prairie at Kirchner Marsh was signaled by a great fall in the *Quercus* curve, combined with rises in NAP curves. Abrupt fluctuations in some of the pollen and seed curves appear to reflect the periodic drying of the lake (Wright et al. 1963; Watts and Winter 1966).

This time of 7200 years ago marked the climax of the climatic trend of increasing warmth and dryness. There are indications of its effects on vegetation as far as northeastern Minnesota, where at Weber Lake a small but significant maximum on all the major NAP curves reflects the approach of prairie openings to a region far within the conifer forest (Fries 1962).

The reverse trend was much slower; prairie openings probably survived in the Jacobson area until 4000 years ago, and true prairie in the Kirchner area until 5500 years ago (Fig. 8). This reversal marked the end of the postglacial period of maximum warmth and dryness, whose bounding dates can therefore be placed at various points along the climatic curves. The figures of 8000 to 4000 years ago serve a good purpose as an average, but the evidence indicates that the peak of the climatic curve was closer to 7000 than to a midpoint of 6000 years ago.

Meanwhile, a new element in the forest composition entered Minnesota—*Pinus strobus*. It may also have had its refuge in the Appalachian Highlands, where it contributes as much as 20 percent of the pine pollen in full-glacial sediments in Virginia (Craig 1969). Its westward migration was much slower than that of the *P. banksiana/resinosa* type. It expanded in southern New England 9000 years ago (Davis 1969b), but it did not persist there long, for the climate became too warm. Its progress through Michigan and Wisconsin is not known, although West's (1961) separation of pine pollen types by size may be valid in showing an abrupt rise of the large type (*P. strobus*, not *P. strobus/resinosa*) at an undated level well above the beginning of the postglacial. At any rate, *P. strobus* reached eastern Minnesota in quantity about 6800 years ago, just a few hundred years after the arrival of the oak savanna from the southwest. It did not extend much farther west at this time, however. Its existence in the Jacobson Lake area may have been marginal during much of the next 2500 years, for the pine pollen values fluctuate strongly between 30 and 60 percent.

The reverse climatic trend toward cooler, wetter conditions can be seen throughout the region. At Pickerel Lake there were the first indications of a fringe of mesic deciduous trees similar to that which now occupies the steep slopes around the basin—with *Tilia americana, Ulmus americana, Fraxinus pennsyl-*

vanica, etc. At Kirchner Marsh, all signs of local prairie were dispelled, and the area became largely an oak forest (in contrast to the more mixed deciduous forest in early postglacial time before the invasion of the prairie). At Jacobson Lake, the changes occurred rather abruptly about 4000 years ago (Wright and Watts 1969). As the decrease in *Quercus* and NAP signaled the end of the oak openings there, conifers increased. Pine values rose from 60 to 80 percent, as *P. strobus* expanded in the region and renewed its migration to the west; it had stopped in eastern Minnesota during the prairie period. The increase in *Picea, Larix, Abies,* Ericaceae, and *Sphagnum* points to the growth of a bog around the lake, a conclusion amply confirmed by stratigraphic macrofossil analyses, which show a great influx of remains of *Picea, Larix, Chamaedaphne, Sphagnum, Vaccinium macrocarpon, V. oxycoccus, Scheuchzeria palustris, Dulichium arundinaceum, Carex angustior,* and similar plants of ericaceous bogs. The upland vegetation, although certainly dominated by *Pinus strobus,* contained more mesic deciduous trees than during the preceding period—*Acer saccharum* and *A. rubrum* were added to the *Tilia, Ulmus,* and *Quercus* communities that survived the xeric interval.

Pinus strobus, which had been halted at the Jacobson Lake area throughout the prairie period, resumed its slow westward migration. It reached the Itasca Park area of northwestern Minnesota about 2700 years ago. Here the prairie and oak savanna, which had dominated the region from 8500 to 4000 years ago, had been replaced by mesic deciduous forest, with the pollen rain dominated by *Betula* and *Ostrya* type.

As the climatic trend continued, the other two pines, *P. banksiana* and *P. resinosa,* followed to the west. During the prairie period these trees probably had their principal refuge in northeasternmost Minnesota; at Weber Lake, for example, the percentages of the *P. strobus* and *P. banksiana/resinosa* pollen types are nearly equal during this interval. The expansion from this refuge may not have extended southward, as it is not recorded at Jacobson Lake. But it reached the Itasca area about 2000 years ago (McAndrews 1969), and now all three pine species are common within 50 km of the prairie border.

The history of growth of the bog around Jacobson Lake in the last 4000 years is probably representative of many lakes in northeastern and northern Minnesota (Wright and Watts 1969). The sequence is seen as far south as Cedar Bog Lake, where an increase in *Larix* and *Sphagnum* occurred about 4200 years ago. It had been preceded there by a rise in *Juniperus/Thuja* pollen, presumably derived from *Thuja occidentalis* (northern white cedar) that gives the bog its name. This is one of the southernmost outliers of coniferous bogs in the state.

Farther north, the great wetlands of the flat glacial lake plains developed about this time. During the prairie period the rainfall was sufficiently slight that surface drainage was adequate even on very gentle slopes, and thus peat did not accumulate. With the climatic reversal that affected this region about 4000 years ago, peat began to form on the plains of glacial lakes Upham, Aitkin, and Agassiz, and today vast ericaceous bogs and, toward the west, patterned sedge fens cover the region, punctuated by raised bogs with *Picea mariana* and *Larix laricina*. Peripheral streams are working back to dissect the bogs and fens, but the area still contains probably the largest continuous conifer wetlands in the world—the Red Lake wetland is an uninterrupted expanse almost 100 km long and 20 km wide.

Although the growth of bogs and fens can be attributed primarily to the climatic change that brought the prairie period to a close, other factors may have been important also. Continued leaching of nutrients from the soils throughout postglacial time must have gradually decreased the base status and pH of the soil water and thus of the surface runoff. When pine and other conifers returned to the slopes at the end of the prairie period, the soil waters that were affected by conifer detritus could have been even more acid. The water accumulating in depressions may therefore have been poor in mineral nutrients and thus favorable for the development of bog plants, especially *Sphagnum*. At the same time, gradual filling of many depressions by sedimentation permitted the peripheral growth of mats and fringes in the very shallow water; these served to screen out mineral nutrients brought by slope wash.

The final major event in the postglacial vegetation history of the Minnesota area resulted from the entrance of white man upon the scene. Extensive cutting of *Pinus strobus* and *P. resinosa* took place in northern Minnesota between about 1880 and 1930. *Betula papyrifera* and *Populus tremuloides* were the principal successors, along with *Pinus banksiana*, in regions that had been burned. The changes in the regional tree pollen rain resulting from these events were not great, however—usually a slight reduction in pine and increase in birch. Of greater significance was the disturbance farther south and west, where agriculture, starting about 1850, converted large areas of deciduous forest and natural prairie into hay and grain fields, with all their agricultural weeds. This event is marked in the pollen sequence by an abrupt rise in the curves for *Ambrosia*, Chenopodiineae, *Iva*, *Xanthium*, and various other weed types, as well as the first occurrence of *Zea mays*, *Avena*, and other cultivated plants.

In the heart of the agricultural region, such as Ohio, the *Ambrosia* values may rise to 40 percent of total pollen (Ogden 1966). Where accelerated soil erosion has accompanied cultivation, the *Ambrosia*-rich sediments may be over a meter thick; more than 4 meters of clay with up to 60 percent *Ambrosia* were found in limestone depressions in Kentucky (Wright et al. 1966). Even in New England and eastern Canada, now largely reforested after abandonment of farms in the mid-nineteenth century, the *Ambrosia* rise is clear.

Although Indians had practiced agriculture locally in the Great Lakes region, starting 1500–2500 years ago, no record of this disturbance has been detected in pollen diagrams.

It might be emphasized that the effects of white man on the vegetation, as severe as they were, had nowhere near so great an effect as did the late-glacial and postglacial climatic changes, at least as far as the pollen record is concerned.

Although the postglacial vegetational changes in the Minnesota area can be traced with some detail, the situation to the east is not so clear, because of too few detailed diagrams. The dozens of pollen diagrams from Wisconsin to Ohio produced by Sears, Potzger, and their students do not include consideration of nonarboreal pollen, have a sampling interval too gross and a pollen sum too small, and are not supplied with radiocarbon dates. The well-dated diagram from Silver Lake in west-central Ohio (Ogden 1966) shows a slight increase of the *Ambrosia* curve for the interval of about 8000 to 4000 years ago, as does the diagram (Fig. 6) for Rogers Lake, Connecticut (Davis 1969b). Both of these may represent long-distance transport of *Ambrosia* pollen from the expanded prairie of mid-postglacial time rather than local openings in the forest (Wright 1968b), for the other major prairie types (Gramineae, *Artemisia*) do not show a comparable expansion, and *Ambrosia* is notorious as a distant disperser of pollen (Bassett and Terasmae 1962). The tree pollen curves for this interval, however, give no indication of a warm, dry period; in fact the traditional interpretation of the pollen zonation from New England to Ohio, with *Quercus/Tsuga* (zone C–1), *Quercus/Carya* (C–2), and *Quercus/Castanea* (C–3), places the warm interval later, i.e. in zone C–2, which is dated to about 4000–1500 years ago. The reasoning is that *Carya* is more common today in the continental interior than it is in the New England area, so its pollen maximum must mark a relatively warm, dry, more continental climate. Uncertainties about the climatic and ecological requirements of many of the New England trees, however, make the climatic interpretation of the pollen sequence as unsure in New England (Davis 1965) as it is in western Europe (Frenzel 1966), which has a comparable oceanic climate. The reason may lie in the fact that the real mid-postglacial xerothermic interval was confined to the prairie border region of the continental interior, which is subject to severe summer droughts such as occurred during the 1930s, when New England remained unaffected (Wright 1968b).

Whatever the changes of forest composition in southern New England during the mid-postglacial time, it seems certain that the last few thousand years have been marked by expansion of conifers and bogs in northern New England and southeastern Canada (Auer 1930; Davis 1965) just as in Minnesota. Although increasing humidity of the climate could explain the change in both areas, one cannot ignore the possible influence of gradual leaching of the soils on forest composition, as well as the shallowing of water by sedimentation to permit the encroachment of fringing bogs.

In southeastern United States most sites have had incomplete sedimentation in postglacial time. Sites in north-central Florida and southern Georgia (Watts 1969, 1971) show thick sediment for the last 8000 years, preceded by an unconformity that probably represents all Wisconsin time as well as the early post-Wisconsin (carbon dates below the unconformity are >35,000 years ago). The post-hiatus pollen sequence (Fig. 3) starts with an assemblage dominated by *Quercus*, along with maxima of *Ambrosia*, Gramineae, Chenopodiineae, and *Artemisia* that rival their mid-postglacial maxima in the prairie border of the Midwest. The chenopod at one of the sites proves to be a semiaquatic *Amaran-*

thus that is present today, but the high values of the other types cannot be so easily explained, and they may indeed represent prairie-like openings in a xeric oak/pine forest. This assemblage gives way about 6000 years ago to one dominated by *Pinus*, which continues to the surface, with increasing representation of swamp margin plants such as *Taxodium* (swamp cypress). This recent development of swamp fringes on lakes resembles the recent growth of bogs in Minnesota and other parts of northeastern North America.

The basic sequence of oak to pine can be identified in postglacial sites in the Southeast as far north as Virginia (Figs. 4, 5), beyond which pine is not significant and the New England type of sequence prevails, involving successively *Tsuga*, *Carya*, and *Castanea* along with *Quercus* (Fig. 6). The past ecotones of vegetation types cannot be located without many more pollen diagrams, nor can the climatic implications be fully understood. With the present meager coverage of postglacial pollen diagrams in the central and eastern United States, it appears that the driest time in the Southeast occurred 10,000–6,000 years ago, in the upper Midwest 8000–4000 years ago, and in the New England area 4000–1500 years ago. Either the climatic interpretations of the pollen assemblages are incorrect, or the climatic fluctuations in the three regions are not in phase. Certainly much more must be learned before either of these explanations is accepted, but at present it appears that the second is more likely. An attempt has been made to provide a climatological rationale for the relations between the Midwest and New England sequences, involving analogs to recent climatic fluctuations, e.g. the droughts of the "dust bowl" in the 1930s had no manifestation whatever in New England (Wright 1968b). Just how the Southeast fits into this general picture is not yet clear, however.

WESTERN NORTH AMERICA

The Southwest

Pre-Wisconsin history is undoubtedly recorded in the deep cores from the San Augustin Plains in western New Mexico, Wilcox Playa in southern Arizona, and other desert basins of the Southwest (Martin and Mehringer 1965), but correlations with glacial and interglacial ages are not very satisfactory, partly because isotope dating does not provide the proper time control, and partly because no obvious interglacial pollen assemblages (presumably like the postglacial) can be identified and thus correlated with the Sangamon and older interglacials. Most of the sites investigated have been dry throughout postglacial time so that pollen-bearing sediments for this time have not accumulated. Possibly the sites were also dry during interglacial intervals, so that interglacial sediments may also be missing.

For Wisconsin time in the Southwest there is little doubt that the climate was pluvial. Geologic evidence indicates that lakes filled many nondesert basins and that glaciers developed or expanded in the mountains. The best-substantiated sequence is from Searles Lake in southeastern California, where literally dozens of radiocarbon dates show that lacustrine clays were deposited in pluvial lakes during the pluvial intervals, with salt beds being formed during interpluvial times (Smith 1968). Lake Bonneville in western Utah also has a history well documented with radiocarbon dates (Morrison 1965). Expansion of the lake to the outlet level occurred about 18,000 years ago, and the lake maintained a generally high level until desiccation started about 12,000 years ago. Although the rise to the outlet level was initially affected by the diversion of a major stream into the basin as a result of volcanic damming, there is little doubt that climatic conditions were primarily responsible for the expansion of the lake.

This confirmation of the traditional glacial/pluvial correlation at Searles Lake, Lake Bonneville, and other sites in the Southwest contrasts with recent results in Mediterranean Europe, where pollen studies indicate that the climate at low latitudes was dry rather than moist at the time of glacial expansion (Wijmstra 1969).

Pollen evidence in the Southwest for a pluvial climate during Wisconsin time comes from the high percentages of *Pinus* and in some cases other montane conifers (*Picea, Abies, Pseudotsuga*) at sites that are now characterized by semi-desert shrubs (Martin and Mehringer 1965). Sites as far east as the west Texas grasslands show great expansion of *Pinus* (Hafsten 1964).

In the mountains of the Southwest the tree line was lowered, so that mountain crests with elevations as low as 2700 meters in northwestern New Mexico were then in the alpine zone, with a pollen rain dominated by *Artemisia*, along with *Picea* and *Pinus* blown up to the crest from the slopes below. As is the case with lowland regions, pollen counts of surface samples in transects across mountain ranges have been useful in providing a basis for paleoecological interpretations (Maher 1963; Bent and Wright 1963), although sources of error are probably rather greater. Thus there is evidence for a kind of telescoping of altitudinal vegetational belts as they were lowered during the cold phase (Wells and Berger 1967; Wright et al., unpublished ms.). Differences in topography may result in differences in the areal coverage (and thus pollen production) of stands of particular species as the vegetational belts were lowered. Nonetheless, the evidence is convincing that the upper tree line itself was about 1000 meters lower during Wisconsin time throughout the region than it is today, and that the rapid transformation to the modern type of vegetation occurred about 12,000 to 10,000 years ago, just as in the Great Lakes region (Mehringer 1967a; Adam 1967).

Subdivisions within Wisconsin time for the Southwest can be suggested from only one locality, the Chuska Mountains of northwestern New Mexico, where changes in the proportions of *Picea* and three species of *Pinus* provide the basis for postulating altitudinal shifts in the vegetation belts between about 35,000 and 12,000 years ago (Wright et al., unpublished ms.). Interestingly enough, one of the pines, *P. flexilis* (limber pine), is no longer present in the area. This indicates the effects of climatic change on the flora of mountain ranges. It may also explain the absence of certain otherwise common trees from certain ranges, for example the absence of *Pinus ponderosa* from several ranges in northwestern Wyoming, despite their presence in nearby ranges.

At no site is there good evidence for a climatic oscillation that might be correlated with that inferred for the Two Creeks/Valders sequence in the Great Lakes region.

For post-Wisconsin time in the Southwest, the pollen sequence has been derived primarily from alluvial deposits rather than lake deposits, mainly because of the paucity of lakes suitable for pollen analysis (Martin 1963; Mehringer 1967b). Alluvium is characteristically discontinuous and frequently contains poorly preserved pollen, so the record is undoubtedly incomplete and imprecise. Major interest centers on the evidence for and against the so-called Altithermal interval, the part of postglacial time during which the climate is supposed to have been warmer and drier than the present. The concept of the Altithermal is based upon geological evidence of erosion and deposition in the alluvial valleys (Antevs 1955; Haynes 1968), but the pollen sequence provides no support for it, according to Martin (1963).

Within the Rocky Mountains proper, all the lakes investigated are glacial lakes, so that the pollen records generally do not begin before late-glacial time. There are enough pollen diagrams to show that the tree line during the glacial period was lowered by an amount comparable to the depression of the snow line, and that the change to the postglacial type of vegetation occurred about 11,000 years ago. Sites in the higher mountains were characterized by a pollen rain with generally more than 50 percent *Artemisia* in late-glacial time—much greater than can be found in any alpine area today, probably an indication of the effect of increased areal coverage of alpine vegetation when the tree line was depressed to mid-elevations on the mountains (Baker 1970). For example, localities now in sagebrush steppe (*Artemisia tridentata*) at 2200-meter elevation at the western base of the Wind River Mountains in northwestern Wyoming were probably surrounded by spruce forest from 11,000 years back to 21,000 years ago (R. C. Bright, unpublished). At a site in southeastern Idaho at 1450 meters in the steppe below the sagebrush belt (Fig. 10), the vegetation before 11,000 years ago was dominated by *Pinus* and *Picea* (Bright 1966). Although in all these cases the pollen stratigraphy (and the altitudinal vegetational belts) are not sharp enough to indicate the exact amount of tree zone depression, a figure of at least 800 meters is reasonable.

The post-Wisconsin pollen sequence in the Rocky Mountain sites indicates slight changes that can be interpreted as the record of the Altithermal interval. The most convincing case is in the Yellowstone Plateau of northwestern Wyoming, where reduction in percentages of *Picea* and *Abies* pollen are interpreted as the result of a rise in the vegetation belts during mid-postglacial time (Baker 1970).

Pacific Northwest and Arctic Alaska

The forested regions from Oregon northward along the cordillera up to the tundra regions of Alaska and the Yukon have been investigated primarily by only two

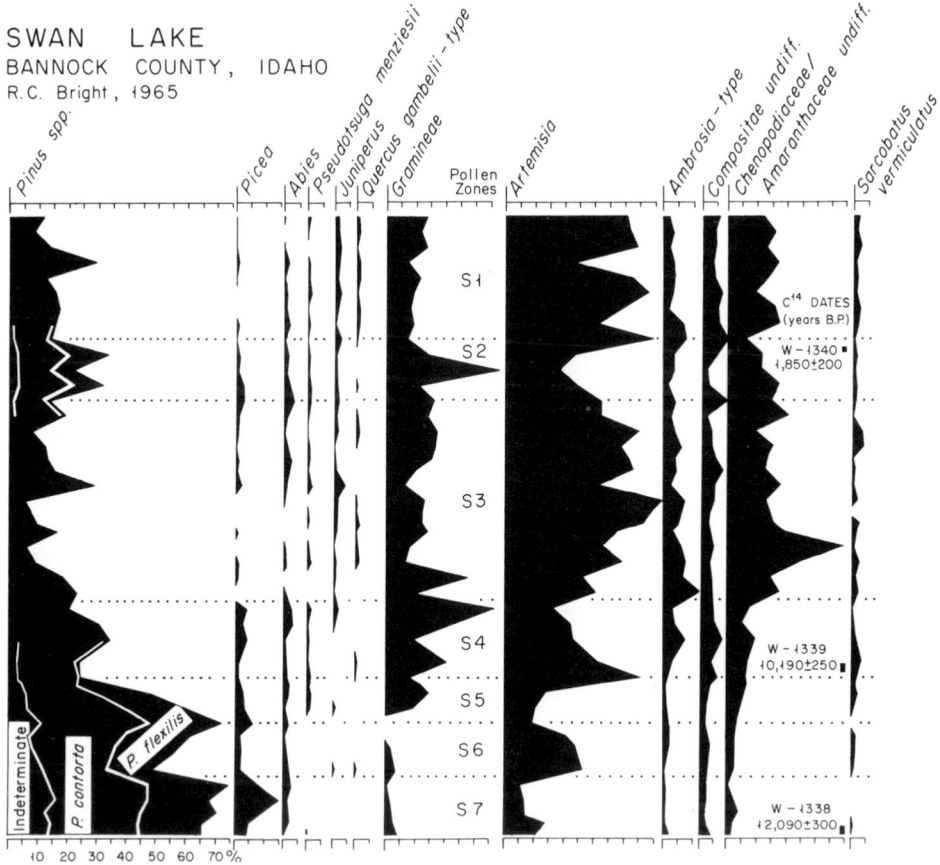

Fig. 10. Pollen diagram for Swan Lake, at the outlet of pluvial Lake Bonneville, southeastern Idaho (redrawn from Bright 1966).

persons; they have produced a host of pollen diagrams as well as extended compilations of the regional vegetation history (Hansen 1947; Heusser 1960). All the important forest trees are conifers, and some of the genera have broad ecological and altitudinal ranges; difficulties in distinguishing pollen types within the genus *Pinus* and the *Larix/Pseudotsuga* group make refinement of the pollen stratigraphy uncertain.

A few of the investigated sites date from intervals prior to the last glaciation, but most of them record late-glacial and postglacial climatic events. Lack of radiocarbon dates makes it difficult to trace the history from place to place in the mountains that characterize the region, but in the southern part the common occurrence of the Glacier Peak volcanic ash (12,000 years old) and the Mount Mazama ash (6600 years) provides stratigraphic markers of great utility.

The general regional picture built by Hansen and Heusser is that with the retreat of the last major glaciers much of the country became dominated by lodgepole pine. As the climate became warmer and drier, the pine forests were succeeded by western hemlock in the moister areas west of the Cascade Mountains, by hemlock and spruce farther north along the coast to the Alaska panhandle, by Douglas-fir and oak in drier intermountain valleys, and even by grasslands in the dry plateaus east of the Cascades (Heusser 1965). Reversal of the climatic trends in late-postglacial time caused further changes in forest composition.

For the modern tundra region of Alaska and adjacent Yukon, the basic pollen sequence was established by Livingstone (1955) from cores of lake sediment in the Brooks Range of northern Alaska, and his interpretation of the sequence made use of surface-sample analyses from different types of herb and shrub tundra at different distances from the spruce forest. A systematic review of this and other investigations in the Alaska tundra has been made by Colinvaux (1967).

The time of Wisconsin glaciation was marked throughout the region by a herb tundra that indicates a colder summer climate than today's, with much greater distance to forest. Subsequent warming caused progressive advances of dwarf-birch tundra and spruce/alder forest. A long core from Imuruk Lake on the Seward Peninsula of western Alaska records older glacial and interglacial intervals, with pollen assemblages similar to those of the Wisconsin and post-Wisconsin zones that top the profile.

An informative pollen diagram has been completed by Rampton (1970) for a site at the east base of the cordillera in the southwestern Yukon, in the northern edge of the *Picea* forest, in an area of permafrost (Fig. 11). The site is a lake on the older of two principal moraine complexes of the region, presumably correlative with the Bull Lake moraine of the Middle Rocky Mountains. The sequence is marked by herb and shrub pollen zones that represent various types of tundra vegetation, with radiocarbon dates from 30,000 to about 8700 years ago. Then follows a spruce/birch zone and, starting about 5700 years ago, a spruce/alder zone, which extends to the surface.

Summary and Conclusions

Studies of vegetational history often have as their principal objective the climatic history of a region, because of all the criteria that can be used to typify the climate of continental areas the gross natural vegetation is the most prominent. In fact the inferred relation is so close that world climatic provinces have been classified and delineated principally according to the vegetation. Because pollen analysis provides a gross rather than a detailed picture of the vegetation, it is a powerful tool in the study of regional vegetational changes. Only after detailed ecological reconstructions, however, can the response to climatic change be properly evaluated. Plant macrofossil analysis can be used to provide some details of local vegetational history. Because many specific identifications are possible, macrofossils provide basic information pertinent to problems of floristic history.

But beyond the problems of climatic history are questions of vegetational

ANTIFREEZE POND
YUKON TERRITORY , CANADA

V. Rampton , 1968

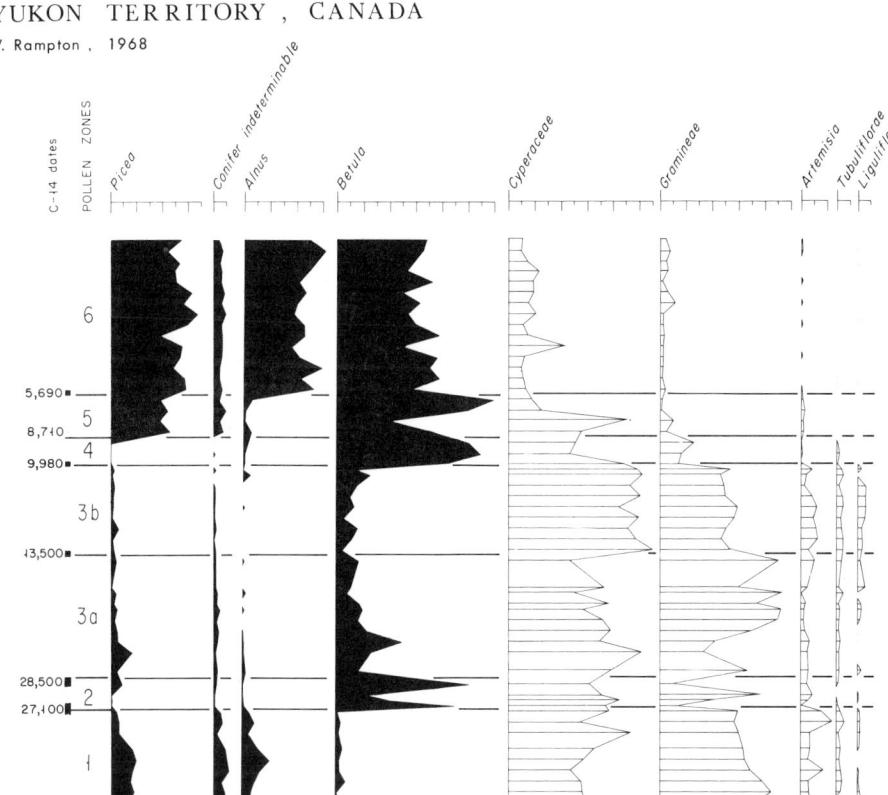

Fig. 11. Pollen diagram for Antifreeze Pond, near the Alaska Highway in southwestern Yukon (redrawn from Rampton 1970).

history, such as the long-range stability of vegetational associations, with or without the influence of climatic change—questions concerned with differential rates of migration of major species in response to controls by soil type, shade tolerance, seed-dispersal mechanisms, and competition.

In North America the vegetational history is becoming so well known for the past 40,000 years, and especially for the last 14,000 years, that it helps fill a gap in climatic history for some areas and supplements or challenges geological criteria in others. Vegetational studies have the great advantage of generally involving stratigraphic succession of organic materials that can be easily dated by radiocarbon analysis, so that the chronology of the sequence can be easily determined, with little danger of missing significant gaps in the record.

Pollen studies have been completed throughout enough of North America to show the major trends in late Quaternary vegetational history. Most information is available for the central and eastern United States, where the abundant lakes and marshes provide numerous sites for study. A beginning has been made on extending the sequence back into early Wisconsin and pre-Wisconsin time.

The last interglacial age (Sangamon) in the central and southeastern United States was marked by vegetation grossly similar to that of today, although many more sites must be studied before the regional patterns can be reconstructed with any certainty. During the early part of the Wisconsin glacial age the vegetation changed drastically, culminating in a boreal forest about 25,000 to 20,000 years ago. This forest at first consisted almost exclusively of *Picea* in northeastern Kansas, *Picea* and *Pinus* from Illinois eastward to the Virginian Appalachians and the coastal plain, and mostly *Pinus banksiana* in the southern Appalachians. As the main Wisconsin glaciation culminated in the Illinois area and the ice began to withdraw, pine was excluded from the forest, and in late Wisconsin time it was totally or essentially absent from the boreal forest for Saskatchewan to Ohio. Tundra may have existed in the higher parts of the northern Appalachians at the time of ice maximum, and it certainly existed as a narrow belt in New England and northern Minnesota at the time of ice retreat.

The climatic change that accelerated the retreat of the Laurentide ice sheet in the Great Lakes region may also have resulted in the changes from boreal *Picea* forest into communities of *Pinus* and *Betula* in the more northerly areas, of *Ulmus* and *Quercus* south of the Great Lakes, of prairie in the west, and of *Quercus* and *Pinus* in the southern Appalachians. These changes in forest composition started about 12,500 years ago in the south and reached northern Minnesota about 10,000 years ago. In a general way, the trend was unidirectional, with no certain reversals that can be correlated with the Two Creeks/Valders ice margin fluctuation or with any other of the numerous glacial oscillations plotted for the Great Lakes region. Here the pollen and geological evidence come into conflict, the resolution of which requires additional study on both sides. The situation may mean that, with the flora available in the area, pollen analysis is not sufficiently sensitive to record other than major vegetational changes, whereas in northwestern Europe, with a different and perhaps more responsive flora, the Allerød/Younger Dryas pollen oscillation is firmly correlated with the central Swedish moraine. On the other hand, the situation may mean that the Valders ice advance in the Lake Michigan lobe did not represent a significant climatic oscillation but instead was the result of a nonclimatic event such as a glacial surge.

Meanwhile, in the Western Cordillera, the tree line during the Wisconsin glaciation was lowered 800 to 1000 meters, and alpine vegetation in the mountains was vastly expanded in areal coverage. The various forest belts were lowered as well, although probably by unequal amounts, and many of the now-desert basins of the Southwest were marked by shrub steppe or even by *Pinus edulis*. At the end of the Wisconsin, as both the Cordilleran ice sheet and the various valley glaciers retreated, the climatic change caused the various vegetation belts to ascend the mountains, changing their composition in the process. The transformation was as abrupt here as it was in the Middle West.

The postglacial vegetational history of North America saw slow change in forest composition, largely in response to climatic change, although the timing of the changes differed from one region to the next. In the Middle West, at least on the north side of the prairie peninsula, the deciduous forest changed to prairie and back again during the general interval 8000 to 4000 years ago. But in the Southeast the inferred dry interval occurred earlier, perhaps about 10,000 to 6,000 years ago, and in the Northeast later—4000 to 1500 years ago. In the Rocky Mountains a mid-postglacial dry interval is only weakly recorded in the pollen sequence and is not well dated, and in the Southwest its very existence has been challenged. The relatively slight vegetational changes of the postglacial, compared to the abrupt change that terminated the Wisconsin, makes paleoclimatic analysis for that interval difficult. The end-Wisconsin climatic change was clearly worldwide and unidirectional in its major effects. But the postglacial changes may have been more localized in their manifestations. Thus the trend toward warmer, drier climate 8000 years ago in the Midwest may have been accompanied by little significant changes in the Northeast, in analogy with the situation during the Midwest drought of the 1930s (Wright 1968b). Slight shifts of air mass distributions in certain regions occurring with greater frequency can have major effects on the vegetation when continued for hundreds of years. The concept of a postglacial xerothermic or altithermal climate for a specific time interval for the entire continent, let alone the world, may not be valid, and we must search instead for regional patterns that correspond to modern climatic provinces delineated by air mass frequencies (Bryson and Wendland 1967).

Although some of the main features of vegetational history for North America are becoming well defined, major gaps still exist. Geographically, perhaps the principal one concerns south-central United States and adjacent Mexico, to determine the southern limit of the boreal forest for the Wisconsin glaciation, as well as the distribution of temperate vegetation belts. But beyond this is the entire subject of vegetation patterns during the Sangamon interglaciation, for which only a very small start has been made. The wealth of information about regional climates and phytogeographic patterns that has come from European studies of literally dozens of interglacial sites indicates the potential in North America for expanding the history backward in time, so that floristic as well as vegetational problems can be examined. The best hope for such expansion comes from the glaciated parts of the Middle West, where the glacial stratigraphy and morphology is established and where interglacial organic sediments can be studied. It will be many years, however, before the European style of early Quaternary biostratigraphy and biogeography can be reviewed.

REFERENCES

Adam, D. P., Late-Pleistocene and Recent palynology in the central Sierra Nevada, California, p. 275 in Cushing and Wright, 1967.

Anderson, S. T., A late-glacial pollen diagram from southern Michigan, *Danmarks Geol. Undersoegelse*, II, no. 80, 140, 1954.

Antevs, Ernst, Geologic-climatic dating in the West, *Am. Antiquity, 20,* 317, 1955.

Argus, G. W., and Margaret B. Davis, Macrofossils from a late-glacial deposit at Cambridge, Massachusetts, *Am. Midland Naturalist, 67,* 106, 1962.

Auer, V., Peat bogs in southeastern Canada, *Geol. Surv. Canada, Mem., 162,* 1, 1930.

Baker, R. G., Late-glacial pollen and plant macrofossils from Spider Creek, southern St. Louis Co., Minnesota, *Bull. Geol. Soc. Am., 76,* 601, 1965.

Baker, R. G., Pollen sequence from late Quaternary sediments in Yellowstone Park, *Science, 168,* 1449, 1970.

Bassett, I. J., and J. Terasmae, Ragweeds, *Ambrosia* species, in Canada and their history in post-glacial time, *Can. J. Bot., 40,* 141, 1962.

Bender, Margaret M., R. A. Bryson, and D. A. Baerreis, University of Wisonsin radiocarbon dates III, *Radiocarbon, 9,* 530, 1967.

Bent, A. M., and H. E. Wright, Jr., Pollen analysis of surface materials and lake sediments from the Chuska Mountains, New Mexico, *Bull. Geol. Soc. Am., 74,* 491, 1963.

Birks, H. J. B., The Late-Weichselian and present vegetation of the Isle of Skye, 440 pp., Ph.D. thesis, University of Cambridge, 1969.

Black, R. F., Ice-wedge casts of Wisconsin, *Wisconsin Acad. Sci., Arts, Letters, 54,* 187, 1965.

Braun, E. Lucy, Deciduous forests of eastern North America, 596 pp., Blakiston, Philadelphia, 1950.

Bright, R. C., Pollen and seed stratigraphy of Swan Lake, southeastern Idaho: Its relation to regional vegetational history and to Lake Bonneville history, J. Idaho State Univ. Mus., *Tebiwa, 9,* 1, 1966.

Brown, C. A., The flora of Pleistocene deposits in the Western Florida Parishes, West Feliciana Parish, and East Baton Rouge Parish, Louisiana, *Louisiana, Dept. Conserv., Geol., Bull., 12,* 59, 1938.

Bryson, R. A., and W. M. Wendland, Tentative climatic patterns for some late-glacial and postglacial episodes in central North America, p. 271 in W. J. Mayer-Oakes, ed., *Life, Land, and Water,* 414 pp., Univ. of Manitoba Press, Winnipeg, 1967.

Colinvaux, P. A., A long pollen record from St. Lawrence Island, Bering Sea (Alaska), *Paleogeograph., Paleoclimatol., Paleoecol., 3,* 29, 1967.

Craig, A. J., Vegetational history of the Shenandoah Valley, Virginia, *Geol. Soc. Am., Spec. Papers, 123,* 283, 1969.

Craig, A. J., Pollen concentration in laminated sediments from northeastern Minnesota, 63 pp., M.Sc. thesis, University of Minnesota, 1970.

Cushing, E. J., Redeposited pollen in late-Wisconsin pollen spectra from east-central Minnesota, *Am. J. Sci., 262,* 1075, 1964.

Cushing, E. J., Problems in the Quaternary phytogeography of the Great Lakes region, p. 403 in Wright and Frey, 1965.

Cushing, E. J., Late-Wisconsin pollen stratigraphy and the glacial sequence in Minnesota, p. 59 in Cushing and Wright, 1967.

Cushing, E. J., and H. E. Wright, Jr., eds., *Quaternary Paleoecology,* 433 pp., Yale University Press, New Haven, 1967.

Davis, Margaret B., Phytogeography and palynology of northeastern United States, p. 377 in Wright and Frey, 1965.

Davis, Margaret B., Pollen accumulation rates at Rogers Lake, Connecticut, during late- and post-glacial time, *Rev. Paleobotany Palynology, 2,* 219, 1967a.

Davis, Margaret B., Late-glacial climate in northern United States: A comparison of New England and the Great Lakes region, p. 11 in Cushing and Wright, 1967b.

Davis, Margaret B., Palynology and environmental history during the Quaternary period, *Am. Scientist*, 57, 317, 1969a.

Davis, Margaret B., Climatic changes in Connecticut recorded by pollen deposition at Rogers Lake, *Ecology*, 50, 409, 1969b.

Deevey, E. S., Jr., Late-glacial and postglacial pollen diagrams from Maine, *Am. J. Sci.*, 249, 177, 1951.

Dreimanis, A., J. Terasmae, and G. D. McKenzie, The Port Talbot interstade of the Wisconsin glaciation, *Can. J. Earth Sci.*, 3, 305, 1966.

Florin, Maj-Britt, and H. E. Wright, Jr., Diatom evidence for the persistence of stagnant glacial ice in Minnesota, *Bull. Geol. Soc. Am.*, 80, 695, 1969.

Frank, A. H. E., Pollen stratigraphy of the Lake of Vico (Central Italy), *Paleogeograph., Paleoclimatol., Paleoecol.*, 6, 67, 1969.

Frenzel, B., Climatic change in the Atlantic/sub-Boreal transition on the Northern Hemisphere: Botanical evidence, in *World Climate from 8000 to 0* B.C., p. 99, Royal Met. Soc., London, 1966.

Fries, Magnus, Pollen profiles of late Pleistocene and Recent sediments at Weber Lake, northeastern Minnesota, *Ecology*, 43, 295, 1962.

Fries, Magnus, Outlines of the late-glacial and postglacial vegetational and climatic history of Sweden, illustrated by three generalized pollen diagrams, p. 55 in H. E. Wright, Jr., and D. G. Frey, eds., International studies on the Quaternary, *Geol. Soc. Am., Spec. Papers, 84*, 565 pp., 1965.

Frye, J. C., and H. B. Willman, Permafrost features near the Wisconsin glacial margin in Illinois, *Am. J. Sci.*, 256, 518, 1958.

Graham, Alan, and Charles Heimsch, Pollen studies of some Texas peat deposits, *Ecology, 41*, 751, 1960.

Grüger, E., The development of the vegetation of southern Illinois since late Illinoian time (preliminary report), *Intern. Congr. Quaternary, 8th, Paris, 1969*, Proc. 1971, in press.

Grüger, J., Vegetation history of the area near Muscotah, northeastern Kansas, unpubl. ms., 1971.

Hafsten, Ulf, A standard pollen diagram for the southern High Plains, USA, covering the period back to the early Wisconsin glaciation, *Intern. Congr. Quaternary, 6th, Warsaw, 1961, Rept., 2*, 407, 1964.

Hansen, H. P., Postglacial forest succession, climate, and chronology in the Pacific Northwest, *Trans. Am. Phil. Soc., 37*, 1, 1947.

Haynes, C. V., Jr., Geochronology of late-Quaternary alluvium, p. 591 in Morrison and Wright, 1968.

Heusser, C. J., Late Pleistocene environments of North Pacific North America, *Am. Geogr. Soc., Spec. Publ., 35*, 308 pp., 1960.

Heusser, C. J., A Pleistocene phytogeographical sketch of the Pacific Northwest and Alaska, p. 469 in Wright and Frey, 1965.

Hibbard, C. W., and D. W. Taylor, Two late Pleistocene faunas from southwestern Kansas, *Univ. Michigan Mus. Paleontol., Contrib., 16*, 1, 1960.

Iltis, H. H., The genus *Gentianopsis* (Gentianaceae): Transfers and phytogeographic comments, *Sida, 2*, 129, 1965.

Janssen, C. R., Recent pollen spectra from the deciduous and coniferous-deciduous forests of northwestern Minnesota: A study in pollen dispersal, *Ecology*, 47, 804, 1966.

Janssen, C. R., Stevens Pond: A postglacial pollen diagram from a small *Typha* swamp in northwestern Minnesota, interpreted from the pollen indicators of surface samples, *Ecol. Monographs*, 37, 145, 1967.

Jelgersma, Saskia, A late-glacial pollen diagram from Madelia, south-central Minnesota, *Am. J. Sci.*, 260, 522, 1962.

Kapp, R. O., and A. M. Gooding, Pleistocene vegetational studies in the Whitewater Basin, southeastern Indiana, *J. Geol.*, 72, 307, 1964.

Kempton, J. P., and J. E. Hackett, The late-Altonian (Wisconsinian) glacial sequence in northern Illinois, p. 535 in Morrison and Wright, 1968.

Lichti-Federovich, S., and J. C. Ritchie, Recent pollen assemblages from the western Interior of Canada, *Rev. Palaeobotan. Palynol.*, 7, 297, 1968.

Livingstone, D. A., Pollen profiles from Arctic Alaska, *Ecology*, 36, 587, 1955.

Maher, L. J., Jr., Pollen analyses of surface materials from the southern San Juan Mountains, Colorado, *Bull. Geol. Soc. Am.*, 74, 1485, 1963.

Maher, L. J., Jr., *Ephedra* pollen in sediments of the Great Lakes region, *Ecology*, 45, 391, 1964.

Martin, P. S., Taiga-tundra and the full-glacial period in Chester County, Penna., *Am. J. Sci.*, 256, 470, 1958.

Martin, P. S., *The last 10,000 years: A fossil pollen record of the American Southwest*, 87 pp., Univ. of Arizona Press, Tucson, 1963.

Martin, P. S., and P. J. Mehringer, Jr., Pleistocene pollen analysis and biogeography of the Southwest, p. 433 in Wright and Frey, 1965.

McAndrews, J. H., Postglacial history of prairie, savanna, and forest in northwestern Minnesota, *Torrey Bot. Club.*, *Mem.*, 22, 72 pp., 1966.

McAndrews, J. H., Paleoecology of the Seminary and Mirror Pool Peat Deposits: Univ. Manitoba Dept. Anthropology, Occ. Pap., 1, 253, 1967.

McAndrews, J. H., Paleobotany of a wild rice lake in Minnesota, *Can. J. Botany*, 47, 1671, 1969.

McAndrews, J. H., R. E. Stewart, Jr., and R. C. Bright, Paleoecology of a prairie pothole: A preliminary report, p. 101 in *Glacial Geology of the Missouri Coteau and Adjacent Areas*, edited by Lee Clayton and T. F. Freers, North Dakota Geol. Survey, Misc. Ser. 30, 170 pp., 1967.

McAndrews, J. H., and H. E. Wright, Jr., Modern pollen rain across the Wyoming basins and the northern Great Plains, *Rev. Palaeobotan. Palynol.*, 9, 17 (1969).

Mehringer, P. J., Jr., The environment of extinction of the late-Pleistocene megafauna in the arid southwestern United States, p. 247 in *Pleistocene Extinctions: The Search for a Cause*, edited by P. S. Martin and H. E. Wright, Jr., 454 pp., Yale Univ. Press, New Haven, 1967a.

Mehringer, P. J., Jr., Pollen analysis of the Tule Springs site, Nevada, p. 130 in *Pleistocene Studies in Southern Nevada*, edited by H. M. Wormington and D. Ellis, 409 pp., Nevada State Museum, Anth. Papers 13, 1967b.

Mehringer, P. J., Jr., C. E. Schweger, R. T. Wood, and R. B. McMillan, Late-Pleistocene boreal forest in the western Ozark highlands? *Ecology*, 49, 567, 1966.

Morrison, R. B., Quaternary geology of the Great Basin, p. 265 in Wright and Frey, 1965.

Morrison, R. B., and H. E. Wright, Jr., Editors, *Means of Correlation of Quaternary Successions*, 631 pp., Univ. of Utah Press, Salt Lake City, 1965.

Ogden, J. G., III, Pleistocene pollen records from eastern North America, *Botan. Rev.*, *31*, 481, 1965.

Ogden, J. G., III, Forest history of Ohio. I. Radiocarbon dates and pollen stratigraphy of Silver Lake, Logan County, Ohio, *Ohio J. Sci.*, *66*, 387, 1966.

Potzger, J. E., Phytosociology of the primeval forest in central-northern Wisconsin and Upper Michigan, and a brief post-glacial history of the lake and forest formation, *Ecol. Monographs*, *16*, 211, 1946.

Potzger, J. E., Nineteen bogs from southern Quebec, *Can. J. Botany*, *31*, 383, 1953.

Potzger, J. E., and B. C. Tharp, Pollen record of Canadian spruce and fir from a Texas bog, *Science*, *98*, 584, 1943.

Rampton, V., Late Quaternary vegetational and climatic history of the Snag-Klutlan area, southwestern Yukon Territory, Canada, *Bull. Geol. Soc. Am.*, 1970, in press.

Ritchie, J. C., Contributions to the Holocene paleoecology of westcentral Canada. I. The Riding Mountain area: *Can. J. Botany*, *42*, 181, 1964.

Ritchie, J. C., and Bernard deVries, Contributions to the Holocene paleoecology of west-central Canada. A late-glacial deposit from the Missouri Coteau: *Can. J. Botany*, *42*, 677, 1964.

Rosendahl, C. O., A contribution to the knowledge of the Pleistocene flora of Minnesota, *Ecology*, *29*, 284, 1948.

Rowe, J. S., Forest regions of Canada, Canada Dept. Northern Affairs Nat. Res., Forestry Branch, *Bull. 123*, 71 pp., 1959.

Shay, C. T., Postglacial vegetation development in northwestern Minnesota, and its implications for prehistoric man: Univ. of Minn., M.S. Thesis, 1965.

Schweger, C. E., Pollen analysis of Iola Bog and paleoecology of the Two Creeks Forest Bed, Wisconsin, *Ecology*, *50*, 859, 1969.

Smith, G. I., 1968, Late-Quaternary geologic and climatic history of Searles Lake, southeastern California, p. 293 in Morrison and Wright, 1968.

Tauber, Henrik, Differential pollen dispersion and the interpretation of pollen diagrams, *Danmarks Geol. Undersoegelse*, Ser. II, no. 89, 69, 1965.

Terasmae, J., A palynological study of Pleistocene interglacial beds at Toronto, Ontario, *Geol. Surv. Canada, Bull.*, *56*, 23, 1960.

Watts, W. A., 1967, Late-glacial plant macrofossils from Minnesota, p. 89 in Cushing and Wright, 1967.

Watts, W. A., A pollen diagram from Mud Lake, Marion County, north-central Florida, *Bull. Geol. Soc. Am.*, *80*, 631, 1969a.

Watts, W. A., The full-glacial vegetation of northwestern Georgia, *Ecology*, *51*, 17, 1970.

Watts, W. A., Interglacial and postglacial vegetation history near Lake Louise (Georgia) and Scott Lake (Florida), *Ecology*, 1971, in press.

Watts, W. A., and R. C. Bright, Pollen, seed, and mollusk analysis of a sediment core from Pickerel Lake, Day County, South Dakota, *Bull. Geol. Soc. Am.*, *79*, 855, 1968.

Watts, W. A., and T. C. Winter, Plant macrofossils from Kirchner Marsh, Minnesota: A paleoecological study, *Bull. Geol. Soc. Am.*, *77*, 1339, 1966.

Watts, W. A., and H. E. Wright, Late-Wisconsin pollen and seed analysis from the Nebraska Sandhills: *Ecology*, *47*, 202, 1966.

Wayne, W. J., Periglacial features and climatic gradient in Illinois, Indiana, and western Ohio, east-central United States, p. 393 in Cushing and Wright, 1967.

Wells, P. V., and R. Berger, Late Pleistocene history of coniferous woodland in the Mohave Desert, *Science, 155,* 1640, 1967.

West, R. G., Late-glacial and postglacial vegetation history in Wisconsin, particularly changes associated with the Valders readvance, *Am. J. Sci., 259,* 766, 1961.

Whitehead, Donald R., Palynology and Pleistocene phytogeography of unglaciated eastern North America, p. 417 in Wright and Frey, 1965.

Whitehead, D. R., Studies of full-glacial vegetation and climate in southeastern United States, p. 237 in Cushing and Wright, 1967.

Wijmstra, T. A., Palynology of the first 30 meters of a 120 m deep section in northern Greece, *Acta Botan. Neerl., 18* (4), 511, 1969.

Wright, H. E., Jr., The roles of pine and spruce in the forest history of Minnesota and adjacent areas, *Ecology, 49,* 937, 1968a.

Wright, H. E., Jr., History of the prairie peninsula, p. 78–88 in *The Quaternary of Illinois,* 179 pp., Univ. Illinois Coll. Agric., Spec. Publ. 14, 1968b.

Wright, H. E., Jr., Glacial fluctuations and the forest succession in the Lake Superior area, *Intern. Assoc. Great Lakes Res., 12th Conf., Ann Arbor, 1969, Proc.,* 397, 1969.

Wright, H. E., Jr., Vegetational history of the Central Plains, in W. Dort and J. K. Jones, eds., Pleistocene and Recent environments of the Central Plains, Univ. Kansas Press, 1970, in press.

Wright, H. E., Jr., and D. G. Frey, eds., *The Quaternary of the United States,* 922 pp., Princeton University Press, Princeton, N.J., 1965.

Wright, H. E., Jr., and R. V. Ruhe, Glaciation of Minnesota and Iowa, p. 15 in Wright and Frey, 1965.

Wright, H. E., Jr., and W. A. Watts, with contributions by Saskia Jelgersma, Jean C. B. Waddington, Junko Ogawa, and T. C. Winter, Glacial and vegetational history of northeastern Minnesota, *Minn. Geol. Surv., Spec. Publ.,* 11, 59 pp., 1969.

Wright, H. E., Jr., T. C. Winter, and H. L. Patten, Two pollen diagrams from southeastern Minnesota: Problems in the regional late-glacial and postglacial vegetational history, *Bull. Geol. Soc. Am., 74,* 1371, 1963.

Wright, H. E., Jr., D. A. Livingstone, and E. J. Cushing, Coring devices for lake sediments, p. 494–520 in B. Kummel and D. M. Raup, eds., *Handbook of paleontological techniques,* 852 pp., W. H. Freeman, San Francisco, 1965.

Wright, H. E., Jr., Barbara Spross, and R. A. Watson, Pollen analyses of the sediment from sinkhole ponds in the Central Kentucky Karst, *Natl. Speleological Soc. Bull., 28,* 185, 1966.

Kazimierz Kowalski

17. THE BIOSTRATIGRAPHY AND PALEO-ECOLOGY OF LATE CENOZOIC MAMMALS OF EUROPE AND ASIA

MAMMALS AS INDICATORS OF PALEOCLIMATE

Mammals are not good indicators of temperature in paleontology. Owing to their homothermia they are capable of tolerating wide ranges of temperatures, using both physiologic and ethologic methods of adaptation, as has been demonstrated in numerous observations and experiments. Suffice it to say that house mice can live and reproduce in cold storage, in which the temperature is kept a few degrees below the freezing point. The great thermal tolerance of mammals is also indicated by the ranges of various species; for example, the wolf occurs in Eurasia from the zone of tundra down to the subtropical territories of India.

Species such as the reindeer, lemmings, and Arctic fox, which are associated with the tundra environment now, are reliable indicators of cool climate in the younger periods of the Pleistocene. In Eurasia some of them occur in middle and upper Pleistocene layers. It is arguable that these species had been associated with the periglacial environment since the beginning of their geological record. However, it seems certain that a group of mammalian species limited in occurrence to the cool zone has existed at least since the Günz glaciation.

The existence of a mammalian fauna of cool environments in earlier periods (i.e. in late Tertiary and Villafranchian) is much more difficult to ascertain. The adaptation of mammals to a cool environment leaves no uniform expression in the structure of the skeleton, the only element we can study in fossil material. It may well be that species occurred, especially among the rapidly evolving rodents, which were associated with tundra areas in the periods of glaciations older than the Günz, but the data obtained so far do not allow their recognition. In general, however, the mammalian groups limited in range to the temperate zone are geologically young; they did not arise before the late Tertiary. This fact suggests that the rise of the set of species associated with tundra took place still later, which by no means implies that the periglacial environments of the oldest Late Cenozoic glaciations lacked mammalian fauna. It should rather be supposed that they supported eurytopic species with a wide range of thermal adaptation.

The only methods we have at our disposal for the paleoclimatic estimation of mammalian faunas in the periods preceding the occurrence of modern arctic species are those of indirect inference. Of these the most important (but so far rarely applied in paleontological studies) are as follows.

1. Adaptations of some mammals are reflected in the structure of their skeletons and teeth and indicate association with definite types of vegetation and, consequently, indirectly with paleoclimatic conditions. For example, primates are associated with a forest environment which provides food the year round and is therefore subtropical or tropical. The development in many groups of ungulates and rodents of hypsodonty and the ability to run fast is an adaptation for grazing and life in open areas. Such connections may sometimes be found even in carnivores, which are generally less dependent on the types of vegetation; e.g. hyenas, typical carrion-eaters, can find sufficient food only in open areas.

2. The number of mammalian species per unit of area is a good paleoclimatic indicator. Ceylon has 83 species of mammals, the British Isles only 41. It can be seen from Simpson's (1963) studies that in North America there is an evident decrease in the number of mammalian species from the tropics toward the pole. The numerical data for Eurasia are incomplete, but this phenomenon is equally clear there. We are rarely well enough acquainted with fossil faunas to determine the exact number of mammalian species of which they are composed, sometimes, however, the estimation may be restricted to one order. Thus in the Pliocene of Poland, 19 species of insectivores have been distinguished, whereas in the present-day fauna of this territory there are only 8 (Kowalski 1964). This is undoubtedly evidence for the existence of a milder climate in the Pliocene; such large numbers of insectivore species are encountered only in subtropical regions nowadays.

3. Morphological changes occurring in particular species of mammals also indicate climatic (especially thermic) changes. However controversial the so-called Bergmann's rule may still be (it deals with the increase in the size of warm-blooded animals in a cool climate), it actually describes the existing trend in many species. For example, Sych (1965) found a gradual increase in the body measurements of the fossil hare *Hypolagus brachygnathus* in the Pliocene and early Pleistocene localities of Poland, indicating a gradual decrease of about 10°C in the mean annual temperature.

Mammalian Faunas of the Neogene of Eurasia

Although we know many localities of fossil mammals of the Neogene in Eurasia, only a few can be used as the basis for paleoclimatic conclusions. The most valuable are long series of sediments that allow the investigation of gradual climatic changes in the same place. One can also draw conclusions about climatic differences by comparing the compositions of the faunas from the same period and examining the differences between them in reference to their geographical situations.

The first method may be applied for the mammals of the Eurasian Neogene in a few areas. One is the Vallés-Penedés region in northern Spain, which displays a rich and long series of continental deposits with mammalian faunas, worked out by Crusafont (1956, 1958).

The lowest layers, which go back to the Burdigalian, provided more than fifty species of mammals. The fauna shows a remarkable endemism in relation to the remaining parts of Europe and suggests the predominance of a damp sylvan environment. The cervids of the genera *Amphitragulus, Palaeomeryx, Lagomeryx,* and *Procervulus* prevail among the ungulates; the cats and weasels, mostly those of the forest genera (*Pseudaelurus, Felis, Stromeriella,* and *Isochyriotis*), prevail among the carnivores. Hydrophilous animals (e.g. the pig, *Anchitherium,* and numerous members of the Sciuridae) form 72 percent of the faunal elements. The whole is an apparent continuation of the Aquitanian fauna.

The Vindobonian is represented by the abundant faunas from Hostalets de Pierola and San Quirze. The mammals go through a distinct evolution, their endemism in relation to the rest of Europe decreases, but the change in the nature of the environment is only slightly marked. Although the forest elements prevail, *Anchitherium* disappears, and so do the Paleomerycidae and Lagomerycidae. The bunodont species of pigs, *Listriodon lockharti,* is replaced by the lophodont species *L. splendens.* The proportion of rodents in the fauna increases at the cost of the artiodactyls and lagomorphs. The diversity of cats and weasels shrinks, while the hyenids, the first antelopes (*Protragoceas* and *Miotragoceras*) and the first saber-toothed cats (*Grivasmilus*) appear. These changes suggest the presence of open areas. The dampness of the environment is shown by the occurrence of otters, beavers, and the like.

The next period represented in the series under discussion, termed the Vallesian by Crusafont (1956), is already marked by the presence of *Hipparion.* This is accompanied by other typical elements of the Hipparion fauna, e.g. *Aceratherium incisivum* and *Euprox dicranoceras;* nevertheless, numerous forms of the previous period persist. A paleoecological analysis shows that the share of forest forms continued to decrease. The hydrophilous elements form about 80 percent of the whole fauna. The numbers of felids and mustelids decrease in favor of hyenids. The bovids begin to dominate over the cervids, and the giraffids increase in number.

The last period of the series corresponds to the Pikermian, according to Crusafont's (1958) division. The makeup of the mammalian fauna indicates a savanna. The change that took place in the fauna between this and the previous period was abrupt. Only 10 percent of the ungulates are forest forms, and hardly 25 percent of the whole fauna may be defined as hydrophilous elements. The total of species is smaller, typical forms of this period being *Hipparion mediterraneus, Crocuta eximia,* gazelles, and numerous giraffes.

The studies carried out by Freudenthal and Sondaar (1964) on the basis of the fauna of rodents confirm the nature of the last two periods. The Cricetodontidae and numerous archaic elements prevail in the first, whereas the second period is marked by a sudden appearance of the Muridae and Cricetidae.

Another European area in which the transformation of the Neogene fauna can be followed continuously is the Viennese Basin (Thenius 1960). Here, the Vindobonian period exhibits faunas in which the deer predominate; at first they indicate evergreen forests and then, perhaps, forests with some deciduous trees. The Tragocerinae are characteristic of these faunas, as were the primates *Pliopithecus* and *Dryopithecus;* and *Zygolophodon lauricensis* and *Bunolophodon angustidens* were important elements of the mastodonts.

The Sarmatian period shows no traces of *Hipparion*, which no doubt lived elsewhere in Europe at that time. In whatever way we interpret this, the fauna does unambiguously indicate the dominance of the savanna-type vegetation. It includes, among other species, *Euprox furcatus, Conohyus simorrensis, Gazella stehlini*, and *Brachypotherium brachypus*.

A further evolutionary stage observed in the Viennese Basin is represented by the Pannonian fauna, which, still a forest fauna, includes *Hipparion*. The forest must have been deciduous, of temperate type. The species that lived then were *Bunolophodon langirostris, Dicerorhinus schleiermacheri, Chalicotherium goldfussi*, etc.

The only Asian region in which, as in the European regions discussed, the gradual changes in the fauna of the Neogene can be observed, is the Siwalik Beds. The Siwalik series do not reach so far back as the European series of sediments. The Kambial zone seems to correspond to the Vindobonian and contains such characteristic species as *Bunolophodon pandionis, Listriodon guptai*, and *Hyaenelurus lahirii*. *Hipparion* appears in the Chinji zone whose fauna, however, retains its fundamental Miocene character and may be regarded as corresponding to the Vallesian. The Nagri zone coincides, at least partly, with the Pikermian, and the Dhok Pathan zone with the Astian. Finally, the Tatrot zone must be referred in great part to the Villafranchian and, therefore, to the Quaternary. In Colbert's (1935) opinion, "The progressive development of the Siwalik faunas was dependent upon a change from a relatively dry flood plain environment to a more moist flood plain and forest environment." Colbert explains this change in climate by the gradual upthrust of the Himalayas, which sheltered their southern slopes from the cold influence of the northern climate and, at the same time, blocked the moist monsoon, causing an increase in the rainfall. However, a striking change, which consisted among other things in a considerable reduction of the number of primate species, occurred between the Dhok Pathan zone and the Tatrot zone.

The Neogene fauna that is best suited for zoogeographical analysis is the so-called Hipparion fauna. The genus *Hipparion*, which no doubt came from America, appeared in Eurasia as early as the Miocene and at places persisted until the Villafranchian; therefore, in the environment of occurrence of the fauna named after this species there were certainly remarkable changes. Moreover, the genus *Hipparion* is not a reliable indicator of the type of vegetation; it is met in associations of the savanna fauna and in those of steppes and temperate forests.

Kurtén (1952) analyzed the Hipparion faunas of China and found that a part of them, i.e. those lying farther southeast in the provinces of Honan and (part of) Shansi, are sylvan in character (gaudryi fauna, called after its characteristic species *Gazella gaudry*), whereas those from the northeastern part of Shansi and Kansu Province are of a steppe nature (dorcadoides fauna after its characteristic species *Gazella dorcadoides*). There is a transitional zone with elements of both faunas which may be interpreted either as an area where these faunas mixed in one period or as evidence of the displacement of their boundaries. At any rate, these studies show that in the period of Hipparion fauna the boundary between the forest and steppe zones ran across the territory of China. According to Orlov (1962), the Hipparion faunas of Siberia and Kazakhstan indicate the diversity of the plant cover and the unquestionable presence of forest. Altan-Teli in Mongolia harbored a steppe fauna, but the presence of a member of Rhizomyidae, *Pararhizomys hipparionum*, which I have described from this locality (Kowalski 1969), suggests a more abundant plant cover than at present. The Crimean faunas of this period include desert elements, e.g. the rodents of the family Dipodidae.

In Europe the fauna from Eppelsheim and those from the neighboring regions of Germany are forestal; the brachydont browsing types of species prevail in them (Thenius 1959). The genus *Giraffa* is characteristic of the giraffes, *Tapirus* is present, and the proboscidians are represented by brachydont *Dinotherium*. The classical Hipparion fauna from Pikermi is more southern in character but in spite of older opinions to the contrary, its forest character is generally supported.

Another rich locality of the Hipparion fauna, in Samos, hardly 250 km away from Pikermi, presents an entirely different type of fauna that approximates that of Maragha in Iran. The hypsodont genera predominate here, the giraffids of the genera *Palaeotragus*, *Samotherium*, and *Helladotherium* are represented in large numbers, but the genus *Giraffa* is wanting. There are few cervids and numerous bovids, including a large number of gazelles.

At the occurrence of the Hipparion fauna in Eurasia the climate-vegetation zones were apparently well developed, indicating a climate which was milder than that of today. The northern, western, southeastern, and southern parts of the continent were covered by moister forest formations, and a vast open tract of steppes and deserts extended across the middle of Eurasia, situated somewhat farther south than the present range of these formations. This made it possible for elements of the fauna to migrate freely from China to western Europe.

If we try to sum up the paleoclimatic data provided by the mammalian faunas of the Eurasian Neogene, we shall see that as early as the beginnings of the Miocene there was a turnover marked by an evident change in the fauna. Although the Aquitanian fauna was for the most part forestal, in relation to the Oligocene faunas it was characterized by a decrease in the occurrence of species of moist tropical forests (which species, however, persisted longer in southern

Asia; Gabunia and Trofimov 1964). The process that caused the climate to become dryer lasted throughout the Miocene and was related to the gradual fall in temperature. These changes are continuous so far as we can judge, and the materials we have at our disposal do not indicate the existence of cyclic changes of the climate. Even when the forest fauna was replaced by the fauna of open areas and then by another forest fauna—as in the Viennese Basin—these changes show the consistent shifting of climatic zones, because the older forest zone was subtropical and the younger one temperate. The direction of climatic changes may have been different at different places, owing to local topographic factors (as in the Siwalik Beds).

Astian

The late Pliocene shows transitional characters between the Tertiary and Quaternary in many respects. The classical faunas of land mammals of this period, from Rousillon and Montpellier in France, show the absence of the genus *Equus*, whose appearance indicates the beginning of the Quaternary, but their stratigraphic equivalents in eastern Europe and Asia already have a fauna of modern character, with the genera *Equus, Leptobos, Camelus,* and *Archidiskodon* (Nikiforova 1964). On this basis the Soviet paleontologists suggest that the Astian should be included in the Quaternary.

The fundamental portion of the Astian fauna in Europe is a continuation of the Pliocene fauna. Here, the observations of the rodent fauna of southern France are particularly instructive (Thaler 1966). The rodents belong to a fast-evolving group which is closely associated with plant cover, and for this reason they may be more useful than large mammals in paleoclimatic studies. In the Pikermian fauna the rodents of southern France are chiefly forestal, and there are many endemic elements among them, including the typical genus *Ruscinomys*. The murids abound here and a few Cricetodontidae have also survived. In the Astian fauna, however, immigrants appear beside the persisting endemic elements; their ancestors should be looked for outside this territory, probably as far distant as Asia. These new elements include *Mimomys*, the oldest vole in western Europe, and *Trilophomys*, the specialized hypsodont genus of cricetids, known also from many localities of eastern Europe. These allochthonous elements next became dominant in the faunas of the Villafranchian.

The locality of Nîmes in southern France is particularly important. It presents many-layered cave deposits, such as are known from numerous Quaternary localities, except that in Nîmes they are of Astian age. In the lower layers of these deposits the allochthonous elements (*Apodemus, Cricetus, Mimomys*) prevail, but the autochthonous fauna of rodents with *Ruscinomys* appears once again in the upper layers. This would be the oldest faunistic evidence of a climatic oscillation—a transitional cool phase. The foregoing facts need further confirmation, which may be provided by the intensive studies now being carried on in small-mammal localities throughout Europe.

Villafranchian

The Villafranchian period is represented by numerous faunas in Eurasia. It began at the same time as the Quaternary and is marked by the appearance of ungulates of the genera *Equus*, *Camelus*, *Leptobos*, and *Archidiskodon*, which also, as has been mentioned, gradually extended their ranges toward the west (Nikiforova 1965). The Villafranchian lasted as long as the Quaternary, and thus a well-known and evident evolution of the fauna occurred during this period. The stratigraphic determination of particular phases of the Villafranchian may be based both on the phenomenon of sequential vanishing of the archaic faunal elements (e.g. mastodonts and *Tapirus*) and on the appearance of new elements. However, the determination of climatic phases of the Villafranchian on the basis of mammalian faunas is more difficult. Kurtén (1963) analyzed the composition of the faunas from numerous localities of this period in western Europe and, having calculated the share of elements associated with water, steppe, and forest, arrived at the conclusion that three pluvial periods and three dryer intervals can be distinguished in the Villafranchian. In his opinion, the pluvial periods may have corresponded to the glacial periods of the late Pleistocene, but their amplitude was smaller.

Eastern Europe has provided a very rich Villafranchian sequence of faunas of small mammals representing the period designated the Villanium by Kretzoi (1961). Their stratigraphic correlation with the particular faunas of large mammals of western Europe has not as yet been completely determined. As a whole they suggest—as a matter of course—a dryer climate than in western Europe, and steppe or even semidesert vegetation. An analysis of the abundance of the mammalian species shows a progressive cooling, but no cyclic fluctuations in temperature or humidity have as yet been demonstrated on a faunistic basis. It is nevertheless characteristic that in the Villanium period a wave of allochthonous small mammals occurred in eastern Europe, analogous with the later waves which no doubt can already be referred to the glaciations. On the whole, however, up to the end of the Villafranchian the climate was evidently warmer all over Europe than it is nowadays.

Middle Pleistocene

The fossil mammalian localities from the first European Quaternary glaciation, the Günz, are known from the western as well as eastern parts of Europe, e.g. from France (Chaline 1969), Germany (Kahlke 1961), Poland (Kowalski 1960), and Romania (Terzea and Jurcsák 1969). These faunas are, as a rule, of steppe character and contain such typical arctic mammals as lemmings, musk-ox, and reindeer. The studies carried out by Kurtén (1960) showed clear fluctuations in the size of several mammalian species, which may be interpreted as resulting from temperature changes.

The Cromerian period, named after the Cromer Forest Beds of England, was correlated with the Günz-Mindel interglacial. The studies by West and Wilson

(1966) show that so far as its climate is concerned the Cromerian of past authors is not a uniform period but consists of at least two warmer phases separated by a cooler interval. Such a detailed division cannot yet be applied for the fauna. It was, generally speaking, a fauna of temperate forests, although it contained some tropical elements such as the macaque and hippopotamus. The composition of the fauna shows distinct changes as compared with the fauna from Tegelen in Holland, referred to the pre-Günz period, and thus concludes the series of Villafranchian faunas. For example, the rodent species *Mimomys pliocaenicus,* dominant at Tegelen, had undergone an evident evolution by that time and in the Cromer Forest Beds is represented by the next stage of this evolutionary line of voles, *M. savini.*

Middle Pleistocene faunal changes in Asia cannot be correlated stratigraphically with the oldest European glaciations, although they must partly coincide with them in time. The oldest period of the middle Pleistocene is represented by the fauna from Ubejdiya in Israel (Haas 1966); in addition to the Villafranchian species of deer, it includes such elements as wild boar (*Sus scrofa*), fallow deer (*Dama mesopotamica*), and *Bison.*

The zoogeographic boundary between the Holarctic and Oriental Provinces had run along the Tsinling Shan Mountains in eastern Asia since as early as the beginnings of the Pleistocene. A faunal association called the *Stegodon-Ailuropoda* complex evolved south of these mountains throughout the Quaternary. Its existence covered the period from the beginnings of the middle Pleistocene until the late Pleistocene (Kahlke 1962); *Gigantopithecus blacki,* a huge member of the primates, was one of its elements in the middle Pleistocene. The oldest faunas of this complex still include the remnants of the Tertiary such as *Mastodon* and Chalicotheriidae. The occurrence of *Palaeoloxodon* and *Megatapirus* is characteristic of the younger faunas, and many elements of this complex, e.g. the giant panda (*Ailuropoda melanoleuca*), have lasted up to now. The older evolutionary stages of the *Stegodon-Ailuropoda* complex unambiguously suggest a warmer climate than that of today; for example, the range of the orangutan reached farther north than it does today. Pei (1963) described interesting changes in size in different members of this faunal complex. They may in part show paleoclimatic changes, or on the other hand result from evolutionary trends occurring in different mammalian groups and not be directly connected with the changes in temperature.

To return to the European situation, it must be stated that a marked cooling of climate occurred toward the end of the Cromerian period, signaling the approach of the next glaciation, the Mindel. The fauna of the middle sands of Mosbach, referred to the end of this interglacial (Kahlke 1961), was of steppe character and included *Elephas trogontherii* as its typical species. The upper sands, overlying the previous layer, contain a periglacial complex of animals including the reindeer.

The most detailed picture of the changes in fauna during the Mindel glaciation is perhaps that presented by the series of sediments from Zlatý Kuň Cave near Prague (Fejfar 1961). It contains an abundant mammalian fauna that allows the recognition of two cool periods separated by a milder phase. The earliest appearance in Europe of lemmings of the genus *Dicrostonyx* has been recorded from this locality.

The period of Mindel glaciation marks a fundamental turn in the history of the European fauna of mammals. After this period the archaic elements appear only sporadically—e.g. those preserved in the islands of the Mediterranean Sea. A complex of arctic mammals, which had probably arisen in the uplands of Asia earlier (Hoffmann and Taber 1967), became dominant in the periglacial zone of Europe at that time, permitting the precise determination of cool periods.

In Eastern Asia the localities stratigraphically corresponding to the Mindel glaciation, though not necessarily representing its coolest phase, have a climate approximating that of the present day or perhaps somewhat warmer. In this group we must number the faunas from Transbaikalia, examined by Erbajeva (1968) and, above all, the fauna from locality I at Choukoutien, accompanying the find of *Sinanthropus pekingensis*. The warm elements of the Choukoutien fauna are confined to the macaque, which among the monkeys shows the greatest climatic tolerance (Kurtén and Vasari 1960). I was in a position to ascertain that the tropical forms of bats described from this site belong, in fact, to species of the temperate zone (Kowalski and Li 1963). The rodents indicate the presence of steppe and forest environments, which agrees more or less with the present character of the Peking region.

The fauna of the Mindel-Riss interglacial resembled the modern fauna in many respects (Kurtén 1968). In England this period seems to have been cooler than the preceding Cromerian and the following Eemian interglacials, for the hippopotamus, recorded from these two interglacials, did not occur in the Mindel-Riss. The fallow deer *Dama clactoniana* was the characteristic form for western Europe.

Late Pleistocene

The uncommonly numerous fossil mammals from the late Pleistocene of Europe and also the more and more numerous finds from the northern part of Asia permit a more detailed reconstruction of life in this period than is possible for the earlier Pleistocene. The rodent-spectrum method enables better recording of even slight fluctuations in the vegetation on the basis of mammalian remains than does any other method. The most important sediments used for it are cave deposits. Because of intense mechanical weathering in cool periods these deposits accumulated quickly, preserving animal remains well. Moreover, many caves provided shelters for owls, whose pellets contain remains of small mammals, chiefly rodents, now found in the deposits. Diet studies of contemporary owls show that these birds hunt within a radius of about 5 km from their shelters and, though they give signs of some food preferences, they take to some extent all species of proper size available. The topographically diversified areas in which the caves occur usually presented a mosaic of different environments—

forest, steppe, tundra, and riverbank. Even slight climatic changes brought about shifts in the ranges of these environments and, consequently, in the numbers of rodents inhabiting them. The shifts eventually found expression in the composition of the remains accumulated in a given cave. In the Polish caves we managed to examine the rodent spectra covering the period from the Riss glaciation, through the last interglacial and the last glaciation with its cool phases and interstadials, up to the present time. The older spectra are difficult to find, though they may be stumbled upon sometimes by happy coincidence.

The Riss glaciation has a typical arctic fauna which hardly differs from that of the last glaciation and therefore contains such elements as the cave bear, wolverine, arctic fox (which appeared for the first time in this period) and, among the ungulates, the mammoth, woolly rhinoceros, and reindeer. The fauna of rodents, despite its rapid evolution, resembles that of the last glaciation, but there are some exceptions. For example, *Microtus nivalis* and *Lagurus lagurus*, unknown from the later deposits, occurred in England in this period (Kowalski 1967).

In the Eemian interglacial, Europe witnessed a warming of climate, which in its optimum provided better weather conditions than those of the present. Of the large mammals, the forest elephant and rhinoceros *Dicerorhinus kirchbergensis* must be mentioned, but the rodent fauna included no extinct species.

The European fauna of the periglacial environment of the Würm is generally known. It must be emphasized that the complex of the then living animals does not correspond to that of the modern tundra of northern Europe and western Asia, even if we allow for the extinct species; an analogy should rather be sought in eastern Asia, where the characteristic constituents of the animal world of steppe and tundra meet. Moreover, this complex was nearly uniform over the vast space of Eurasia (including Alaska) at that time. After the withdrawal of the ice sheet, the old elements of this complex were either extinct or preserved in eastern Asia (steppe lemming *Lagurus lagurus* and *Microtus gregalis*), in the Eurasian tundra (lemmings and reindeer), or in steppes (saiga antelope, ground squirrels, and hamsters).

Additional material for the determination of climatic changes during the Late Cenozoic is provided by the intercontinental migrations of mammals between Eurasia and America (Repenning 1967). At least four periods of intense exchange between the faunas of the Old and the New World can be distinguished in the interval between the present day and the middle Pliocene (Hemphillian in the American stratigraphy). Without going into the paleontological details one can say that the first wave of migration indicates the occurrence of a moist and warm forest environment in the Bering Land Bridge region. The migration wave corresponding to the Villafranchian (or the Blancan in the New World) suggests the presence of forest vegetation on the route, but with open areas and a temperate climate. The deterioration of the climate on the migration route of the mammals continued, and the great wave of migration in the late Pleistocene included only arctic species, inhabitants of steppes, tundra and, at the most, the northern zone of taiga. As time lapsed, the faunal exchange was more and more limited to one direction only. The peak was reached in the late Pleistocene; in this period 23 mammalian species passed from Eurasia to North America and none migrated from the opposite direction.

These data add evidence that the formation of the arctic complex of mammalian faunas was a unique phenomenon in the history of this group, owing to which the Pleistocene stands out in relief against all the preceding epochs of the Cenozoic. It had certainly been preceded by the gradual evolution of the mammalian faunas of the temperate zone lasting throughout the Neogene, as the general climate deteriorated on the continents of the northern hemisphere.

There are different approaches to the subject of the Late Cenozoic mammalian faunas. We may treat them as valuable material for the stratigraphic correlation of continental deposits, or look into the exciting problem of evolution and speciation of particular groups or, finally, examine the mammals as the background of the evolution of man. I have been concerned with them chiefly as indicators of paleoclimate, being aware of all the limits inherent in the material under study. In this respect, also, I recognize the fact that, although the Pleistocene saw the culmination of changes that began many millions of years earlier, it was an exceptional phenomenon in the more recent history of the earth.

REFERENCES

Chaline, J., Les rongeurs du pleistocène moyen et supérieur de France, Thesis, Université de Dijon, 1969.

Colbert, E. H., Siwalik mammals in the American Museum of Natural History, *Trans. Am. Phil. Soc.*, n.s., 26, 1935.

Crusafont Pairó, M., Análisis bioestadístico de las faunas de mamíferos fósiles del Vallés-Penedés, *Cursillos Conf. Inst. "Lucas Mallada,"* 3, 73, 1956.

Crusafont Pairó, M., Endemism and paneuropeism in Spanish fossil mammalian faunas, with special regard to the Miocene, *Comment. Biol. Soc. Sc. Fennica*, 8, 1, 1958.

Erbajeva, M. A., Taphonomie der Fundstellen von Kleinsäugerresten des Anthropogens in Westtransbaikalien, *Ber. Deut. Ges. Geol. Wiss.*, A, 13, 335, 1968.

Fejfar, O., Review of Quaternary vertebrata in Czechoslovakia, *Prace Inst. Geol. Warszawa*, 34, 109, 1961.

Freudenthal, M., and P. Y. Sondaar, Les faunes à Hipparion de Daroca (Espagne) et leur valeur pour la stratigraphie du Neogéne de l'Europe, *Proc. Koninkl. Ned. Akad. Wetenschap.*, B, 67, 473, 1964.

Gabunia, E. K., and B. A. Trofimov, Connections between Tertiary mammalian faunas of Europe and Asia (in Russian, with English summary), in *Tertiary Mammals*, edited by J. A. Orlov, p. 7, Nauka, Moskva, 1964.

Haas, G., On the vertebrate fauna of the lower Pleistocene site Ubeidiya, in *The Lower Pleistocene of the Central Jordan Valley*, edited by M. Stekelis, p. 1, The Israeli Academy of Sciences, Jerusalem, 1966.

Hoffmann, R. S., and R. D. Taber, Origin and history of holarctic tundra ecosystems, with special reference to their vertebrate faunas, in *Arctic and Alpine Environments*, edited by H. E. Wright, Jr., and W. H. Osburn, p. 143, Indiana University Press, 1967.

Kahlke, H. D., Revision der Säugetierfaunen der klassischen deutschen Pleistozän-Fundstellen von Süssenborn, Mosbach und Taubach, *Geologie*, Berlin, *10*, 493, 1961.

Kahlke, H. D., Zur relativen Chronologie ostasiatische Mittleistozän-Fundstellen und Hominoidea-Funde, in *Evolution und Hominisation*, edited by G. Kurth, p. 84, Fischer, Stuttgart, 1962.

Kowalski K., An early Pleistocene fauna of small mammals from Kamyk (Poland), *Folia Quaternaria*, 1, 1960.

Kowalski K., Paleoecology of mammals from the Pliocene and Early Pleistocene of Poland (in Polish, with English summary), *Acta Theriologica*, 8, 73, 1964.

Kowalski K., *Lagurus lagurus* (Pallas, 1773) and *Cricetus cricetus* (Linnaeus, 1758) (Rodentia, Mammalia) in the Pleistocene of England, *Acta Zool. Cracov.*, *12*, 111, 1967.

Kowalski K., *Pararhizomys hipparionum* Teilhard & Young, 1931 (Rodentia) from the Pliocene of Altan Teli, Western Mongolia, *Paleontologia Polonica*, *19*, 163, 1969.

Kowalski K., and C. K. Li, Remarks on the fauna of bats (Chiroptera) from Locality 1 at Choukoutien, *Vertebrata Palasiatica*, 7, 144, 1963.

Kretzoi, M., Stratigraphie und Chronologie, *Prace Inst. Geol. Warszawa*, *34*, 313, 1961.

Kurtén, B., The Chinese Hipparion fauna, *Comment. Biol. Soc. Sc. Fennica*, *13*, 4, 1952.

Kurtén, B., Chronology and faunal evolution of the earlier European glaciations, *Comment. Biol. Soc. Sc. Fennica*, *21*, 5, 1960.

Kurtén, B., Villafranchian faunal evolution, *Comment. Biol. Soc. Sc. Fennica*, *23*, 3, 1963.

Kurtén, B., *Pleistocene Mammals of Europe*, 317 pp., Weidenfeld and Nicolson, London, 1968.

Kurtén, B., and Y. Vasari, On the data of Peking man, *Comment. Biol. Soc. Sc. Fennica*, *23*, 7, 1960.

Nikiforova, K. V., On the stratigraphic position of the Astian, *Report VIth Intern. Congr. on Quaternary*, 2, 547, 1964.

Nikiforova, K. V., Stratigraphische Equivalente des Villafranchiens in der Sovietunion, *Proc. Koninkl. Ned. Adad. Wetenschap.*, B, *68*, 237, 1965.

Orlov, J. A., Quelques données sur la faune de l'Hipparion de la Sibérie et du Kazachstan, *Colloq. Intern. Centre Natl. Rech. Sci. (Paris)*, *104*, 319, 1962.

Pei, W. C., On the problem of the changes of body size in Quaternary mammals, *Sci. Sinica (Peking)*, *12*, 231, 1963.

Repenning, C. A., Palearctic-Nearctic mammalian dispersal in the Late Cenozoic, in *The Bering Land Bridge*, edited by D. M. Hopkins, p. 288, Stanford University Press, Stanford, 1967.

Simpson, G. G., Species density of North American Recent mammals, *Systematic Zoology*, *12*, 57, 1963.

Sych, L., Fossil Leporidae from the Pliocene and Pleistocene of Poland, *Acta Zool. Cracov.*, *10*, 1, 1965.

Terzea, E. and T. Jurcsák, Contribution à la connaissance des faunes pléistocènes moyennes de Betfia (Roumanie) (in Roumanian, with French summary), *Lucrarile Inst. Speol.*, 8, 201, 1969.

Thaler, L., Les rongeurs fossiles du Bas-Languedoc dans leurs rapports avec l'histoire des faunas et la stratigraphic du tertiaire d'Europe, *Mem. Museum Nat. Hist. Nat. (Paris)*, n. s., 17, 1966.

Thenius, E., Wirbeltierfaunen, in *Tertiär*, edited by A. Papp and E. Thenius, 328 pp., Enke, Stuttgart, 1959.

Thenius, E., Die jungtertiären Wirbeltierfaunen und Landfloren des Wiener Beckens und ihre Bedeutung für die Neogenstratigraphie, *Mitt. Geol. Ges. Wien*, 52, 203, 1960.

West, R. C., and D. G. Wilson, Cromer Forest Bed series, *Nature*, 209, 497, 1966.

C. C. Flerow

18. THE EVOLUTION OF CERTAIN MAM-
MALS DURING THE LATE CENOZOIC

Extensive information has been accumulated on the faunas of the northern hemisphere, and on the patterns of their evolution and dispersal during intervals of
the Pliocene and Quaternary periods. This knowledge is based on a study of
abundant factual data discussed in numerous papers on the systematics, paleoecology, phylogeny, and stratigraphy of mammals for the period under question
(see the figures at the end of this paper). There are many articles and summaries on the continents of Europe, Asia, and North America which give a more or
less clear idea of the composition and distribution of mammals in the northern
part of both hemispheres. On the basis of these researches we are now able to
reconstruct the fundamental stages in the development of Quaternary and Recent forms of mammals.

First it is necessary to show the role played by the changes in the fauna itself and in the living conditions of the mammals during the Pliocene and Pleistocene on the developmental processes of Recent complexes. The composition and
general nature of mammalian faunas in tropical and subtropical zones had taken
their present form by the end of the Pliocene. The population of African savannas,
as indicated by the history of its development, represents in composition and
origin the descendants of the so-called Hipparion faunas, which were widely
developed during the Neogene in the present tropical and temperate zones of
Europe, Asia, and Africa. It is known that areas which are now characterized
by temperate zone conditions had a much warmer climate and, correspondingly,
a different vegetational cover during the period of the Hipparion fauna. Mammals
included in this fauna belong to thermophilic species mostly inhabiting forest-
steppes and savannas and partly humid plains and vast river valleys with forest
thickets along the banks. Their composition is well known and need not be discussed here. During the Pliocene this complex was subjected to certain changes
and by the early Quaternary had become virtually the present savanna type.

Similarly, mammals inhabiting the Indo-Malayan area have their roots in
the Pliocene, but they are characterized by a predominance of forest and swamp
species. Typical for the fauna of this area are Indian elephants and rhinoceroses,

tapirs, wild boars, tragulides, deer, and oxen; among the carnivores are the various bears and big cats, tree-climbing rodents, and primitive primates—squirrel shrews, true apes, etc.

In genera, and to a great extent even in species, this entire association is apparently a direct successor of the Pliocene complex. Especially striking confirmations of this fact are the tapirs, tragulides, deer, and bears. Axis deer and samburs, some of the oldest representatives of the family Cervidae, have been known in Asia and Europe since the middle Pliocene. Although in the Palaearctic region of the Eurasian continent these groups, like the muntjacs (Cervulinae), were not preserved in their original state and disappeared at the end of the Pliocene and early Pleistocene, having left numerous descendants represented by more specialized genera and subgenera (*Elaphurus, Przewaliskum*), in the Indo-Malayan area they continue to exist even now. Especially striking is the pattern of changes in the former dispersal of various chevrotains—tragulides of the families Gelocidae and Tragulidae. Tragulides are living now in forest and swamp areas of tropical Asia and Africa (genera *Tragulus, Moschiola, Hyaemoschus*); in the Palaearctic region they disappeared completely during the second half of the Pliocene, whereas the Oligocene and Miocene are characterized by a wide development in Europe and in the northern half of Asia of their numerous representatives—*Lophiomeryx, Miomeryx, Prodremotherium, Dorcatherium*, and many others. During the post-Tertiary the place of chevrotains in the north is occupied by their descendants the musk deer (Moschidae), artiodactyls that stand close to them in many morphological features, but are much more highly adapted to cold climatic conditions and have specific traits for a life in low-temperature areas having a snow cover and providing a boreal or alpine vegetation for their food. An analogous picture can be observed for many other groups of mammals.

In contrast to African savanna faunas, India is characterized by many typically forest animals that are virtually absent in Africa. The reason for this difference lies in the physical and geographical conditions of the time, when Hipparion faunas formed on the Asian continent migrated to Africa. There they undoubtedly encountered obstacles that served as insurmountable barriers to the dispersal of forest animals—like vast expanses of steppes that can be easily crossed by horses, steppe rhinoceroses, elephants, antelopes, wild boars, giraffes, and big cats. Later many of these produced an abundant community species of steppe dwellers, like antelopes for instance. In India, on the contrary, real steppe mammals, like horses and the majority of antelopes, are almost completely absent.

In contrast to the south, the composition of mammals inhabiting the Holarctic sharply differs from both the Pliocene and early Pleistocene complexes. In preglacial Pleistocene the animals here included many thermophilic species; the glacial and postglacial population of the Holarctic consists of species to a greater or lesser degree adapted for a life under cold conditions with a snow cover. With the beginning of a colder spell many Pliocene species of mammals became extinct, others retreated to the south, and the rest were sufficiently adaptable to adjust themselves quickly to new conditions, in some cases undergoing fundamental morphological and physiological changes. Certain adaptive traits have appeared: cold-resistance, seasonally changing hair, an alteration of coloring, and the winter hair of the mature animal always being different from that of the young animals as well as from species of the same genera living in warmer regions. Various adaptations appeared in the limbs for movement on snow and ice.

The developmental period of the Quaternary fauna of the Holarctic is characterized by a rapid extinction of some genera widely developed in the Pliocene. At the end of the Pliocene, *Hipparion* had become extinct because of shrinkage of extensive humid plains, and tapirs completely disappeared in Europe and Palaearctic Asia; on the other hand, a substantial number of the genera remained and in a changed state became parts of newly formed complexes. Rhinoceroses and elephants, for instance, acquired a peculiar appearance. At the same time new species and genera were appearing—*Elasmotherium*, southern elephants (*Archidiscodon meridionalis*), big-horned deer (*Megaloceros*), *Alces latifrons*, and many others.

Presently available data on the Quaternary fauna indicate that the climate not only became colder but fluctuated considerably, which caused complicated changes in the entire physico-geographical medium and in the animal kingdom throughout this period. New species emerged, some species adapted to new conditions and others migrated, and new biocoenoses were formed.

At the present time we cannot trace details of this complicated sequence of events over the territory that has been inhabited by boreal faunas. It is most distinct in the Asian and especially in the European regions during the upper Paleolithic. One can assume that during the Quaternary period the vast areas for the various components of the early Quaternary faunistic complexes were gradually shrinking. There was more pronounced discontinuity in the distributions of many faunal species. This was a typical condition for a Recent fauna.

Let us summarize the main events and moments of especially distinct changes in the history of mammalian fauna that took place from the end of the Pliocene up to the Recent.

1. At the end of the Pliocene a number of species became extinct. The areal distribution of many mammals shrank considerably. Thermophilic forms—tragulides, southern species of deer (muntjacs, axis deer, samburs), tapirs, the majority of rhinoceroses and elephants, mastodonts, apes, etc.—disappeared completely in the northern parts of Asia and Europe. At the same time, during the late Pliocene and early Pleistocene, a number of new groups appeared, typical of the Pleistocene and Holocene (*Elasmotherium*, gigantic deer, true deer—*Cervus elaphus* L. and *C. nippon* Temminek—and others). In the U.S.S.R. the Taman and Tiraspol faunistic complexes correspond to this period of time.

2. At the beginning of the middle Pleistocene there was extensive development in the northern parts of Asia, Europe, and North America of truly

arctic and subarctic genera and species like the reindeer, snow sheep, musk-ox, long-horned bison, woolly rhinoceros, mammoth, polar fox, polar bear, wolverine, various lemmings, etc. These events are characteristic of the upper Paleolithic faunistic complex.

3. At the end of the Pleistocene there was an extinction of species adapted to life under glacial conditions: woolly rhinoceros, mammoth, and to a lesser degree the musk-ox and long-horned bison. The last two having survived to our time, they were gradually restricted in distribution, retreating to regions in which prevailing conditions are associated with ice; they became extinct in Europe and Asia but are preserved in North America and in Greenland.

This time is characterized by the shrinkage or, to be more exact, a northward displacement of the habitat of the majority of arctic species that had previously been displaced from the areas covered by masses of continental ice and had migrated during the second half of the Pleistocene (during the period of maximum glaciation) far to the south. Examples are the reindeer, polar fox, lemmings, wolverines, and others. It should be remarked that some species apparently migrated during postglacial time farther north than their present distribution, which indicates that in many areas the climate of the early Holocene was warmer than at present.

It should be stated that the formation of recent zoogeographic groups in the Holarctic must have taken place during the last stages of the geological and paleogeographical history of Asia, Europe, and North America. We should regard the Holocene as the time when typically arctic forms were retreating to the northeast, inasmuch as their life was associated with conditions created by the presence of continental ice. The last representatives of a truly glacial fauna—the musk-ox and the wood bison—are living testimonials of this phenomenon.

All the above evidence proves beyond dispute that a knowledge of Pliocene and early Pleistocene faunas and floras is of decisive importance for the understanding of the recent development of mammals (as well as of many other animals and certainly of plants) in the tropical and subtropical zones; to appreciate their distribution in the temperate and arctic zones one has to know the faunas and floras of the Pleistocene and early Holocene.

Some influence pertaining to the effect of man upon the changes in mammalian fauna was undoubtedly felt most in places of lengthy habitation—the Tigris-Euphrates interfluve, the Nile Valley, India, China, and Europe. The influence of man proceeded in two ways—first in a passive way determined by the development of territories in the expansion of agriculture, felling of forests, and plowing. This resulted in a displacement of, mainly, large species, a destruction of harmful animals, a transformation of the fauna by domestication, and the development of animal breeding. Second, man effects changes in the fauna by the active extermination of animals. The latter became especially effective with the invention of firearms and the development of hunting techniques.

We know that in Asia and in northern regions of America bisons became extinct in the early Holocene. In this case, apparently, we have a pattern analogous to the development of the musk-ox which became extinct all over the Palaearctic but was preserved in North America and Greenland. In other words it survived in places where climatic and general landscape conditions are associated with the presence of continental ice. Even in North America the musk-ox retreated gradually from the western bank of the Mackenzie River toward the east, closer to glacial territories. With the recession of the ice and the beginning of a warmer spell in Europe and northern Asia, followed by a change in vegetation, the retreat of the musk-ox began. Like the musk-ox, the forest bison also retreated gradually eastward, its grazing area shrinking until it found its last refuge in the forests of the Great Slave Lake area.

What can explain this extinction over a vast territory of Siberia and the northern parts of North America? Many scientists believe that the exterminating activity of man was one of the main reasons for the extinction of large Pleistocene mammals. There is no doubt that it was extensive (during the last two to three millennia especially). However, hunting was not sufficient during Paleolithic and Neolithic times to reduce substantially the number of such animals as mammoths, woolly rhinoceroses, cave bears, and lions. It has been said that elephants and rhinoceroses are still present in Africa because the population there has not been dense, but southern Asia, India, and Indochina have been very densely populated and yet elephants, rhinoceroses, and many other large mammals are still living there.

The ideas of population density of Europe and areas of Siberia held by those who believe man to have been the main cause for the extinction of large mammals will not stand up to scrutiny. According to their reasoning, Europe must have been very sparsely populated—or at least much less densely than eastern and western Siberia—because big-horned deer, aurochs, and (until recently) wisents are still preserved there, whereas they had been exterminated all over Siberia by man. This approach to the anthropic factor is demonstrably false.

It is generally known that in North America bisons flourished until the Union Pacific Railroad was built, despite the fact that the Indians had always hunted them. When the railway was completed, the Americans shot nearly all the millions of bisons. The same can be said about elephants, rhinoceroses, and quaggas in Africa, where these species have been killed by Europeans. But men of the Paleolithic, Neolithic, and even of the Bronze Age were certainly, absolutely, unable to destroy completely the populations of large animals.

What, then, are the actual causes for the extinction of Pleistocene mammals? First it should be said that extinction is determined by a *set* of causes. For different species, however, different factors are of decisive importance.

We know that in western Siberia the elk (moose) does not live in all areas having a deep snow cover, and the northern limit of its distribution beyond the Urals sharply descends southward along the Ob River to the extreme southern margins of the forest zone, rising again far to the north in eastern Siberia where the snow cover is much thinner.

The American wood bison lives in the very severe conditions of Athabasca and the Great Slave Lake area. The latter remains frozen for seven months of the year. The temperatures range to –54°C. The average temperature in January is about –24°C. Winter begins in the middle of October and lasts until the end of April.

All this, however, does not interfere with the normal existence of the wood bisons. What is the reason? It happens that the snow cover in Athabasca does not exceed an average depth of half a meter. Permanently frozen ground is at a depth of 10 cm. There is also no permanently frozen ground on the European plain, but the snow cover is thicker and the European bisons retreated for this reason to the west and south into regions with less snow. During the winter the American bison can be satisfied with snow alone and does not need water. The European bison, on the contrary, needs drinking water. In the steppes of the Transbaikal area, in Manchuria, and in Mongolia there is often no snow in winter. This fact alone created unfavorable conditions for the existence of the bisons in western Siberia and Alaska which are characterized, as is well known, by the deepest snow covers.

The European bison and the wood bison need leaves and thin branches for their food, from such trees as willow, aspen, oak, lime, mountain ash, etc. In addition, both these animals cannot live without definite types of grass fodder which are completely absent in northern Siberia and Alaska. Moreover, both species try to avoid the vast sphagnum peat muskeg and its associated plant assemblages; these muskegs and plants are widely distributed in Siberia.

The Evolution of Bisons

The earliest bisons are known from the late Pliocene of the Siwaliks and China. They represent a separate, morphologically well-defined group of the oldest bisons of the *Eobison* subgenus.

Bisons probably originated in southern Asia during the late Pliocene. Their ancestors stand close to *Leptobos*. Real *Bison* appeared in the early Pleistocene and was widely disseminated in the temperate zones of Asia and Europe. Southern Europe up to the Caucasian plains in the east was inhabited by short-horned forms—*B. tamanensis*, later *B. voigtstedtensis*, *B. lagenocornis*, and *B. schoetensacki*. By the beginning of middle Pleistocene *Bison* disappeared in southern Asia, but was widely distributed in the Holarctic (*B. schoetensacki*). During the middle Pleistocene, Europe, Asia, and the northern half of North America were inhabited by long-horned bisons: *B. priscus priscus*, *B. priscus crassicornis*, *B. priscus gigas*.

Landscape conditions throughout these territories were similar, and bisons occupied the entire area. From Asia the bisons dispersed into America (*B. chaneyi*) before the maximum glaciation (Illinoian). During the Illinoian glaciation of Canada they were forced southward and inhabited only the territory of the present United States, developing there into the gigantic *B. latifrons*. During

Wisconsin time the southern population was completely isolated from the northern population that continued to live in Asia, Beringia, and Alaska. During this time in the United States the ancestors of steppe bisons (*B. alleni*) and the later steppe bisons (*B. bison*) were developed. After the disappearance of the Canadian glacial sheet bisons again penetrated into Canada from the northwest and Alaska.

As the ice sheet disappeared the bisons began to get smaller. The long-horned *B. priscus priscus* and *B. priscus crassicornis* were replaced by short-horned *B. priscus mediator* and *B. priscus athabascae*. They also inhabited a very extensive area. At the end of the Würm that area was broken up into parts and by the beginning of the Holocene it was completely separated. Bison were preserved only in Europe (excepting its northern part), eastern Siberia, Alaska, Canada, and the United States. Independent populations were being formed and the bisons began to differentiate. An endemic form originated in Europe—*B. bonasus*.

A large *B. priscus athabascae* lived in eastern Siberia, Alaska, and Canada, which soon became extinct in Asia and is now preserved only in Canada.

By the beginning of the Holocene three geographically isolated populations had originated, with independent lines of historical development, very different ecologically and morphologically and well adapted to different conditions: *B. bonasus* inhabiting broad-leaved and mixed forests of Europe with a temperate climate and thin snow cover; the plains bison (*B. bison*) lived in the open expanses of the southern half of North America; and the woods bison (*B. priscus athabascae*) lived in the severe climate of the northern taiga forests of Canada.

Stratigraphical Ranges of the Main Mammalian Groups
of Europe and Northern Asia

						Zygolophodon borsoni	Anancus arvernensis	Archidiscodon			Mammuthus		Palaeoloxodon	
					Faunistic complexes			gromovi	meridionalis	wüsti *	chosaricus	primigenius	antiquus	namadicus
RECENT					Living									
PLEISTOCENE	W				Upper Paleolithic							Late Form		
	R-W											early form		?
	R	Kha za rian												
	M-R	Singili- an												
	M	Tiraspo- lian												
	G-M													
VILLAFRANCHIAN	G	Tamanian												
	D	Khaprovian												
	pre-D	Moldavian												

* *Mammuthus trogontherii trogontherii*

	VILLAFRANCHIAN				PLEISTOCENE					RECENT
	pre-D	D	G	G-M	M	M-R	R	R-W	W	
	Moldavian	Khaprovian	Tamanian		Tiraspolian	Singili-an	Kha-za-rian	Upper Paleolithic		Living
Faunistic complexes										
Dicerorhinus etruscus										
Kirkhbergensis										
hemitoechus							?			
Hanxatherium										
Coelodonta early species										
tologoijensis					?					
antiquitatis										
Elasmotherium										

VILLAFRANCHIAN				PLEISTOCENE					RECENT	Faunistic complexes
pre-D	D	G	G-M	M	M-R	R	R-W	W	Living	
Moldavian	Khaprovian	Tamanian		Tiraspolian	Singilian	Khazarian	Upper Paleolithic		Living	
										Hipparion
										Probosci-dipparion
										Plesippus — stenonis-group
										süssenbornensis-gr.
										E.sivalensis-sanmeniensis-group
										Equus — Simionescui
										mosbachensis-group
										caballus

sivalensis · sanmeniensis · early form · late form

large · small

Faunistic complexes	VILLAFRANCHIAN			PLEISTOCENE						RECENT
	pre-D	D	G	G-M	M	M-R	R	R-W	W	
	Moldavian	Khaprovian	Tamanian		Tiraspolian	Singi-lian	Khazarian		Upper Paleolithic	Living
Paraca-melus	╎	╎	╎							
Camelus					?	╎	╎	╎	╎	
Euctenoce-ros, Euclado-ceros		‖								
Proeme-gaceros		?	‖	‖	‖					
Praeda-ma					‖					
Megalo-ceros						‖	‖	‖	‖	‖
Libral-ces			?	?	‖					
Alces loti-frons		╎	╎	╎	╎ A.l. postremus	╎	╎	╎	╎	╎
Alces alces					‖	‖	‖	‖	‖	‖
Rangifer					‖ ?	‖	‖	‖	‖	‖

Bison (Bison)

Faunistic complexes · Soergelia · Boopsis · Praeovibos · Ovibos · Bison (Eobison) · voigtstedtensis · schoetensacki · priscus · bonasus · Bos primigenius

RECENT — Living
PLEISTOCENE — W · R-W · R (Khazarian) · M-R (Singilian) · M (Tiraspolian) · G-M
VILLAFRANCHIAN — G (Tamanian) · D (Khaprovian) · pre-D (Moldavian)

Upper Paleolithic

B.p. occidentalis
B.p. mediator **
B. priscus crassicornis
B priscus priscus **
recticornis pallantis
süssenbornensis
sivalensis · palaeosinensis · tamanensis

* In European Mindel I - B. schoetensacki lagenocornis,
in the end of Mindel I and Mindel II B.s. schoetensacki.
** In Europe and West Siberia.
*** In East Siberia.
Note: in Rissian of Southern Siberia and Mongolia-Bison
priscus gigas.

	VILLAFRANCHIAN			PLEISTOCENE						RECENT	Faunistic complexes
	pre-D	D	G	G-M	M	M-R	R	R-W	W	Living	
	Moldavian	Khaprovian	Tamanian		Tiraspo-lian	Singi-lian	Kha-za-rian	Upper Paleolithic		Living	
Ursus etruscus											
Ursus deningeri		?									
Ursus spelaeus						?	?				
Ursus arctos											

Legend

—··—··— Northern Asia

———— Eastern Europe (within U.S.S.R.)

══════ Western Europe

Compiled by C. C. Flerow and A. V. Sher after their original work and after papers by: W. Gromova, *Brief Review of Quaternary Mammals of Europe* (in Russian), 143 pp., Moscow, 1965; and E. Vangengeim and V. Zazhigin, in *The Main Problems of Anthropogene Geology in Eurasia*, pp. 47–59, Moscow, 1969. W = Würm, R = Riss, M = Mindel, G = Günz, D = Donau.

Walter William Bishop

19. THE LATE CENOZOIC HISTORY OF EAST AFRICA IN RELATION TO HOMINOID EVOLUTION

It is not possible to understand the literature that records details of the Late Cenozoic history of East Africa without an appreciation of the principal phases through which studies in the area have developed since 1890. A brief history of investigations over the last eighty years is outlined in Figure 1.

Geological exploration commenced with the classic foot safari of J. W. Gregory to the Kenya Rift valley which now bears his name. This journey may be said to have opened the first "normal" stage of Cenozoic investigation. Rift valley environments were described for the first time in relation to tectonic and volcanic histories while the existence of former wet basins was given recognition where relevant.

In 1914 the influence of scientists, concerned with the pattern of climatic change on a worldwide basis, led to broader studies. Various hypotheses were put forward which inferred the presence of tropical pluvials as the low-latitude correlatives of temperate glaciations. For over twenty years field observations in East Africa were to be colored by theoretical climatic expectations. The end points of this phase of investigation, dominated by climatological theory, are seen in the papers of Brooks (1914) and Simpson (1934). In East Africa, evidence of the former greater extent of present lakes was sought and found.

The value of this period of development was to focus interest upon the study of depositional histories in low-lying lacustrine basins. Its danger was that pluvial explanations were applied to virtually all situations where strandlines or lacustrine sediments were found above present lake levels. Some normal geological research continued through this period in Kenya (Gregory 1921) and Uganda (Wayland 1926, 1934), but in general the effects of tectonic instability and contemporaneous volcanic activity were ignored.

In 1939 two significant events were the onset of a war-enforced moratorium in field studies and the publication of a paper by Solomon (1939) criticizing the pluvial-interpluvial concept. Solomon had been associated with Leakey in inves-

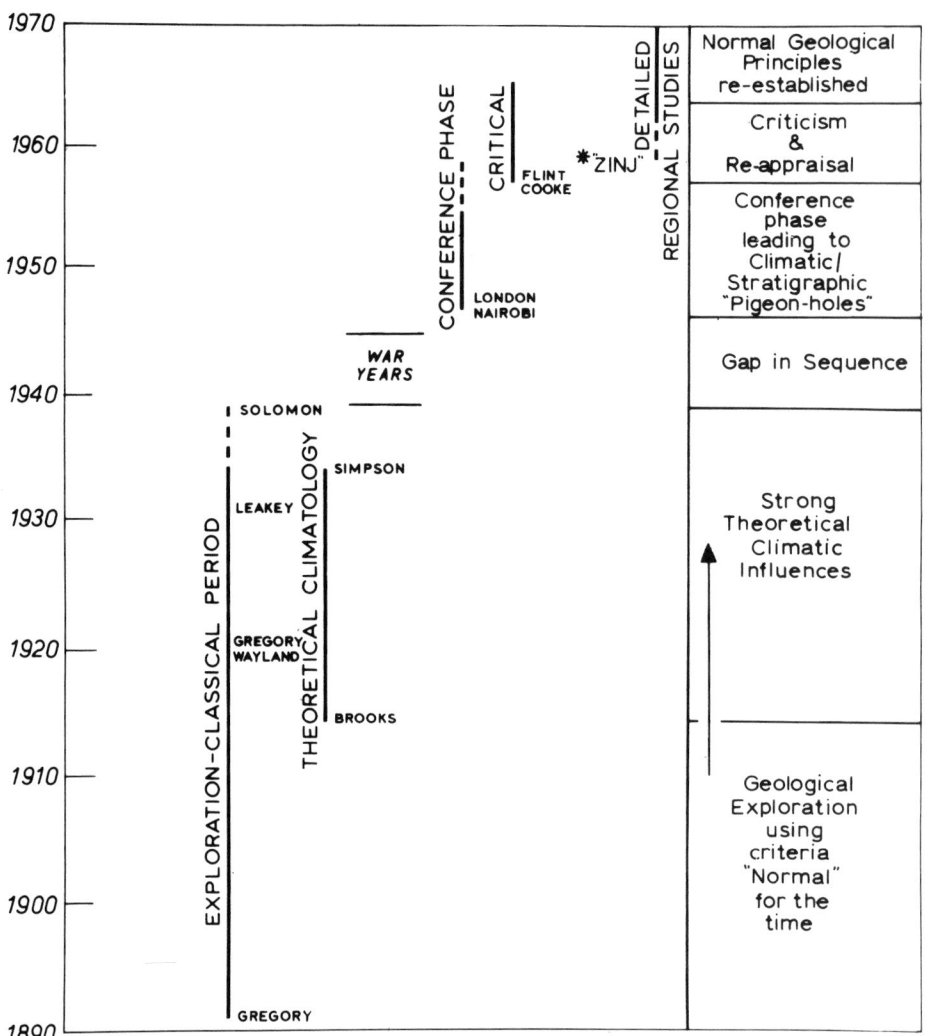

Fig. 1. History of Late Cenozoic geological investigations in East Africa.

tigating the type area of the Gamblian pluvial (Leakey and Solomon 1929; Leakey 1931; Nilsson 1932). However, further work in Uganda with O'Brien (1939) convinced him that "the Pluvial Hypothesis rests on very slender foundations."

Because of the war years this first salvo of criticism passed virtually unnoticed. Indeed the postwar period saw a veneer of conference respectability added to the so-called stratigraphic–climatic terminology. After an initial meeting in Nairobi in 1947 (Leakey 1952) and a subsequent report to the International Geological Congress held in London in 1948 (Sandford and Blondel 1951), succeeding Pan African congresses on prehistory (Clark 1957; Mortelmans and

Nenquin 1962) engraved the pluvial terminology ever more deeply upon the Pleistocene stratigraphic record. These all too attractive terminological pigeon-holes were extrapolated from their type localities in East Africa and used (or, more correctly, misused) over the whole of the African continent.

This situation continued until Cooke (1958) and Flint (1959) voiced strong criticisms which led to reappraisal of the evidence at each of the alleged type localities. This revealed the unsure sedimentary foundations upon which the whole sequence of inferred climatic changes had been erected. The "critical" phase continued until 1965 when it culminated in a conference which was unanimous in recommending a return to normal stratigraphic principles (Bishop and Clark 1967) as illustrated in various national stratigraphic codes.

The year 1959 was important not only in the publication of Flint's critical review paper but also for the discovery by Mary Leakey of the skull of the homi-nid *Australopithecus* (*Zinjanthropus*) *boisei* at Olduvai Gorge in northern Tan-zania. This proved to be the first of a steadily increasing number of specimens belonging to the australopithecine/hominine "solid solution series" which East Africa has yielded during the last ten years. These finds of potential ancestors of *Homo sapiens* have been the fruits of diligent investigations initiated by Dr. and Mrs. Leakey but more recently carried out by numerous workers. The recent phase of research was triggered off by the finding of *A. boisei* ("Zinj") which led to increased input of funds to support further geological and archaeological ex-plorations, often in difficult terrain. It will be suggested below that unusual com-binations of paleoenvironmental conditions have contributed to the survival of this unrivaled historical record of the morphology and activities of man's forebears and of their mammalian contemporaries.

The aim of this paper is to outline the stratigraphic time scale against which the discoveries of hominoid fossils must be viewed, and to review in detail only the research of the last ten years. For the purposes of the paper East Africa is restricted to Kenya, Tanzania, and Uganda with brief mention of the Omo Valley in southern Ethiopia. However, so much research has been carried out, with a great deal of it still continuing, and so abundant have been the finds that it is possible to discuss only six areas. Five of these, labeled A, B, D, E, and F in Fig-ure 2, occur in Kenya and northern Tanzania within or near the eastern or Gregory Rift valley. Area C, in the western or Albertine Rift in Uganda is referred to only briefly.

AREA A. THE FOSSILIFEROUS MIOCENE VOLCANICS OF THE KENYA-UGANDA BORDER

To assist in defining a time base for the Late Cenozoic, East Africa provides a sharp stratigraphic break approximately at the base of the Miocene ($\pm$ 25 m.y.). Numerous potassium-argon ages have been established for rocks bracketing or in-terbedded with Miocene fossil mammal assemblages (Bishop et al. 1969). These suggest that the earliest mammalian and hominoid assemblage in Africa south of the Sahara is that lying beneath the volcanics of Mount Elgon, Uganda (Fig. 2;

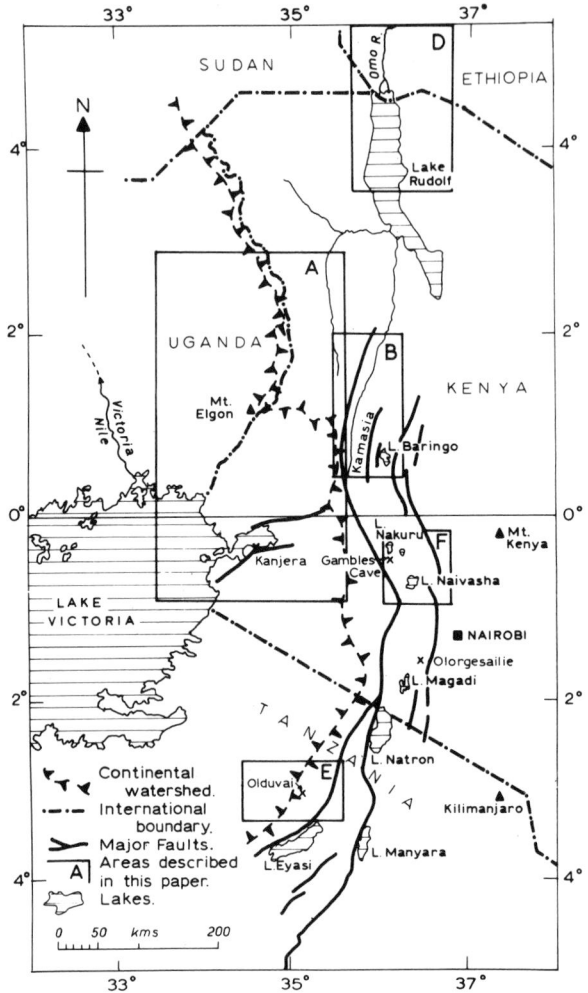

Fig. 2. Locality map showing the principal areas of East Africa discussed in this paper.

Walker et al. 1969). Potassium-argon ages of about 22± and 24± m.y. have been obtained for lavas above and below the fossiliferous sediments (Bishop et al. 1969). The mammalian fauna from this locality includes the hylobatine genus *Pliopithecus*. The only earlier African mammalian assemblages for which isotopic dates are available are from the Fayum Beds north of the Sahara. Here a potassium-argon age for lava overlying the sedimentary sequence suggests that the fossil primates from these localities are probably of an age in excess of 25 m.y. and perhaps more than 30 m.y.

Figure 3 illustrates diagrammatically the probable range through Miocene time of activity from the five central vent volcanoes of Rangwa, Tinderet, Elgon, Napak, and Moroto. These cones, now in varying stages of dissection, straddle the intercontinental watershed along the Kenya-Uganda border (Area A, Fig. 2).

The five volcanoes are associated with important Miocene fossil mammal localities as follows (for detailed references see Bishop 1963, 1967, 1968; Van Couvering and Miller 1969):

Rangwa center	Rusinga, Mfwanganu, and Karungu localities
Tinderet area	Songhor, Koru, and Fort Ternan localities
Mt. Elgon center	Bukwa locality
Napak center	Napak I, IV, V, and IX and several other minor localities
Moroto center	Localities Moroto I and II

As seen from Figure 3, although activity associated with the Rangwa volcano may have commenced as early as 22.5 m.y. in the vicinity of the Karungu fossil locality, all the remaining earlier Miocene fossil mammal- and hominoid-bearing localities fall within the two-million year bracket from 20 to 18 m.y. Volcanicity at Moroto may have continued until 14+ m.y., but the mammalian assemblages peripheral to the base of the Moroto volcanic pile are similar to those from Napak and are probably of an equivalent age.

There are numbers of other fossiliferous localities (Bishop 1963, 1967) which are not shown in Figure 3. They have not been firmly dated and have yielded only limited mammalian fossils and insignificant numbers of hominoids compared with the localities illustrated. Up to 1963 the total published Miocene hominoid fossils from East African localities was 355. Of these the Rangwa volcano had yielded 198, mainly from Rusinga Island, and the Tinderet area had provided a further 132, leaving a total of only 25 from all other localities. Further finds have been made from the Rangwa volcano and Tinderet area localities but are unpublished in detail. The finds from the Napak volcano since the first discovery of fossils in 1958 now total 48 hominoid fragments and from Moroto since 1961 numerous fragments representing three individuals. Very few additional hominoid fragments have been recovered from other localities.

The hominoids obtained from the earlier Miocene of Kenya and Uganda since the first discoveries (Hopwood 1933) include abundant remains of *Dryopithecus* (*Proconsul*) *major*, *D.* (*P.*) *nyanzae* and *D.* (*P.*) *africanus*, and of the hylobatine genus *Pliopithecus* (*Limnopithecus*) *legetet* and *P.* (*L.*) *macinesi*. Even without the additional figures for unpublished hominoid finds from the Rangwa and Tinderet volcanoes, the concentration of well-preserved fossils in the time span of 18 to 20 m.y. is very clear, particularly following Van Couvering's remapping of the Rusinga Island localities (Van Couvering and Miller 1969).

The occurrence of this concentration, and indeed of virtually all the important mammalian localities, in association with the five carbonatite-nephelinite volcanoes appears to argue some relationship between the volcanic deposits and the existence of conditions suitable for fossilization. Some fossils occur in conglomerates and gravels, interpreted as representing fluviatile channels on the Basement Complex surface beneath the volcanic sequences. However, the bulk of the finds and all the best-preserved specimens come from subaerial tuffs or, more locally, tuffaceous lacustrine deposits. These accumulated on the flanks of the intermit-

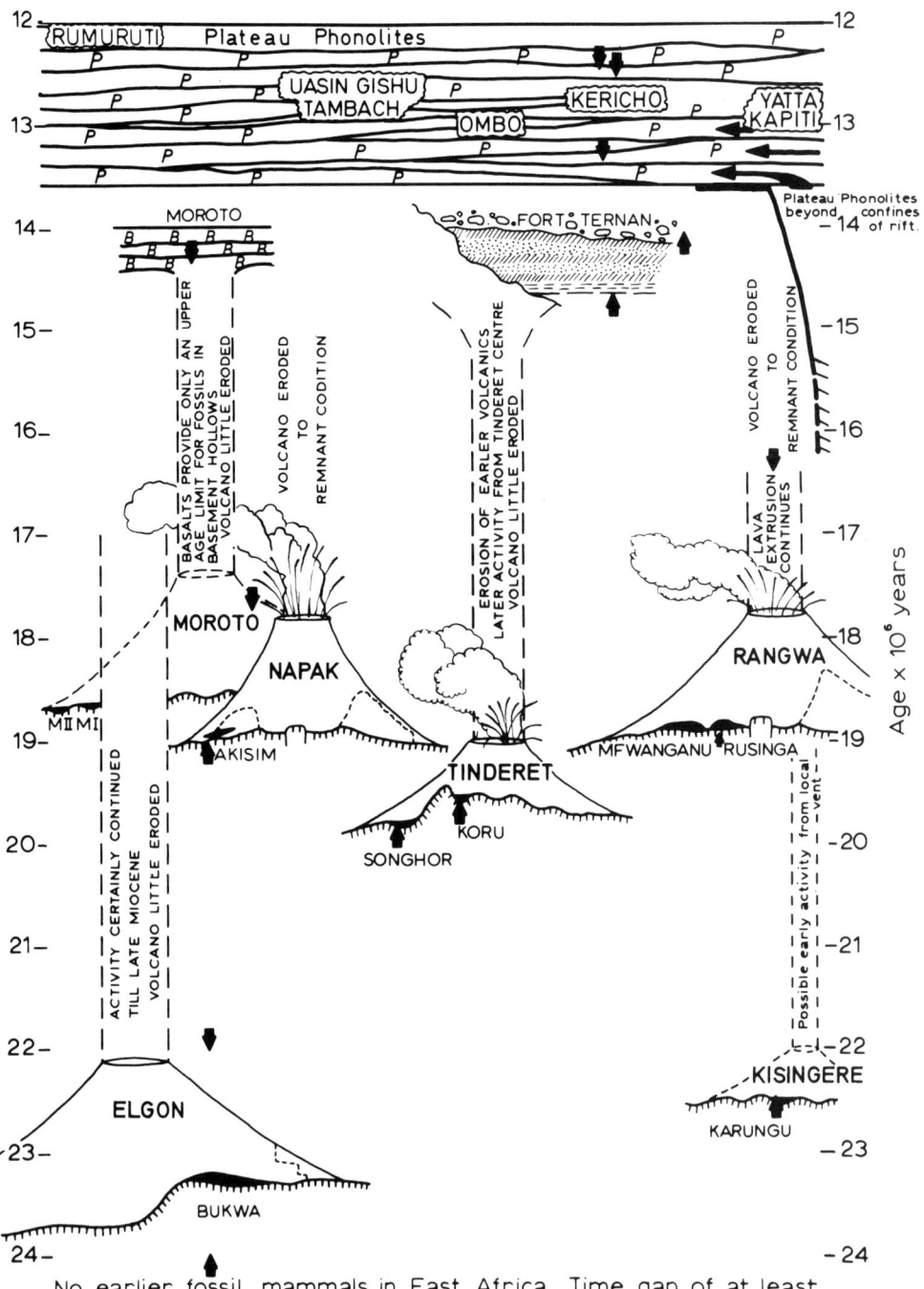

Fig. 3. Volcanic centers and fossiliferous localities of the Kenya-Uganda border through Miocene time. Arrows indicate important isotopic dates. Fossiliferous sediments shown in black but cross-hatched for Fort Ternan.

tently active volcanoes and became interbedded with the growing piles of volcanic debris.

In the last ten years investigations at newly discovered localities have concentrated upon trying to understand the nature of the total fossil assemblages. Prior to the discovery of the first fossils from Napak in 1958, each Miocene locality was represented only by a faunal list. This was usually based solely upon identifications of dental and cranial material and gave no idea of relative abundance of different species or groups. Frequently, specimens were preselected as "identifiable" and "not identifiable" and the latter, consisting mainly of postcranial fragments, were not collected. The total analysis of the Napak and Moroto assemblages (Bishop 1958, 1968; Bishop and Whyte 1962) and the as yet unpublished analysis of the Fort Ternan fauna by Gentry, have revealed the dangers of subjective and partial collecting. They also underline the value of additional information that can be obtained from the total investigation of natural associations of bone and tooth.

Table 1 gives figures for an analysis of 7876 specimens of fossils collected at four localities within one lithological unit at Napak. These figures were obtained by combining the individual details from each of the four localities.

Detailed conclusions to be drawn from these and similar figures must await completion of comparative studies. However, certain factors are immediately evident—for instance, the consistently high percentage of skulls, teeth, and jaws (14.2) in the total assemblages; the numerical dominance of rodents in the death assemblage; and the remarkably high percentage of primates when considered against the normal scant occurrence of these animals in the fossil record. Some preliminary discussions of these problems and others suggested by the extremely par-

Table 1. Total Analysis of Specimens from Napak Fossiliferous Tuffs

Napak I, IV, V, IX; total mammalian pieces, 7876

Indeterminate fragments	69.6%
"Better" bones (probably identifiable)	16.2%
Skulls, teeth, and jaws	14.2%

Analysis of skulls, teeth, and jaws, 1119 pieces comprising 14.2% of total

Rodent	64.0%
Proboscidea	10.9%
Artiodactyl	9.3%
Primate	6.0%
Others	9.8%

Analysis of a sample of 305 postcranial "better" bones

Vertebrae	39.3%	Calcaneum	6.3%
Phalanges	16.7%	Metapodials	5.9%
Femur fragments	9.2%	Humerus	4.9%
Astragalus	9.2%	Others	8.5%

tial nature of the survival record of certain bones from the postcranial skeleton have been touched upon by Bishop (1968).

Here it must suffice to note several conclusions that have resulted from such attempts to collect complete associations of fossil bone and to study them in relation to their environmental setting. They include:

1. Recognition of the consistent pattern of selective destruction of skeletal material prior to burial or stratification. This may result from weathering or removal by erosion on land surfaces and/or the action of predators.

2. Indication of the role of carbonatite volcanics in providing suitable chemical environments for the preservation of bone and tooth as suggested by studies of trace elements (Dawson 1962, 1964; Bishop 1968).

3. Recognition of the need to consider fossils as an integral part of the raw material of the sediments in which they occur. This leads naturally to the study of the sedimentology of the fossiliferous deposits and of their inferred environments of deposition.

4. Appreciation of the lack of comparative data from present-day situations and particularly of how modern carcasses and skeletons survive weathering processes and predator activities in being converted from potential into actual fossils. Studies in taphonomy are currently being undertaken in Kenya and Uganda game parks. These should provide information to allow more accurate interpretation of environments of death. Such information is of crucial importance before any attempt can be made to reconstruct possible environments occupied by man and animals during life.

5. Recognition of the varied environmental niches occurring in East Africa, particularly within the rift system, which has led to geomorphological studies of present-day processes in selected areas in a deliberate attempt to understand and reconstruct Cenozoic environments.

The faunal sequence is extended upward, into the later Miocene, from the concentration of evidence falling in the 18- to 20-m.y. age span, by the rich mammalian assemblage of Fort Ternan. This is at present the most accurately dated Miocene fossil locality in East Africa at 14± m.y. This results from the close correspondence between K^{40}/Ar^{40} ages, established at two laboratories, for biotite from a horizon immediately below the fossiliferous strata, and a sequence of whole-rock ages for overlying phonolite lavas (Bishop et al. 1969). The situation of the locality is illustrated diagrammatically in Figure 3.

The discovery by a Kenya farmer, Mr. F. Wicker, of this well-dated mammalian assemblage from subaerial tuffs was a valuable addition to the East African Miocene record (Leakey 1962; Bishop and Whyte 1962). The hominoids represented include a hylobatine, a dryopithecine, and an early hominid—the much-discussed *Ramapithecus* [= *Kenyapithecus*] *wickeri*. The Fort Ternan assemblage stands rather on its own, being separated from the earlier Miocene assemblages by at least 3 million years which are represented locally only by erosion of the earlier nepheline-bearing lavas and tuffs. The almost complete faunal turnover between the 18± m.y. and 14± m.y. assemblages appears to confirm the time gap indicated by the potassium-argon ages.

The lavas which effectively seal the Fort Ternan fossiliferous sediments are part of a widespread occurrence of plateau phonolites. Although comprised of flows from various sources they form a convenient stratigraphical datum of later Miocene age. Potassium-argon ages have been established for samples of these petrologically similar phonolite flows from localities up to 200 miles apart. They yield sixteen tightly grouped dates between 12.0 ± 0.3 and 13.6 ± 0.6 m.y. (Bishop et al. 1969). This suggests that the phonolites may be a convenient indicator of the uppermost Miocene in Kenya.

AREA B. THE FORMER "PLIOCENE GAP"

Area B of Figure 2 lies to the east of Area A and is situated between the midline of the Gregory Rift, approximately through Lake Baringo, and its western boundary fault here marked by the fall of the Elgeyo escarpment from 9000 to 4000 feet. The principal relief feature in the area is the tilt block of the Kamasia or, more correctly, the Tugen Hills. On the east, faults step down strata from the crest of the range at 8500 feet to below Lake Baringo (3500 feet). On the west, a dissected dip slope descends into the Kerio Valley (4000 feet). The range was the type area of the so-called Kamasian pluvial and the region in which Gregory during his exploratory journey recognized the sediments which he called Kamasia Lake Beds. These sediments were raised to the formal status of a pluvial by the 1947 Pan African Congress (Leakey 1952). This decision was ratified by the 1948 International Geological Congress despite criticism by Shackleton that as the area was not mapped geologically it was unsuitable for designation as a type locality (Sandford and Blondel 1951).

It remained unmapped until officers of the Kenya Geological Survey commenced reconnaissance mapping of the southern part of Area B early in the 1960s (McCall et al. 1967). In 1965 a more detailed mapping program of the area was undertaken by the East African Geological Research Unit based at Bedford College, University of London, under the direction of Professor B. C. King.

The investigations of research students engaged in the mapping project are still largely unpublished. However, preliminary descriptions of sediments and faunas of Plio-Pleistocene age have been given by Martyn (1967, 1969, 1970) and of earlier Pliocene age by Bishop and Chapman (1970). In most areas of the Gregory Rift all earlier strata are obscured by mid- to later Pleistocene sediments and lavas. However, along the eastern face of the Tugen (Kamasia) Hills faultarch, a long sequence is exposed from Miocene through Pliocene into Pleistocene. This succession rests upon Basement Complex rocks seen to be over 1000 feet in thickness at the base of the escarpment. Almost 10,000 feet of later Cenozoic volcanics and sediments overlying the Basement can be inferred for this portion of the rift. The older 6000± feet of this succession outcrop in the upper part of the main escarpment and are capped by the Kabarnet trachyte, aged about 7.0± m.y.

The younger rocks are exposed in the grid-faulted region between the scarp foot and Lake Baringo (Fig. 2).

As mapping is still in progress, only preliminary details of lavas, sediments, faunal content, and isotopic ages are given in Table 2, together with indications of the major erosional and tectonic episodes. It is evident from the nature of the six sedimentary units described in Table 2 that deposition has occurred in this part of the rift, as seen in a sectional view through geological time, in a pattern identical with that viewed in plan from an aircraft or satellite above the Gregory Rift at the present day. Sedimentary basins have formed from time to time as the result of interplay of faulting and volcanicity. Their infill reflects the local hydrographic situation with volcanicity playing an intermittent role in supplying detritus. Basins eventually cease to hold water, or to be catchments for sediments, as the local erosional, tectonic, or volcanic situation changes.

Against such a background it is not possible to see the more subtle changes initiated by climatic change until comparative stability is attained, albeit only temporarily, in the later Pleistocene. Thus the Kokwob Beds at the upper terminus of the stratigraphic column may prove to correlate with climatically induced changes of lake level in other East African wet basins (see below).

The type section of the former Kamasian pluvial is seen to contain six sedimentary episodes ranging in age from 10 or 11± m.y. to later Pleistocene. There is no evidence that can be interpreted as support for a major earlier Pleistocene pluvial.

The interbedded sequence of lavas and sediments spans, with breaks, the whole of Pliocene time. At the base of the sequence the strata are the time equivalent of Fort Ternan, although no fauna of this age has yet been found from Area B. The upper three sedimentary units, of Pliocene to late Pleistocene age, are contemporary with the well-dated sequences from Areas D, E, and elsewhere which will be discussed below.

The value of the Tugen Hills–Baringo area sequence lies in:

1. Bridging the "Pliocene" gap in the sedimentary sequence and fossil mammal record in Africa south of the Sahara. The gap formerly ranged from the Fort Ternan fossiliferous sediments (14± m.y.) to the base of the Omo fossiliferous sediments and their equivalents (4± m.y.).

2. Providing, in the form of the crown of a single upper molar from the Ngorora formation (between 9 and 12 m.y.), the first specimen from an approximately 10-million-year gap in the hominoid fossil record (Bishop and Chapman 1970).

3. Providing a sequence of mammalian associations which although they do not contain abundant fossils are already sufficient to show, when calibrated by potassium-argon dates, that many assemblages containing fossils which were originally lumped as Villafranchian and placed in the earlier Pleistocene are in fact dated at 7.0± m.y., or 5.0± m.y., or 2 to 3± m.y. The

Table 2. Simplified Succession for the Tugen (Kamasia) Hills and the
Area West of Lake Baringo (Area B, Figure 2)

Relative age	Lithological unit	Preliminary details
Later Pleistocene	Kokwob Beds	Max. thickness 9 ft; raised strandlines of Lake Baringo from 60 to 12 ft above present lake level containing abundant bivalves and ostracods with some fish and occasional mammalian bones; isotopic ages awaited but probably < 10,000 yrs B.P.

Faulting and Erosion

Relative age	Lithological unit	Preliminary details
Later Pleistocene	Kapthurin formation	Max. 300 ft; the fluviatile and tuffaceous sediments divided into five members (Martyn 1970); two fossiliferous localities, one yielding mammals (Fuchs 1950) and the other artifacts plus mammals, including a hominid mandible (M. Leakey 1970)

Faulting and Erosion

Relative age	Lithological unit	Preliminary details
Earlier to mid-Pleistocene	Volcanics	Lavas (phonolite, trachyte, mugearite, etc.) extruded over a long time span; the potassium-argon dates established through the lavas range upward from 2.0± m.y. near the base
Later Pliocene	Chemeron formation	Max. 700 ft; the fluviatile to lacustrine sediments divided into five members (Martyn 1967, 1969); twelve major fossiliferous localities known), yielding mammals, reptiles, fish, a few mollusca, and fossil wood; the fauna is concentrated at three stratigraphic levels, the youngest yielding an australopithecine temporal fragment
		(a) *Kipcherere Beds;* broadly the time equivalent of the Chemeron formation (Martyn 1969) with four fossiliferous localities along the outcrop of the lowest of the three members
		(b) *Kaperyon Beds;* the time equivalent of the Chemeron formation at least for part of their 600-ft thickness but have yielded only a limited mammalian fauna

Faulting and Deep Weathering

Relative age	Lithological unit	Preliminary details
Later Pliocene	Basaltic volcanics	Max. 1500 ft; basaltic lava flows with some interbedded tuffs
Later Pliocene	Lukeino Beds	Max. 200–250 ft; fluvio-lacustrine sediments have yielded a sparse mammalian fauna

(*cont.*)

Table 2 (*cont.*)

Relative age	Lithological unit	Preliminary details
		Faulting and Erosion
Mid-Pliocene	Kabarnet trachyte	Max. 800 ft; trachyte lava flows; three potassium-argon dates established for lower flows: 6.7 ± .03, 6.8 ± 0.2, 7.2 ± 0.3 m.y.
	Mpesida Beds	Max. 100 ft; local pockets of sediment between Kabarnet trachyte flows; several mammalian localities are known but deposits are only sparsely fossiliferous
		Faulting, Weathering and Erosion *Angular Unconformity*
Earlier Pliocene	Volcanics	Basaltic lava flows, 120 and 45 ft in various episodes; phonolitic lava flows 800–1000 ft; lowest flow gives a preliminary potassium-argon date of approximately 9± m.y.
Earlier Pliocene	Ngorora formation	Max. 1200+ ft; Ngorora sediments contain numerous channel conglomerates concentrated at two stratigraphic levels yielding fossil mammals and other vertebrates (Bishop and Chapman 1970); lateral change of facies occurs into sediments predominantly of shale and tuffaceous sandstone lithology; shales contain well-preserved fish (*Tilapia* sp.)
Earlier Pliocene and late to mid-Miocene	Volcanics	Max. 4000+ ft; a thick sequence of largely phonolitic lava flows (some basanites, tephrites, and mugearites) with interbedded tuffs, shales, and grits; uppermost phonolite flows yielded preliminary potassium-argon dates of approximately 12± m.y.
		Phonolites slightly lower in the sequence are probably lateral equivalent of Uasin Gishu Plateau phonolites dated at 12.0 ± 0.3 to 13.6 ± 0.6 m.y. (Bishop et al. 1969); dates for phonolites earlier in sequence show that these lavas range back at least to 16± m.y.

older faunas should undoubtedly be assigned to the Pliocene, despite some uncertainty over the definition of the Plio-Pleistocene boundary.

4. The fact that many of the lavas and pyroclastic rocks in the sequence contain fresh potash feldspars makes them eminently suitable for K^{40}/Ar^{40} dating. This, combined with the use of a portable fluxgate magnetometer to

establish the paleomagnetic polarity of individual flow units, gives the possibility of recognizing and calibrating major polarity epochs earlier than the Matuyama-Gauss-Gilbert, reversed-normal-reversed sequences. The establishing of precise boundaries between epochs and the failure to recognize shorter "events" is limited by the size of the probable error of the potassium-argon ages in relation to the length of the period of reversed or normal polarity.

5. Providing an intermittent sequence of sedimentary units spanning the Pliocene for the first time in Africa south of the Sahara. Similar sequences could possibly be established elsewhere in the rift by boring beneath the mid- to later Pleistocene cover rocks. The Pliocene strata record a change from the largely carbonatite-nephelinite central vent volcanoes in which the Miocene fossils are preserved. These succeeding strata indicate characteristically fluvio-lacustrine environments, with pumice tuffs, trachytic and phonolitic lavas, and subsidiary basalt flows and their weathering products providing the characteristic debris which infills sedimentary basins. Further search will undoubtedly increase the number of fossils from the Pliocene localities. However, it seems likely that the basic change in the conditions of preservation of fossils will be reflected in the different proportions of mammals represented. The change from conditions favoring fossilization of small fauna in general and hominoids in particular during the Miocene may be indicated by the molar crown which is at present the sole surviving representative of 9 to 10 million years of hominoid evolution.

AREA C. THE ALBERTINE RIFT VALLEY

The Uganda western rift is not illustrated in Figure 2 and only brief reference will be made to this area. The general setting of this branch of the rift is a mirror image, to the west of Lake Victoria, of the Gregory Rift to the east. The similarities between the eastern and western branches of the rift system end with this broad symmetry of layout.

The Albertine Rift is a classic graben, as illustrated in Holmes (1965, Figs. 766, 768) and in various other textbooks. The simplicity of physical setting is reflected in the sedimentary record which has been typically that of a wet, fault-bounded basin since Miocene time. Evidence of volcanicity is absent, and tectonic activity appears to have been restricted to repeated movements of the boundary faults. Grid faulting on the floor of the graben has been limited to minor movements of later Pleistocene age. Lake Albert is deeper than the small fault-bounded lake basins characteristic of the Gregory Rift (Fig. 2). Along much of its shoreline Lake Albert laps directly against the escarpment walls which delimit the graben. The sediments accumulating on the present lake bed represent the uppermost stratum of a long record of lacustrine deposition.

Up-to-date accounts of the stratigraphy of the Albertine Rift have been given elsewhere (Bishop 1965, 1969) and for the Congo sections of the western rift by

Heinzelin (1963) and Gautier (1965, 1967). Faunal evidence has been discussed by Hooijer (1963) and Cooke and Coryndon (1970). Geophysical investigations suggest a maximum thickness of over 8000 feet of deposits in the center of the graben (Brown 1956) and a bore hole penetrated over 4000 feet of sediments before running into the rift wall (Harris et al. 1956). The best-exposed sequence occurs at the southwest end of the lake and is summarized in Table 3.

As might be expected in a persistently lacustrine environment, the Albertine Rift has yielded no evidence of hominoid fossils or artifacts earlier than the mid- and later Pleistocene. It is proposed to discuss here only the Kaiso formation and in particular the fossiliferous ironstones in view of the climatic inferences which were read into them by early workers.

The ironstone bands are thickest and usually oolitic when they are unfossiliferous or contain only gastropods. They are more sandy and thin or ill-defined in the areas that yield abundant mammalia. The ironstones contain coarser quartz grit and become virtually unfossiliferous as the boundary escarpment is approached.

The fossiliferous ironstones are lenses and probably represent former shallow basins or swampy lagoons, with abundant vegetation to provide food for the gastropods, joined by sluggish rivers. From time to time individual basins virtually dried out, killing the freshwater mollusca and fish. Later inundation and a fresh supply of detritus sealed the horizon, and comparatively rapid sedimentation pro-

Table 3. Stratigraphic Succession Southwest of Lake Albert, Uganda

Lithology	*Interpretation*
5. Recent swamp deposits and low level strandlines	Recent variations of lake level yield swamps and strands
MINOR GRID FAULTS	
4. Wasa Beds (and plateau gravels) thickness 100+ ft	*Mid to later Pleistocene:* Coarse fluvio-lacustrine sediments resulting from rejuvenation of boundary faults
MAJOR MOVEMENT OF RIFT BOUNDARY FAULTS	
3. Kaiso formation: Rhythmic beds of drab gray clays, buff silts, and fine sands with ironstones which are sometimes fossiliferous; thickness 2700 ft	*Later Pliocene:* Rhythmic deepening of rift yields cyclothems in lacustrine sediments
2. Passage Beds: red, blue, and gray clays with subsidiary sandstones; thickness 650 ft	*Earlier Pliocene:* Stable deeper water lacustrine conditions
1. Kisegi Beds: Coarse grits, current-bedded sandstones, subordinate clays, and lignites; thickness 900 ft	*Miocene:* Fluvio-lacustrine deposits; fast streams from newly exposed scarps
MAJOR MOVEMENT OF RIFT BOUNDARY FAULTS	

vided conditions suitable for the burial and eventual preservation of bones of any mammalia or crocodiles which may have died at or near the margins of the shallow basins or swamps.

The consistent chemical, lithological, faunal, and stratigraphical nature of the Kaiso formation wherever it occurs in Uganda and the Congo raises the problem of the wider environmental setting in which the deposits were formed. The shallow-water nature was not questioned by earlier workers. Wayland (1926, p. 8) envisaged sedimentation keeping pace with rift subsidence and sometimes exceeding it but considered the fossiliferous ironstones to be evidence of desiccation. Fuchs (1934, p. 146) assigned the Kaiso sediments to a first pluvial and considered that toward the end of deposition the climate became more arid, while Wayland (1934, p. 347) refers the "Kaiso bone beds" to an interpluvial period.

In contrast Solomon (1939, p. 27) wrote, "The fossiliferous horizons do not necessarily indicate a marked climatic change." Heinzelin (1955) suggested that the deposits gave evidence of a humid climate, with a period of aridity marked by beds of gypsum in the Congo. Despite the diversity of the above views, the upper fossiliferous horizons of the Kaiso formation were assigned to a first interpluvial by the Pan African Congress (Clark 1957, p. xxxi).

Of the above views I agree with Solomon and believe that climatic interpretations are not warranted, in view of the rift valley location of the deposits. The nature of the climate prevailing at the time of deposition of the formation is not proven, although intermittent emergence undoubtedly took place (Flint 1959).

The characteristic lithologies of the Kaiso deposits indicate lacustrine environments into which little coarse or unweathered detritus was brought. In sharp contrast the succeeding coarse, current-bedded Semliki deposits contain abundant quartz and fresh feldspar as also do the Kisegi Beds which precede the Passage Beds and the Kaiso formation. Both the Kisegi and Semliki lithologies suggest major movements of the rift valley boundary faults, producing pronounced escarpments. Repeated minor movements of the boundary faults would account for the rhythmic sedimentation seen in the Kaiso formation and also for the input of coarser detritus from the escarpment and the onset of more rapid sedimentation associated with the ironstones.

It is suggested that such tectonic mainsprings would explain the consistent nature of the Kaiso environments. Far from ruling out tectonic causes, the repeated conditions necessary to produce the numerous ironstone horizons through 2700 feet of sediment are most convincingly accounted for as the result of successive small fault movements. A parallel might be sought in the rhythmic sedimentation seen in Carboniferous strata. A sequence of clay, succeeded by sand, gravel, and oolite grading up again into fine sands, silts, and clays, would constitute a typical Kaiso cyclothem. Identifications of mammalian fossils by Cooke and Coryndon (1970) allow one to suggest that the lower Kaiso assemblage may be as old as $5\pm$ m.y., by analogy with faunas for which isotopic ages have been established in the Gregory Rift. Similarly, the assemblages of mammals that occur toward the top of the formation may be as young as $3\pm$ or $2\pm$ m.y.

Contrasting Histories of the Gregory and Albertine Rifts

If the rift stratigraphic record is considered as an indicator of tectonic activity, a well-dated sequence may be used to construct a pseudoseismic record. This is attempted in Figure 4 as a means of contrasting the tectonic history of the Albertine Rift with that for Area B of the Gregory Rift over the last 12 million years.

It cannot be pretended that the sequence for the Lake Albert basin is as well calibrated as that for the Baringo area. The date of 10 to 11 m.y. for the fault movement which initiated deposition of the Kisegi Beds rests solely upon the evidence of the fragmentary mammalian fauna described by Hooijer (1963). This fauna is Miocene in character but may be much earlier than the minimal later Miocene–early Pliocene age assigned here. This would necessitate the drawing out of the Lake Albert record with a much longer time span assigned to the clays of the Passage Beds. However, such an amendment would not alter the obvious contrast between the Albert and Baringo basins in the tectonic pattern as reflected in the sediments.

The Baringo sequence is well calibrated by potassium-argon dates. They are sufficient to emphasize the pulsatory and repetitive nature of the tectonic history through 12 million years. The relative amplitudes of the oscillations of the trace used to indicate major fault movements can be based only upon subjective estimation. It must also be admitted that fault activity probably is virtually continuous and that the oscillations on the Baringo trace only indicate periods of more accelerated movement. The curves on the Kaiso section of the Lake Albert trace represent diagrammatically the more insidious movements recorded in the sediment as cyclothems.

Nevertheless, the differences in tectonic expression and the absence from the Albertine Rift of the omnipresent volcanicity characteristic of the Gregory Rift (for details of the various volcanic associations see Williams 1969) have to be taken into account in any geophysical model for East Africa. Attempts have been made recently to explain the African rift system by analogy with midoceanic ridges and rifts, plate tectonics, and the breakup of continents. It is unfortunate that the latest of these makes use of crude dating evidence to establish periods of faulting and volcanicity (Girdler et al. 1969, Figs. 3, 4). The authors conclude that attenuation of the crust has taken place and figures are quoted of up to 2 cm yr^{-1} for the rate of "stretching."

The figures and arguments are not repeated in detail here. It is obvious that the pattern of repeated fault movement and volcanicity seen in Figure 4, coupled with the more accurate isotopic and mammalian controlled ages which have become available in the last four years, must supersede the data quoted by Girdler and colleagues. In particular, these authors ignore the fact that the lateral extension that can be assessed from observed faults does not account for any appreciable amount of the stretching suggested. Dikes, which might also take up some of the geophysically deduced slack, are virtually absent from the rift floor. The suggested extension must be considered in the light of the occurrence of substan-

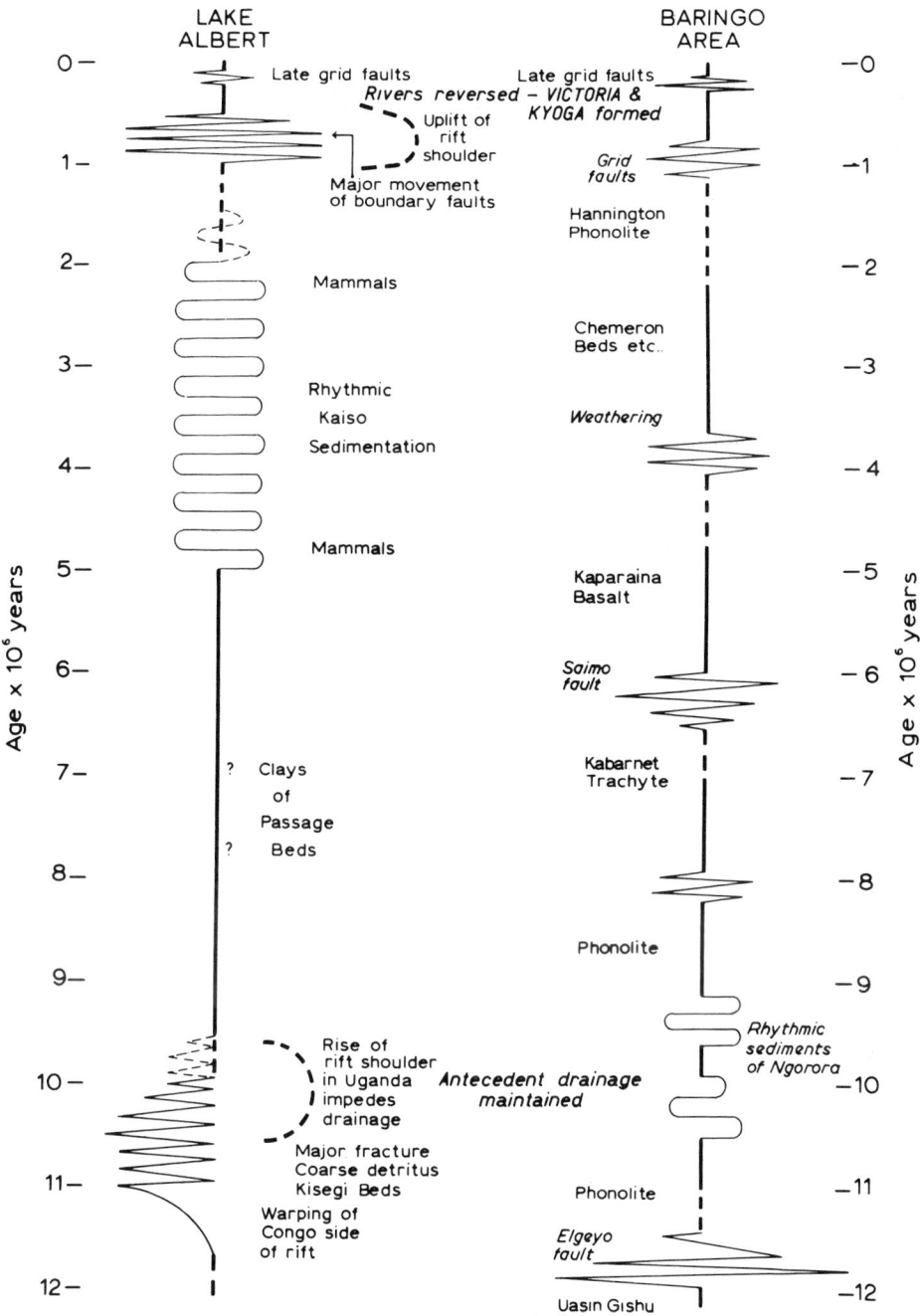

Fig. 4. Pseudoseismic traces derived from the stratigraphic and sedimentary record, illustrating the contrast in tectonic history between the Lake Albert Basin and the Baringo district of the western and eastern rifts respectively.

tial "Basement" outcrops within the rift. The pattern of older faults and volcanicity toward the margins of the rift is to be expected and can be simply explained as a result of continued accumulation of later sediments and volcanics on the floor of the graben with eventual burial of earlier strata and all but the most recently active volcanoes and fault lines.

One can only speculate as to a possible cause for the observed contrasts between the eastern and western rifts. However, it is possibly relevant to observe that the Albertine Rift cleaves across an area of former westerly flowing drainage. Local updoming, associated with the rise to the rift shoulder, disturbed erosion surfaces whose planar extent suggests a previous period of prolonged stability on the site of this rift. In contrast, the Gregory Rift dome with its attendant volcanicity strikes at an acute angle across a line of early established culmination or swell. This receives geomorphological expression in the form of the intercontinental watershed with remnant monadnock-like mountain blocks. These rise above the pediment bevels which mark the inland termini of planation surfaces. Although a departure from previous thinking, a possible relationship might be sought between crustal thinning with associated volcanicity and lines of prolonged cymatogenic uplift and erosion.

AREA D. THE LAKE RUDOLF BASIN AND OMO VALLEY

Area D as delimited on Figure 2 includes at its northern end the valley of the Omo River in southern Ethiopia which received pioneer investigation in 1932–33 (Arambourg 1947). The Omo Basin has been the subject of detailed international study since 1966. Investigation is continuing, but preliminary stratigraphical and faunal results have been published on behalf of the United States team (Howell 1968, 1969; Butzer and Thurber 1969), the French group (Arambourg and Coppens 1967; Arambourg et al. 1967, 1969), and the Kenya party (R. E. F. Leakey et al. 1969).

The difficult terrain in Kenya to the east of Lake Rudolf (Fig. 2) has been investigated since 1968 by a team under the leadership of R. E. F. Leakey. After an exploratory season in 1968, preliminary geological investigation in 1969 established the stratigraphical succession and confirmed the richness of the area in mammalian fossils (R. E. F. Leakey et al. 1970).

In addition to these two important regions adjacent to each other to the north and east of Lake Rudolf, Area D of Figure 2 might have been extended to take in the strip of country immediately southwest of the lake. This includes three important inliers of lava and sediment rising above the plain underlain by Recent sediments. These three localities, named in order from north to south Lothagam, Ekora, and Kanapoi, have been investigated recently by a team from Harvard University (Patterson 1966; Patterson et al. 1970).

Lothagam, Kanapoi, and Ekora

In the vicinity of Lothagam Hill an isolated, westward-tilted fault block exposes a succession almost 3000 feet thick of which sediments, comprising three distinct lithological units, account for well over 2000 feet. The oldest unit, Lothagam 1, is divided into three fluviatile members which are interpreted as clastic infill at the margins of a rapidly sinking basin. This unit contains the lower mammalian fauna (lower fossiliferous beds of Maglio 1970). This fauna is suggested by Patterson and colleagues (1970) as being older than early Kaiso but possibly equivalent to the Kaperyon Beds which are probably older than 5± m.y. (Area B, Fig. 2, and Table 2).

A potassium-argon date of 3.71 ± 0.23 m.y. for a sill intruded at the top of Lothagam 1 gives only a minimum age for the fauna. Lothagam 2 is much thinner, consisting of silts, clays, and bedded tuffs, and is interpreted as being of lacustrine origin. Vertebrate fossils are rare but beds of gastropods are common. Lothagam 3 (upper fossiliferous beds of Maglio 1970), marks a return to fluviatile conditions. The scanty fauna shows remarkable similarity with that of Kanapoi where an overlying basalt has yielded three potassium-argon dates ranging between 2.9 ± 0.3 and 2.5 ± 0.2 m.y. This gives an upper age limit for this fauna and, on a basis of its similarity to other dated faunas, an age of about 4± m.y. is suggested for Kanapoi (Patterson et al. 1970) which also shows similarities to the Chemeron formation fauna (Table 2). The Ekora fauna is shown to be younger than that of Kanapoi, and the elephants suggest that the assemblage may be rather earlier than the oldest part of the main Omo succession.

Thus the three localities provide a later Pliocene sequence which at the top is rather earlier than the base of the Omo fossiliferous sequence about 4 m.y. It extends downward to overlap in time the Chemeron formation in the Baringo-Tugen Hills succession. The óldest Lothagam strata are possibly similar in age to the Kaperyon Beds at perhaps 5.5 m.y.

A specimen of *Australopithecus* sp. cf. *A. africanus*, from near the top of Lothagam 1, and a specimen of cf. *Australopithecus* from Kanapoi extend the hominid record back in time from the lowest Omo levels. These tentatively dated specimens of *Australopithecus* are the earliest assigned to the genus. The sequence will be an extremely important one in the search for further early hominids. I am grateful to Professor Bryan Patterson and to A. K. Behrensmeyer and V. J. Maglio for helpful discussion of this key area.

The Omo Valley

I have not visited the Omo area and am indebted to Professor Clark Howell for discussing the succession and fauna. The outline that follows can be amplified by reference to Howell (1968, 1969) and Butzer and Thurber (1969). The classical Omo Beds are now formally designated as the Shungura formation. They consist of over 1500 feet of clays, silts, sands, and tuffs, predominantly of deltaic origin, in a series of cyclic units. Ten tuff bands are useful marker horizons and have been labeled in ascending order from A to J. Six potassium-argon determinations on feldspar-bearing pumices between horizons B and I_2 range in age from 3.75 m.y. to 1.84 m.y. (personal communication, F. H. Brown). All the dates have low probable errors and occur in correct order of stratification, thus providing a unique calibration for this fossiliferous sedimentary sequence. Ten hominid localities occur in the Shungura formation between tuffs C and H, spanning an age range from approximately 1.9 m.y. to 3.0 m.y.

Full details are not yet published for the hominid finds made by the French expedition, although they include specimens from seventeen localities which indicate that robust and gracile australopithecines existed simultaneously through virtually the whole time range of the Shungura formation (Arambourg et al. 1969). Arambourg and Coppens (1967) described a mandible lacking tooth crowns from below tuff D with an age in excess of 2.37–2.56 m.y. This specimen was assigned to a new genus, *Paraustralopithecus aethiopicus*. In the absence of tooth crowns it seems preferable to accept it as an australopithecine but without assigning it to a particular taxon until a detailed description of this specimen and of the other Omo hominid material has been published.

The hominid material found by the United States team includes a mandible and a hemimandible with robust australopithecine affinities. The more complete specimen is tentatively referred to *Australopithecus* cf. *boisei* (Howell 1969). The age of these mandibles is rather less than 2.0 m.y., but isolated teeth found by both the French and United States parties cause the fossil record of robust australopithecines located in the Shungura formation to extend back to at least 2.5 m.y. B.P. Seventeen isolated teeth from nine other localities are referred tentatively to a small australopithecine, *Australopithecus* cf. *africanus*. They range in age from < 2 m.y. to > 3 m.y.

At White Sands, another locality in the Omo Valley, the Usno formation is exposed. The deposits are silts and clays representing deltaic or floodplain deposits and sands and gravels of fluvial origin. The Usno formation is probably the lateral equivalent of part of the Shungura formation. A K^{40}/Ar^{40} age of 3.1± m.y. for a basalt underlying the sequence provides a maximum age for the commencement of deposition. The White Sands locality has yielded eight hominid teeth. A nearby locality of similar age, Brown Sands, has yielded eleven teeth. Howell (1969) refers all these teeth to a hominid species which "resembles the Sterkfontein sample referred to A. *africanus*."

Two other lithological units older than the Shungura and Usno formations have been recognized in the Omo Valley and termed the Nkalabong and Mursi formations. The type area of the latter is at the Yellow Sands locality some 50 miles north of the present shore of Lake Rudolf. The unit, which constitutes the oldest sedimentary formation in the area, consists of over 450 feet of deltaic and prodeltaic sediments. These have been divided into three members and are overlain by a basalt member. The basalt has yielded potassium-argon ages of 4.05 m.y. and 4.25 m.y.

The Nkalabong formation rests unconformably upon the weathered and eroded surface of the basalt. Deltaic beds are virtually absent but fluviatile current-bedded silts and clays indicate former courses of the Omo River. A lapilli tuff in member II has yielded a potassium-argon date of 3.95± m.y. Thus the two formations extend the Omo sequence a little farther back in time from the earliest deposits of the Shungura formation.

Another younger unit, the Kibish formation, occurs as widely distributed horizontal delta and delta plain sediments throughout the lower Omo Basin. The deposits are discussed in detail by Butzer and Thurber (1969) and shown to be of later Pleistocene age. The upper part of the sequence is calibrated by thirteen radiocarbon dates ranging from greater than 37,000 years to 3250 ± 150 years B.P. Three skulls and some skeletal material ascribed by Day (1969) to an early representative of *Homo sapiens* are of later Pleistocene age but older than 37,000 years B.P. The evidence for periods of high lake level during the deposition of the Kibish formation will be discussed below in relation to that from the Elmenteita, Nakuru, and Naivasha basins which lie in the rift valley some 400 miles to the south (Fig. 2).

East Rudolf; The Koobi Fora Sequence

The fossiliferous sediments outcropping east of Lake Rudolf have received preliminary description by Behrensmeyer (1970) under the name of Koobi Fora I, II, and III. The three units have a total thickness of at least 1000 feet. Koobi Fora I consists of thin-bedded limonitic clays and sands containing fish and mollusca and is interpreted as indicating a lacustrine environment. The overlying Koobi Fora II marks a change to fluviatile conditions, indicated by sandy clays, cross-bedded sandstones, and pebble conglomerates, and locally yields abundant mammals. This division has been prospected in some detail and the mammalian fossils, including hominids, are described by R. E. F. Leakey et al. (1970).

The five hominid specimens include a complete skull, lacking teeth crowns, which has many of the characteristics of the robust australopithecine *A. boisei*. This occurs in the lower part of Koobi Fora II and thus would seem to be slightly older than the tuff horizon which in another section has yielded a potassium-argon age of 2.61 ± 0.26 m.y. (Miller and Fitch 1970). Two robust mandibular fragments and a weathered maxilla from the East Rudolf area also seem attributable to robust australopithecines. A second hominid cranium was recovered in 1969 about a mile away from the locality that yielded the robust skull, in an area that also lies in the lower part of the Koobi Fora II unit. Only the parietals and basioccipital region have been recovered, but the morphology suggests that it is "either a gracile species of the genus *Australopithecus* or else an early form attributable to the genus *Homo*" (R. E. F. Leakey et al. 1970).

In several localities stone artifacts lie on the surface of the Koobi Fora beds, sometimes in association with bone fragments and more complete mammalian fossils. A preliminary excavation at one site has shown that the flake artifacts on the surface originate from the upper part of tuffaceous channel fill. This is the horizon which at a nearby locality yielded the pumice samples for which K^{40}/Ar^{40} ages and Ar^{40}/Ar^{39} ages were established by Miller and Fitch (1970).

The age of 2.61 ± 0.26 m.y. awaits corroboration by dates for other horizons. However, this technically "good" age, for sanidine-feldspars from pumice cobbles, which are assumed to be broadly contemporary with the deposition of the tuff, makes these the oldest artifacts for which dates have been established and pushes back the record of tool-making activities by about three-quarters of a million years. I am indebted to Richard Leakey for conducting me over this potentially most important area in September 1969.

The prime importance of the Lake Rudolf Basin is that the localities yield collectively a series of isotopically dated fossiliferous sediments which link the largely Pliocene sequence already described for Area B with the well-known hominid-bearing sequences at Olduvai Gorge, Tanzania.

Area E. Olduvai Gorge and Its Environs

So much has been written of this unique locality that this paper will only outline the history of geological investigation and underline the important developments which have taken place since 1959.

In common with other parts of East Africa the Olduvai deposits were initially described in lithological terms (Reck 1914) and only later in the *climate of opinion* which invoked pluvial and interpluvial interpretations (Leakey 1951, 1952; Clark 1957). The return to a normal approach commenced with geological mapping over a wider area by Pickering (1958). However, it was the work of Hay (1963a, 1963b, 1967) and Mrs. Leakey that finally changed the oversimplified interpretation of Beds I, II, and IV as indicating lacustrine (pluvial) conditions with the red bed, Bed III, as an interpluvial phase.

The detailed stratigraphical observations of Hay and Mrs. Leakey provide a firm geometry for the numerous fossiliferous and archaeological occurrences and illustrate the true nature of the environmental setting. During its early history the Olduvai sedimentary basin was dominated by a volcanic highland to the east from which intermittent pyroclastic activity contributed deposits ranging from those typical of *nuées ardentes* to derived volcanic debris. Fault movements took place within the basin of sedimentation from Bed II times. The extent and form of the lake on the basin floor was in part fault-controlled from that time. For much of its history it was a shallow saline lake or a playa. Bed III was reddened and cemented by zeolites as the result of oxidation of ferrous iron and chemical reaction in a saline environment with a high pH (Hay 1967, p. 224). Hay concluded that throughout the deposition of the 350 feet of Pleistocene sediments the climate had been "relatively dry and not unlike the present," although some climatic fluctuations seemed to have taken place.

If a modern analog is required for the pattern of sedimentation within the shallow Olduvai basin represented by Beds I and II, it may be supplied by the floor of the nearby Ngorongoro Caldera at the present day. Here a shallow and annually fluctuating saline lake is surrounded by extensive alluvial flats traversed by watercourses marked by gallery forests which give way to swamp, approaching the lake. Carbonatitic volcanic activity still occurs intermittently from Ol Doinyo Lengai, an active cone only 25 miles northeast of Ngorongoro. The analogy breaks down in that the Lengai debris does not discharge along rivers that debouch into the Ngorongoro lake. However, abundant plains game and attendant carnivores which inhabit the caldera floor complete a picture remarkably similar to that recorded by the Olduvai sediments and faunal assemblages. Man is still present, albeit in rather a different role, and the only notable absence is active fault movement. Against the realistic environmental background outlined by Hay, other important recent developments at Olduvai can be summarized as follows.

1. Detailed field investigations and archaeological excavations under the direction of Dr. and Mrs. Leakey of the well-exposed stratified sequences yield a unique record of hominid fossils and associated artifacts, together with abundant mammalian and other vertebrate fossils. This work was concentrated initially on Beds I and II but is now being continued in the later strata.

Sixteen major hominid finds are known from Beds I and II, including almost complete skulls, fragments of skull, and mandibles. The forms occurring include *Australopithecus* (*Zinjanthropus*) *boisei* from Bed I. The more controversial taxa, whose relationships were referred to above as resembling a solid solution series, range from *Australopithecus* sp. cf. *A. africanus* to *Homo habilis* and cf. *Homo erectus* (the last refers to a specimen from near the top of Bed II).

Four principal horizons in Bed I have yielded assemblages of Oldowan artifacts while Bed II is typified by a similar number of levels with more developed Oldowan industries. Interbedded with these, at levels near the middle of Bed II, are several horizons yielding early Acheulean material (M. D. Leakey 1967). A circular concentration of basalt blocks just above the Basalt Member in Bed I may mark the location of a windbreak or primitive habitation. Few hominid fossils have been recovered from Beds III and IV, but large assemblages of Acheulian artifacts have been excavated. The lower member of Bed V contains Kenya Capsian artifacts.

2. The application of potassium-argon dating, and to a minor degree the fission track method, to calibrate the fossil and archaeological record at Olduvai has been a tremendous stimulus to studies of the later Cenozoic in East Africa (Evernden et al. 1964; Evernden and Curtis 1965; Fleischer et al. 1965). Despite a total of over fifty dates listed for Olduvai Gorge in the above publications, only a few have proved reliable when tested against stratigraphical data or on a basis of mutual consistency (Hay 1967; Isaac 1969). Firmly dated strata at Olduvai are only four in number and restricted to Bed I. They are based upon eleven potassium-argon runs as follows:

Age m.y.	Locality etc.
1.57–1.66	Three runs (1045, 1050, 1062*), plagioclase from tuff near top of Bed I (marker Bed A)
1.70–1.76	Five runs (1043, 1053, 1057, 1058, 1179); anorthoclase from *nuée ardente* horizon in Bed I
1.92	Plagioclase and augite (1100) from basalt lava within Bed I
1.85–1.91	Two runs (1080, 1088); anorthoclase from tuffs 8 feet below the Basalt Member in Bed I

* The numbers in parentheses refer to samples quoted in Evernden and Curtis (1965).

This series of dated horizons is of extreme importance because of the impetus given by this pioneer research to the application of potassium-argon dating to other early Pleistocene rocks. The Olduvai sequence fits perfectly onto the upper end of the set of more recently established ages for the Omo sequence (Howell 1969). It should be noted that no unequivocal ages now exist for the critical time spans of Bed II and Bed IV. The absence of dates for Bed II is probably to be accounted for by the lack of good primary tuffs (Evernden and Curtis 1965, p. 351). Bed IV in addition to this factor lies squarely in the umbra of the isotopic dating blank which at present separates the upper and lower range limits of the potassium-argon and radiocarbon methods respectively.

The date for the Basalt Member in Bed I receives some support from paleomagnetic stratigraphy in the form of the Olduvai Normal Event. Finally it should be stressed that the Plio-Pleistocene potassium-argon ages for hominid-bearing strata in East Africa are made more valuable by the virtual absence of similar chronometric data elsewhere in Africa.

HOMINOID FOSSILS THROUGH LATE CENOZOIC TIME

Although it is not proposed to pursue the historical sequence further by discussing other localities, the important work of Isaac on the Peninj Group, Lake Natron (Isaac 1965, 1967) and in the Olorgesailie Basin (Fig. 2) must be noted (Isaac 1966). The Olorgesailie sediments have yielded only artifacts and mammalian fossils, but the Peninj finds include a robust australopithecine mandible in addition to a lower Acheulian industry.

The details of the numerous other localities in East Africa which have yielded artifacts of later Pleistocene age, and the comparative few which have provided skulls of other hominids including *Homo sapiens*, are perhaps a more respectable study for an archaeologist than a historical geologist. Also, it is desirable to ignore them in order to avoid the logarithmic time scale which would become essential if they were included.

To reveal the density of hominoid finds through 25 million years and to appreciate the tempo of evolution, Figure 5 is presented against a constant time scale. According to the isotopic and stratigraphic dating evidence discussed above, hominoid finds are indicated by black circles (and one doubtful occurrence by an open circle). To allow comparisons to be made, fossil monkeys and galagids are also indicated on the same time base. The expansion of artifact occurrences through the last two million years is indicated by X's.

It should be noted that the circles do not indicate numbers of individuals but merely record the occurrence of hominoids of one or more types at a particular time line. Thus a circle may indicate a few or up to 200 dryopithecine fossils in the Miocene, a single tooth crown in the Pliocene, a skull, or several parts of a skeleton from the Pleistocene.

It can be seen that the record obtained may be divided into eight phases as follows:

Phase A 25 to 20 m.y.	Only the Bukwa locality, Mount Elgon (between 22 and 24 m.y.) occurs in this phase. It has yielded *Pliopithecus*.
Phase B 20 to 18 m.y.	A remarkable concentration of localities (e.g. Rusinga, Mfwanganu, Songhor, Koru, Napak, and probably also Moroto) yield large assemblages of hominoids. Dryopithecines occur (robust forms probably ancestral to the gorilla and more lightly built chimpanzee-like forms), and it has been suggested that the genus *Kenyapithecus* may extend back to this level. Hylobatines (two species), monkeys (colobine and cercopithecine), and galagids are also present.
Phase C 18 to 14 m.y.	This phase marks a gap in the hominoid record. Lack of dating evidence is contributory but even if all the known hominoid fossils, principally from rift valley localities, that may belong in this phase were firmly dated there would be only a sparse fossil record compared with Phase B.
Phase D 14 ± m.y.	The single Fort Ternan locality samples only a narrow time band but has yielded the hominid *Ramapithecus* [= *Kenyapithecus*] *wickeri* and also a possible dryopithecine.
Phase E 14 to 5.5 m.y.	This represents a major gap in the hominoid record, yielding up to the present only a single molar crown from the Ngorora formation (between 9 and 12 m.y.).
Phase F 5.5 to 1.5 m.y.	Several well-dated sequences (Lothagam, Kanapoi, Chemeron, East Rudolf, Omo, and Olduvai) yield abundant australopithecines. The fossils occur only sparingly at first but from 3.5 to less than 1.5 m.y. abundant finds occur, representing robust and gracile forms living as contemporaries. Stone artifacts are first recognized just before 2.5 m.y. From an initial meager record artifacts become progressively more abundant after about 2.0 m.y. The genus *Homo*, represented by *H. habilis*, appears at about 1.75 m.y.
Phase G 1.5 to 0.05 m.y.	Hominid fossils of more advanced type are represented in the Olduvai Bed II sequence, but firm dating evidence is lacking. Artifacts continue to increase in abundance and become more diverse in typology.
Phase H 0.05 to 0 m.y.	Abundant artifacts of many specialist industries together with limited numbers of hominid fossils lie within the range of radiocarbon dating.

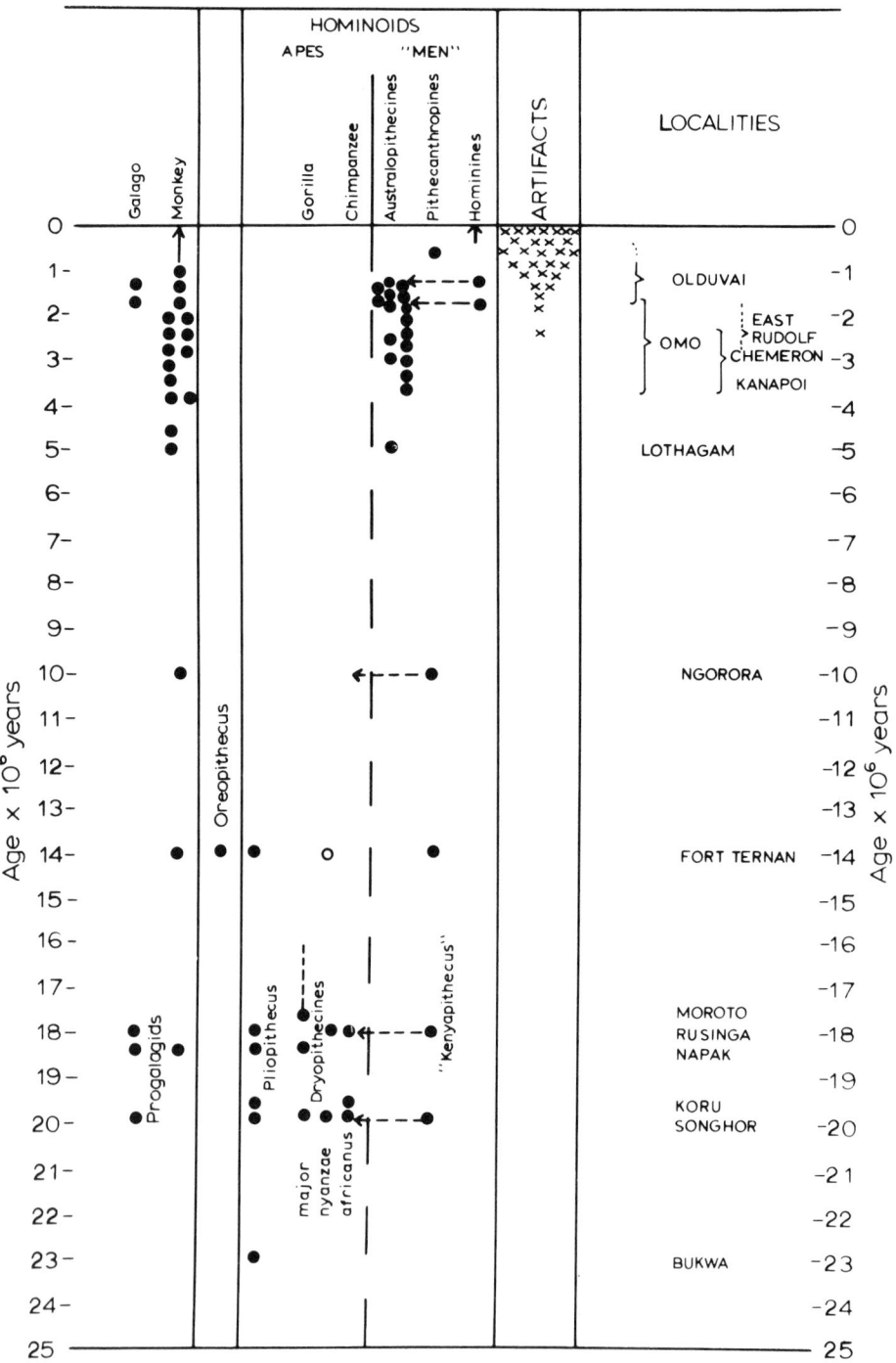

Fig. 5. East African hominoid fossils plotted against a constant time base through 25 million years of Late Cenozoic time. Arrows suggest possible second positions for some taxa.

Two periods in this sequence stand out as yielding particularly rich concentrations of fossils: Phase B (20 to 18 m.y.) and the last two million years of Phase F (i.e. from 3 or 3.5 m.y. to 1.5 m.y. and ranging on into Phase G). These are obviously much richer in hominoid and primate fossils than other sections of the record. Indeed, Figure 5 demonstrates graphically the great blanks which still characterize even the rich East African fossil hominoid record. It stresses the need for evidence to bridge these gaps.

In addition, it is important to continue to study the richer fossiliferous localities in an attempt to isolate any peculiar factors that will aid in understanding these natural "funeral parlor" concentrations of hominoid fossils. The present evidence indicates that the most important environments were the flanks of intermittently active carbonatitic volcanoes during restricted phases of Miocene time. The other concentration of evidence is in broadly deltaic environments, including overbank and channel situations, spanning two to three million years of Plio-Pleistocene time. The occurrence of volcanic activity within these lacustrine basins, around the margins of which the hominoid fossils were stratified, cannot be ignored. In the Olduvai Basin the volcanicity was again of carbonatitic affinity.

EVIDENCE OF CLIMATIC CHANGE

From this review of the history of various sedimentary units spanning the last 25 million years it is clear that tectonic activity, combined to a greater or less degree with volcanicity, has been the principal factor governing local environmental change as reflected in the sediments.

Climatic changes have certainly taken place in East Africa during this period, but the isolation of their effects in the depositional record has been rendered impossible by the presence of other variables. In the course of this paper several type localities of the former stratigraphic/climatic terminology have been elucidated. They are the Kamasia Hills (Area B), as the alleged type area for the Kamasian (or second) pluvial, and the Kaiso formation (Area C); and Bed III at Olduvai Gorge (Area E), representing respectively the first and second interpluvials. Reinvestigation of the Kagera Valley in Uganda (Bishop 1969) shows that there is no stratigraphic evidence to support a Kageran (or first) pluvial. The Kanjeran (or third) pluvial was not accepted as a formal stratigraphic/climatic unit by the 18th Geological Congress held in London in 1948 (Sandford and Blondel 1951).

With the removal of this terminology and its associated climatic pigeonholes it is possible to look back over the record for acceptable evidence of climatic change. With virtually no evidence remaining in support of the first three pluvial periods or the first two interpluvials it is pertinent to examine the last interpluvial and its successor the Gamblian, or last, pluvial. Some paleontological, geomorphological, or sedimentary evidence of former climatic conditions may have survived in basins formed during the later Pleistocene which have remained stable since that time.

The "MN horizon," an artifact-bearing rubble in the Kagera Valley in southwest Uganda, suggests a climatic oscillation toward drier conditions about 40,000 to 50,000+ years B.P. (Bishop 1969). The evidence requires a major recession of the western margin of Lake Victoria, accompanied by a later transgression. Unfortunately, this is an area where later Pleistocene tilting toward the east is demonstrable for the period between approximately 25,000 and 15,000 years B.P. (Bishop and Cole 1967). Although "see-saw tectonics" (Solomon 1939) would seem unlikely as an explanation for the recession and return of the lake waters, it cannot be ruled out. Also the dating of the events referred to above is unsatisfactory as it involves "secondhand" radiocarbon ages extrapolated, on a basis of similarity of archaeological assemblages, from actual dates obtained at Kalambo Falls in Zambia (Clark 1962).

The notion of a last interpluvial based upon the Kagera Valley evidence must be placed in a suspense account until firm dates and supporting information of lowered lake levels are forthcoming.

The Gamblian pluvial was named from Gambles Cave near Lake Elmenteita in Kenya (letter G, Fig. 6) and was based upon descriptions of strandlines in the basins of lakes Nakuru, Elmenteita, and Naivasha (Leakey 1931; Nilsson 1932). The physical setting of this highest part of the Gregory Rift is seen in Figure 6, which illustrates the manner in which three volcanic centers (the crater of Menengai, the mass of Eburru, and the cone of Longonot) divide the rift into two basins.

The northern basin was investigated geomorphologically by Washbourn (1967). Accurate leveling revealed the existence of an undeformed shoreline at 6370 feet above sea level, or 600 feet above the present surface of Lake Nakuru. The probable extent of this lake, which had an outlet toward the north, is indicated on Figure 7. The spectacular increase in water volume can be appreciated when the 600-foot-deep lake in a steep-sided basin is compared with the remnant shallow saline lakes Elmenteita and Nakuru.

Washbourn located a patch of diatomaceous silts yielding freshwater shells, including *Corbicula africana* and *Melanoides tuberculata* in abundance, 400 feet above Lake Nakuru. The radiocarbon age established by Thurber for shell sample L.1201 was 9650 ± 250 years (Kamau 1970).

There is no firm evidence to relate these mollusca to the 600-foot strandline. However, *Melanoides tuberculata* and *Corbicula africana* were recorded from the beach gravel in Gambles Cave II (Leakey 1931), which lies on the level of Washbourn's 600-foot beach. Three radiocarbon dates on charcoal from between 0 and 60 cm in the occupation debris immediately overlying the beach yielded ages between 8000 and 8500 years B.P. (G. Isaac, personal communication). This is in accord with the Capsian industry from these levels as similar radiocarbon dates spanning this period have been obtained for Capsian industries in North Africa. Also, the dates are in keeping with the evidence of a bone harpoon from Gambles Cave II (Oakley 1961).

Figure 6 shows the probable extent of a similar pluvial lake in the Naivasha

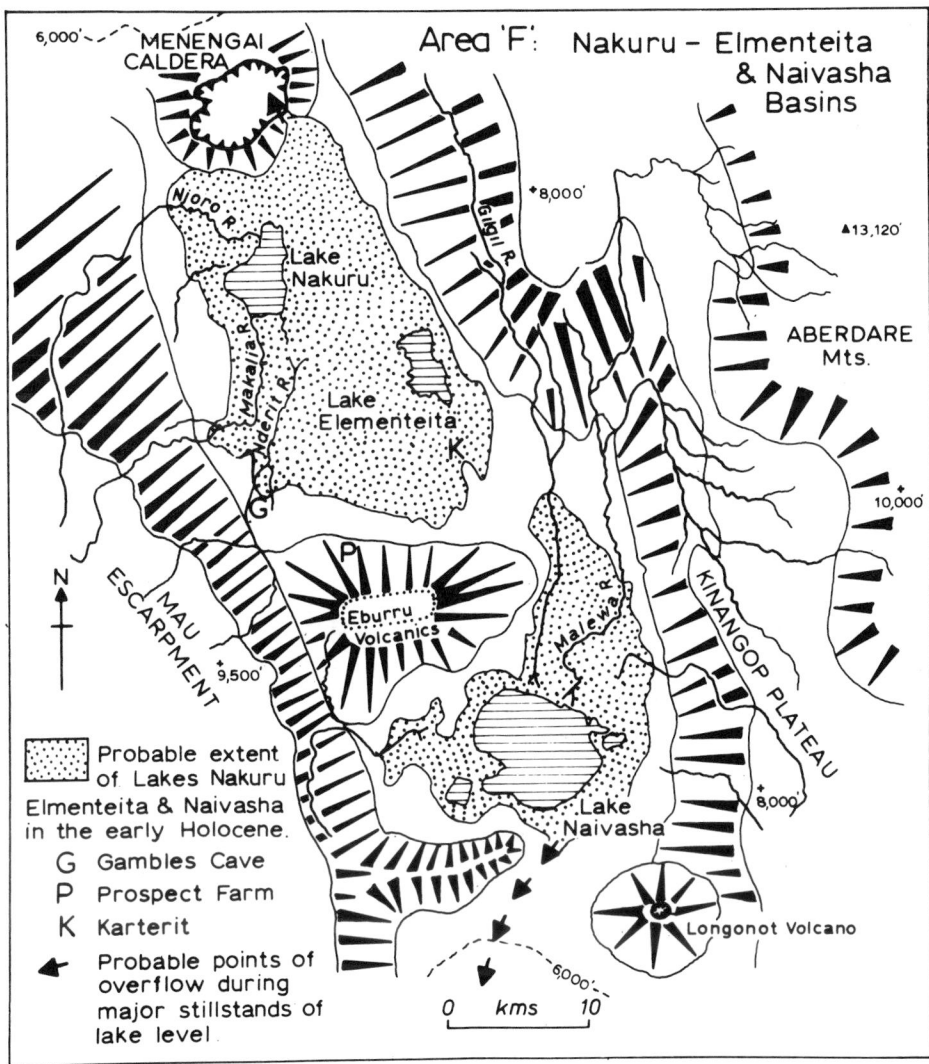

Fig. 6. Major relief features and extent of former lakes in Area F (Fig. 2), the Na-kuru-Elmenteita and Naivasha basins.

Basin. Richardson (1966) adopted a different approach by taking a core from sediments underlying Lake Naivasha. Paleolimnology, supported by radiocarbon dates, yielded a sequence of:

	Years B.P.
Base of core; large lake greater in extent and depth than today	9200 ± 160
Beginning of lake's decline in size	5650 ± 120
Just after complete drying up	3000 ± 60

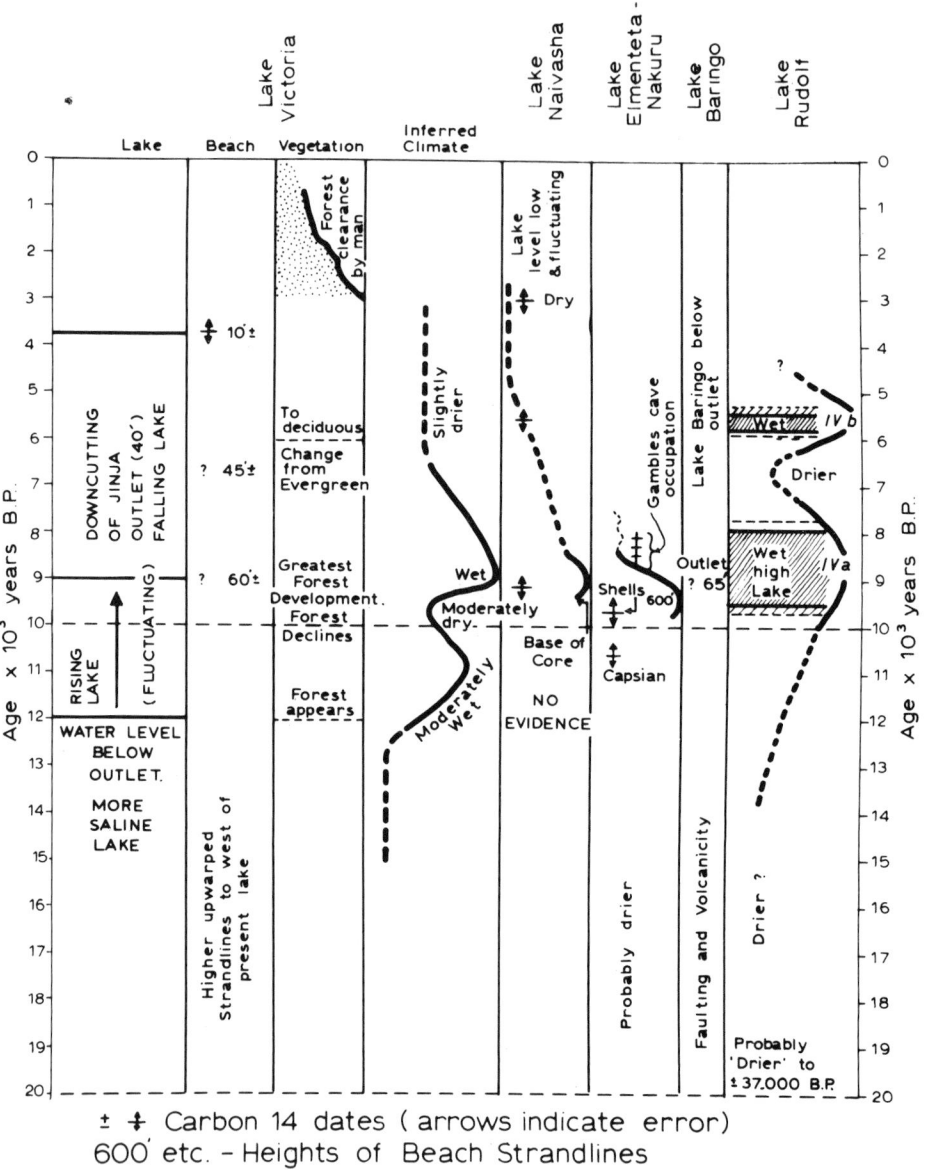

Fig. 7. Changes in lake level and inferred variations in climate through the last 15,000 years for several East African lake basins. Details for Lake Victoria from Kendall (1969), Lake Naivasha from Richardson (1966), Lake Elmenteita-Nakuru from Washbourn (1967) and Kamau (née Washbourn—1970), Lake Rudolf from Butzer and Thurber (1969).

This offers independent evidence for a high lake level at about 9000+ years ago. As seen from Figure 6, Lake Naivasha obtains the bulk of its inflow from tributaries originating on plateaus northeast of the rift. In contrast, flow into the Nakuru-Elmenteita Basin derives from the Mau Highlands southwest of the rift. The virtually synchronous high levels in the two basins lend support to a climatic origin rather than to a local change within a single catchment area.

The published evidence for change in lake level in the two basins is summarized in Figure 7. It is compared with sequences having a similar time span, established for the Lake Victoria and Lake Rudolf basins. Evidence from three sublake cores taken near the northern shore of Lake Victoria was given by Kendall (1969). Chemical history, algae, diatoms, and pollen stratigraphy were studied. Calibration was established by 28 radiocarbon dates at regular intervals through 18 meters of one core and by an additional date for each of the other two cores. The history of Lake Victoria deduced by Kendall from this evidence is summarized in Figure 7.

A sequence of shallow lake and strandline deposits was described for the younger members of the Kibish formation of the Omo Valley by Butzer and Thurber (1969). Thirteen radiocarbon dates suggested two high stands of Lake Rudolf, the earlier between 9500 and 7900 years B.P. and the later 5750 to 5450 years B.P.

From Figure 7 there seems to be a broad agreement between oscillations of lake level in the four widely diverse basins. This suggests an underlying climatic cause. One of the maxima of the former Gamblian pluvial, far from being the time equivalent of any part of the last glaciation, falls within the Holocene.

Further detailed observations are urgently required. They will eventually reveal a relationship, although possibly not on a simple one-to-one basis, between wet phases in tropical East Africa and the well-documented pluvial oscillations and continental glaciations of more temperate latitudes.

Acknowledgment: I thank the numerous colleagues who assisted with the field investigations described here. I am grateful to those who kindly gave their time to illustrate their own research areas and to discuss specific problems, also to those from whose publications I have quoted in writing this review. My best thanks go to Dick and Peggy Flint for happy memories of a stimulating safari through Uganda in May 1957. At that time many germs of ideas were implanted which have gradually developed through the intervening years.

REFERENCES

Arambourg, C., Mission scientific de l'Omo. *Museum Nat. Hist. Paris,* fascs. 1–3, 1943–47.

Arambourg, C., and Y. Coppens, Sur la découverte dans le pléistocène inférieur de la vallée de l'Omo (Ethiopie) d'une mandibule d'Australopithécien, *Compt. Rend. Acad. Sci. Paris,* 265 (D), 589, 1967.

Arambourg, C., J. Chavaillon, and Y. Coppens, Premiers résultats de la nouvelle mission de l'Omo (1967), *Compt. Rend. Acad. Sci. Paris, 265* (D), 1891, 1967.

Arambourg, C., J. Chavaillon, and Y. Coppens, Résultats de la nouvelle mission de l'Omo (2ᵉ campagne 1968), *Compt. Rend. Acad. Sci. Paris, 268* (D), 759, 1969.

Behrensmeyer, A. K., Preliminary geological interpretation of a new hominid site in the Lake Rudolf Basin, *Nature, 226,* 225, 1970.

Bishop, W. W., Miocene mammalia from the Napak volcanics, Karamoja, Uganda, *Nature, 182,* 1480, 1958.

Bishop, W. W. The later Tertiary and Pleistocene in eastern equatorial Africa, in *African Ecology and Human Evolution,* edited by F. C. Howell and F. Bourliere, p. 246, Aldine, Chicago, 1963.

Bishop, W. W., Quaternary geology and geomorphology in the Albertine rift valley, Uganda, in *International Studies of the Quaternary,* edited by H. E. Wright, Jr., *Geol. Soc. Am., Spec. Papers, 84,* 1965.

Bishop, W. W., The later Tertiary in East Africa—volcanics, sediments and faunal inventory, in *Background to Evolution in Africa,* edited by W. W. Bishop and J. D. Clark, p. 31, University of Chicago Press, Chicago, 1967.

Bishop, W. W., The evolution of fossil environments in East Africa, *Trans. Leicester Lit. Phil. Soc., 62,* 22, 1968.

Bishop, W. W., Pleistocene stratigraphy in Uganda, 128 pp., *Geol. Surv. Uganda, Mem., 10,* Entebbe, 1969.

Bishop, W. W., and F. Whyte, Tertiary mammalian faunas and sediments in Karamoja and Kavirondo, East Africa, *Nature, 196,* 1283, 1962.

Bishop, W. W., and J. D. Clark, eds., *Background to Evolution in Africa,* 935 pp., University of Chicago Press, Chicago, 1967.

Bishop, W. W., and G. H. Cole, Revised stratigraphical nomenclature for the Kagera and Orichinga valleys, Uganda, in *Background to Evolution in Africa,* edited by W. W. Bishop and J. D. Clark, p. 522, Chicago University Press, Chicago, 1967.

Bishop, W. W., and G. R. Chapman, Early Pliocene sediments and fossils from the northern Kenya rift valley, *Nature, 226,* 914, 1970.

Bishop, W. W., J. A. Miller, and F. J. Fitch, New potassium-argon age determinations relevant to the Miocene fossil mammal sequence in East Africa, *Am. J. Sci., 267,* 669, 1969.

Brooks, C. E. P., The meteorological conditions of an ice sheet and their bearing on the desiccation of the globe, *Quart. J. Roy. Meteorol. Soc., 40,* 1914.

Brown, J. M., Geophysics, in *Oil in Uganda,* N. Harris et al., Geol. Surv. Uganda, *Mem.,* 9, 1956.

Butzer, K. W., and D. L. Thurber, Some late Cenozoic sedimentary formations of the lower Omo Basin, *Nature, 222* (5199), 1132, 1969.

Clark, J. D., ed., *Proceedings 3rd Pan.-Afr. Cong. on Prehistory,* Chatto and Windus, London, 1957.

Clark, J. D., Carbon 14 chronology in Africa south of the Sahara, *Proc. 4th Pan-Afr. Cong. on Prehistory,* Leopoldville, 303, 1962.

Cooke, H. B. S., Observations relating to Quaternary environment in east and southern Africa, 73 pp., Du Toit Memorial Lecture No. 5, *Geol. Soc. S. Africa, Annex. 60,* 1958.

Cooke, H. B. S., and S. Coryndon, Pleistocene mammals from the Kaiso formation and other related deposits, in *Fossil Vertebrates of Africa*, vol. 2, Academic Press, London, 1970, in press.

Dawson, J. B., Carbonatitic volcanic ashes in northern Tanganyika, *Bull. Volcanol.*, 27, 81, 1964.

Day, M. H., Omo human skeletal remains, *Nature*, 222 (5199), 1135, 1969.

Evernden, J. F., and G. H. Curtis, The potassium-argon dating of late Cenozoic rocks in East Africa and Italy, *Current Anthropology*, 6 (4), 343, 1965.

Evernden, J. F., D. E. Savage, G. H. Curtis, and G. T. James, Potassium-argon dates and the Cenozoic mammalian chronology of North America, *Am. J. Sci.*, 262, 145, 1964.

Fleischer, R. L., L. S. B. Leakey, P. B. Price, and R. M. Walker, Fission track dating of Bed I, Olduvai Gorge, *Science*, 148, 72, 1965.

Flint, R. F., On the basis of Pleistocene correlation in East Africa, *Geol. Mag.*, 96 (4), 265, 1959.

Fuchs, V. E., The geological work of the Cambridge expedition to the East African lakes 1930–31, *Geol. Mag.*, 71, 837, 1934.

Fuchs, V. E., Pleistocene events in the Baringo Basin, Kenya Colony, *Geol. Mag.*, 87, 149, 1950.

Gautier, A., Relative dating of peneplains and sediments in the Lake Albert Rift area, *Am. J. Sci.*, 263, 537, 1965.

Gautier, A., New observations on the later Tertiary and early Quaternary in the western rift: The stratigraphic and palaeontological evidence, in *Background to Evolution in Africa*, edited by W. W. Bishop and J. D. Clark, p. 73, University of Chicago Press, Chicago, 1967.

Girdler, R. W., J. D. Fairhead, R. C. Searle, and W. T. C. Sowerbutts, Evolution of rifting in Africa, *Nature*, 224, 1178, 1969.

Gregory, J. W., *The Rift Valleys and Geology of East Africa*, 479 pp., Seely, London, 1921.

Harris, N., J. W. Pallister, and J. M. Brown, *Oil in Uganda*, 33 pp., Geol. Surv. Uganda, *Mem.*, 9, 1956.

Hay, R. L., Stratigraphy of Beds I through IV, Olduvai Gorge, Tanganyika, *Science*, 139 (3557), 829, 1963a.

Hay, R. L., Zeolitic weathering in Olduvai Gorge, Tanganyika, *Bull. Geol. Soc. Am.*, 74, 1281, 1963b.

Hay, R. L., Revised stratigraphy of Olduvai Gorge, in *Background to Evolution in Africa*, edited by W. W. Bishop and J. D. Clark, p. 221, University of Chicago Press, Chicago, 1967.

Heinzelin, J. de, Le fossé tectonique sous le parallèle d'Ishango, *Mission J. de Heinzelin*, Vol. 1, Brussels (Inst. Parcs, Natl. Congro Belge), 1955.

Heinzelin, J. de, Palaeoecological conditions of the Lake Albert–Lake Edward rift, in *African Ecology and Human Evolution*, edited by F. C. Howell and F. Bourliere, p. 276, Aldine, Chicago, 1963.

Holmes, A., *Principles of Physical Geology*, 1058 pp., Nelson, London, 1965.

Hooijer, D. A., Miocene mammalia of Congo, *Ann. Mus. Roy. Afrique centrale* (Tervuren) *Sci. Geol.*, 46, 1963.

Hopwood, A. T., Miocene primates from Kenya, *J. Linn. Soc.* (*Zool.*) *London*, 38, 437, 1933.

Howell, F. C., Omo Research Expedition, *Nature, 219* (5154), 567, 1968.

Howell, F. C., Remains of Hominidae from Pliocene/Pleistocene formations in the lower Omo Basin, Ethiopia, *Nature, 223* (5212), 1234, 1969.

Isaac, G. L., The stratigraphy of the Peninj Beds and the provenance of the Natron Australopithecine mandible, *Quaternaria, 7,* 101, 1965.

Isaac, G. L., The geological history of the Olorgesailie area, *5th Pan-Afr. Cong. on Prehistory,* Tenerife, 125, 1966.

Isaac, G. L., The stratigraphy of the Peninj group—Early middle Pleistocene formations west of Lake Natron, Tanzania, in *Background to Evolution in Africa,* edited by W. W. Bishop and J. D. Clark, p. 229, University of Chicago Press, Chicago, 1967.

Isaac, G. L., Studies of early culture in East Africa, *World Arch., 1* (1), 1, 1969.

Kendall, R. L., An ecological history of the Lake Victoria Basin, *Ecol. Monographs, 39,* 121, 1969.

Leakey, L. S. B., *The Stone Age Cultures of Kenya Colony,* Cambridge University Press, 1931.

Leakey, L. S. B., *Olduvai Gorge,* Cambridge University Press, 1951.

Leakey, L. S. B., editor, *Proceedings of the First Pan-African Congress on Prehistory,* Nairobi, 1947, Blackwell, Oxford, 1952.

Leakey, L. S. B., A new lower Pliocene fossil primate from Kenya, *Ann. Mag. Nat. Hist., 4* (47), 689, 1962.

Leakey, L. S. B., and J. D. Solomon, East African archaeology, *Nature, 124,* 9, 1929.

Leakey, M. D., Preliminary survey of the cultural material from Beds I and II, Olduvai Gorge, Tanzania, in *Background to Evolution in Africa,* edited by W. W. Bishop and J. D. Clark, p. 417, University of Chicago Press, Chicago, 1967.

Leakey, M., An Acheulian industry with prepared core technique and the discovery of a contemporary hominid mandible at Lake Baringo, Kenya, *Proc. Prehistoric Society,* Cambridge, 1970.

Leakey, R. E .F., K .W. Butzer, and M. H. Day, Early *Homo sapiens* remains from the Omo River region of south-west Ethiopia, *Nature, 222* (5199), 1132, 1969.

Leakey, R. E. F., A. K. Behrensmeyer, F. J. Fitch, J. A. Miller, and M.D. Leakey, New hominid remains and early artifacts from northern Kenya, *Nature, 226,* 223, 1970.

McCall, G. J. H., B. H. Baker, and J. Walsh, Late Tertiary and Quaternary sediments of the Kenya rift valley, in *Background to Evolution in Africa,* edited by W. W. Bishop and J. D. Clark, p. 191, University of Chicago Press, Chicago, 1967.

Maglio, V. J., Early Elephantidae of Africa and a tentative correlation of African Plio-Pleistocene deposits, *Nature, 225,* 328, 1970.

Martyn, J. E., Pleistocene deposits and new fossil localities in Kenya, *Nature, 215* (5100), 476, 1967.

Martyn, J. E., Unpublished Ph.D. thesis, University of London, 1969.

Martyn, J. E., Notes on the geology of the Kapthurin Beds, in M. Leakey, *An Acheulian Industry with Prepared Core Technique etc.,* Proc. Prehistoric Soc., Cambridge, 1970.

Miller, J. A. and F. J. Fitch, Radioisotopic age determinations of Lake Rudolf artefact site, *Nature, 226,* 226, 1970.

Mortelmans, G., and J. Nenquin, eds., *Actes du IVe Congrès panafricain de Préhistoire et de l'Etude du Quaternaire* (Tervuren), 1962.

Nilsson, E., Quaternary glaciations and pluvial lakes in British East Africa, *Geog. Ann.* (*Stockholm*), *13*, 249, 1932.

Oakley, K. P., *Antiquaries J.*, *41*, 86, 1961.

O'Brien, T. P., *The Prehistory of the Uganda Protectorate*, Cambridge University Press, 1939.

Patterson, B., A new locality for early Pleistocene fossils in north-western Kenya, *Nature*, *212*, 577, 1966.

Patterson, B., A. K. Behrensmeyer, and W. D. Sill, Geology and fauna of a new Pliocene locality in north-western Kenya, *Nature*, *226*, 918, 1970.

Pickering, R., *Degree Sheet 12, S.W. Quarter*, Geol. Surv. Tanganyika, Dodoma, 1958.

Reck, H., Erste vorlaufige Mitteilung über den Fund eines fossilen Menschenskelettes aus Zentral-Afrika, *S. B. Ges. Naturf. Fr. Berlin*, 3, 1914.

Richardson, J. L., Changes in level of Lake Naivasha, Kenya, during post-glacial times, *Nature*, *209*, 290, 1966.

Sandford, K. S., and F. Blondel, Proceedings 18th Internat. Geol. Congress (London 1948) Part 14, *Assoc. Services Geol. Africains Proc.*, 19, 1951.

Simpson, G. C., Studies in world climate, *Quart. J. Roy. Meteorol. Soc.*, *60*, 425, 1934.

Solomon, J. D., The Pleistocene succession in Uganda, in T. P. O'Brien, *The Prehistory of the Uganda Protectorate*, p. 15, Cambridge University Press, Cambridge, 1939.

Van Couvering, J. A., and J. A. Miller, Miocene stratigraphy and age determinations, Rusinga Island, Kenya, *Nature*, *221* (5181), 628, 1969.

Walker, A. C., P. G. W. Brock, and R. A. Macdonald, Fossil mammal locality on Mount Elgon, Eastern Uganda, *Nature*, *223*, 591, 1961.

Washbourn, C. K., Late Quaternary chronology of the Nakuru-Elmenteita Basin, Kenya, *Nature*, 243, 1967.

Washbourn-Kamau, C. K., Lake levels and Quaternary climates in the Eastern Rift Valley of Kenya, *Nature*, *216*, 672, 1967.

Wayland, E. J., The geology and palaeontology of the Kaiso Bone-beds, *Geol. Surv. Uganda, Occ. Paper*, *2*, 5, 1926.

Wayland, E. J., Rifts, rivers, rains and early man in Uganda, *J. Roy. Anthrop. Inst.*, *64*, 333, 1934.

Williams, L. A. J., Volcanic associations in the Gregory rift valley, East Africa, *Nature*, *224* (5214), 61, 1969.

William R. Farrand

20. LATE QUATERNARY PALEOCLIMATES OF
THE EASTERN MEDITERRANEAN AREA

The eastern Mediterranean region is, and has long been, of prime interest to historians and archaeologists, both classicists and prehistorians. It includes the cradle of civilization and the area of emergence of agriculture, upon which civilization was based, and is the birthplace of our modern Western culture that has sprung from the Greco-Roman tradition. Furthermore, this area has long been of great interest to students of the Quaternary because geographically and climatologically it provides the transition from the glaciated northern continents to the pluviated tropical lands. In spite of this long tradition, which for example goes back to the studies of the Dead Sea by Lartet (1865), Quaternary stratigraphy and paleoclimatology are still.very poorly understood. Much of the work in the past has been done in connection with archaeological excavations, which in itself is good, but these studies have in general been limited to isolated spot localities, which is bad. Regional synthesis of the facts collected has been largely based on broad-scale comparisons and facile correlation, supported in the main by the use of human artifacts as guide fossils. Unfortunately, superficial similarities can be misleading on the one hand and, on the other, nothing can ever be learned about the details of evolution and migration of prehistoric man and his culture if one accepts, *a priori,* that all artifacts or artifact assemblages of the same kind are of the same age. Regional synthesis based on careful stratigraphic tracing from area to area and on the establishment of local chronologies free from the implications that necessarily follow when one applies a name characteristic of a distant chronology, such as "Würm glaciation," is still very much needed throughout the Near East.

Even Quaternary paleoclimate and climatic change in this area is not yet well understood. There are conflicting interpretations that have been derived from different kinds of data; for example, snowline lowering and deep-sea cores suggest a rather strong temperature decrease during the last continental glaciation of northern Europe, while pollen studies suggest that there was not much change. Moreover, data from the study of fossil rodents is interpreted as showing a steady, unidirectional trend toward desiccation from the last interglaciation through the last glaciation and on into postglacial time. Tectonic activity in the Jordan Rift valley and changes in sea level along the Mediterranean coast have had local influences on the climate recorded in those areas; the regional climatic trends may therefore be masked by local effects and can be sorted out only with difficulty.

Relatively recent summaries of Quaternary stratigraphy and climates in the eastern Mediterranean area (Butzer 1958; Howell 1959) are available, but much work has been done in the last decade in many domains—glaciology, cave sedimentology, marine littoral and deep-sea stratigraphy, fluvial geomorphology, palynology, paleontology, etc. The purpose of the present article is to update in summary fashion our knowledge of late Quaternary paleoclimates in the eastern Mediterranean.

In order to be reasonably comprehensive in depth of coverage, this summary will be restricted temporally and geographically. My primary goal is the evaluation of the amplitude of climatic change through a glacial-interglacial cycle. Therefore, the paper will be limited to the *last* glacial-interglacial cycle, for which quite a lot of information is now available.

The terms "last glaciation" and "last interglaciation" will be used in the sense of geologic-climate units that are based on climatic events, namely continental glaciation, in more northerly latitudes. Radiocarbon dating provides the necessary verification of time equivalence between events in the eastern Mediterranean and northern Europe for approximately the last half of the period being surveyed— the last 40,000 years or so. Correlation with the beginning of the last glaciation is reasonably well assured by means of marine deposits of the transgression that is related to the last interglaciation. These marine deposits, which are generally found at 5 to 8 meters above present sea level, have been called Neotyrrhenian or Tyrrhenian III and are correlated with the Eemian interglaciation. (However, see the discussion of G. A. Wright 1967, and Butzer 1967.) These deposits have been dated in the western Mediterranean at 75,000 to 90,000 years old (Stearns and Thurber 1965).

The geographic limits of this summary (Fig. 1) will be the maritime littoral of the Mediterranean east of the longitude of the Greek peninsula (about 21°E). The continental hinterlands generally lying 100 to 200 kilometers inland from the coasts are excluded since they represent quite different climatic and vegetational zones that, in the present state of our knowledge, can only confuse the picture seen in the maritime areas. Thus, our area comprises the cave of Haua Fteah (Cyrenaican Libya) and the long pollen core of the Tenaghi Philippon (northeastern Greece) near the western extremity and the Dead Sea Rift and the Syrian caves of Yabrud and Jerf 'Ajla near the eastern limit.

DEEP-SEA CORES

Two deep-sea cores from the eastern Mediterranean (Fig. 1, nos. 2 and 3) have been studied and reported in detail; the first of these lies about midway between

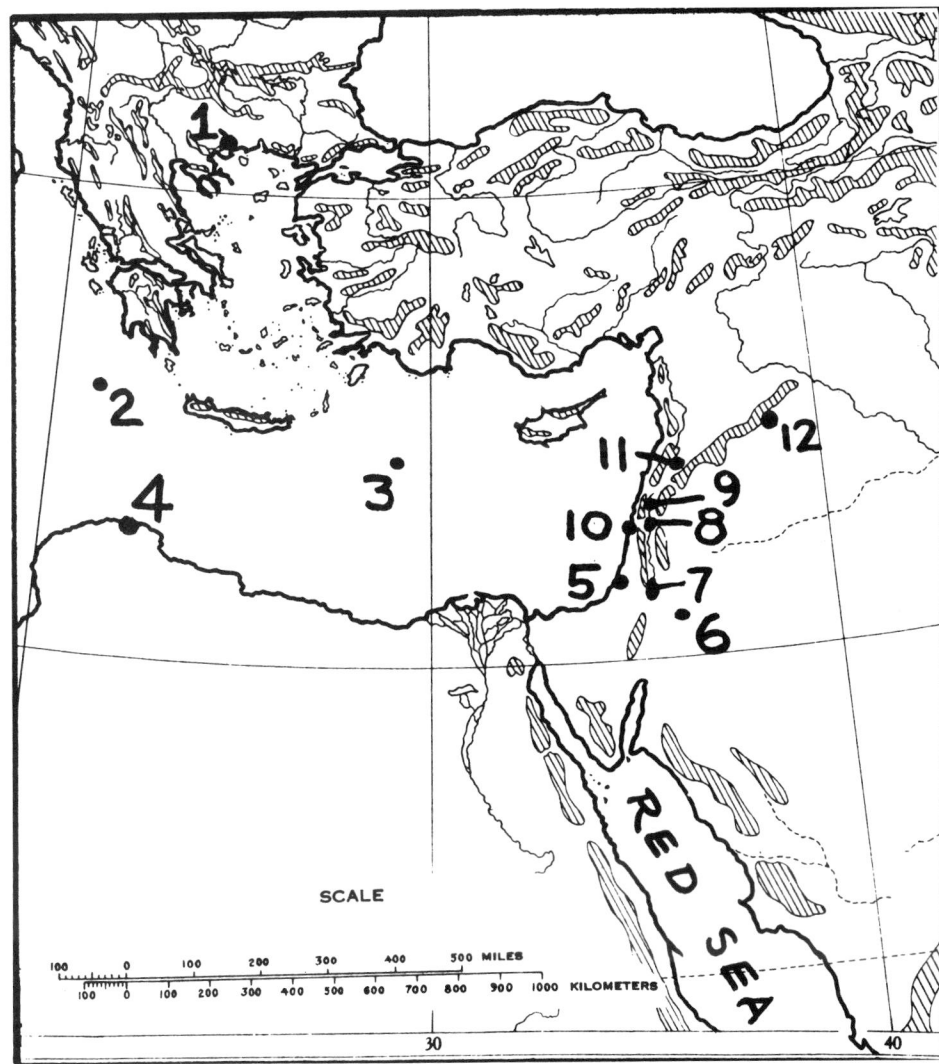

Fig. 1. Index map of localities discussed in text. 1, Tenaghi Philippon; 2, core V 10–67; 3, core 189; 4, Haua Fteah; 5, Ashdod, Israel; 6, El Jafr basin; 7, Dead Sea; 8, Lake Tiberias; 9, Lake Hula; 10, Mount Carmel and Haifa; 11, Yabrud; 12, Jerf 'Ajla.

Crete and Cyprus (Emiliani 1955) and the second (core V 10–67) is west of Crete at about longitude 20°E (Vergnaud-Grazzini and Herman-Rosenberg 1969; Herman et al. 1969). The latter study is by far the more complete, and in general it verifies the earlier interpretations made by Emiliani. Climatic interpretation of core V 10–67 (Fig. 2) is based on a complete faunal analysis and oxygen-isotope analysis, and it spans some 100,000 years of late Quaternary time. The outstanding results of this analysis are as follows.

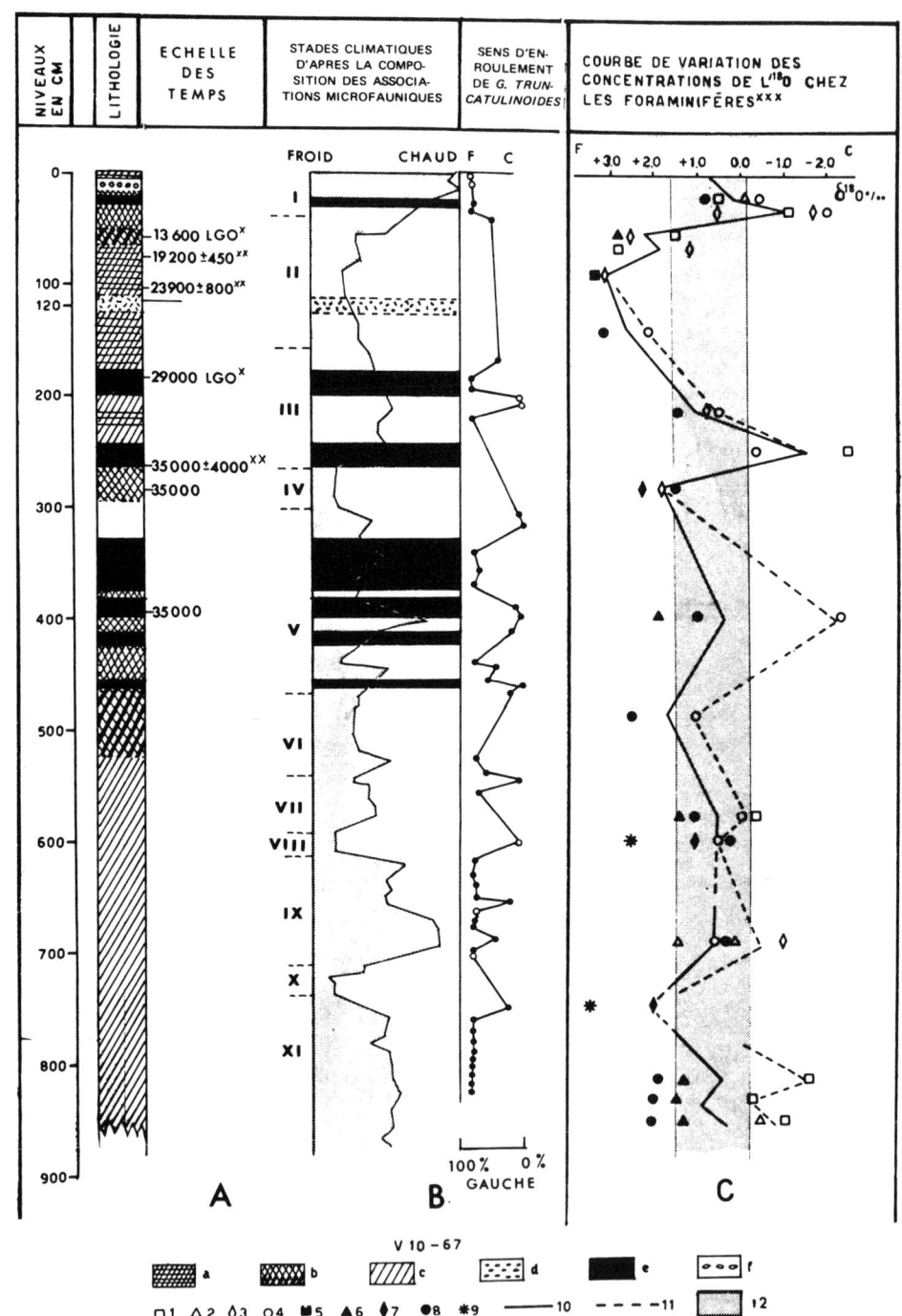

1. The alternation of warm-water and cool-water faunas (foraminifera and pteropods) reinforces the interpretation of the oxygen isotopes, and in particular the coiling direction of *Globorotalia truncatulinoides* differentiates clearly the warm postglacial from the cooler conditions of the major part of the core and suggests, along with the oxygen isotopes, that the lower part of the core (below about 625 cm) reflects the climate of the last interglaciation.

2. The oxygen-isotope analysis shows a range of variation in the lower half of the core that is about the same as that in modern planktonic fauna in the same area. However, in the part of the core younger than about 35,000 years the shells of both foraminifera and pteropods are strongly enriched in oxygen-18, indicating the effects of strong dilution of Mediterranean waters as a result of the influx of glacial meltwaters via the Black Sea. According to the authors, this change in the oxygen-isotopic ratio indicates that the waters of the eastern Mediterranean were more diluted than today, with a salinity decrease of 1 to 5 per mil, and that the temperature decrease was 5 to 10°C during the period between 30,000 and 12,000 years ago.

3. The upper half of the core is marked by repeated layers of black muds resulting from periods of stagnation of the bottom waters. Stagnation apparently became a common phenomenon after about 50,000 years ago, being thus coincident with periods of low sea level that were tied to maxima of continental glaciation. During these very low sea levels the eastern Mediterranean basin must have been nearly cut off from the western basin and the Atlantic Ocean. The Straits of Sicily would have been reduced to a narrow passage hardly more than 30 km wide at times of sea level minima.

Emiliani's (1955) study was restricted to oxygen-isotope analysis, and he reached conclusions similar to those of Vergnaud-Grazzini and Herman-Rosenberg (1969). The latter, however, estimated the range of paleotemperatures to be rather smaller than that suggested by Emiliani. Emiliani (1955, Fig. 1) estimated the temperature minimum of the last glaciation to be about 6°C and the maximum

---

Fig. 2. Analysis of core V 10–67 from the eastern Mediterranean (from Vergnaud-Grazzini and Herman-Rosenberg 1969). Column A. Lithology: a = beige sediment; b = gray sediment with reduced horizons; c = gray sediment; d = tephra; e = black sediment of stagnant conditions; f = pteropod beds. LGO° = dates from Lamont Geological Observatory; °° = dates from Centre Scientifique de Monaco (Thommeret). Column B. Paleotemperatures determined by faunal analysis and coiling direction of *Globorotalia truncatulinoides*: froid = cold, chaud = warm; F = left coiling, C = right coiling. Column C. Oxygen isotope determinations: 1, *Globigerinoides rubra;* 2, *G. trilobus sacculifer;* 3, immature forms and *Limacina* sp.; 4, *Orbulina universa;* 5, *Globigerina pachyderma;* 6, *Globorotalia truncatulinoides;* 7, *Globigerina bulloides;* 8, *Globorotalia inflata;* 9, benthonic species; 10, average isotopic composition; 11, isotopic variation in the genus *Globigerinoides;* 12, range of variation of isotopic composition of living planktonic species; °°° = determinations by the Laboratory of Nuclear Geology at Pisa.

of the preceding interglaciation about 21°C; in other words, the temperature decrease for the last cycle of glaciation was about 15°, rather than 5–10°, as suggested by the more recent study. The difference between these two estimates stems primarily from the corrections applied by the different authors for the storage of water in the form of ice sheets during a glaciation. In light of recent knowledge of the isotopic ratios found in the Greenland ice sheet, the estimate of Vergnaud-Grazzini and Herman-Rosenberg (1969) should be considered the more reasonable.

Deep cores from the Red Sea have also been studied (Herman 1968) and have led to the recognition of four climatic phases there that parallel those in the eastern Mediterranean as well as those of the open oceans (Fig. 3). As in the previously cited study, the climatic interpretation is based on a study of the entire planktonic fauna, foraminifera and pteropods, rather than a single indicator species, although the figure reproduced here is based on *Globigerinoides sacculifera* alone. Phases I and III in the Red Sea represent conditions very similar to those of the present day and probably represent postglacial and last interglacial times respectively. Phase II, dated from 65,000 to 12,000 years ago, was characterized by a cooler and more humid climate, "as evidenced by the planktonic fauna and the larger amounts of terrigenous material" in the cores. No estimate of the quantitative change in temperature is given in this study except the qualification that the temperature south of latitude 23°50′N must not have been lower than 18°C during Phase II, based on the presence of *Globorotalia tumida*.

Pollen Studies in Coastal Areas

Moving from the deep sea into littoral areas we can next review the paleoclimates that have been interpreted from palynology. In the northwestern sector of the area under discussion, in the Tenaghi Philippon in northeastern Greece (Fig. 1, no. 1), a 120-meter-deep bore has brought to light a long, continuous section including perhaps the latter half of the Quaternary (see Van der Hammen, this volume). The uppermost 30 meters of this core, described and interpreted by

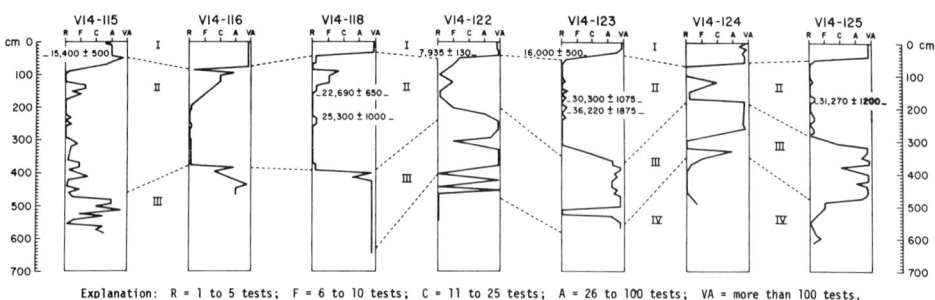

Explanation: R = 1 to 5 tests; F = 6 to 10 tests; C = 11 to 25 tests; A = 26 to 100 tests; VA = more than 100 tests.

Fig. 3. Correlation of seven cores from the Red Sea based on the variation in frequency of *Globigerinoides sacculifera* (warm stenothermal species; after Herman 1968).

Wijmstra (1969), spans the entirety of postglacial and last-glacial time and reaches into the last (Eemian) interglaciation. The Tenaghi Philippon is a structural basin only 40 meters above sea level and only a few kilometers from the Aegean coast, from which it is separated by a low mountain barrier. This core has been analyzed in great detail, but only a summary diagram based on samples taken every 100 cm is presented here (Fig. 4). Paleoclimatic interpretation is based primarily on the variations in the ratio of arboreal to herbaceous vegetation, which shows a clear dominance of herbs over trees throughout the time equivalent to the last European glaciation. This is interpreted as a period that was both cooler and drier than the present climate of this area. Radiocarbon dating of this core reveals a very good correlation with stadials and interstadials known from western Europe.

The differentiation of interglaciations and interstadials in the Tenaghi Philippon core is made possible by associations of arboreal vegetation that are not constant throughout. At different levels in the core the herbaceous vegetation gives way to one of three arboreal complexes: (1) open *Quercus ilex–Pistacia–Juniperus;* (2) *Quercus sessiliflora–Q. pubescens;* (3) *Pinus nigra–P. sylvestris.* The first of these associations is not very tolerant of frost; however, the primary difference between the two oak associations (1 and 2) is one of humidity, the second group requiring greater moisture. On the other hand, the pine association differs from the others mainly in its greater tolerance for low temperatures. Therefore, during the interstadials of the last glaciation it is the pine association that interrupts the herbaceous vegetation, whereas in postglacial and interglacial times oak and pistachio dominate over pine.

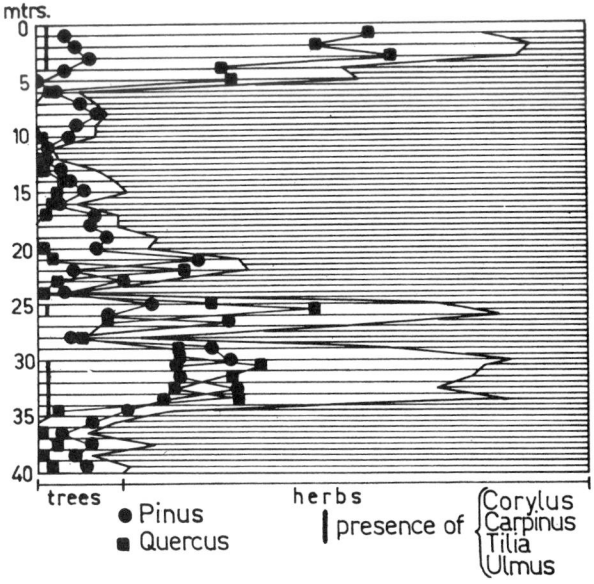

Fig. 4. Generalized pollen diagram from the Tenaghi Philippon, northeastern Greece, based on a sampling interval of 1 meter (from Wijmstra 1969, Fig. 4).

In Figure 4 the gradual deterioration of forest vegetation during the transition from interglaciation to glaciation contrasts distinctly with the rapid change (at about 5-meter depth) in the opposite sense at the end of the last glaciation.

Quantitative estimates of climatic change are difficult to assess here, and Wijmstra (1969) does not attempt it. Changes in both temperature and precipitation are indicated, and they are, as usual, difficult to untangle. One indication of the temperature change, however, may be suggested on the basis of the altitude of the transition from oak to pine associations found at the present time on the uplands surrounding the Tenaghi Philippon. *Pinus nigra* first appears at and above 580 meters above sea level (Wijmstra 1969), but only somewhat higher does it become the dominant species. Pine then continues in dominance up to the tree line, which lies at 1600 meters on the north side of the Tenaghi Philippon. The dominance of the pine association during interstadials of the last glaciation, therefore, suggests a lowering of the oak-pine transition by roughly 600 meters relative to the present condition. If this transition is controlled primarily by temperature rather than by humidity, as Wijmstra suggests, and if one applies the normal lapse rate of temperature, 0.65°C per 100 meters, then a decrease of 3.9° is suggested. However, this should be considered only as a rough indication of the real change, and since this figure applies to interstadials, the stadials were presumably even cooler. On the other hand, the alternations of herbaceous vegetation with pine forests during the last glaciation may have been more strongly dependent on changes in humidity than in temperature.

Therefore, one can say cautiously that the temperature difference during the last glacial-interglacial cycle in northeastern Greece was about 4°C. With some greater degree of certainty, one can say that the vegetation indicates that the climate in this area during the last glaciation was distinctly drier than at present.

Palynology of a truly littoral zone has been carried out in borings on the coastal plain of Israel by Rossignol (1961, 1962, 1963). The stratigraphy underlying the coastal plain is now well known by means of a great number of wells drilled for ground-water exploration and has recently been summarized by Issar (1968). An excellent three-dimensional reconstruction of the alternations of deep-water, littoral, and continental sediments is now available (Fig. 5) and has been correlated with raised marine strandlines along the Levantine coast by means of the fossil faunas (e.g. Issar and Picard 1969).

Rossignol's (1961, 1962) primary study was that of a core 189 meters long recovered near Ashdod on the southern coast of Israel (Fig. 1, no. 5). The microfossil assemblage consisted of autochthonous pollen from plants now growing in Israel, allochthonous pollen and spores transported by longshore currents along with sediments from the Nile, and marine planktonic hystrichospheres (Fig. 6). The allochthonous pollen and hystrichospheres were very useful in distinguishing marine from nonmarine sediments and in evaluating the effects of reworking of the sediments by waves and currents. Among the autochthonous pollen Rossignol (1962) recognized three assemblages: (1) Graminae–Cyperaceae; (2) Compositae–

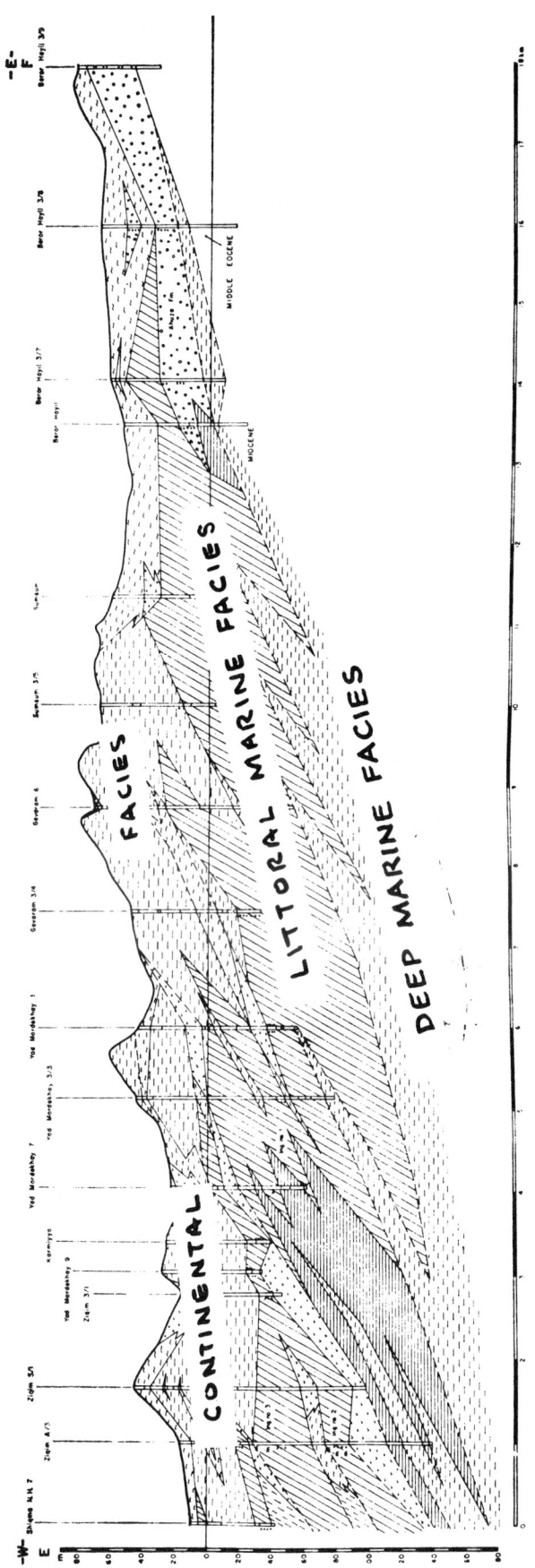

Fig. 5. West–east geologic section through the Israel coastal plain (from Issar 1968, Fig. 4).

Fig. 6. Pollen diagram from the Ashdod 15/0 bore hole (from Rossignol 1962).

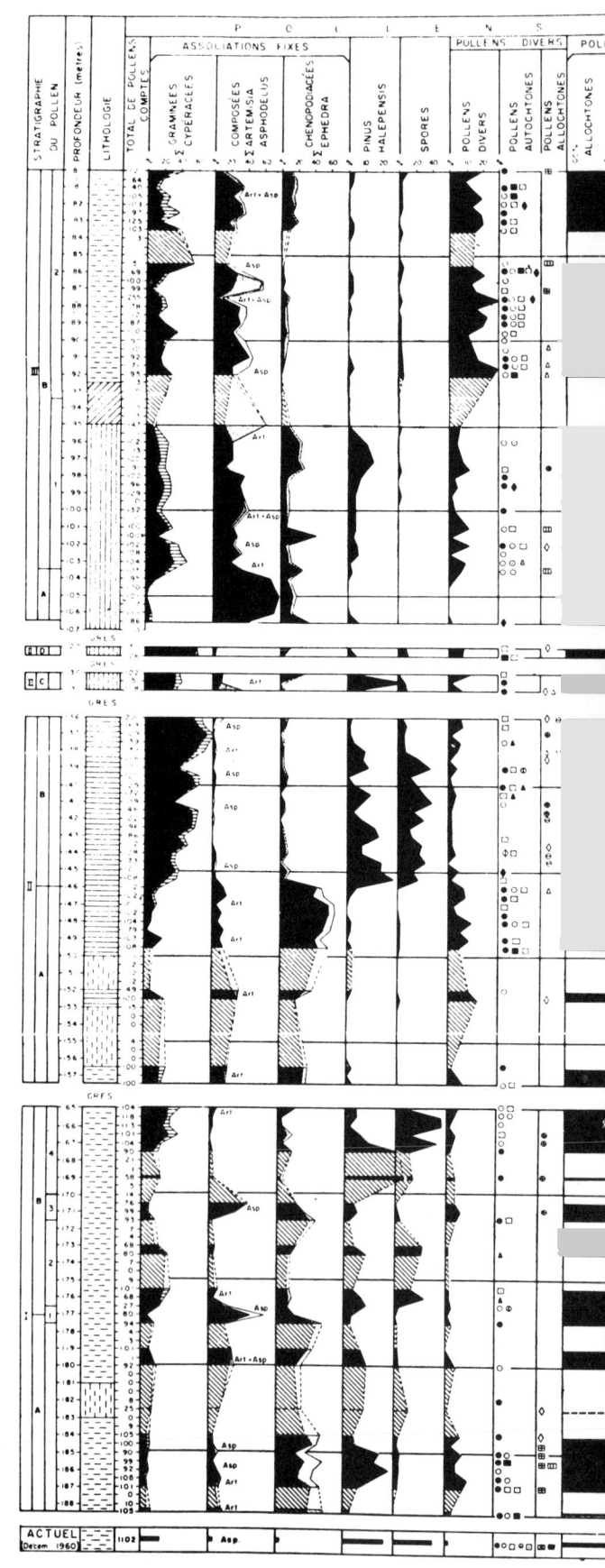

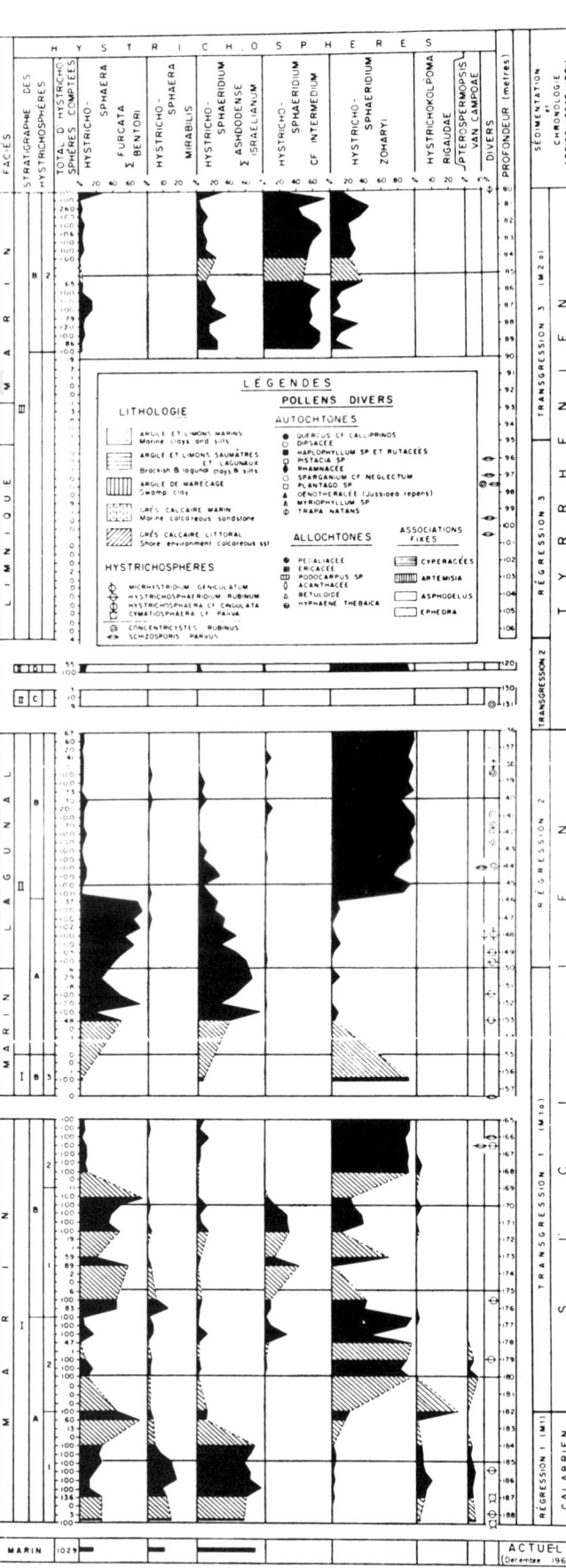

LÉGENDES

LITHOLOGIE

ARGILE ET LIMONS MARINS
Marine clays and silts

ARGILE ET LIMONS SAUMÂTRES
ET LAGUNAIRE
Brackish & lagunal clays & silts

ARGILE DE MARÉCAGE
Swamp clay

GRÈS CALCAIRE MARIN
Marine calcareous sandstone

GRÈS CALCAIRE LITTORAL
Shore environment calcareous sst

HYSTRICHOSPHÈRES

MICRHYSTRIDIUM GENICULATUM
HYSTRICHOSPHAERIDIUM RUBINUM
HYSTRICHOSPHAERA CF CINGULATA
CYMATIOSPHAERA CF PARVA
CONCENTRICYSTES RUBINUS
SCHIZOSPORIS PARVUS

POLLENS DIVERS

AUTOCHTONES

● QUERCUS CF CALLIPRINOS
○ DIPSACÉE
■ HAPLOPHYLLUM SP ET RUTACÉES
□ PISTACIA SP
◆ RHAMNACÉE
◇ SPARGANIUM CF NEGLECTUM
□ PLANTAGO SP
▲ OENOTHÉRALÉE (Jussioea repens)
▲ MYRIOPHYLLUM SP
⊕ TRAPA NATANS

ALLOCHTONES

● PÉDALIACÉE
● ÉRICACÉE
▦ PODOCARPUS SP
◊ ACANTHACÉE
△ BÉTULOIDE
● HYPHAENE THEBAICA

ASSOCIATIONS FIXES

▦ CYPERACÉES
▥ ARTEMISIA
ASPHODELUS
EPHEDRA

Artemisia–Asphodelus; (3) Chenopodiaceae–*Ephedra*. The first of these is considered to indicate humid conditions; the other two are indicators of dry climates. A very distinct alternation is apparent throughout the core between the Gramineae-Cyperaceae assemblage and that of the chenopod group, indicating that these two groups were competing for the same terrain under changing climatic conditions. The relations of the Compositae assemblage is not so clear, suggesting in this case that there was no direct competition between the Compositae and the grasses and sedges. An independent variable is the occurrence of *Pinus halepensis*, the Aleppo pine, which now grows on the Judaean uplands; it shows a strong direct correlation with the humid grass-sedge group.

The reconstruction of the former vegetation envisaged by Rossignol on the basis of her pollen study is that of a coastal-plain landscape alternately occupied by a grass-sedge association during times of relatively great humidity and by a halophytic chenopod association under an arid climate. By an examination of the facies relationships known in the coastal-plain area, she has determined that the transition from dry to humid conditions usually occurs rather early during a regression of the sea, and that the humid conditions prevail from then until the very end of the following transgression, at which time dry conditions set in again. Thus a direct correlation between high sea level (= interglacial) and coastal swamps with grasses and sedges (= humid) is excluded, and one can conclude, in fact, that regional climatic change must be the explanation for the changes in vegetation types. The interpretation of regional climatic change is reinforced by the increased frequency of *Pinus halepensis,* indicating greater degree of forestation on the Judaean Hills, and by a strong increase in allochthonous spores from the Nile, indicating greater runoff in the upper Nile Valley, coincident with Gramineae-Cyperaceae maxima in the sediments.

Rossignol (1963) has also investigated the pollen in five cores in the vicinity of Haifa Bay on the northern part of the Israeli coastal plain (Fig. 1, no. 10). The results here complement those of the Ashdod area by adding post-Tyrrhenian oscillations to the section. The latter are represented at Ashdod only by sterile sands. Similar fluctuations of the pollen types are seen here, although the chenopod association is much less abundant than at Ashdod, *P. halepensis* is infrequent, and allochthonous spores are nearly absent, the distance from the Nile being much greater. The story is not so clear as at Ashdod. The vertical distance between samples in the cores is rather great, and the relationship to regressions and transgressions is not very clear. Nevertheless, Rossignol (1963) concludes that "dans la région de Haifa, l'évolution récente du climat est indiquée par celle de la végétation: le climat est mediterranéen, et ses fluctuations ont une faible amplitude: d'abord un peu plus humide, puis plus sec, enfin plus humide de nouveau à la fin de la période envisagée." It will be interesting to compare this trend toward greater humidity during post-Tyrrhenian time to the simultaneous trend toward greater desiccation shown by mammalian microfauna, which we shall discuss later.

PLUVIAL LAKES

The concept of a pluvial lake originated in the Near East when Lartet (1865) related the past changes in level of the Dead Sea to periods of glaciation. This is an area that is quite sensitive to slight shifts in climate since it lies, as does the southwestern part of the United States, in the tension zone between the humid, temperate belt of westerlies and the dry, subtropical high-pressure belt. Many pluvial lakes are known throughout this region, especially in the Nile Valley, in Anatolia, and in the Dead Sea Rift (see Butzer 1958). We shall concentrate on the Dead Sea Rift and the immediately surrounding area because recent studies there are most directly pertinent to the question of late Quaternary climate.

The Dead Sea Rift valley (Fig. 1, nos. 7, 8, 9) extends from the northeastern corner of the State of Israel to the Gulf of Aqaba on the east side of the Sinai Peninsula, a distance of more than 400 km. There are three contemporary water bodies in the rift. Lake Hula in the north, drained about 30 years ago for peat exploitation, was a relatively shallow, small, swampy lake in historic time and apparently throughout much of the Quaternary; Lake Tiberias (Yam Kinneret or the Sea of Galilee), 15 to 20 km farther south, is still a freshwater lake; and the Dead Sea in the middle sector of the rift is a highly saline basin now and was so throughout middle and late Quaternary time. The Jordan River has several heads north and east of the Hula depression; it flows into Lake Hula at about 72 meters above sea level. South of the Hula the Jordan is incised into the Korazim cover basalts, and it drops to 209 meters below sea level at the point where it enters Lake Tiberias. South of Lake Tiberias the Jordan traverses a lowland underlain primarily by chalky lacustrine deposits of the ancient pluvial Lisan Lake and debouches into the Dead Sea at about —400 meters. The drainage from the south into the Dead Sea is via the Wadi Arava, which contains no perennial streams.

The Dead Sea Rift valley is a well-recognized structural trough; its main period of taphrogenesis (graben tectonics) occurred during Quaternary time. Both strike-slip movement along its border faults and subsidence of the rift floor have taken place discontinuously and have occurred as late as latest Pleistocene or even Holocene time (Zak and Freund 1966; Neev and Emery 1967). Therefore, the role of tectonics must be recognized in the interpretation of the high-level lakes that existed there during the Quaternary.

A quick review of the development of the Dead Sea Rift will be useful in providing a background against which late Quaternary events can be viewed. This review depends heavily on the recent summary by Neev and Emery (1967). The first sedimentary unit that is restricted to the rift in its spatial distribution is the Rock Salt unit (in the Mount Sodom area), which contains Plio-Pleistocene pollen and spores. Thus it is thought that the formation of the graben was initiated at about the beginning of the Quaternary period. In early Quaternary time, and perhaps as late as middle Quaternary time, one or more connections may have persisted with the open sea, either in the south to the Gulf of Aqaba, to the west through a fossil channel to the Mediterranean in the vicinity of Gaza, or to the northwest via the Plain of Esdrelon to Haifa Bay. Further tectonic movements isolated the rift valley from the sea, and a lacustrine regime ensued wherein gypsiferous and calcareous marl deposition prevailed in the central parts of the valley, and alluvial fans spread outward from the valley walls. This is the "Lisan stage" of Neev and Emery (1967). The Lisan marl itself constitutes the rock unit of the latter part of this stage and is up to 40 meters thick. The Hamarmar formation (clays and brines) constitutes a transitional unit between the Salt unit and the Lisan marl and is a little thicker than the Lisan; earlier Quaternary sediments in the rift valley, including the Salt unit, are up to 4000 meters in total thickness. The Lisan stage is followed by the Dead Sea stage during which the configuration of the Dead Sea and Lake Tiberias was essentially the same as we see today. Apparently about 20,000 years ago there was a coincidence of strong subsidence of the graben floor and climatic desiccation at the time of the transition from the Lisan stage to the Dead Sea stage. This transition was rapid, occurring within about 1000 years, during which the level of the Lisan Lake dropped some 190 meters as a result of the combined tectonic–climatic interaction. At this time the Dead Sea, Lake Tiberias, and an intermediate lake in the Bet She'an area came into existence. The Bet She'an lake existed until about 5000 years ago, when it was drained by headward erosion of the Jordan River. Moreover, early in the Dead Sea stage the level of the Dead Sea was considerably lower than it is today; its north and south basins were at times separated from each other in the area of the Lisan Peninsula to such an extent that even in historic time caravans could traverse the rift valley at this point. The Dead Sea has experienced rising water levels throughout the last 1000 years; the north and south basins were joined again only some 400–500 years ago, and the maximum level of the present Dea Sea was reached only about 50 years ago. On the other hand, Lake Tiberias has persisted essentially unchanged since the beginning of the Dead Sea stage (Ben-Arieh 1967).

Looking more closely at the upper Quaternary Lisan Lake, we find that the land–water relationships then must have been very different from those that prevail today. As determined by the distribution of the Lisan marls and shoreline tufa deposits, the Lisan Lake extended continuously from the south shore of the present Lake Tiberias to a point some 35 km south of the south shore of today's Dead Sea (Neev and Emery 1967, Fig. 15); see Figure 7. In fact, the entire Tiberias basin was probably part of the Lisan Lake (Ben-Arieh, 1967). Its north–south dimension was thus about 220 km and its maximum width about 17 km; its highest shoreline lay at −180 meters (compared to −400 meters for the Dead Sea), and its water depth was at least 190 meters. At that time the water volume would have been about 325 km³ and the salinity about 135 grams per liter, "too great for most aqueous life forms other than diatoms" (Neev and Emery 1967, p. 104). The present Dead Sea, with a volume of 136 km³, has a salinity of 322 grams per liter.

Rossignol (1969) has investigated the fossil pollen in Lisan Lake sediments. The Lisan marl itself is sterile, apparently because the lithology of the sediments is not conducive to the preservation of pollen, but the upper part of the underly-

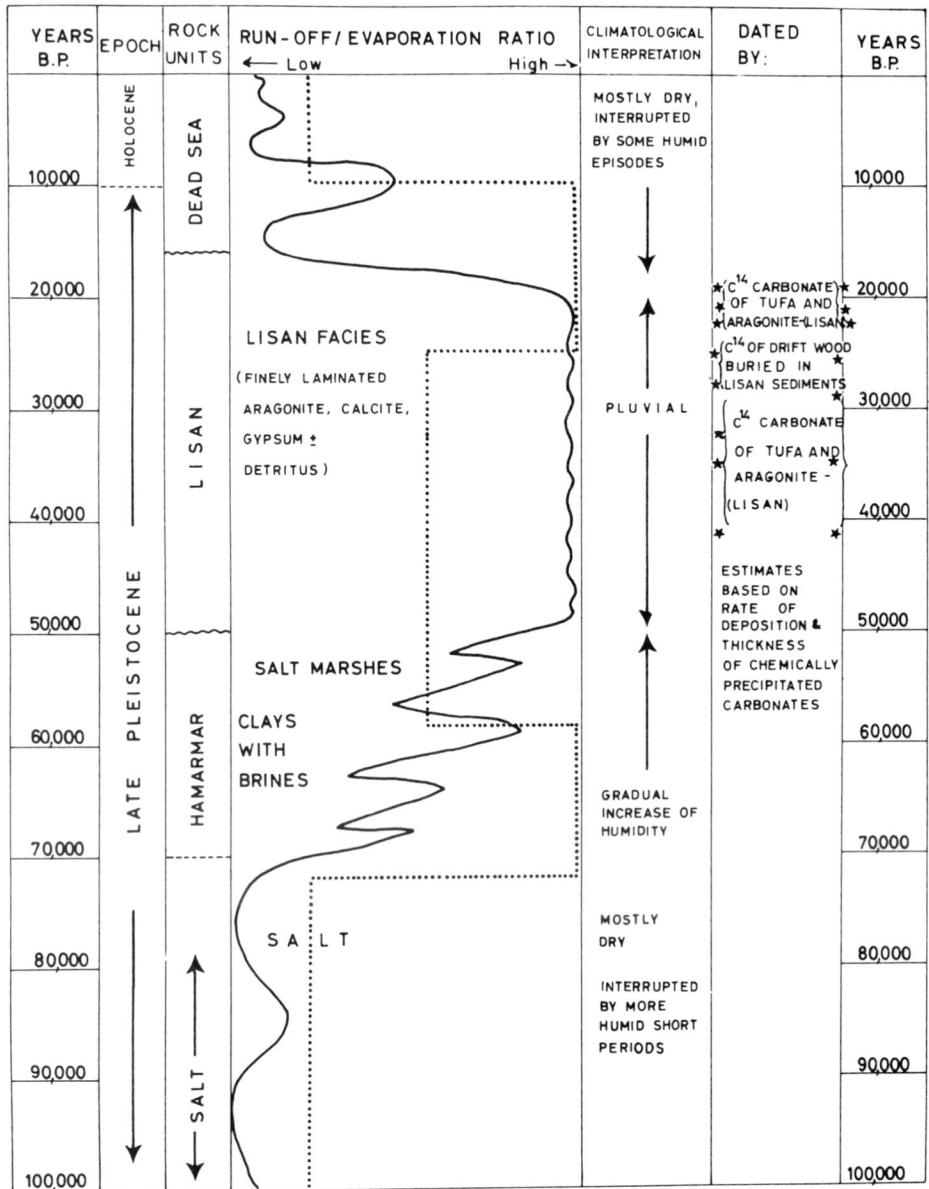

.... IDEALISED LEVEL-CHRONOLOGY FOR GREAT BASIN LAKES
BY BROECKER & WALTON (1959)

Fig. 7. Generalized climatic fluctuations during the past 100,000 years for the Dead Sea region compared with similar data from the Great Basin of the United States (dotted line; after Neev and Emery 1967, Fig. 16).

ing transitional clayey Hamarmar sediments has yielded enough pollen for an interpretation of vegetational history. These sediments should date from the period 100,000 to 70,000 years ago (Neev and Emery 1967). In the top 16 meters of the Hamarmar Beds Rossignol finds the following:

(bottom)	Zone I	Aleppo pine dominant, Graminae rather abundant, few Chenopodiaceae, Compositae, *Ephedra*
	Zone II	Aleppo pine and Graminae decrease strongly, Chenopodiaceae, Compositae, *Artemisia*, and *Ephedra* increase
(top)	Zone III	Rise and dominance of Aleppo pine, decrease of Chenopodiaceae, slight decrease of *Artemisia*, increase of *Ephedra*, no change in Compositae

Rossignol concludes that the climate of zones I and III was more humid than today, although still biseasonal Mediterranean, with perhaps 800 mm rainfall at 850 meters above sea level; zone II was drier with about 400 mm rainfall on the uplands. The pollen assemblages suggest a vegetational source with an altitudinal gradient from −200 to +850 meters above sea level. For comparison, the site sampled is now about 400 meters below sea level in an area with 50 mm or less rainfall per year. The Judaean Hills, on the windward side of the Dead Sea graben, rise to 700 to 1000 meters and have an annual rainfall of about 400 mm at 800–900 meters above sea level.

The areal extent of the Lisan Lake was perhaps the most important determinant for late Quaternary climate in the Palestine-Transjordan area. The lake must have had a surface area roughly twice that of the present Dead Sea and Lake Tiberias combined. Moreover, the altitudinal difference between the lake surface and the surrounding uplands was about 200 meters less than at present. Therefore, evaporation must have been taking place from a much greater surface area during late Quaternary time, and the distance of transport of the water vapor from the lake to the surrounding uplands was much shorter than under present conditions. Furthermore, the pluvial El Jafr Lake (discussed below), just to the east of the Dead Sea Rift and contemporary with the Lisan Lake, must also have added a considerable amount of humidity to the atmosphere of the surrounding region.

The demise of the Lisan Lake was, as mentioned above, rapid and was conditioned both by tectonic subsidence and by climatic change about 20,000 years ago. Radiocarbon dating of shoreline tufas shows that the lake level dropped from −180 to about −370 meters in less than 1000 years. A lake with a shoreline at −370 meters and the altitude of its floor the same as that of the present Dead Sea could have held only about half the water of the former Lisan Lake, even allowing for subsequent sedimentation that has partially filled in the Dead Sea since the time of subsidence. Thus, climatic desiccation must also be invoked. The time of this rapid transition has been dated by radiocarbon analysis of shoreline tufas, of aragonite of the Lisan marl, and of pieces of driftwood recovered from the lacustrine sediments (Broecker *in* Neev and Emery 1967, p. 25). The lower part of the Lisan sediments is beyond the range of radiocarbon, but Neev and Emery (1967. p. 26) estimate that the Lisan stage began some 70,000 to 100,000 years ago.

The Lisan marl is in general very little disturbed by tectonic movements, but it is not entirely free from such influence. Zak and Freund (1966) estimate that some 150 meters of strike-slip movement along border faults of the rift valley has occurred since the demise of the Lisan Lake. Furthermore, Langozky (1963) finds Hamarmar-type sediments, i.e. characteristic of the lake that immediately preceded the Lisan Lake in the rift valley, 85 meters above the uppermost Lisan sediments. He attributes this occurrence to pre-Lisan faulting and correlates it with several unconformities recognized between the Lisan and Hamarmar (or "Samra") beds. Some of these occurrences are situated as high as 60 meters *above* sea level. Thus it appears that progressive taphrogenesis throughout the Quaternary has continued to lower the level of the Dead Sea graben and its lakes, at least relative to the surrounding uplands if not relative to present sea level.

Post-Lisan history in the vicinity of the Dead Sea is recorded by terraces both in the tributary wadis and along the Jordan itself. Vita-Finzi (1964) recognizes the following sequence in the tributaries to the lower Jordan.

Lower wadi terrace	Roman and Iron Age artifacts
Upper wadi terrace	Rolled middle and Upper Paleolithic artifacts; numerous, but not contemporary with terrace
Tufas and cemented gravel	"Pre-Neolithic"
Lisan formation (gravels)	One Upper Paleolithic-type artifact that could be "pre-Aurignacian" (i.e. Amudian)

The well-bedded, generally fine-grained character of the lower terrace sediments suggests to Vita-Finzi that rainfall at the time of its accumulation was more uniformly distributed than today. Moreover, he attributes the alluviation in general to a strong increase in frost action in the upper reaches of these wadis, some 600 meters above sea level, and he finds support for such an interpretation in the observations by himself and others in Libya and elsewhere in North Africa that suggest as much as a 20°F (11°C) decrease in January temperature.

The northern part of the Jordan Valley, i.e. north of Lake Tiberias, presents a different kind of story, although with similar conclusions concerning paleoclimates. Immediately north of Lake Tiberias the Korazim cover basalts constitute a topographic barrier that separates the Hula basin to the north from Lake Tiberias (Fig. 8). This barrier has persisted throughout middle and late Quaternary time, and therefore the Hula basin has developed independently of the Lake Tiberias-Dead Sea, or Lisan Lake, area. The Quaternary of the northern Jordan Valley has recently been summarized by Picard (1965), who draws heavily on his own earlier studies in this area (especially Picard 1963).

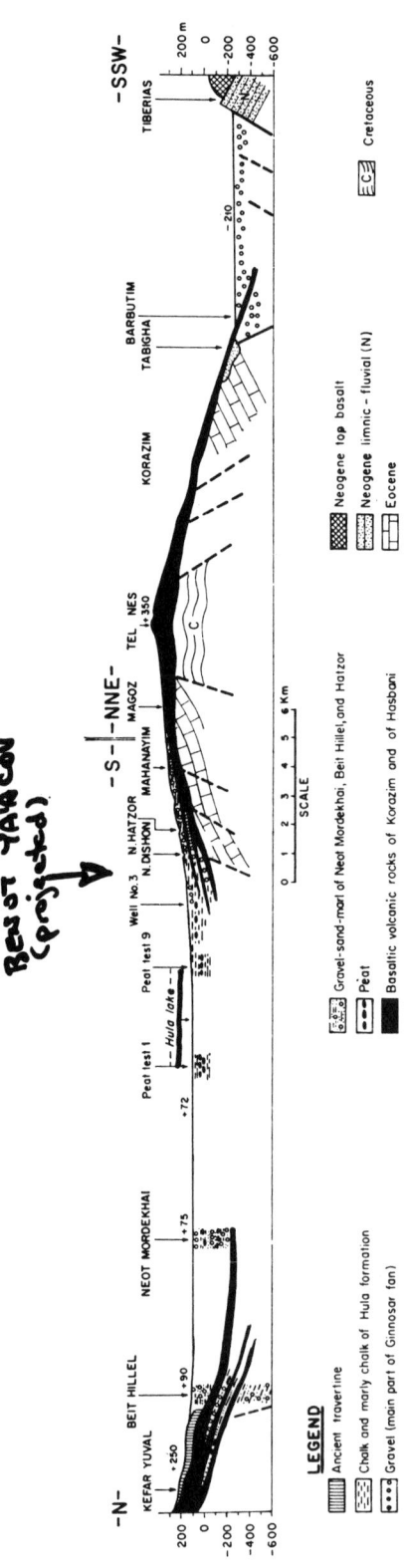

Fig. 8. North–south cross section through the Hula and Tiberias sectors of the Jordan graben, showing stratigraphy and structure (from Picard 1965).

The brief outline of the development of the northern part of the Jordan Valley includes the following major steps (Picard 1965).

1. Upper Villafranchian deposition (Ubeidiya formation) in an extensive shallow lake alternating with a floodplain environment; they include the important archaeological site at Ubeidiya with artifacts of pebble culture and "Abbevillian" affinity along with a hominid once referred to the *Sinanthropus* group. However, Tobias (1966) finds that "the only decidedly archaic feature seems to be the great thickness of the parietal bone," and he prefers to classify it as *Homo spec. indet.*

2. A period of moderately strong tectonic activity in which the Ubeidiya formation was folded and truncated by erosion.

3. Accumulation of Naharayim fluvial clastics in the area just south of Lake Tiberias and of terrace deposits of the Jordan with Acheulian artifacts at Benot Yaaqov in the southeastern corner of the Hula basin.

4. Extensive middle Quaternary volcanism then follows, according to Picard (1965), including the Korazim and Yarda basalts (up to 200 meters thick) between the Lake Tiberias and Lake Hula basins and the Yarmuk basalt just south of Lake Tiberias. However, Schulman (1967) is of the opinion that the Korazim and Yarda basalts are part of the much more extensive Pliocene "cover basalts"; he places the thin basalts that are found in the Benot Yaaqov terrace section and interbedded in the Hula lacustrine series (see below) in post-cover-basalt time. Picard had considered them to be tongues from the main Korazim lavas. The Yarmuk basalt, however, remains in its middle Quaternary position since it can clearly be seen in stratigraphic position between the Naharayim and Lisan formations.

5. A period of lacustrine sedimentation in both the Hula and the Tiberias basins, which were formed or at least accentuated during the preceding period of middle Quaternary volcanism. The Lisan Lake reached into the Tiberias basin, as we have seen previously, and Lisan marl was deposited directly upon the Yarmuk basalt. In the Hula basin Picard (1963, 1965) recognizes a lower Lacustrine unit, at least 20 to 30 meters thick, a Main Peat unit that reaches a maximum of 50 meters in thickness, and an upper Lacustrine unit of 25 to 50 meters in thickness. Thin basalts are interbedded into these lacustrine beds on the flanks of the Hula basin, as mentioned in section 4 above, and because of his views on the correlation of these basalts with those at Benot Yaaqov, which he correlates with the Yarda (= Korazim) basalt, Picard (1965) considers the lower Lacustrine and the Main Peat units of the Hula formation to be middle Quaternary in age. However, as we shall see below, the Hula formation apparently is entirely of late Quaternary age.

Lacustrine sedimentation continued into Holocene time throughout this area. As mentioned earlier, Lake Hula was drained artificially within very recent time, and a thin Holocene peat overlies the upper Lacustrine unit recognized by Picard. On the other hand, the Bet She'an basin, just south of the Tiberias basin, was occupied by a lake in which Lisan-type marls were accumulating as recently as about 5000 years ago (Neev and Emery 1967).

Since we are primarily concerned with late Quaternary climates, it will be necessary to look more closely at the Hula formation. J. M. Remy (*in* Picard 1963) was the first to investigate the pollen in these lacustrine beds and found the following succession, beginning at the top:

A4	"humid-warm"
A3	"dry": slight maximum of Compositae with Oleaceae of short duration
A2	"humid-warm": *Quercus* maximum
A1	?
B4	Again "dry": Compositae maximum, no Oleaceae
B3	Less "dry": more *Cedrus*, some *Picea*, Graminae maximum, frequent Oleaceae; some climatic change
B2	again "dry": Compositae maximum
B1	transition: reduction of Compositae, few *Cedrus* and Oleaceae
C4	"dry-warm": reduction of *Cedrus*, no Picea or Oleaceae, but 24 per cent Compositae
C3	"dry-warm": much *Quercus*
C2	"humid-warm": 16 per cent *Cedrus*, some *Picea*, few Compositae or Oleaceae; "climatic optimum"
C1	"humid-warm": *Cedrus* increase, more Compositae than C2

Remy interprets the transition from the Main Peat (B) to the upper Lacustrine unit (A) to be sharper than the transition from C to B, and Picard (1963, p. 9) correlates this sharper transition with the extinction of two mollusks, *Viviparus apamea* and *Melanopsis aaronsohni*, which characterize both the lower Lacustrine unit and the chalky interbeds in the Main Peat. These two mollusks are also abundant in the *Viviparus* beds of the old Jordan terrace at Benot Yaaqov and are found in the Ubeidiya formation (Villafranchian). These extinctions reinforce Picard's (1963, 1965) placement of the upper Lacustrine unit in the upper Quaternary and the lower part of the Hula formation in the middle Quaternary.

However, some recently determined radiocarbon dates suggest that both the upper Lacustrine and the Main Peat units are of late Quaternary age. A 54-meter core from the center of Lake Hula is being studied by G. E. Hutchinson and U. Cowgill (see Stuiver 1969, pp. 591–92). The uppermost 41.65 meters of this core constitute a "main lacustrine period" that is dated between 2480 ± 100 (Y–2424) and 22,400 ± 400 years ago (Y–2514). The lower 13 meters penetrated by the core is divided into three periods of very shallow water interrupted by transitory lacustrine episodes. The lowest date reported, 29,700 ± 800 years ago (Y–2517), comes from a depth of 49.70 meters in the core, although slightly higher, at 48.48 meters, there is an older age of 32,900 ± 800 (Y–2426). Stuiver (1969, p. 592) cautions that some of the dates are probably too old because of the incorporation of old carbonate, and this perhaps explains the inversion of dates just mentioned. If, therefore, the "main lacustrine period" recognized here is identical with the upper Lacustrine unit of Picard (1963), then that unit falls into the last major subdivision of the last glaciation of Europe and North America (main Würm and Woodfordian respectively), and it is most reasonable that the lower Lacustrine unit of the Hula formation should be assigned to the early part of the last glaciation.

Another C[14] date has recently been reported by Horowitz (1968) from the top of the first peat encountered in a bore hole (K–JAM) in the Hula basin. This peat underlines 30 meters of lacustrine chalk, Picard's Upper Lacustrine Unit, and its age was determined to be 18,800 ± 195 years (Hv 1725), in reasonable agreement with the date of 22,400 years mentioned in the preceding paragraph. Horowitz's (1968, 1969) palynological study of the K–JAM (Hula) core leads him to the conclusion that the climate there during the last interglaciation ("Eemian") was warm and very dry and that the time of the last glaciation, which he dates between 60,000 and 11,500 years ago, was cooler and humid and included two warm, humid interstades. He reports a very close relationship between the climatic events in the northern Jordan Valley and in western Europe.

Just 100 km southeast of the southeastern corner of the Dead Sea lies the basin of El Jafr (Fig. 1, no. 6), which contained a pluvial lake in late Quaternary time at least the size of Lake Constance. Like the Dead Sea, this basin lies in an area of less than 50 mm rainfall per year. Huckriede and Wiesemann (1968) have recently restudied this basin, and they report that the El Jafr pluvial lake must have had an area of 1000 to 1800 km², that is, approximately the same surface area as the Lisan Lake. Furthermore, radiocarbon dating (Huckriede and Wiesemann, 1968, p. 81) shows that the El Jafr lake was a contemporary of the Lisan Lake: the upper part of the El Jafr lacustrine limestone dates 26,400 ± 870 years ago (Hv 1719). (These authors also report additional dates from the Lisan marl itself that confirm those previously reported in Neev and Emery, 1967, and mentioned above.)

The series of events interpreted by Huckriede and Wiesemann (1968) are as follows.

1. "Main Würm pluvial," during which up to 25 meters of limestone and marl rich in ostracodes and mollusks were deposited. The shorelines of this lake are abundantly littered with Levalloiso-Mousterian artifacts, along with some cruder types attributed to the Levalloisian or upper Acheulian. The demise of this lake was somewhat more recent than 26,000 years ago.

2. A period of progressive desiccation during which the lake becomes brackish, followed by a time when sheets of gravel spread across the basin. Levalloiso-Mousterian artifacts are found *on* this gravel surface; they are heavily patinated and wind polished. The end of this period, called the "post-Lisan arid phase," is characterized by extreme aridity and sand storms more intense than at present.

3. Moist conditions return in the "upper Würm" pluvial phase, and the increased humidity causes solution of part of the older pluvial limestone and converts the basin into a mud-flat playa. Upper Paleolithic artifacts of a mid-

dle Paleolithic character (called the "Matakhium industry" by the authors) occur on the surface of this playa and on the surface of a delta belonging to this young moist phase.

Thus we see that the occurrence of markedly pluvial conditions in the Palestine-Transjordan area during late Quaternary time is confirmed in three separate basins, the Dead Sea-Tiberias basin, the Hula basin, and the El Jafr basin, all three of which have yielded identical radiocarbon dates for the final phases of their pluvial lakes. The climate of the Dead Sea Rift and of western Jordan must, therefore, have been much more pleasant during that time. This explains to a great extent the relative abundance of prehistoric artifacts in areas that are essentially uninhabitable today, such as the El Jafr Playa and the Judaean Desert on the west bank of the Dead Sea (Neuville 1951).

GLACIATION AND PERIGLACIAL ACTIVITY IN THE NEAR EASTERN MOUNTAINS

Since the publication of Butzer's review (1958) there have been several investigations of Quaternary glaciation, snowlines, and periglacial phenomena in the Mediterranean area, primarily the work of German and Swiss scientists (Klaer 1962; Kaiser 1963, 1965; Messerli 1966a, 1966b, 1967). The points of view of these different workers have not been the same in all cases, and their results diverge somewhat, but there is unanimity among them that the climatic snowline was distinctly lower during the Quaternary than it is today. Although some of the higher mountains in Turkey have given clear evidence of glacier expansion in late Quaternary time and are snowcovered at present, much discussion has centered on the question of Quaternary glaciation in the lower mountains of the Levant—the Lebanon, the Anti-Lebanon, and Mount Hermon. There now seems to be a consensus that small glaciers did appear on these ranges, and Messerli (1967) considers two morainic systems at about 2500 meters on the south slope of Mount Hermon to be the best glacial features in all the Lebanese mountain area. An example of differences of opinion, on the other hand, is provided by interpretation of the cirque-like Cedars of Lebanon Valley. Impressive moraine-like hills occur on the floor of this "cirque," and Kaiser (1963) interprets them as the results of both Riss and Würm glaciation. However, both Klaer (1962) and Messerli (1966a) interpret these features as the results of mass movement, in particular, slippage on a lower Cretaceous glide surface, and they find no evidence of glacial activity in the Cedars valley.

Opinions on the depression of climatic snowline during the Quaternary do not differ so radically. Both Kaiser (1963) and Messerli (1967) conclude that snowline was lowered about 1000–1200 meters in Asia Minor (Fig. 9). If such depression is due entirely to a temperature change, then a decrease of some 6 to 7°C would have occurred in the Levant. Klaer (1962), on the other hand, recognizes only about 500 meters of snowline lowering in the Lebanon, and a temperature decrease of about 3.5°C, if due to temperature alone, or only about 2.5° if there

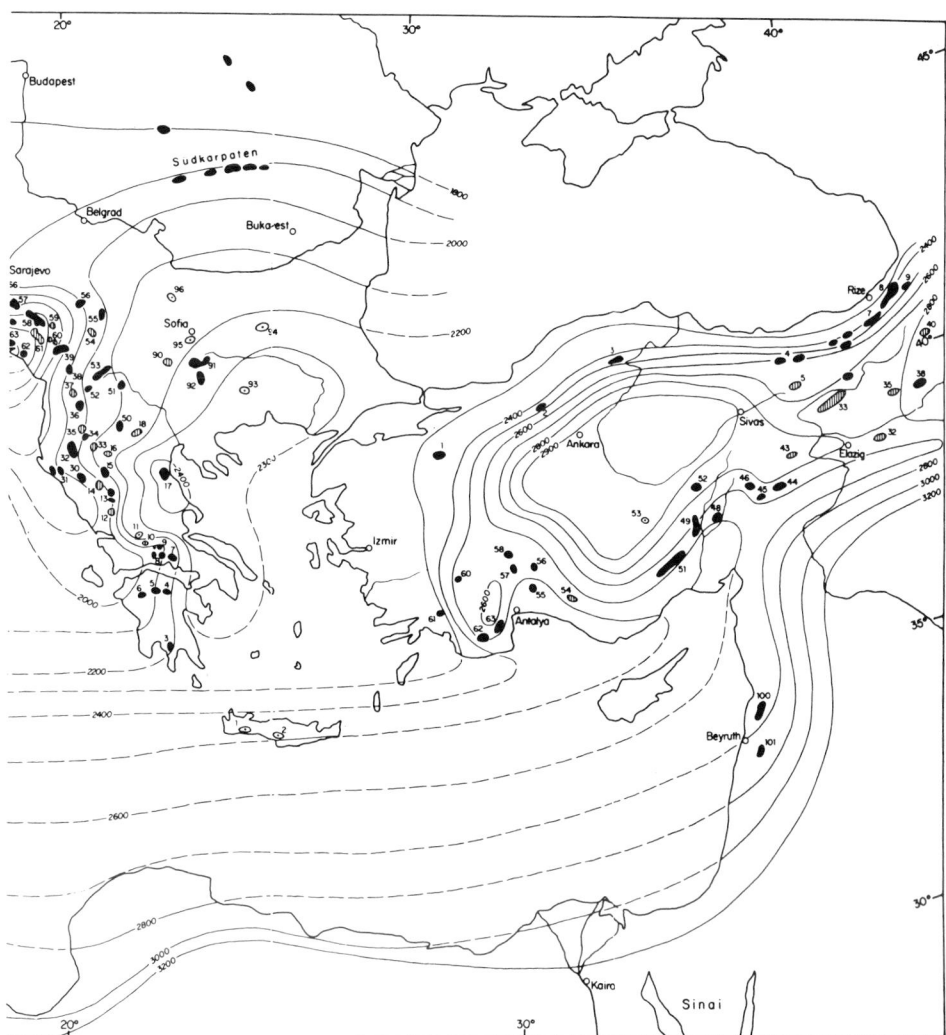

Fig. 9. Würm-age mountain glaciation (numbered spots) and snowline configuration in the eastern Mediterranean area (from Messerli 1967).

was a concomitant increase in available atmospheric moisture. The principal difference of opinion in the case of snowlines comes from the interpretation of the present-day snowline in the Lebanon. Klaer (1962) believes that it must lie very near the summit of the Lebanon (Qornet es Saouda, 3088 meters) because of the occurrence there of isolated patches of perennial snow. Messerli (1967), on the contrary, interprets these snow patches as strictly local phenomena, and he places the modern snowline well above the Qornet es Saouda at the level of the 4.5°C July isotherm, which he believes corresponds closely to the position of the perennial snowline elsewhere.

Kaiser (1963) also discusses the lowering of the "periglacial" zone in the Lebanese mountain area, finding that zone to have been lowered by 700–850 meters below its present position during the last glaciation. Kaiser's views are summarized in Figure 10.

Other evidence of periglacial activity, although not so intense as permafrost or solifluction, has been reported from the eastern Mediterranean littoral (e.g. De Vaumas 1968). The evidence is scattered and weak in many cases, and certain forms (e.g. slope breccias) are open to alternative interpretations and are therefore controversial. Dresch (1967), for example, saw no certain evidence of protracted frost activity (frost wedges, cryoturbation, etc.) below about 600 meters above sea level, but Nir (1963) and De Vaumas (1968) affirm that frost weathering has been important not only on the Judaean and Galilean uplands but also all the way down to sea level—in the Levant as well as on the island of Cyprus (De Vaumas 1964). At present snow does fall occasionally on the Palestinian uplands, and bare soil at 750 meters near Jerusalem is reported to freeze (Nir 1963). At lower altitudes in the Levant frost is extremely rare today. Therefore, frost activity at sea level, which De Vaumas (1968) reports as far south as the Negev at least, must imply a significant cooling of the climate in these areas.

Late Quaternary (Lisan and post-Lisan) alluviation in the Dead Sea wadis, discussed above, has likewise been attributed to this increased frequency of freeze–thaw cycles in the Palestinian and Jordanian uplands (Vita-Finzi 1964).

Another indication of frost activity is found in prehistoric cave sites around the eastern Mediterranean. Small angular limestone fragments generally interpreted as the result of freeze–thaw alternations characterize the bulk of the coarse

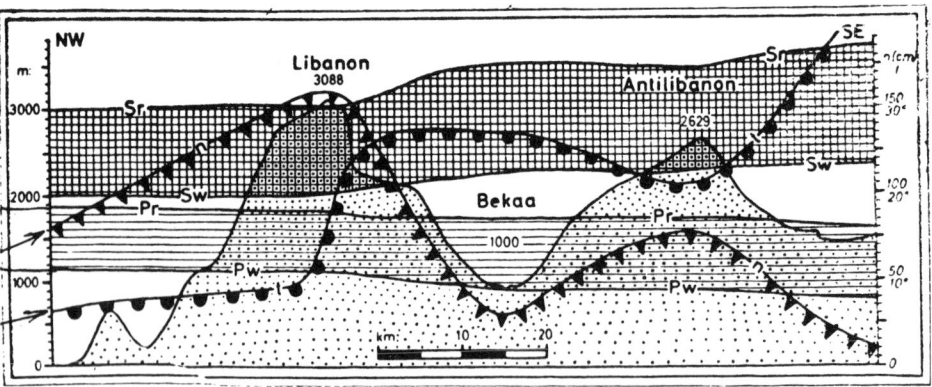

Fig. 10. Schematic profile of the altitudinal position of the present and ice-age climatic snowline and the lower limit of periglacial phenomena in the Syrio-Lebanese mountains (from Kaiser 1963, Fig. 7). Sr = recent snowline; Sw = Würm-age snowline; Pr = recent lower periglacial limit; Pw = Würm-age lower periglacial limit; m = present-day mean annual precipitation; t = present-day temperature difference between mean monthly temperatures of warmest and coldest months (°C).

fraction of the sediments in the rock shelters at Yabrud and Jerf 'Ajla in Syria (Farrand 1965; Goldberg 1968), in the Haua Fteah in Libya (McBurney 1967), and apparently also in Ksar 'Akil in Lebanon (H. E. Wright 1962) as well as in the caves of the Judaean Desert (Neuville 1951). Furthermore, recent observations by the writer have shown that similar limestone debris occurs in the cave of Jebel Qafzeh at Nazareth in Galilee (about 350 meters above sea level). The case for frost weathering in the Carmel caves on the Israeli coastal plain is as yet much less clear. Freezing temperatures are rare or absent in all these areas under the present climate. The sites with the coolest climate at present are Yabrud and Jerf 'Ajla where the mean January temperature is positive and even the normal minimum daily temperatures are above 0°C; extreme temperatures in January occasionally drop at low as $-7°C$, thus making freeze–thaw cycles few and far between in these areas (Syrian Arab Republic 1964). Therefore, the prevalence of frost-derived debris in these caves indicates a significant lowering of the mean winter temperatures in these areas during the period of the last glaciation of Europe.

Cryoclastic activity requires both subfreezing temperatures and the presence of moisture, but the question of how much moisture is yet to be resolved. The presence of pluvial lakes in these same areas indicates certainly that there was an increase in the precipitation/evaporation ratio, but it does not necessarily require an increase in the absolute amount of rainfall. General limits on the amount of increased rainfall in the Levant have been set by De Vaumas (1968), who sees an increase in cryoclastism and solifluction, both involving an increase in the effective atmospheric moisture, but an absence of linear erosion by running water. Thus there was an increase in the effective precipitation, but not a very great increase. This conclusion is in line with those of Rossignol (1969) and Picard (1965), mentioned above, that climatic change in the Near East was not very marked during late Quaternary time.

Prehistoric cave sites have yielded other indications of paleoclimate, both from their sediments and from the fossils contained in the strata. (The mammalian faunas will be discussed below.) In the Haua Fteah the conclusion that a cooler climate prevailed there during the last 50,000 years or more is based not only on the presence of cryoclastic limestone rubble but also on the dominance of caprine goats (an indicator of cool climate), and on isotopic temperature analysis of marine mollusks that constituted part of the food supply of the prehistoric inhabitants of the cave (Emiliani in McBurney 1967, p. 58). The average temperature during the last glaciation was about 15°C, whereas mollusks from the interglacial (?) and postglacial sediments indicate average temperatures of about 20 to 22°C, thus suggesting a change in the sea surface temperature along the north coast of Libya of about 5 to 7°C during the last glacial-interglacial cycle. This temperature shift is quite similar to that recognized in Mediterranean deep-sea cores (5–10°C) and from lowering of the snowline (6–7°C), discussed previously.

Furthermore, changes in effective moisture have also been interpreted from these cave sediments. In general, a greater degree of chemical corrosion (leading to the rounding of rock fragments in situ), cementation by secondary calcium carbonate, and accumulation of phosphates in the sediments have been taken as indications of periods of increased atmospheric moisture (Neuville 1951; Goldberg 1968; Farrand, in press). In general, moister conditions prevailed during the Levalloiso-Mousterian habitations of Jerf 'Ajla, Yabrud, et-Tabun, Qafzeh, and Oumm-Qatafa (Farrand, in press), as well as at the Haua Fteah (McBurney 1967), roughly between 30,000 and 70,000 years ago.

Mammalian faunal analysis, of both megafauna and microfauna, comes primarily from archaeological sites in the area under consideration. The megafaunas from eastern Mediterranean cave sites have been recently reviewed by Higgs (in McBurney 1967) in connection with his own work at the Haua Fteah. In general, Higgs finds that a high frequency of bovine remains correlates with warm, dry conditions throughout the circum-Mediterranean area. Cool, moist conditions, on the other hand, are indicated by relatively great numbers of caprine goats. Furthermore, Higgs believes he can demonstrate that selection by the human inhabitants of these caves has not played an overwhelming role in the determination of the fossil assemblage; that is, selection by man has not strongly biased the animal types represented. The megafauna recovered at the Haua Fteah in sediments spanning the last 85,000 years or more is shown in Figure 11 (from Higgs in McBurney 1967).

Additional information on the megafauna comes from the studies, as yet largely unpublished, of Bouchud at Qafzeh. Bouchud (1969) concludes that the terminal upper Paleolithic at Qafzeh, characterized by dominance of *Gazella*, was very dry, whereas the Mousterian habitation fell into a humid phase in which Cervidae were abundant along with *Capra ibex*, which suggests a cool climate, and horses, suggesting the proximity of steppe conditions.

The upper Pleistocene rodent faunas, on the other hand, have been published in detail recently by Tchernov (1968). These faunas were likewise recovered from archaeological cave sites: Oumm-Qatafa, Kebara, Fallah, and Abu Usba. The last two sites are Natufian-Neolithic in age, and thus postglacial or postpluvial, in terms of the general climatic chronology, but Oumm-Qatafa in the Judaean Desert and Kebara on the coast together apparently span the time of the last interglaciation and the last glaciation. On the whole, the climatic implications of these rodent faunas bear only on the humidity factor and give little or no information on paleotemperatures. Moreover, Tchernov sees a unidirectional trend from relatively humid conditions to arid conditions throughout the time period studied. Specifically, the lower sediments of Oumm-Qatafa yield tropical rodent faunas dependent on the proximity to water bodies in a "swampy-steppe" environment, although this cave is situated in the midst of a very dry, rocky desert. These tropical elements decrease in abundance (Fig. 12) during Acheulian and lower Levalloiso-Mousterian times, but a forest biotope persists until the upper Paleolithic, during which time the forest environment begins to disappear, leaving the Oumm-Qatafa area with only maquis[1] vegetation and bare rock slopes. According

1. Dense shrubby vegetation characterizing much of the Mediterranean coastal area.

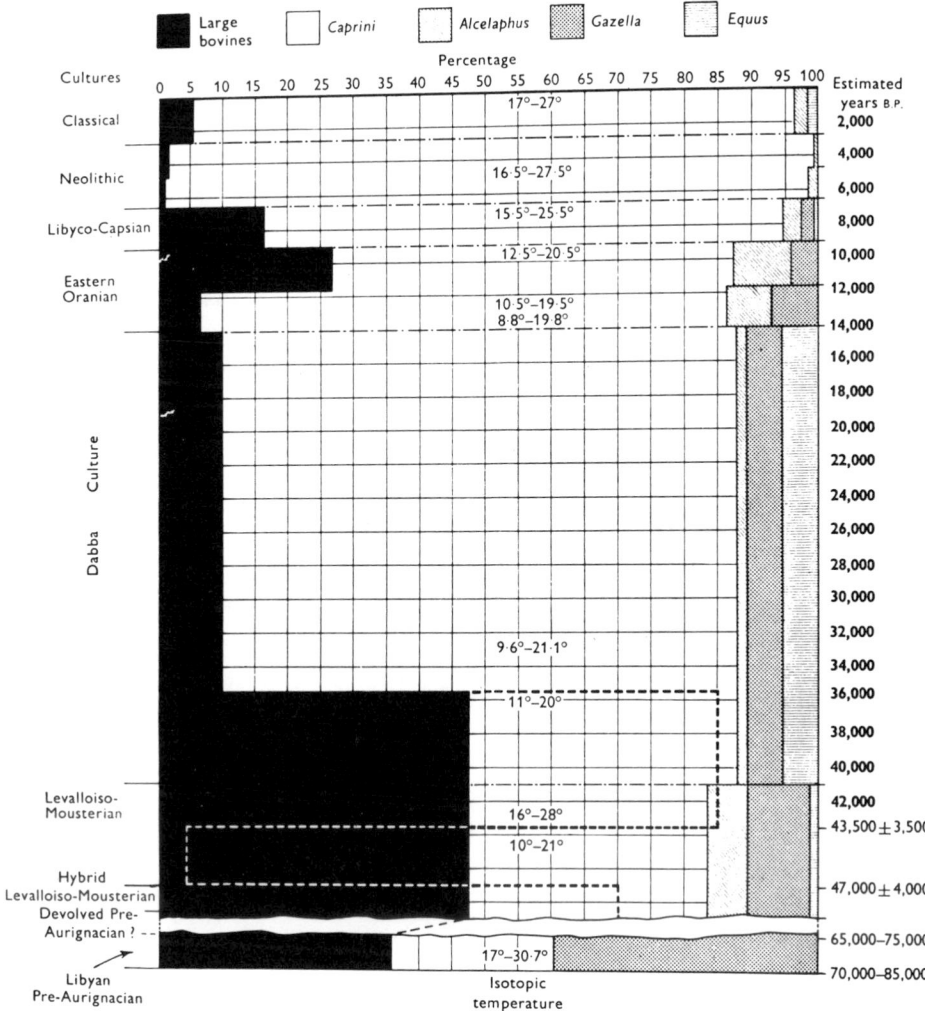

Fig. 11. Variations in the main constituents of the mammalian fauna at the Haua Fteah compared with isotopic temperature readings and plotted against a C¹⁴ time scale. The two earlier dates are regarded as minimum ages (From McBurney 1967, Fig. II.6).

to the time scale used by Tchernov, and generally accepted by others, Acheulian industries are dated from the time of the last interglaciation (warm/dry?) in the Near East, while the Levalloiso-Mousterian peoples inhabited this area during the height of the last glaciation (cool, moist?). Therefore, rodent faunas that show a moisture gradient from moist to dry over this same time span indicate a trend exactly opposite to that expected. The only indication of paleotemperature that Tchernov finds is through the application of Bergman's rule to changes in size of anatomical characters in *Spalax ehrenbergi*, which supports other estimates of

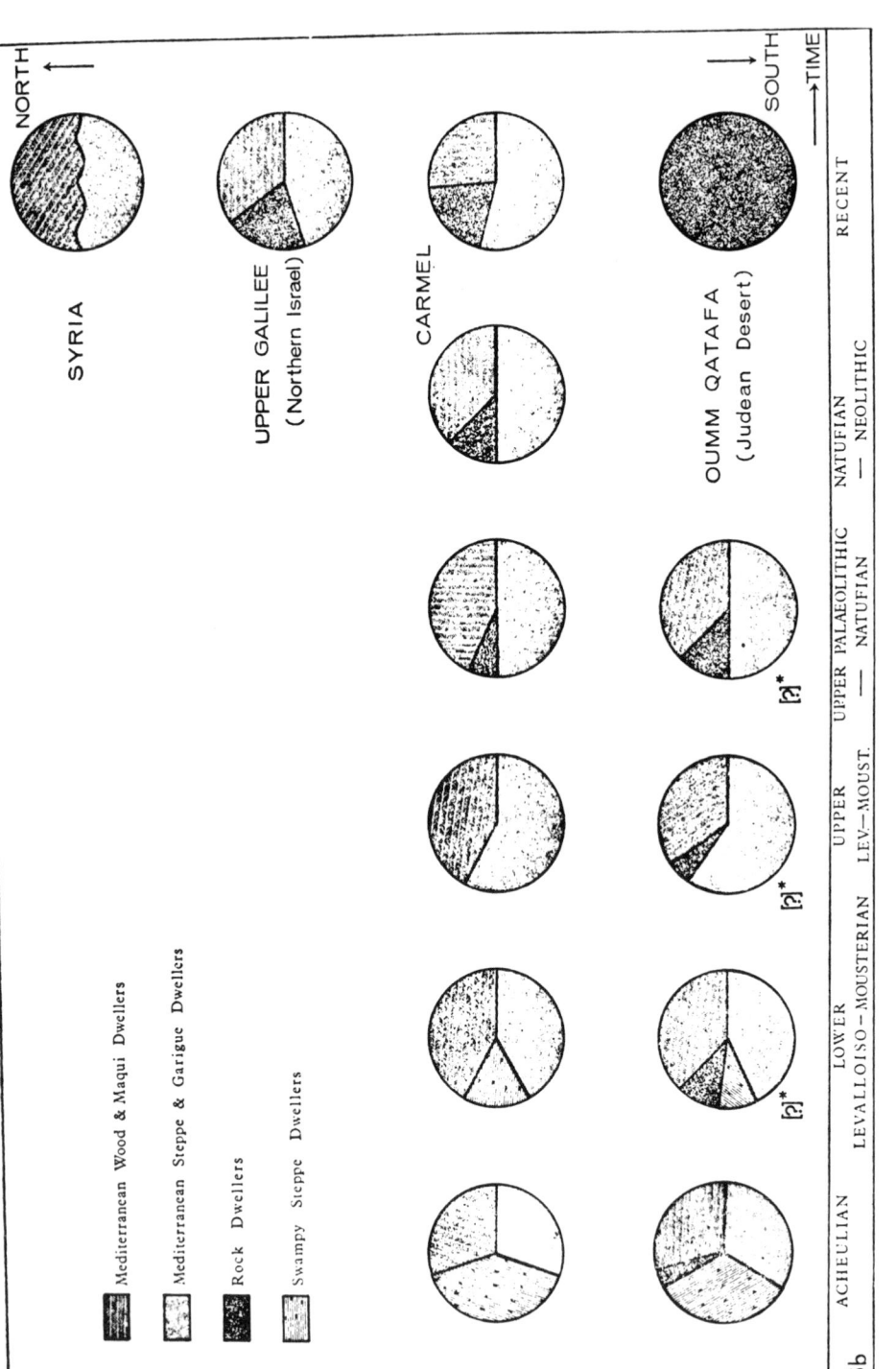

Fig. 12. Schematic changes in biotopes throughout the upper Pleistocene in parts of Israel and surrounding regions, as determined by rodent faunas (from Tchernov 1968).

a 5°-latitude shift of climatic belts in upper Pleistocene time, with the "coldest" temperatures falling at about 20,000 years ago (Tchernov 1968, p. 141).

Tchernov discusses the apparent paradox of a swampy, steppe habitat in the vicinity of Oumm-Qatafa during the last interglacial while the Dead Sea, only some 20 km away, was undergoing desiccation. However, his conclusions would have been different if he had used the modified interpretation of Dead Sea history (Picard 1965; Neev and Emery 1967) rather than that of Picard dating from 1943. In 1943 Picard had placed the Lisan Lake in middle Quaternary time and its desiccation at the time of the last interpluvial (interglaciation). As discussed above, the Lisan Lake is now known to have prevailed throughout most of the last glaciation, i.e. until about 20,000 years ago, and its demise was brought about partly by tectonic subsidence of the rift valley. Tchernov (1968, p. 134) was looking for such subsidence to explain the apparent paradox, but he presumably was not aware of the current interpretation of Dead Sea history. He would have liked to document a change in lake level of about 400 meters during late Quaternary time in order to explain the desiccation he recorded in the rodent fauna. Now it is known that there was a change of more than 200 meters from the maximum of the Lisan Lake to the minimum level of the Dead Sea, and this locally abrupt tectonic change may well have been more important in its effect on the fauna than were the subtle climatic changes associated with glaciation or pluviation.

Conclusions

The data and arguments reviewed in this paper are summarized in Table 1. All the lines of evidence lead to the conclusion that during the last glaciation the climate of the eastern Mediterranean area was definitely cooler and moister than now. Quantitative estimates point to a decrease of 5 to 7°C in mean temperature. Certainly there must have been an increase in the effective precipitation in this area due to the independently documented temperature decrease (thus, less evaporation), but evidence for an absolute increase in precipitation is not so clear. Increased intensity of alluviation in the Dead Sea wadis (Vita-Finzi 1964), however, points in this direction. Similar evidence. not discussed in this paper, comes from the middle Nile Valley where Butzer and Hansen (1968) have documented rather heavy alluviation in wadis draining from the Red Sea Hills westward into the Nile during the time of the last glaciation. They attribute the increased precipitation to a southward shift of the Mediterranean winter rain belt. Such a shift should have also benefited the eastern Mediterranean region and the Levant, although the increase may not have been very great. Butzer and Hansen (1968, p. 152) estimate an increase of only some 50 to 100 mm per year in the Red Sea Hills area.

Furthermore, there are several suggestions of at least one humid–less humid–more humid cycle within the time of the last glaciation. Rossignol (1969) has documented such a cycle in the Hamarmar formation of the Dead Sea basin and, according to the time scale of Neev and Emery (1967), this should have occurred

Table 1. Climatic Change in the Eastern Mediterranean Area During the Time of the Last European Glaciation

Type of information	Climate of the "last glaciation" Temperature (°C)	Humidity	References
Deep-sea cores			
West of Crete	5–10° cooler		Vergnaud-Grazzini (1969)
East of Crete	15° (?) cooler		Emiliani (1955)
Red Sea	cooler	increased	Herman (1968)
Palynology			
Tenaghi Philippon	cooler	decreased	Wijmstra (1969)
Israel coast		increased	Rossignol (1962)
Dead Sea		humid–dry–humid	Rossignol (1969)
Hula basin		humid–dry–humid	Picard (1965)
Pluvial lakes			
Dead Sea		humid–dry–humid	Neev & Emery (1967)
Hula basin		humid–dry–humid	Picard (1963)
El Jafr		humid–dry–humid	Huckriede & Wiesemann (1968)
Snowlines			
Lebanese mountains	6–7° cooler		Kaiser (1963), Messerli (1967)
Lebanese mountains	2.5–3.5° cooler		Klaer (1962)
Frost action			
Galilee, Cyprus	colder		Nir (1963), De Vaumas (1968)
Dead Sea wadis	colder	increased	Vita-Finzi (1964)
Caves (Syria, Israel)	colder	increased	Farrand (1965), Goldberg (1968)
Caves (Libya)	colder	increased	McBurney (1967)
Faunas			
Mollusks (Libya)	5–7° cooler	increased	McBurney (1967)
Mammals (Libya)	cooler	increased, then decreased	McBurney (1967)
Mammals (Israel)	cooler		Bouchud (1969)
Rodents (Israel)	cooler (?)	progressively drier	Tchernov (1968)

some 50,000 to 70,000 years ago, thus early in the last glaciation. Somewhat later the Dead Sea basin experienced relatively moist conditions that brought into existence the Lisan Lake; then desiccation (as well as tectonics) brought about the demise of the Lisan Lake, but this dry episode was followed by a moister period near the end of the last glaciation that caused alluviation in Dead Sea wadis (Vita-Finzi 1964). Therefore, a sequence of humid–dry–humid–dry–humid conditions appears to be substantiated for the Dead Sea basin.

Just to the north, in the Hula basin, both pollen and sediments indicate moisture fluctuations (Picard 1965). The sequence of lacustrine marl followed by shallow-water peat and in turn followed by more lacustrine marl can be examined in greater detail in the pollen record. Horowitz (1968, 1969) situates a humid ("pluvial") period of about 50,000 years' duration between two dry periods that correspond to last interglacial and postglacial times. Moreover, the "pluvial" episode was interrupted by two relatively warm although still humid interstades, the second of which began at about the same time as the demise of the Lisan Lake, about 20,000 years ago.

Furthermore, lacustrine marl deposition in the El Jafr basin just southeast of the Dead Sea appears to have been contemporary with the Lisan pluvial (Huckriede and Wiesemann 1968). This humid interval in the El Jafr basin was also followed by desiccation and then by another humid interval with alluviation, all of which were apparently in phase with similar events in the Dead Sea Rift valley.

The sediments of the cave at Jebel Qafzeh near Nazareth, Israel, have been interpreted by Neuville (1951) as representing a sequence of cold and humid climates followed by temperate conditions and in turn followed by a cold and dry climate. The temperate phase corresponds to final Mousterian time. Faunal study of this cave by Bouchud (1969) confirms this conclusion to the extent that it documents a change from humid conditions during Mousterian times to very dry conditions in the terminal upper Paleolithic. Moreover, McBurney (1967) reaches the conclusion that there was a minor warm, dry interval during final Mousterian time in the Haua Fteah, roughly 45,000 to 33,000 years ago.

Without more precise absolute or stratigraphic dating it is impossible to determine at present whether the evidence of warmer, drier intervals that have just been discussed represent a single simultaneous period of climatic amelioration in the eastern Mediterranean during the last glaciation. Some of the dates discussed above suggest a lack of synchroneity or else a multiplicity of mild intervals, but the chronological controls presently available are not adequate for reaching a conclusion on this point. In contrast, Wijmstra (1969) is able to see ten interstadials within the last glaciation as recorded by pollen in the Tenaghi Philippon in Greece. The last six of these are controlled by a number of radiocarbon dates so that Wijmstra can conclude that there is good agreement between his interstadials and those recognized in northwestern Europe. It must be admitted, however, that Wijmstra's study is unique for this region in its degree of detail and wealth of radiocarbon dates. Furthermore, it is rather far removed geographically from most of the other sites discussed; moreover, Wijmstra concludes that the climate

Table 2. Geological Correlation of Near Eastern Archaeological Sites*

(McBurney 1967; Farrand, in press)

Climate	Jerf 'Ajla	Yabrud	Tabun	Qafzeh	Oumm Qatafa	Haua Fteah
dry, cold	Upper (U.P.) (L.M.)			C (U.P.) D (U.P.)		ii (U.P.) —33,000
minor, warm	Interstade			E, F, G (L.M.)		iii —45,000
cold, moist	Lower (L.M.)	I-A (L.M.) I-B (YAB-Pre-A) (MIC)	B (L.M.) C (L.M.) D (L.M.)	H, I (L.M.) J (L.M.) K (L.M.) L (L.M.)	B (L.M.) C (L.M.)	iv (L.M.) —60,000 (?)
warm, dry (5–8 m shoreline)		I-C (YAB)	E (MIC-Pre-A) F (ACH) G (TAY)	M (—)	D (MIC) (ACH)	v (Pre-A)
cold, moist		I-D (YAB-ACH) IV (TAY)			E (ACH) (TAY)	
warm, dry					F (TAY) G (TAY) H-J (—)	

* Key to industries:
U.P. = Upper Paleolithic
L.M. = Levalloiso-Mousterian
YAB = Yabrudian
Pre-A. = Pre-Aurignacian, or Amudian

MIC = Micoquian
ACH = Tayacian, or Tabunian
TAY = Acheulian

throughout the last glaciation there was cooler and *drier* relative to the present climate. This conclusion is exactly opposite that of all the other areas discussed here and perhaps reflects the fact that northern Greece lay under the influence of dry continental European air masses during the last glaciation, whereas the other areas were still subject to a Mediterranean climate, although with an increase in winter rainfall during that time.

A final conclusion concerns the apparent position of prehistoric cultures with respect to climatic change. I have previously attempted such a correlation (Farrand, in press) for five sites in the Levant, and McBurney (1967) and his colleagues have drawn similar conclusions from the Haua Fteah. My previous conclusions were drawn primarily from the comparison of sedimentary sequences in cave sites. McBurney was able to use sediments, fauna, and oxygen istotopes, and in addition he has a series of consistent radiocarbon dates at his disposal. It must be emphasized that our correlations (Table 2) are based strictly on nonarchaeological arguments. The names of the industries that are found in given strata of a particular site were added to this table after the correlations based on sediment study had been made. It is only in this way, I believe, that real temporal and geographical relations between different groups of prehistoric men can ever be deciphered. This table marks only a first step. Archaeological conclusions drawn from it will be only as good as the archaeological information that was available; at the present time such information is uneven, erratic, and maybe even erroneous for certain sites. Some of the sites are being restudied, and others should be restudied. After such a re-evaluation it will be interesting to look at this table again.

Summary. Data pertinent to climatic change—taken from deep Mediterranean and Red Sea cores, pollen studies in Greece and Israel, pluvial lakes in the Jordan graben and in Jordan, determinations of snowline lowering, and prehistoric cave habitations—all lead to the conclusion that the climate of the eastern Mediterranean during the last continental glaciation was both cooler and more humid than at present. The mean temperature was lowered by some 5 to 7°C, and there was a definite increase in effective moisture. The absolute amount of atmospheric precipitation may also have been increased. Several lines of evidence lead to the identification of at least one humid–dry–humid cycle within the last glaciation; other evidence points to progressively drier conditions throughout this period of time. Tectonic activity in the Jordan graben may have influenced the local climate more strongly than did the distant glaciations.

Acknowledgments. My interest in this subject has grown largely from my participation in several archaeological excavations in the Near East. Three years of study of the et-Tabun cave of Mount Carmel directed by A. J. Jelinek, University of Arizona, have been sponsored by the Foreign Currency Program of the Smithsonian Institution and by the National Science Foundation. Further support and encouragement have come from Bernard Vandermeersch, Faculté des Sciences, Paris, who invited me to study the sediments at the Jebel Qafzeh; this work is funded by the RCP no. 50 of the Centre National de la Recherche Scientifique (France). Moreover, the manuscript was read by Dr. A. Issar of the Geological Survey of Israel, who offered valuable criticism.

REFERENCES

Ben-Arieh, J. Last stage in the formation of Lake Kinneret (Tiberias) and the oscillations of lake level (abstract), *Israel J. Earth Sci., 16,* 48, 1967.

Bouchud, Jean, Etude paléontologique de la faune du Djébel Qafzeh, Israël (abstract), *Intern. Quaternary Assoc., 8th Congr., Résumés,* Paris, 118, 1969.

Broecker, W. S., and A. Walton, The geochemistry of C^{14} in freshwater systems, *Geochim. Cosmochim. Acta, 16,* 15, 1959.

Butzer, K. W., Quaternary stratigraphy and climate in the Near East, 157 pp., *Bonner Geographischen Abhandlungen, 24,* 1958.

Butzer, K. W., Reply: On late Pleistocene chronology, *Current Anthropology, 8,* 353, 1967.

Butzer, K. W., and C. L. Hansen, *Desert and River in Nubia,* 562 pp., University of Wisconsin Press, Madison, 1968.

Dresch, J., Questions de géomorphologie en Israël, *Bull. Assoc. Geograph. Franc.,* 350–51, 2, 1967.

Emiliani, C., Pleistocene temperature variations in the Mediterranean, *Quaternaria, 2,* 87, 1955.

Farrand, W. R., Geology, climate and chronology of Yabrud rockshelter I, *Ann. Archéologiques de Syrie, 15,* 35, 1965.

Farrand, W. R., Geological correlation of prehistoric sites in the Levant, *UNESCO Symposium on Environmental Changes and the Origin of Modern Man, Proceedings,* in press.

Goldberg, P., Sediment analysis of two prehistoric rockshelters in Syria, 69 pp., unpublished masters thesis, University of Michigan, Ann Arbor, 1968.

Herman, Yvonne, Evidence of climatic changes in Red Sea cores, in *Means of Correlation of Quaternary Successions,* edited by R. B. Morrison and H. E. Wright, Jr., *Intern. Quaternary Assoc., 7th Congr., Proc., 8,* 325, 1968.

Herman, Y., Y. Thommeret, and C. Vergnaud-Grazzini, Micropaleontology, paleotemperatures and radiocarbon dates of Quaternary Mediterranean deep-sea cores (abstract), *Intern. Quaternary Assoc., 8th Congr., Résumés,* Paris, 174 bis, 1969.

Horowitz, A., *Upper Pleistocene-Holocene Climate and Vegetation of the Northern Jordan Valley (Israel),* Geol. Survey Israel, Paleontology Division, Report P/2/68, 123 pp. (in Hebrew with English summary), 1968.

Horowitz, A., Les changements climatiques en Israël pendant le Pleistocene supérieur et l'Holocène (abstract), *Intern. Quaternary Assoc., 8th Congr., Résumés,* Paris, 153, 1969.

Howell, F. C., Upper Pleistocene stratigraphy and early man in the Levant, 65 pp., *Proc. Am. Phil. Soc., 103,* 1959.

Huckriede, R., and G. Wiesemann, Der jungpleistozäne Pluvial-See von El Jafr und weitere Daten zum Quartär Jordaniens, *Geol. Palaeontol.* (Marburg), 2, 73, 1968.

Issar, A., Geology of the central coastal plain of Israel, *Israel J. Earth Sci., 17,* 16, 1968.

Issar, A., and L. Picard, Sur le Tyrrhénien des côtes d'Israël et du Liban, *Bull. Assoc. Franc. Etude Quaternaire, 6e annee,* no. 18, 35, 1969.

Kaiser, K., Die Ausdehnung der Vergletcherungen und "periglazialen" Erscheinungen während der Kaltzeiten des quartären Eiszeitalters innerhalb der Syrisch-Libanesischen Gebrige und die Lage der klimatischen Schneegrenze zur Würmeiszeit in östlichen Mittelmeergebiet, *Intern. Quaternary Assoc., Reports, 3* (Warsaw, 1961), 127, 1963.

Kaiser, K., Ein Beitrag zur Frage der Solifluktionsgrenze in den Gebirgen Vorderasiens, *Z. Geomorphol., 9*, 460, 1965.

Klaer, W., Untersuchungen zur klimagenetischen Geomorphologie in den Hochgebirgen Vorderasiens, *Heidelberger Geograph. Arb., 11*, 135 pp., 1962.

Langozky, Y., High level lacustrine sediments in the Rift Valley at Sdom, *Israel J. Earth Sci., 12*, 17, 1963.

Lartet, L., Note sur la formation du bassin de la Mer Morte ou lac Asphaltite et sur les changements survenus dans le niveau de ce lac, *Bull. Soc. Geol. France*, Ser. 2, *22*, 420, 1865.

McBurney, C. B. M., *The Haua Fteah (Cyrenaica) and the Stone Age of the Southeast Mediterranean*, 387 pp., University Press, Cambridge, 1967.

Messerli, B., Das Problem der eiszeitlichen Vergletscherung am Libanon und Hermon, *Z. Geomorphol., 10*, 37, 1966a.

Messerli, B., Die Schneegrenzhöhen in den ariden Zonen und das Problem Glazialzeit-Pluvialzeit, *Mitt. Naturforsch. Ges. Bern*, n. s., *23*, 117, 1966b.

Messerli, B., Die eiszeitliche und die gegenwärtige Vergletscherung in Mittelmeerraum, *Geograph. Helv., 22*, 105, 1967.

Neev, D., and K. O. Emery, The Dead Sea, *Bull. Geol. Surv. Israel, 41*, 147 pp., 1967.

Neuville, R., Le Paléolithique et le Mésolithique du désert de Judée, 270 pp., *Arch. Inst. Paleontol. Humaine, Mem., 24*, 1951.

Nir, D., Indices de gélivation récente et pleistocène en Israël, *Biul. Peryglacjalny, 12*, 161, 1963.

Picard, L., The Quaternary in the northern Jordan valley, 34 pp., *Israel Acad. Sci. Humanities, Proc., 1*, 1963.

Picard, L., The geological evolution of the Quaternary in the central-northern Jordan graben, Israel, *Geol. Soc. Am., Spec. Papers, 84*, 337, 1965.

Rossignol, M., Analyse pollinique de sédiments marins quaternaires en Israël, I: Sédiments récents, *Pollen Spores, 3*, 303, 1961.

Rossignol, M., Analyse pollinique de sédiments marins quaternaires en Israël, II: Sédiments pleistocènes, *Pollen Spores, 4*, 121, 1962.

Rossignol, M., Analyse pollinique de sédiments quaternaires dans le plaine de Haifa-Israël, *Israel J. Earth Sci., 12*, 207, 1963.

Rossignol, M., An Upper Pleistocene Dead Sea area climatic sequence, Israel (abstract), *Intern. Quaternary Assoc., 8th Cong., Résumés*, Paris, 113, 1969.

Schulman, N., Remarks on the Quaternary in the northern Jordan valley, *Israel J. Earth Sci., 16*, 104, 1967.

Stearns, C. E., and D. L. Thurber, $Th^{230}-U^{234}$ dates of late Pleistocene marine fossils from the Mediterranean and Moroccan littorals, *Quaternaria, 7*, 29, 1965.

Stuiver, M., Yale natural radiocarbon measurements IX, *Radiocarbon, 11*, 545, 1969.

Syrian Arab Republic, Ministry of Planning, Directorate of Statistics, *Statistical Abstract for 1963*, 421 pp., Government Press, Damascus, 1964.

Tchernov, E., *Succession of Rodent Faunas during the Upper Pleistocene of Israel*, 152 pp., Paul Parey, Hamburg and Berlin, 1968.

Tobias, P. V., A member of the genus *Homo* from 'Ubeidiya, *Israel Acad. Sci. Humanities*, 12 pp., 1966.

Vaumas, E. de, Phénomènes cryogèniques et systèmes morphogénétiques en Méditerranée orientale (Chypre, Galilée), *Rev. Geograph. Phys. Geol. Dyn.*, 6, 291, 1964.

Vaumas, E. de, Questions de Géomorphologie en Israël, *Bull. Assoc. Geograph. Franc.*, 362–63, 167, 1968.

Vergnaud-Grazzini, C., and Herman-Rosenberg, Y., Etude paléoclimatique d'une carotte de Méditerranée orientale, *Rev. Geograph. Phys. Geol. Dyn.*, 11, 279, 1969.

Vita-Finzi, C., Observations on the late Quaternary of Jordan, *Palestine Exploration Quarterly, 96th yr*, 19, 1964.

Wijmstra, T. A., Palynology of the first 30 m of a 120 m deep section in northern Greece, *Acta Botan. Neerl.*, 18, 511, 1969.

Wright, G. A., On late Pleistocene chronology, *Current Anthropology*, 8, 353, 1967.

Wright, H. E. Jr., Late Pleistocene geology of coastal Lebanon, *Quaternaria*, 6, 525, 1962.

Zak, I., and R. Freund, Recent strike slip movements along the Dead Sea rift, *Israel J. Earth Sci.*, 15, 33, 1966.

Maurice Ewing

21. THE LATE CENOZOIC HISTORY OF THE ATLANTIC BASIN AND ITS BEARING ON THE CAUSE OF THE ICE AGES

In the sixties we have seen lively advances in knowledge of the Late Cenozoic history of the ocean floor. We have obtained an extensive body of data and completely new points of view about a wide variety of phenomena that bear directly on the subject of the Late Cenozoic glacial ages. Seismic refraction and reflection measurements have taught us much about the nature of the underlying crustal and mantle rocks and about stratification and thickness of the sediments. They have also guided coring efforts to outcrops, so that we could sample the older sediments more effectively. Magnetics, gravity, and thermograd studies supplied other facets of our knowledge, and together these have guided the drilling campaign of *Glomar Challenger,* and enhanced the utility of that operation. All these results have provided us with a useful picture of the evolution of the Atlantic Ocean basin during the Cenozoic.

THE ATLANTIC OCEAN BASIN AND LATE CENOZOIC GLACIATIONS

In barest outline, the Atlantic was formed early in the Mesozoic era as a narrow rift in the lithospheric plate. This rift split the Americas from Africa and Europe and has been progressively widened by sea floor spreading about an axis now marked by the mid-Atlantic ridge. The bordering continents were carried on the rigid plates without major deformations. Conservation of surface area is provided by destruction of lithospheric plates by underthrusting from the oceanic side at deep-sea trenches.

Knowledge of the schedule of magnetic field reversals for the past several million years and of the axial distances of the associated anomaly strips has given us rates of separation of the lithospheric plates amounting to several centimeters per year.

Thus, as indicated in Figure 1, a broad strip of new ocean bottom running the length of the Atlantic was created in Cenozoic time at rates now fairly well

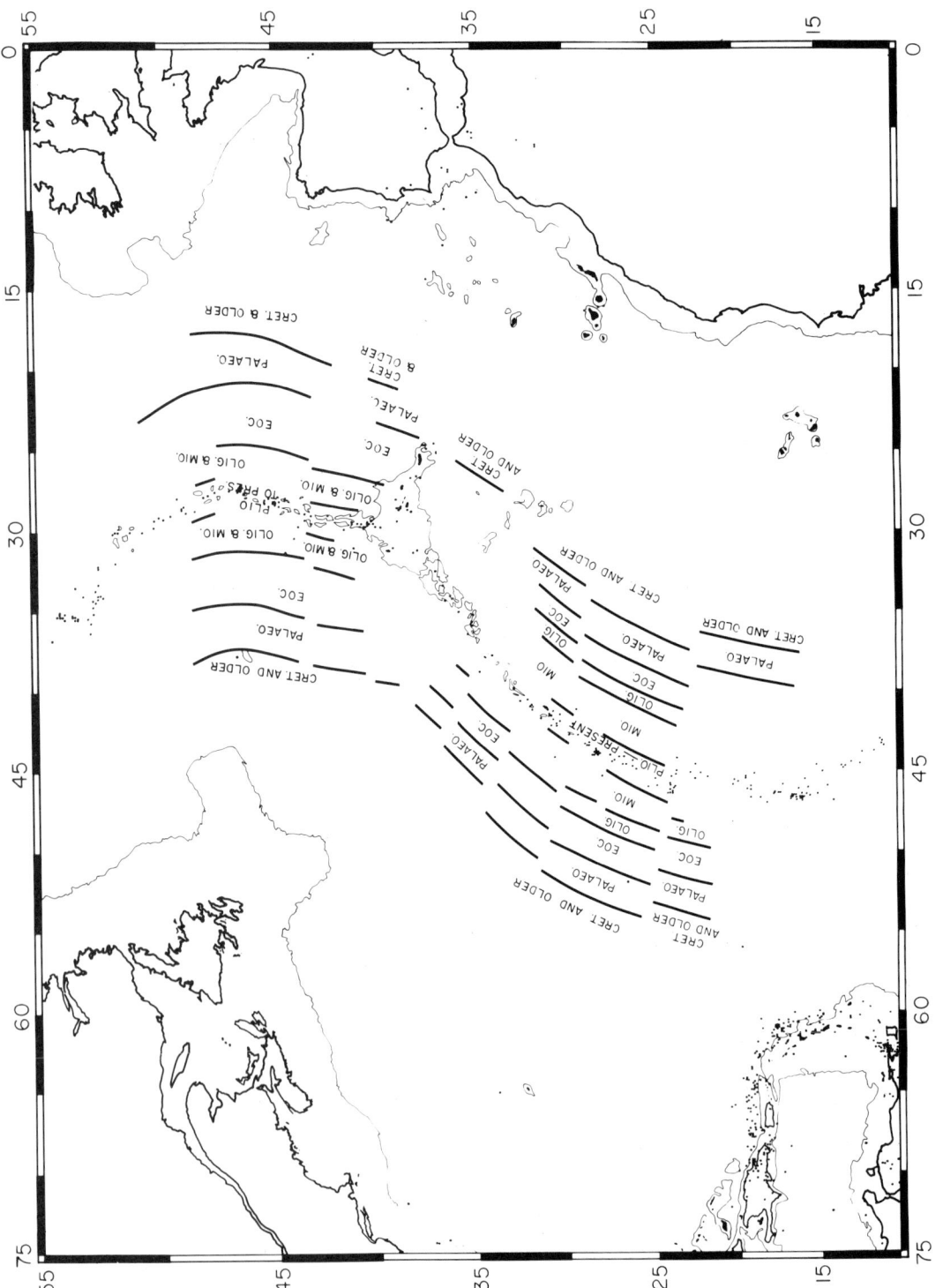

Fig. 1. Cenozoic sea floor spreading in the North Atlantic. The isochrons show the extent of spreading at various Cenozoic epochs, as deduced from the pattern of magnetic anomalies (courtesy of W. C. Pitman).

known. By means of marine seismic surveys, the thickness and stratification of the sediment on the deep-sea floor also is being thoroughly mapped. Although only the topmost 10 to 20 meters are extensively sampled by piston coring, with the aid of a system that uses acoustic reflections to locate outcrops of older strata, we have sampled pre-Pleistocene sediments effectively despite this limitation. The Lamont-Doherty core collection includes hundreds of pre-Pleistocene cores, tens of Cretaceous cores, and a fair number of Jurassic ones. These cores demonstrate that a great variety of processes control the transportation, deposition, and erosion of deep-sea sediment.

The absence of any samples older than late Jurassic, or even of any geophysical evidence of extensive areas where they might be found by drilling, is strong support for the corollary of sea floor spreading, systematic destruction of oceanic crust and overlying sediment.

There are some specific results relating to the details of ocean floor spreading that must be considered in coupling the Cenozoic history of the Atlantic Ocean basin to the ice ages.

The discontinuity in sediment thickness at the ridge crest, observed on many reflection traverses and apparently not attributable to topographic control of deposition, suggested "stop-and-go" spreading of the sea floor, with a far greater rate during the past 10 million years than before that time (Ewing and Ewing 1967). The *Glomar Challenger* cores, however, according to preliminary interpretation, seem to indicate a constant rate of spreading (Peterson et al. 1970). Further study is necessary to resolve the question of possible variations in spreading rate, a question of utmost importance in the study of the Late Cenozoic ice age. It remains to be seen whether the geophysical observations will ultimately be explained by variation in spreading rate or by some complication of the process of sedimentation.

Another feature of the combination of reflection traverse data and the results of the deep-sediment coring program has some bearing on our view of the development of the geometry of the ocean basins with time. An extensive, nearly level sub-bottom reflector called Horizon A has been mapped in the low- and middle-latitude basins of the Atlantic and its age estimated as late Cretaceous to early Cenozoic. Some *Glomar Challenger* cores (Ewing et al. 1969) show it to be a chert sequence terminated in mid-Eocene. The date of the beginning of chert deposition is not yet known. This reflector maintains an approximately level attitude and is terminated against the slope of basement on the ridge flank at a point where the age of the basement, judged on the basis of accepted spreading rates, is found to be not much greater than that of the sediments. This suggests that the ridge was very low in mid-Eocene time and that each parcel of sea floor formed at the ridge axis may remain permanently at about the same elevation it occupied at the time of its formation. Other facts about the distribution of sediment suggest that a topographic ridge may not in all cases be present at the locus of an axis of spreading.

At 2 centimeters per year (a spreading rate that is compatible with the results of LePichon 1968) the width of a rift such as that between Spitsbergen and Greenland would incerease by 20 km per m.y., or about 200 km in 10 m.y., and may thus have been very narrow at the beginning of the Late Cenozoic ice age. Any process of climatic control that is dependent on interchange of sea water through such a strait assumes a vastly different aspect when the facts of sea floor spreading are taken into account.

Finally, both on the basis of evidence from deep-sea sediments and continental glacial deposits, major glaciation on the earth seems to have commenced as early as the middle Miocene and possibly even earlier. Many of these results are discussed elsewhere in this volume, and some relevant papers are listed in the bibliography.

RELATION TO THE EWING-DONN THEORY OF ICE AGES

It is of interest to re-examine—in terms of these current ideas about sea floor spreading, pole wandering, and chronology of ice ages—the geographical theory which Donn and I propounded several years ago (Ewing and Donn 1956; Donn and Ewing 1966). We proposed that an ice age results from movement of the poles to thermally isolated positions, a proposal which has received additional support as more information about the high-latitude areas became available. From the start, the theory consisted of two main parts. The first was offered to explain the simple fact of occurrence of ice ages, widely separated in time as they are in Late Cenozoic, Paleozoic, and Precambrian time. The second part was an attempt to account for the oscillatory fluctuations between glacial and interglacial stages, which were so striking a feature of the Late Cenozoic ice age.

In the first part we emphasized the fact that the effectiveness of ocean currents in producing an equable distribution of climate by convection of heat from low to high latitudes was critically dependent upon the situation of the poles in open ocean areas in which there was also good access to currents from the equatorial area. The equable pattern of climate during Mesozoic and much of Cenozoic time was attributed to pole positions in the open ocean, which was even then being suggested by paleomagnetic evidence for pole wandering. There is every indication—under the assumption of a regime of solar radiation like that prevailing now—that no permanent ice cover could form around a pole in the open ocean. The fundamental point is that the Cenozoic ice age resulted from the movement of both poles into thermally isolated positions—the south pole into the Antarctic continent, and the north pole into the almost landlocked Arctic Ocean.

At some stage, the Arctic Ocean froze over, increasing the extent of thermal isolation. Formation of an extensive circumpolar ice cover, either on land or on landlocked ocean, would certainly have the additional effect of some lowering of the mean global temperature (as has been discussed by many writers), owing to the increase in albedo of the ice-covered areas. Lowering of temperatures of high-latitude areas and formation of the ice sheets was attributed primarily to the thermal isolation and secondarily to the increased albedo.

The importance of an adequate supply of moisture is the other requirement for formation of an ice sheet. This point was illustrated by comparing the present conditions on Greenland and Antarctica in relation to moisture supply from nearby open ocean areas to that of the desert lands which now border the Arctic Ocean. This observation strongly influenced us in our belief that an ice-free Arctic Ocean accompanied the initiation of the continental ice sheets of the Late Cenozoic glacial stages in the northern hemisphere. The second part of our hypothesis also initially made use of the concept of alternation between ice-free and ice-bound conditions of the Arctic Ocean, and of throttling the Artic–Atlantic interchange, in the effort to explain the oscillations in the system. This feature, however, is only a detail, and one for which several alternatives are possible.

The main theme that ice ages are caused by the thermal isolation of circumpolar areas—which may occur either as an accident of geography in the course of continental drift or as a result of pole wandering, or a combination of the two—may be tested for applicability to the late Paleozoic and earlier ice ages as well as to the present one. The principal objection to the hypothesis that the current ice age may be attributed to the thermal isolation of the Late Cenozoic pole positions in the landlocked Arctic Ocean and the Antarctic continent has been that the poles moved into their present positions in the early Cenozoic, so early that the interval between occurrence of a favorable geographic environment and the initiation of glaciation is unacceptably great. But continuing motion of a few centimeters per year between almost any two of the continents involved is now taken to be almost axiomatic; so, to answer a question such as the degree of isolation of the Arctic Ocean from the Atlantic Ocean a few million years ago, it is necessary to have data on the rate and direction of motion obtained from magnetic anomaly patterns to supplement conclusions from the paleomagnetism from rocks of various ages from the adjacent continents. One must even consider the changes which have taken place in the channel for interchange of water between the Arctic and Pacific oceans during the time since the poles reached their present positions.

Irving and Robertson (1968) discussed these aspects of our hypothesis several years ago. They concluded that, "No case in support of the Ewing-Donn hypothesis for the initiation of the current ice age can be made on the basis of the present incomplete data, but it is clear that the paleomagnetic observations are not necessarily inconsistent with it, as may appear at first sight." They point out that if our hypothesis is correct, then a high concentration of land would be expected in high latitudes during ice ages. They show latitudinal distribution of continental crust, calculated from paleomagnetic results, and find evidence to support a suitable concentration of land in Antarctic regions in late Mesozoic and in Arctic regions in the Late Cenozoic. The results of Smith and Hallam (1970) yield many of the same conclusions about the grouping of the Gondwana continents that were expressed by Du Toit (1937), whose work, being based on the very limited data available at that time, was described by Smith and Hallam as "a triumph of imaginative synthesis." The reconstruction of the southern continents offered by Smith and Hallam (1970) supports very strongly the conclusion of Irving and Robertson (1968) by showing a high concentration of land through conjunction of the Gondwanic continents in late Mesozoic time. This, in turn, suggests that the hypothesis that thermal isolation of polar regions is the cause of ice ages is applicable to the Permo-Carboniferous ice age as well as to that in the Late Cenozoic.

Briden (1967) made the observation that there appear to be three stable Paleozoic positions for the paleomagnetic pole relative to the five Gondwanic continents, separated by short periods of pole movement in the Devonian and Carboniferous. The stable position for the interval from upper Carboniferous into Permian is of interest with respect to the thermal isolation at the time of the Permo-Carboniferous ice age, and the move to it may relate to initiation of the ice age.

The recent reconstruction of Gondwanaland by McElhinny and Luck (1970), based on paleomagnetic measurements alone, places the continents in an essentially unambiguous configuration and traces a path followed by the pole from northwest Africa in Cambrian to the border between South Australia and Antarctica in Permo-Carboniferous time. They showed support for this configuration by matching across continental boundaries of cratons (ages at least 2000 m.y.) with their surrounding younger metamorphic belts and geosynclines (ages 450 to 650, maximum 1100 m.y.). It differs only in minor detail from that of Smith and Hallam (1970), except for the separation of Australia and Antarctica (which they say might well be narrowed when further Paleozoic data from Australia become available).

The McElhinny and Luck reconstruction provides very satisfactory explanation of the Permo-Carboniferous glaciation in terms of thermal isolation of the south pole while it was in the critical area. The migration they show for this pole illuminates the remark made by M. F. Glaessner (1967, personal communication) during a demonstration of evidence for this ice age that it was *almost, but apparently not exactly* synchronous in the several Gondwanic continents. Fairbridge (1969, 1970) reports evidence for a continental ice sheet over 4000 km of North Africa during the Ordovician and Silurian, which extends the remark by Glaessner and offers additional evidence for production of an ice age by thermal isolation of the poles.

The second part of our hypothesis related to the oscillatory fluctuations between glacial and interglacial stages which characterized the Late Cenozoic ice age. The original proposal was that these fluctuations resulted from alternation between ice-free and ice-covered condition of the Arctic Ocean, the alternation from thawing to freezing being dependent on the degree of throttling of the Arctic–Atlantic interchange when sea level was lowered by removal of water to form the continental ice caps. The objection to this view at the time it was offered, and in the subsequent years when more information about the bathymetry of the connecting straits became available, was that the straits were so deep that the Pleistocene lowering of sea level would be inadequate to throttle the flow. We consider that this objection is largely removed by the modern results about sea floor spreading (for instance LePichon 1968; and Barazangi and Dorman, 1970) as shown in Figure 2, together with the lengthened time scale allowed by various results mentioned earlier in this paper. The results of Hays et al. (1969) appear to confirm the opinion that has gradually been forming that there were far more than four glacial stages and that the period of oscillation was 70,000 to 120,000 years, which is not in agreement with the predictions of the Milankovitch theory.

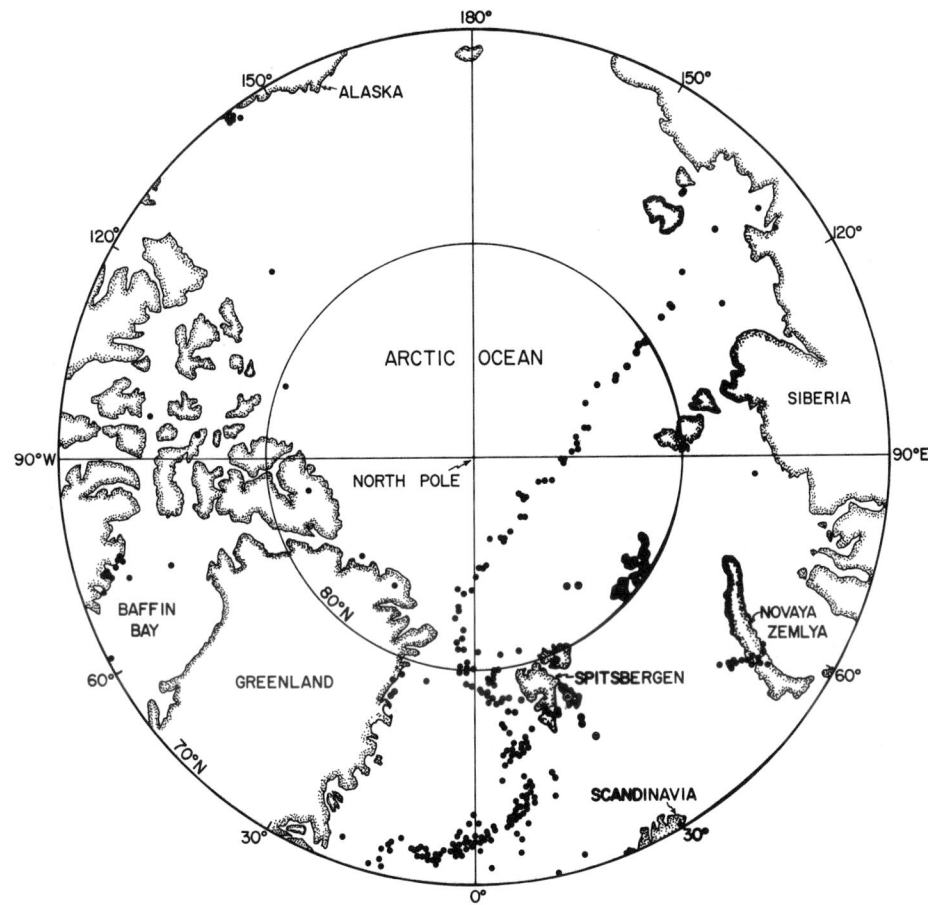

SEISMICITY OF THE ARCTIC, 1961-SEPT. 1969, ESSA, CGS EPICENTERS

Fig. 2. Seismicity of the Arctic region, delineating the axis of spreading which opens the channel between Spitsbergen and Greenland. At 80°N latitude, the present distance between 300-meter isobaths is about 300 km. At 2 cm per year spreading, this separation would have been only 200 km 5 m.y. ago, and 100 km 10 m.y. ago. With the probable accompanying depth changes, this would have provided a far more severe throttling of the Arctic–Atlantic interchange (from Barazangi and Dorman 1970).

Conclusions

Information gathered during the last decade or two from a variety of sources, but much of it from the oceanic areas of the world for which little information was available before then, has led to wide acceptance of the belief that the concept of sea floor spreading must be at the heart of any discussion of the Late Cenozoic history of the Atlantic basin. This concept, of course, implies continuous pole wandering and continental drift, and data have now been provided for good estimates of the rates at which both processes are occurring. Studies of seismicity reconfirm the opinion that such changes are occurring in areas which controlled the exchange of water between the Arctic Ocean and the Atlantic and Pacific during the period in which the Late Cenozoic ice age occurred. Explorations in Antarctica and in the sediments of the adjacent deep-sea floor appear to have placed the commencement of Antarctic glaciation into the Miocene, and to have lengthened considerably the time available for these changes to occur.

The correlation of carbonate content in deep-sea sediment with paleontologic and isotopic indications and with magnetic changes appears to demonstrate, if one grants that each carbonate maximum corresponds to a glacial stage, that the initial stages of glaciation are synchronous in the northern and southern hemispheres, that the number of these cycles is considerably greater than had been estimated previously, and that the period appears to be somewhat irregular, ranging from 70,000 to 100,000 years.

The results of Shaw and Donn (in press) from calculations of the heat budget are interpreted to indicate that either ice-free or ice-covered conditions of Arctic climate would be stable. The need is to discover the mechanism that would cause oscillation between the two states, each of which appears to be metastable. In the opinion of the writer, the best hope is the effect of throttling of the Arctic–Atlantic interchange by lowering of sea level during glacial stages. The expansion of the ice cap into West Antarctica through grounding of the ice in areas such as the Ross Sea (Hollin 1962; Calkin et al. 1970) serves to increase the fluctuation of sea level computed from estimated volume of ice.

The large number of glacial stages indicated may seem surprising, but may perhaps be less so if the problem of estimating the number of glaciations from evidence on continents is contrasted with the problem of estimating the number of glaciations from evidence in deep-sea sediments. The first may be compared in complexity to estimating the number of times the blackboard has been erased; while the second may be compared to finding the number of times the wall has been painted.

Improved knowledge about details of continental drift and pole wandering appear to favor the hypothesis of thermal isolation of polar regions as a general cause of ice ages.

Acknowledgments. Financial support for this work was provided by the Office of Naval Research under contract N00014-67-A-0108-0004 and by National Science Foundation grant GA-1615.

Lamont-Doherty Geological Observatory Contribution no. 1598.

REFERENCES

Barazangi, M., and H. J. Dorman, Seismicity map of the Arctic compiled from ESSA Coast and Geodetic Survey epicenter data, January 1961–September 1969, *Seism. Soc. Am. Bull., 60,* no. 5, 1970.

Briden, J. C., Recurrent continental drift of Gondwanaland, *Nature, 215,* 1334, 1967.

Calkin, P. E., R. E. Behling, and C. Bull, Glacial history of Wright Valley, southern Victoria Land, Antarctica, *Antarctic J., 5,* 22, 1970.

Donn, W. L., and M. Ewing, A theory of ice ages III, *Science, 152,* 1706, 1966.

Du Toit, A. L., *Our Wandering Continents,* Oliver and Boyd, Edinburgh, 1937.

Ewing, J., and M. Ewing, Sediment distribution on the mid-ocean ridges with respect to spreading of the sea floor, *Science, 156,* no. 3782, 1590, 1967.

Ewing, M., and W. L. Donn, A theory of ice ages, *Science, 123,* 1061, 1956.

Ewing, M., J. L. Worzel, and C. A. Burk, Introduction in Ewing et al., 1969, *Initial Reports of the Deep Sea Drilling Project,* vol. 1, p. 3, U.S. Government Printing Office, Washington, 1969.

Fairbridge, R. W., Early Paleozoic south pole in northwest África, *Bull. Geol. Soc. Am., 80,* 113, 1969.

Fairbridge, R. W., South pole reaches the Sahara, *Science, 168,* 878, 1970.

Hays, J. D., T. Saito, N. Opdyke, and L. Burckle, Pliocene-Pleistocene sediments of the equatorial Pacific: Their paleomagnetic, biostratigraphic, and climatic record, *Bull. Geol. Soc. Am., 80,* 1481, 1969.

Hollin, J. T., On the glacial history of Antarctica, *J. Glaciology, 4,* 173, 1962.

Irving, E., and W. A. Robertson, The distribution of continental crust and its relation to ice ages, in R. A. Phinney, ed., *The History of the Earth's Crust,* p. 168, Princeton Univ. Press, Princeton, N.J., 1968.

LePichon, X., Sea-floor spreading and continental drift, *J. Geophys. Res., 73,* 3661, 1968.

McElhinny, M. W., and G. R. Luck, Paleomagnetism and Gondwanaland, *Science, 168,* 830, 1970.

Peterson, M. N. A., N. T. Edgar, C. C. von der Borch, and R. W. Rex, Cruise leg summary and synthesis, in Peterson et al., *Initial Reports of the Deep Sea Drilling Project,* vol. 2, p. 413, U.S. Government Printing Office, Washington, 1970.

Shaw, D., and W. L. Donn, A thermodynamic study of Arctic Ocean control of climate, in press.

Smith, A. G., and A. Hallam, The fit of the southern continents, *Nature, 225,* 139, 1970.